THE BRAHMAPUTRA BASIN WATE

Water Science and Technology Library

VOLUME 47



*The titles published in this series are listed at the end of this volume.*

# THE BRAHMAPUTRA BASIN WATER RESOURCES

edited by

VIJAY P. SINGH

*Department of Civil and Environmental Engineering,*
*Louisiana State University, Baton Rouge, U.S.A.*

NAYAN SHARMA

*Water Resources Development Training Centre,*
*Indian Institute of Technology, Roorkee, Uttaranchal, India*

and

C. SHEKHAR P. OJHA

*Department of Civil Engineering,*
*Indian Institute of Technology, Roorkee, Uttaranchal, India*

KLUWER ACADEMIC PUBLISHERS

DORDRECHT / BOSTON / LONDON

A C.I.P. Catalogue record for this book is available from the Library of Congress.

ISBN 1-4020-1737-5

Published by Kluwer Academic Publishers,
P.O. Box 17, 3300 AA Dordrecht, The Netherlands.

Sold and distributed in North, Central and South America
by Kluwer Academic Publishers,
101 Philip Drive, Norwell, MA 02061, U.S.A.

In all other countries, sold and distributed
by Kluwer Academic Publishers,
P.O. Box 322, 3300 AH Dordrecht, The Netherlands.

*Printed on acid-free paper*



Printed in the Netherlands.

**Dedicated to**

**the people of**

**Northeast India**

# TABLE OF CONTENTS

# PREFACE

The Brahmaputra River is a precious gift of Nature to India and especially to its northeast region. The meaning of this gift becomes evident immediately when one considers the vastness of water resources which the Brahmaputra River basin possesses. Nearly 30% of India's water resources potential and 41% of the country's total hydropower potential are found in this basin. In terms of the average discharge, the Brahmaputra River is the fifth largest river in the world. Traversing a distance of about 2,900 kilometers from Tibet at an elevation of 5,150 meters to the Bay of Bengal and draining an area of nearly 1.6 million square kilometres shared by Tibet, India and Bangladesh, it has shaped the social, economic, cultural, religious, political, and linguistic fabric of the northeast region of India from time immemorial. Its majestic disposition is nothing short of poetic. The northeast region has a colourful ethnic and non-ethnic population and represents almost a miniature India. Its biodiversity of flora and fauna and other natural ecosystems is unsurpassable.

The northeast region of India is endowed with enormous natural resources and the huge potential of the Brahmaputra River basin. These resources notwithstanding, this part of India remains woefully underdeveloped. The current state of low development of the northeast region can arguably be attributed to a multitude of factors. Regardless of the debate about these factors, what is abundantly clear is that a sustainable development of this region cannot be achieved without the water resources development of the Brahmaputra River basin. To that end an integrated basinwide approach is needed. This approach must combine social, political, economic, administrative, cultural, and legal considerations with scientific and technological paradigms. It must involve people's participation and must strive to promote the larger interest of the region rather than the good of a few. It is these considerations which motivated the undertaking of this book project. The theme of this book is vast and no single individual has the expertise, the in-depth knowledge, or the experience of all aspects of water resources of the Brahmaputra basin. Therefore the editors solicited contributions from the people representing different disciplines bearing on the development of water resources and who are well known in their speciality areas. It is hoped that this book will stimulate further interest and development in the Brahmaputra River basin.

The subject matter of this book is divided into four sections encompassing 30 chapters. The first section, comprising seven chapters, presents preliminaries needed for planning and developing a water resources project. These relate to the atmosphere, lithology, and pedology. Introducing the theme of the book in chapter 1, the weather and climate of the Brahmaputra valley are discussed in chapter 2, followed by a short treatment of hydrometeorology in chapter 3. Soils and agriculture of the Brahmaputra basin are presented in chapter 4. The next three chapters deal with various aspects of geology of this basin. Specifically, chapter 5 presents an overview of the Brahmaputra River, chapter 6 fluvial geomorphology, and chapter 7 hydrogeology of the Brahmaputra basin.

Section 2 deals with hydrology and hydraulics that constitute the engineering foundation of any water resources project. This section comprises 6 chapters. The hydrology of the Brahmaputra basin is treated in chapter 8. The next five chapters discuss hydraulic issues. Chapter 9 provides a discussion of channel processes, followed by a discussion of spatio-temporal morphological features in chapter 10. Mathematical modelling and braid indicators are treated in chapter 11 and hydraulic structures of the Brahmaputra basin in chapter 12. The section is concluded with a discussion of scour around spurs with particular reference to the Gumi site in the basin in chapter 13.

For the planning of water resources a number of constraints must be taken into account. Section 3, encompassing 9 chapters, is devoted to a discussion of some of these constraints. Chapter 14 reflects on preliminary considerations required in water resources planning. Chapter 15 dwells on the problems arising owing to flooding in the Brahmaputra basin and the prospects for addressing them from an administrative perspective. The role of the Brahmaputra in Assam's economy is examined in chapter 16, followed by a close look at the financial constraint to development of Assam in chapter 17. The social and cultural fabric of the Brahmaputra basin is analysed in chapter 18. The remaining three chapters deal with ecological and environmental issues. The ecology of the Brahmaputra basin is discussed in chapter 19, followed by a discussion on wetlands and their role in geo-environment in chapter 20. Water quality, sediment biogeochemistry, and mineral transport are presented in chapter 21. The section is concluded with a discussion of cropping systems in the Brahmaputra basin in chapter 22.

The concluding section 4 is comprised of 8 chapters and deals with water resources development. Chapter 23 discusses groundwater development, followed by a discussion of irrigation development in chapter 24. A discussion of power development is presented in chapter 25. Water resources development may play a key role in the transport of goods. Chapter 26 discusses navigation and chapter 27 presents transport infrastructural scenarios. Flooding is an annual problem in the Brahmaputra basin. Flood management is the theme of chapter 28. For effective flood management modern tools for data acquisition, such as remote sensing, and processing of such data through geographical information systems (GIS) are needed and these are discussed in chapter 29. This section is concluded with a discussion of seismotectonic and earthquake design parameters in chapter 30, for the Brahmaputra basin is one of the most seismically active areas of the world and this information is vital for design of hydraulic structures.

The book will be useful to engineers, agricultural scientists, environmentalists, economists, planners, managers, politicians, and administrators who are concerned with development of water resources in the Brahmaputra basin. The people of the northeast are the best symbol of the current state of development of their region in general and what this mighty river has done and can do for them. The book is, therefore, dedicated to them. The editors would like to take this opportunity to express their deep gratitude to the contributors who took time from their busy schedules to write their contributions. The book symbolizes their hard work. Dr. S. Jain, National Institute of Hydrology, provided invaluable assistance with the final preparation of the book manuscript. His assistance is most gratefully acknowledged. Our families were supportive of this book project from the beginning and without their support the book would not have come to fruition.

V.P. Singh
Baton Rouge, Louisiana, U. S. A.

N. Sharma and C.S.P. Ojha
Roorkee, India

# CONTRIBUTORS

Dr. D.N. Barthakur
Swapnalaya Apartments,
Guwahati 781 024, Assam, India

Dr. M. Barthakur
White House,
National Highway No. 37
P.O. Gauhati University,
Guwahati 781 014, Assam, India

Dr. D. Bhattacharya
Depatment of Soil Science
Assam Agricultural University
Barbheta, Jorhat 785 013, Assam, India

Mr. B. K. Bhattacharyya
Step By Step English Medium School,
Housing Road, Hengra Bari,
Guwahati 781 006, Assam, India

Dr. A. K. Bora
Department of Geography
Gauhati University
Gopinath Bordoloi Nagar,
Guwahati 781 014, Assam, India

Mr. R. Chakraborty
Public Works Department
Nijarapar, P. O. Chandmari
Guwahati 781003, Assam, India

Mr. R.M. Das
Inland Water Transport Department
Assam
Ulubari, Guwahati 781 007
Assam, India

Mr. B. Datta
"Upendralaya" Professor Para,
K.B. Road, Guwahati 781 008
Assam, India

Dr. P. S. Goswami
Department of Economics
Cotton College,
Panbazar, Guwahati 781 001
Assam, India

Dr. A. K. Sarma
Civil Engineering Department
Indian Institute of Technology (IIT)
North Guwahati, Guwahati
Assam, India

Dr. S. Hazarika
Department of Soil Science
Assam Agricultural University
Barbheta, Jorhat 785 013
Assam, India

Dr. M. Hussain
Department of Political Science
Gauhati University
P.O. Gauhati University,
Guwahati 781014,
Assam, India

Mr. S. Krishna
Assam Sachivalaya, Govt. of Assam
Dispur, Guwahati 781 006
Assam, India

Dr. Chandan Mahanta
Civil Engineering Department
Indian Institute of Technology (I.I.T.)
P. O. North Guwahati
Guwahati, Assam, India

Mr. S. B. Medhi
Special RAPP
National Human Rights Commission
West of Nabgraha Temple
P. O. Kharguli
Huwahati 781004, Assam, India

Mr. A. K. Mitra
Flood Control Department, Assam
Chandmari, Guwahati 781 003

Assam, India

Dr. C. S. P. Ojha
Civil Engineering Department
Indian Institute of Technology
Roorkee 247667, Uttaranchal, India

Late Dr. A. B. Palaniappan
Nnational Institute of Hydrology
Jal Vigyan Bhawan
Roorkee 247667, Uttaranchal, India.

Mr. S.N. Phukan
B-304, Swapnalaya Apartments
Narikalbari, Zoo Narengi Road
Guwahati 781 024, Assam, India

Dr. V. Prakash
Earthquake Engineering Department
Indian Institute of Technology
Roorkee 247 667, Uttaranchal, India

Dr. B. Purkait
Petrology Division, Marine Wing
Geological Survey of India
63, N. S. C. Bose Road, Calcutta
700040, India

Dr. A. Sharma
NEDFI
IDBI Building, G.S. Road
Guwahati 781 005, Assam, India

Dr. N. Sharma
Water Resources Development
Training Centre
Indian Institute of Technology
Roorkee, Uttaranchal 247 667
India

Dr. P. Sarma
Department of Geography,
Cotton College,
Panbazar, Guwahati - 781 001
Assam, India

Professor V. P. Singh
Department of Civil & Environmental
Engineering
Louisiana State University
Baton Rouge, LA 70803-6405
U. S. A.

Dr. A. Sinvhal
Earthquake Engineering Department
Indian Institute of Technology
Roorkee 247 667, Uttaranchal, Indi

Dr. V. Subramaniam
Asstt. Professor
Deptt. of Civil Engineering
Indian Institute of Technology
P. O. North Guwahati
Guwahati, Assam, India

Mr. V. K. Talwar
Director, Gates (E & NE) Directorate
Sewa Bhawan, C.W.C, R. K. Puram
New Delhi 110066, India.

Dr. B. K. Talukdar
"EVER GREEN"
Samanwoy Path (Survey)
PO: Beltola, Guwahati 781 028
Assam, INDIA

Dr. N. C. Talukdar
Department of Soil Science
Assam Agricultural University
Barbheta, Jorhat ?785 013
Assam India

# CHAPTER 1

# INTRODUCTION

C.S.P. OJHA & V.P. SINGH

Planning, development, and management of water resources are interdisciplinary and require an integrated approach. This approach involves both science and humanities. Sometimes humanities are facetiously termed 'soft' sciences, and sciences including physical and biological sciences, as well as their engineering offsprings, as 'hard' sciences. The science includes climatology, geology, soil science, agriculture, civil and agricultural engineering, environmental science and ecology; whereas the humanities include society, social structure, politics, economics, administration, culture, religion, languages, and so on. This chapter introduces the Brahmaputra River basin and summarizes the various aspects of water resources planning and development in the basin.

## 1.1 The Brahmaputra Drainage Basin

The variation in the availability of water in the different river basins of India is huge. The average annual runoff of all the rivers of the country has been assessed as approximately 1869 $km^3$. Of this amount, the utilizable portion by conventional storage and diversion is estimated to be about 690 $km^3$. In addition there is substantial replenishable ground water potential in the country estimated at 432 $km^3$. The per capita availability of water at the national level has dropped from about 5177 $m^3$ in the year 1951 to 1869 $m^3$ at present (India, 2001). With burgeoning population, rising standard of living, increasing thrust of industries for water, and growing vulnerability of water to pollution, it is expected that the per capita availability of water is not likely to increase, at least not in the near future. Thus it has become increasingly important to develop a systematic approach to the study of different river basins in the country. For effective planning and management of water resources of river basins this approach is vital.

Among different river basins of India, Brahmaputra River basin enjoys a special status for its resources and majesty. From the standpoint of its water resources, it can be characterized as a flowing sea. It forms the lifeline of the north eastern region of India, which includes Arunachal Pradesh, Assam, Meghalaya, Manipur, Nagaland, and Tripura. The drainage area of the basin is estimated as 580,000 sq. km, of which China, India, Bangladesh, and Bhutan share 50.5%, 33.6%, 8.1% and 7.8 %, respectively. The drainage area as a percentage of the basin within India for different states is: Arunachal (41.88 %), Assam (36.33%), Meghalaya (6.00%), Nagaland (5.55%), Sikkim (3.75%) and West Bengal (6.47%). The Brahmaputra River originates in South China (in the



*V.P. Singh et al. (eds.), The Brahmaputra Basin Water Resources*, 1-16.

Kailash range) at an elevation of about 5300 m above mean sea level (MSL) and exhibits a wide range of slope variations from 4 --17 m/km (gorge section), 27 cm/km (Pasighat), and 10 cm/km (Guwahati) in its total length of about 2880 km, which is comprised of 1625 km in Tibet, 918 km in India and 337 km in Bangladesh (Goswami, 2001).

The water resources of the Brahmaputra River basin are huge and yet so little development has occurred to date. An idea of the availability of water resources and utilisation in respect of this basin can be obtained from Table 1.1 (Goswami (2001). Other relevant statistics for this system are given in Table 1.2 along with several other river basins. Needless to say, the water resources potential of the Brahmaputra system is the highest among all the river basins in the Indian subcontinent. Other related statistics about the Brahmaputra River basin can be also gleaned from different chapters of the book.

Table 1.1 Water resources: availability and utilisation (after Goswami, 2001).

| | |
|---|---|
| Total water resources potential | 30% of the country's total |
| Per capita and per hectare water availability | Highest in the country |
| Hydropower potential | About 41% of the country's total |
| Potential developed so far | Only about 3% |
| Irrigation potential | 4.26 million hectare (m.ha) |
| Present coverage of irrigation | 0.85 million hectare (20% of existing potential against the national average of 56.4%) |
| Groundwater potential | Very high in valley areas (at shallow depth) |
| Potential developed so far | 4.3% (against the national average of 32%) |

Recent satellite data provide an indication of the total basin land use pattern (Goswami, 2001): the forest cover (14.5%), grass land (44.0%), agricultural land (14.0%), crop land/natural vegetation mosaic (12.8%), barren/sparsely vegetated land (2.5%), water bodies (1.8%), snow and ice (1.1%), and urban built-up land (0.02 %).

Hussain (1994) describes the Brahmaputra drainage system as unique in the world, as it provides the only example in the world where the drainage pattern runs in a diametrically opposite direction. For example, in Tibet it flows from west to east but immediately south of Tibet it flows from east to west in Assam. Furthermore, the Himalayas are considered to be younger in age than the Brahmaputra River. Considering the energy crisis in many developing countries of the world today, there is a lot of awareness of harnessing hydropower potential. In fact, hydropower is superior to other sources of energy and nearly 25 percent of the total power of the world is supplied by hydropower stations (Prasad, 2000). In India the economically exploitable hydropower potential is about 84,044 MW at a 60% load factor, equivalent to an installed capacity of about 150,000 MW which will yield an annual energy of 442 billion units of electricity. The importance of the Brahmaputra River basin becomes immediately clear

Table 1.2 Water Resources Potential in River Basins of India (after CWC, 1996).

| S. No. | River Basin | Catchment Area, sq. km | Average Annual Potential m.ham | Estimated Utilisable flow m.ham | Estimated present use of surface water, m.ham | Present stage of utilization percent |
|---|---|---|---|---|---|---|
| 1 | Indus (up to border) | 321,289* | 7.33 | 4.60 | 4.0 | 87 |
| 2a. | Ganga | 861,452* | 52.50 | 25.0 | data classified | |
| b. | Brahmaputra | 231,136* | 58.56 | 2.4 | | |
| 3. | Sabarmati | 21,674 | 0.38 | 0.19 | 0.18 | 93 |
| 4. | Mahi | 34,842 | 1.10 | 0.31 | 0.25 | 81 |
| 5. | Narmada | 98,796 | 4.56 | 3.45 | 0.80 | 23 |
| 6. | Tapi | 65,145 | 1.49 | 1.45 | N.A. | |
| 7. | Brahmmni-Baitarni | 39,033 | 2.85 | 1.83 | N.A. | |
| 8. | Subernrekha | -- | 1.24 | 0.68 | N.A. | |
| 9. | Mahanadi | 141,589 | 6.69 | 5.00 | 1.70 | 34 |
| 10. | Godavari | 312,812 | 11.05 | 7.63 | 3.80 | 50 |
| 11. | Krishna | 258,948 | 7.81 | 5.80 | 4.70 | 81 |
| 12. | Pennar | 55,213 | 0.63 | 0.69 | 0.50 | 79 |
| 13. | Kaveri | 81,155 | 2.14 | 1.90 | 1.80 | 95 |
| 14. | Other East flowing rivers between Mahanadi and Pennar | -- | 2.25 | 1.31 | N.A. | |
| 15. | East flowing rivers between Pennar and Kanya Kumari | -- | 1.65 | 1.67 | N.A. | |
| 16. | West flowing rivers of Kachch, Saurashtra | -- | 1.51 | 1.50 | N.A. | |
| 17 | West flowing rivers, south of Tapi | -- | 20.09 | 3.62 | N.A. | |
| 18. | Minor river basins draining into Bangla-desh, Burma | -- | 3.10 | -- | -- | |
| | | | 186.93 | 69.03 | -- | |

* Areas in Indian territory; N.A.: not applicable

by noting that out of 84,044 MW hydropower potential at a 60% load factor, the Brahmaputra accounts for as much as 34,920 MW. However, only 1% of this potential has been utilised so far. Sadly enough, the Brahmaputra water resources also happen to be the least developed water resources in India. It is noted that there is some difference between the developed potential of hydropower as reported by Prasad (2000) and that reported by Goswami (2001). The very low values of hydropower development in the basin are indicated nevertheless.

For development of the Brahmaputra River basin, it is imperative to examine the factors which have hampered the development of its huge water resources. To that end, a brief glimpse of the history of the north eastern region is given, followed by the efforts of Government of India in developing the water resources of the country. The different components required for a systematic study of water resources are also identified and these are described in the context of the Brahmaputra River.

## 1.2 History of Development

The history of development of the Brahmaputra River basin is closely linked with that of the north eastern region of India. Most parts of the river basin lie in the states of Assam, Arunachal Pradesh, and Nagaland, followed by a large portion of it in Bangladesh. Prior to the independence of India in 1947, all these states were a part of Assam. The advent of Assam is traced to as early as 1228 AD (Hussain, 1994). This state was then conquered by the invaders from southeast Asia, who ultimately ruled the state for nearly six hundred years. In the history of Assam, this rule of six hundred years is also known as the era of Ahoms. In the early part of the 19$^{th}$ century Ahom rulers had to face Burmese invasion. To safeguard the state from the Burmese invasion the help of the British was requested by the rulers. The British established a treaty with Burma. This treaty is known as the YANDABU TREATY. Although this treaty helped save the Ahom rulers from the Burmese invasion, it also paved an easy way for the British to enter the state and establish their rule by the year 1826. Thus it is interesting to observe that the state of Assam came into the possession of the British at a much later stage, and thus the process of development could be initiated only after the integration of the state with the British Colonial rule. Unfortunately, the time after 1826 was too short for the British to do much for the development, as in 1857 they had to face the first revolt by Indian patriots. Much awareness for the freedom got an impetus from this struggle and the period up to 1947 could no longer remain conducive for a healthy development of this region which had very poor infrastructure, inaccessibility to many regions because of the presence of dense forests and hilly terrains, coupled with too wide a stretch of water of the Brahmaputra River and its tributaries. This was further compounded by the complex demographic status in terms of the presence of many tribes, language and cultural diversity. The history of Assam is full of many incidents which indicate very little or almost negligible interaction with a large portion of masses in the region. It is worthwhile to mention that in Independent India the complex dynamics of these diversities has also been responsible for the division of the State of Assam into many small sister states, i.e., Assam, Arunachal Pradesh, Manipur, Mizoram, Meghalaya,

Nagaland, and Tripura. Even today these dynamics continue, and the demands for the further disintegration of many of these states are now in vogue.

A significant part of the Brahmaputra River basin also lies in Bangladesh. There is evidence of illegal migration of a large number of people from Bangladesh to these sister states. In fact, the history of Assam after independence concentrates mainly around the illegal immigrants and their role in the sister states. After independence a lot of time has been lost in resolving the issue of illegal immigration. The immigrants are a determinant factor in elections, and hence their stay or deportation from Assam has become a political issue. The period from 1979 to 1982 is marked with many mass movements with wide local support. This also led to an accord with the Government of India and is widely known as the Assam accord. Although the objectives of this accord include the overall development of Assam, in addition to resolving the problem of illegal immigration, the stalemate continues even to the present day.

### 1.2.1 DEVELOPMENT OF WATER RESOURCES IN INDEPENDENT INDIA

It is relevant here to first address the issue of development of water resources in India after independence in 1947. Table 1.3 provides certain statistics for various plan periods. Thus it can be seen from this table that the efforts by the Government of India are very much on the right track, and the importance attached to the water sector has continued in every plan. In fact, several collaborative projects with the assistance of different countries have also been initiated in the last decade.

Water is vital for realising the full potential of the agricultural sector and the country's development. Optimum development and efficient utilization of water resources therefore assume greater significance. The National Water Policy, adopted in September 1987, stresses that "water is a primal natural resource, a basic human need and precious national asset." Since the adoption of the national water policy in 1987 a number of problems and challenges have emerged in the development and management of the water resources sector.

In 1991 the National Watershed Development Project for Rainfed Areas (NWDPRA) was launched in 25 states and 2 union territories. The project continues even in the present Ninth Five Year Plan with the purpose of increasing agricultural productivity through sustainable use of natural resources. The program measures under NWDPRA include treatment of arable and nonarable land, treatment of drainage lines, and production systems in arable and nonarable lands with emphasis on multiple/ mixed/ strip cropping preferably of legume components, organic farming, dry land horticulture, agro forestry, silvi pastures, live stock management and household production systems (India, 2001). In addition, the World Bank aided watershed development projects have been also initiated in the region. Assam Rural Infrastructure and Agriculture Support Project is one such example of a World Bank aided watershed development project, which was implemented in 1995 for a period of eight years. The objective of this project is to alleviate poverty and accelerate agricultural growth through improved use of resources. Alternative farming is replacing Jhum farming in the northeast hill region and attempts have been also made at nutrient management (India, 2001).

Table 1.3 Planwise position of irrigation potential created and utilized (in million hectares).

**Note**: S.W.[1]: Surface water; G.W.[2]: Ground water; N.A.: Not available

| Plan | Potential created Major & Medium | Minor S.W[1]. | Minor G.W[2]. | Minor Total | Total | Potential Utilised Major & Medium | Minor S.W[1]. | Minor G.W[2]. | Minor Total | Total |
|---|---|---|---|---|---|---|---|---|---|---|
| 1 | 2 | 3 | 4 | 5 | 6 | 7 | 8 | 9 | 10 | 11 |
| At the end of I Plan (1951--56) | 12.20 | 6.43 | 7.63 | 14.06 | 26.26 | 10.98 | 6.43 | 7.63 | 14.06 | 25.04 |
| II Plan (1956--61) | 14.33 | 6.45 | 8.30 | 14.75 | 29.08 | 13.05 | 6.45 | 8.30 | 14.75 | 27.80 |
| III Plan (1961--66 ) | 16.57 | 6.48 | 10.52 | 17.00 | 33.57 | 15.17 | 6.48 | 10.52 | 17.00 | 32.17 |
| Annual Plans (1966--69) | 18.10 | 6.50 | 12.50 | 19.00 | 37.10 | 16.75 | 6.50 | 12.50 | 19.00 | 35.75 |
| IV Plan (1969--74) | 20.70 | 7.0 | 16.50 | 23.50 | 44.20 | 18.69 | 7.00 | 16.50 | 23.50 | 42.19 |
| V Plan (1974--78) | 24.72 | 7.50 | 19.80 | 27.30 | 52.02 | 21.16 | 7.50 | 19.80 | 27.30 | 48.46 |
| Annual Plans (1978--80) | 26.61 | 8.00 | 22.00 | 30.00 | 56.61 | 22.64 | 8.00 | 22.00 | 30.00 | 52.64 |
| VI Plan (1980--85) | 27.70 | 9.70 | 27.82 | 37.52 | 65.22 | 23.57 | 9.01 | 26.24 | 35.25 | 58.82 |
| VII Plan (1985--90) | 29.92 | 10.99 | 35.62 | 46.61 | 76.53 | 25.47 | 9.97 | 33.15 | 43.12 | 68.59 |
| Annual Plans (1990--92) | 30.74 | 11.46 | 38.89 | 50.35 | 81.09 | 26.32 | 10.29 | 36.25 | 46.54 | 72.86 |
| VIII Plan (1992--97) Provisional | 32.96 | N.A. | N.A. | 56.60 | 89.56 | 28.44 | N.A. | N.A. | 52.32 | 80.76 |
| Annual Plan (1997--98) | 33.64 | N.A. | N.A. | 57.62 | 91.26 | 28.93 | N.A. | N.A. | 53.05 | 81.98 |
| Annual Plan (1998--99) | 34.52 | N.A. | N.A. | 58.80 | 93.32 | 29.81 | N.A. | N.A. | 53.78 | 83.59 |
| Annual Plan (1999--00) | 35.35 | N.A. | N.A. | 59.38 | 94.73 | 30.47 | N.A. | N.A. | 54.23 | 84.70 |

Although some progress has been made in the irrigation sector, much has not been achieved in hydropower development. All India power shortage at present stands at approximately 13% (Prasad, 2000). An initiative was taken by the Government of India in 1991 to attract private entrepreneurs to bring additional resources, but the response of the private sector has been poor. The investment decision for many new projects and implementation of many projects get held up on account of nonavailibility of environment and/or forest clearance from the Ministry of Environment and Forests.

Recently, environmental activists have directed their sustained attacks on large hydro plants involving storage dams. The environmental lobby is against major dams and associated mega projects, e.g., Tehri and Sardar Sarovar projects. Under the present condition the development of small/mini/micro hydel schemes (capacity up to 15 MW) could be one of the ways to mitigate power shortages to some extent. These schemes can be completed within two to three years and they have the least impact on the natural environment. In India the small hydropower potential is estimated to be of the order of 10,000 MW out of which 5,000 MW would come from canal falls and another 5,000 MW from hilly regions (Prasad, 2000). This would mean that as most of the Brahmaputra River basin has a hilly terrain there is considerable potential for developing small hydropower in the region.

## 1.3 Considerations for Water Resources Development

For the study of water resources of the Brahmaputra River basin it is essential to consider the various issues that may be relevant to the technicians, planners and political leaders. Some such issues may be: (1) geology of the basin; (2) climatic conditions; (3) hydrology; (4) soil types and their distribution; (5) agriculture, cropping pattern and irrigation; (6) hydraulics of the river; (7) environmental aspects; (8) hydropower development; (9) social, financial and administrative, political and judicial issues; and (10) navigation and transport. It will be appropriate here to address these issues briefly. For economy of space and the convenience of description, these issues are arranged in different but interrelated groups.

### 1.3.1 GEOLOGY AND SEISMO-TECTONICS

The Brahmaputra River system lies in one of the most seismically active regions of the world. Thus seismo-tectonics and geology of the basin needs an important coverage. In fact, recent times have witnessed an increased seismic activity all around the globe. Any controversy regarding seismic safety of the project may upset the construction schedule in view of the nature of the earthquake occurrence and ground motion being probabilistic. Seismic parameters have a great impact on the safety of the structure and social implications in the event of any failure. Recently many hydropower projects, e.g., Tehri have suffered on this account (Prasad, 2000). As the Brahmaputra basin lies in one of the most seismically active regions of the world, an exclusive coverage to the seismic aspects has been given in the book. This is expected to be helpful for the construction of various engineering structures.

The importance of geology in the context of the Brahmaputra is described by Rao (1975), who emphasizes a scientific and intensive study of the valley, including (i) geological base, (ii) tectonic events, (iii) geomorphology including river morphology, (iv) sedimentation and fluvial processes, (v) geophysics, (vi) geohydrology, and (vii) engineering geology. An attempt on these lines was initiated by the Geological Survey of India at the instance of K.L. Rao in 1970--1972. This study helped with a better understanding of the basin characteristics, some of which find coverage in this book. Hussain (1994) also provides a useful compilation of the rocks in the basin. Granite and gneisses are the principal rocks in the Maghalaya Plateau. At the foot of hills are frequently found beds of conglomerates. In between these two ranges lies the narrow Brahmaputra valley, which is alluvial in origin and consists of sand, sandstones, pebbles, clay and sometimes a mixture of sand and clay with decomposed vegetable matter. The fragile rocks in Arunachal Pradesh are prone to severe erosion. Also, the occurrence of land slides leads to an increased input of sediment load into the river. It is well known that sediment laden streams are prone to a phenomenon of braiding, which, in a simple way, is the division of a stream into many segments with the formation of islands in between. In fact, the Brahmaputra River possesses one of the largest islands of the world, i.e. Majuli Island. The deposition of sediments in the river beds raises the flood level each year, resulting to an increasing loss of life and property. Geological risks can not be foreseen at the Detailed Project Report stage itself in the Himalayas as in the Nathpa Jhakri Project where a heavy landslide took place at the dam site and which could not be predicted in advance in spite of the association of some of the world's renowned engineering geologists with the project (Prasad, 2000). Thus geology dictates to a large extent the success of any development project.

### 1.3.2 HYDROLOGICAL AND CLIMATOLOGICAL ASPECTS

The climate of any basin influences soil types. Climatic conditions are often characterized by temperature and rainfall variations, and the level of humidity. The whole Brahmaputra basin is divided into four micro-climatic zones. Basically, the climate of the Brahmaputra basin is humid subtropical, and high rainfall and high humidity are the main features thereof. The number of rainy days may sometimes equal three weeks in a month. Thus rainfall is the most dominating component of the climate of the Brahmaputra basin. During rainy and summer seasons water needs are less than the available water, and this results in the moist soil or wet areas for nearly two thirds of a year. To know the details of variation of temperature, rainfall, and humidity in the Brahmaputra basin is essential to understand the presence of a wide variety of soils in the basin. The climatology coupled with the hydrology can also provide a basis for estimating surface runoff and the erosion potential of different sites in the basin. For this reason a detailed coverage to the climatological and hydrological aspects is covered in the book.

### 1.3.3 SOILS PROFILES

For agricultural development a knowledge of soil profiles is essential, as soils provide essential minerals for the growth of crops and plants and at the same time they work as a storage to supply moisture for healthy development of plants. Soil profiles representing major soil orders of the Brahmaputra basin are: Entisol, Inceptisol, and Alfisol. The basin also encounters the presence of poorly drenched soils, and the nature of soil is such that the growth of paddy in the basin is practiced at a very large scale. To fully understand the characteristic of the soil it is essential to examine the site characteristics and morphological features of soil profiles, the structure and texture of soils, the presence of sand, silt and clay, bulk density, saturation percentage and moisture holding capacity, pH, cation exchange capacity, exchangeable cations, percent base saturation, etc. In addition, the mineralogy of soils which includes sand, silt and clay mineralogy, is also equally important for studying the suitability of the soil for a specific crop. Details pertaining to some of these characteristics of soils along with the productivity potential, which is a function of the soil moisture content, drainage, effective depth of soil, texture and structure of root zone, average content of a horizon, nature of clay and nutrient reserve, are given for different zones of the Brahmaputra basin. In addition, the microbial status of Brahmaputra basin soils, fertility status of Tea soils and cropping systems, and crop productivity are covered in the present work. In most of the basin, soil degradation has occurred and specific mention of the degradation of the soil of the valley by effluent of coal mining and oil drilling industry is included.

### 1.3.4 HYDRAULICS OF THE BRAHMAPUTRA RIVER BASIN

For development of civilisations rivers have been often used by engineers to create storage and diversion of river waters to meet specific needs of the society. The use of hydraulic structures finds a lot of attention in the British regime, particularly in the north-western region of the country. Years after India's independence have also witnessed the construction of hydraulic structures in different parts of the Brahmaputra basin, but the pace of development has been less than encouraging. For design of hydraulic structures the knowledge of probable maximum flood along with a suitable resistance relationship are very much needed. Unlike other rivers in the country, the Brahmaputra is well known for its braiding characteristics, and many stretches of this river have been subjected to out flanking. Attempts have been also made in the past to train the river by using a series of spurs. All these issues related to hydraulic structures, resistance relationships, and river training, along with the modelling aspects of braided river find an exclusive coverage in the book. It is pertinent to mention here that for studying the resistance characteristics of the Brahmaputra River, an extensive data base available in Bangladesh is utilised. Similarly, the performance of spurs at the Gumy site is included to enrich the experiences in training of this mighty river. The Brahmaputra valley of Assam is in a state of perpetual flux owing to the rapid geomorphological changes (Rao, 1975) and hence the geomorphology also finds coverage in the book.

### 1.3.5 ENVIRONMENTAL ASPECTS

In modern times environmental issues cannot be ignored. It is essential to identify the stretches in the stream where the entry of pollutants takes place. The influence of discharge of such pollutants in relation to the dilution capacity of the stream is to be assessed. Also the wetlands of the Brahmaputra basin, being an important component of the environment, need to be included in describing the overall situation of the environmental health of the basin. For this reason a chapter is exclusively devoted to emphasizing the water quality status of the stream in a selected stretch of the river and offers a valuable insight into the waste handling capacity of the stream. Similarly a discussion of wastelands of the Brahmaputra basin is provided to strengthen the environmental aspects. It is also emphasized here that certain actions of the Ministry of Environment and Forests have also come under criticism in recent years. For example, the treatment of very highly degraded areas of free draining catchment, i.e., up to the next project upstream at the project cost. Prasad (2000) considers that it is not justified to charge the full cost of catchment area treatment to hydro projects only and argues that the cost should be proportioned amongst various beneficiary sectors such as agriculture, transport, irrigation, flood control, etc. along with hydropower projects. Many of the projects in India are held up because of environmental considerations. Thus the development of a basin needs to be commensurate with a healthy environment and the environmental issues form a very important part of water resources development.

### 1.3.6 SOCIAL, FINANCIAL, AND ADMINISTRATIVE ISSUES

The availability of water and societal development are closely linked. Historical evidence clearly indicates the existence of civilizations in river basins. In fact the social fabric of a society is closely linked with the water resources. Table1.4, based on a study done in 1988 by Rao et al. (Singh, 1999), clearly indicates the linkage of social status with the water resources development. Thus living standards of the society can be improved through management of water resources. This, in turn, involves administrative as well as financial issues.

Table 1.4 Linkage between irrigation and social status.

| Percent Gross Area with Irrigation | Percentage of people below poverty line |
|---|---|
| Below 10 | 68.8 |
| 20 to 30 | 45.6 |
| Above 50 | 26.5 |

Administrative issues gain a lot of importance when the basin states are affected by insurgent and militant elements. Abduction of innocent people working at a project, extortion of money from them, and constant threat to life works as the greatest impediment in the project's execution. Furthermore, the cost of security arrangements add to the project's cost and cost overruns due to insurgency. The Doyang (Nagaland) and Loktak Hydroelectric project in Manipur are the examples of projects which have suffered on account of insurgency (Prasad, 2000). Also, during the project execution liaison with local authorities, district administration, law and order enforcing authorities, revenue authorities, licensing departments, and ministries is required to be maintained. If people are not able to appreciate the project requirements in the right perspective owing to varied interests, undue procedural delays may occur, thus emphasizing the role of administration in the process of development.

To emphasize the financial aspects it is appropriate to look at some of the available statistics related to the cost of creating additional irrigation potential in order to improve the social status. From 1978 onwards the cost had been increasing on an average of Rs. 29,600 per ha till 1992 (Hassim, 1999). The costs would be even higher today. Such increases are mainly on account of harvesting of water from increasingly difficult situations. Despite such high costs, the pricing system in India is far from being competitive; prices do not cover even the maintenance costs of the delivery system (Hassim, 1999). In the hydro power sector too, financial and structural reforms are very much needed for the development of hydropower, as many of the State Electricity Boards in India are facing financial constraints on account of transmission and distribution losses, illegal connections, and corruption (Prasad, 2000). Thus, consideration of financial aspects is equally important for the development of water resources.

### 1.3.7 FLOOD MANAGEMENT

Floods are natural phenomena resulting from unusual or heavy precipitation. Tropical countries with monsoon climates, such as India are more prone to them, although floods in the relatively small Red River in Canada and the U.S.A. in April-May, 1997 have shown that developed, temperate zone countries are not immune from them (Singh, 1997). Engineering measures such as storage, marginal embankments, improvement of waterways, etc., can and do reduce flood intensity and flood damage. Similarly, non-engineering measures comprising watershed treatment and control, flood forecasting and warning system, and flood plain zoning can also play an effective role in the mitigation of flood damage, provided that they are implemented and reinforced (Singh, 1997).

Hydroelectric projects are the worst sufferers of natural calamities such as floods. With a very few exceptions, most of the projects have been delayed owing to flash and unprecedented floods, which either washed away or overtopped the cofferdams. In some cases the flood water entered the tunnels, accumulating therein a lot of silt and slush whose removal added to time and cost overruns (Prasad, 2000). The floods also caused heavy loss of material and machinery.

The occurrence of floods in the Brahmaputra River basin is an annual feature. A severe flood occurred in 2000. Each year floods lead to a considerable loss of life, property, and agriculture. Thus attempts have been made to manage the flood. For this reason two chapters are included in the book which highlight the damage owing to floods and the effective steps undertaken so far to manage the floods.

### 1.3.8 AGRICULTURE, CROPPING PATTERN, AND IRRIGATION

Agriculture plays an important role in the Indian economy. Its share in GDP is about 30 % and its share in employment is much larger, say around 65% (Hassim, 1999). The availability of land is expected to be a major constraint on the growth of agriculture in India. Owing to the increased demand for land for housing and the limitations of compensating for this demand by converting forests into land, it is being realised that increase in land is possible only through reclamation of waste land. Apart from the availability of land, other issues are also important to be considered in the agriculture sector. For example, the United States of America, despite its gross land area of 937 M. ha and annual river flows approximately the same as that of India, irrigates only about 20 M. ha, yet its food grain production is over 250 million tons. In comparison, the mean annual yield of irrigated food grains from irrigated cultivation is still less than 2 ton per hectare (Thomas, 1999).

Considering the limited net sown area in any system, it is reasonable to augment the availability of land through increased cropping intensity. However, enhancement of cropping intensity requires expansion of irrigation facilities and demands higher efficiency of the irrigation. At the national level the ultimate irrigation potential is estimated to be about 113 million hectares. With this irrigation potential and the current cropping intensity coefficient with respect to irrigation, the cropping intensity can go up to 149 as against the current level of 134.20 (Hassim, 1999).

Table 1.5 Annual water demand in m ha-m for years 2000 and 2025 (Hussain, 1994).

| Purpose | Year | |
|---|---|---|
| | 2000 | 2025 |
| Domestic use | 3.3 | 5.2 |
| Industrial | 3.0 | 12.0 |
| Energy | 2.7 | 7.1 |
| Irrigation | 63.0 | 77.0 |
| Others | 3.0 | 3.70 |
| Total | 75.0 | 105.0 |

In the Indian context, irrigation happens to dominate the annual water demand, as is seen from Table 1.5. The efficiency of irrigation in India is not that high. Water use efficiency in India is estimated to be only between 38 to 40% for canal irrigation and about 60% for ground water schemes. The optimal efficiency can be as high as 60% for canal schemes and about 75% for ground water schemes. Thus, irrigation has received the attention in India. This is obvious from Table 1.6 compiled by Chaturvedi (1985). Although the statistics reported in this table are not up to date, however, these signify the future of irrigation in India.

Table 1.6. Developments and future of irrigation in India (after Chaturvedi, 1985).

| Year | Net area sown (m ha) | Net area irrigated (m ha) | % of net sown area irrigated |
|---|---|---|---|
| 1950 -- 51 | 110.75 | 20.85 | 17.56 |
| 1955 -- 56 | 129.16 | 22.75 | 17.62 |
| 1960 -- 61 | 133.20 | 24.66 | 18.51 |
| 1967 -- 68 | 139.70 | 27.52 | 19.70 |
| 1973 -- 74 | 142.71 | 32.60 | 22.85 |
| 1977 -- 78 | 149.93 | 40.00 | 26.68 |
| 1982 -- 83 | 155.17 | 52.40 | 33.77 |

### 1.3.9 NAVIGATION AND TRANSPORT

The use of streams and rivers to transport goods from one place to another has been widely used in India since the pre-independence era. Also the transport from one side of the bank to the other side through boats and ships is a part of daily life in Assam. The Brahmaputra River is known for its large width which may vary from a several hundred meters to a few kilometers. Some of the stretches of the river are as wide as twenty kilometers. The construction of bridges in such reaches is highly uneconomical. Even in other reaches, as the width of the river is fairly large, the use of surface water transport has received importance. The navigation industry also employs many people and thus influences the financial and social aspects of life in the basin as a whole. In addition to the navigation industry, transport by rail and road is equally important, as it helps in the exchange of resources within, as well as across the basin boundaries. Both of these transport sectors are equally important. In India, owing to saturation of train and road corridors, the proposal to increase transport of cargo by navigation received attention in the recent past. One example of this is the proposal to develop the Ganga water way, which alone could, by 2015, help in the carriage of 20 to 30 million tonnes of containerised and other cargo, with mechanised country boats operating feeder services (Verghese, 1999). With this in view, the development of these two sectors is highlighted in this book.

### 1.3.10 HYDROPOWER

One of the possible uses of water has been through the generation of hydropower. The available statistics on the Brahmaputra water resources indicate that there is tremendous potential for hydropower generation and its utilisation. Table 1.7 supports this fact. In fact, studies on interbasin transfers in India indicate that it is economically unfeasible to transport water from Brahmaputra to other states and the only way the river can be beneficial to the country is by harnessing the huge hydropower potential and diverting this to other states. Attempts to harness this potential in the past as well as plans to harness this in the future are expected to prove useful for deriving the benefits of the basin.

Table 1.7 Hydro-energy potential and utilization.

| River Basin | Power | Percent |
|---|---|---|
| Indus | 19,988 | 12.90 |
| Ganga | 10,715 | 17.07 |
| Central Indian rivers | 2,740 | 19.00 |
| West flowing peninsular | 6,149 | 51.84 |
| East flowing peninsular | 9,532 | 37.0 |
| Brahmaputra | 34,920 | 1.06 |
| Total | 84,044 | -- |

As per a recent survey, operational hydroelectric schemes in India account for only 15% whilst those under execution for only 7% of the total potential. Thus the bulk of the potential amounting to 78%, remains to be developed (Prasad, 2000). With this in view, a coverage of this important topic is presented in this book.

## 1.4 Development of Water Resources: Problems and Challenges

The statistics presented so far indicate that both the sectors including irrigation and hydropower have a considerable scope for the development. However, these developments have been hampered by rapid geomorphological changes, occurrence of annual floods, migration of channels, erosion of land, disruption of road and rail communications, flooding of human habitat and cultivable lands and finally aggravated by the disastrous earthquakes that occur in this valley which is a geologically unstable zone (Rao, 1975). The situation is further complicated by political factors which have been responsible for the upsurge of terrorist activity in certain parts of the basin. The issue of illegal immigrants has disrupted the normal life of Assam from time to time, and even the assembly elections of May, 2001, also witnessed an upsurge of terrorist activities. The lack of technical institutions in the region is also important. It is, however, encouraging that in recent years one of the Indian Institutes of Technology has been established in Guwahati. The influence of this can be appreciated from a few chapters in this book having been contributed by faculty of this Institution. The establishment of a technical institution is bound to have a far reaching scientific and technical impact on the Brahmaputra basin at different fronts and might help achieve the objectives of development. At this point it will be also appropriate to mention the establishment of the Brahmaputra Board by the Government of India in 1982. The objectives of this board also include the overall development of the basin. It is anticipated that the future will embrace developments in the region despite the challenges being faced so far.

## 1.5 An International Perspective

As the Brahmaputra River basin includes a part of China, Bhutan, India, and Bangladesh, it is an international river. There is no documentation of sharing of the potential of river water between India and China. However, there exists a possibility of cooperation between India and China. In fact, at the place where the river enters India from Tibet, the river drops from the Tibetan Plateau from 3,350 m to 800 m in India, giving a head of nearly 2500 m (Rao, 1975). The minimum discharge in the river at this point of diversion is 1,000 cubic meters per second (cumecs). Corresponding to this there exists a possibility of diverting the run of the river from the plateau to join straight into the river again by a 20 km long tunnel. The valley characteristics are such that the river at the point of diversion is in China, and the site of the power house is in India. Therefore it requires the cooperative effort of both countries to generate energy from this colossal 'Power Store House' (Rao, 1975).

The river has been also in the news between Bangladesh and India. Bangladesh receives the inflows from two major rivers including Ganges and Brahmaputra, which finally pass through it before discharging their waters into the Bay of Bengal. For its development and to meet the demand for water, Bangladesh has been asking for an increased share of the Ganges' water from the Government of India. However, the requisite demand of diversion of water into Ganges cannot be met by India because of its own needs of water. Thus for the development of region surrounding the course of the Ganges River in Bangladesh, the issue of transferring a part of the Brahmaputra water into the Ganges water has been receiving attention in Bangladesh. Although this proposal requires a construction of nearly 320 km of canal, it has not been materialised owing to financial constraints. If this diversion comes into operation it can also safeguard a lot of land from the fury of floods in certain regions of Bangladesh. Thus the Brahmaputra River has received an international recognition and is considered to build a more friendly and stable relationship between India and Bangladesh.

## 1.6 Concluding Remarks

One of the objectives of this chapter has been to introduce different facets of this mighty river basin of India. Most of the statistics have been avoided in this chapter, since they will be a repetition of those given in different chapters. One of the reasons for the poor development of this basin has been the lack of information on different fronts. For the overall development of water resources it is essential to have a complete description of different components, which may encompass a wide spectrum of issues including geology, hydrology, climatology, environmental and hydraulic aspects, social, financial, and administrative issues, seismo-tectonics, soil types, agriculture, crops, irrigation, ground water potential, etc. Thus a large number of such components were identified to collect information. In some of the places information has not been up to date for the difficulties in procuring data. However, the new information generated from the present compilation is expected to serve the purpose of the development of the Brahmaputra Basin in a significant way despite the challenges it encounters against its development.

## References

Central Water Commission (CWC) (1996). Water and related statistics.

Chaturvedi, M.C. (1985). Water resources of India: an overview, Sadhana, 8(29), pt. I.

Chitale, M.A. (1999). "Water and environment." Special Issue of Dr. Kanwar Sain Memorial Lectures, Journal of Indian Water Resources Society, pp. 16--23.

Gowankar, V. (1999). "Evolution of meteorology as the water resources experts may like to see." Special Issue of Dr. Kanwar Sain Memorial Lectures, Journal of Indian Water Resources Society, pp. 12--15.

Hashim, S.R. (1999). "Water as a social and economic good." Special Issue of Dr. Kanwar Sain Memorial Lectures, Journal of Indian Water Resources Society, pp. 69--74.

Hussain, M. (1994). Encyclopaedia of India: Assam, RIMA Publishing House, New Delhi.

India (2001). A reference manual, compiled and edited by Research, Reference and Training Department, Publication Division, Ministry of Information and Broadcasting, Government of India, New Delhi.

Mehta, M.C. (1999). "The role of judiciary and citizens in environment protection and sustainable development." Special Issue of Dr. Kanwar Sain Memorial Lectures, Journal of Indian Water Resources Society, pp. 75--83.

Menon, P.G. (1999). "Water planning for the country : Application of management sciences." Special Issue of Dr. Kanwar Sain Memorial Lectures, Journal of Indian Water Resources Society, pp. 62--68.

Prasad, Y. (2000). "An overview on hydropower", Indian Journal of Power and River Valley Development, 50 (2), 181--186.

Shah, R.B. (1999). "Water resources – Myths and Realities." Special Issue of Dr. Kanwar Sain Memorial Lectures, Journal of Indian Water Resources Society, pp. 50--61.

Singh, B. (1997). "Water resource development in India – a perspective." Proc. Int. Symp. On 'Emerging trends in hydrology', edited by D.C. Singhal, N.K. Goel, D.K. Srivastava, R. Singh, H. Joshi and B.S. Mathur., Sept. 25-27, University of Roorkee, Roorkee, India, pp. 1--17.

Singh, B. (1999). "Development of water resources in Ganga Brahmaputra basin." Special Issue of Dr. Kanwar Sain Memorial Lectures, Journal of Indian Water Resources Society, pp. 41--49.
Thomas, K.C. (1999). "Water resources of India and its development prospects." Special Issue of Dr. Kanwar Sain Memorial Lectures, Journal of Indian Water Resources Society, pp. 1--4.
Verghese, B.G. (1999). "Himalaya-Ganga development and cooperation for a billion people." Special Issue of Dr. Kanwar Sain Memorial Lectures, Journal of Indian Water Resources Society, pp. 5--11.

# CHAPTER 2

# WEATHER AND CLIMATE

M. BARTHAKUR

The Brahmaputra valley enjoys an atypical climate, incomparable with any other part of India. This chapter presents a brief discussion of the valley's climate and the factors that control it. Also discussed are different seasons that the people of this valley enjoy.

## 2.1 Climate

Climate is not a singular phenomenon developed solitarily and independently at the confines of a particular area. Climate as such is a broad atmospheric phenomenon created by the combined influence of many factors. The Brahmaputra valley is a part of North-East India which, in turn, is an integral part of sub-tropical South-East Asia. Being a part of the sub-tropical belt, obviously its climate is akin to the South-East Asiatic monsoon. The local physical conditions, of course, modify the general characteristics of South-East monsoon to a certain extent. This is what we actually observe in the valley.

The physiographic configuration, enormous water bodies, absence of North-East monsoon and upper air circulation, etc., have moulded the climatic personality of the Brahmaputra valley. Structurally, the valley is like a winnowing fan, shut in the North, North-east and South by a high mountain rim and table land, but wide open to the South-west. The periodic South west trades (called monsoon streams) enter into the valley through two inlets, one from the Eastern part of the lower Gangetic valley and the other through Barak valley and hill gaps of the Borail ranges. The monsoon wind basically moulds the weather phenomenon. However, to understand the weather and climate of the Brahmaputra valley one must have a general knowledge of the physiographic make up of the valley. In fact, it is a ramp valley bordered by lofty hills of Arunachal on the north, the Patkai range of the Eastern Himalayas on the East and abutted by the Karbi and Meghalaya plateau on the south. The area, in general, forms an integral part of the south Asiatic monsoon land, but its peculiar orographic configuration plays a dominant role in causing deviations from normal weather conditions. According to W. Kopper, this region as a whole, excluding the high mountain barriers, enjoys 'Cwg' or humid mesothermal Gangetic type of climate.



*V.P. Singh et al. (eds.), The Brahmaputra Basin Water Resources*, 17-23.

## 2.2 Climatic Controls

The major climatic controls of the Brahmaputra valley are: (a) orography; (b) presence of alternating pressure calls of North west and Bay of Bengal; (c) predominance of tropical maritime humid air masses; (d) the roving periodic and occasional western disturbances; and (e) the local mountain and valley winds. The minor factors or controls are: (i) sub-tropical location; and (ii) occasional development of local depression, reduction of thermal difference by the extensive forest etc., deviate from the normal Gangetic type of humid mesothermal climate.

### 2.2.1 OROGRAPHIC CONTROL

The Eastern Himalayan high lands on the north, their syntaxial bend and parallel ranges in the east always push the southwest moist air up the slope and result in rapid moist adiabatic cooling that causes condensation and rainfall. This is an annual phenomenon particularly during April to September. Being obstructed, the air unable to cross the high Himalayan range is divided into two branches, one branch penetrates into the syntaxial parallel ranges in the north and the other branch to the south west that causes the precipitating weather in part of Arunachal, Nagaland and Manipur. Besides the innumerable rivers, streams and marshy lands in the valley supply enormous evaporated water even during the dry period from October to Februray. The lofty mountain rim in the east always repels the north east trades from the South China Sea and South Pacific Ocean. The incoming air mass whirls down the Irrawadi and Salween valley without any influence in the valley.

### 2.2.2 PRESSURE CELL CONTROL

During the period of summer from March to August low pressure develops in western India as well as over the Bay of Bengal. These cells are the dominating factors so far as incursion of the moist tropical air mass is concerned. The Bay of Bengal cell has its influence zone all over the Brahmaputra valley and its surroundings. Causally, a weak low also forms in the upper air layers, causing strong attractions to the tropical moist air masses. Similarly, with the shift of the thermal zone to the south, northwest India as well as southern Bengal develop high pressure cells and come under the continental cool stable air masses from Tibet or central Siberia, causing severe cold over the northern Gangetic plains, including the mountain and sub-mountain periphery of the northern Brahmaputra valley. However, the southern tongue of this high pressure cell does not extend beyond the hills, for which the severity of winter is much less than in the Gangetic plains in the west.

### 2.2.3 PREDOMINANCE OF MT AIR MASSES

The entire India is dominated by maritime tropical air masses for the whole year, except for a small spell during December and January. The maritime tropical unstable air masses become active with the intensification of the low pressure cells and move on a southeast-northwest trajectory that creates weather throughout the Gangetic plains and the northeast India.

## 2.2.4 LOCAL MOUNTAIN AND VALLEY WINDS

Owing to the peculiar fan shaped valley structure and development of local weak low pressure cells over the plains, the dry mountain winds slide down the southern slope of northern high lands. While blowing over the innumerable rivers, streams and marshy lands, enormous moisture is picked up which creates a dense fog on the southern part of the valley. This is a special feature of the valley weather in winter.

The meteorologist refers to the extension of the easterly jet stream and upper air westerlies to the extreme northeast Assam. The jet streams often pull down the western disturbances to the north of the valley and develop longer rainy days during the later part of the winter season. Owing to prolonged sunny weather during January and February or early March, rapid evaporation takes place over the entire valley. Rising temperatures often help the growth of local pressure centers in the valley. The supply of moisture and rising temperature stimulate growth of thunderstorms and hail storms, followed by heavy showers. This is another feature of the weather of the Brahmaputra valley. Still another feature of valley weather is the complete absence of the northeast monsoon.

## 2.2.5 LOCAL MOUNTAIN AND VALLEY WINDS

Local mountains and winds exert profound effects on the weather of the Brahmaputra valley, especially during the post winter season. A wide thermal gap develops between the valley and the surrounding high lands. The diurnal thermal fluctuation becomes more prominent owing to sunny days followed by continuous evaporation from the valley grounds, which were left saturated with the monsoon water. During the day time the ground is warmed and the warm air moves up towards the high land and at night the process is reversed. The reversal of circulation in the valley is responsible in formation of the morning fog. The mighty Brahmaputra River and its tributaries supply moisture which, in contact with the cool ground surface, helps formation of dense fogs, especially in the morning hours. At times the fog rises up and reduces the visibility to a considerable extent. Under the impact of the northern winds, the fog spreads over the karbi and Meghalaya plateau also.

## 2.2.6 LOCAL DEPRESSION

The temperature in the valley gradually rises from 13 °C in late January to 28 °C in April. The rising temperature, coupled with upper air low, develops surface depression and stormy weather follows. In the case in which the western depression reaches the valley, it develops prolonged stormy and cloudy weather in the basin. The presence of enormous evaporated water augments the growth of thermodynamic local storms. In the aftermath of these storms, the valley experiences cold spells. The severity of the cold winter is felt when the southern tongue of the Tibetan high pressure cell extends over the northern high land of the valley. These are some of the distinct departure of the normal tropical monsoon regime.

## 2.2.7 SEASONAL WEATHER

The climate of the valley may be divided into four primary seasons: (a) winter; (b) pre-monsoon; (c) monsoon or rainy; and (d) retreating monsoon. Locally the year is divided into six seasons: (a) summer; (b) rainy; (c) autumn; (d) hemanta; (e) winter; and (f) spring. Each of the seasons is constituted of two months, each exhibits different weather and changes in the natural environment.

### *2.2.7.1 Winter Season*

The winter season begins in late November and continues up to January. January is the coldest month and records the lowest temperature, as shown in Table 2.1. Cool gentle northerly and northeasterly winds blow during January and February. The cool, fair and pleasant weather is interrupted by driving showers, associated with western disturbances which lower the temperature and bring cold spells. Ground fog is very common during winter on the southern bank of the Brahmaputra. The northerly or north easterly winds drive the fog towards the southern plateaus of Karbi hills and Meghalaya. On the northern bank, especially in the north Lakhimpur district, fog is found only occasionally. The cool mountain winds always keep the weather clean and sunny. On an average the foggy days in the south bank vary between 60 to 100 days. Rainfall is lowest in this season and December in the driest month. Occasional driving showers, bright sun, the cool gentle northerly or month easterly winds, the moist fog, winter dews, etc., constitute the winter, delightful and pleasant weather of the valley.

### *2.2.7.2 Pre-monsoon Season*

Pre-monsoon is a transitional season between dry winter and wet monsoon. Pre-monsoon is a comparatively short period. The duration of the season varies from year to year. At times the pre-monsoon becomes difficult to distinguish from monsoon or dry winter season especially in matters of rainfall. The rain starts from middle of March and continues up to August. Usually the month of March and April, constitute the pre-monsoon season. From the month of March the temperature gradually rises with higher diurnal range. In the upper ranges the average diurnal temperature varies between 19.3 °C to 20.6 °C whereas it is slightly higher in the lower valley (between 21.3 °C and 20.11 °C). However, with the march of the sun north toward of the equator, the rising temperature forms local lows and casual touch to the weather. Such rains occur in association with western disturbances and may lead to thundery weather. However, in late April the western disturbances totally disappear. During this period hail storms are not uncommon. In general the weather remains partially rainless for most part of the season and westerly winds dominate. The winds become strong in the early hours, particularly in the western part of the valley that causes dusty weather. In Guwahati and its suburbs, dust storms are very common in the month of March. The pre-monsoon season is not a very pleasant season. The rapid changes of the atmospheric temperature make the climate trying and uncomfortable.

### *2.2.7.3 Monsoon*

Monsoon is the busiest season of the Brahmaputra valley. The agricultural activities begin and the people become very active. The season actually starts in the later half of the month of May and continues till September. This is the longest season, during which the rising temperature is brought down considerably by intermittent rains. Rainy

Table 2.1 Temperature, humidity and rainfall at different locations in the Brahmaputra valley.

| SEASONS | | WINTER | | PRE-MONSOON | | RAINY MONSOON | | | | | RETREATING MONSOON | | WINTER | |
|---|---|---|---|---|---|---|---|---|---|---|---|---|---|---|
| Major Statio-ns | Climatic Variables | Jan | Feb | Mar | April | May | June | July | Aug. | Sept. | Oct. | Nov. | Dec. | Av. Annual |
| Dibrugarh | Temperature in °C | 13.33 | 16.00 | 19.83 | 22.22 | 24.83 | 26.28 | 26.67 | 27.00 | 26.28 | 24.33 | 19.83 | 14.22 | |
| | Rainfall in centimeter. | 3.78 | 6.20 | 16.51 | 24.10 | 30.71 | 49.96 | 53.64 | 45.11 | 35.18 | 15.21 | 3.35 | 1.60 | 285.35 |
| | Humidity in percentage | 94 | 88 | 79 | 82 | 86 | 90 | 90 | 90 | 89 | 95 | 95 | 94 | |
| Sibsagar | Temperature in °C | 13.17 | 15.67 | 19.39 | 22.22 | 25.11 | 27.17 | 27.56 | 27.59 | 26.83 | 24.28 | 19.17 | 14.11 | |
| | Rainfall in centimeter. | 3.07 | 5.13 | 11.18 | 25.70 | 30.63 | 38.42 | 45.69 | 41.58 | 30.05 | 13.13 | 3.10 | 1.47 | 249.16 |
| | Humidity in percentage | 96 | 91 | 84 | 84 | 87 | 88 | 90 | 89 | 90 | 91 | 92 | 97 | |
| Tezpur | Temperature in °C | 14.00 | 16.33 | 20.61 | 23.17 | 25.22 | 26.94 | 27.33 | 27.50 | 27.44 | 24.72 | 20.22 | 15.17 | |
| | Rainfall in centimeter. | 1.47 | 2.74 | 4.88 | 15.27 | 27.05 | 30.84 | 34.77 | 33.07 | 20.98 | 10.41 | 2.29 | 0.64 | 184.41 |
| | Humidity in percentage | 91 | 84 | 74 | 77 | 84 | 87 | 89 | 90 | 89 | 85 | 84 | 91 | |
| Guwahati | Temperature in °C | 14.17 | 16.44 | 20.83 | 24.00 | 25.83 | 27.44 | 28.11 | 28.11 | 27.56 | 25.06 | 20.00 | 15.28 | |
| | Rainfall in centimeter. | 0.97 | 2.97 | 5.05 | 14.50 | 23.60 | 31.24 | 31.19 | 26.06 | 16.74 | 7.06 | 1.40 | 0.41 | 161.19 |
| | Humidity in percentage | 91 | 82 | 73 | 76 | 82 | 85 | 86 | 86 | 85 | 86 | 89 | 92 | |
| Dhubri | Temperature in °C | 14.56 | 16.67 | 21.39 | 24.39 | 25.44 | 26.50 | 27.50 | 27.56 | 27.00 | 25.11 | 20.79 | 16.27 | |
| | Rainfall in centimeter. | 0.97 | 1.88 | 4.22 | 13.77 | 40.11 | 61.23 | 43.66 | 33.76 | 35.61 | 12.45 | 1.02 | 0.18 | 284.68 |
| | Humidity in percentage | 87 | 79 | 69 | 74 | 85 | 82 | 88 | 88 | 88 | 85 | 82 | 86 | |

weather, sometimes, continues for several days, and June and July are the rainiest months. Rainfall is mostly associated with storms. These two months usually record the highest amount of rainfall. The humidity in the air increases beyond 90% and the temperature remains as high as 27 °C or more, as shown in Table 2.1. High humidity and high temperature make the weather sultry oppressive and sweltering. Monsoon is the most trying season at the same time, it is the busiest season of the valley. The melting snow over the eastern Himalayan region, coupled with enormous rain water, incapacitates the river channels and swelling waters spread over the flood plains causing devastating floods. The floods usually occur in two or three waves, leaving behind irreparable destruction to man, animal and the crop. The aftermath of floods leaves a trail of misery, epidemic, and poverty in rural areas.

*2.2.7.4 Retreating Monsoon*

Towards the end of September the south west monsoon withdraws from the Bay of Bengal and monsoonal disturbances in the valley abruptly retreat. With this retreat of monsoon wind the temperature falls and morning mist and fogs appear. Rainfall is negligible, the weather progressively clears up, and fair sunny weather prevails, weak cool north wind blows giving a pleasant winter touch. The bright sunny weather with a tendency to raise atmospheric heat in the afternoon and rapid fall of temperature at night, makes the weather the most endurable and pleasant period of the year. This is the hilarious season when most of the social merry-making, festivals are held. In fact, this is the festive season of the valley.

## 2.3 Concluding Remarks

The Brahmaputra valley or the state of Assam including the adjoining regions, such as Arunachal Pradesh, Nagaland, Meghalya, Mizoram, Tripura and Manipur, have a typical climatic personality, incomparable with any other part of India. The area in general forms an integral part of the south east Asiatic monsoon land, but its peculiar, high land guarded orography plays a dominant role in creating local weather phenomenon and climatic individuality. According to Mr. W. Koppen, this region, excluding the high mountain barriers, falls under 'Cwg' or typical Cwg type. In the valley about 60% of the rainfall is associated with storms, which is not an usual phenomenon in the Ganga valley. Besides, the mountain and valley winds act like land sea breeze and minimize the thermal ranges. Fog is another local weather character rare in the Ganga plain. Therefore the climate of Brahmaputra valley can be distinguished from 'Cwg' in many respects. The climate of the valley may be called Humid Mesothermal Brahmaputra type or 'Cwb' type. The summer season of the rest of the country is not at all prominent in the valley.

## References

Barthakur, M. (1968). Some Aspects of Weather in the Brahmaputra valley. Assam Science Society Vol. XI.
Barthakur. M. (1970). Climate of Assam, Geography of Assam. H.P. Das. N.B.T.I New Delhi.
Barthakur. M. (1978). Climate of Assam – Extension Lecture Session –2 N.E.I.G.S.
Chatterjee. S.B. (1965). The Trends of Indian Weather and Climate: Past and Present. International Quartenary Congress. U.S.A.

Mukherjee, A.K. (1969). Meterologist–Barjhar Observatory, Gauhati, Notes on Assam's weather (unpublished)

Ray. T.K. Occurrence of an unusual hailstrom over Brahmauptra Valley. Vayu Mandal – I.M.S. Vol.I, No. 4. New Delhi.

Sarkar . M.B. and Gaikwad. S.D. (1980). Continental effect over India and neighbourhood. Vayu mandal. I.M.S. Vol. 10, No. 324.

Menon. P.A. (1979). North East Monsoon.Vayu Mandal, I.M.S. Vol 9. No.324, New Delhi.

# CHAPTER 3

# HYDROMETEOROLOGY

B. PURKAIT

The Brahmaputra valley predominantly receives southwest tropical monsoons during April through October and occasional winter showers. The monsoon rains account for 90% of the annual rainfall with an average between 2500 and 3200 mm. The extremely dynamic regime coupled with the unique physiographic setting of the Brahmaputra basin with its orographic position in the upper catchment areas contributes to frequent flooding in the valley. This chapter provides a short discussion of hydrometeorological aspects of the basin.

### 3.1 Introduction

Lying to the north of the tropic of Cancer and bounded by high hill ranges in the north, east and south, the Brahmaputra valley experiences the predominant influence of the southwest tropical monsoon which prevails from April to October with occasional winter showers. The approach of the monsoon is marked by strong winds, overcast skies and occasional thundershowers, hailstorms and cyclones during April and May and it starts its full play of heavy showers beginning with June.

Past environments and climate to certain extent, as also the past geomorphic history, depended on the movements involved with the emergence of the Himalayas. The environment in the Himalayas varied from a shallow marine in early Tertiary times to that of subaerial subject to orogenic forces. Mountain building activity continued into Pleistocene also (Krishnan, 1968). Bengal basin, which is to the south and southwest of the Brahmaputra basin, was a scene of marine transgression in post Tertiary times (Sengupta, 1966). The southern part of the Brahmaputra had been a scene of intense constructional activity with the penecontemporaneous structural deformation.

### 3.2 Atmospheric Variables

The Brahmaputra basin, as shown in Figure 3.1, experiences a typical tropical humid climate. Three distinct seasons, such as (i) winter, (ii) spring and (iii) monsoon (summer), prevail in the area. The rainy season starts from May and continues up to October, but actually the rainfall is distributed from March itself. Although the main monsoon rains begin in May, the pre-monsoon showers start in mid-April often accompanied by hailstorms. The mean annual monsoon rainfall varies from 145 cm to



*V.P. Singh et al. (eds.), The Brahmaputra Basin Water Resources*, 24-34.

188 cm, while the average monthly rainfall recorded at Tezpur is 30.84 cm, 34.77 cm, 33.07 cm and 20.98 cm in the months of June, July, August and September, respectively. The rainfall in the basin increases from south to north and also from west to east. However, the maximum precipitation takes place during June and July. The

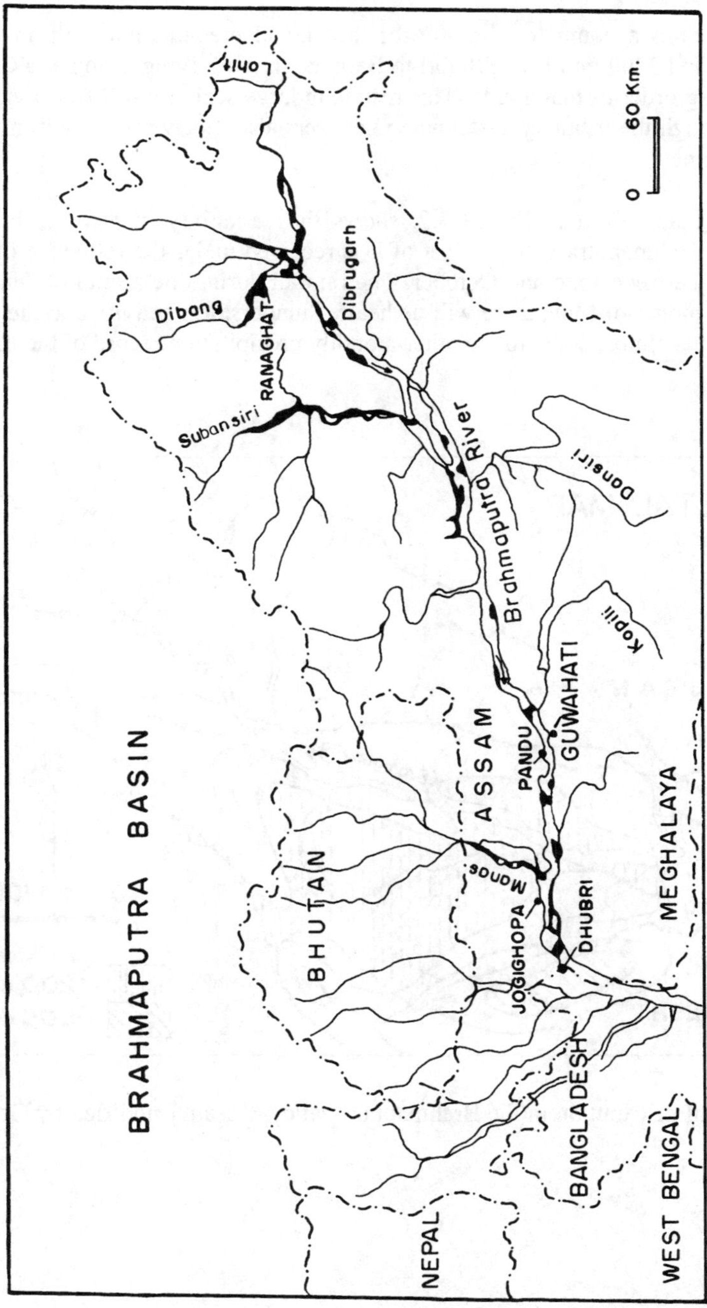


Fig. 3.1 The Brahmaputra basin.

minimum temperature comes down to 9° C during December while the temperature shoots up to about 37° C in the peak summer month of July. Humidity is quite high through the year, but more so in the wet season.

### 3.3 Precipitation Characteristics

The monsoon rains account for the 90% of the annual precipitation with an average between 2500 and 3200 mm from Dhubri to Barpeta, through Bongaigaon and Goalpara in the increasing order of magnitude. This results in large scale run-off by way of flood discharges through the tributary systems. Some rivers also receive snowmelt in varying degrees in summer.

The isohyetal map, given in Figure 3.2, shows the variability in the distribution of rainfall of the Brahmaputra valley. Most of it is received under the influence of south-west monsoon between June and October. The rainfall during the month of June is the highest. In the month of May, there will be heavy thunderstorm activity and the amount of precipitation is almost equal to the other monthly precipitation record of the monsoon months.

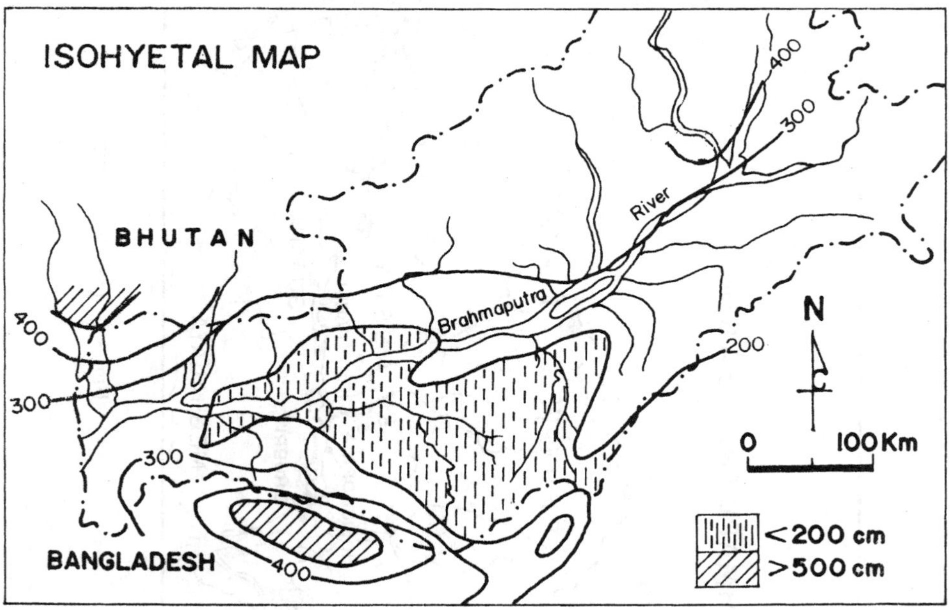


Fig. 3.2 Rainfall distribution in the Brahmaputra valley, Assam (modified by Goswami, 1985).

Owing to orographic effects, the rainfall in the sub-Himalayan zone is usually higher than that over the plains. The plot of Figure 3.3, owing to Hoffer (1998), reveals that the two graphs of precipitation are not parallel. In the years 1954, 1958, 1959 --1967 and 1978, they even follow an opposite trend. The precipitation patterns in the mountains from and in the plains often differ each other and perhaps are not even caused by the same climatological events. The annually flooded area in West Bengal is well correlated with the precipitation of the plain itself. Widespread floods usually coincide with high monsoonal rainfall over the plain (1956, 1959, 1968, 1971 and 1978). This indicates that at least some of the floods in West Bengal are 'home made', that is, they are mainly caused by the precipitation on the plain itself and not by the rains in the hills and mountains. This, again, is a very interesting example that demonstrates that the hydrological input from the Himalayas is not a decisive factor for producing floods further down stream and that more emphasis ought to be given to the regional rainfall input over the flood-affected areas themselves.

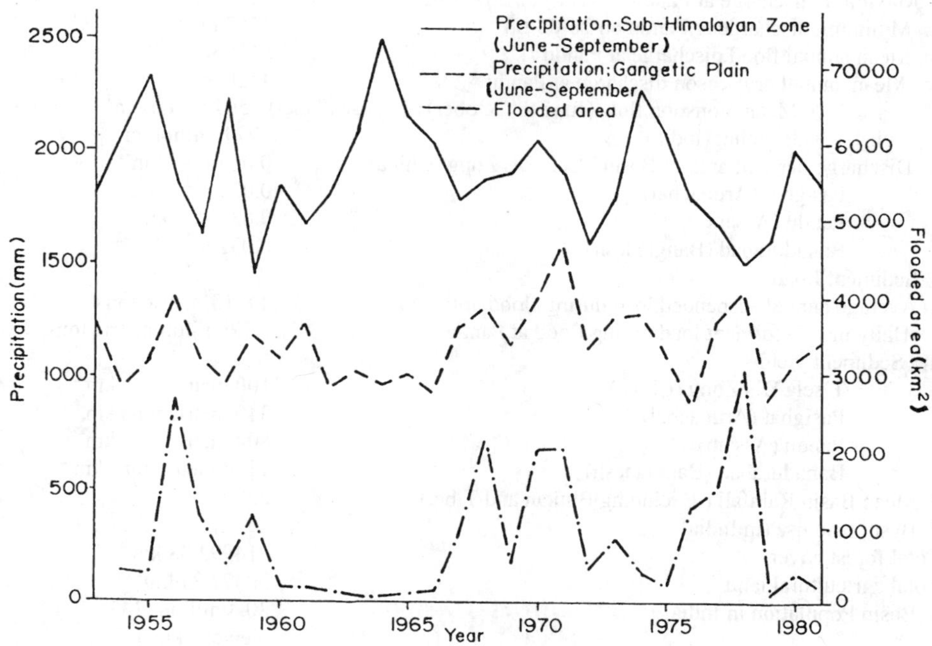


Fig. 3.3 Monsoonal precipitation in the sub-Himalayan zone and Ganga plain (West Bengal) compared with the flood-affected area in West Bengal from 1953—1981 (Source: Chaphekar and Mhatre, 1985; Parthsarthy et al., 1987; Hoffer, 1998).

Table 3.1 Salient features of the Brahmaputra basin.

| | |
|---|---|
| 1 Total catchment area | 580000 $km^2$ |
| i. Catchment area within China | 293000 $km^2$ |
| ii. Catchment area within India | 195000 $km^2$ |
| iii. Catchment area within Bhutan | 45000 $km^2$ |
| iv. Catchment area within Bangladesh | 47000 $km^2$ |
| 2 Length from its source to confluence with the Ganga | 2880 km |
| i. Length within Tibet (China) | 1625 km |
| ii. Length within India | 918 km |
| iii. Length within Bangladesh upto the confluence with the Ganga | 337 km |
| 3 Gradient | |
| i. Reach within Tibet | 0.00260 |
| ii. Reach between Indo-China border and Kobo in India | 0.00190 |
| iii. Reach between Kobo and Dhubri | 0.00014 |
| iv. Reach within Bangladesh First 60 km from India Border | 0.00009 |
| Next 106 km reach | 0.00008 |
| Next 92 km reach | 0.00004 |
| Next 79 km reach | 0.00003 |
| 4 Discharge characteristics | |
| i. Maximum discharge at Pandu (Assam) on 23-08-62 | 72794 $m^3$ $s^{-1}$ |
| ii. Minimum discharge at Pandu on 20-02-68 | 1757 $m^3$ $s^{-1}$ |
| iii. Mean annual flood discharge at Pandu | 51156 $m^3$ $s^{-1}$ |
| iv. Mean annual dry season discharge at Pandu | 4420 $m^3$ $s^{-1}$ |
| v. Mean monsoon flow (June to October) Shigatse (Tibet) | 507 million $m^3$ |
| vi. Pasighat (India) | 3979 million $m^3$ |
| v. Discharge per unit area of Basin T'sela D's Zong (China) | 0.01 $m^3$ $s^{-1}$ $km^{-2}$ |
| Pasighat (Arunachal) | 0.023 $m^3s^{-1}$ $km^{-2}$ |
| Pandu (Assam) | 0.03 $m^3s^{-1}$ $km^{-2}$ |
| Bahadurabad (Bangladesh) | 0.032 $m^3s^{-1}km^{-2}$ |
| 5 Sediment Load | |
| i. Average annual suspended load during flood at Pandu | $4 \times 10^8$metric tons |
| ii. Daily mean sediment load during flood at Pandu | 2.12million metric tons |
| iii. Sediment yield | |
| T`sela D`s Zong (China) | 100 metric tons $km^{-2}$ |
| Pasighat (Arunachal) | 340 metric tons $km^{-2}$ |
| Pandu (Assam) | 804 metric tons $km^{-2}$ |
| Bahadurabad (Bangladesh) | 1128 metric tons $km^{-2}$ |
| 6 Mean Basin Rainfall ( Excluding Bhutan and Tibet) | 230 cm |
| 7 Basin land use (in India) | |
| Total forest cover | 114992.08 $km^2$ |
| Total agricultural land | 50473.84 $km^2$ |
| 8 Basin Population in India | 30.4million (143 persons/$km^2$) |

The main reason for high flood is that the rivers in northern and eastern regions originate from the Himalayan ranges and have snow fed catchments and lie in high rainfall zone. It is observed that out of the total water resources of 1674 million cubic meters of India, 872 million cubic meters is contributed by the Ganga and the Brahmaputra River systems alone. Of this the Brahmaputra basin contributes 23% of the total water resources of the country. The Brahmaputra basin in India receives an annual rainfall of about 230 cm. The salient features of the Brahmaputra basin are shown in Table 3.1 (Goswami-1998).

In the Brahmaputra catchment area of heavy precipitation and easily erodible rocks it is imperative to check the erosion of the hill slopes, and thus the accumulation of sand and silt in the main river. The rate of infiltration into the ground water reservoir depends on the intensity and duration of the storm precipitation. The alluvial zone and the piedmont plains have a heavy intake capacity. It is observed that the 'Older Alluvium' has got a lower infiltration rate compared to the 'Newer Alluvium.' The permeability of the former is also to be correspondingly less, perhaps owing to a greater degree of cementation or induration. Consequently these contrasting parameters will induce sectional imbalance in groundwater movement both vertically and horizontally

### 3.4 Precipitation Formation

A number of rivers, streams, and a good number of natural lakes or ponds, locally known as bils are dynamic and static sources of surface water. They receive supplies from the large annual rainfall, directly or indirectly. Some of the important tributary rivers of the Brahmaputra from east to west are Manas, Champamati, Gaurang and Sankosh. The Gangadhar is the only important river near the border of Bangladesh in the west. All these rivers have headwaters in the Bhutan Himalayas. The Manas River receives the snowmelt in summer, even though the major flood discharges originate from the rainfall in the beginning of the southwest monsoon with peak discharge during July to September. All these rivers mentioned above pass through alluvial plains except the Champamati which drains the inselberg zone. All these rivers are perennial in nature. As such the precipitation formation seeping underground within the thick alluvial downslow sections is quite high and adequate to provide a sizeable base flow for all these rivers. From the groundwater studies it is also confirmed that all the run-off systems are affluent round the year though during peak monsoon, some of these rivers like Manas, Beki, etc become temporaily influent.

To the east and west of the Barpeta area, a group of bils occupy shallow depressions which may possibly indicate the interwedged back swamps of the Manas-Beki River system and also of the main Brahmaputra River. The local precipitation and small perennial channels help sustain these bils all through the year. Some bils are cut-off oxbows of the older channels and are sustained by groundwater flow as well as by the small rivers/streams emerging from the peripheral area. In the inselberg zone, some large bils occur in 5 to 10 m depressions in the Chapar Formation. Besides the receiving of much surface runoff from the stream of the inselberg zone these bils also have a direct contact with the water table. Thus both the surface and the subsurface flows have an equilibrium outlet through streams feeding the main Brahmaputra river. During high floods these outlets also act as channels of back flooding. The progressive reduction of these bils through reclamation and conversion to paddy lands in the proximity of Manas, Beki and Brahmaputra causes marginal transpirational loss of waters which ultimately may upset the long term balance of the runoff systems.

### 3.5 Storms and their Movements

The Brahmaputra is a veritable source of disaster in the entire north-east India. The valley is long and narrow being only about 80 km wide on an average. The broadest part of the valley is where the river divides the districts of Sibsagar and Lakimpur. The rainfall in the valley ranges from 175 cm in Kamrup to about 640 cm in the northeast frontier regions. Some parts of Nowgong district in the north of the Khasi and the Jaiyanti hills lie in a shadow, the annual rainfall being 110 cm. The frequent changes of the course of the Brahmaputra and the braiding of channels associated with the erosion of river banks become a major problem of the area during floods. The occurrence of cyclones also aggravates the situation. The cyclonic storms are accompanied by heavy rainfall and high speed winds. During the monsoon months (June to September) the Brahmaputra and its tributaries attain high discharge with frequent inundations. The flow regime of the river responds to the seasonal rhythm of monsoons and the freeze-thaw cycle of Himalayan snow. Discharge per unit area of the basin is very high (Table 3.1) and is comparable to that of any other large rivers of the world.

The Brahmaputra basin represents a powerful monsoon rainfall regime under wet humid conditions, receiving an annual rainfall of about 230 cm. There is a marked variability in the distribution of precipitation over the catchment (Figure 3.2). For example, the Jiadhol catchment on the northeast receives 410 cm of rainfall while the Kopili catchment in the south central part receives only 175 cm. Rainfall in the Himalayan sector amounts to 500 cm per year with the lower ranges receiving more than the higher areas. For example, the annual rainfall at Dibrugarh in the eastern part of the valley is 285 cm, whereas at Pasighat, located in the foothill region, it is 507 cm, and at Tuting, located further up in the Himalaya, it is 274 cm. Monsoon rains from June to September account for 60--70% of the annual rainfall. These rains which contribute a large portion of the runoff in the Brahmaputra and its tributaries are primarily controlled by the position of a belt of depression called the monsoon trough extending from the northeast of India to the head of the Bay of Bengal. In the course of its north--south oscillations in summer, when this axis moves to the foothills of the Himalayas heavy precipitation is caused in Assam and adjoining highlands. The pre-monsoon season covering March, April, and May produces 20--25 % of annual rainfall caused primarily by depressions moving from the west and by local convectional storms. Precipitation in the Meghalayan Plateau to the south of the valley is of the order of 1,000 cm per year.

### 3.6 Hydrologic Characteristics

The gradient of the Brahmaputra is 4.3 to 16.8 m/km in the gorge section upstream of Pasighat, and 0.1 m/km near Guwahati (Goswami, 1985). Details of the gradient of the Brahmapputra basin at different reaches are shown in Table 3.1.

Soils derived from the Tertiary sandstones of the lower Himalayan zone are friable in nature. They consist mainly of sands with admixtures of cobbles and boulders having a lower threshold of erosion. The Brahmaputra valley is made up mainly of alluvial soils. These are called 'Newer Alluvium' and form the present day river deposits. A few isolated pockets of Pleistocene deposits of 'Older Alluvium' occur within the valley as well as along the foothill region whereas laterites and red loams are found in the hilly areas to the south of the Brahmaputra.

The hydrographs of annual flow and sediment discharge of the Brahmaputra river reflect a repetitive patterns of rise and fall of flow and sediment load in the river corresponding to the seasonal variation of the monsoonal precipitation and the freeze--thaw cycle of the Himalayan snow (Figure 3.4). The daily discharge hydrographs at two sections (Jogighopa and Pandu) indicate the discharge fluctuations and different waves of floods during the monsoon period (Figure 3.5). The average discharge at the mouth of the Brahmaputra is the fourth ($19.83 \times 10^3$ $m^3$ $s^{-1}$) as compared to other large rivers of the world e.g. Amazon ($99.15 \times 10^3$ $m^3$ $s^{-1}$), Congo ($39.66 \times 10^3$ $m^3$ $s^{-1}$), Yangtze ($21.80 \times 10^3$ $m^3$ $s^{-1}$), Hwang Ho ($19.83 \times 10^3$ $m^3$ $s^{-1}$) (Goswami, 1998).

From an analysis of the frequency distributions of annual floods, it was observed that the mean annual flood has a recurrence interval of 2.1 years, whereas in the case of the highest ever recorded flood of 72,748 $m^3$ $s^{-1}$, it is 100 years. The most devastating floods were recorded in 1954, 1962, 1966, 1972, 1973, 1974, 1977, 1978, 1983, 1984, 1986 and 1988 during a span from the years 1952 to 1996.

The extremely dynamic monsoon regime vis-à-vis the unique physiographic setting of the basin with its orographic position at the upper catchment areas attribute to the frequent occurrence of flood in this region. The water yield of the basin is one of the highest among the large rivers of the world. High rates of yield together with the limited width of the valley and the greatly flattened gradient lead to tremendous drainage congestion and the resultant flooding. Because of the heavy silting, the bed levels of the Brahmaputra and some of its tributaries have risen considerably, reducing the height of banks and the carrying capacity of the channels. As a result the water of the main river and its tributaries submerge the banks during summer high flows and inundate the surrounding lowlands. The rank of sediment yield of the Brahmaputra is the second highest (1,128 tons $km^{-2}$ $yr^{-1}$ at Bahadurabad, Banladesh and 804 tons $km^{-2}$ $yr^{-1}$ at Pandu, Assam) in the world after the Yellow River (1,403 tons $km^{-2}$ $yr^{-1}$ ) (Goswami, 1998).

Encroachment of the large number of wetlands that serve as natural reservoirs such as the bils, swamps and marshes in the flood plain zone has also reduced the retention capacity of the drainage system causing the flood levels to rise.

The flood problems of the Brahmaputra valley may be viewed from different angles such as (a) the problem of inundation of riverine areas owing to submersion of the banks by the main river and its tributaries, (b) the problem of drainage congestion, especially near the outfall of the tributaries during high stages of rivers, and (c) the problem of bank erosion and channel instability during spilling of flood waters over the banks due to excessive supply of sediment load hindering its carrying capacity.

Strengthening of flood forecasting and warning system through application of sophisticated technology like satellite remote sensing, computer based mathematical watershed modeling and telemetric methods of real-time data collection and transmission etc. are essential for meeting the challenge posed by flood hazard. Flood

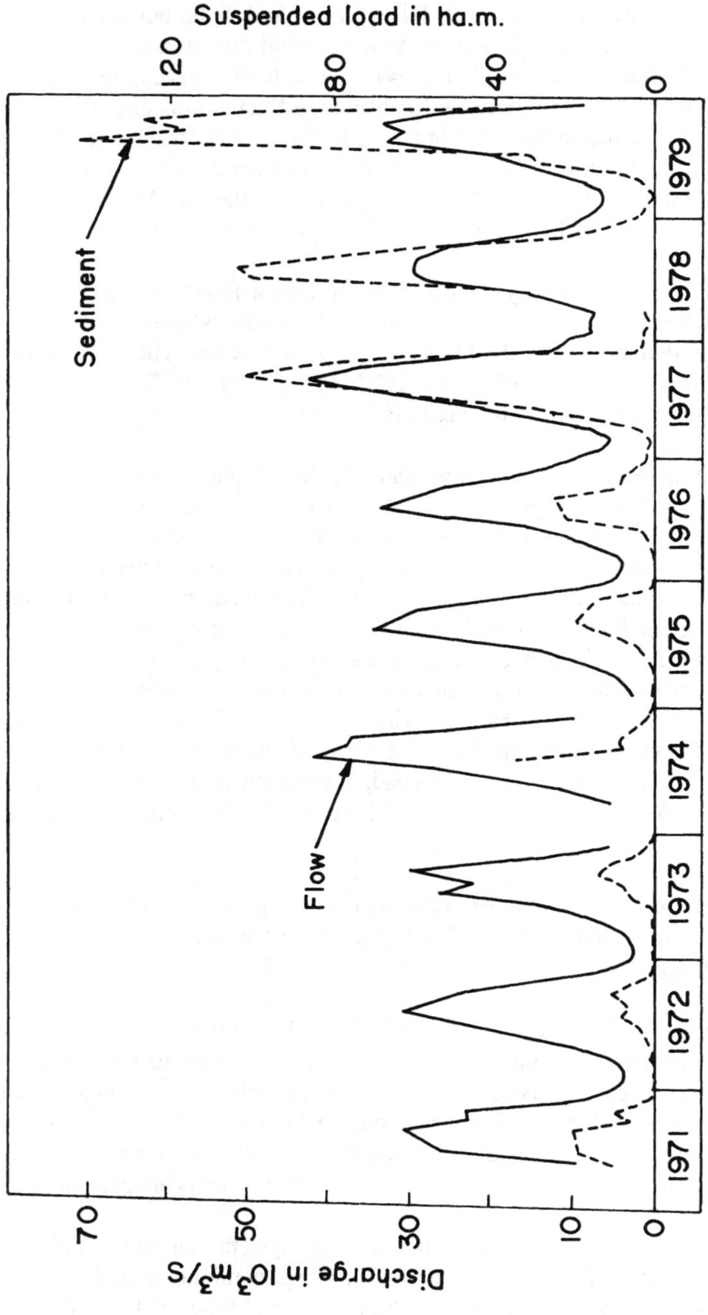


Fig. 3.4 Annual flow and sediment discharge hydrographs of the Brahmaputra River (Source: Goswami, 1985).

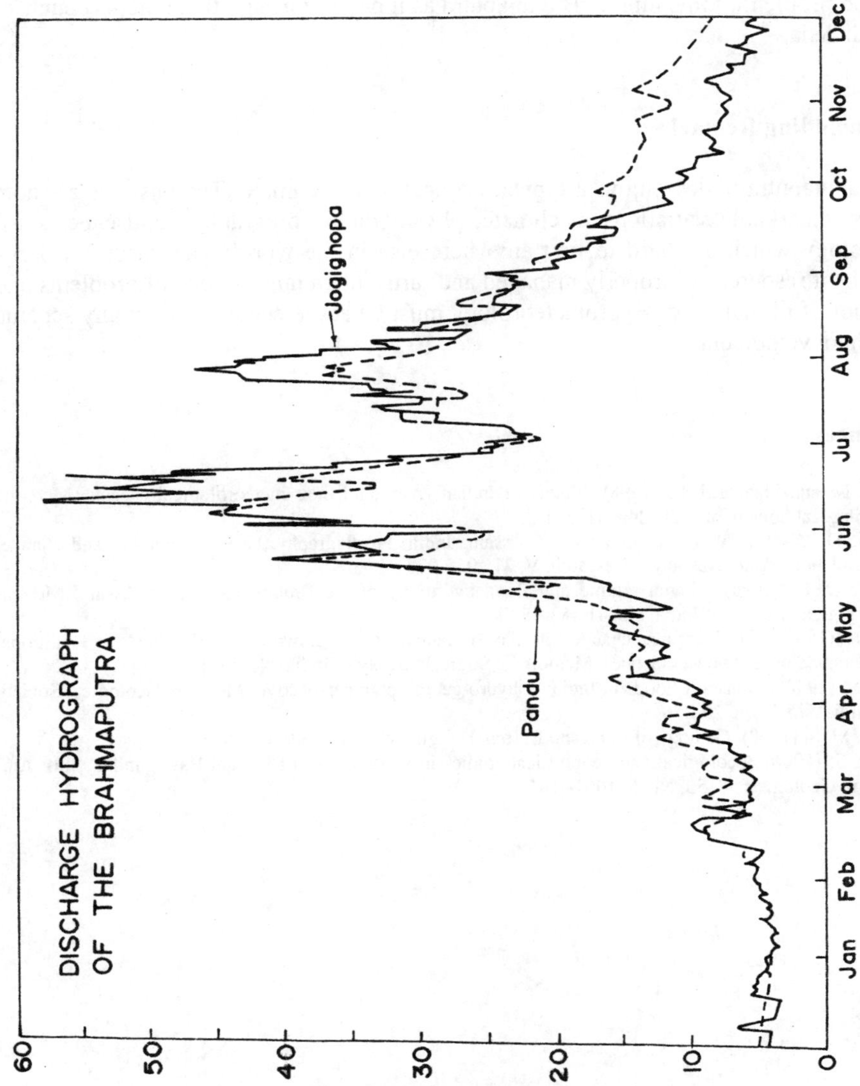


Fig. 3.5 Daily discharge of the Brahmaputra River at Pandu and Jogighopa.

plain zoning based on remotely sensed data using modern techniques such as the GIS provide an effective tool for management of floods. The reliable and up-to-date hydrological data base is required to overcome the formidable effect of flood hazards. Adequate regional as well as international cooperation and concern is required to tackle the flood problem of this mighty Brahmaputra as it passes through three major countries of South Asia.

## 3.7 Concluding Remarks

The Brahmaputra valley enjoys a typical tropical humid climate. The basin as a whole exhibits remarkable variations in climate, physiography, orography, landscape, soils, and geology which are hard to find anywhere else in the world. The basin is a large reservoir of resources if properly managed and turns into a huge source of problems and frustration if ill-managed. Hydrometeorology must be at the center-stage in any scheme for basin development.

## References

Dhar, O. N. and Nandargi, S. (1998). Floods in Indian rivers and their meteorological aspects. Memoirs Geological Society of India. No. 41. p. 1-25

Goswami, D. C. (1985a). Brahmaputra river, Assam, India: Physiography, basin denudation and channel aggradation. Water Resources Research, V. 21, .959-978

Goswami, D. C. (1998). Fluvial regime and flood hydrology of the Brahmaputra river, Assam. Memoir Geological Society of India. No. 41. 53-75

Hoffer, T. (1998). Do land use changes in the Himalayas affect downstream flooding? - Traditional understanding and new evidences. Memoir Geological Society of India. No. 41, p.119-141

Kale, V. S. (1998). Monsoon floods in India: A hydro-geomorphic perspective. Memoirs Geological Society of India. No.41, p.229-256

Krishnan, M. S. (1968). Geology of India and Burma. Hogginbothams, Madras, p. 604.

Sengupta, S. (1966). Geological and geophysical studies in western part of Bengal Basin, India. Bull. Am. Ass. Petrol. Geologists. V. 50, No. 5. 1001-1017.

# CHAPTER 4

# SOILS AND AGRICULTURE

N. C. TALUKDAR, D. BHATTACHARYYA AND S. HAZARIKA

The soils of the Brahmaputra River basin are similar to the soils found in the Indo--Gangetic plains. These soils are classified according to the soil taxonomy and represent widely divergent soil groups. Climate has been the most dominant factor in formation and development of soils of the basin. More specifically, the rainfall and temperature have exercised the most conspicuous influence. Introducing the characteristics of these soils, the land classification is described. Then their influence on agricultural productivity is treated.

## 4.1 Introduction

The Brahmaputra valley, with an area of about 56,000 km$^2$, constitutes more than 50% of the geographical area of Assam. The valley extends about 700 km from east to west with varying width ranging from 65 to 100 km. It is one of the greatest of the world's river systems. The entire valley area is interspersed with small hillocks and swampy low land where flood recurs annually. In fact, occurrence of *beels*, oxbow lakes and huge marshy tracks is very frequent in this valley. Innumerable *chars* and *chaparis* are also found in the valley amongst which Majuli (929 km$^2$ area) is reported to be the world's largest river island. Soils of the valley are of varied types and their characteristics reflect the influence of both the parent material (geology) and the peculiar climate and vegetation of the region. Land use pattern, crop productivity and crop response to nutrient management in the Brahmaputra basin are greatly controlled by its landforms and soil types and climate of the valley.

The whole Brahmaputra basin is broadly delineated into the four macroclimatic zones: (a) North Bank Plain (NBPZ); (b) Upper Brahmaputra Valley (UBVZ); (c) Central Brahmaputra Valley (CBVZ) and (d) Lower Brahmaputra Valley (LBVZ). Each of these zones has been further divided into different belts. The climatic characteristics of these zones are distinct, as shown in Table 4.1.

## 4.2 Geology of the Brahmaputra Basin

Brahmaputra basin was formed during the Pleistocene and in recent periods from the sediments derived from the Assam plateau in the south and the Assam Himalayas in the north and brought down by the Brahmaputra River and its tributaries (Wadia, 1975).



*V.P. Singh et al. (eds.), The Brahmaputra Basin Water Resources*, 35-71.

The core of the Assam Plateau is composed largely of gneisses and granites of Archean age. However, the Assam Himalayas have quite different stratigraphical and structural constitution containing fossiliferous records of successive marine geographical system representing all ages from Cambrian to early Tertiary. The formation of the basin is comparatively younger than the Assam plateau and consists mainly of gneisses, schists and sand stones (Wadia, 1975). Thus the alluvium of the Brahmaputra basin derived from the Assam plateau and Assam Himalayas manifest wide diversity. The geological nature of the alluvium brought down by the Brahmaputra river is quite different from that deposited by its 35 tributaries flowing down from the foothills located on both sides resulting in different types of alluvium.

Table 4.1 Climatic characteristics of the four zones of the Brahmaputra basin.

| **Zone** | **Average rainfall (mm)** | **Humidity (%)** | **Temperature (°C)** | |
|---|---|---|---|---|
| | | | Maximum July-(August) | Minimum. (January) |
| NBPZ | 1,000 | 80 | 37 | 5 |
| UBVZ | >2,000 | >80 | 37 | 5 |
| CBVZ* | 1,600 | <80 | 38 | 8 |
| LBVZ | 1,700 | 80 | 31 | 10 |

* include 30% of area which receive less than 600 mm rainfall.

### 4.3 Climate

The climate of the Brahmaputra basin is humid sub-tropical characterized by high rainfall and high humidity. It is mainly influenced by both the south west monsoon from the Bay of Bengal and the surrounding hills of the Assam Himalayas, lower Himalayas, and the Assam plateau. The climate is characterized by hot and wet summer and dry and cool winter and are divided into four distinct seasons viz., pre-monsoon, monsoon, post monsoon and winter.

The pre-monsoon (March to May) weather is characterized by a gradual rise in the temperature, vanishing of fog, occasional thunder showers, cool morning and hot afternoon winds which occasionally become strong causing dust storms. The rainfall during the period varies from 20% to 30% of the total precipitation. The monsoon season (June--August) is characterized by cloudy weather, high humidity and weak variable surface winds. Temperature and rainfall increase with advancement of the season. The number of rainy days varies between 18--20 days in a month. August is the warmest month. The total precipitation varies between 60%--70% of the total annual rainfall. The post monsoon season (September--November) is characterized by fair weather with the fall in temperature and rainfall. The number of rainy days reduced to about three by the month of November. The winter (December--February) is characterized by the fall in temperature and rainfall with occasional cool breeze from north and morning fog.

Climatologically, the Brahmaputra basin can be divided into three broad regions viz., eastern, western and middle. Rainfall is the most dominating component of climate of the Brahmaputra basin. The basin enjoys a fairly good amount of rainfall during five months from May to September. The eastern region is characterized by the high rainfall with mean annual rainfall (MAR) of 2,452 mm followed by the western and middle region with MAR of 1,982 and 1,527 mm, respectively (Bhattacharyya and Sidhu, 1984). The representative mean monthly climatic data is presented in the Figure 4.1. The average minimum temperature of the valley is about 10 °C that attain in December/January and the maximum is about 32°C which falls in July/August. The mean annual temperature (MAT) of eastern, western, and middle regions are 23.5°C, 24.5°C and 24°C, respectively, with temperatures more than 20°C for 9 months in a year in the entire basin. During rainy and summer seasons the water supply exceeds the water need. The soil, therefore, remains moist for 7 to 9 months and qualifies for udic moisture regime. In this period the low lying areas remain saturated and ground water levels remain high. Such localized wet areas qualify for aquic moisture regime. The relative humidity (RH) on an average exceeds 80% for the entire basin. Even in dry months of the winter season the average RH does not fall below 75%.

## 4.4 Soils of the Brahmaputra Basin and their Characteristics

The alluvium derived soils of the Brahmaputra valley have been grouped into new and old alluvial and forest soils according to the conventional system of soil classification. These ill-defined groups include widely divergent soils which are placed together simply because of the common mode of formation of parent material or their existing land use (Chakravarty *et al.,* 1981). However, at present these soils are classified according to the Soil Taxonomy (Soil Survey Staff, 1975). The major groups of alluvium-derived soils are Entisols, Inceptisols and Alfisols (Chakravarty *et al.*, 1981; Karmakar, 1985; Bhattacharyya, 1991; Dutta and Karmakar, 1995; Chamuah *et al.*, 1996; Das *et al.*, 1997: Dey and Sehgal, 1997a; Bhattacharyya *et al.*, 1998) which are similar to soils as observed in the Indo Gangatic Plains (Chakravarty *et al.,* 1981). The representative soil profiles of these soil orders are shown in Figure 4.2. As stated earlier, climate has the most dominant role in formation and development of soils of the Brahmaputra basin. Of the climatic components the rainfall and temperature have the most conspicuous influence (Chakravarty *et al.,* 1980; Karmakar, 1985). Topography of the land form modify the influence of climate on these soils. The duration to which soil forming processes are in action also give rise to variations in the profile characteristics of the upland, old flood plain and recent flood plain soils. The upland soils are more developed due to the origin of the alluvium in pleistocene age, whereas other soils are less developed due to their origin on younger alluvium.

### 4.4.1 MORPHOLOGICAL CHARACTERISTICS

Morphological properties of some representative soils of Brahmaputra basin are given in Figure 4.3.

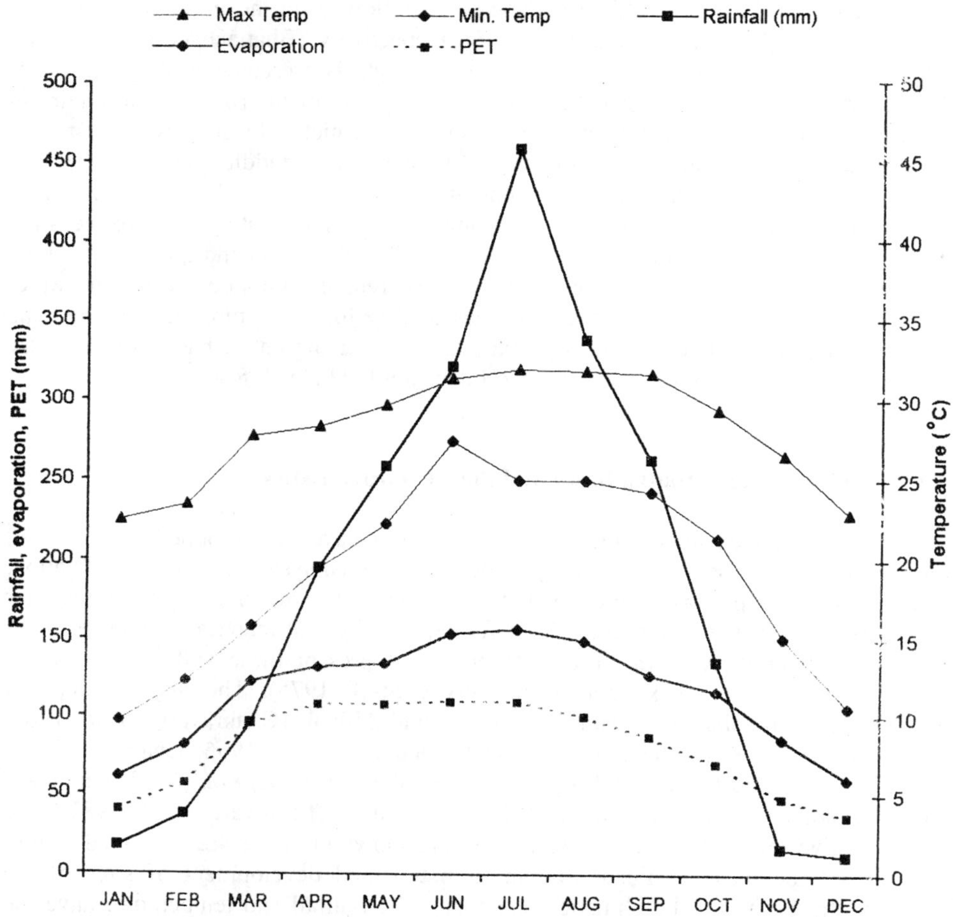


Fig. 4.1 Mean monthly rainfall, evaporation, potential evapotranspiration (PET) and temperature (maximum and minimum) at Jorhat, Assam (1974--1996). (Source: Annual Report 1996--97, AICRP on Agrometeorology, AAU, Jorjhat).

Fig. 4.2 Soil profiles representing major soil orders of the Brahmputrra basin. A Entisol (Fluvaquent), B. Inceptisol (Haplaquept), and C. Alfisol (Peleudolf).

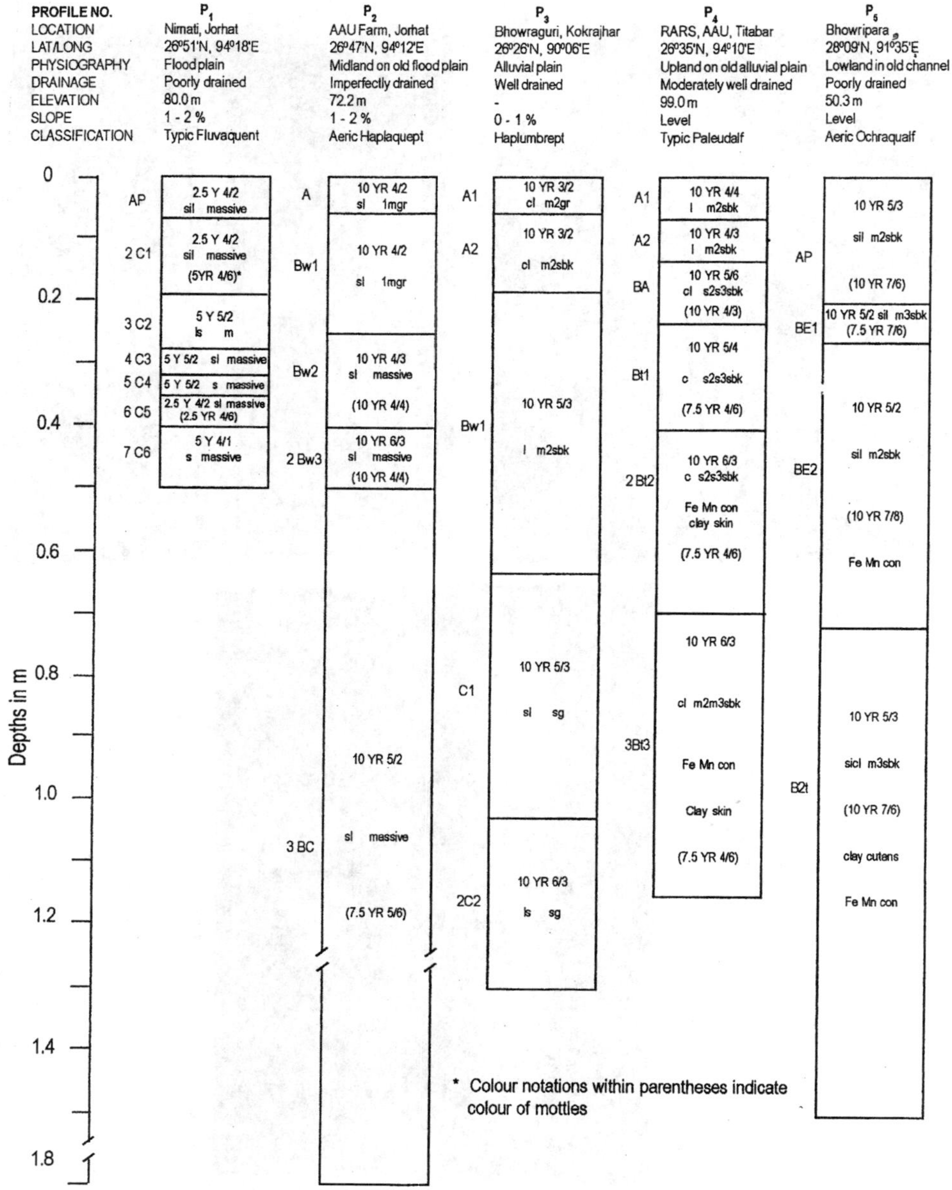


Fig. 4.3 Site characteristics and morphological features of the representative soil profiles of the Brahamaputra basin.

*4.4.1.1 Colour*

The dominant hue of the colour of the soils of the Brahmaputra basin is 10 YR. The change of grayness of the soils of the eastern part of the basin to yellowish brown along with reddish mottles in the western part of the basin is a characteristic feature of the soil colour of Brahmaputra basin (Chakravarty *et al.*, 1978). Generally, soils of uplands are dark brown to yellowish brown with values ranging from 4 to 6 and chroma 3 to 8. These colours of the soils indicate their better drainage conditions (Chakravarty *et al.* 1978; Bhattacharyya and Siddhu, 1984; Dutta and Karmakar, 1995; Chamuah *et al.*, 1996). Some of these soils also contain mottles with high chroma at variable depths of the profile. The fluctuations in the depth of water tables in the profile causes the variation in the depth of mottles as the soils remain saturated temporally during rainy season and subsequently oxidized during dry periods.

The recent and old flood plain soils, the channel soils, and low lying soils of upland have characteristic gleyed colours associated with wetness (Chakravarty *et al.*, 1978; Bhattacharyya and Siddhu, 1984; Chamuah *et al.*, 1996; Das *et al.*, 1997; Dey and Sehgal, 1997a; Bhattacharyya *et al.*, 1998). The poorly drained soils have dark gray to pale brown colour with chromas of 1 and 2 within a depth of about 75 cm from the soil surface. These soils reduce conditions during the rainy season leading to development of low chroma to qualify them for aquic moisture regime. Some of these soils are also subjected to water logging due to their low lying positions and remain under flood every year. The chemical processes associated with reduction and mobilization of iron and manganese (gleying) under saturated conditions responsible for lowering of chroma of soil colour and their subsequent oxidation and precipitation under oxidized conditions results in higher chroma of soil colour.

The paddy growing soils of the basin also develop characteristics of grayish matrix colour (Chroma $<2$) throughout the profiles (Dey and Shegal, 1997; Gangopadhya et al., 1998). Colours of surface horizons of these soils are dark gray to gray (10 YR, 6/1 to 5 YR, 5/1). Distinct mottles with colours varying from red/yellowish red to strong brown (5 YR/25YR, 5/8 to 7.5 YR, 5/6) are present in these soils throughout the profiles. The uniform greyish matrix colour of most of the paddy soils of Brahmaputra basin indicates a strong influence of characteristic hydrology of paddy cultivation on the soil colour of the Brahmaputra basin (Dey and Sehgal, 1997).

*4.4.1.2 Fe--Mn and $CaCO_3$ concretions*

Iron--manganese (Fe--Mn) concretions/nodules are formed in the well developed soils of the eastern and western part of the basin (Chakravarty, 1977). The Fe--Mn concretions are irregular to sub-rounded in shape and very dark gray to brown (7.5 YR, 5/0 and 7.5 YR, 5/2) in colour. Some soils of the middle region, mostly soils of Nowgong district of the basin contains fine to coarse, irregular, white (5 YR, 8/1) $CaCO_3$ concretions in the subsurface horizons of the soils profile. The reduction of Fe and Mn and their subsequent movement and precipitation resulted in the formation of ferruginous concretions, while the $CaCO_3$ nodules are of primary origin (Chakravarty *et al.*, 1982).

*4.4.1.3 Structure and Texture*

The structure of recent and old flood plain soils is weak and granular to sub-angular blocky in surface horizons and massive to sub-angular blocky in sub-surface horizons. The structure of upland soils is moderate to strong, sub-angular to angular blocky at lower horizons. The soils develop moderate to strong blocky structures due to alternate wet and dry seasons prevailing in the basin along with considerable illuviation of clay in subsurface horizons. Structural development of the soils of the Brahmaputra valley also have relation to the stage of their pedogenic development (Chakravarty, 1977; Das *et al.*, 1997).

The soils of the Brahmaputra valley have wide textural variations ranging from sand to clay (Chakravarty *et al.*, 1978; Bhattacharyya and Sidhu, 1984; Dutta and Karmakar, 1995; Das *et al.*, 1997; Dey and Sehgal, 1997a; Gangopadhyay *et al.*, 1998). The flood plain soils are predominantly silty loam to sand. The upland soils are loam to clay loam at the surface horizons and clay loam to clay at the lower horizons. The alluvium near the Brahmaputra River is newer and stratified. In these flood plain soils, coarser materials are deposited first followed by deposition of finer materials at the surface. The soils towards Assam plateau and Assam Himalayas are older in age and finer in texture even in the lower horizons due to their being at more advanced stage of weathering and considerable translocation of clay in lower depths.

The morphological properties of the soils show that the thickness of the solum and degree of profile development of the soils increase as one moves from the recent flood plains on the bank of the Brahmaputra River towards the old flood plains and then to the upland soils (Chakravarty *et al.*, 1978; Chamuah *et al.*, 1996).

4.4.2 PHYSICO-CHEMICAL CHARACTERISTICS

Soils of the Brahmaputra basin have great diversity in their physical and chemical properties. Some important physicochemical properties of few representative soil profiles of this basin is presented in Table 4.2.

*4.4.2.1 Sand, Silt and Clay Content*

Flood plain soils of the Brahmaputra basin contain a higher percentage of sand at lower depths (Chakravarty *et al.*, 1978; Bhattacharyya and Siddhu, 1984; Karmakar, 1985; Bhattacharyya *et al.*, 1998). Sand content of upland soils is less than 28 per cent at the surface and it decreases further at lower depths. The fine and very fine fractions of sand together constitute the major portion of the total sand. The difference in sand content and the proportion of fine to very fine sand fractions also confirms the general observations of the soils of the Brahmaputra basin recorded from the morphological features. Soils near the Brahmaputra River are coarser in texture and those away from the river are finer in texture.

The silt content of the soils of the Brahmaputra basin ranges from 30--75 per cent. The depth-wise distribution of silt fraction in the profiles of flood plain soils of this basin is irregular. However, a gradual decrease of silt in upland soils is observed down the profiles. In most of the soils, the content of medium silt exceeds that of coarse and fine silt (Chakravarty *et al.*, 1978). Lithological discontinuity is observed between two

Table 4.2. Physico-chemical characteristics of some soils of Brahmaputra basin

| Depth | Sand | Silt | Clay | OM | PH | $Fe_2O_3$ | $Al_2O_3$ | $CaCO_3$ | Ex. Cataions (cmol ($P^+$) $kg^{-1}$ | | | | CEC (cmol ($P^+$) $kg^{-1}$ | B.S |
|---|---|---|---|---|---|---|---|---|---|---|---|---|---|---|
| (cm) | | % | | | | | % | | $CA^{2+}$ | $Mg^{2+}$ | $K^{2+}$ | $Na^+$ | | |
| **P1 Typic fluvaquent** | | | | | | | | | | | | | | |
| 0-7 | 27.7 | 67.1 | 5.2 | 1.3 | 8.0 | 0.71 | 1.45 | 1.85 | 2.5 | 1.3 | 0.4 | 0.2 | 6.7 | 65.0 |
| 7-19 | 40.5 | 54.9 | 4.6 | 1.0 | 8.0 | 0.57 | 1.70 | 0.98 | 2.3 | 1.4 | 0.3 | 0.1 | 6.7 | 60.5 |
| 19-28 | 84.0 | 30.2 | 2.8 | 0.1 | 8.2 | 0.39 | 1.79 | 0.94 | 1.2 | 1.4 | 0.1 | 0.1 | 2.9 | 62.1 |
| 28-31 | 67.3 | 29.5 | 5.0 | 0.2 | 8.1 | 0.86 | 1.80 | 0.68 | 1.8 | 0.7 | 0.2 | 0.3 | 4.8 | 62.0 |
| 31-35 | 88.7 | 9.5 | 1.8 | 0.1 | 8.2 | 0.57 | 1.68 | 0.25 | 1.7 | 0.8 | 0.2 | 0.3 | 3.8 | 78.9 |
| 35-40 | 48.6 | 45.6 | 5.8 | 0.4 | 8.1 | 0.67 | 2.08 | 1.23 | 2.0 | 0.9 | 0.2 | 0.1 | 5.8 | 55.3 |
| 40-48 | 92.2 | 4.2 | 3.6 | 0.2 | 8.3 | 0.39 | 1.86 | 0.08 | 1.3 | 0.5 | 0.1 | 0.1 | 2.9 | 69.0 |
| **P2 Aeric Haplaquept** | | | | | | | | | | | | | | |
| 0-6 | 55.1 | 33.8 | 9.2 | 1.0 | 4.4 | 0.90 | 0.40 | ND | 1.5 | 0.8 | 0.2 | 0.1 | 7.1 | 34.5 |
| 6-25 | 53.1 | 34.9 | 9.7 | 0.9 | 4.7 | 0.70 | 0.30 | ND | 1.2 | 0.8 | 0.1 | 0.1 | 6.1 | 36.0 |
| 25-40 | 68.1 | 19.8 | 9.9 | 0.6 | 4.7 | 0.70 | 0.60 | ND | 0.7 | 1.1 | 0.1 | - | 5.0 | 38.8 |
| 40-50 | 69.0 | 20.3 | 8.1 | 0.6 | 5.1 | 0.30 | 0.80 | ND | 1.1 | 1.2 | - | - | 5.5 | 41.8 |
| 50-180 | 70.7 | 18.4 | 9.2 | 0.3 | 5.4 | 1.20 | 0.90 | ND | 0.3 | 1.1 | - | - | 3.6 | 41.0 |
| **P3 Umbric Dystrochrept** | | | | | | | | | | | | | | |
| 0-6 | 38.2 | 32.8 | 29.0 | 5.28 | 4.1 | 1.61 | 0.85 | ND | 4.0 | 3.4 | 0.5 | 0.10 | 22.5 | 35.6 |
| 6-18 | 39.7 | 33.1 | 27.2 | 4.55 | 4.0 | 1.22 | 0.87 | ND | 3.7 | 2.8 | 0.2 | 0.06 | 20.3 | 33.2 |
| 18-63 | 35.2 | 42.3 | 22.4 | 0.83 | 4.0 | 1.34 | 0.72 | ND | 3.2 | 2.7 | 0.1 | 0.05 | 10.5 | 56.8 |
| 63-103 | 79.3 | 7.2 | 13.5 | 0.18 | 4.3 | 0.77 | 0.37 | ND | 1.2 | 0.6 | 0.2 | 0.06 | 6.0 | 34.5 |
| 103-130 | 85.2 | 6.9 | 7.9 | 0.05 | 4.7 | 0.66 | 0.38 | ND | 1.6 | 1.6 | 0.1 | 0.04 | 5.1 | 65.8 |
| **P4 Typic Paleudalf** | | | | | | | | | | | | | | |
| 0-7 | 38.6 | 39.6 | 21.8 | 1.5 | 5.8 | 1.21 | 2.50 | ND | 2.6 | 1.8 | 0.1 | 0.2 | 8.6 | 55.2 |
| 7-13 | 39.6 | 41.2 | 19.2 | 1.5 | 5.4 | 1.2 | 1.6 | ND | 2.6 | 1.4 | 0.1 | 0.3 | 9.6 | 46.0 |
| 13-23 | 38.0 | 39.4 | 22.6 | 1.2 | 5.6 | 1.5 | 2.0 | ND | 2.1 | 3.1 | 0.2 | 0.3 | 11.5 | 49.6 |
| 23-40 | 35.0 | 37.6 | 27.4 | 1.0 | 5.7 | 1.7 | 2.8 | ND | 2.4 | 3.5 | 0.1 | 0.3 | 14.4 | 44.4 |
| 40-115 | 34.0 | 30.4 | 35.6 | 0.6 | 5.8 | 2.1 | 2.1 | ND | 2.5 | 5.6 | 0.2 | 0.4 | 18.3 | 48.0 |
| **P5 Aeric Occhraqualf** | | | | | | | | | | | | | | |
| 0-20 | 5.9 | 65.8 | 26.3 | 1.3 | 4.8 | 1.7 | 0.3 | ND | 2.3 | 1.5 | 0.1 | 0.1 | 7.9 | 51.0 |
| 20-26 | 4.0 | 67.5 | 24.8 | 0.8 | 5.9 | 2.1 | 0.8 | ND | 2.7 | 2.3 | 0.1 | 0.6 | 8.5 | 77.6 |
| 26-72 | 7.6 | 68.2 | 20.7 | 0.3 | 6.3 | 2.2 | 0.8 | ND | 3.1 | 1.6 | 0.2 | 0.5 | 7.3 | 84.0 |
| 72-150 | 5.2 | 60.6 | 30.5 | 0.1 | 6.2 | 2.8 | 1.4 | ND | 3.5 | 1.5 | 0.2 | 0.7 | 7.1 | 83.9 |

adjacent horizons of flood plain soils of this basin as the differences of sand/silt ratios between these horizons are 0.2 or more (Chakravarty *et al.*, 1978; Bhattacharyya and Sidhu, 1984). Lithological discontinuity indicates polygenic evolution of the flood plain soils owing to periodic fluviatile action that keeps the alluvium stratified.

The soils developed on uplands, old flood plain, lower horizons of some recent flood plain and upper horizon of old channel soils contain higher amount of clay (21-58%). Clay increases with depth in upland and old flood plain soils. Moreover, there is an increase in the ratios of fine clay to coarse clay (1.00--3.47) and fine clay to total clay (0.50--0.77) in the lower horizons of the profiles indicating the presence of argellic horizons (Chakravarty *et al.,* 1978). A significant negative correlation ($r = -0.64$) between silt and clay contents on the upland profiles and old flood plain profiles indicates that clay is formed by the weathering of silt. The movement of clay to lower depths in profiles of the upland soils of the Brahmaputra valley depends on the rainfall pattern of the basin. The clay content increases from a depth of 28 to 49 cm in soils of middle, western and eastern parts of the basin with the increase in rainfall. Higher MAT of the western region increases the intensity of weathering of the soils which results in more amount of clay (28.45%) of the upland soils compared to that (23.30%) of the soils of the eastern region (Chakravarty *et al.,* 1978; Bhattacharyya and Sidhu, 1984).

*4.4.2.2 Bulk Density, Saturation Percentage and Moisture Holding Capacity*

The bulk density of the soils of the Brahmaputra basin ranges from 1.36 to 2.10 $gm/cm^3$ (Chakravarty, 1977). Relatively more developed upland soils of the basin have a higher bulk density than other soils. The bulk density of soils depends on their clay content. The saturation percentage of the soils varies from 25.7 to 54.7 per cent and the field capacity varies from 3.1 to 29.2 per cent. The field capacity of the soils is significantly correlated with the silt ($r = 0.6$) and clay content ($r=0.5$) of the soils. Therefore the silt and clay content of the soils of the Brahmaputra basin determine the moisture retention capacity of the soils (Chakravarty *et al.,* 1978).

*4.4.2.3 Organic Matter*

The organic matter content in the surface horizon of the soils of the Brahmputra basin varies from 0.60 to 2.9 per cent and its distribution decreases with depth (Chakravarty *et al.,* 1978; Bhattacharyya and Siddhu, 1984; Das *et al.,* 1997; Dey and Sehgal, 1997a; Bhattacharyya, *et al.*, 1998; Gangopadhyay *et al.*, 1998). Some recent flood plain soils near the Brahmaputra River have an irregular depthwise distribution of organic matter. This has been attributed to the youthful and stratified nature of these soils (Bhattacharyya and Siddhu, 1984). The organic matter content of the soils of Brahmputra basin increases with the increase in the amount of rainfall and decreases with the increase in the temperature of the basin (Chakravarty *et al.,* 1978).

*4.4.2.4 Free $CaCO_3$, $Fe_2O_3$ and $Al_2O_3$ Content*

Soils of the middle region of Brahmaputra basin contain free $CaCO_3$ and carbonate nodules at lower depths (below 139cm). The source of these carbonate is the lime rich argillaceous alluvium from the North Cachar and Karbi Anglong of Assam plateau (Chakravarty *et al.,* 1978; Das *et al.,* 1997). The flood plain soils contain carbonate throughout the profile indicating a lack of carbonate leaching and minimum profile development in these soils (Bhattacharyya and Siddhu, 1984). The calcium carbonate content of these soils is higher at the surface horizons which indicates recent deposition of Brahmaputra alluvium derived from calcium rich rocks. But $CaCO_3$ is completely washed out of the profiles of the well developed soils on other landforms of the Brahmaputra basin. Low pH, high rainfall, and moderately well drained to well drained conditions of these soils enhance dissolution and leaching of carbonates. Therefore the electrical conductivity (EC) of the Brahmaputra basin soils is low. Although the EC of the recent flood plain soils are high compared to other well-developed soils of uplands and old flood plains, the EC of these soils are also within the limits (< 0.8 $dsm^{-1}$) of normal soils. This eliminates the salinity problems in the soils of the Brahmaputra basin (Bhattacharyya and Siddhu, 1984).

The free iron oxides ($Fe_2O_3$) content of the soils of the Brahmaputra basin ranges from 0.57 to 5.28% and that of the aluminum oxides ($Al_2O_3$) ranges from 0.31 to 3.95 per cent. These constituents of soils have been found to increase gradually with age of the alluvium from flood plain soils to upland soils indicating that the rate of weathering is also increasing in the same manner (Chakravarty *et al.*, 1978; Bhattacharyya and Sidhu, 1984).

*4.4.2.5 pH*

The pH of the recent flood plain soils of the Brahmaputra basin ranges from 7.6 to 8.0 and that of the old channel soils, old flood plain soils and upland soils ranges from 5.5 to 8.4, 4.4 to 7.5 and 4.8 to 6.4, respectively (Chakravarty *et al.,* 1978; Bhattacharyya and Sidhu, 1984). The pH of these soils decreases with increase in the intensity of rainfall. The pH of the soils also decreased with the increase in age and the degree of pedogenesis of the soils from the river Brahmaputra towards the Assam range and Assam Himalayas. Moreover, soils with high ground water table have a higher pH at lower horizons, and also the pH of soils containing free $CaCO_3$ is higher.

*4.4.2.6 Cation Exchange Capacity (CEC)*

The CEC of the flood plain soils of the Brahmaputra basin varies from 3.0 to 12.0 cmol. (p+) $kg^{-1}$ of soil. The CEC of the upland soils of the basin are higher (11.0 -- 23.3 cmol (p+) $kg^{-1}$). The CEC of upland soils increases regularly with depth but the depth distribution of CEC in flood plain soils is irregular. In these soils highly significant and positive correlation is observed between clay and CEC (r=0.84). Organic carbon content of the soils is also positively correlation with CEC. But the contribution of clay content of the soils of the Brahmaputra basin to CEC of the soils is about 94 per cent (Chakravarty, 1977).

*4.4.2.7 Exchangeable Cations*

Exchangeable hydrogen ($H^+$) is the dominant cation in the exchange complex of the soils of Brahmaputra basin. Except in the soils of the recent flood plain, exchangeable H+ generally tends to decrease down the profile of other soils of the basin. The profile distribution pattern of exchangeable $H^+$ is similar to that of the organic matter distribution in these soils and these parameters are correlated significantly with r values of 0.70. However, the dominant basic cations of the exchange complex of the soils of Brahmaputra basin are calcium ($Ca^{2+}$) and magnesium ($Mg^{2+}$). Their distribution is generally irregular in flood plain soils of the basin. The content of $Mg^{2+}$ is generally higher in old alluvial soils of the basin.

*4.4.2.8 Percent Base Saturation (PBS)*

The percent base saturation (PBS) is an indication of the extent to which exchangeable basic cations occupy the active site of the exchange complex of the soils. In the soils of the Brahmaputra basin, PBS is mostly higher than 50 per cent. The PBS of the upland and old flood plain soils increases regularly down the profiles. But the distribution of PBS in recent flood plain and old channel soils of the Brahmaputra valley is irregular. This is a reflection of extensive leaching of the basic cations not having occurred. Alternately, under alternate wet and dry seasons, the minerals are weathered to release bases that contribute towards the replenishment of the bases. As the soils are subjected to leaching during rainy season there is movement of bases down the profiles of the soils. The rate of accumulation of bases also depends on the heaviness (texture) of the soils and their positions in the landscape.

*4.4.2.9 Total Elemental Composition*

Higher rainfall and temperature increase base status and the Fe and Al oxides and decrease the silica/sesquioxide ratio of the soils (Chakravarty, 1977; Karmakar, 1985; Dey, 1999). In general the silica/sesquioxide ratio decreases with depth in most of the upland soils and this is indicative of more advanced stages of peodogenic development of these soils. As the soils of eastern region of the basin experience a higher rate of weathering (owing to its higher rainfall than in the middle region of the basin) the soils contain a higher amount of iron and aluminum and lower silica/sesquinoide ratios than the soils of the middle region. The soils of the western region contain the highest amount of iron and aluminium and the lowest amount of silica as well as silica/sesquioxide ratios indicating the highest rate of weathering of the soils of this part (Chakravarty, 1977). This may be attributed to the temperature effect which is the highest among different regions. The recent flood plain soils of the valley also contain a higher amount of Fe and Al as the alluvium contains a higher amount of feromagnesium minerals (Bhattacharyya and Sidhu, 1987).

4.4.3 MINERALOGY OF SOILS

*4.4.3.1 Sand Mineralogy*

The light sand fraction of the soils of the Brahmaputra basin constitute about 93.0--99.9 per cent of the total sand (Chakravarty, 1977). The important minerals present in this fraction of sand are the light sand minerals (sp. qr. < 2.57), quartz, mica ( muscovite ) and feldspars (Bhattacharyya an Sidhu, 1987; Chakravarty *et al.,* 1979; Chakravarty, 1977). Quartz is the most dominant sand mineral and its content varies from 57--88 per cent in these soils. The muscovite content in light fractions of sand varies from 25 to 29.2 per cent. In most of the soils of this basin muscovite has a general tendency to increase in lower depths of the profiles. This mineral is present in greater amounts in flood plain soils than in the upland soils. The muscovite flakes belong to $2M_1$ polymorph and were formed under the conditions of high temperature and pressure, i.e., under igneous and/or high grade metamorphic environments (Bhattacharyya and Sidhu, 1987).

The feldspars constituted 3.0 to 23.0 per cent of light sand fractions. The feldspar minerals present are mainly K-feldspars. The K-feldspars are more abundant in recent flood plain soils than in well developed upland soils of the basin. The amount of triclinic K-feldspar, i.e., microcline decreases from flood plain soils to upland soils whereas the amount of monoclinic K-feldspar increases from upland soils to recent flood plain soils of the Brahmaputra basin (Bhattacharyya and sidhu, 1987). The value of triclinicity of K-feldspar minerals of these soils is 0.95, irrespective of the differences in their pedogenesis.

The heavy minerals (sp.gr. > 2.5) in sand fractions of the soils ranges from 0.1 to 6.1 per cent (Chakravarty *et al.,* 1979). This group of sand minerals includes varieties of ferromagnesium, translucent and opaque minerals represented by zircon, biotite, chlorite, amphibole, tourmaline, pyroxene, rutile, garnet and kyanite in the decreasing order of abundance. The chlorite, biotite, amphibole and pyroxene combined together constitute the major proportion of heavy minerals of sand fractions of soils of the basin. Sand mineralogy of the soils of the Brahmaputra basin indicates that the percentage of resistant minerals, i.e., quartz, zircon and tourmaline, decreases and that of weatherable minerals, i.e., feldspars, micas and chlorite increases with increase in the depth of these soils

*4.4.3.2 Silt Mineralogy*

Silt fractions of the flood plain soils have quartz, feldspars, micas, amphiboles, calcite, chlorite and kaolinite. In the well developed upland soils of the basin silt fractions do not contain calcite and amphiboles. Moreover, comparatively lower amounts of calcite, micas and feldspar are present in silt fractions than in sand fractions of the upland soils of the Brahmaputra basin (Bhattachrayya and Sidhu, 1987; Chakravarty, 1977).

*4.4.3.3 Clay Mineralogy*

Clay mica (illite) and kaolinite are the major minerals present in clay fractions of the soils of the Brahmaputra basin (Karmakar, 1985; Bhattacharyya and Sidhu, 1987; Chakravarty *et al.*, 1992; Bhattacharyya, 1997; Dey and Sehgal, 1997b). Clay mica is the dominant mineral in clay fractions of flood plain soils (53-61%). The very sharp 10 Å peak of the clays of these soils indicates good crystallinity of clay illites (Bhattacharya and siddhu, 1987). They are mostly dioctahedral in nature. Random powder X-ray diffraction pattern of illite obtained after concentrating this mineral by dissolving other 14 Å components in 2M HCl indicates that illite in clay fractions of the soils of Brahmaputra basin is the same mineral that occurred in sand and silt fractions indicating their origin from the same parent rocks (Bhattachrayya and siddhu, 1984; Dey and Schgal, 1997).

Chlorite constitutes 8 to 10 per cent of the clay fractions of the flood plain soils of the basin. It is mostly absent in well developed upland soils. The 060 reflections of the X-ray difractogram indicate that chlorite of the Brahmaputra basin is trioctahedral in nature. Kaolinite is another dominant clay mineral occurs in soils of the Brahmaputra basin constituting 36--40 per cent of the clays. The occurrence of small amounts of goethite is also reported in clays of some of these soils (Bhattachryya, 1997). Other non-phyllosilicate minerals present in clays of the Brahmaputra basin are quartz and feldspars.

The characteristic hydrology associated with paddy cultivation altered the clay mineralogy of the paddy growing soils of the Brahmaputra basin (Dey and Sehgal,

1997). Generally, the mica content of paddy soils is lowered by 22 to 59 per cent than that of the associated non paddy soils. The 10 Å reflection of mica peaks of paddy soils is relatively more asymetrically broadened towards higher d-spacing. This indicates greater replacement of K by hydrated cations in paddy soils of the basin. The decrease in the mica content in paddy soils is accompanied by the pronounced increase in the content of kaolinite and to a lesser extent that of vermiculite, smectite and chlorite. In this submerged subtropical environment, depotassication of mica takes place and thus mica is transformed to smectite through a vermiculite intermediate phase in the presence of Ca and Mg rich environment. Under the poor drainage conditions of paddy growing soils smectite and vermiculite are more stable in paddy soils than the associated non-paddy soils (Dey and Sehgal, 1997).

*4.4.3.4 Soils Orders and their Distribution*

The major groups of the alluvium-derived soils of the Brahmaputra basin are entisol, inceptisol and alfisol (Karmakar, 1985; Chakravarty *et al.,* 1981; Bhattacharyya, 1991; Dey and Sehgal, 1997a). At presently the data on the distribution of soils under these orders in the Brahmaputra basin is not available. However, the situation for whole of the state of Assam indicates that majority of soils of Assam are inceptisols comprising of 45% of the soils followed by entisols and alfisols which represent 36.6 and 12.3 per cent of the soils of Assam, respectively (Table 4.3 and Figure 4.4).

Table 4.3 Distribution of soils under different orders and sub-orders.

| Sl. No. | Order / Sub order | Percent of soils | Per cent of TGA | Area × 1000 ha |
|---|---|---|---|---|
| 1 | Inceptisols | 45.0 | 41.4 | 3245.3 |
| | Aquepts | 12.5 | | |
| | Ochrepts | 32.5 | | |
| 2 | Entisols | 36.6 | 33.6 | 2640.1 |
| | Aquents | 29.2 | | |
| | Psaments | 0.8 | | |
| | Fluvents | 5.8 | | |
| | Orthents | 0.8 | | |
| 3 | Alfisols | 12.3 | 11.3 | 886.9 |
| | Udalfs | 12.3 | | |

Source: NBSS & LUP, regional office, Jorhat, Assam (1994)

*4.4.3.5 Land Capability Classification*

The land Capability Classification of soils (U.S.D.A.) is a system designed primarily for soil conservation and is intended for determining the maximum intensity of land use consistent with low erosion risks and sustained productivity. The basic USDA land capability classification allocates land suitable for cultivation to classes I to IV, followed by land suitable for grazing to classes V--VI and forestry to class VII, leaving class VIII for wild life and recreation. These classes may be divided into sub classes based on the number and severity of several limitations that downgrade the land from class I. These include erosion risk (e), wetness (w), rooting zone (soils) limitation (s) and climatic limitations (e). Depending on the severity of the limitations, such as erosion and wetness, land units 1, 2, 3 may also be used (Dent and Young, 1981).

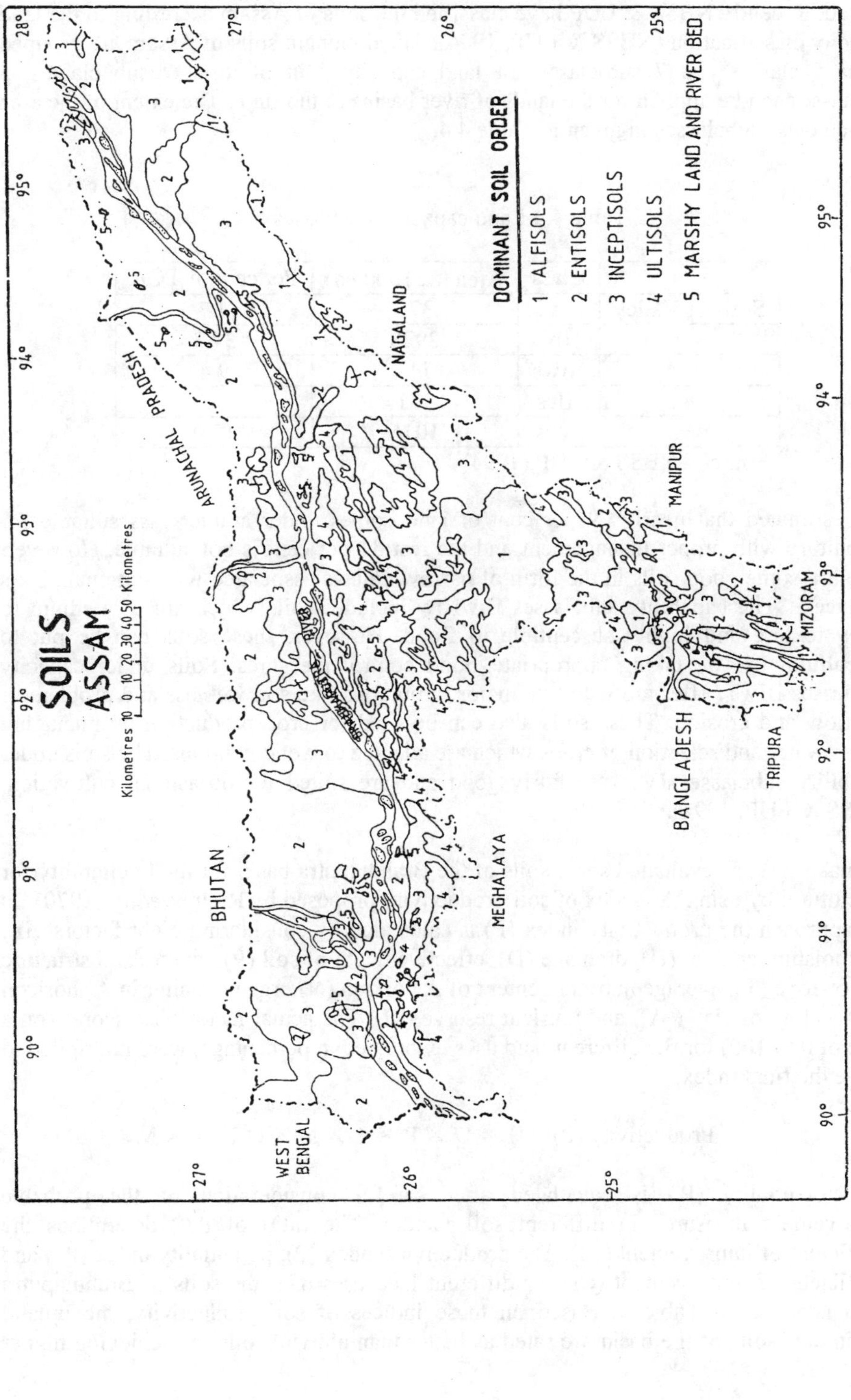


Fig. 4.4 Distribution of domain soil orders in Assam.

The information on land capability classes of the soils of the Brahmaputra basin are limited. Recently NBSS & LUP have classified the soils of Assam according to the land capacity classification (NBSS & LUP, 1994). The dominant soils of Assam are grouped under 5 classes and 7 sub-classes of land capacity. Out of these 7 sub classes, 5 subclasses can be applied for the lands of river basins of the state. The extent of the area in each class/subclasses in given in Table 4.4.

Table 4.4 Land capacity subclasses.

| | Class | Area ( $\times$ 1000 ha) | Per cent of TGA |
|---|---|---|---|
| Soils of Valley | IIw | 372 | 4.7 |
| | IIIw | 2928 | 37.3 |
| | IIIws | 744 | 9.4 |
| | IIIes | 331 | 4.2 |
| | IVes | 1016 | 13.0 |

Source : NBSS & LUP (1994)

It is estimated that nearly 69 per cent of land comes under arable class suitable for agriculture with proper management and the rest 31 per cent is not suitable. However, there are some good soils in the form of narrow valleys associated with dominant soils in places. The capability subclasses IIw/IIIw includes soils which are of medium to heavy texture and highly susceptible to flood. However, these soils can be put to agricultural use by taking appropriate flood control measures. Soils under capacity subclasses IIIws / IIIes are with two major limitations like soil wetness and problem of overflow and erosion. These soils also can be used for crop production by managing excess water and selection of crops which are adapted to wet conditions. The soils under capability subclasses IVes are fairly good and are suited for occasional cultivation. (NBSS & LUP, 1994).

Karmakar (1995) evaluated some soils of the Brahmaputra basin for their suitability for agricultural by using the index of soil productivity proposed by Riquier *et al.* (1970). In this approach the productivity index (P) is calculated by considering eight factors viz., soil moisture content (H), drainage (D), effective depth of soil (P), texture and structure of root zone (T), average nutrient content of A horizon (N),organic matter in $A_1$ horizon ( O ), nature of clay ( A ) and nutrient reserve (M). The actual factor wise scores (on a scale of 0 to 100) for the different land uses, expressed in percentage, were multiplied to derive the final index:

$$\text{Productivity (P)} = H \times D \times P \times T \times N \times O \times A \times M$$

The potentiality ($P'$) is calculated after careful consideration of the probable improvement measures in different soil factors. The ratio of $P'/P$ determines the coefficient of improvement (CI). The productivity index (P), potentiality index ($P'$ ) and coefficient of improvement (CI) for different land uses of some soils of Brahmaputra basin is given in Table 5. Based on these indices of soil productivity, the upland (piedmont) soils of the basin are rated as better than alluvial soils for achieving higher crop yields. The soils of the flood plains and alluvial fans are rated as poor. However, the improvement of productivity of these soils will require adoption of measures, such as drainage, timely irrigation, and application of nutrient inputs. Among the soil groups evaluated by productivity indices, the Udispamment of alluvial fans are the poorest in productivity and these soils are not likely to respond favourably to improved management for producing satisfactory crop yield level. The very poor structures and coarse texture of these soils exercise an overriding influence in making them the poorest productive soils.

Table 4.5 Productivity (P), potentiality (P′) and coefficient of improvement (CI) of some soils of the Barhmaputra basin (Karmakar, 1995).

| Soils | Physiographic Units | Crops | | | Pasture | | | Commercial trees | | |
|---|---|---|---|---|---|---|---|---|---|---|
| | | P | P′ | CI | P | P′ | CI | P | P′ | CI |
| Dystrochyst | Lower piedmont plain | 48.7 (2) | 65.0 (I) | 1.33 | 46.3 (2) | 58 (II) | 1.25 | 73.1 (1) | 81.2 (I) | 1.11 |
| Udic spamment | Alluvial fan plain | 8.2 (4) | 17.4 (II) | 2.12 | 10.0 (4) | 20.8 (III) | 2.08 | 3.3 (5) | 10.3 (IV) | 3.12 |
| Epiauept | Alluvial plain | 27.1 (3) | 61.4 (II) | 2.26 | 64.0 (1) | 81.2 (I) | 1.24 | 8.1 (4) | 36.1 (II) | 4.46 |
| Udifluvent | Flood plain | 18.7 (4) | 58.1 (II) | 3.11 | 36.9 (2) | 69.2 (I) | 1.87 | 3.5 (5) | 34.2 (III) | 9.7 |

The figures in the parentheses indicate productivity (P), potentiality (P′), classes as follows:

| Rating | P | P′ | Classes |
|---|---|---|---|
| 65--100 | 1 | I | Excellent |
| 35--64 | 2 | II | Good |
| 20--34 | 3 | III | Average |
| 8--19 | 4 | IV | Poor |
| 0--7 | 5 | V | Extremely poor |

## 4.5 Fertility Status of Agricultural Soils

The soils of the Brahmaputra alluvium are partly new or recent and partly old. These soils show wide variations in their physical, chemical and biological properties. The nutrient status of these soils also varies considerably. These variations can be traced back to geological origin on which soils have developed. The alluvium deposits have been brought down by the Brahmaputra and its tributaries which originate from the foot hills on both the banks of the mighty river.

4.5.1 ORGANIC MATTER CONTENT

The organic matter reserve of the Brahmaputra valley soils ranges from low to high. It is particularly high in *bheel and* virgin soils. In continuously cultivated soils, organic content has decreased to an alarming level. The average organic matter content of new alluvium, old alluvium and upland soils are 0.97%, 1.02% and 2.70%, respectively. The distribution of organic matter in the profiles shows a gradual decrease with depth (Chakravarty *et al.,* 1987) The light, medium and heavy textured soils of CBVZ of Assam contain low to high level of organic C with values ranging from 0.39 to 1.11% (Moriom,1994). Surface soils collected from cultivated fields of different locations representing seven agro--ecological situations in UBVZ contained low organic C (0.19 to 1.14%) (Tamuli, 1997; Medhi, 1999). The organic C content of alluvial soils of Guwahati, Tezpur, Nagaon and Jorhat varies from 0.87% and 1.10% (Chakravarty *et al.,* 1984).

Sarma (1990) carried out fractionation of humus in inceptisols and alfisols of the Brahmaputra valley. By and large the content of fulvic acid fraction predominated over humic acid fractions in almost all soils of the valley studied. Alfisol of the valley contained higher amount of humus carbon than humus carbon of inceptisol. The content of C, H and N in the humuic acid fraction was more than their content in fulvic acid fraction. The oxygen content of fulvic acid was higher than the humic acid and thus the O : H ratio in the fulvic acid was higher and the C:H ratio was lower than the humic acid. This was the indication of the weak aromatization of the fulvic acid fraction. The total acidity, equivalent acidity of -COOH and phenolic–OH group of fulvic acids were higher than those of humic acids.

4.5.2 NITROGEN CONTENT

The N content of soils exhibits variations between sites and layers of the same soil profile. The total N content in the top soils of old alluvial vary from higher to low, while the new alluvial soils contain medium level of total N. The average value of total N content in the valley was reported to be 0.1% (Raychaudhury *et al.*, 1963). Dutta (1974) reported that the amount of the total N in new alluvial soils of Brahmaputra valley varied between 0.013% and 0.135%. On the other hand, the N content of the old alluvial soil is reported to be within the range 0.034 to 0.165%. The total N content of some rice field soils of Brahmaputra valley varied between 0.049% and 0.217% (Barthakur, 1975). The C/N ratios of new and old alluvial soils are 10.7 to 60.0 and 6.7 to 28.0, respectively. Surface soils, representing seven agro-ecological situations of Jorhat and Golaghat district in the UBVZ were low to medium in available N content with values ranging from 173.8 to 425.5 kg/ha (Medhi, 1999). The general trend of distribution of the total N in soil profiles is that the soil horizons below the surface layer exhibit a gradual decrease.

4.5.3 PHOSPHATE CONTENT

Top soils of the old alluvium have low to deficient levels of phosphate available for plant use. In general, available $P_2O_5$ content in old alluvial soils has been reported to be less than 0.01% (1% citric acid soluble). The top soils of new alluvium are relatively rich in phosphate where available phosphate may be as high as 0.055% (Raychaudhury *et al*., 1963). Soil acidity is the main reason of the low level of phosphorus available for plants in these soils. The details of the status of available $P_2O_5$ in few districts of the North and South bank of the Brahmaputra have been reported by Raychaudhury *et al.* (1963). The average content of $P_2O_5$ in the new alluvium soils of Lakhimpur, Nagaon, and Kamrup districts was 0.025%, 0.020% and 0.015%, respectively. Various studies indicated that the total amount of P in acid soils of the valley was quite high but the fraction of P in available form decreases with maturity and acidity of soils (Chakravarty and Majumder, 1970; Chakravarty and Majumder, 1971). The total P content in some rice field soils of the Sibsagar district ranged between 83.5 ppm to 550 ppm (Barthakur, 1975). Organic forms of P constitute a significant fraction of the total P. A fractionation study of soil P showed that Fe--P and occluded P constituted 34.0% and 37.3% of the total inorganic P. Al--P and Fe--P fraction control to a great extent the available P content as positive correlations are obtained between Al-P and Fe-P fractions and available P content of the soils. The pH and organic matter content of the soils of the basin also exercise some control on P availability to crops. Tewari *et al.* (1969a) reported that 81% to 93% of the soluble P applied to soil could be retained by the soil through fixation. The P fixing capacity of surface layer soils collected from profiles of old flood plains of southern Brahmaputra valley ranged between 71.5 to 81.4% (Ali, 1992). In another study Medhi (1999) found available $P_2O_5$ in the surface soils of the Jorhat and Golaghat districts of UBVZ in the range between 2.24 and 23.60 kg/ha.

### 4.5.4 POTASH CONTENT

Top soils of the new alluvium are richer in available potash compared with top soils of the old alluvium. The average values of available $K_2O$ content of new alluvial soils of Lakhimpur, Nagoan and Kamrup districts were 0.006%, 0.014% and 0.002%, respectively (Raychaudhury *et al*, 1963). A study on forms of the potassium reserve of alluvial soils indicated that the total K, exchangeable K water soluble K and $HNO_3$-soluble K varied between 1,500 to 19,500 ppm, 22.07 to 188.60 ppm, 1 to 10 ppm and 28.40 to 406.70 ppm, respectively (Nath, 1978). On the other hand, Ali (1992) reported that the contribution of K from non-exchangeable K to the total plant uptake was maximum in coarse texture alluvial soils of Assam compared to fine textured alluvial soils. Surface soils of the old flood plain of the southern Brahmaputra valley contained 0.78 to 11.37 ppm of exchangeable K (Ali, 1992). From a thorough study Medhi (1999) reported that the available K content of the representative soils of seven agroecological situations in the UBVZ of Assam ranged between 50 to 187.5 ppm.

### 4.5.5 MAGNESIUM AND CALCIUM CONTENT

In the Brahmputra valley soils the percentage of total (acid soluble) MgO is relatively greater than that of CaO, but available (replaceable) Mg is lower than Ca. This is also true for all the soils of Assam (Raychaudhury *et al.*, 1963). In recent alluvium and low-lying soils exchangeable $Ca^{++}$ and $Mg^{++}$ are dominant. Cations of clay surface and their content increase down the profile of upland soils. In old alluvium derived soils, only exchangeable $Mg^{++}$ increases with depth of profile. The presence of higher amount of $Mg^{++}$ in older soils of Assam has been reported by Raychaudhury *et al.* (1963) and Mandal *et al.,* (1975). Status of exchangeable Ca and Mg in the surface soils of

Brahmaputra valley soils of Nagoan, Tezpur, Guwahati and Jorhat is low with values ranging between 73 to 148 and 3.6 to 39, respectively. The Ca:Mg ratio of these soils varied between 1.41 and 12.17 (Chakravarty *et al.*, 1984). Exchangeable $Ca^{++}$ and $Mg^{++}$ in some old flood plain soils of southern Brahmaputra valley was reported to be 28.4 to 98 and 3.6 to 24.96 ppm, respectively (Ali, 1992).

4.5.6 STATUS OF MICRONUTRIENTS

The distribution of different forms of manganese in acid alluvial soils of Assam was reported by Das *et al,* (1971). Highly significant positive correlations were observed between clay and easily reducible Mn content of soils. Similar correlations between organic carbon and total Mn has also seen reported. The average values of various forms of Mn, viz., total, easily reducible, water soluble, exchangeable and available Mn in the new and old alluvial soils of Assam were reported to be 974.1 and 439.4, 37.5 and 42.7, 3.33 and 2.62, 13.55 and 19.44, 15.90 and 22.1 ppm, respectively. The available Mn content of these soils is reduced with increase in soil pH.

The available Cu ($NH_4$-acetate extractable) in the alluvial soils varied between 0.4 and 1.0 ppm (Tewari and Chakravarty, 1972). Acid soils of the alluvial origin are generally deficient in available Mo ($NH_4$-oxalate soluble) and it was reported to be 0.02 mg $kg^{-1}$ soils (Tewari, 1970). The amount of available Fe in Jorhat soils of UBVZ was as high as 732 ppm (Tewari, 1969b). The total Zn content in Assam soils is quite high (42.5 to 250.0 ppm). New alluvial soils contain lower amounts of Zn than do the old alluvial soils (Nath and Das, 1980). The total Zn content showed highly significant positive correlation (r=0.657) with clay content of soils of the valley.

## 4.6 Crop Response to Applied Nutrients and Lime

Field experiments have shown that crop yields in most of the arable soils of the Brahmaputra Valley can be considerably increased by adequate manuring combined with proper cultural practices. Most of the old alluvial soils are required to be adequately manured, particularly with nitrogenous and phosphatic fertilizers to increase yield of crops. Organic manures, such as cowdung, compost, oil cake and green manures, have been found effective for most crops in these soils. However, the available quantities of these manures are not sufficient to meet the crop demand for nutrients.

New alluvial arable soils of the Brahmaputra valley are less acidic in reaction as compared to old alluvial soils and are relatively rich in phosphate, potash and calcium. But the N content of these soils are medium. Organic manures and nitrogenous fertilizers have been effective in increasing crop yields in these soils. A significant response of rice to NPK fertilizers on different soils types of Brahmaputra valley has been obtained. However, the magnitude of response varies depending on these types of soils, crop varieties and climatic conditions. The response of rice to applied NPK fertilizer in representative soils of UBVZ, CBVZ and NBPZ of Assam is shown in the Table 4.6. The results of experiments carried out in the Regional Agricultural Research Station, Lakhimpur (NBPZ) showed that the response of rice to applied NPK in the sandy loam soil under high rainfall conditions is very less. A major portion of the fertilizer nutrients specifically N is leached through the sandy loam soil. The response of crops to phosphatic fertilizers grown in acidic soils of the Brahmaputra valley is also invariably low. A high fixation capacity of these soils is one of the major constraints of low response of crop to applied P. As mentioned before, phosphate fixation capacity of these soils have been reported to vary between 71.5 and 93% (Tewari *et al.,* 1969; Ali, 1992).

Short duration rice varieties grown in the cultivators' fields of Kamrup in the LBVZ and Nagaon in the CBVZ responded significantly to fertilizer N. Application of N @ 50 to 75 kg N/ha is recommended for soils of Kamrup district in order to obtain the maximum net return from per rupee investment on fertilizer N on rice. The recommendation for the Nagoan district is 25 to 50 kg N/ha is (Barthakur and Sharma, 1981).

Recent estimates show that the extent of occurrence of acid soil is as high as 96.0 per cent in the UBVZ and 91.6% in the CBVZ and LBVZ in Assam (Figure 4.5). Soil acidity is one of the important factors of low response of crops to fertilizers in the alluvial acid soils of Assam. However, the research work (Harmine, 1968; Chakravarty, 1975; Chaliha, 1971; Pathak, 1975; Singh, 1975, Barthakur, *et al.,* 1981) conducted in the state suggested that liming acid soils could increase the response of crops to applied nutrients tremendously. The lime requirement of different acid soils of Assam varies from 1.75 to 8.79 t/ha of $CaCO_3$ (Bora, 1967). The lime requirement of some acid rice field soils of Barhamaputra Valley varies from 1.8 to 6.8 t/ha. in the form of $CaCO_3$ as liming material (Barthakur and Majumder, 1967).

Table 4.6 Response of rice crop to applied nutrients.

| Location | Soil Characteristics | Response range (kg grain/kg NPK) | Referene |
|---|---|---|---|
| 1. UBVZ | Acidic clay laom with Av. N = 460 kg/ha Av. $P_2O_5$ 10 kg/ha Av. $K_2O$ = 179 | 5.2--25.50 | Bhattacharyya *et al* ., 1994 |
| 2. UBVZ | Acidic sandy loam with Av. N = 303 kg/ha Av. $P_2O_5$= 15.78 Av. $K_2O$ = 146.20 | 6.78--14.55 | Ahmed and Baroova, 1992 |
| 3. CBVZ | Acid soil (farmers field) | 8.4 -- 22.4 | Annual report 1989-90, 1990-91 & 1991-92, RARS, AAU Shillongani, Nagaon Assam |
| 4. NBPZ | Acidic sandy loam with Av. N = 10.67kg/ha Av. $K_2O$ = 104.53 | 1.37--9.20 | Talukdar and Chakravarty, 1988. |

## 4.7 Microbial Status of the Brahmaputra Soils and Response of Crops to Microbial Inocula

Data on population and distribution of bacteria and fungi in different soils of Brahmaputra basin is limited. In rice growing soils of pH 4.9, bacteria and fungi population were found to be $2.84 \times 10^9$ and $8.9 \times 10^5$/g soil (Barua 1999). The microbial biomass carbon in soils of the valley ranges from 461 to 3016 kg/ha (Hazarika *et al.,*2000). The microbial biomass C in a 50 year cultivated soils was 76% less than that in adjacent fallow field soils (Talukdar and Hazarika, 1998).

Soils of the valley harbour a rich diversity of beneficial soil microorganisms, such as *Rhizobium, Azospirillum,* phosphate solubilizing bacteria and vesicular arbuscular mycorhizal fungi (VAMF) (Das, 1990, Talukdar and Barthakur, 1981; Singh, 1997; Talukdar, 1999). Das isolated 350 *Rhizobium* strains from root nodules of common pulses and legumes grown in the valley. Phosphate solubilising bacteria and fungi are widespread in the rice rhizosphere soils of the valley (Talukdar and Barthakur, 1981; Talukdar, 2000a). Inoculation of pulses grown in acidic soils of Assam by efficient *Rhizobium* strains resulted in significant yield increase over the uninoculated control. There was an increase of yield by 3.3, 5.6, 2.7 and 4.4 q/ha in green gram (*Vigna radiata*), black gram (*Phaseolus aureas*), Lentil (*Lens esculenta*) and pea (*Pisum sativum*), respectively due to inoculation with the compatible and efficient strains of *Rhizobium* (Das, 1990). Application of *Azosprillium* and PSB biofertilizer along with compost resulted in 6.0 q/ha increase in rice yield over the yield (23.5 q/ha) of uninoculated control treatment (Talukdar *et al.,* 2000). In low P soils of the valley, application of PSB biofertilizer with insoluble rock phosphate enhanced crop yield by more than 100% (Figure 4.6) (Lahar *et al*., 2000). Vesicular arbuscular mycorrhizal fungi spores in the genera *Glomus, Gigaspora, Acaulospora, Enthrophosphora* and *Scutellospora* in the acidic soils of Brahmaputra basin have been reported (Venkataraman *et al.,* 1985; Talukdar, 1999). The beneficial effect of inoculum of VAMF on growth of agricultural crops and forest tree species (*Tectona grandis* and *Malina arborea* grown in degraded forest soils of the valley have been demonstrated (Talukdar, 2000b).

## 4.8 Fertility Status of Tea Soils of the Brahmaputra Valley

In the Brahmaputra valley tea is grown on both banks of the river on land with elevation between 50 to 120 m above mean sea level (Baruah, 1989). The tea soils of the Brahmaputra valley, which occupy 196,173 ha, are mostly alluvial and vary widely in texture. Some of the soils on the north bank are usually very sandy and poor in plant nutrients. There are small areas of old red soils which are mostly loam in texture and suitable for tea. Mann and Gokhale (1960) analysed a few typical tea soils of Brahmaputra valley and reported the analyses data as pH 5.2 -- 5.7, organic carbon 1.2%--2.2%, total N 0.071--0.106%, available phosphorus 0.003%--0.012% and available potash 0.004% -- 0.009 %.

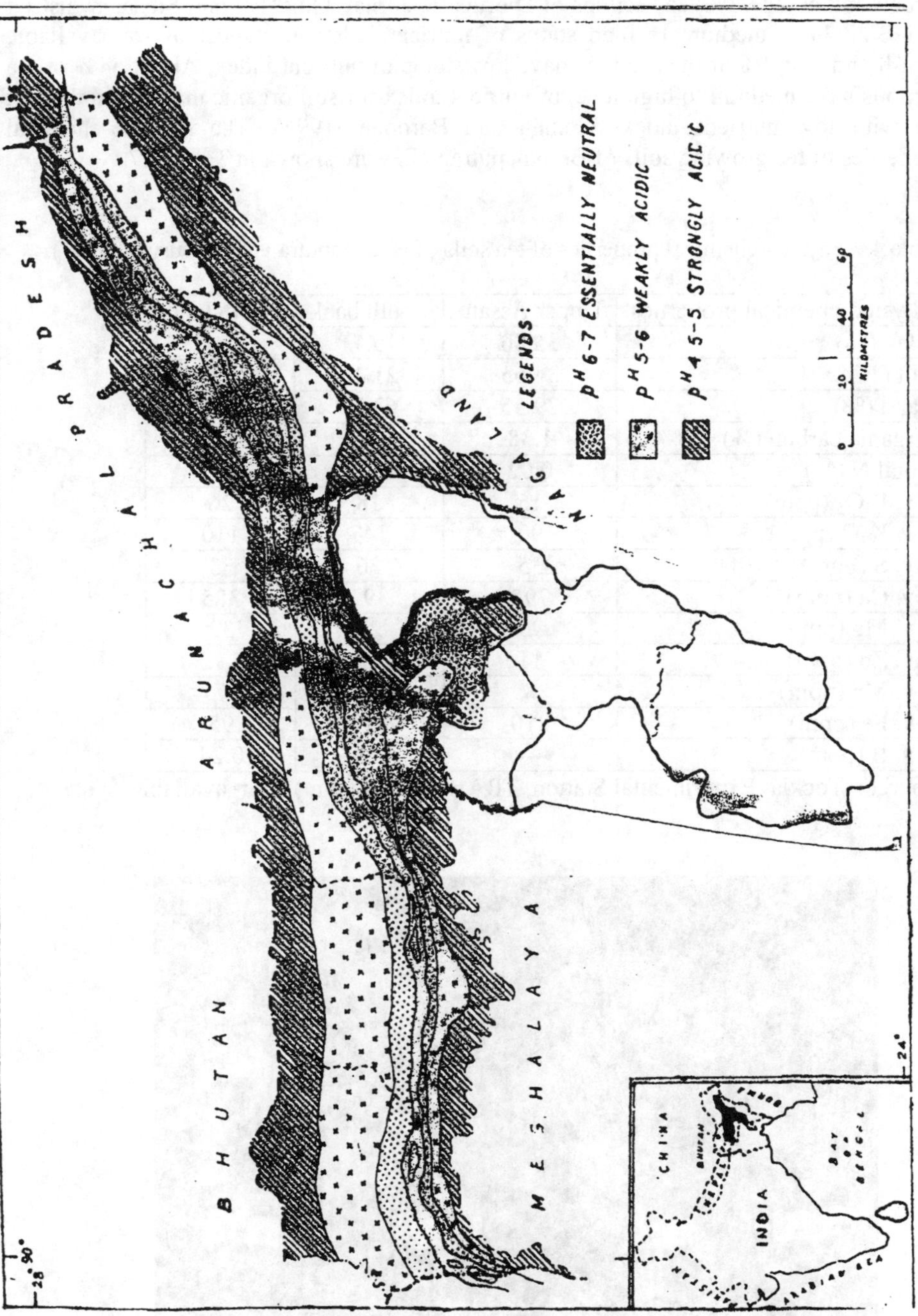


Fig. 4.5 Variation in pH of soils of the Brahamaputra basin.

Most of the tea soils of Brahmaputra valley have low to medium status of organic carbon and about 93% tea sections of Dibrugarh, Jorhat, Golaghat and Sibsagar district of UBVZ have medium to high status of nutrient index in respect of soil available potash and rest 7% of the sections have low status of nutrient index. About 54% of tea sections have medium to high level of nutrient index for soil organic matter and the rest are with low nutrient index (Gohain and Barooah, 1996). The physico-chemical properties of tea growing soils of Brahmaputra valley are shown in Table 4.7.

Table 4.7 Physico-chemical properties of tea soils of Brahmaputra valley (Survey, 1991).

| Physico-chemical properties | Upper Assam | South bank | North bank |
|---|---|---|---|
| Clay (%) | 35.80 | 15.11 | 12.47 |
| Silt (%) | 34.65 | 21.16 | 11.52 |
| Sand (%) | 29.55 | 63.73 | 76.01 |
| Organic carbon (%) | 1.38 | 0.72 | 0.55 |
| Total N (%) | 0.02 | 0.13 | 0.09 |
| Av. $P_2O_5$ (ppm) | 15 | 18 | 26 |
| Av. $K_2O$ (ppm) | 90 | 147 | 110 |
| Av. S (ppm) | 35 | 26 | 33 |
| Av. Ca (ppm) | 795 | 392 | 385 |
| Av. Mg (ppm) | 23 | 23 | 32 |
| Av. Zn (ppm) | 11 | 2 | 3 |
| Av. Mn (ppm) | 8 | 10 | 7 |
| Av. Fe (ppm) | 210 | 119 | 95 |
| Av. B (ppm) | --- | 0.20 | 0.24 |

(Source: Tocklai Experimental Station, TRA, Jorhat, Assam) Av.= available

Fig. 4.6 Effect of phosphate of solubilizing bacteria on transplanted rice (variety: Ranjit).

### 4.9 Land Use Pattern in the Brahmaputra Valley

Crop fields mostly under rice and surrounding human settlement (villages), forest cover, tea plantations, grazing land and water logged swampy areas with a huge network of tributaries are commonly observed components of the land surface of the Brahmaputra and Barak valley of Assam. The scenario and magnitude of shift with respect to use of lands, surface area for forestry, agriculture, pasture and grazing over a period of almost 50 years from 1951--52 to 1997--98 are shown in Table 4.8. There has been a reduction of 27.5% in forest cover and an increase of 71.5% in cropped area within 50 years since 1951--52. The cropping intensity has increased from 116.7% in 1951 to 143.9% in 1997--98. The best forest patches of upper Assam are confined to Dibrugarh and Tinsukia districts of UBVZ.

### 4.10 Land Tenure and Farm Size of the Brahmaputra Valley

The land holdings in the Brahmaputra valley are small (average size 1.47 ha). A number of socio-economic problems of the farmers of the valley have linked to the larger number of the fragmented, small and marginal holdings. The share tenancy is quite wide spread. As per the world agricultural census, 16 per cent of the holdings are yet wholly rented, while another 9 per cent are partially rented. The 59.66 per cent of the land holdings are of size 1 hectare or less, 22.62% of size 1--2 ha and 9.04% is of 2--3 hectare (Mandal et al. 1981).

Table 4.8 Area under different types of land use in Brahmaputra and Barak valley of Assam from 1951--52 to 1997--98.

| Type of land use | Area (1000 ha) | | | |
|---|---|---|---|---|
| | 1951--52 | 1960--61 | 1975--76 | 1997--98 |
| Forest | 2403 | 1745 | 1659 | 1742 |
| Area unavailable for cultivation | 1064 | 1416 | 1364 | --- |
| Other uncultivated area * | 436 | 704 | 598 | --- |
| Fallow land | 4603 | 395 | 244 | --- |
| Tea plantations | --- | 162 | --- | 227 |
| Net sown area | 1973 | 2003 | 2664 | 2769 |
| Area sown more than once | 312 | 396 | 632 | 1215 |
| Total cropped area | 2285 | 2373 | 3025 | 3984 |

* fallow area not included

### 4.11 Cropping Systems

Cultivation of cereals is the basic features of cropping in the Brahmaputra basin. Among cereals, rice is the dominant crop covering almost 80 per cent of the net cropped area. Out of the three factors, namely, soil, rainfall, and temperature as determinants of the cropping pattern of a particular area, rainfall has exercised an overriding influence in evolution of rice cropping systems of Brahmaputra valley. The annual average rainfall (2,000 mm) provides adequate standing water for growing rice throughout the year. Therefore rice monoculture is the predominant cropping system in Assam covering about 1800,000 hectare of land as *kharif* rice. The rice-rice cropping system is followed in approximately 15% of the area followed by rice--mustard (10%). Thus (a) rice--rice and (b) rice--mustard are the two predominant cropping systems of Assam (Annual report of All India Co-ordinators Research Project on cropping systems submitted by AAU Centre, Directorate of Research, AAU, 1999). Along with rice the other crops such as, jute, pea, potato, wheat, and vegetables are also grown in a limited area depending upon land situations. Common cropping systems of the Brahmaputra valley are presented in Table 4.9. In UBVZ, the uplands are mostly occupied by tea cultivation.

Rice is grown in the Brahmaputra valley in three seasons (i) *ahu* (Autumn-March to June), (ii) *Sali* and *bao* (Winter-June to November/December) and (iii) *Boro* (summer or Spring--November--April). *Ahu* (autumn rice) is grown both as direct seeded and transplanted crops in upland and irrigated fields. In irrigated fields *ahu* is grown mainly as transplanted crops. *Boro* (spring rice) was earlier grown in marshy land in the *rabi* season but in recent years *boro* rice is also grown in lowlands with an irrigation facility. Seed bed for *boro* rice is raised in November--December and transplanting is done in January--February and the crop is harvested in April--May. The cultivation of the *Boro* rice is mainly confined to Nagaon, Kamrup and Goalpara districts of the Brahmaputra valley. Because the cumulative sunshine hour during November-December to February--May is maximum, *boro* rice productivity is the highest among the three types of rice cultivated in the valley. With increase in irrigation facility the area under *boro* rice in the valley has increased from 30,500 ha in 1988--89 to 175, 000 ha in 1997-98.

### 4.12 Crop Productivity

The crop productivity of different crops in the Brahmaputra valley is presented Table 4.10. The crop productivity in the valley is low. The low productivity of the soils cannot be directly attributed to low fertility of soils. But rather the productivity is the result of socio-economic constraints, including power and slow rate of adoption of improved techniques of crop production. The slight jump in productivity over a period of one decade (Table 4.10) can be attributed to the increase in use of inputs, such as irrigation, fertilizer, and HYV of crops, particularly rice by farmers. The fertilizer consumption in the state has increased from 7 kg/ha in 1980 to 17.8 kg/ha in 1997--98 and irrigation facilities have increased from 290,000 ha in 1980 to 480,000 ha in 1997--98. The total area under HYV rice was 125,000 ha in 1997-98 against 89,000 ha in 1970.

### 4.13 Degradation and Restoration of Soils of the Brahmaputra Basin

As per the FAO's (1969) definition of soil degradation (process which lowers the current and/or potential capability of soils to produce qualitatively and/or quantitatively, goods or services) a major portion of soils of Brahmaputra basin is under degradation hazards. Out of 7,852,000 ha of the geographical area of Assam, 6,029,600 ha is degraded owing to flooding, waterlogging and acidification, loss of nutrients and organic matter (Annual Report of NE regional center NBSS & LUP, Jorhat, Assam). A major portion of these degraded soils is confined to the Brahmaputra basin, although the exact estimate for the valley is not available.

Table 4.9 Common cropping systems followed in different agro-climatic zones under different land situations.

| Land situation | Cropping systems | | |
|---|---|---|---|
| | Kharif | Rabi | Summer |
| Rainfed medium lands | Rice | Fallow | Fallow |
| | Rice | Vegetables | Jute/rice |
| | Rice | Pea/toria/potato | Rice |
| | Rice | Chilli/brinjal/wheat/rice | Rice/Jute/Fallow |
| Rainfed low lands | Rice | Fallow | Fallow |
| | Rice | Fallow | Rice |
| | Rice | Fallow | Jute |
| | Rice | Fallow | Jute |
| | Rice | Vegetable/toria/wheat | Rice/Jute |
| Chronically flooded areas | Fallow | Toria/wheat/vegetables | Rice |
| | Short duration Post flood rice | Fallow | Preflood rice |
| | Jute (Capsularis) | Toria/wheat/pea/french bean | Short duration rice |
| Deep water areas | *Bao* rice | | |
| | *ahu* and *boro* as mixed crop | | |
| Rain fed uplands | Vegetable | Wheat/pea/toria/lentil/vegetable | Vegetable |
| | green gram/ black gram | Toria/niger | Summer rice. |

## 4.13.1 DEGRADATION OF AGRICULTURAL SOILS

The agricultural soils of the basin are subjected mainly to chemical and biological degradation. Chemical degradation is the accumulated negative impact of chemicals and chemical processes on those properties that regulate the life process in soil. The problem of chemical deterioration of soil is severe in the Brahmaputra valley, particularly in the central and lower Brahmaputra Valley zones where the population density is high and the land is cultivated with a relatively high cropping intensity without following adequate fertilizer input and residue management practices. For example, the organic carbon content of farmers' fields under rice cultivation was 45.6% lower than that of the adjacent undisturbed fields under native vegetation cover. Similarly, plant nutrient contents were also much lower in the continuously cultivated soils (Hazarika *et al.,* 2000). The decrease in organic carbon content is also associated with decrease in microbial biomass C of these soils. The decrease in organic matter content and microbial biomass may also cause a decline in the productivity of soils. Such evidence was obtained from the results of experiment conducted in sandy loam soils containing 0.32% organic carbon and 369.6 kg microbial biomass C/ha. The rice yield in such soils was increased by 6.3 q/ha owing to inoculation with *Azospirillum* and phosphate solubilizing bacterial culture in a compost base over the yield of the uninoculated control plot. The organic matter content and microbial biomass C of the inoculated plot were also enhanced (Talukdar *et al.,* 2000). The inorganic fertilizer could increase the yield only by 1.77 q/ha over the no fertilizer control treatment. In restoration of soils of rice cultivated fields of the valley, organic amendments and microbial inoculum will play a very useful role.

Table 4.10 Productivity of rice, jute, wheat pulses and oil seeds in 1978--79 and in 1997--98.

| | 1978-79 | | | 1997-98 | | |
|---|---|---|---|---|---|---|
| Crop | Area (×1000 ha) | Production (×1000 tonnes) | Yield (kg/ha) | Area (×1000 ha) | Production (×1000 tonnes) | Yield (kg/ha) |
| Rice | 2241.3 | 2172.3 | 969 | 2523.6 | 3382.8 | 1359 |
| Jute* | 114.8 | 899.0 | 1409 | 94.5 | 9040.9 | 1720 |
| Wheat | 68.1 | 69.2 | 1015 | 84.6 | 110.0 | 1299 |
| Pulses | 103.5 | 42.4 | 410 | 128.9 | 71.1 | 552 |
| Rape and mustard | 196.3 | 85.8 | 437 | 279.2 | 154.5 | 553 |

*Figures indicate the number of thousands of bales

### 4.13.2 DEGRADATION OF RICE SOILS OF THE VALLEY BY EFFLUENT OF COAL MINE AND OIL DRILLING INDUSTRY

The impact of mining operations, such as oil exploration (involving drilling for mineral oil and pumping operations) and coal mining, on soil quality in certain patches of the valley has been recognized in recent years. Oil exploration is confined mainly to the upper Brahmaputra valley Zone comprising the Tinsukia, Dibrugarh, Sibsagar and Jorhat districts. The crude oil from the production wells flow through an individual 10 cm flow line to the group gathering stations (GGS)/oil collecting stations (OCS) where crude oil is processed into separate associated chemicals which are called effluent. This effluent is disposed of in the environment through different types of technology. A major amount of this is released to the environment with or without treatment. Normally the crude content in untreated effluent ranges from 3000 to 5000 ppm. A case study indicated that the effluents traveled a distance of 5 km along the water way and damaged rice crops and soils. The pH of the effluent contaminated standing water ranged from 9.4 near OCS to 8.3 at a distance of 5.0 km. Similarly, the soil pH ranged from 8.65 to 6.3. The pH of the normal rice soil of this zone ranges from 4.5 to 5.5 (Barthakur, 1995; Chaudhury, 1996). A continuous deposition of effluents from drilling stations in rice fields has caused severe damage to the quality and productivity of soils.

Rice field soils of the Tinsukia districts are affected by acid mine waste. The coal reserve of the Patkai forest range contains 4% elemental sulfur and a substantial amount of this material gets oxidized in the overburden dumps of the coal mine and washed off to the surrounding rice field soils by heavy rain during monsoon seasons. The pH of the rice field water surrounding coal fields ranged from 2.02 to 2.30. Likewise the soil pH is also reduced. The paddy field soils near the colliery also receive the deposition of the coal dust which affects soil quality (Barthakur, 1988).

### 4.13.3 DEGRADATION OF SOILS UNDER FOREST AND TEA PLANTATION

The forest area of the Brahmaputra valley zone has been shrinking (see Table 4.8). Encroachment, logging, and mining in forests result in removal of the forest area, and consequently the forest soils of such areas are also degraded. Felling of trees from the forest paves the way for free passage of stray animals and human beings and soil surface layer become very compact. Natural regeneration of forest trees in such compact soil becomes difficult, especially in the sal (*Shorea robusta*) forest area of the valley. This difficulty of natural regeneration has caused disappearance of forests at an accelerated pace. The organic matter and other nutrient content of the degraded forest soils are also affected (Table 4.11). A study on degraded forest soils of the valley indicated that conspicuous changes occurred in physico-chemical properties of forest soils under varying state of degradation (Table 4.11). The view of soil profiles of few forest sites of the valley along with the above ground vegetation cover is presented in Figures 4.7 and 4.8.

Table 4.11 Few selected chemical properties of soils under intact and degraded forests of Brahmaputra basin.

| Site | Forest cover | Soil depth (cm) | Soil chemical properties | | | | |
|---|---|---|---|---|---|---|---|
| | | | pH | OC (%) | Available nutrients (kg/ha) | | |
| | | | | | N | $P_2O_5$ | $K_2O$ |
| Akhbandha, Tinsukia (UBVZ) | Mixed evergreen forest intact | 0-25 | 4.5 | 1.87 | 452 | 59.4 | 107 |
| Akhbandha, Tinsukia (UBVZ) | Forest cover removed 10 years ago | 0-25 | 5.3 | 0.92 | 316 | 54.2 | 103 |
| Hoograjuli, Charduar (NBPZ) | Mixed deciduous forest intact | 0-15 | 5.8 | 0.95 | 293 | 32.1 | 175 |
| Hoograjuli, Charduar (NBPZ) | Vegetation removed five years ago | 0-15 | 5.7 | 0.47 | 247 | 32.6 | 67.2 |

Fig. 4.7 Soil profile (B) under intact evergreen forest (A) and soil profile (D) under adjacent degraded forest (C) of Upper Dihing Reserve forest of Digboi Forest Division of UBVZ.

Fig. 4.8 Soil profile (B) under intact evergreen forest (A) and soil profile (D) under adjacent degraded forest (C) of Charduar Reserve Forest under Sonitpur Forest Division of NBPZ.

The performance of tree seedlings in degraded soils is also affected owing to reduced level of organic carbon, soil compaction, and poor nutrient status, etc. However, the growth of seedlings could be improved by proper treatment. For example, the growth seedlings of two important tree species, *Tectona grandis* and *Malina arborea* could be improved by application of compost and VAMF inoculum in pot culture experiment (Figure 4.9). Compost and VAMF inoculum manifested their beneficial effects on seedling growth by improving P nutrition (Talukdar, 2000b).

Tea soils of the Brahmaputra basin have undergone chemical and biological degradation. Heavy use of chemical inputs such as pesticides and fertilizers reduced biological activity. A recent study indicated that soils which are put under tea cultivation for a long period experienced a reduction in microbial biomass and dehydrogenous activity (Table 4.12). Although the organic matter content of soil did not vary much, dehydrogenase activity was reduced significantly. The dehydrogenase activity is a reflection of microbial activity in soils. In general, an increase in the age of plantation is associated with a higher level of cumulative pesticides application and residues in soil. (Annual Report of ACRIP on pesticide residue, Department of Entomology, A. A. U.).

Table 4.12 Soil organic carbon, microbial biomass C and dehydrogenase activity of soils under tea cultivation over varying period.

| Years under tea | OC (%) | MB-C (μg/g soil) | Dehydrogenase activity (μg TPF/g soil) |
|---|---|---|---|
| Virgin soil | 0.91 | 164.1 | 1.08 |
| 8 | 0.91 | 91.3 | 0.67 |
| 20 | 0.88 | 69.1 | 0.58 |
| 43 | 0.77 | 76.7 | 0.33 |

Fig. 4.9 Effect of organic matter and VAMF inoculation on the growth of teak (Tectona Grandis) seedling grown in degraded forest soils.

The tea-cultivated soils are also reported to experience physical and chemical degradation. From a comparative study, Dey et al. (1967-68) reported that young, medium, and old age tea soils had much lower values of soil aggregates compared to those of adjacent virgin soils. Similarly, tea cultivation reduced soil pH and the amount of exchangeable bases substantially (Dey et al. 1967-68).

### 4.13.4 DEGRADATION OF WETLANDS OF BRAHMAPUTRA VALLEY

An area of 102.31 $km^2$ of the Brahmaputra plains is demarcated as wetlands which have geographical importance as storage basins during floods. Besides, these wetlands are rich in biodiversity of vegetable fodder, fruits, medicinal plants, fish, snails, frogs, crabs, eggs, and insects. Overall there are 3,513 freshwater wetlands in Assam and the majority of these is in the Brahmaputra basin. These wetlands support an appreciable assemblage of rare, vulnerable species or sub species of plants and animals and have an appreciable number of individuals of any one of these species. They also support more than 20,000 birds and a substantial number of water fowl. This is indicative of the value, productivity, and diversity of wetlands of the basin (Bhattacharyya, 1998). However, in recent years many wetlands of the basin have been degraded and large numbers are under serious threat of degradation. The wetlands are getting degraded owing to various factors, such as (i) inorganic and organic deposits which include sand, silt, clay, pebbles and other inorganic materials; this results in the reduced depth of wetland; (ii) the blockage of feeder channels and roads; (iii) construction of embankments, irrigation channels and roads etc.; (iv) reduced depth of many swampy area have paved the way for enchroachment and cultivation. These factors have contributed to the degradation of wetlands by way of shallowing or fragmentation leading to an increase in the number of wetlands with less water spread area or decrease in the depth of wetlands with more water spread areas. This is clearly observed in the case of wetlands in the Kaziranga National Park (KNP) where the total number of wetlands was 155 in 1967. The wetlands in Kaziranga national park has increased to 257 in 1997. However, by area wetlands in KNP have been decreased to 27.4 $km^2$ in 1997 from the original area of 36.6 $km^2$ in 1967 (Baruah *et al.,* 1998).

A systematic study in a few famous beels such as Deepar, Urpad and Elenga beel, has shown that soil and water quality of these beels are severely deteriorated. The concentration of nutrients in the water of these beels has increased resulting in the growth of aquatic plants and algae which subsequently have affected aquatic life. The phosphate content has been reported to be as high as 0.05 mg $L^{-1}$. Solid contents of these waters are also high (Kakati and Bhattacharyya, 1989). The concentration of trace elements, such as Zn, Fe were also higher in soils of the Deepar beel. The pH of the Deepar beel soil is 4.1--5.0. Similarly, the organic C content of the beel soil is reported to be high (Deka and Goswami, 1992).

The Elenga beel is polluted by the effluent of paper industries, Jagiroad, Assam which generates huge loads of sodium sulphite, sodium carbonate, calcium hypochlorite, magnesium bisulphite. The Nagaon paper Mill, Assam, produces 2,270 $M^3$ of effluent daily. These effluents ultimately find their way into the Elenga beel. The soils of the beel area are heavy in texture with very low infiltration rate ranging from 0.15 to 0.25 cm/hr. There are accumulations of soluble salts in soils where rice has been grown by irrigating with effluent containing water of Elenga beel. The irrigated soils had 3 times greater EC values than those of adjacent non-irrigated soils (0.06 ds/m) (Barthakur and Bhattacharyya, 1995).

## 4.14 Concluding Remarks

The soils of the Brahmaputra valley are alluvium derived and are classified into old alluvial and forest soils. These soils encompass a wide range, when viewed in terms of their physical, chemical, and biological characteristics. These variations can be traced back to their geological origin. Most of the arable soils of the Brahmaputra valley need to be adequately manured for increased agricultural productivity. New alluvial arable soils are less acidic in reaction than are old soils.

## 4.15 References

Ahmed, A.A. and Baroova, S.R. (1992). Efficiency of nitrogen sources in rice (*Oryza sativa*) and their residual effect on wheat (*T. aestivum*). *Indian J. Agron.* **37** (1): 55--59.

Ali, B. (1992). Pedological evaluation of some alluvium derived soils of southern Brahmaputra valley of Assam. Unpublished M.Sc. (Agri) Thesis, Assam Agricultural University, Jorhat, Assam.

Barthakur, H. P. and Majumder, N.N. (1967). A laboratory study on lime requirement of some acid soils of Sibsagar district of Assam. *J. Assam Sci. Soc.* **19**: 1--4.

Barthakur, H.P. and Sharma, K.C. (1981). Summary of Research Findings (1975--76 to 1979--80). Agric. Chem. Res. Unit, Directorate of Research, Assam Agric. Univ., Jorhat, Assam.

Barthakur, H.P.(1975). Transformation of inorganic phosphorus and its availability in rice field soils in relation to chemical and microbiological activities. Unpublished Ph. D. Thesis, Guwahati University, Guwahati, Assam.

Barthakur, H.P., Ahmed, S. and Barthakur, A (1981). Summary of Research Findings (1975-76 to 1979-80), Agric. Chem. Res. Unit., Directorate of Research, Assam Agric. Univ. Assam.

Barthakur, H.P. 1988. Report on environmental impact of open cast mining of coal in patkai ranges by M/S Coal India submitted by the Directorate of Research, AAU, Jorhat, Assam.

Barthakur, H.P. 1995. Environmental impact of oil exploration in Assam. *Ecol. Environ and Cons.* **1**: 109--113.

Barthakur, H.P. and Bhattacharya, D. 1995. Possibility of growing paddy irrigating with Nagaon Paper Mill effluent. *J. Agric. Sci. Soc. NE India.* **8(2)**: 162-168.

Barua, R., Talukdar, H, and Talukdar, N.C. 1999. Soil chemical properties as influenced by the application of fertilizers and Farm Yard manure (FYM). *Intern. J. Trop. Agric.* Vol. 171.

Baruah, D.N. (1989). Science and Practices in Tea Culture. pp. 509. Tea Research Association, Calcutta and Jorhat.

Baruah, P., Sharma, P. and Goswami, D.C. 1998. A database for wetlands of Assam: A study using remote sensing technique. Proc. NAGI's (Eastern region) Nat. Con. Environment and Sustainable Development, Oct 13--14, Vo 1. pp 45--52.

Bayan, N. and Talukdar, N.C. 1997. Effect of co-inoculation with Gigaspora and Bacillus in green gram (*Vigna radiata*). *Indian Journal of Microbiology*, **37** : 103--104.

Bhattacharya, P.C. (1998). Wetlands of Assam-taken for granted. The Hindu Survey of Environment, 83-87.

Bhattacharyya , H.C. Saud, R. K. and Pathak, A.K. (1994). Response of promising long duration rice varieties of levels of NPK fertilizers. *JASS* 7(1): 1-4.

Bhattacharyya, D. (1991). Pedogenesis and classification of some soils of Assam. *Intern. J. Trop.Agric.* **9** : 1--8.

Bhattacharyya, D. (1997). Clay mineral assemblage in some soils of Assam. *J.Argic. Sci. Soc. N.E. Indian.* **10** :109-- 112.

Bhattacharyya, D. and Sidhu, P.S. (1984). Morphology and physiochemical characteristics of some soils of Assam. *J. Res. Assam agric. Univ.* **5**: 20-29.

Bhattacharyya, D. and Sidhu, P.S. (1987). Mineralogy of some soils of Assam. *Clay Research.* **6**: 23-30.

Bhattacharyya, T, Dubey, P.N., Das, T.K., Baruah, U.,Gangopadhyay, S.K. and Kumar Dilip. (1998). Soil formation as influenced by Geomorphic processes in the Brahmaputra flood plain of Assam. *J. Indian Soc. Soil Sci.* **56**: 647 -- 651.

Bhattacharyya, T, Baruah, U, Gangopadhyay, S.K.and Kumar Dilip. (1997). Characterization of some Aquepts occurring in Assam valley. *J Indian Soc. Soil Sci.* **45** : 791 -- 796.

Bora, P.K. (1967).A study of some aspects of fertility status of Assam soils. Unpublished M.Sc. (Agri) Thesis, Dibrugarh University, Dibrugarh, Assam.

Bordoloi, L.J. and Talukdar, N.C. and Talukdar, H. 2000. Effect of slurry method of P applicationa and phosphate solubilizing bacteria on growth of transplanted rice. Communicated to J. Indian Society of Soil Science.

Chakravarty, D.N. and Majumder, N.N. (1970). A study on phosphate status of Assam soils, I. Total, organic, inorganic and available phosphorus and fixation of phosphorus. *Indian J. Agric. Chem.* **3**: 87--94.

Chakravarty, D.N. and Majumder, N.N. (1971). A study on phosphate status of Assam soils. II. Fractionation of inorganic phosphorus. *Indian J. Agric. Chem.* **4**: 52--55.

Chakravarty, D.N., Tewari, S.N. and Barthakur, B.C. (1987). *Assam Soils and Their Management. Research Bull.* No. **8**, Directorate of research, Assam Agricultural University, Jorhat, Assam, pp 55.

Chakravarty, D.N. (1977). Influence of climatic and topography on the pedogenesis of alluvium derived soils in Assam. Ph. D. thesis ,PAU , Ladhuiana.

Chakravarty, D.N., Sehgal, J.L. and Dev, G. (1980). Pedogenesis of some soils at varying stages of development in the southern bank of river Brahmaputra in Assam, factors influencing. *J. Res. Assam agric. Univ.* **1 and 2**: 153--159.

Chakravarty, D.N., Sehgal, J.L. and Dev, G. (1978). Influence of climte and topography on the pedogenesis of alluvium-derived soils of Assam. *Indian J. agric. Chem.* **11**: 77--97.

Chakravarty, D.N., Sehgal, J.L. and Dev, G. (1979). Sand mineralogy of the alluvium derived soils of southern bank of the river Brahmaputra in Assam. *J. Indian Soc. Soil. Sci.* **27** : 417-426.

Chakravarty, D.N., Sehgal, J.L. and Dev, G. (1982). Characterization and pedogenic development of Fe-Mn gleabules in the alluvium-derived soils of Assam, Indian. *Pedologie*. **32**: 39--51.

Chakravarty, D.N., Tewari, S.N. and Borthakur, B.C. (1987). *Assam soils and their fertility management. Research Bulletin No. 8.* Assam Agril. Univ. Jorhat.

Chakravaryty, S.K., Sinha, H. and Mathur, B.S. (1984). Morphological and physicochemical properties of some alluvial soils of Assam. *J. Indian Soc. Soil Sci.* **33** : 1328--36.

Chakraverty, D.N., Sehgal, J.L., Pal D.K. and Dev. G. (1981). Soil Taxonomy of some alluvium-derived soils of Assam. *J. Indian Soc. Soil Sci.* **29** : 125-- 126.

Chakraverty, D.N., Sehgal, J.L., Pal D.K. and Dev. G. (1992). Clay mineralogy of Assam soils developed on alluvium. *Agropedology*. **2**: 45--49.

Chaliha, A.(1971). Response of three dwarf wheat varieties to different levels of NPK fertilizer application with and without lime. Unpublished M.Sc. (Agri) Thesis, Assam Agricultural University, Jorhat, Assam.

Chamuah, G.S.; Gangopadhyay, S.K.; Walia, C.S, and Baruah, U. (1996). Soils of Jorhat district: Physiography Relationship. *Agropedology* . **6** : 29--36.

Chaudhury, M. 1996. Characterization of soils of Rudrasagar oil field in the vicinity of Group Gathering Station (GGS) in relation to affluent contamination. M. Sc. thesis submitted to the Department of Soil Science, Assam agricultural University, Jorhat, Assam.

Das, K.K. 1990. Evaluation of efficient *Rhizobium* strains for common pulses grown in Assam. Ph. D. Thesis, AAU, Jorhat.

Das, K.N., Borah, N. Bordoloi, P.K. and Sharma, P.K. (1997). Soils of rain shadow belt of Central Brahmaputra Valley of Assam ( India ) – their characterization and classification. *Intern J. Trop. Agric.* **15** : 149-- 158.

Das, M. C., Bora, P. K. Sarma, S. C. and Sarmah, K. C. (1971). Manganese status of Assam soils. *Fert. News* **16** : 42-44.

Deka, S.K. and Goswami, D.C. 1992. Hydrology, sediment characteristics and depositional environment of wetlands. A case study of Deepar beel of Assam. *J. Assam Sci. Soc.* **34(2)**:62-84.

Dent, D. and Young A. (1981). Soil Survey and Land Evaluation. London.

Dey (1999). Elemental composition and association in some Inceptisols and Alfisols of Assam under varying land use. *J. Indian soc. Soil Sci.* **47**: 305--310.

Dey, J.K. (1987). Pedogenesis and classification of some submerged soils of Assam and their suitability for rice cultivation. Ph. D. thesis, PAU, Ludhiana.

Dey, J.K. and Sehgal J.L.(1997a) Characterization and classification of some alluvium-derived paddy and associated non-paddy soils of Assam. Agropedology. 7: 22--31.

Dey, J.K. and Sehgal, J.L. (1997b). Clay mineralogy of some alluvial soils of Assam under long term paddy cultivation. *Agropedology* **7** : 48--53.

Dey, S.K., Bhattacharyya, N.G., Sengupta, A.K., Barua, D.N. 1967--68. Soil fertility survey. Part 1: Soils of Jorhat district. Toklai Occassional scientific papers No.1. Tea Research Association, Jorhat-13, Assam, India.

Dutta S. and Karmakar. R.M. 1995. Characterization and classification of soils of Instructional-cum-Research farm of Assam Agricultural University. *JASS*. **8** : 40 -- 46.

Dutta, M.N. (1974). Studies on profiles of some characteristic soil groups of Assam. Unpublished M.Sc. (Agri) Thesis, Assam Agricultural University, Jorhat, Assam.

Gangopadhyay, S.K., Walia, C.S., Chamuah, G.S. and Baruah, U. (1998). Rice growing soils of Upper Brahmaputra Valley of Assam – their characteristics and suitability. *J. Indian Soc. Soil Sci.* **46** : 103--109.

Gaswami, D.N. (1960). Geology of Assam., Department of Publication, Univ. Gauhati, Assam.

Gohain K.K. and Barooah, A.K. (1996). Nutrient indices of Assam estate tea soils with reference to organic matter and soil available potash. *Two and a Bud* **43** (2) : 18-19.

Harmine, A (1968). Agronomic studies on response to sugarcane variety CO 419 to application of phosphatic and potassic fertilizers in presence and absence of lime. Unpublished M.Sc. (Agri) Thesis, Dibrugarh University, Dibrugarh, Assam.

Hazarika, S., Medhi, B.K., Talukdar, N.C. and Thakuria, D. 2000. Effect of long-term cropping with or without fertilization on physico-chemical and biological properties of an aeric haplaquept soil of Assam, India. In Extended summaries vol. 2. Voluntary papers-Natural Resources, International Conference on managing natural resources for sustainable agricultural production in the 21st century. New Delhi, India, February 14-18, pp 193--194.

Kakoti, G.N. and Bhattacharya, K.G. 1989. Physicochemical properties of the deepar beel water. *J. Assam Sci. Soc.* **31 (3):**15--20.

Karmakar, R.M. (1985). Genesis and classification of soils of Northern Brahmaputra valley of Assam. Ph. D. Thesis ; IARI, New Delhi.

Karmakar. R.M. (1995). Productivity and potentiality evaluation of some soils of lower Brahmaputra Valley Zone of Assam. In *Proc. Sem. Agric. Sci. Soc.* NE India, pp 199-203, Nov. 27--28, 1995.

Mandal, S.C., Sinha, M.K. and Sinha, H. (1975). *Acid soils of India and Liming*, ICAR *Tech. Bull.* (Agric) No. **51**, ICAR, New Delhi.

Mandal, S.C., Pande, H.K., Mahanta, K.C. and Borthakur, B.C. and Singh, P. 1981. Report of the ICAR research review committee for Assam Agricultural University, Jorhat, Assam. Indian Council of Agricultural Research, Krishi Bhavan, New Delhi.

Mann, H. H. and Gokhale, N. G. (1960). Soils of tea growing tracts of India. *J. Indian Soc. Soil Sci.* **8** : 191--200.

Medhi, M. (1999). Changes in available nutrients in soils of UBVZ of Assam as affected by liming. Unpublished M.Sc. (Agri) Thesis, Assam Agricultural University, Jorhat, Assam.

Moriom, G.A. (1994). Phosphate adsorption and desorption in acid soils of central Brahmaputra valley zone (CBVZ) of Assam. Unpublished M.Sc. (Agri) Thesis, Assam Agricultural University, Jorhat, Assam.

Nath, A.K. and Das, P. (1980). A study on zinc status of soils of Assam and its availability. *Indian J. Agric. Chem.* **13**: 93--100.

NBSS & LUP (1994). Soils of Assam (India) for land use planning NBSS & LUP, Nagpur-10 .

Pathak, J. C. (1975). Response of sugarcane (*Saccharum officinarum*) variety CO 740 to rates of NPK and lime. Unpublished M.Sc. (Agri) Thesis, Assam Agricultural University, Jorhat, Assam.

Riquier, J., Bramao, D.L. and Carnet, J.P. (1970). A new system of soil appraisal in terms of actual and potential productivity. (AGL/TESR/70/6) FAO, Rome.

Raychaudhuri, S.P., Agarwal, R.R., Dutta Biswas, N.R., Gupta, S.P. and Thomas, P.K. (1963). *Soils of India*, ICAR, New Delhi, pp. 20-35.

Sarmah, A.C. (1990). Characterization of organic matter in two major soil orders of Assam. Unpublished Ph. D. Thesis, Assam Agricultural University, Jorhat, Assam.

Singh, G. (1975). Response of wheat varieties to different levels of NPK fertilizer application in presence and absence of lime. Unpublished M.Sc. (Agri) Thesis, Assam Agricultural University, Jorhat, Assam.

Singh, A.K. (1997). Studies on characterization of four phosphate solubilizing bacterial (PSB) isolates and their effect on green gram (*Vigna radiata* L. Wilczek). A. M. Sc. Thesis submitted to the Department of Agricultural Biotechnology, Assam Agricultural University, AAU, Jorhat, Assam.

Talukdar, H. and Barthakur, H.P. (1981). Solubilization of rock phosphate by root suspension inoculum of certain rice varieties grown in Assam. *J. Res. Assam Agric. Univ.* **25(1)**: 46--51.

Talukdar, N.C. 2000a. Isolation and characterization of Azospirillum and phosphate solubilizing bacteria from rhizosphere soils of rice grown in acidic soil of Assam. Communicated to the Indian Journal of Microbiology.

Talukdar, N.C. and Chakravarty, D.N. (1988). Effect of varying levels of N, P and K on grain yield and nutrient uptake of *ahu* rice in a light textured soil under high annual rainfall. *Ann. Agric. Res.* **9** (2): 159--164.

Talukdar, N.C. and Hazarika, S. (1998). Comparision of Rhizobium nodule number and yield of green gram grown in fallow and fifity year cultivated soils amended with farm yard manure. Paper presented in the International Conference on Environment and Agriculture held at Kathmandu, Nepal. Nov 1--3, 1998.

Talukdar, N.C. (1999). Vesicular-arbuscular mycorrhizal fungi (VAMF) biofertilizer-propspects and potential in N.E. regions of India. In *Use and Development of Microbial Inoculants* (eds. T. Singh, A.K. Yadav, S. Roychaudhury). Published by Regional Biofertilizer Development center, Imphal. pp 85-98.

Talukdar, N.C. (2000b). Third annual report of the adhoc project "Growth of two timber yielding trees in degraded forest sites as affected by their symbiont mycorrhizal fungi" submitted to the Indian Council of Forestry Research and Education, Dehra Dun from the Soil Microbiology laboratory, Department of Soil Science, Assam Agricultural University, Jorhat, Assam.

Talukdar, N.C., Mannan, A.M., Thakuria, D. and Barua, P.P. (2000). Effect of inoculation of N-fixing and P-solubilizing bacteria in compost base to rice and green gram (*Vigna radiata*). Comunicated to the Indian Journal of Microbiology.

Tamuli, A. (1997). Behaviour of potassium in some soils of upper Brahmaputra valley zone (UBVZ), Assam. Unpublished M.Sc. (Agri) Thesis, Assam Agricultural University, Jorhat, Assam.

Tewari, S. N, Dutta, I. And Deka, A. K. (1969b). HF-extractable ammonium and ammonium fixing capacity in some soils of Assam. *Indian J. Agric. Chem.* **2**: 18--21.

Tewari, S. N. (1970). Molybdenum nutrition in plants as influenced by lime and phosphate application. *J. Inst. Chemist.* (India) **62** : 29--32.

Tewari, S. N. and Chakravarty, D. N (1972). Copper status of Assam soils. *Assam Agric.* **2**: 73--77.

Tewari, S. N., Chakravarty, D. N. and Bora, P. K. (1969a) Retension and transformation of soluble phosphate added to Assam (India) soils. *J. Inst. Chemist.* (India) **61**: 174--176.

Venkataraman, M.K., Kakoty, R. and Singh, H. (1990). Occurrence and distribution of vesicular-arbuscular mycorrhizal fungi in acid soils of North Eastern India. In (Jalali, B.L. and Chand, H. eds) Current trend in mycorrhizal research. Proceedings of the national conference on mycorrhiza, HAU, Hissar. pp 29--30.

Wadia, D.N. (1975). Geology of India. Tata McGraw Hill, New Delhi, pp. 73--88.

# CHAPTER 5

# AN OVERVIEW OF THE BRAHMAPUTRA RIVER SYSTEM

J. N. SARMA

The Brahmaputra River is an antecedent, snow fed, and large trans-Himalayan river originating in southern Tibet (China). It runs for a length of 2880 km and its river system drains parts of China, Bhutan, India and Bangladesh. Considerable variations in width, gradient, discharge and channel pattern occur throughout its course. The system is described here by dividing the course into ten unequal reaches. The name of each reach and number of notable tributaries joining it are given sequentially as follows: from the source to Pindzoling-8, Pindzoling to Gytsa Dzong-8, Gytsa Dzong to Monku-4, Monku to Pasighat-38, Pasighat to Laikaghat-6, Laikaghat to Dihingmukh-15, Dihingmukh to Dhansirimukh-14, Dhansirimukh to Pandu-33, Pandu to Dhubri-17 and Dhubri to Goalundo-7. The course through the plains of India and Bangladesh possess several distributary channels. The tributaries contributing a significant part of the discharge of the Brahmaputra are: the Kyi, Chu in Tibet; the Dihang, Lohit, Subansiri, Burhi Dihing, Kopili, Jia Bhareli, Manas and Gangadhar in India; and the Tista in Bangladesh.

## 5.1 Introduction

The Brahmaputra River, termed a moving ocean, is an antecedent snow fed river which flows across the rising young Himalayan Range. Geologically, the Brahmaputra is the youngest of the major rivers of the world and unique in many respects. It happens to be a major river for three countries, viz., China, India and Bangladesh. The river basin of the Brahmaputra is bounded on the north by the Kailash and Nyen-Chen-Tanghla ranges of mountains; on the east by the Salween River basin and the Patkai range running along the Indo-Myanmar border; on the south by the Nepal Himalayas, the Naga and Barail ranges and the Meghalaya Plateau; and on the west by the Ganga river basin. The maximum meridional extent of the basin is 1540 km along $29^0 30'$ N latitude and maximum latitudinal extent is 780 km along $90^0$ E longitude.

The total length of the river is 2,880 km, out of which 1,625 km fall in China (Tibet), 918 km in India (Arunachal Pradesh and Assam) and the rest 337 km in Bangladesh. Several tributaries join the river all along its length. The drainage basin of the Brahmaputra extends over an area of 520,000 km$^2$. Of this 293,000 km$^2$ are in China,



*V.P. Singh et al. (eds.), The Brahmaputra Basin Water Resources*, 72-87.

45,000 $km^2$ in Bhutan, 195,000 $km^2$ in India, and 47000 $km^2$ in Bangladesh. The average annual runoff of the Brahmaputra at Pasighat and Pandu in India as well as Bahadurabad in Bangladesh is 186,290, 494,357 and 589,000 million cubic metres, respectively. The monsoonal flow of the Brahmaputra at Tesla Dzong in Tibet is 36.27% of the flow at Pasighat (WAPCOS, 1993).

## 5.2 The River Course

The river finds its origin in Tibet and flows through its southern part in the name of 'Tsangpo' in the easterly direction almost parallel to the Himalayas for about 1,100 km. At places it has a wide channel and flows at a moderate gradient. Then, it turns north-east and then north to traverse through a series of cascades and rapids within the Himalayan range; thereafter, it turns south and south-west and thus traversing about 525 km at a steeper gradient, enters Arunachal Pradesh of India in the name of the Dihang or the Siang. After flowing about 226 km through the Himalayas in Arunachal Pradesh it enters into the plains near Pasighat. About 52 km downstream from Pasighat, the Dihang is joined by two other Trans-Himalayan rivers, viz., the Dibang and the Lohit near Laikaghat. From here the combined flow of the three rivers is called the Brahmaputra. The Brahmaputra flows through a narrow valley which is known as the Assam valley or the Brahmaputra valley in about east--west (E--W) direction for 640 km with a very low gradient. The valley is like a winnowing fan bounded on its three sides by hill ranges and a plateau on the south. The average width of the valley is about 86 km and of the width of the valley the river itself at places occupies 15--19 km. Throughout its course in the Assam valley the Brahmaputra is braided with some well defined nodal points where the river is narrow and restricted by stable banks. In this valley it is joined by several tributaries from both sides. On the west the valley is open and beyond Assam it widens into a broad low lying deltaic plain of Bangladesh. The Brahmaputra, after traversing the spurs of the Meghalaya Plateau, turns south and enters Bangladesh with the name of 'Jamuna'. The river flows across the plains of Bangladesh for 337 km before joining the Ganga at Goalundo.

The drainage map of the Brahmaputra, showing its major tributaries, is given in Figure 5.1. Figure 5.2 shows the longitudinal profile of the river. It is apparent from the profile that a sudden and large drop in the height of the river bed occurs just before the Indo-China border.

## 5.3 Earlier Work

According to ancient literature, e.g., the 82nd and 83rd chapters of the Kalika Purana, and the 55th chapter of the Padmapurana the source of the Brahmaputra lies within the Tibetan Himalyas. The river is regarded as the son of lord Brahma; hence the Brahmaputra is a masculine river. But the general contention was that the river Lohit, the tributary coming from east, is the main course of the Brahmaputra. Therefore, the name Luit or Lohitya has also been used for the Brahmaputra in some parts of Assam and in Sanskrit literature.

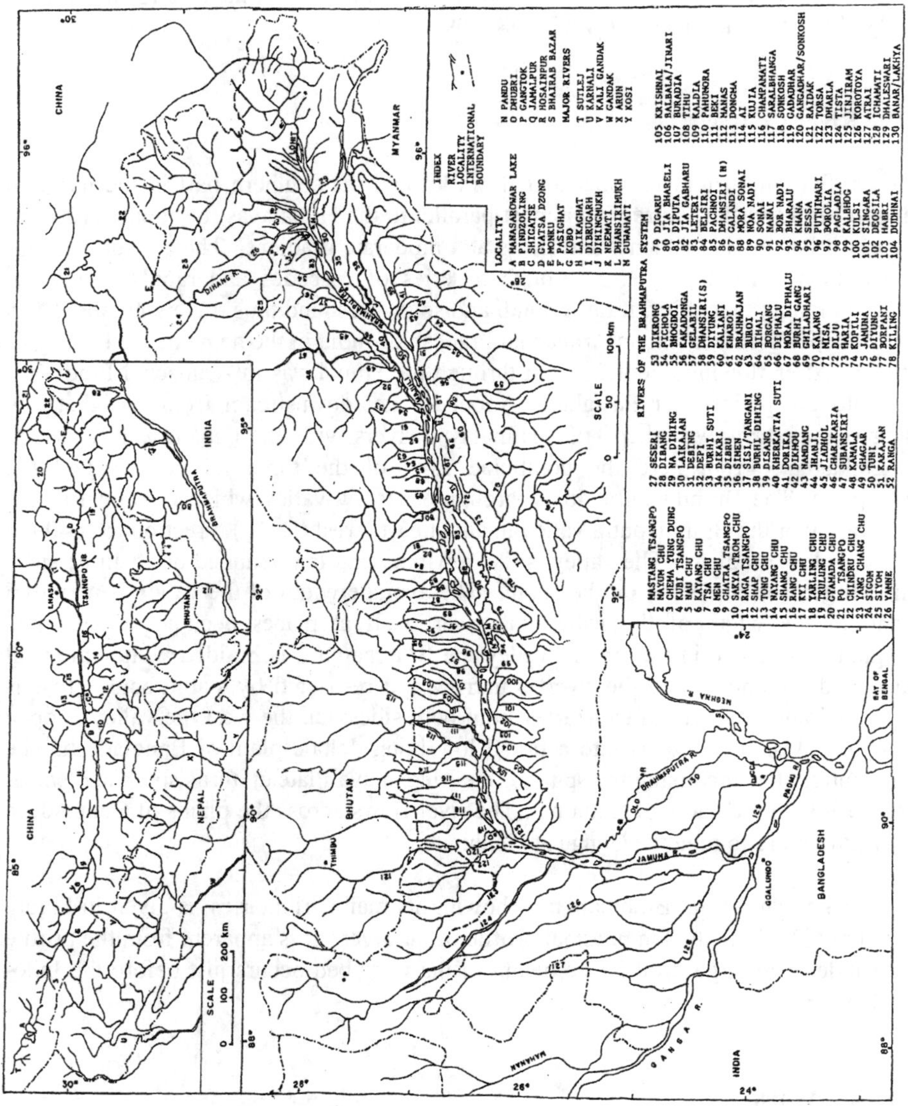


Fig. 5.1 The map of the Brahamaputra River system. The Tsangpo is shown in the inset.

Index for Figure 5.1

| | | | | | | | |
|---|---|---|---|---|---|---|---|
| A | Manasarowar Lake | 15 | Shang Chu | 54 | Pichola | 93 | Bharalu |
| B | Pindzoling | 16 | Rang Chu | 55 | Bhogdoi | 94 | Khana |
| C | Shigatse | 17 | Kyi Chi | 56 | Kakadonga | 95 | Sessa |
| D | Gyatsa Dzong | 18 | Yarling Chu | 57 | Gelabil | 96 | Puthimari |
| E | Monku | 19 | Trulung Chu | 58 | Dhansiri(S) | 97 | Borolia |
| F | Pasighat | 20 | Gyamada Chu | 59 | Diyung | 98 | Pagladia |
| G | Kobo | 21 | Po Tsangpo | 60 | Kaliani | 99 | Kalbhog |
| H | Laikaghat | 22 | Chindru Chu | 61 | Kharoi | 100 | Kulsi |
| I | Dibrugarh | 23 | Yang Chang Chu | 62 | Brahmajan | 101 | Singara |
| J | Dihingmukh | 24 | Sigom | 63 | Buroi | 102 | Deosila |
| K | Neemati | 25 | Siyom | 64 | Bihali | 103 | Habri |
| L | Dhansirimukh | 26 | Yamne | 65 | Borgang | 104 | Dudhnai |
| M | Guwahati | 27 | Seseri | 66 | Diphalu | 105 | Krishnai |
| N | Pandu | 28 | Dibang | 67 | Mora Diphalu | 106 | Balbala/Jinari |
| O | Dhubri | 29 | Na Dihing | 68 | Burhi Gang | 107 | Buradia |
| P | Gangtok | 30 | Laikajan | 69 | Ghiladhari | 108 | Tihu |
| Q | Jamalpur | 31 | Debing | 70 | Kalang | 109 | Kaldia |
| R | Hosainpur | 32 | Depi | 71 | Misa | 110 | Pahumora |
| S | Bhairab Bazar | 33 | Burhi Suti | 72 | Diju | 111 | Beki |
| T | Shutlez | 34 | Dikari | 73 | Haria | 112 | Manas |
| U | Karnali | 35 | Dibru | 74 | Kopili | 113 | Dongma |
| V | Kali Gandak | 36 | Simen | 75 | Jamuna | 114 | Ai |
| W | Gandak | 37 | Sisi/Tangani | 76 | Diyung | 115 | Kujia |
| X | Arun | 38 | Burhi Dihing | 77 | Borpani | 116 | Champamati |
| Y | Kosi | 39 | Disang | 78 | Kiling | 117 | Saralbhanga |
| 1 | Mastang Tsangpo | 40 | Kherkatia Suti | 79 | Digaru | 118 | Sonkosh |
| 2 | Mayum Chu | 41 | Dorika | 80 | Jia Bhareli | 119 | Gadadhar |
| 3 | Chema Yung Dung | 42 | Dikhou | 81 | Diputa | 120 | Gangadhar/Sankosh |
| 4 | Kubi Tsangpo | 43 | Namdang | 82 | Jia Gabharu | 121 | Raidak |
| 5 | Nau Chu | 44 | Jhanji | 83 | Leteri | 122 | Torsa |
| 6 | Kayang Chu | 45 | Jiadhol | 84 | Belsiri | 123 | Dharla |
| 7 | Tsa Chu | 46 | Charikaria | 85 | Pachnoi | 124 | Tista |
| 8 | Men Chu | 47 | Subansiri | 86 | Dhansiri (N) | 125 | Jinjiram |
| 9 | Chatra Tsangpo | 48 | Kamala | 87 | Galandi | 126 | Korotoya |
| 10 | Sakya Trom Chu | 49 | Ghagar | 88 | Mora Sonai | 127 | Atrai |
| 11 | Raga Tsangpo | 50 | Tuni | 89 | Noa Nadi | 128 | Ichamati |
| 12 | Shap Chu | 51 | Kakajan | 90 | Sonai | 129 | Dhaleswari |
| 13 | Tong Chu | 52 | Ranga | 91 | Nanai | 130 | Banar/Lakhya |
| 14 | Nayang Chu | 53 | Dikrong | 92 | Bor Nadi | | |

The course of the Brahmaputra was not completely surveyed till as recently as the first quarter of the twentieth century. Several workers attempted to identify the actual course of the river and drew various conclusions. Wade (1927) wrote the first geographical account of the rivers of Assam. He also accepted the Lohit as the main flow of the Brahmaputra. British explorers later established that the Dihang was the main course of the Brahmaputra. In the early nineteenth century several British officers, such as Captain Wilcox, Captain Neufville, Captain Bedford, etc., tried to explore the upper reaches of the Dihang without much success. Later, Krick and Bourry, two French missionaries, were murdered while making attempts to explore the upper reaches of the river. Colonel Woodthorpe and Captain Harman also made attempts, but were unsuccessful. However, Harman was able to measure the discharges of the Dihang, and Dibang and Lohit near Sadiya in 1980. Three pandits, Nain Singh, Nem Singh and Kishen Singh (1870-80), distinguished as Buddhist monks, ascertained the course of the Tsangpo between Shigatse, Tse-tang, and Gyala. The observations made by the explorer Kinthup in 1884, who was sent by Captain Harman, were not believed as he was illiterate. Exploring the upper course of the Dihang in 1886 Captain Needham could dispel most of the doubts that the Tsangpo is flowing into India as the Dihang. About 1913 Lt. Col. Bailey and Captain H.T. Marshead established the course of the Dihang when they went upstream from Sadiya. However, Captain Kingdon Ward and Lord Cawdar (1924) established the course along the upper reaches. Later Swami Pranavananda visited the source of the Brahmaputra on June 17 and 18, 1937 and gave an account of the same in the book 'Kailas Manasarowar'. It is noteworthy that various modern studies on the Brahmaputra have been confined mostly to the part of the river system within the territory of India, and mostly the Assam valley.

## 5.4 The Brahmaputra River System

Taking into consideration certain features of the river such as channel width and type, gradient, river flow and junction of major tributaries, an attempt is made here to describe the entire Brahmaputra River system part by part by dividing its course into ten equal reaches. The major features of the river system in each reach, starting from its origin to the confluence point, are described sequentially in what follows.

### 5.4.1 THE REACH FROM THE SOURCE TO PINDZOLING

From the source to the place named Pindzoling (29° 25'N, 88° 02'E), this initial reach of the Brahmaputra in Tibet is about 650 km in length. The river is known as the Tsangpo in Tibet and near its origin it is called the Mastang Tsangpo or the Tamchok Khambab. According to the Tibetans the source of this river lies in the Kanglung Kang glacier at an elevation of 5,300 m (82° 10'E, 30° 30'N) near Lake Konggyu Tso, where it is called the Tamchok Khambab Kangri. Many snow fed streams join the river from the passes of Maryum-La (5,150 m) and Marnyak-La (5,303 m). These passes are about 60 km southeast of Lake Manasarowar from where two other big rivers, viz., the Indus and the Sutlej have their sources.

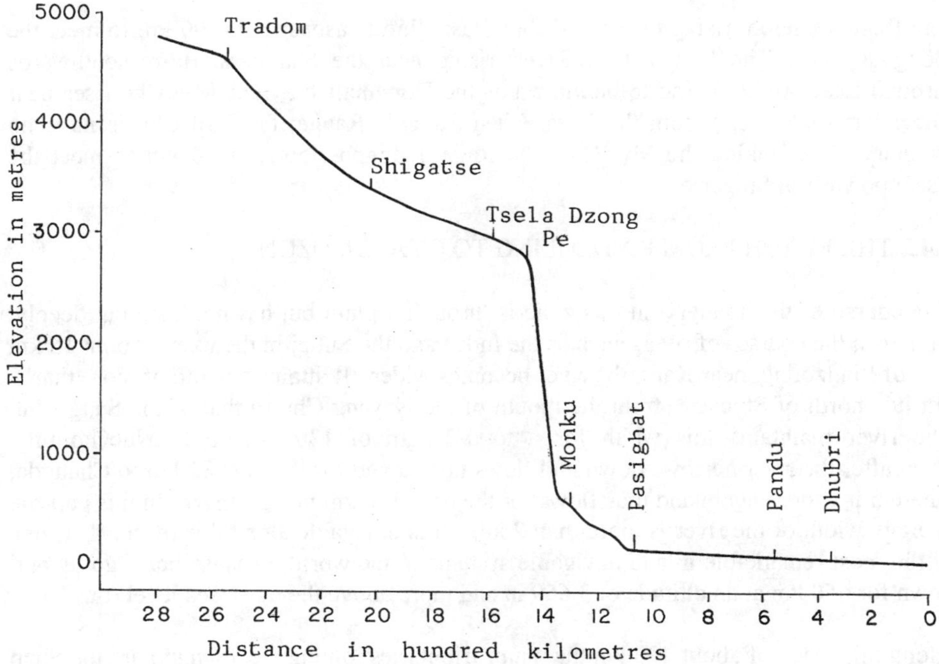


Fig. 5.2 The longitudinal profile of the Brahamaputra.

After flowing for 80 km in an easterly direction, the Tsangpo meets two homologous rivers, viz., the Mayum Chu and the Chema Yung Dung. The Tsangpo continues its easterly journey through two stupendous mountain ranges, viz., the Lapchung and the Nyen Chen Tanghla in the north and the Himalayas in the south. The northern boundary of the basin is formed by several glaciers located in the various ranges from which the river drains water. From west to east these are the Kailash range, 6,100 m (longitude 82°--84° E), the Lunkar range, 6,000 m (longitude 84° -84° 05'E) and the Nyen Chen Tanghla range, 7,088 m (long. 86° 25' -- 95° 00'E). The southern limit of the Tsangpo basin is formed by the Himalayas. The peculiarity of the Himalayas in this reach is that they do not form a watershed. The main river is about 160 km north of the Himalayas. The drainage area spreads to a maximum of 80 km to the south and 135 km to the north from the Tsangpo when measured across Tradom (29° 40' N, 84° 10' E). The major tributaries joining the reach are the Nau Chu. Tsa Chu, Men Chu, Chatra Tsangpo, and Raga Tsangpo in the left bank and the Kubi, Kayang and Sakya Trom Chu in the right bank besides numerous unnamed streams. Of these the Nau Chu, Chatra Tsangpo, and Raga Tsangpo are large tributaries.

The Chatra Tsangpo rises in the Sangmo Perlic Pass (5,818 m). It flows south to Lake Lapchung Tso and from there it is known as the Chaktak Tsangpo. Thereafter, it flows west of Saka and takes its principal tributary, the Rukshok Tsangpo, on the right bank.

Glaciers from the Lapchung, Kangchung Kangri and Kailash ranges feed this river.

The Raga Tsangpo, rising near the Gebuk Pass, flows eastwards for 90 km to meet the Dongma Chu. The Ksham Chu River, rising near the Sha Pass, flows southwards through Lake Amkyok Tso to be known as the Dongma Chu. The Myi Chu rises near Pubchenri (6,200 m ) from the Nyen Chen Tanghla Range; the Bup Chu is its main tributary. After taking the Myi Chu, the Raga Tsangpo flows for 50 km to meet the Tsangpo west of Shigatse.

### 5.4.2 THE REACH FROM PINDZOLING TO GYATSA DZONG

The course of the Tsangpo in this zone is through a plain but has not been cut deeply, similar to the courses of rivers such as the Indus and the Sutlej in the west. About 15 km east of Pindzoling, near Karu, the river becomes wider. It attains a width of more than 3 km just north of Shigatse, from the mouth of the Nayang Chu to that of the Sang Chu. The river maintains this width for a total length of 130 km up to Nubchoilung. Thereafter the river narrows down and flows unchanged for the next 85 km to Changda, wherein it widens again and thus flows for the next 170 km up to Sangri. In this part the average width of the river is more than 2 km. It is navigable along this reach. It is one of the most remarkable inland navigable systems in the world, because boats go up and down for 650 km at an altitude of 3,659 m and more above the mean sea level (m.s.l.)

Along this reach of about 490 km the major tributaries joining the Tsangpo are the Shap Chu, Nayang Chu, Rang Chu, Yarling Chu and 33 more unnamed notable ones on the right bank as well as the Tong Chu, Shang Chu, Kyi Chu and 36 unnamed ones on the left bank. Out of these the Nayang Chu and Kyi Chu are much bigger. The precursor of the Nayang Chu is the Tumbayung Chu which rises from Lake Yu Tso, lying west of Toma La Pass in the Central Himalayan Range north of Chomu Lhari (7,314 m). On its way it receives the Kangrupo Chu on the left and the Tran Chume Chu on the right bank. After flowing past Gyangtse Dzong, it is known as the Nayang Chu. Thereafter it flows in a northwesterly direction and takes the Laraphu Chu on the left and passes east of the great trade centre of Shigatse to meet the Tsangpo. The Nayang Chu is a snow fed river. The Kyi Chu or the Gya Chu is the largest tributary of the Tsangpo in Tibet. It is about 300 km in length and the capital city of Lhasa on its bank. The Miggi Tsangpo, rising near southeast of Lake Dong Tso (31° 0' N, 91° 0' E) at the Shang Slung Pass (6,410 m) in the Nyen Chen Tanglha range, is the precursor of the Kyi Chu. Along its southerly course it receives the Nab Chu and the Dugong on the left and the Uri Rong on the right and thereafter is known as the Gya Chu. After passing by the side of Lasha it receives the Tobing Chu, a snowfed tributary, on its right bank before meeting the Tsangpo.

### 5.4.3 THE REACH FROM GYATSA DZONG TO MONKU

The town of Gyatsa Dzong, lying near the longitude 93° E, is situated on the left bank of the Tsangpo; from the longitude 92° 30' E the Tsangpo enters into the most difficult reach in a gorge with rocky banks thousands of metres high. The reach from Gyatsa Dzong to Monku (29° 12' N, 95° 0' E), situated near the Indo-China border, is about 485 km long. The riddle of this part remained unsolved till 1924. It is quite surprising that the Tsangpo which continues a north-easterly journey along the northern slope of the Himalayas, appears to be untraceable within the stupendous mountainous gorge between two adjacent high peaks, viz., the Gyala Peri (7,150m) and the Namcha Barwa (7,755 m). The river takes a hairpin bend to turn south to cut across the Himalayas and south-west through the same. Along this reach 40 notable tributaries including the Trulung Chu join the river on its right bank and 24 notable tributaries together with the Giamda Chu, Po Tsangpo and Chindru Chu join the Tsangpo on its left bank.

From Gyatsa Dzong to Tsela Dzong (near 94° E ), the southern side of the reach is dotted with numerous glaciers which partly feed the Tsangpo and partly the Subansiri, Jia Bhareli and Manas rivers which join the Brahmaputra in India. On the northern side also the river drains many glaciers in a wide reach from the Nyen Chen Tanghla Range and the Nyen Chen Tanghla Chomo Range (long. 90° -- 97° E).

The river changes its direction from east to northeast at about 50 km before reaching Tsela Dzong. Further east, at Pe (2950 m), the river is still a broad placid stretch of water about 600 m wide. Then turning abruptly to the northeast and north, it makes its way by a succession of stupendous gorges between the huge mountain masses of the Gyala Peri and the Namcha Barwa. Through these deep narrow gorges the Tsangpo rushes down in a series of cascades and rapids. At Gompone it takes from north the Po Tsangpo. Thereafter the river turns progressively eastward, southward and finally southwestward to reach Monku. About 1.6 km before the border there is a fall of about 24.40 m on the main river. This is the only known fall in this river. The length of the river between Tsela Dzong and Indian border is about 260 km and its drop is by about 2,200 m. Within this reach the maximum drop is found in between Pe and 50 km upstream of Indo-China border.

The Giamda Chu joins the Tsangpo at Tsela Dzong. Its precursor is called the Kam Chu which passes by Giamda Dzong to be known as the Giamda Chu. Its chief tributary is the Drukla Chu, which is larger than the main river. The Drukla Chu rises as the Song Chu in the Nyen Chen Tanglha range and meets the Giamda Chu at Tro. The combined flow runs for about 100 km south-east to meet the Tsangpo. The Po Tsangpo is a hilly river which flows in a direction just opposite to that of the Tsangpo and meets the latter as a 70 m wide swift torrent. The Ngagong Chu and the Poto Chu join at Khata to create the Po Tsangpo of which the latter is longer and originates from the glaciers of the southern slope of the Nyimo Chhomo Range (95° 30' -- 96° 40' E). The Cho Dong Chu and the Yu Chu are tributaries of the Ngagong Chu and the Poto Chu, respectively. The Nok Chu and the Sya Chu, both rising from the Nyen Chen Tanghla range, join the Po from the Alado Chu. The Alado Chu, after passing through Lake Yigrong Tso, is known as the Yigrong Chu which joins the Po Tsangpo from north. The Rong Chu is another tributary joining the Po Tsangpo from west at about 15 km downstream of the Yigrong Chu.

Although the Tsangpo has an easterly course throughout the north of the Himalayas, a large number of its small and large tributaries flow in the westerly direction, i.e., meeting the Tsangpo by flowing from the opposite direction, thereby developing a barbed type drainage pattern. Among these major tributaries are the Sakya Trom Chu,

Rhe Chu, Nayang Chu, Rang Chu, and Yarling Tsu. This feature has led to the speculation that the Tsangpo might originally have flowed westwards. Which route it chose is not certain for it may have followed what is now the course of the Kali Gandak or the Karnali or the Sutlej. According to Burrard et al. (1932) there is some evidence that the Kali Gandak (a tributary of the Ganga in Northern India) might be the most likely contender, because the Phuto Pass, separating the Kali Gandak and the Tsangpo, is an extraordinary depression barely 75 m higher than the Tsangpo valley. Moreover, the gorge of the Kali Gandak through the main Himalayan range is also too deep to have been cut by it which has a very small catchment above the gorge. Subsequently the Dihang captured the Tsangpo by its headward erosion and reversed the flow of the latter.

### 5.4.4 THE REACH FROM MONKU TO PASIGHAT

The Tsangpo enters the Indo-China border near Monku at an elevation of 660 m and flows for 5 km as the international boundary to arrive at Korbo in Arunachal Pradesh of India. In Arunachal Pradesh the river is known as the Siang or Dihang. The lengths of the river reaches from Korbo to its mouth and to Pasighat are 918 km and 226 km, respectively. The Dihang flows through the Himalayas of Arunachal Pradesh at first in the southwest direction but after meeting the Sigong River on its right bank takes a right angled turn to flow southeastwards and finally southwards to cut across the strike of the geological formations. The average gradient of the river in this reach is 1 in 515. Between Rotung and Rengging, just upstream of Pasighat, the river meanders for a distance of 15 km. River terraces are also noticed along the river stretch between Yinkiong and Pasighat. Some of the terraces are at a height of 250 m above the present bed of the river. At Pasighat the Dihang has a maximum discharge of 29,643 $m^3 sec^{-1}$ and minimum of 1076 $m^3 sec^{-1}$.

A large number of hilly tributaries join the Dihang from both sides depicting a higher drainage density on the southern slope of the Himalayas. From Monku the tributaries joining the Dihing successively downstream are the Siyi, Rinsong, Rebuing, Siyi, Nidgong, Angong, Sibot, Sipu, Siring, Sipo, Sirap, Sisam, Sirnyuk, Mebung, Shimong, Siyom or Yombung, Yembung, Siding, Sireng, Sirpa Korong and Sille Korong on the right bank. Likewise, the left bank tributaries from Monku are the Yang Sang Chu, Siri, Sirin, Sipat, Sijar, Sikar, Sikem, Ebung, Pyabung, Rebung Yumbung, Sileng, Yapin ( Yamne ), Sibot Korong, Sopa Korong and Siku. Among these the Sigon, Siyom and Yang Sang Chu are much larger than the rest.

### 5.4.5 THE REACH FROM PASIGHAT TO LAIKAGHAT

Although this reach is only 56 km in length, this is where the river attains its characteristic braided pattern. The Dihang descends from the hilly terrain into the Assam Valley and is joined by the combined flow of two majestic rivers, viz., the Lohit and the Debang, near east of Kobo and north of Laikaghat. Only from here the Dihang is known as the Brahmaputra. The Dihang enters the plains near Pasighat and flows in the east southeast (ESE) direction as a braided river. Its flow is divided into two unequal channels. The major one is the Dihang proper and the other, known as the

Sibia, flows southeasterly at about 4 -- 5 km east of the former. This has resulted in formation of alluvial islands of enormous size, viz., Pekar, Lali and Kobo. The Sibia takes a small river named Seseri and meets the Lohit at the outfall of the main Dibang after flowing for another 23 km. On the downward journey the Dihang receives tributaries like the Kemi Korong, Ripiang Korong, Sille and Leko and also shoots off a branch, called Burhi Suti, on its right bank before meeting the Lohit at a distance of 20 km west of the outfall of the Sibia. At the tri-junction of the Dihang, Lohit and Debang the contribution from the Dihang is found to be about 31.63% of the total discharge of the Brahmaputra.

In ancient times the Lohit was regarded as the main Brahmaputra. According to an Indian tradition the Brahmaputra originated from the celebrated shrine named Brahmakund, which is situated at the foothills of the Mishimi Hills on the bed of the Lohit. The Lohit River finds its origin as the Yayul Chu from Mount Yoku (6,614 m) ( 29°30' N, 97°15' E) of China. Flowing south-easterly for 180 km the Lohit enters India and thereafter it flows for 190 km to meet the Dihang. Near the Indian border it is called the Tellu. Its important tributaries are the Rongtu Chu, Tiding, Kundil, Dibang and Sibia on its right bank. The tributaries joining its left bank are the Chai, Ghalum, Lati, Lang, Kerom, Tengapani, Noa Dihing and Chaikhowa. In the plain it has a braided channel infested with large channel bars and islands resulting in width varying from 2 to 6 km. The Lohit and many of its tributaries are perennially snow fed rivers. The Lohit contributes about 7.8% of the total flow of the Brahmaputra.

The Dibang is another major river of this reach and its entire basin lies in India. It rises as the river Adzon near a mount (5,355 m) at the northernmost point of Arunachal Pradesh. It has a linear southerly course. The major tributaries of Dibang are the Angi, Tajan, Ethun and Deopani,on its left bank and the Emra, Ahui, Sirki and Ishi on its right bank. Its lower 45 km course is extremely braided where it bifurcates; the western branch named Gango meets the Lohit about 2 km upstream of the outfall of its main branch. The Dihang contributes about 6.64% of total flow of the Brahmaputra.

### 5.4.6 THE REACH FROM LAIKAGHAT TO DIHINGMUKH

This reach is only about 97 km in length. The river has a westerly course up to the outfall of the Dibru River and then it has a southwesterly course to a little beyond Dihingmukh. Its channel is braided, and there are many channel bars and islands. The channel islands have their maximum and average lengths of 14.00 km and 3.75 km, respectively. The gradient of the river is 1 in 3700 between Laikaghat to Dibrugarh and 1 in 5595 from Dibrugarh to Neemati.

The river is navigable throughout the reach and has the important townships of Tinsukia and Dibrugarh on its South bank. The river shoots off a distributary named Burhi Suti from its north side at about 14 km downstream of Laikaghat and it joins the main river 60 km downstream. The Sisi and Tangani are two rivers joining the Brahmaputra directly on its north bank. On the south bank it receives small rivers, such as the Laikajan, Mohmorajan, Dibru, Maijan and the biggest tributary of this reach, the Burhi Dihing. The Burhi Dihing is formed by the Namphuk, Namchik, Megantion, Khaikhe,

Tirap, Digboi, Tingrai, and Sessa. It has a length of 360 km and the drainage basin of about 6,000 $km^2$ contributes 1.87% of the total discharge of the Brahmaputra.

A characteristic feature of the Brahmaputra in the Assam valley is the presence of distributaries (local name 'Suti'). The Buri Suti, mentioned above, flows almost parallel to the Brahmaputra maintaining a difference of 3 to 5 km. Many small streams, viz., the Yagalong, Rayang, Je, Debing, Gali, Depi, Nate, Dikari, Narad, Somkhong, and Simen join this Suti progressively from the east. Moreover, on the way the Burthi Suti also shoots off three branches to meet the Brahmaputra; amongst them the Lakhmi Suti is the largest. All the tributaries feeding this Suti find their origins in the lowermost slope of the Arunachal Himalayas. In this area the change in the courses of these rivers is very frequent resulting in interlacing of many streams and an intricate drainage network.

### 5.4.7 THE REACH FROM DIHINGMUKH TO DHANSIRIMUKH

This reach is about 130 km in length. The river has a braided channel with large channel bars. Here the river attains a width of about 15 km during flood period. The important townships are North Lakhimpur in the north and Sibsagar, Jorhat, and Golaghat on the South.

The major north bank tributaries of the river in this part are the Jia Dhol, and great Subansiri; along with the latter a combined flow of the Ranga, Pabha, Meneha and Digrong also meets the Brahmaputra. The Kachikata, Mora Dikrong and Pinchola are other smaller rivers of the north bank. On the south bank the major tributaries such as the Disang, Dikhow, Jhanji and Dhansiri along with smaller ones like the Balama, Dorika and Kikila join the Brahmaputra in this reach.

The Jia Dhol finds its origin in the lowermost Himalayas. It is famous for floods and frequent diversion of its channel. The Subansiri is the largest tributary of the Brahmaputra. Gold was mined commercially from its placer deposits till the recent past, hence it derived its name from gold (Sanskrit: Subarna = Gold). It is a perennially snowfed Trans-Himalayan river which is formed by the association of the Lokong Chu (Char Chu), Chayal Chu and Tsdri Chu in Tibet. After traversing the Himalayas and passing through the Miri Hills, it enters the plains and flows through them for 90 km before meeting the Brahmaputra. The Kamala, Ghagar and Sampara are its major tributaries in India. The total length of the river is 442 km. Draining an area of 37,700 $km^2$, the Subansiri observed a maximum discharge of 18,799 $m^3\ sec^{-1}$ and minimum of 131 $m^3\ sec^{-1}$. It contributes 7.92% of total yield of the Brahmaputra.

The Disang River originates from the mount of Patkai Bum in Arunachal Pradesh and flows 230 km. The Dikhow, the most famous holy river of ancient times, and the Jhanji both find their origins from the Naga Hills. The Yangmun and the Namdang are major tributaries of the Dikhow and the Mitong is the main tributary of the Jhanji. Some small rivers, viz., the Teok, Mudoijan, and Kakojan all originating in the plains flow collectively through the Kokila to the Brahmaputra.

The Dhansiri (south) originates in the southwest corner of the Laishong Peak (1868 m) of the Barali range. Its major tributaries are the Diyung near the foothills and the Kaliani and Gelabil near its mouth. The Bhogdoi and the Kakodong are two large tributaries flowing through the Gelabil to the Dhansiri. The Dhansiri has a length of 354 km and draining an area of 12,250 $km^2$, it contributes 1.03% of the total discharge of the Brahmaputra.

The most remarkable feature of this reach is the Majuli island which has an area of 1,250 $km^2$ and is considered to be the biggest river island of the world. It is formed by bifurcation of the Brahmaputra into two branches, viz., the Kherkatiya Suti or the Luit Suti on the north and the Brahmaputra (Dihing) on the south. Majuli is not a normal alluvial island formed by the Brahmaputra. It was a piece of land known as Majali, on the south bank of the Brahmaputra (Luit) situated north of the tributary of the Dihing River (Burhi Dihing) till the year 1,750. An extreme flood of that year caused the Brahmaputra (Luit) to divert its course southwards on the east of Majali, thereby joining it with the Dihing River. Subsequently the Brahmaputra diverted its main flow through the lower reach of the Dihing, and the land Majali in between the two rivers resulted in the island Majali which is now called Majuli. The Tuni is the main river in Majuli.

### 5.4.8 THE REACH FROM DHANSIRIMUKH TO PANDU

The reach is about 210 km in length. The river flows in this reach through alluvium with inselbergs at some places. The gradient of this river in this zone from Neemati to Pandu is 1 in 6571. The river has a braided channel except at Pandu where it is constricted into a single channel of 1.2 km wide. This is the narrowest point in the entire reach. The river is navigable in this reach throughout the year and has the important townships of Tezpur on its north bank and Naogaon and Guwahati on the south. A great number of tributaries of various magnitudes meet the river along this reach. On the north bank progressively from east are the Kharoi, Brahmajan, Buroi, Bihali, Jinjia, Borgang, Burhigang, Sadharu, Ghiladhari, Jia, Bharali, Deputa, Jia, Gabharu, Belsiri, Panchnoi, Dhansiri (north), Doipham, Mora Dhansiri, Dhansiri, Galandi, Noanadi, Na Nai, and Barnadi. The biggest amongst them is the Jia Bhareli which is a snow fed trans-Himalayan river. It enters India as the Kameng River. It has a length of 264 km in India and drains a basin of 14,450 $km^2$. The Tenga and the Dikrai are the two principal tributaries of the Jia Bhareli. The Jia Bhareli contributes about 4.9% of total discharge of the Brahmaputra.

On the south bank of this reach the Brahmaputra is met progressively from east by the Diphalu and Mora Diphalu and tiny Borghop. Thereafter, the river shoots off a branch called Kalang Suti near Kukurakata hill. Although the Kalang now happens to be a branch of the Brahmaputra, it does not depict any characteristics of the main braided river. On the contrary it is a perfectly meandering river. There is ample evidence to regard it as the lower reach of the Dhansiri, detached from the latter owing to erosion of the Brahmaputra. The Kalang has a length of 160 km, it receives small rivers, such as the Nanai, Misa, Diju, Haria and Digaru and also a big river, the Kopili. Within the enclave of the Kalang and the Brahmaputra the Kadam, Leteri, Sonai, and Mora Sonai join the Brahmaputra. Beyond the outfall of the Kalang a small river, the Amchang, joins the Brahmaputra on the east of Guwahati city and another small river, the Bharalu, joins it within the city.

The Kopili finds its origin in North Cachar Hills of Assam at an elevation of 1,630 m. Its total length is 297 km and it drains a basin of 15868 $km^2$. Its major tributaries are the Diyung, Jamuna, Borpani and Kiling. The Kalang along with Kopili contributes 1.47% of the total flow of the Brahmaputra. In the past the Brahmaputra flowed in a northerly course from Tezpur to Biswanath, whereas it flowed through a southerly course through the Sonai in the Naogaon and Morigaon districts of Assam.

### 5.4.9 THE REACH FROM PANDU TO DHUBRI

This reach has a length of 189 km. The average width of the river in this reach is 7.93 km. The minimum width is 2.4 km at Soalkuchi. During flood the river here is one sheet of water from bank to bank and has a maximum width of 18.5 km. The gradient of the river from Pandu to Goalpara (115 km) is 1 in 8875 and from Goalpara to Dhubri is 1 in 14,650.

Eleven tributaries, viz., the Puthimari, Pagladia, Manas Beki, Kujia, Champamati, Saralbhang (Gaurang), Sonkosh, Gadadhar, Sulibhangi and Gangadhar (Sonkosh) join the river from north. Except the Kujia, Chulibhangi and Gadadhar and others find their origins in the Bhutan Himalayas. There are also many important sub-tributaries. The Sessa and Boroloia are two notable rivers, being tributaries of the Puthimari and Pagaldia, respectively. A branch of the Pagaldia, called the Buradia, is joined successively by the Mora Pagaldia, Tihu, Kaldia, Pahumora and Palla. This unified flow meets the Beki which is a branch of the River Manas. The Manas is the biggest tributary of this reach. It finds its origin as the Kuru Chu from a glacier in the northern slope of Mount Kula Kangri (28° 15' N, 90° 35' E) in Tibet. After flowing for 90 km through Tibet and 140 km through Bhutan it bifurcates near the Indian border to create the Beki. The Manas and the Beki flow individually for 112 and 198 km respectively through the Assam Valley before meeting the Brahmaputra. The Manas is known as the Kar in Bhutan where it is joined by its major tributary the Dongma. The Manas has a length of 375 km and drains a basin of 38176 $km^2$ with a maximum discharge of 7,641 $m^3\ s^{-1}$. It contributes 5.48% of total discharge of the Brahmaputra. The Ai is the main tributary of the Manas in India. Another big river of this reach is the Gangadhar (or Sonkosh ), the westernmost river in Assam, whose principal tributary is the Raidak. It has a length of about 240 km and contributes 2.81% of total discharge of the Brahmaputra.

Compared to the north bank tributaries the south bank tributaries of this reach are smaller because of the closeness of the Meghalaya Plateau on the south. The tributaries of this region are the Khana, Kalmani, Kalbhog, Kulsi, Deosila, Habri, Krishnai and the Balbala (Jinari). Another notable river, the Jinjiram, drains its basin from this reach. It divides itself into two branches, one of which joins the Jamuna near Indo-Bangladesh border and the other meets the old Brahmaputra near Divanganj in Bangladesh. The Kulsi and the Krishnai are two notable rivers in this reach. The Rani is the major tributary of the Kulsi and the Dudhnoi that of the Krishnai.

The important townships of this reach are Goalpara on the south bank and Dhubri on the north. The river is navigable throughout the year in this reach. Prior to the sixteenth century A.D. the Brahmaputra had a course on the north of Hajo which is at present 6 km north of the present channel.

### 5.4.10 THE REACH FROM DHUBRI TO GOALUNDO

The Brahmaputra turns southwards and divides itself into three branches, viz., the Sialdoh, Boalmari, and the main stream about 12 km west of Dhubri and then runs for 30 km as the Indo-Bangladesh border. From Dhubri it traverses nearly 337 km, taking a new name, the Jamuna, in Bangladesh, to join the Ganga at Goalundo. From Goalundo the combined flow of the two majestic rivers is known by its dual name, the Padma or the Ganga. The Padma comes down for about 105 km in a south-easterly direction before it is joined by a major drainage system from northeast which starts with the Barak in the hills of Manipur in India and traversing its valley in Assam and Bangladesh culminates as the Maghna at its confluence with the Padma near Chandpur. Thereafter the Padma is called Maghna which flows southwards for 50 km to debouch into the Bay of Bengal. But till the eighteenth century the Brahmaputra flowed through Bangladesh in a different channel from south of the Garo Hills of the Meghalaya Plateau in a southeasterly direction through Jamalpur, Nasirabad and Hosainpur to meet the Meghna directly near Bhairab Bazar. At that time there was no existence of the Jamuna. The earlier course of the Brahmaputra is known as the old Brahmaputra, which is now a major distributary channel of the main river in Bangladesh. It has a distributary called the Banar or Lakhya.

From Dhubri to Goalundo the only notable tributary joining the Jamuna on its left bank is the Jinjiram, whereas a few distributary channels shoot off from the river from this bank. In addition to the old Brahmaputra another very large distributary is the Dhaleswari which flows through Manikganj and Narayanganj to meet the Meghna. On the other hand, a number of tributaries join the Jamuna on its right bank; these are successively the Torsa (Dhudhkumar or Raidak), Jaldhaka (Dharla), Tista, Artai, Ghaghat and Korotoya.

The Torsa joins the Jamuna at 15 km downstream from the last Indian tributary, the Gangadhar. It rises below the Tang Pass (roughly 27° 49' N, 89° 11' E) and travels through the Chumbi Valley of Bhutan for a length of 145 km where it is called Amo Chu before entering India near Rangamati. The river travels for a distance of 100 km through North Bengal as the Torsa and after flowing past Balampur it joins the Jamuna. Its main tributaries are the Kaljani, Jayanti, and the Raidak.

The Jaldhaka or Dharla rises in Sikkim and enters the plains of India after flowing southerly for about 50 km. It traverses a total distance of 186 km passing through Bhutan, India and Bangladesh till it finally joins the Jamuna at about 15 km downstream from the outfall of Torsa near Alipur in Bangladesh. Its tributaries are the Dudhuya and a few distributaries of the Torsa. It has a basin area of 4,883 $km^2$.

The Tista joins the Jamuna 35 km downstream from the Dharla outfall. It is the largest of the rivers joining the Jamuna. Two rivers, viz., the Lhonak and Lachung rising in the Himalayas join about 80 km north of Gangtok to from the Tista. The main one rises in the Chitamu Lake ($28^0$ 2' N, $88^0$ 44' E) in Tibet. Running through narrow gorges for nearly 138 km it enters the plains at Sivok in India. It flows in a steady southeasterly course to Jalpaiguri, beyond which it records frequent changes. It has a number of tributaries, many of which are mountain torrents. The most notable of them are the Rangni, Rangpo, Great Rangit, Reli, Lish, Gish, and Chel. It has a drainage area of 12,540 $km^2$ and after traversing a distance of 309 km joins the Jamuna below Rangpur town in Bangladesh. Up to the close of the eighteenth century it flowed into the Ganga, but in a great flood of 1,787 it suddenly turned eastwards and joined the present Jamuna.

The Atrai rises near Siliguri in India, and flows southwards for 80 km to Debiganj. Then it takes a south-westerly course for 16 km as the Atrai up to Mandla where it shoots off a channel called the Sib which joins the Ganga directly. Thereafter turning southeast it shoots off another identical channel called the Baranai which also joins the Ganga. Along its course of the next 130 km it receives three tributaries, viz., the Jamuna, Korotoya, and Ichamati. The Artai meets the Jamuna about 20 km downstream from the Dheleswari bifurcation and 90 km downstream from the outfall of the Tista.

## 5.5 Discussion and Concluding Remarks

The river system of the Brahmaputra drains a vast area with varying climatic and geomorphic conditions. A majority of the tributaries constituting this river system are small rivers fed by rain and spring water, whereas its major tributaries are of snow fed type. Although hydrological data of the river in its Tibetan part are not available, according to a report of WAPCOS 1993 the monsoonal flow at Shigatse in Tibet is about 12.74% of that at Pasighat, which increases to 36.27% at Tsela Dzong. The only large tributary joining this reach is the great Kyi Chu. The contribution of the Kyi Chu appears to be greater than that of the Tsangpo.

The contribution of the Dihang at Pasighat is 31.63% of total discharge of the Brahmaputra. In Indian territory the major tributaries and their percent of contribution are: Subansiri (7.92), Lohit (7.90), Debang (6.64), Manas (5.48), Jia Bareli (4.90), Sonkosh / Gangadhar (2.81), Burhi Dihing (1.87), Kopili (1.47), Dhansiri South (1.05), Disang (0.85), Dikhou (0.60) and Dhansiri North (0.39). In the Bangladesh sector the principal tributary joining the river is the Tista. These sub-systems constitute a major part of the main Brahmaputra system.

Most of the tributaries in Tibet are of snow fed type, having their catchment areas over high mountains. They exhibit least variation in their river system. The tributaries of the alluvial valley of Assam may be divided into two categories. The tributaries joining from north rise in the Arunachal Himalayas and bring more sediment. As a result the north bank tributaries choke up their bed leading to diversion of channel, river capture, as well as channel bifurcation to modify their river systems more frequently. In contrast to this the south bank tributaries have a lower rate of sediment yield and their drainage systems exhibit a more or less conservative nature.

A considerable amount of change in the river has been observed in the last 200 years in the Bangladesh sector. Leaving behind the earlier route, i.e., the old Brahmaputra, the river started its southward journey through the present Jamuna only after a great flood of 1787. Prior to that there was a small channel, named Jenni, which shot off from the old Brahmaputra near the then Mallansa to join the 'Deleswari' in the south near Jaffirgung of those days. The' Deleswari' was a small channel in 1779. Before 200 years ago there was no existence of the majestic Teesta, instead in 1799 there was a small creek known as Teesta Creek, which joined the Ganga. Similary the Korotova was a tributary of the Ganga in 1799 and in ancient times it had united with the Mahananda and Kosi of west to flow together directly into the Bay of Bengal.

It is apparent from the overall study and the discussion that the river system changes with time. The change may be so minor that it may be unnoticed but the major changes have resulted in considerable modification of the geomorphology and hydrological characteristics of the river at different times of the past. A proper understanding of the entire Brahmaputra river system is quite essential for various hydrological and hydrogeological studies as well as for harnessing its potential tributary rivers.

## 5.6 References

Burrard, S.G., Hayden, H.H. and Heron, A.M. (1932). Geography and Geology of the Himalayan Mountains and Tibet (2nd Ed.) Dehradun.

Das, H.P. (1969). The Brahmaputra drainage system. Jour. N.E.I.G.S., Vol. 1.

Dutta Chaudhury, A.N. (1978). The Brahmaputra and the valley of his blessings. Assam Science Society. pp. 28-47.

Goswami, D.C. (1985). Brahmaputra river, Assam India: Physiography, basin denundation and channel aggradation. Water Resources Research, 21(7), pp. 959-978.

Harman, H.J. (1879). On the operation for obtaining discharge of large rivers in Assam 1877-88. Assam Valley Series, Survey of India. pp. 4-36.

James Rennell's memoir of map Hindustan or the Mughal Empire and his Bengal Atlas (1779).

Kedia, D. (1978). The course of the Brahmaputra. Souvenir, Assam Science Society. pp. 17-23.

Law, B.C. (1968). Mountains and rivers of India. 21st Int. Georg. Cong.

Rawat, I.S. (1975). Indian explorer of the 19th century. Pub. Div. Govt. of India.

Sharma, J.N. (1992). Asomar nad nadi. Asom Sahitya Sabha. Jorhat, 337p.

Swami, P. (1949). Kailas Manasarowar. Motilal Banarasidas, Delhi.

Wade, J.P. (1927). An Account of Assam 1800. Lakhimpur. 372p.

WAPCOS. (1993). Morphological studies of River Brahmaputra. Kailash, New Delhi.

# CHAPTER 6

# FLUVIAL GEOMORPHOLOGY

A. K. BORA

Geology, geomorphology, physiography, climate, and soils of the Brahmaputra basin exhibit marked variations. An understanding of these variations is vital for the development of the basin. This chapter presents a short treatment of different aspects of fluvial geomorphology of the Brahmaputra River basin.

## 6.1 The Brahmaputra Basin

The Brahmaputra River is one of the biggest rivers in the world. It is known as the Tsangpo in Tibet (China), the Siang or Dihang in Arunachal Pradesh (India), the Brahmaputra in Assam (India) and the Jamuna in Bangladesh. Originating at an altitude of 5,300 m just south of the lake Konggyu Tso about 63 km south-east of the Manasarowar lake in Tibet, the river traverses a total distance of 2,906 km through the Himalayan mountains, hills and plains until reaching the Bay of Bengal in Bangladesh. The river, with its Tibetan name Tsangpo in the uppermost reach, flows through southern Tibet for about 1,625 km eastward and parallel to tributaries, viz., the Nau Chhu, the Tsa Chhu, the Men Chhu, the Charta Tsangpo, the Raga Tsangpo, the Tong Chhu, the Shang chhu, the Gya Chhu, the Giamda Chhu, the Po Tsangpo and the Chindru Chhu and the right bank tributaries, viz. the Kubi, the Kyang, the Sakya Trom Chhu, the Rhe Chhu, the Rang Chhu, the Nyang Chhu, the Yarlang Chhu, and the Trulung Chhu join the river along its uppermost reach. At the extreme eastern end of its course in Tibet the Tsangpo suddenly enters a deep narrow gorge at Pe where in the gorge section the river has a gradient ranging from about 4.3 to 16.8 m/km (Figure 6.1).

Traversing through deep narrow gorges of the Himalayan terrain the Tsangpo takes a southward turn and enters Indian territory at an elevation of 660 m. After crossing the Indo-China border the Tsangpo changes its name to the Siang of Dihang in the State of Arunachal Pradesh (India) and flows in a southernly course for a distance of 278 km up to Kobo. The river then enters the State of Assam (India) taking two important tributaries the Dibang and the Luhit. At the exit of the gorge the slope of the river is only 0.27 m/km. After taking the Dibang and the Luhit the Dihang is then known as the Brahmaputra, flowing initially southwestward and then westward through the Assam Valley. At the head of the valley near Dibrugarh the river has a gradient of 0.09--0.17 m/km which is further reduced to about 0.1 m/km near Pandu (Figure 6.2). The mighty Brahmaputra rolls down the valley from east to west for a distance of 640 km up to the Bangladesh border (Table 6.1).



*V.P. Singh et al. (eds.), The Brahmaputra Basin Water Resources*, 88-112.

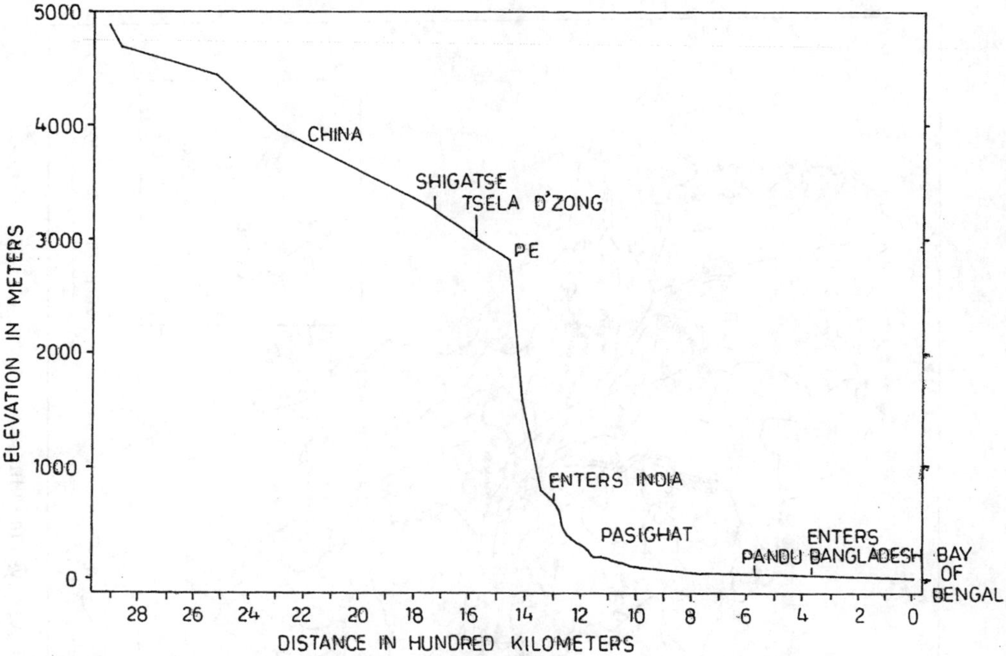


Fig. 6.1 Longitudinal profile of the Brahmaputra River (after Gosawami, 1985).

Near Dhubri (India) the Brahmaputra turns southwards and enters Bangladesh with its name the Jamuna. In Bangladesh the Jamuna meets the Padma near Goalundo, traversing a total distance of 363 km from the Indo-Bangladesh border it finally pours into the Bay of Bengal. The Brahmaputra basin covers an area of about 580,000 sq. km extending from 82°E to 97° 50′ E longitudes and 25° 10′ to 31° 30′N latitudes. The basin spans over an area of 293,000 sq.km in Tibet (China), 45,000 sq. km. in Bhutan, 194,413 sq.km. in India and 47,000 sq.km. in Bangladesh. In India, the basin lies in the states of Arunachal Pradesh, Assam, Nagaland, Meghallaya, Sikkim and West Bengal (Figure 6.3). The statewise distribution within India and the countrywise distribution of the drainage area are presented in Table 4.2. The Indian part of the basin has a maximum east--west length of about 1,540 km and a maximum north--south width of about 682 km along 93° east longitude. The basin is characterised by large variations in relief slope, landforms, climate, vegetation and landuse. The upper catchment lying in Tibet (China) and in the eastern Himalayas of Arunachal Pradesh, Sikkim and Nagaland comprise mostly mountain ranges and narrow valleys and the channels are restricted within steep and narrow valleys in the mountains. In Assam and Meghalaya the basin consists of hills, plateaus and plains covered by forests, tea gardens, agricultural lands and built-up areas. In West Bengal also the basin covers hills and valleys dominated by forests, tea gardens, agricultural lands and built-up areas. The lower portion of the basin in Bangladesh consists largely of fertile plains and delta areas.

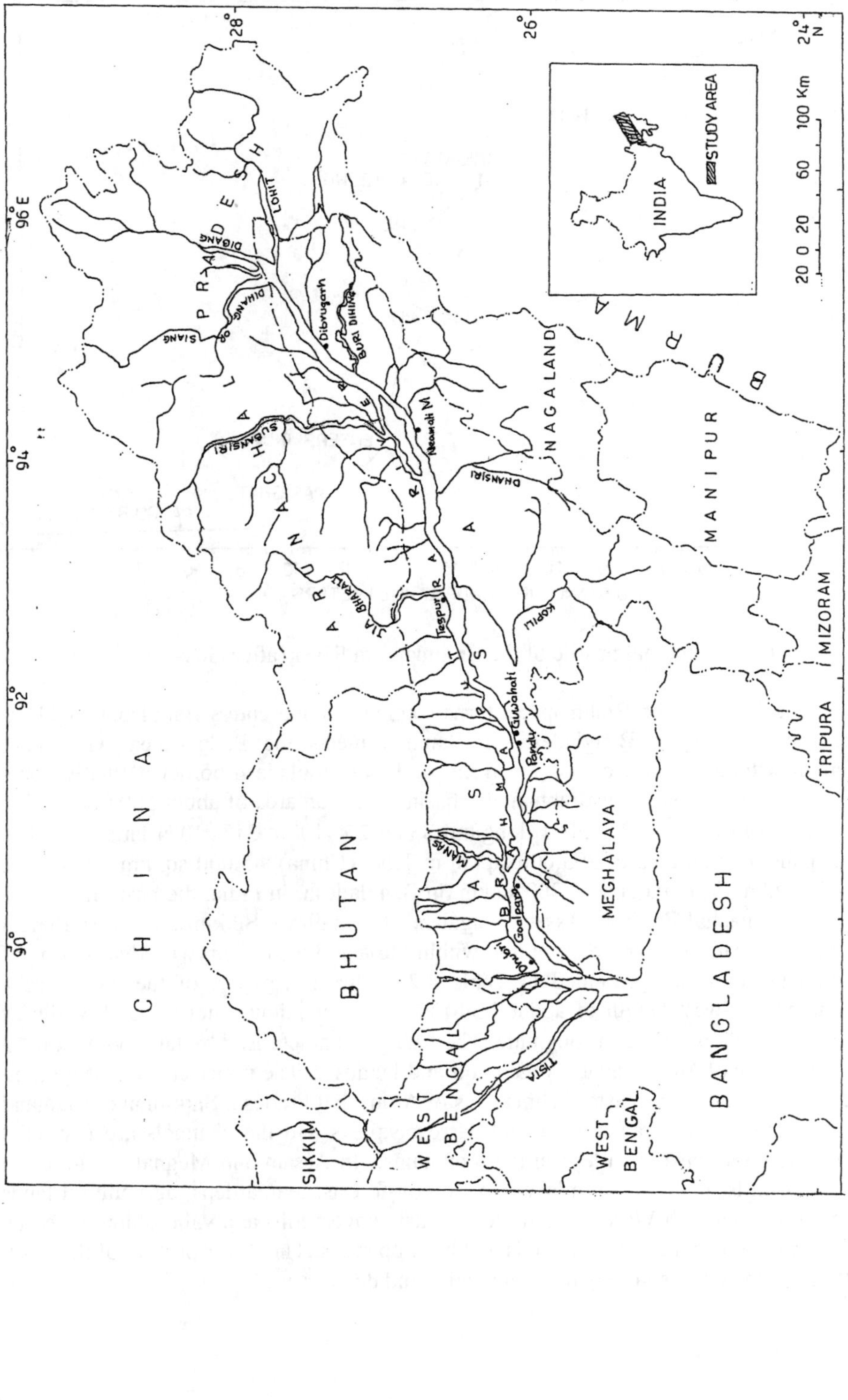


Fig. 6.2 The Brahmaputra River system in India.

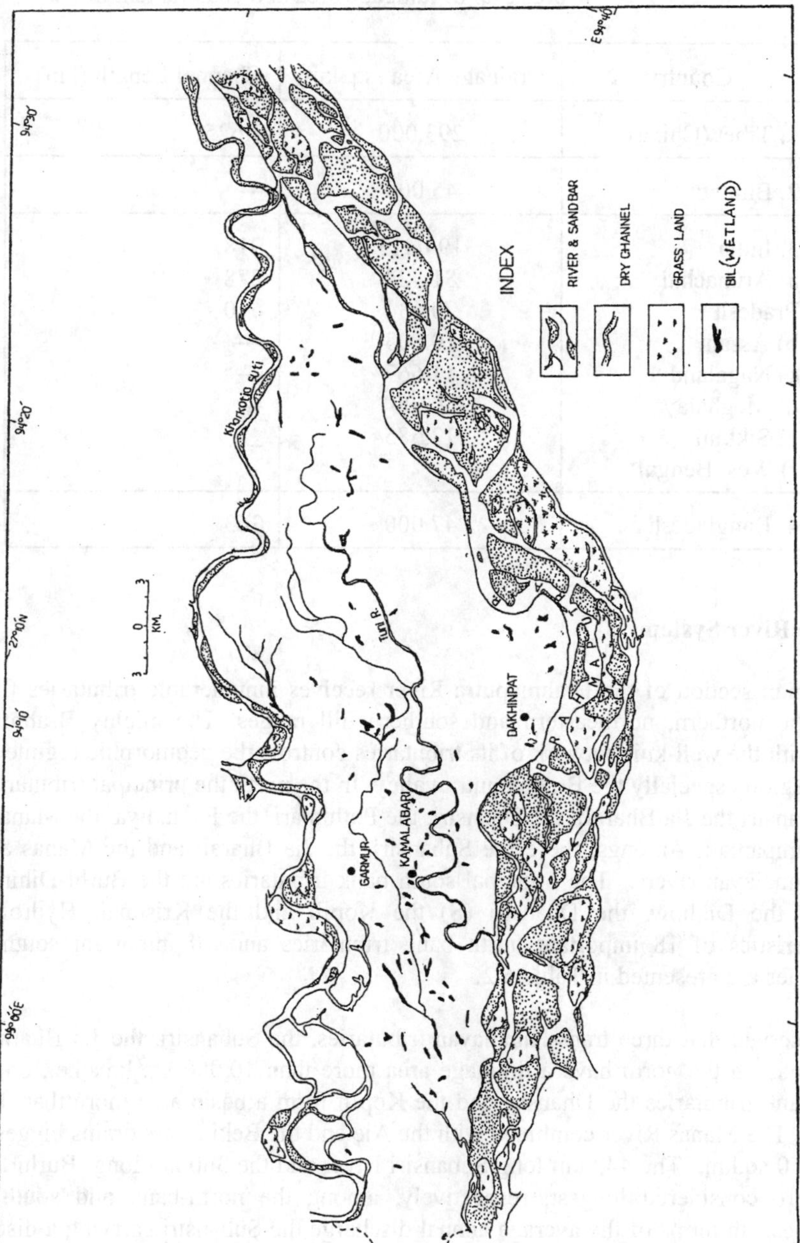


Fig. 6.3 Majuli River island in 1915.

Table 6.1 The Brahmaputra River: Country and Indian state-wise break-up of raingage area and channel length.

| Country | Drainage Area (sq. km) | Channel Length (km) |
|---|---|---|
| 1. Tibet (China) | 293,000 | 1,625 |
| 2. Bhutan | 45,000 | --- |
| 3. India<br>a) Arunachal<br>Pradesh<br>b) Assam<br>c) Nagaland<br>d) Meghalaya<br>e) Sikkim<br>f) West Bengal | 194,413<br>81,424<br>70,634<br>10,803<br>11,667<br>7,300<br>12,585 | 918<br>278<br>640<br>---<br>---<br>---<br>--- |
| 4. Bangladesh | 47,000 | 363 |

**6.2 The River System**

The Indian section of the Brahmaputra River receives innumerable tributaries flowing down the northern, northeastern and southern hill ranges. The mighty Brahmaputra along with the well-knit network of its tributaries controls the geomorphic regime of the entire region especially the Brahmaputra valley. In the north the principal tributaries are the Subansiri the Jia Bharali, the Dhansiri, the Puthimari, the Pagladiya, the Manas, and the Champamati. Amongst these the Subansiri, the Jia Bharali and the Manas are the trans-Himalayan rivers. The principal south bank tributaries are the Burhi-Dihing, the Disang, the Dikhow, the Dhansiri (S) the Kopili and the Krishnai. Hydrological characteristics of 18 important north bank tributaries and 10 important south bank tribjutaries are presented in Table 6.2.

It is observed that three trans-Himalayan tributaries, the Subansiri, the Jia Bharali and the Manas on the north have a drainage area more than 10,000 sq. km, i.e., only two south bank tributaries the Dhansiri and the Kopili form a basin area more than 10,000 sq. km. The Manas River combined with the Aie and the Beki rivers drains biggest area of 41,350 sq.km. The 442 km long Subansiri River and the 360 km long Burhi Dihing River are considered longest, respectively, among the north-bank and south bank tributaries. In terms of the average annual discharge the Subansiri carrying a discharge of 755,771 $m^3 s^{-1}$ ranks first among all the 28 important tributaries. The Jia Bharali and the Manas in the north carrying an average annual suspended sediment load of 2,013 ha.m. and 2,166 ha.m., respectively, are the leading rivers in the case of sediment discharge. Of all the north and south bank tributaries as many as fourteen have sediment yields in excess of 500 tons/$km^2$/year, the highest being 4,721 tons/$km^2$/year from the

Jia Bharali basin (Table 6.2). Besides the Jia Bharali, the other two trans-Himalayan tributary basins, viz., the Subansiri and the Manas yield 959 and 1,581 tons/km$^2$ /year respectively.

Table 6.2 Hydrological characteristics of some important north-bank and south bank tributaries of the Brahmaputra River.

| Rivers | Catchmet Area (sq.km) | Length (km) | Average Annual Discharges ($m^3s^{-1}$) | Avearage Annual Suspended Load (ha.m) | Sediment Yield (ton/km$^2$)/ year) |
|---|---|---|---|---|---|
| Northern Tributaries | | | | | |
| 1.Subansiri | 28,000 | 442 | 755,771 | 992 | 959 |
| 2. Ranganadi | 2,941 | 150 | 74,309 | 186 | 1,598 |
| 3. Buroi | 791 | 64 | 20,800 | 16 | 529 |
| 4. Bargang | 550 | 42 | 16,000 | 27 | 1,749 |
| 5. Jia Bharali | 11,716 | 247 | 349,487 | 2013 | 4721 |
| 6. Gabharu | 577 | 61 | 8450 | 11 | 520 |
| 7. Belsiri | 751 | 110 | 9300 | 9 | 477 |
| 8. Dhansiri (N) | 1,657 | 123 | 26,577 | 29 | 463 |
| 9. Noa Nadi | 907 | 75 | 4450 | 6 | 166 |
| 10. Nanoi | 860 | 105 | 10,281 | 5 | 228 |
| 11. Barnadi | 739 | 112 | 5756 | 9 | 323 |
| 12. Puthimari | 1,787 | 190 | 26,324 | 195 | 2,887 |
| 13. Pagladiya | 1674 | 197 | 15201 | 27 | 1,883 |
| 14. Manas-Aie-Beki | 41,350 | 215 | 307,947 | 2,166 | 1,581 |
| 15. Champamati | 1,038 | 135 | 32,548 | 13 | 386 |
| 16. Gaurang | 1379 | 98 | 22,263 | 26 | 506 |
| 17. Tipkai | 1,364 | 108 | 61,786 | 31 | 598 |
| 18. Godadhar | 610 | 50 | 7,000 | 0.21 | 272 |
| Southern Tributaries | | | | | |
| 1. Burhi Dihing | 8,730 | 360 | 1411,539 | 210 | 1,129 |
| 2. Disang | 3,950 | 230 | 55,101 | 93 | 622 |
| 3. Dikhow | 3,610 | 200 | 41,892 | 34 | 252 |
| 4. Jhanzi | 1,130 | 108 | 8,797 | 16 | 366 |
| 5. Bhogdoi | 920 | 160 | 6072 | 15 | 639 |
| 6. Dhansiri (S) | 10,242 | 352 | 68,746 | 147 | 379 |
| 7. Kopili | 13,556 | 297 | 90,046 | 118 | 230 |
| 8. Kulsi | 400 | 93 | 11,643 | 0.6 | 135 |
| 9. Krishnai | 1,615 | 81 | 22,452 | 10 | 131 |
| 10. Jinari | 594 | 60 | 7,783 | 3 | 96 |

The north bank and south bank tributaries of the Brahmaputra show marked differences among them in terms of their fluvio-geomorphic characteristics. The variations in environmental settings including geology, geomorphology, physiography, relief, precipitation and soils of the two regions belonging to the north bank and south bank river basins bring about notable differences between these two groups of rivers. A systematic study reveals the following points of differences.

1. The northern tributaries have very steep channel gradient but the southern tributaries have comparatively flatter gradients.
2. The northern tributaries have shallow braided channels while the southern ones have their meandering channels over the plains.
3. In the case of northern tributaries the long section of the river course is in the hilly terrain and a short distance is over the plains whereas the southern tributaries have their long courses over the plains.
4. The northern tributaries have generally coarse sandy beds with occasional gravel beds up to some distance from the foothills. But the southern tributaries have beds and banks composed of fine alluvial soils.
5. The northern tributaries carry an enormous sediment load as compared to the southern tributaries. On the average the sediment yield of the north-bank tributaries is three times higher than that of the south bank tributaries coming out of the Naga, Mikir hills and the Meghalaya plateau.
6. The northern tributaries generally have flash floods but the southern ones have comparatively less flash floods.
7. The northern tributaries are characterised by frequent shifting of their channels during floods but the southern tributaries change their courses less frequently. As revealed by the study the northern tributaries have peculiar channel shifting patterns. The Subansiri and all other eastern rivers shift their channels westward while the rivers between the Pagladiya and Subansiri shift eastwards. Again from the Manas up to the river Sonkosh in the west all the rivers migrate westward.

## 6.3 Geology, Geomorphology, and Tectonics

The Brahmaputra basin in India comprising varying geologic and geomorphic characteristics represents its peculiar physiographic make-up. The basin is bounded by the Eastern Himalayas on the north and east, the Naga-Patkai ranges on the northeast and Meghalaya Plateau and Mikir hills on the south. The region can be geologically and tectonically divided into four major zones, viz. the Himalayan folded and Tertiary hills and mountains, the Naga-Patkai ranges, the Meghalaya Plateau and Mikir hills and the Brahmaputra valley in Assam (Figure 6.4). The Himalayan zone comprises three topographic units that rise progressively to the north. The lowermost ranges called sub-Himalayas with an average elevation of 1,000 m consist mainly of Tertiary sand stones and are conspicuous by the presence of many raised, relatively young terraces (Gansser, 1964). The Middle Himalayas, having an average elevation of 4,000 m are underlain by lower Gondwana (Palaeozoic) deposits comprising shales, slates, and phyllites overlain by a thick horizon of basaltic rocks. The Greater Himalayas with an average elevation of 6,000 m consist primarily of granites and gneisses (Goswami, 1985). The Himalayan mountains with their syntaxial N--E bends originated out of the Tethyan Geosynline (Wadia, 1968) and are essentially composed of loose sedimentary rocks. The sub-Himalayas and the lower Himalayas are characterised by piedmont zones, low discontinuous ridges, low linear ridges, high rugged hills and upland valley depressions. The Patkai-Naga ranges stand on the eastern and southeastern border of the

Brahmaputra valley in Assam. These ranges with an average elevation of 1,000 m are composed of Tertiary sediments and characterised by the presence of a large number of active faults. This zone consists of piedmont plains, anticlinal ridges and synlinal valleys with terraced alluvial fills, undifferentiated sharp ridges and narrow valleys, upland valley depressions and plateau remnants. The Meghalaya plateau and the Mikir hills attaining an elevation ranging from 600 m to 1800 m are made up primarily of gneisses and schists. This part, being a rigid mass, belongs to the Daccan plateau of the stable Indian peninsular block of Pre-cambrian age. It is characterised by plateau remnants, inselbergs, deeply dissected uplands with faulted monoclines of Tertiary cover, denuded hills, basement controlled structural ridges covered with Tertiary rocks and upland valley depressions.

Table 6.3 Rainfall, temperature and humidity conditions at some selected stations of the Brahmaputra basin.

| Station | Average Annual Rainfall (cm) | Annual Normals of Temperature Maximum (°C) | Minimum (°C) | Annual Normal of Relative Humidity (in percentage) |
|---|---|---|---|---|
| 1. Cherrapunji | 1042 | 20.5 | 14.2 | 73 |
| 2. Dibrugarh | 215 | 27.3 | 18.5 | 85 |
| 3. Sibsagar | 190 | 27.5 | 18.8 | 87 |
| 4. Guwahati | 175 | 29.3 | 19.2 | 81 |
| 5. Shillong | 260 | 21.1 | 11.9 | 69 |
| 6. Tezpur | 180 | 28.7 | 19.5 | 82 |
| 7. Darjeeling | 310 | 15.7 | 9.8 | 74 |
| 8. Jalpaiguri | 172 | 29.2 | 19.1 | 82 |

Table 6.4. Widths of the nodal points and the downstream braided reaches of the Brahmaputra.

| Nodal Points | Width at the Nodal Point (km) | Width of the Braided Reach downstream of the nodal points (km) |
|---|---|---|
| 1. Murkongselek | 4.8 | 9.6 |
| 2. Desangmukh | 5.1 | 10.0 |
| 3. Downstream of Jhanjimukh (S) | 3.75 | 8.4 |
| 4. Upstream of Dhansiri (N) | 4.0 | 7.5 |
| 5.Downstream of Dhansirimakh (S) | 4.4 | 5.2 |
| 6. Upstream of Tezpur | 3.6 | 12.3 |
| 7. Pandu, Guwahati | 1.2 | 7.5 |
| 8. Soalkuchi | 2.4 | 10.5 |
| 9. Pancharatna | 2.4 | 10.5 |

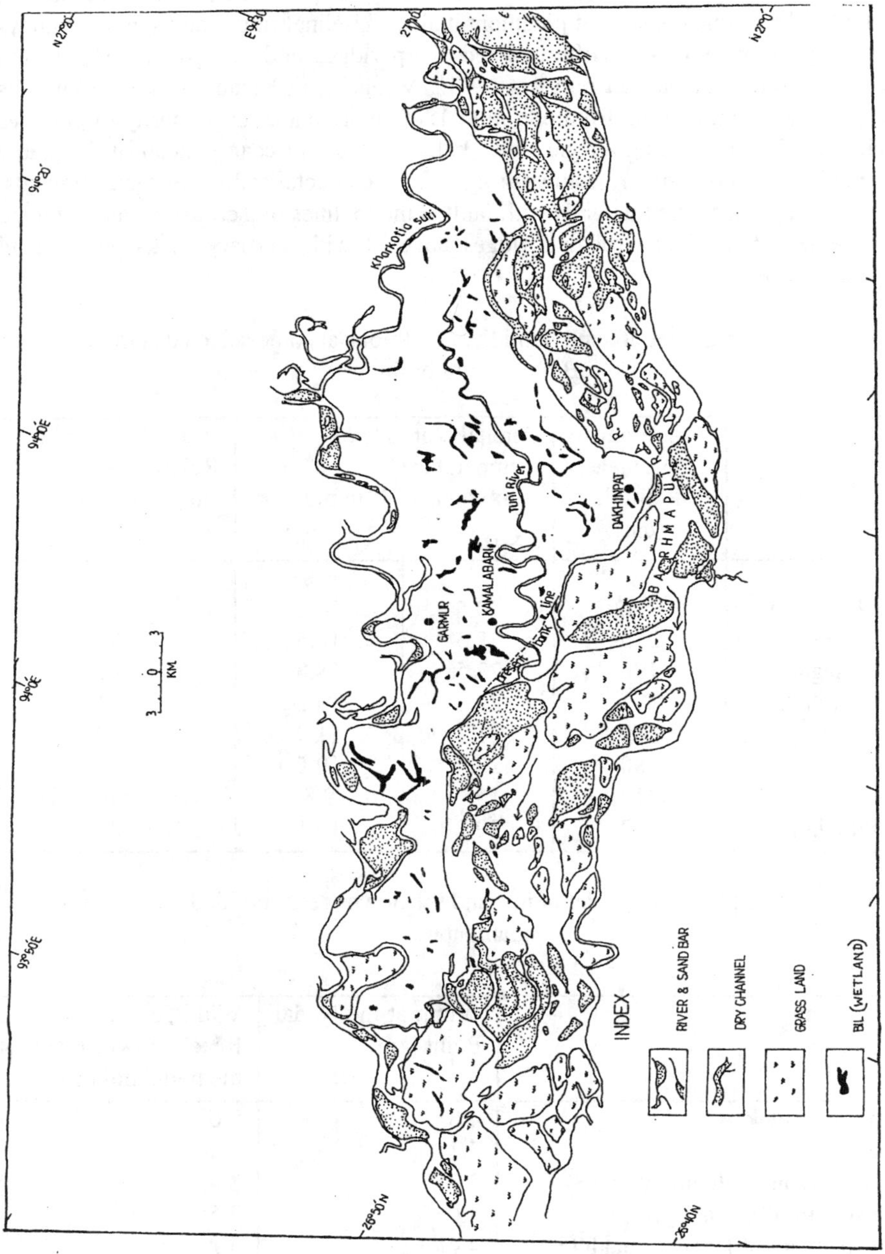


Fig. 6.4 Majuli River island in 1995.

The Brahmaputra valley in Assam, on the other hand, is underlain by recent alluvium approximately 200--300 m thick, consisting of clay, silt, sand, and pebbles (Geological Survey of India, 1974). The valley developed over the foredeep in between the Peninsular mass and the Tethyan Geosyncline. The foredeep characterised by some complicted tectonic features represents a series of faults and thrust extending in the NE--SW direction from the eastern margin of the Meghalaya plateau across the North Cachar Hills to Tirap District of Arunachal Pradesh. These thrusts originated at the time of the late Himalayan—Patkai--Naga Hills orogeny and pushed the tertiary deposits into folds and faults. The foredeep is believed to be under the sea till the sub-recent period received deposits during all the periods of the tertiary and quaternary ages. The tertiary deposits consist mainly of sandstones, shale, grit, conglomerate and lime stones. Towards the close of the Pleistocene period, alluvium began to be deposited in the from of sand, pebbles and gravels especially along the northern foothills of the Brahmaputra valley. These valley deposits of reddish brown sandy clay with some pockets of unassorted pebble, cobble, sand and silt have been identified as older alluvium. The tertiary beds of the valley are overlain by a thick layer of newer alluvium composed of sand, silt and clay which are being brought down from the rising Himalayas in the north, the Patkai Naga ranges in the east and southeast and the Meghalaya plateau in the south by numerous tributaries of the Brahmaputra.

The characteristic geological and tectonic framework coupled with structural complexities has rendered the Brahmaputra basin geomorphologically a most complicated one. A variety of landform under varied climatic conditions have formed over the geologic and tectonic base of the region. The peri-glacial, glacio-fluvial, and fluvial processes are dominantly operative in the basin at varying altitudes. The higher elevations of the Himalayas experience periglacial and glacio-fluvial erosion and deposition. The bare relief of the sub-Himalayas and Greater Himalayas suffer from immense sheet erosion owing to periglacial solifluction. The low hill ranges with hot and humid climate and heavy rainfall concentrated to a few months of the year experience solifluction, sheet erosion and landsides. The incidence of landslides is high in the Himalayan foothills where heavy rainfall, high seismicity and toe cutting of hill slopes by the streams are most frequent. Heavy rains often loosen soil and the soft rocks of the young Himalayan ranges. Rain water percolates through joints, fractures, foliations, and pores of rocks and soils and finally makes them loose and heavy, which cause heavy slope failure. Fluvial processes are, on the other hand, significantly dominant on the valley bottoms and plains where alluvial deposition takes place due to erosion of the higher surface by rivers and floodings in the valleys. The erosional and depositional processes conspicuously intensified by copious rainfall and frequent seismic movements, however, play a dominant role in creating various fluvio-geomorphic environments in the basin.

### 6.4 Climate and Soils

In a year four distinct seasons, viz, winter, summer monsoon and autumn or post-monsoon are experienced in the Indian portion of the basin. The winter season with a mean temperature varying from 15 °C to 17.5 °C begins in December and continues to the end of February. Light northeasterly winds blow down the Brahmaputra valley in Assam and light northerly to north-westerly winds in West Bengal. The weather is occasionally changed by the passage of western disturbances across the region. Light rainfall occurs in January and February along the hills, increasing towards north-east of the Brahmaputra valley. From March onwards, the hot weather starts and continues upto May with a mean temperature between 25 °C and 27.5 °C in the lower Brahmaputra valley. But the sub-Himalayan regions record temperatures lower than 25 °C in the period. In this season the basin is under the sway of three air streams---a northwesterly current from Uttar Pradesh and Bihar, a southerly stream from the head of the Bay of Bengal, and a stream of north-easterly to easterly winds from the Brahmaputra valley blowing over the plains of North Bengal and Bihar. Being influenced by these three air streams, the season is marked with atmospheric disturbances sometime preceded by dust-raising squalls. Rainfall increases both in amount and frequency as the season advances, being generally associated with afternoon or evening thunderstorms and squalls. They are generally more concentrated in the upper Brahmaputra valley. The monsoon starts by the last week of May or in early June following depressions developed in the Bay of Bengal. Subsequently a series of such depressions, forming at the head of the Bay and moving inland leads to spells of continuous and moderate to heavy rain generally over the region. July and August are the rainiest months with a mean temperature varying from 27.5 °C to 30 °C. Rain is most frequent and heavy along the Himalayan foothills in the north and north east. The monsoon withdraws in the last week of September or the first week of October. After the withdrawal of the monsoon winds, light unsteady winds are experienced by the middle of October when a mean temperature over the basin varies from 25 °C to 27.5 °C.

The basin with varied terrain characteristics and being an integral part of the monsoonal regime of southeast Asia shows a marked spatial variation in the distribution of precipitation. The rainfall in the Tista valley varies from 164 cm in the south to 395 cm in the north. The average annual rainfall in the lower Brahmaputra valley is 213 cm while the same in the north-eastern foothill belt is 414 cm. The basin as a whole has the average annual rainfall of 230 cm with a variability of 15--20 per cent. The Himalayan sector receives 500 cm of rainfall per year, the lower ranges receiving more than the higher areas (Goswami, 1985). During the monsoon, months of May to October receive about 12 percent of the annual total. The average annual rainfall, annual normals of temperature and relative humidity at some selected places located within the basin are presented in Table 4.5.

In the sub-Himalayan belt soils with little depth developed over the Tertiary sand stones generally belong to red loam, laterite, and brown hill soil type with admixtures of cobbles and boulders. The greater part of the Brahmaputra valley is made up of new alluvium of recent deposition overlying Tertiary, Mesozoic and Archaean bedrocks. Along the piedmont zone there occurs some patches of older alluvium extending along the interfluves of the tributaries flowing from the Himalayan foothills. The soils of the Meghalaya plateau and the Mikir Hills in the south are of laterite and loamy silt and fine silt types.

## 6.5 Channel Processes

The Brahmaputra River in India forms a complex river system characterised by the most dynamic and unique water and sediment transport pattern. The Brahmaputra is the fourth largest river in the world in terms of the average discharge at the mouth with a flow of 19,830 $m^3 s^{-1}$(Goswami, 1982). The water yield from per unit drainage area of the basin rivers is among the highest of the major rivers of the world. The Jia Bharali River, a major tributary for example carries a mean annual discharge in the order of 0.0891 $m^3 s^{-1} km^{-2}$ (Bora, 1990). As estimated by Goswami (1982) the Brahmaputra yields 0.0306 $m^3 s^{-1} km^{-2}$ at Pandu. As regards sediment transport, the river has also set records in carrying alarming volumes of sediment. The high intensity of monsoonal rains, easily erodible rocks, steep slopes, and high seismicity contribute a lot by rendering the river a heavily sediment-laden one. Thus the Brahmaputra becomes one of the leading sediment carrying rivers of the world. Amongst the large rivers of the world it is second only to the Yellow River in China in the amount of sediment transport per unit of drainage area (Goswami, 1985). At Pandu the river carries an average suspended load of 402 million metric tons. A river with such gigantic water and sediment discharge magnitudes represents its most dynamic fluvial regime. Its large alluvial channel having a width of 6 to 10 km is, therefore, marked by intense braiding, rapid aggradation and drastic bankline changes.

The Brahmaputra is a uniquely braided river of the world. Although braiding seems to be best developed in rivers flowing over glacier outwash plains or alluvial fans, perfect braiding is also found to occur in large alluvial rivers having low slope, such as the Brahmaputra in Assam (India) and Bangladesh or the Yellow River in China. The Assam section of the Brahmaputra River is, in fact, highly braided and characterised by the presence of numerous lateral as well as midchannel bars and islands. The high degree of braiding of the Brahmaputra channel near Dibrugarh and downstream of Guwahati is indicated by the calculated braiding indices of 5.3 and 6.7, respectively, for the two reaches (Goswami, 1985) following the method suggested by Brice (1964). A braiding Index of 4.8 for the entire Assam section of the river calculated by the author on the basis of satellite data of 1993 also suggests a high degree of braiding of the Brahmaputra channel.

In general, braiding in the Brahmaputra follows the mechanism of central bar type of braid formation. During high flow a central bar is deposited in the channel and gradually the bar accrets vertically to the level of the floodplain. It also builds on the downstream end through deposition of bedload material due to the slack water occurring behind the bar. The bar growth causes a decrease in total cross-sectional area leading, thereby, to the instability of the channel. Lateral erosion then follows on one or both the banks. Through repetition of this process in the divided reach, a well developed braided reach with multiple sandbars and islands is produced. However, in the study of the Bangladesh portion of the Brahmaputra river, Coleman (1969) suggests that in addition to the central bar type of braid formation, there is another mechanism in which narrow, stable banks of low erodibility initiate the braiding pricess. In the Assam section of the river also, the presence of such nodes of stable banks is found to effect the formation and location of the bars. There are nine nodal reaches of narrow constriction at various locations along

Table 6.5 Flood damage during major floods in the Brahmaputra valley.

| Years of Major Floods | Area Affected (mha) | Cropped Area Damaged (mha) | Population Affected (m) |
|---|---|---|---|
| 1954 | 2.90 | 0.30 | 1.30 |
| 1955 | 1.35 | 0.07 | 0.18 |
| 1958 | 1.23 | 0.06 | 0.40. |
| 1962 | 1.59 | 0.36 | 3.91 |
| 1966 | 1.51 | 0.06 | 3.62 |
| 1969 | 1.06 | 0.07 | 0.89 |
| 1972 | 1.01 | 0.36 | 2.95 |
| 1973 | 2.31 | 0.17 | 1.84 |
| 1977 | 1.02 | N.A+ | 4.54 |
| 1980 | 1.14 | 0.26 | 3.35 |
| 1984 | 1.49 | 0.14 | 5.61 |
| 1987 | 1.62 | 0.43 | N.A.+ |
| 1988 | 3.82 | 1.12 | 8.31 |
| 1993 | 1.25 | 0.22 | 5.30 |
| 1996 | 1.33 | 0.28 | 5.12 |

+N.A. stands for Not Available.

the Brahmaputra which are at Murkongselek (4.8 km), Disangmukh (5.10 km), Downstream of Jhanjimukh (3.75), upstream of Dhansiri north (4.0 km), downstream of Dhansirimukh (4.4 km), upstream of Tezpur (3.6 km), Pandu, Guwahati (1.2 km), Soalkuchi (2.4 km) and Pancharatna (2.4 km) (Table 6.6). Since banks are relatively stable in these reaches the river scours deeper to accommodate the flood discharge. The scoured debris are then deposited in the channel immediately downstream from the narrow section. As a result, the channel becomes wider and shoaler and bars and islands are produced. Formation of bars causes reduction in cross sectional area and the river, therefore, cuts its banks laterally to accommodate the discharge. Thus the downstream of the nodes intense braiding develops resulting in channel widening through continuous migration of both banks of the Brahmaputra.

As reported from the studies carried out on braided rivers of the world, the major factors thought to be responsible for braiding and bar formation are steep channel gradient, high erodibility of bank materials, great variability in discharge, overabundance of load, and aggradation of the channel bed. In the case of the Brahmaputra River in Assam bar formation and channel division are owing to a combination of factors like high variability in discharge, excessive sediment transport, easily erodible bank materials and aggradation of the channel.

Being the fourth largest river in the world with an average discharge of 19,830 $m^3 s^{-1}$ at its mouth (Goswami, 1982), the Brahmaputra carries 82 per cent of its annual flow at Pandu (Assam) only during the rainy season from May to October. The mean annual maximum and minimum flows in the river during 1955--97 are 48,161 $m^3 s^{-1}$ and 3072

$m^3 s^{-1}$, respectively. On average, therefore, the maximum flow is more than fifteen times the minimum. The flow in the river is less than or equal to the mean annual flow about 51% of the time. High variability in discharge of the river is mainly caused by seasonal rhythm of the monsoon and the freeze--thaw cycle of the Himalayan snow. As regards the pattern of sediment transport, the river has the record of carrying excessive sediment load which is believed to be one of the important factors responsible for braiding.

Table 6.6 Maximum water level of the Brahmaputra river at different locations.

| Site | Date | Maximum Water Level (m) | Danger Level (m) |
|---|---|---|---|
| Dibrugarh | 29.7.82 | 107.95 | 104.24 |
| Neamati | 14.8.87 | 87.79 | 85.04 |
| Bechamara | 26.8.88 | 88.23 | 85.32 |
| Tezpur | 27.8.88 | 66.59 | 65.23 |
| DC's Court (Guwahati) | 29.8.88 | 51.37 | 49.75 |
| Pandu | 29.8.88 | 49.76 | 48.77 |
| Goalpara | 29.8.88 | 37.29 | 36.27 |
| Dhubri | 29.8.88 | 30.36 | 28.65 |

During the monsoon month from May to October the Brahmaputra carries an average daily sediment load of 2.12 million metric tons at Pandu (Assam). Besides carrying a high volume of suspended load of the order of 402 million metric tons on average annual basis at Pandu, the river stands second only to the Yellow River in China in terms of sediment yield per unit of drainage area. In the case of the composition of bed and sediments the river is found to exhibit a high degree of erodibility of sediments along its Assam section except the narrow reaches of relatively rigid banks. Bank materials of the Brahmaputra are mostly composed of dominant proportions of fine sand and silt with the only occasional presence of clay being less than 5%. The median diameter of bank and bed sediments of the river varies from 0.08 mm to 0.22 mm of which the finer size belongs to bed sediments (NEC, 1993). It shows that the sediments, especially of the banks, are mainly in the fine sand and silt fractions which largely contribute to the progress of the braiding process in the channel. Besides, aggradation of its channel plays no less an important role in inducing development of braided channel patterns in the Brahmaputra River. The Assam part of the river has, in general, an aggrading type of bed which is found to be having a secular trend of aggradation. A suspended load budget indicates that at least 70 per cent of the incoming suspended load is trapped in the Assam section of the river, resulting in an overall aggradation of the channel by about 16cm. The aggradational rates are estimated to be higher in years following the big earthquakes (Goswami, 1985).

The wide alluvial bed of the Brahmaputra in the Assam valley is, however, dotted with more than 600 small and big sandbars locally called *chars* which are the characteristic features of a braided river. These *chars* with varying sizes and degree of permanancy are separated by several distributary channels. The fully developed braided charactor of the Brahmaputra can well be viewed especially in the winter season. During summer most of the *chars* are submerged by flood waters and they change their locations, sizes, and physical parameters. It is to be noted that only a small fraction of the total number of *chars* are more or less permanent in existence and among them only a few remain visible even during high floods. With all these characteristics, the chars of the Brahmaputra present a peculiar geomorphic make-up quite distinct from all other physical units of the Brahmaputra valleys of Assam. Some *chars* being endowed with fertile alluvial top soil presently support a good number of permanent, semi-permanent, and quite temporary human settlements. The cores of the bars (*chars)* which are relatively high and least affected by floods support permanent settlements. Away from the cores with the decrease in height, the effect of flood gets increased rendering settlements a semi-permanent character. On the other hand, the fringes of the *chars* are highly susceptible to erosion and frequent inundation and consequently less attractive for human settlements. These *chars* have now assumed much importance from the conservation of the riverine ecology point of view.

Unlike the numerous bars and islands formed as a result of intense braiding processes of the Brahmaputra, Majuli, the largest riverine island of the world, has a peculiar history of its origin. The Majuli was formed as a result of the southward migration of the Brahmaputra which is another important fluvio-geomorphic process of the river. In fact, Majuli was an integral part of the present Jorhat District of Assam on the south bank of the Brahmaputra. As revealed from the historical records, southward shifting of the Brahmaputra by way of capturing the lower course of its tributary Burhi Dihing flowing parallel to the master stream had caused Majuli to emerge as a riverine island during the first half of the 18th century (Bora, and Das, 1998). Being situated within the river Brahmaputra between 26° 45′ and 27° 15′N latitudes and 93° 39′ and 94° 35′ E longitudes the lenticular shaped island is flanked by a minor channel on the north and a major channel on the south. Covering presently an area of 645 $km^2$, it has a total population of 135,378 (1991 census) inhabiting in 155 villages. The island is now the administrative sub-division of Jorhat District of Assam.

In recent years, specially after the great earthquake of 1950, recurring floods and severe erosion are causing havoc in the island, inundating and eroding large areas and a good number of human settlements. The 1950 earthquake, mostly affecting the upper Brahmaputra valley, was the most severe one causing tremendous changes of the riverine topography and sudden rise of river beds. The bed of the Brahmaputra near Dibrugarh was significantly raised, which resulted in an abnormal rise or water level up to 3.00m. Such a phenomenon was highly responsible for severe floods in the Dibrugarh-Neamati reach along the Majuli island. The aggressive nature of the flood in Majuli thus made a quantum leap, especially after the great 1950 earthquake. It is important to note that becasue of its peculiar location amidst the active floodplain of the Brahmaputra, Majuli gets inundated not only in severe floods but also in normal floods. The recent flood occurrence scenario of the island reveals that the nature of floods in terms of frequency of occurrence has been changing. It has now become a common feature that floods of different order occur at least 3--4 times every year. Moreover, the most dominant impacts of these floods is on the agricultural sector, which happens to be the mainstay of the people of Majuli. As regards erosion havocs, the island is also facing serious threat to its existence in recent years. Majuli is a highly erosion-prone riverine landmass under the constant forces for erosion from four sides. It is being eroded by the Brahmaputra on the east, west, and southern frontage and by the Subansiri River, near its outfall on the northwest. Available records show that erosion in Majuli attained serious proportions after the flood of 1954, and thus the first serious attack of erosion was experienced during 1955--56 on its eastern side and thereafter the island has fallen prey to severe forces of constant erosion. A spatio-temporal study of the erosional scenario of Majuli done by the author reveals that in 1915 the island occupied an area of 12,459 $km^2$ (Figure 6.5) and in 1971 the area was reduced to 924.6 $km^2$ accounting for an erosional rate of 5.63 $km^2$ /year. In 1984 its area was again reduced to 752 $km^2$ following considerable shrinking in the order 12.33 $km^2$ /year. The area of the island was further reduced to 645.49 $km^2$ in 1995 (Figure 6.6), thereby maintaining a high erosion rate of 8.88 $km^2$ /year during the period. Under the forces of erosion there has been a total disappearance of 58 villages from Majuli and about 1000 families have already migrated, particularly to the neighbouring districts of Lakhimpur, Jorhat and Golaghat.

In the case of an alluvial braided river like Brahmaputra it is an agreeable fact that the rate of rise and fall of the river water, the number and position of major channels active during floods, the formation and movement of large bedforms, cohesion and variability in composition of bank materials, and intensity of bank slumping are the important factors responsible for bank erosion and bank instability. Coleman (1969) observed along the Bangladesh section of the river Brahmaputra that the movement of thaw is high during the rising stages in May--June, relatively less during the peak of the flood in July--August, most erratic during the falling stage during September--October, and very little during the low flow stage during November--March. The most significant bankline modifications take place during the falling stage when excess sediment is deposited as bars within the channel, causing a change in local flow direction and migration of the thalweg. During floods, because of certain changes in river configuration, the channel starts shifting at some vulnerable sites, which in turn induces bank erosion. Moreover, the impinging flood waters raise the pore pressure of the bank wall, which results in a reverse flow of sand and silt into the river during the falling stages, leading eventually to a subaqueous bank failure. Under these situations the banks of the river near Nagajuli, Maijan, Majuli, Bhairabpur, Balikuchi, Kaziranga, Howlighat and Palasbari experience the threat of severe bank erosion during floods. The channel deposits, thus acquired through bank failure and lateral movement of the channel, lead to channel aggradation. The secular aggradational trend of the bed of the Brahmaputra since the 1950 flood (Goswami, 1988) combined with intense braiding, large discharge and heavy sediment load has been the key factor in causing the river extremely unstable at some vulnerable points.

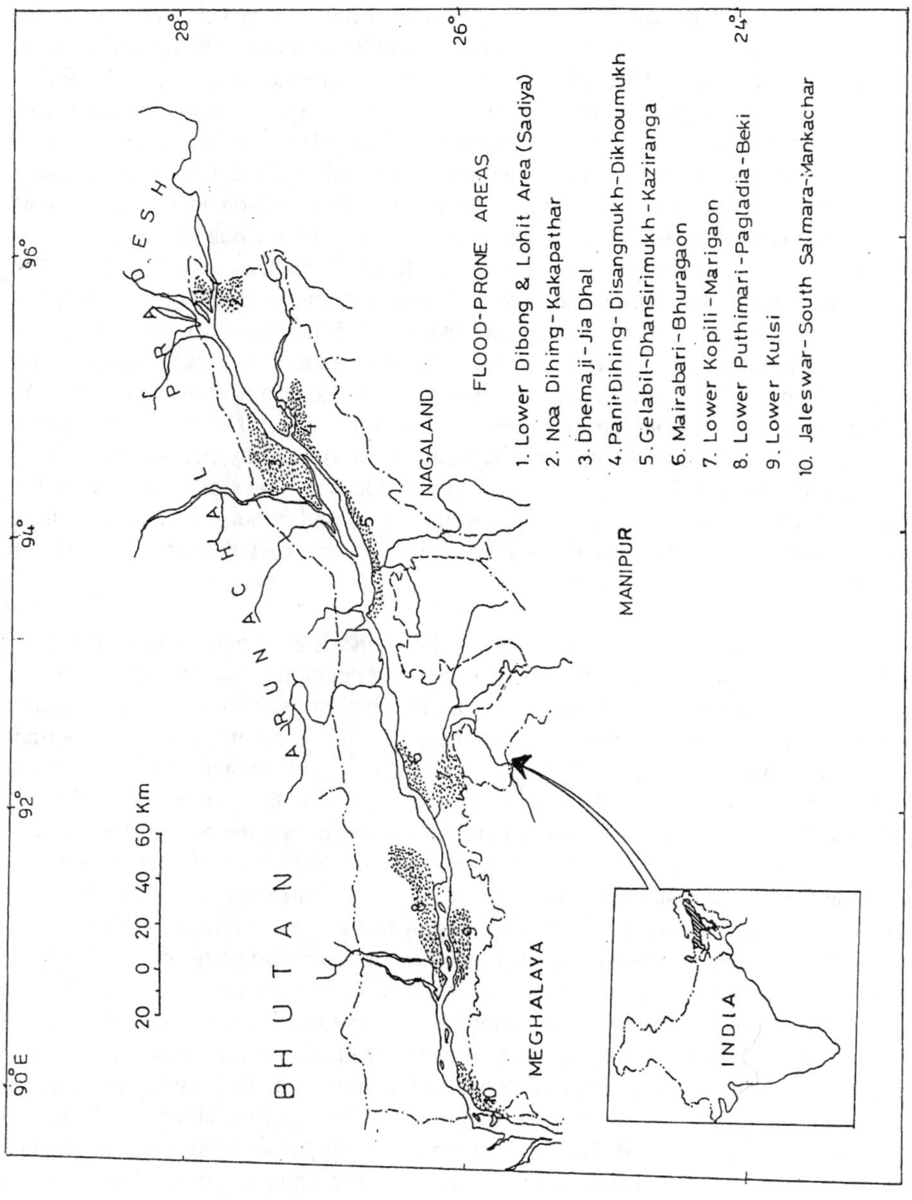


Fig. 6.5 Areas chronically inundated by tributary floods in the Brahamaputra valley, Assam, India.

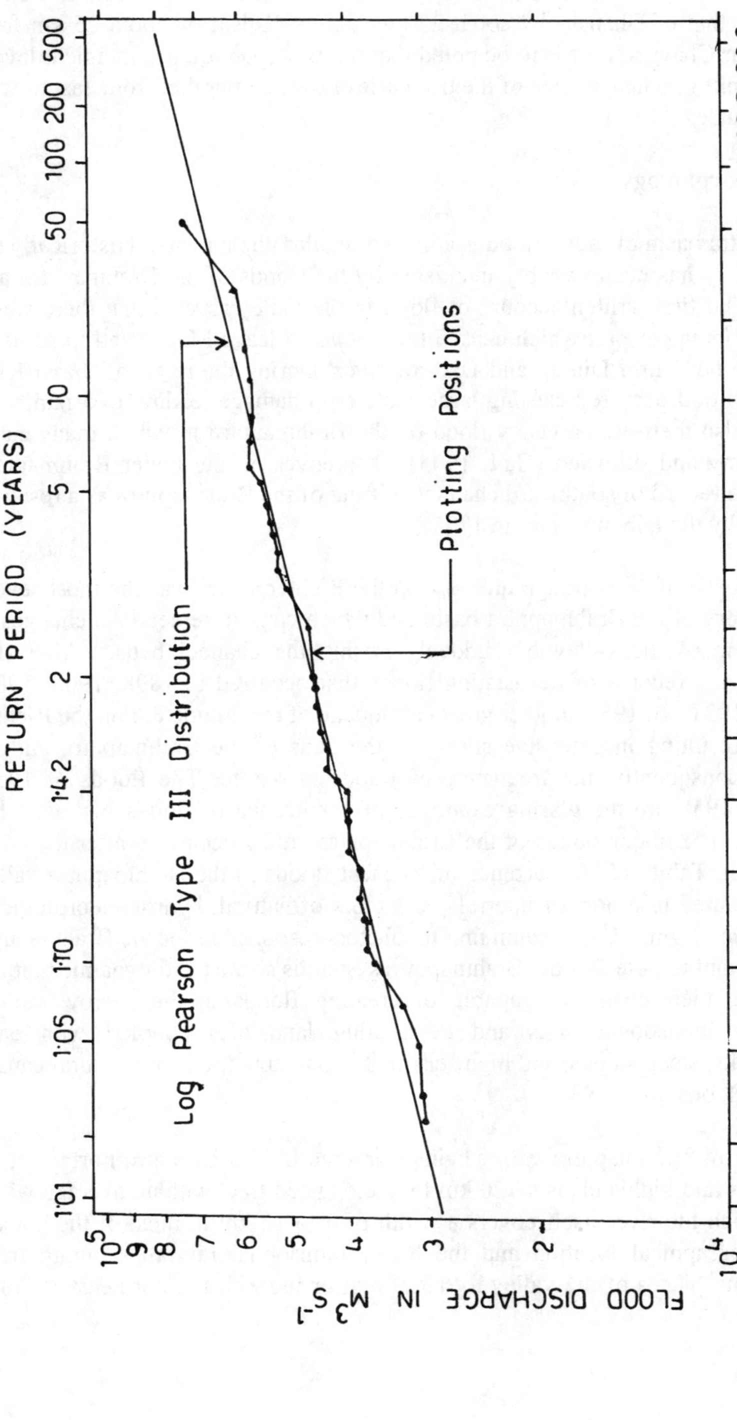


Fig. 6.6 Flood probability analysis of the Brahamaputra at Pandu, Assam, India, during 1955-97 using log-Pearson type 3 distribution.

In general, the Brahmaputra is migrating laterally southward. The southward erosion of the left bank varying from 2 to 4 km has been observed. It is clear from the eastward migration of the confluences of the south bank tributaries of the river. The southward migration of the Brahmaputra engulfed the lowermost parts of the south bank tributaries, and their confluences are thus migrating eastward. There is evidence that the confluences of Dibru moved northeastward by about 10 km, that of Disang moved east by about 30 km, that of Dhansiri by about 25 km, that of Kulshi by about 25 km from Nagarbera to near Chaygaon. It is to be noted that the Brahmaputra has incipient levees on both of its banks and hence most of the tributaries flow sub-parallel from east to west in their lower courses.

## 6.6 Flood Geomorphology

In the Brahmaputra basin of India, flood is almost a regular phenomena. Historically the Brahmaputra valley has been variably inundated by the floods of the Brahmaputra and its tributaries. The first written account of flood in the valley reveals that there was a flood in 1241 in its upper part which caused the *Ahoms* to leave Majuli and to settle in the area between the Burhi Dihing and Dikhow river. During the reign of *Ahom* King Sukapha a great flood occurred causing large scale crop damage leading to a famine in 1570. In 1642 also there was a heavy flood on the Brahmaputra in which many cattle were washed away and drowned (Gait, 1905). Moreover, in the upper Brahmaputra valley there is the record of southward channel shifting of the Brahmaputra as a result of the flood caused by the Dibong River in 1735.

The great earthquake of 1897 (magnitude 8.7 on the Richter scale) was the most severe quake in the history of the Brahmaputra basin and which caused tremendous changes in the fluvial regime of the valley by suddenly raising the channel beds. After this earthquake there are records of devastating floods that occurred in 1898, 1905, 1907, 1916, 1921 and 1931. In 1950 another great earthquake of magnitude 8.6 on the Richter scale occurred, resulting in extensive silting in the beds of the Brahmaputra and its tributaries, and consquently, the frequency of floods increased. The floods of 1954, 1955, 1956 and 1958 are the glaring examples of occurrence of floods just after the 1950 earthquake. The major floods of the Brahmaputra valley occurring after the earth quake are listed in Table 6.5. Occurrence of frequent floods in the Brahmaputra valley can well be attributed to a host of interrelated factors of natural, hydrometeorological, and anthropogenic origin. While examining the factors responsible for the Brahmaputra floods it is important to note that the Brahmaputra system is so vast and dynamic that the natural processes themselves are capable of creating floods in the narrow valley. However, heavy monsoonal rains and devastating landslides coupled with easy erodibility of rocks, steep slopes, and high seismicity constitute the major natural causes of Brahmaputra floods.

The riverine narrow Brahmaputra valley being bordered by the northern, north-eastern and southern hills and highlands is a 720 km long elongated tract with an average width of 80 km, of which the river itself covers a width of 6 to 10 km in most of the places. The adverse geographical location and the heavy monsoonal rainfall rainging from 2,480 mm in some places of the valley to 6,350 mm or more in the northeastern hills, nearly 60--70 percent of which occur during the monsoon period from June to September are responsible for frequent and damaging floods. High intensity of rainfall along the foothills bordering the valley and peaking of tributaries at the confluences within a very short period of time eventually give rise to high flood level of the Brahmaputra. Synchronising of hydrometeorological factors with flood timing in the Brahmaputra and its tributaries also aggravate the flood situation.

The Brahmaputra basin lies in an area of acute seismic activity. The earthquakes of 1897, disturbing the topography of the lower Brahmaputra valley, and of 1950, disturbing mainly the topography of the upper part of the valley, caused extensive landslips and rockfalls in the hills and subsidence and fissuring of the ground in the valley. The aggressive nature of flooding in the Brahmaputra valley made a quantum leap, especially after the great 1950 earthquake (Bhattacharyya and Bora, 1997). The anomaly caused by this earthquake in the river system and the surrounding landmass has naturally increased the frequency and intensity of floods. Changes in river course and accompanying erosion have together accelerated silt deposition on the river bed. The abnormal rise of the bed of the Brahmaputra near Dibrugarh and downstream was recorded in the order of 2.5 to 3.00 meters (Flood Control Department, Assam 1989). This has resulted in spilling of the river on both the banks up to the outfall of the Subansiri River in the reach of more than 200 km. In the central reach of the river in the Brahmaputra valley, a number of spill channels have been thrown up on both the banks. The uncontrolled discharge and subsequent spilling of these channels give rise to a serious flood problem in the area. The lower reach of the river along the districts of Kamrup and Goalpara is also subjected to spill and resultant flooding. In 1988 corresponding highest water levels had been experienced at different locations along the Brahmaputra River. As evident from Table 6.6, the upper reach extending from Bechamara (near Majuli) through the middle and lowermost reach upto Dhubri in the Brahmputra valley recorded the highest water levels during 26--29 August, 1988. This fact well explains the devastating nature of the 1988 flood.

The floods of the tributaries of the Brahmaputra having characteristic features also contribute significantly towards aggravating the overall flood situation of the valley. The tributaries in their lower reaches up to the confluence with the Brahmaputra are generally found to have inadequate drainage capacity of the channels which have to evacuate the rain water as well as the spills from the sub-tributaries. The situation further worsens when the Brahmaputra is in high flood, as the flood waters then usually back up into the tributaries and make them spill creating floods in the lower reaches of the tributaries. In this context reference may be made to Figure 6.7 showing the locations of ten flood-prone areas in the Brahmaputra valley which are chronically affected by the combined inundation of the Brahmaputra and its tributaries.

Besides the natural and hydrometeorological factors, anthropogenic factors are also largely responsible for severe floods in the valley. In many cases man's intervention in the river system, including the embankment system, has caused severe floods. Large scale deforestation in hilly catchments and practice of shifting cultivation on the hill slopes of the basin have led to heavy soil erosion and consequently, enormous quantities of sediments are generated and continuously deposited on the riverbeds. It is estimated that about 2.5 mm of top soil is being annually eroded by the river system and carried to be deposited on the riverbeds. Moreover the high growth of human settlements in the floodplains, destruction of natural wetlands and poorly managed exbankment network are the other notable contributory factors of flooding in the valley. With the increase of population and construction of man-made structures like roads, railways and buildings, etc., the free movement of run-off in the floodplain is hindered, causing artificial floods even in many new area previously considered flood-free.

Wetlands, constituting an important component of the fluvio-geomorphic regime of the Brahmaputra valley in Assam, have their major impacts on flood dynamics. The riverine valley provides favourable geo-ecological conditions for the development and growth of wetlands. The genesis and growth of the wetlands are directly associated with the geo-physical conditions and tectonic evolution of the region, hydrological and fluvio-geomorphic behaviours of the Brahmaputra and its tributaries, and the pattern of land use and human occupancy in the valley. As revealed from the satellite data analysis, the Brahmaputra valley, with its eastwest extension of about 720 km and the average width of 80 km, has as many as 2,167 natural wetlands above the areal size limit of 2.5 hactares. These wetlands of riverine origin categorised as natural ones include fresh water lakes (locally called beels), ox bow lakes, marshy tracts, and seasonally waterlogged areas. The fresh water lakes or beels are mostly located along the Brahmaputra as they generally originate owing to channel shifting of the river. The ox-bow lakes are the abandoned channel or channel cut-offs formed owing to shifting of the tributary channels. On the other hand, the swampy and marshy areas, which constitute a major portion, are the waterlogged areas directly or indirectly connected with the Brahmaputra and its tributaries. However, the wetlands of the valley bear immense significance from geohydrologic, fluvio-geomorphic, economic, and biodiversity points of view which can be realised as follows: (i) In ground water recharge and discharge the wetlands play a significant role, thereby providing water for human use in different purposes; (ii) the wetlands help in reducing flood height and thus flood damage as they act as buffer zones for rivers and other water ways during the period of heavy rains; (iii) the wetlands improve the water quality by decreasing water pollution owing to the presence of certain species of algae; (iv) the wetlands provide suitable sites for the conservation of bio-diversity as they support a variety of wild life and endangered species, birds, insects, and other flora and fauna during different seasons and at different depths; (v) the wetlands act as an economic resource base especially in terms of water for irrigation and fish, etc., for human consumption. In spite of all the well recognised importance of wetlands, the wetland eco-system in the valley suffers from the problem of serious ecodegradation.

Rapid eutrophication caused by debris from the floating vegetation and silt carried by surface runoff as well as streams, large scale human encroachment, growth of human activities like agriculture, cattle rearing, industrial set-up on the banks, extensive fishing and above all mismanagement of wetlands have all contributed much to the destruction of the wetlands system of the Brahmaputra valley. Thus there is a need to conserve these wetlands and protect their unique bio-diversity. Besides being a resource base, the wetlands, if properly managed, may play a key role in reducing flood peaks to a considerable extent by retaining excess water during high floods.

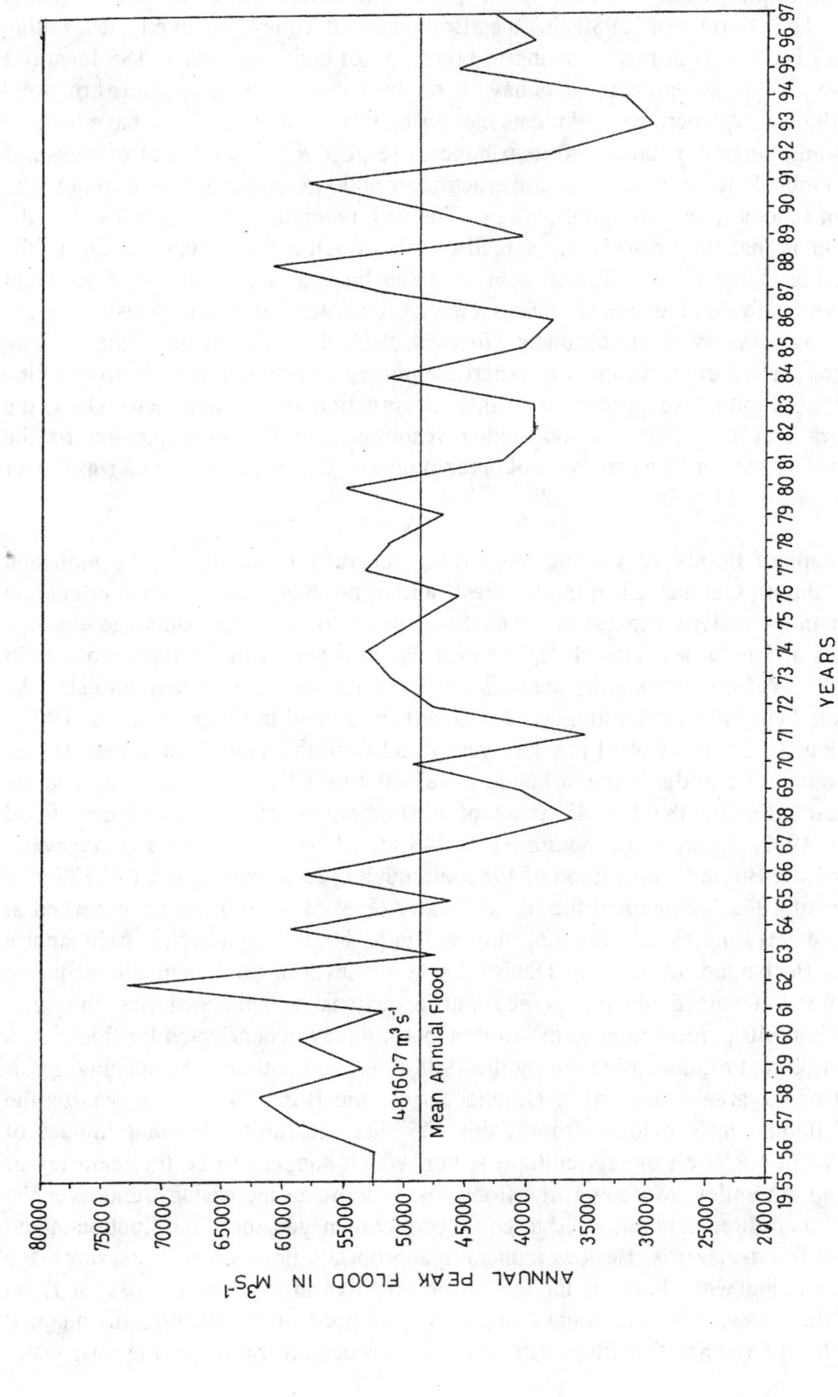


Fig. 6.7 Annual peak floods about the mean annual flood of the Brahamaputra at Pandu, Assam, India, 1955-97.

Embankments, on the other hand, often aggravate the flood problems, as the floodwater coming across them as a result of overtopping or breaching, eventually remains stagnant in a locality for some days and causes heavy damage. Moreover, during the time of the breach catastrophic flash flooding takes place and extensive areas are suddenly inundated. The floods of 1988 in the Brahmaputra valley assumed devastating dimensions (Table 6.5) primarily owing to breaching of embankments at 185 locations in the valley. Till today embankments have been the widely adopted measures of flood control in the valley, where embankments measuring 984 km and 2,567 km have been so far built along the Brahmaputra and its tributaries respectively. Flood control measures other than embankments include the construction of 600 km long drainage channel, 421 anti-erosion schemes, and 56 major sluices. The well recognised adverse effects of the embankment is that they restrict the spread of silt over the flood plain rendering the plains infertile in the course of time. On the other hand, a huge amount of sediment confined within the embankments causes the riverbeds to rise, thereby lessening the water carrying capacity of the channels. In view of all these ill effects of the existing embankment system in the valley, the expert committee constituted in 1989 to examine the causes of flooding suggested no further construction of embankments along the Brahmaputra and its tributaries, but rather recommended the strengthening of the existing ones by keeping the provision of break points so that excess water can spill over at the predetermined points.

The Brahmaputra floods, of varying magnitude, generally occur during the monsoon months of May to October when the heaviest rainfall produces the strongest effects on the Brahmaputra and its tributaries. The flood series of the river obtained through observations at Pandu near Guwahati, however, cover a period of 43 years from 1955 through 1997. A flood probability analysis for the series has been carried out using the Log-Pearson Type III Distribution, and the same is presented in Figure 6.8. The 100 yr flood is about 1.50 times that of the 2.33 year flood from the annual maximum series. The mean annual flood discharge at Pandu is 481,60.7 $m^3$ $s^{-1}$ with a recurrence interval of 2.60 years. During the last 43 years of obsrvations at Pandu the highest flood occurred in 1962 (Figure 6.9), measuring 72,748 $m^3$ $s^{-1}$ with the corresponding water level of 49.35m. But the recent flood of 1988, although of lesser magnitude (47,737 $m^3s^{-1}$) than that of 1962, maintained the highest water level of 49.76m so far observed at Pandu. It is interesting to note that not only at Pandu, but throughout the Brahmaputra valley from Bechamara (Majuli) to Dhubri, the water level of the Brahmaputra during the 1988 flood remained higher, thereby causing extensive damage during the year. Normally about 40 percent area of the Brahmaputra valley gets affected by floods in a year. But the extent of inundation during the 1988 flood was unthinkable, accounting for 62 percent of the area of the valley (Bhattacharyya and Bora, 1997). As regards the impact of floods, it is evident from Table 6.5 that the most dominant impact of Brahmaputra floods is on the agricultural sector, which happens to be the mainstay of the Brahmaputra valley. Moreover, the flood losses show an increasing trend over the years. It is, therefore, an urgent need to go for proper management of the floodplain and reduction of flood- hazards. Besides launching appropriate flood control measures, the approach of living with floods is thought to be effective in averting the risk of flood damage in the valley. The traditional cropping system needs to be scientifically adapted to the rhythm of regular flooding. However, the proper utilization of the vast water resources of the Brahmaputra with a hydropower potential of about 41,000 MW, which is only nominally harnessed to date, can certainly be the key factor for overall socio-economic development of the Brahmaputra valley.

## 6.7 Concluding Remarks

The Brahmaputra basin in India shows marked differences in geology, geomorphology, physiography, climate and soils. The Brahmaputra system with a dense network of its innumerable tributaries is one of the most gigantic and dynamic river systems of the world. The high intensity of monsoonal rains, easily erodible tertiary rocks of the Himalayan ranges, steep slopes, high incidence of landslides and high seismicity in the basin have rendered the river a highly water-carrying, sediment-yielding and heavily sediment laden one. The river, representing a high seasonal variability of flow and sediment transport, and characterised by the locations of some nodal points along its course through the plains, exhibits the intensely braided channel pattern which is largely responsible for drastic bankline changes and severe bank erosion. The narrow and elongated Brahmaputra valley of Assam, with its peculiar geographic locations amidst the hill ranges and highlands in the northern, northeastern, and southern sides coupled with strongest effects of the hydrometeorological factors, is obviously subject to natural flooding. The recently growing aggressive nature of floods in the valley may be attributed to the changes in the river course and morphology caused by the 1950 earthquake, and the anthropogenic factors like high growth of human settlements in the floodplain, destruction of wetlands and poorly managed embankment network. The vast water resource of the river needs to be properly utilised for hydel power generation, irrigation, and allied developmental purposes which will, in turn, considerably reduce the magnitudes of flood hazards in the Brahmaputra valley.

## 6.8 References

Bhattacharyya, N.N. and Bora, A.K. (1997). Floods of the Brahmaputra River in India, Water International, IWRA, The University of New Maxico, USA, Vol. 22, No.4.

Bora, A.K. (1990). Jia Bharali River of Assam: A study in Fluvial Geomorphology, unpublished Ph.D. Thesis, Guwahati University. Assam (India).

Bora, A.K. and Das, D. (1998). Flood and Erosion Hazards in the Riverine Island of Majuli, Assam, Paper presented at the IXth conference of the Indian Institute of Geomorphologists, University of Delhi, January 30-Feb.1.

Brice, J.C. (1964). Channel patterns and Terraces of the Loup Rivers in Nebraska, U.S. Geol. Surv. Prof. Paper, 422-D, 27.

Coleman, J.M. (1969). Brahmaputra River : Channeal Processes and sedimentation , sediment, Geol, 3(2/3), 129--239.

Flood Control Department, Assam (1989). Report of the Committee for Examining the Causes of flood occurred in Assam in 1988, August, 1989, pp. 25 .

Gait, E. (1905). A History of Assam, Lawer's Book stall publication, Guwahati, Assam, Reprinted in 1984.

Gansser, A. (1964). Geology of the Himalayas, John Wiley, New York, pp. 222--231.

Geological Survey of India, (1974): Geology and Mineral Resources of the States of India, Misc. Publ. 30, Geol. Surv. of India, Delhi, pp. 29--56.

Goswami, D.C. (1982). Brahmaputra River, Assam (India): Suspended Sediment Transport, Valley Aggradation and Basin Denudation, unpublished Ph.D. Thesis, Johns Hopkins University, Baltimore, Maryland,

Goswami, D.C. (1985). Brahmaputra River, Assam, India: Physiography, Basin Denudation and Channel

Aggradation, Water Resources Research, 21, pp. 959--978.

Goswami, D.C. (1988). Estimation of Bedload Transport in the Brahmaputra River, Assam, India. Journal of Earth Sciences, Vol. 1, No. 1, pp. 14--26.

North Eastern Council (NEC), (1993). Morphorlogical Studies of River Brahmaputra, Govt. of India, New Delhi

Wadia, D.N. (1968). The Himalayan Mountains: Its origin and Geographical Relations, Mountains and Rivers of India, p.42, 21st Int. Geog. Cong. Net. Comm. Geogr. Calcutta.

# CHAPTER 7

# HYDROGEOLOGY

B. PURKAIT

The sediment deposition in the Brahmaputra basin is of recent origin. Therefore sediments contain a high water content and are loosely compacted. The watersheds are subject to high erosion and huge amounts of sediment supplied to the river channel cause the river to adjust its channel pattern in response to differing flow regimes. This chapter provides a short discussion of hydrogeological aspects of the Brahmaputra basin.

## 7.1 Introduction

The Ganga--Brahmaputra River system together forms one of the largest deltas in the world comprising some 59,570 sq. km.. Swami Pranamananda during his stay (1928--1938) in the Himalayas as a pilgrim observed that Brahmaputra River originates from the Chemayun-dung glaciers at an elevation of about 5,000 m (m.s.l.) near the Manasaravar-Rakhas lakes between Kailash and Kanglung Kangri ranges, east of Manasaravar (Figure 7.1). It is called Matsang Tsangpo in the early reaches and Tsanpo in Tibet (locally Tsanpo means 'purifier') with a long eastward course of about 1,600km up to Namcha Barwa peak (7,710m) keeping parallel with the main Himalayan ranges. It has a sharp hairpin bend near Namcha Barwa and then it turns southerly. The altitude variations are from 5,200m, 4,600m, 3,965m and 2,957m till Namcha Barwa. Near Namcha Barwa bend the altitude variations are 2,835m (near Gyala), 1,615m (near Gompo Neh-confluence of Yogrong chu) and 792m (near Kapu confluence of Chindru chu), in a river length of about 128 km. The main river has different names in its different parts with a total length of about 2,890 km, of which about 885 km is in India, 405 km is in Bangladesh and 1,600 km is in Tibet. Out of 885 km in the Indian Territory 720 km lies in Assam and 165 km lies in Arunachal Pradesh. On entering the Arunachal Pradesh after going round the Namcha Barwa peak of the Himalayas, the river is known as the Siang in its upper reaches and the Dihang in its lower reaches along Passighat. Near Sadiya the river is joined by two other trans-Himalayan tributaries---the Dibang and the Lohit. Below the confluence of these tributaries the river is known as the Brahmaputra, which, in Indian mythology, means the son of the creator, 'Lord Brahma.' Between Namcha Barwa and its confluence with Dihang, the river has a discent of about 2,200m and its water power resources (average discharge at mouth) have been presumed to be the fourth biggest in the world coming after the Amazon ($99.15 \times 10^3 m^3 s^{-1}$), Congo ($39.66 \times 10^3 m^3 s^{-1}$),Yangtze ($21.80 \times 10^3 m^3 s^{-1}$)



*V.P. Singh et al. (eds.), The Brahmaputra Basin Water Resources*, 113-138.

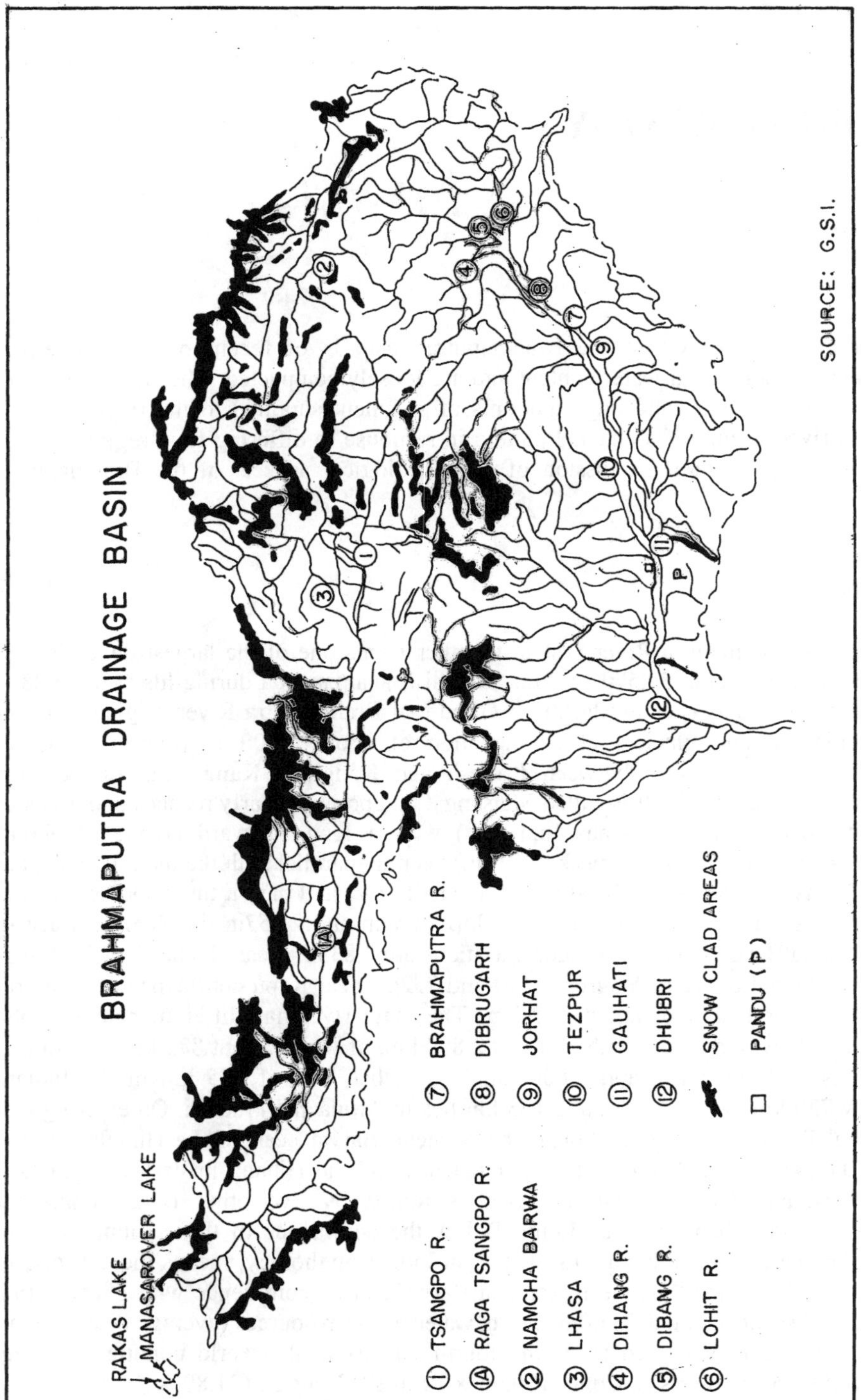


Fig. 7.1 The Brahmaputra drainage basin.

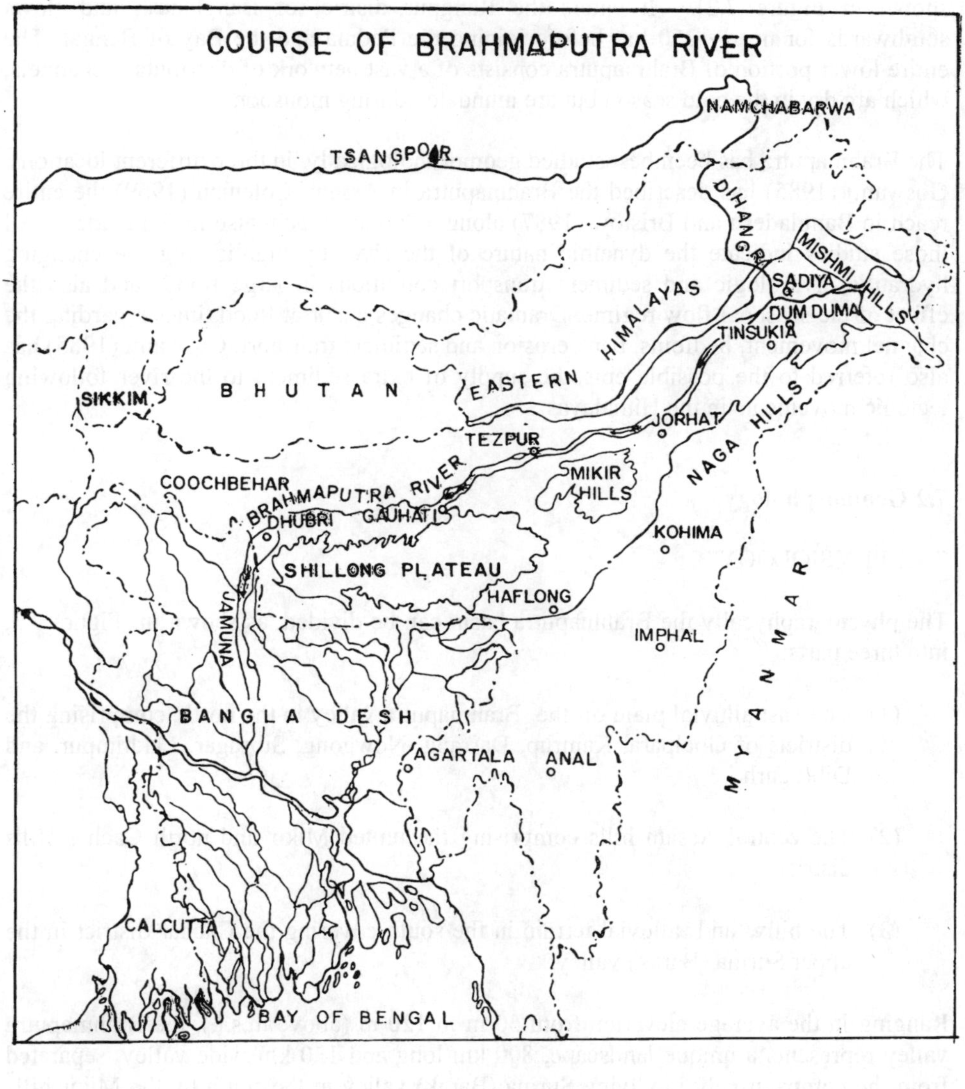


Fig. 7.2 The course of the Brahmaputra River.

and Brahmaputra ($19.83 \times 10^3 m^3 s^{-1}$).The first identity of Brahmaputra with Tsangpo was established through the expedition of Kinthup in 1884 and J.F. Needtham in1886.

The final course of the Brahmaputra lies in the low lying country of Bangladesh, as shown in Figure 7.2). It enters the Rangpur district of Bangladesh and flows southwards for nearly 250 km before joining the Padma and the Bay of Bengal. The entire lower portion of Brahmaputra consists of a vast network of distributary channels, which are dry in the cold season but are inundated during monsoon.

The Brahmaputra has been best studied geomorphologically, in three different locations. Goswami (1985) has described the Brahmaputra in Assam, Coleman (1969) the entire reach in Bangladesh and Bristow (1987) along a 200 km reach, also in Bangladesh. All those studies indicate the dynamic nature of the river by highlighting the changing hydraulic, hydrologic and sediment transport conditions in large floods and also the effect of the seasonal flow regime. Dramatic changes occur at flood times regarding the channel movement, bedfoms, bank erosion and sediment transport. Goswami (1985) has also referred to the possible episodic supply of extra sediment to the river following tectonic movements in the Himalayas.

## 7.2 Geomorphology

### 7.2.1 PHYSIOLOGY

The physiographycally the Brahmaputra basin can be divided, as shown in Figure 7.3, into three parts:

(1) The vast alluvial plain of the Brahmaputra valley in the north, comprising the districts of Goalpara, Kamrup, Darrang, Nowgong, Sibsagar, Lakhimpur, and Dibrugarh.

(2) The central Assam hills comprising the united Mikir and north Cachar Hills district.

(3) The hilly and alluvial terrain in the south covering the Cachar district in the upper Surma (Barak) valley.

Ranging in the average elevation from 50 m to 120 m (above m.s.l.), the Brahmaputra valley represents a unique landscape, 800 km long and 130 km wide valley, separated from the comparatively low lying Surma (Barak) valley in the south by the Mikir hills and Barail range (North Cacher Hills) in the central part. A few granitoid inselbergs and bornhardts occur south and southwest of the Brahmaputra basin (Figure 7.4). The general elevation of this granitoid ridges varies from 200 to 700 m above m.s.l. There are number of channel cut-offs or bils, oxbows and marshy tracts which occupy part of the flood plain deposits. The general elevation of the alluvial plain varies from 60--110m above m.s.l. with a general northerly gradient.

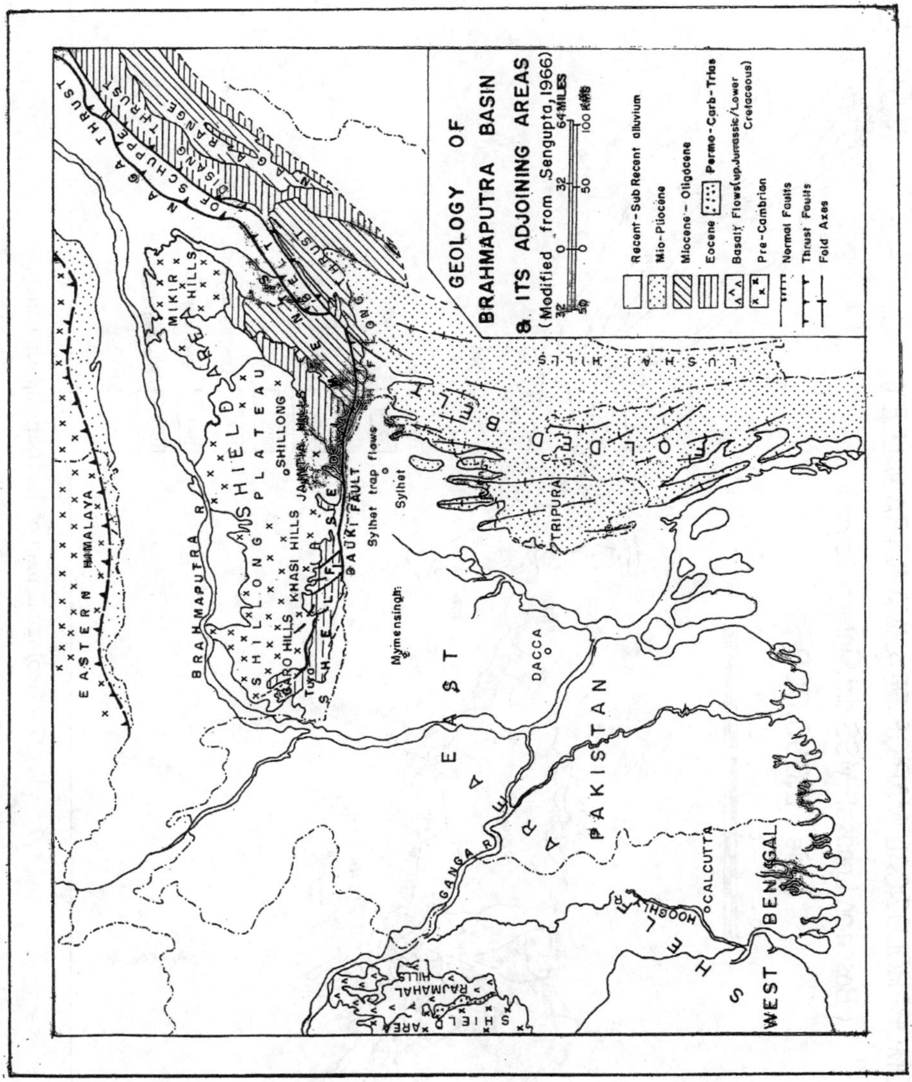


Fig. 7.3 The geology of the Brahmaputra basin.

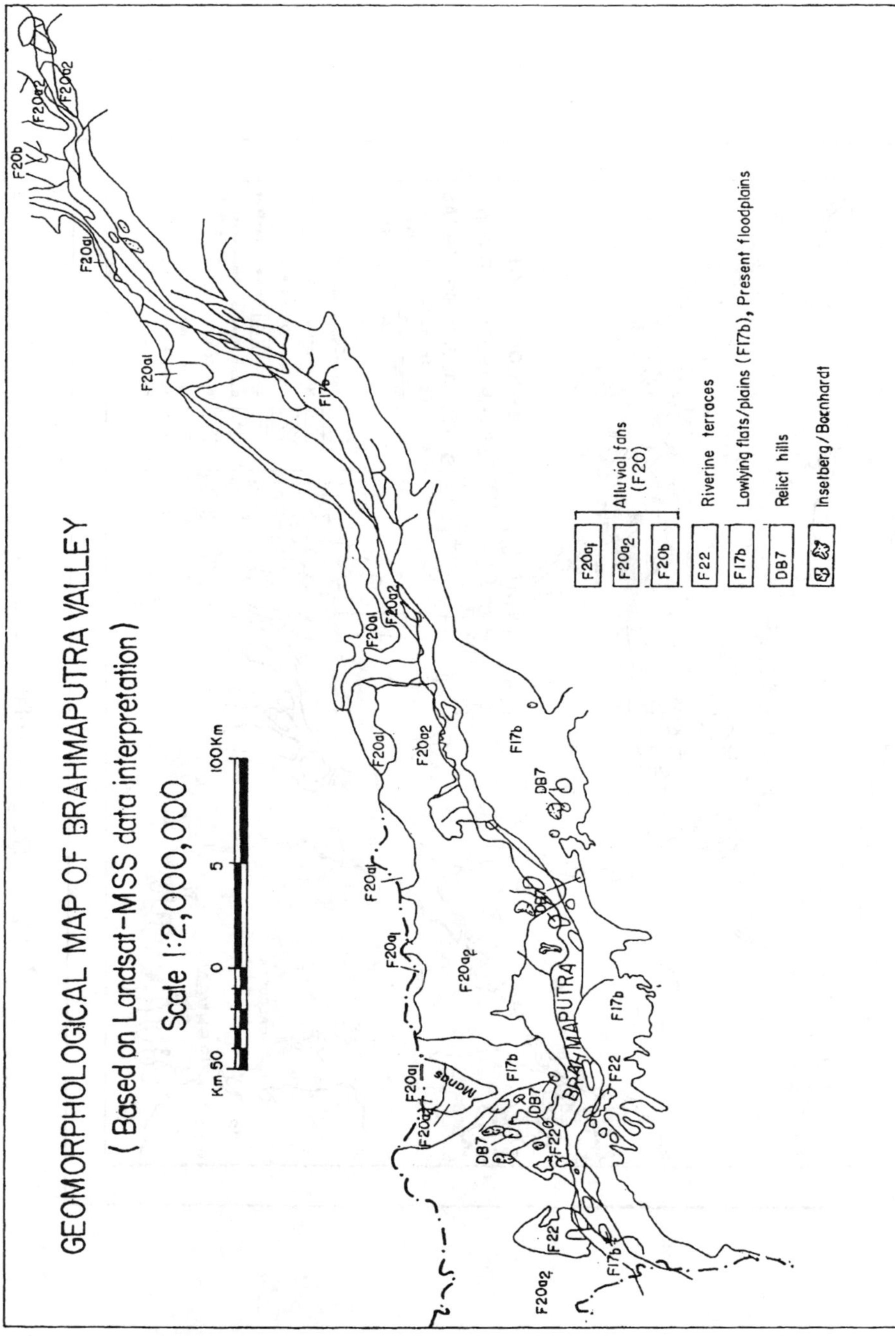


Fig. 7.4 Geological map of the Brahmaputra valley.

The Brahmaputra valley of Assam near Gauhati can be divided into an upper Assam valley having an ENE SW trend whereas the lower Assam valley has a trend of E W. The upper Assam valley is a 'Rift valley', bordered on the north by the easterly continuation of the Himalayan thrust sheets and on the south by the Schuppen belt of the Naga hills. The upper Assam valley is a monotonously plain area, filled up by a thick pile (~5,000m) of Cretaceous, Tertiary and Recent sediments. The lower Assam valley is a 'Ram valley' or sag in the crust formed tectonically during Miocene period along with the Gangetic depression in front of the rising Himalayas. North of the Shillong plateau, there are some Archaean hills as remnants of the northerly sag of plateau, only 30--40 km from the Himalyan foothills. The similarity of the strike of foliations in the crystallines of these isolated Archaean hills with those of the northerly sloping Shillong plateau and the geomorphology of the region suggests a recently drowned landscape (Gansser, 1964).

## 7.2.2 DRAINAGE

The catchment area of the entire river is about 580,000 square kilometers, out of which 195,000 sq. km. lies in India. The maximum discharge of the Brahmaputra as measured at Pandu (Figure 7.1) in 1962 (23-8-62) was of the order of 72,794 $m^3s^{-1}$, whilst the minimum discharge was 1,757$m^3s^{-1}$ in 1968 (on 20-02-68) (Goswami, 1998).The drainage can be elaborated below under (i) river system as well as (ii) drainage pattern.

### *7.2.2.1 River System*

The river is joined by as many as 30 tributaries from the north and 20 tributaries from the south. The catchments of the north bank tributaries lie in the Himalayan range and both in size and height they are bigger and receive higher rainfall than the tributaries flow from the Assam range of hills. The north bank tributaries have very steep slope and shallow braided channels for a considerable distance from the foothills in some cases right up to the outfall. They are subjected to flash floods and carry heavy silt charge. On the other hand, the south bank tributaries have comparatively flatter slopes, deep meandering channels almost from the foothills, bed and banks of fine alluvial soil and comparatively low silt charge. The important north bank tributaries are Subansiri, Dhansiri, Puthimari, Pagladiya, Manas. The south bank tributaries are Noa-Dehang, Burhi-Dehang, Dikhu, Kopili, etc..

### *7.2.2.2 Drainage Pattern*

The drainage pattern of the Brahmaputra network (within India), in general, may be described as dendritic in the northeast and northwest, subparallel in the south ,and reticulate or anastomosing in the main course (Figure 7.1). The drainage pattern in the valley is of antecedent type. The Yazoo drainage pattern is also most significant and predominantly seen over the composite flood plain to the south of the Brahmaputra where the tributaries such as Gola bil, Mara Disai, Mara Dhansiri, etc., flow from east to west parallel to the Brahmaputra for longer distances before debouching into it.

In the southern periphery (Karbi Anglong hills) the patterns, such as angular, sub-angular-dendritic, subangular-trellis, etc., are structurally controlled. The presence of (a) ENE WSW, (b) NE SW, (c) NW SE, and (d) NS trending lineaments /fractures in the

Karbi-Anglong areas could be inferred from these. The ENE WSW trending lineaments are predominantly seen in the southern foothill areas.

## 7.3 River Morphology

The Brahmaputra river, which is under dynamic equilibrium with fluvial as well as seismotectonic processes, does not display a stable river morphology. The alluvial plains of the Brahmaputra measure nearly 96 to 112 km wide and 644 to 805 km long. The bed is about 10 km wide. The depth of the channel of the Brahmaputra is 4,523 m at Tradon, 3,609 m at Shigatse, 135 m at Sadiya, 47 m at Gauhati and 25 m at Dhubri. The bed slope is very gentle, varying from 1 in 6,600 to 1 in 9,900, and in places the river bed lies below the mean sea level.

The channel oscillates from side to side leaving large islands in between. Thus it displays a braided pattern in plan view, and short term channel migration is quite drastic, with rates of movements as high as 870 m a year being common.The most significant bank line modifications take place during the falling river stage, when excess sediment is deposited as bars within the channel, causing a change in local flow direction and migration of the thalweg. The Brahmaputra followed a route some 950 km to the east of its present course to meet the Meghna River in Bangladesh only 200 years ago indicating its preferred migration westward. This movement is probably controlled by major faults or fractures in the earth's crust.

The channel bars or braids with large areal extent (2--8 sq km) are classified into stable bars (tapus in local terminology) having tall grass and unstable bars without any grass cover Expatriate settlements are located on these tapus. Most of these tapus are either washed off completely or remain as small remnants owing to successive torrential flows.

The left bank of the Brahmaputra channel is very steep, its height varies from 3 to 7 m and is unstable. Owing to its southerly migration, the left bank line is also shifting to the south causing inundation or erosion of the neighboring low lying areas.

The Dhansiri, one of the major tributaries of the Brahmaputra, is a highly meandering channel with a sinuosity ratio greater than 1.8. Numerous meander scars are seen scattered over the meander belt of the Dhansiri. Several point bars of limited areal extents (a few hundred sq m) are seen over the slip off slope side of the Dhansiri channel where the active cultivation is going on. The height of the Dhansiri channel banks varies from less than 1 m. to 8 m.

The geological history of the Brahmaputra valley indicates that the Brahmaputra River is a very young river and its present configuration took shape only during the Pleistocene and Recent times.

The bank line position of the Brahmaputra and its channel metamorphosis during the last 150 years (approximately), the active depositional and erosional processes and the active seismicity indicate that the Quaternary endogenic and exogenic processes responsible for the development of the alluvial landscape are still continuing during the present time.

### 7.3.1 GEOMORPHIC SET-UP AND MORPHO-STRATIGRAPHIC UNITS

The Brahmaputra valley has a physical structure comparable to a winnowing fan abutted on the north by the lofty eastern Himalayan ranges, on the south by the Meghalaya plateau and the Mikir Hills, on the east by the Naga Hills, the western side being open and continuous with the Gangetic valley. The numbers of morpho-stratigraphic units are developed in different sub-basins of the Brahmaputra valley. They are identified on the basis of geomorphological, lithological, and pedological parameters, namely, morphology, relief, break of slope drainage pattern, degree of oxidation, nature of sedimentary materials, etc. Basinwise, these units are correlated (Table 7.1). Besides these, different geomorphic features recently developed in the Brahmaputra valley, have been identified based on the Landsat-MSS data interpretation (Figure 7.4).

## 7.4 Geology

### 7.4.1 GENERAL LITHOLOGY

The Brahmaputra valley, excepting a few isolated sedimentary hills in upper Assam, inselberges/ bornhardt of gneissic hills in the Darrang, Kamrup, and Goalpara districts, and a few in lying patches of Older Alluvium in the Darrang and Goalpara districts, is covered by the recent alluvium throughout its stretch.

Situated south of Assam, the Bengal basin is primarily of Quaternary alluvial deposits laid down by the Ganges, Brahmaputra and Meghna rivers (Figure 7.3). The Lower Jurassic trap rocks border on the west and north west of the Bengal basin and the Shillong hills predominantly composed of Eocene Sandstones and Numulitic Limestones occur on its northeast. The Tripura and Chittagong Hills border its eastern boundary. The sediment belongs to Paleocene to Pliocene and are lifted and folded giving an irregular topographical configuration of the area. The Bay of Bengal is its southern boundary.

Being recently deposited, the sediments contain a high water content and are loosely compacted. The predominantly finer sediments and the huge amount of material imposed on the channels by the watershed cause the rivers constantly to adjust their channel pattern to differing flow regimes.

Table 7.1 Correlation of major morphostratigraphic (quarternary landforms and formations) units with their broad lithology of different sub-basins of the Brahmaputra valley.

| | | | | | | Assam | | | | |
|---|---|---|---|---|---|---|---|---|---|---|
| Geological age | Manas sub-basin | Pagladiya sub-basin | Jia-Bhareli sub-basin | Subansiri sub-basin | Lower Brahmaputra basin | South of Brahmaputra river | North of Brahmaputra river | Dhansiri sub-basin | Saralbhang a-Sankosh sub-basin; | Arunachal Pradesh |
| Present Day to Late Holocene | Recent Flood Plain/Barpeta Formation | Recent Flood Plain//Nalbari Formation | Kuruwani Surface /Formation | Recent Flood Plain and deposits | Barpeta I & II surface /Formation | Recent Flood Plain and deposit | Recent Flood Plain,Fan & deposit | Recent Flood Plain deposit /Formation | Recent Flood Plain Formation | Recent Flood Plain,Fan & Formation |
| Late Holocene to Middle Holocene | Hauli Surface/ Formation | Dhamdhama Surface/ Formation | Seijosa (Solabari) Surface/ Formation | Jaihing-North Lakhimpur Surface/ North Lakhimpur Formation | Hauli Surface/ Formation | Dangori Surface/ Formation | Hauli Surface/ Formation | Mangaldai Formation | Uttar Narayanpur Formation | – |
| Early Holocene to Late Pleistocene | Sorbhog Surface/ Formation<br><br>Kukulong Surface/ Formation | Sorbhog Surface / Formation<br><br>Subankhatta Terrace / Kukulong Formation | Itakhola Surface/ Formation<br><br>Balipara (Rangapara) Surface/ Formation | –<br><br>Harmoti Surface/Formation | Sorbhog Surface/ Formation<br><br>Chapar Surface/ Formation | Doom Dooma Surface/ Formation<br><br>Chapar Surface/ Formation | Sorbhog Surface/ Formation<br><br>Chapar Surface/ Formation | Kharupetia Formation<br><br>– | Kechugaon Formation Pekua Formation Phisu Formation<br><br>– | Pasighat Surface/ Formation (Fan)<br><br>Roing Surface/ Formation (Fan) |
| Late Pleistocene | – | – | – | – | – | Jairampur Surface / Formation | – | – | – | Jairampur Surface / Formation |

Plio – Pleistocene Siwalik Pediment / Group

Precambrian Inselbergs (Gneissic Complex) Basement rocks

7.4.2 GEOLOGY OF THE BRAHMAPUTRA BASIN

The geological formations in the Brahmaputra basin covering the state of Assam are summarized below into the following stratigraphic sequences:

| | | | | |
|---|---|---|---|---|
| Quaternary | Recent | Unclassified | Newer or Low Alluvium | Clay, sand, silt and shingle |
| ----------Unconformity---------- | | | | |
| | Pleistocene | Unclassified | Older or High level Alluvium | Clay, coarse sand, shingle, gravel and boulder deposit |
| ----------Unconformity---------- | | | | |
| Tertiary | Pliocene | Dihing Group | Dihing Formation (900m, not divided) | Pebble bed, sandy clay, clay, conglomerate, grit and sandstone |
| ----------Unconformity---------- | | | | |
| | Miocene to Pliocene | Dupi Tila Group | Dupi Tila Formation (Surma valley: 3,300m) and Namsang Formation (Upper Assam: 800m) | Sandstone, mottled clay, grit and conglomerate with coal beds at places |
| ----------Unconformity---------- | | | | |
| | Miocene | Tipam Group | Girujan Formation (1800m) | Mottled clay, sandy shales and subordinate gritty sandstone |
| | | | Tipam Sandstone Formation (2300m) | Bluish grey to greenish, coarse to gritty, false bedded, ferruginous sandstone, clay, shale and conglomerate |

| | | | |
|---|---|---|---|
| | Surma Group | Boka Bil Formation (900m to 1,800m) | Shale, sandy shale,siltstone,mudstone,and lenticular, coarse, ferruginous sandstone |
| | | Bhuban Formation (1,400m to 2,400m) | Alternations of sandstone, sandy shale and thin conglomerate |
| ---------Unconformity--------- | | | |
| Oligocene | Barail Group | Renji Formation ( 600m to 1,000m) | Massive and bedded Sandstone (=the Tikak Parbat Formation in upper Assam) |
| | | Jenum Formation (1,000m to 3300m) | Shale, sandy shale and carbonaceous shale (= the Baragolai Formation in Upper Assam ) |
| | | Laisong Formation (2,000m to 2,500m) | Well bedded,compact, flaggy sandstone and subordinate shale(= the Naogaon Formation in Upper Assam) |
| ---------Unconformity--------- | | | |
| Eocene | Geosynclinal: Disang Group ( Not divided) | | Splintery, dark grey shale and thin sandstone interbeds |
| | Shelf: Jaintia Group | Kopili Formation | Shale, sandstone and marl |
| | | Shell Formation | Sylhet Limestone Member (Fossiliferous) |
| | | | Sylhet Sandstone Member |
| ---------Unconformity--------- | | | |
| Precambrian | Shillong Group (Not classified) | | Quartzite, phyllite and schist |
| ---------Unconformity--------- | | | |
| Archaean | Archaean Group (Not classified) | | Complex metamorphic Group of gneisses and schists, metasediments, later intruded by basic rocks. |

**7.5 Description of Different Rock Units**

7.5.1 ARCHAEANS

The Archaean Group of rocks comprise mainly biotite–hornblende, sillimanite, gneisses, and schists associated with feldspathic biotite granulites, pyroxene-hornblende-granulite, calc-granulites, aplites intruded at places by massive porphyritic, coarse grained granite, pegmatite, quartz veins and basic intrusives. They fringe the southern borders of Goalpara and Kamrup districts, and occur in the northern part of the North Cachar Hills, eastern part of the Nowgong district and as isolated inselbergs in the Brahmaputra valley.

7.5.2 PRECAMBRIAN

The Precambrian Shillong Group of rocks comprising quartzite, phyllite and schist, unconformably overlie the Archeans in the northern part of the North Cachar Hills, western flank of the Mikir Hills across the Kopili valley of Assam. In the central Mikir Hills, occurrences of weathered trap of Sylhet suite (Jurassic) of Meghalaya are reported. Regionally, the foliation trends EW with frequent change of direction from NE SW with steep to vertical dip.

7.5.3 TERTIARIES

The rocks of Tertiary unconformably overly the Precambian Group of rocks and are represented by Jaintia (shelf) and Dishang (geosynclinal) groups of rocks of Eocene age of the Lower Tertiary sequence. These are again overlain by Barail (Oligocene), Surma (Lr Miocene), Tipam (Up Miocene), Dupi Tila (Mio-Pliocene) and Dihing (Pliocene) groups of Upper Tertiary sequence (both shelf and geosynclinal facies). The Jaintia Group is devided into basal Shella (including the former Therria Sandstone and overlying Sylhet Limestone) Formation and the Upper Kopili Formation. These are exposed in the Garampani area of North Cachar Hills. The Shella Formation is once again divided into basal Sylhet Sandstone Member of about 60m thick comprising sandstone, clay with carbonaceous shale and a thin coal seam. The Upper Sylhet Limestone Member attains a thickness of upto 120m. It is a thick foraminiferal limestone with marl bands, upper part being ferruginous and argillaceous in nature having subhorizontal dip. The Kopili Formation represents greyish ferruginous splintery shale with sandstone and calcarious marl. It attains a thickness of about 500m. The Jaintia Group of rocks extends along the southern slopes of the Mikir Hills from Selvetta in the west through Dilai Parbat in the east of Doigrung and further NE. The Geosynclinal facies of Jaintia Group is designated as Disang Group and is a monotonous sequence of dark grey,splintery shale interbedded with thin sandstone. They are exposed to the south of Haflong Disang thrust in the central part of North Cachar Hills from Jaintia valley eastward upto the headwaters of Dhansiri. The overlying Barail Group (Oligocene age) of Surma valley and North Cachar Hills, successively includes the Laisong, Jenum and Renji formations and represents geosynclinal facies. The Laisong (2,000m to 2,500m thick) comprises well bedded, compact, flaggy sandstone with subordinate shale and streaks of coal. It gradually passes up into argillaceous Jenum Formation and represents shale, sandy shale, carbonaceous shale with streaks of coal and interbedded sandstone. The overlying Renji Formation comprises mainly massive and bedded sandstone. The rocks of the above formations are exposed along two different strips covering (a) south-eastern part of the

North Cachar Hills to the south of Haflong-Disang thrust and (b) part of north Cachar and Mikir Hills to the north of Haflong Disang thrust. The Barail Group in the Upper Assam oil fields is characterized by numbers of oil sand horizon within a depth range of 40m to 370m.

The Lower Miocene Surma Group of rocks unconformably rest on Barails and cover a large area in the Surma valley and the North Cachar Hills. It is further divided into basal Bhuban Formation and the upper Boka Bil Formation. The Bhuban Formation represents mainly the arenaceous facies consisting of sandstone, sandy shale and conglomerate, whereas the overlying Boka Bil Formation represents mainly the argillaceous facies consisting mainly of shale, sandy shale, and siltstone.

The Surma Group of rocks are exposed as inliers in the southern part of the Surma Valley and as a strip in the northern part of the valley. About 12 oil sand horizons are reported within the Bhuban Formation of the Cachar district.

The overlying Tipam Group (Upper Miocene age) consists of a lower arenaceous facies, the Tipam Sandstone Formation and an upper argillaceous facies, the Girujan Formation. The Tipam Sandstone Formation represents coarse to gritty, ferruginous sandstone, interbedded with shale, sandy shale, clay and conglomerate. The rocks of Girujan Formation consist mainly of lacustrine, mottled claystone, sandy shale with subordinate mottled, coarse to gritty, ferruginous sandstones.

The Tipams are exposed, in general, in many parts of the Surma valley and also include several oil sand horizons of the upper Assam. Both Tipams and Surmas include fragments of silicified and semi-carbonised fossil wood. The Tipam Group is unconformably overlain by the Dupi Tila Group of rocks belonging to Mio-Pliocene age. It consists mainly of coarse, loose ferruginous sandstone, mottled clay, and poorly consolidated sandstone with layers and pockets of pebbles. They are well exposed in the southern part of Surma valley. In the upper Assam, the Dupi Tila is represented by the fluviatile Namsang Formation.

The Dupi Tila is unconformably overlain by the fluviatile Dihing Group of rocks consisting of alternating pebble beds with coarse sandstones, clay, grit and conglomerate, and partly decomposed plant remains at places. These are well exposed along the Dihing river section near Jaipur in upper Assam.

### 7.5.4 QUARTERNARY DEPOSITS

The Quaternary deposits are unconformably overlying the Dihing Group and are designated variously as 'Terrace deposits', 'Older or High level Alluvium', 'Unstratified Drifts' or 'Red Bank soils'. Sometimes it is difficult to differentiate these deposits from the underlying Dihings. The Quaternary deposits represent indurated yellowish, brownish or reddish clay mixed up with sands, shingle, gravel, and boulder deposits. The pebbly beds of the Older Alluvium at various levels as revealed by borehole logs, indicate cyclic nature of sedimentation. A major part of the Brahmaputra valley is covered by Recent alluvium including the flood plain deposits and is grouped as Newer or Low Level Alluvium. These deposits comprise clay, coarse sand, shingle, gravel and boulder and attain a thickness of approximately 200 to 300 m. In general, the proportion of finer clastics increases towards the south.

### 7.5.5 STRUCTURE AND TECTONICS

The present configuration of the Brahmaputra valley is the result of uplift and subsidence of the Precambrian crystalline land masses, the remnant of which is now represented by Mikir Hills (Karbi-Anglong Hills) of Assam and Shillong plateau of Meghalaya. They form a 'Foreland spur' which has been over thrust from the northwest by the Eastern Himalayas, from the northeast by the Mishmi Hills and from the southwest by the Naga-Patkai range during the Tertiary geotectonic cycle (Figures 7.5a and 7.5b).

Four geotectectonic provinces have been recognized in the northeast India through which the Brahmaputra flows:

i) The stable shield area of Shillong plateau and Mikir Hills.
ii) Platform area peripheral to the shield, now covering the North Cachar Hills, Bangladesh plains and the Brahmaputra valley.
iii) The Naga-Patkai and Eastern Himalayan mobile geosynclinal belts and
iv) The Transitional zones between the platform and geosyncline, probably with narrow pericratonic downwarps marginal to the shield

The above four geotectonic provinces are bounded by major tectonic lineament that include:

a) The basement EW trending Dauki fault along the southern margin of the Shillong plateau, extending upto Haflong, a suspected EW fault along the Brahmaputra valley which has probably a tectonic control of the river in the state of Assam, numbers of NW--SE trending faults located in the SW of the Shillong plateau, the Mikir Hills and its east.
b) The 'belt of Schuppen,' a NE--SW trending structural features of imbricate thrusts over the northern part of the Naga-Patkai range.
c) The frontal Himalayan thrust belt trending EW to NE--SW.
d) The NW-SE trending Mishmi thrust along the Lohit foothills.
e) Probable northeasterly extension of NE--SW Calcutta-Mymensing gravity high to the south of the North Cacher Hills through Cacher district.

On the southern boundary of the Shillong massif occur Upper Cretaceous and Tertiary sediments, overlying basalt trap flows. Thick alluvium deposits of the Brahmaputra and the Ganges rivers cover the whole area between the peninsular shield and the folded belts of the Naga-Lushai Hills.

Subsurface geology as revealed by deep drilling indicates that the Tertiary sediments overlying the basement complex are gently folded into domal and antiformal structures with low to moderately dipping limbs affected by numbers of fault, mainly trending ENE--WSW, NNE--SSW, NE--SW and EW with throw ranging up to 200m.

The course of the Ganges and Brahmaputra rivers along a northwest--southeast zone of fractures within the Garo-Rajmahal gap may be owing to the large-scale strike slip movements of the Dauki fault. Evans (1964) suggested the very large scale strike slip movements of the Dauki fault, causing eastward movements of the whole Shillong plateau relative to the Surma valley region, and postulated that this movement was the principal cause of the existing gap between the Garo and Rajmahal hills.

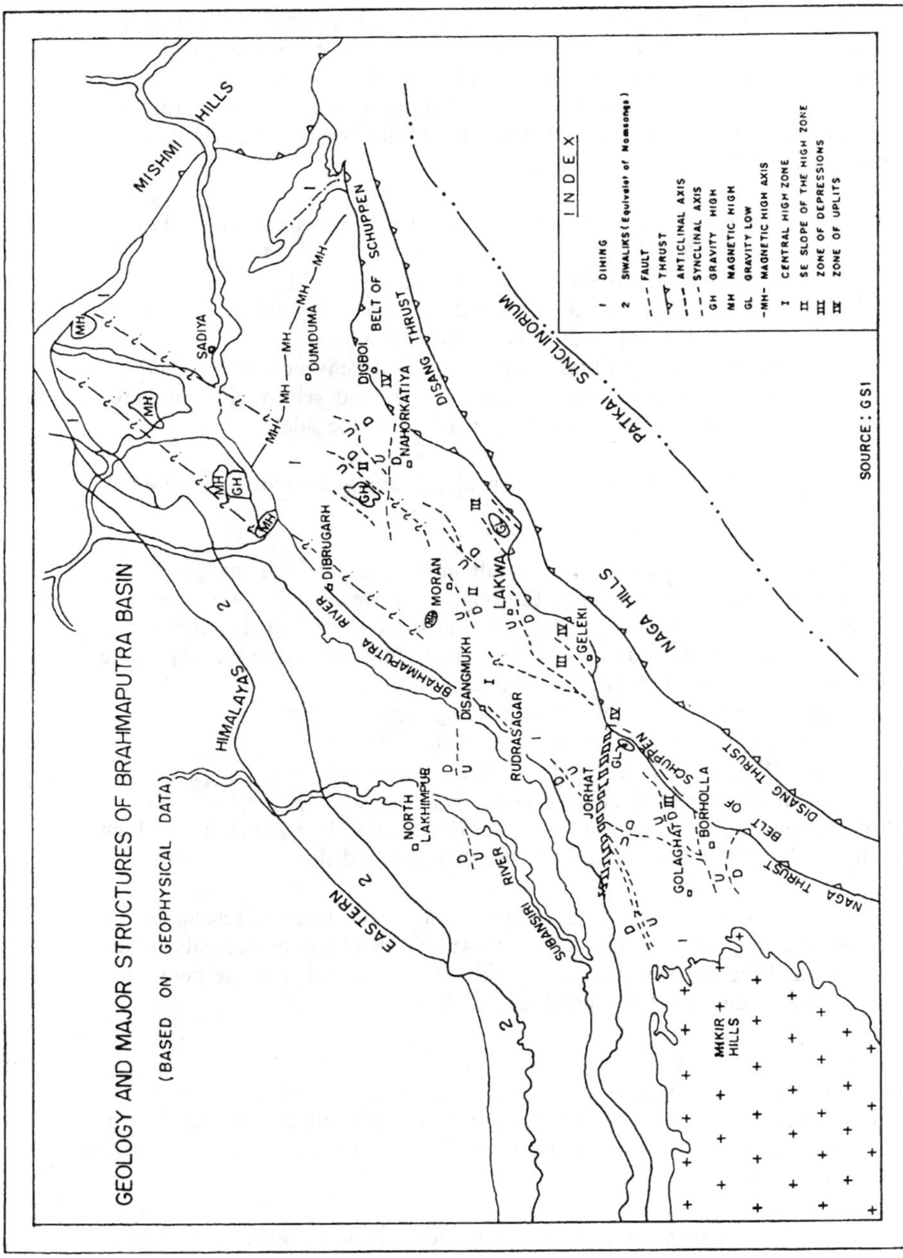


Fig. 7.5a Geology and major structures of the Brahmaputra basin.

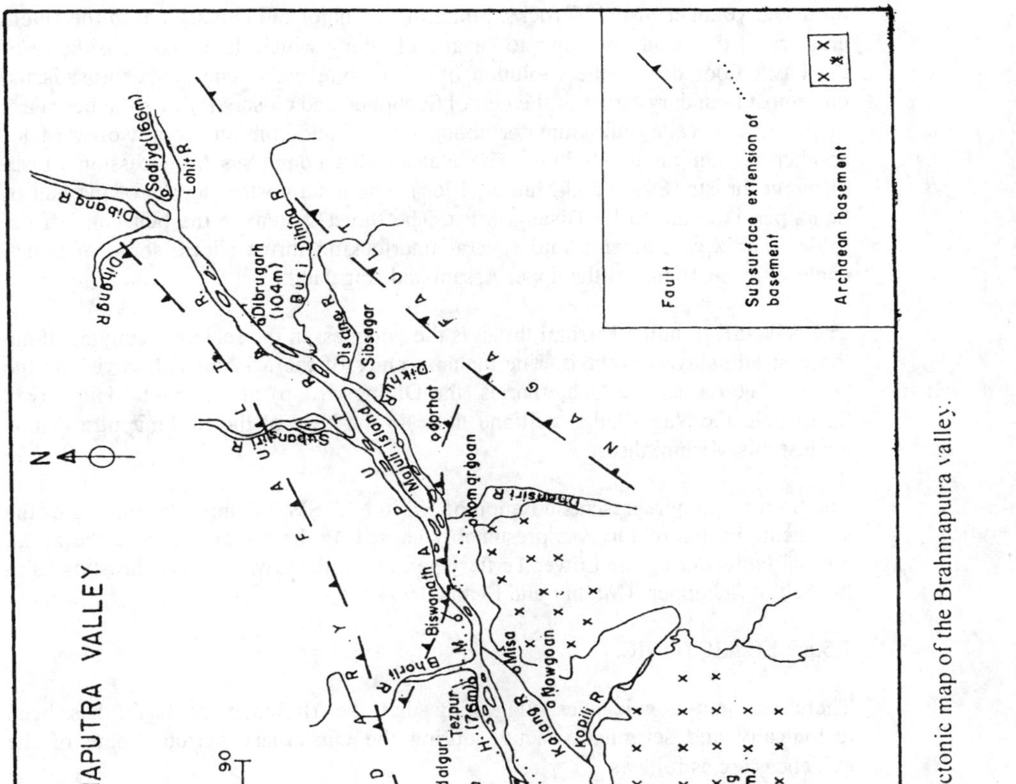


Fig. 7.5b Tectonic map of the Brahmaputra valley.

The faulting continued intermittently throughout the sedimentation from Eocene to Pleistocene, some of the faults affect only the Oligocene rocks, and others extend up to the younger group of rocks. Some of the major faults marginal to the shield area are of the nature of down to basin fault along which there was subsidence of the basin floor during the evolution of the mobile belts. The Naga thrust is the outermost boundary thrust of the belt of Schuppen and closely follows the boundary of the Assam valley alluvium for about 350 km and continue southwestward for another 50 km up to Haflong. The Naga thrust comprises a succession of six different thrusts (Evans and Mathur, 1964). The innermost boundary of the belt of Schuppen is bounded by Disang thrust. The thrust system in the belt cuts off the folded Tertiary sediments into several interlocking thrust slices, some of which enclose the coalfields of the upper Assam and Nagaland.

The NW--SE trending Mishmi thrust is the youngest in the region occupying along the east Himalayan syntaxis. The metamorphics of the Mishmi Hills override the younger Tertiaries and Quaternaries (the Dihing Group) of the frontal Himalayan thrust belt, the Naga-Patkai belt and those in the head of the Brahmaputra valley, against this Mishmi thrust.

The Surma valley represents a region of NS to NE--SW folding. The folding of the sediments in this region was presumably caused by the minor, down to the basin, normal faults during the Lower Tertiary period but there was no overthrusting as in the belt of Schuppen (Mathur and Evans, 1964 ).

### 7.5.6 NEOTECTONIC

There are ample evidences suggesting that the Brahmaputra basin has been tectonically and seismically active during the Quaternary period. Some of the evidences are as follows:

#### *7.5.6.1 Geomorphic Evidences*

Step-like pattern of the alluvial terraces/geomorphic units (quaternary depositional surfaces) indicates progressive down cutting due to uplift.

i) Unpaired terraces occupying different elevations of the same age .

ii) Presence of younger sediments in upstream areas followed by older sediments in some streams.

iii) Longitudinal profile of the Bihora geomorphic surface indicates differential warping. The 5--12 m high vertical scarp of this surface aligned parallel to the ENE--WNW lineament (in the Archaean Group of crystalline) may be a fault line scarp.

iv) Numbers of meander cut–offs, numerous bils or flood plains back swamps indicate the transient nature of the streams. The subsidence of the Brahmaputra valley floor is indicated by these back swamps.

v) The rapid southerly migration of the Brahmaputra and its channel metamorphosis also supports the subsiding nature of the valley floor.

vi) Localized reversal of surface water flood and gently upstream dips of sediments (e.g., Pishugaon area).

*7.5.6.2 Structural Evidences*

(1) The straight course of most of the streams and the drainage patterns like rectangular, subangular, dendritic, etc., directly show the NE--SW, NW--SE, ENE--WSW and NS trending linear segments within the Quaternary alluvium. They indicate the presence of fractures / faults along these directions which are either reactivated during the Quaternary or created due to rupture in recent times. The Kalong and the Kopili rivers are aligned along the major lineaments of the area (NW--SE lineament).

(2) The ENE-WSW and NE-SW trending fractures, present in Archaean Group of crystallines, are seen extending into the alluvium, e.g., the NE-SW trending fractures within the Karbi-Anglong crystallines extend into the alluvium where a segment of the Mara Diphlu Nadi is aligned parallel to this fracture.

(3) The ENE--WSW and NE--SW trending fractures/faults within the Archaean Group of crystalline of Karbi-Anglong are found to be still active as they are associated with thermal springs.

*7.5.6.3 Seismic Evidences*

(1) The Indian Meteorology Department has compiled the major earthquake intensity data indicating the presence of a few earthquake epicenters well within or below the alluvium. These earthquake epicenters are indicative of the seismic zone of this area.

(2) The analysis of microearthquake data recorded during 1982 to 1985 in the NE India led to the identification of a high level of seismicity in some areas (e.g. Nowgong and Karbi-Anglong). The trends of this seismicity are incompatible with ENE--WSW and NE--SW lineaments.

(3) The effect of severe earthquakes is the presence of sand dykes cutting through some formations of Holocene age (e.g., Pathsala, Tangla).The sand dykes upto 15 cm in thickness show a preferred orientation indicating remobilization of sand along fracture planes from an underlying, water saturated sand bed, under the effect of pressure generated by severe seismic impulses during 1897 and 1950 Assam earthquakes.

(4) The Brahmaputra basin falls well within the seismotectonically disturbed zone. The river shifted its course many times in the recent past owing to the effects of the earthquakes in the area during the years 1889, 1897, 1953 and 1988.

Thus the neotectonic activity has definitely affected the morphology and physiographic features of this area.

## 7.6 Hydrogeology

Hydrogeologically, the Brahmaputra basin can be divided into two distinct categories, viz., (a) dissected alluvial plain and (b) the inselberg zone. The first category is represented in the flood plain extending from south of sub-Himalayan piedmont fan zone in the north to right up to the main rock promontory of Garo Hills and Shillong plateau. The sediments of this vast alluvial plain have been brought down by the past and present Himalayan rivers. There are a large number of buried channels consisting of gravel and sands, natural levees of sands and silts, back swamps / bils of silts and clays. Borehole data indicate that there is a general coarsening of sediments at depth with two to three zones of cyclothemic gravels within depth of 90–100m, indicating a cyclic nature of sedimentation. There is also a coarsening of grain size noted on the present surface away from the main channel towards the higher altitudes. The inselberg zone is characterized by fractured, jointed, and weathered ancient crystalline rocks with inter-hill narrow valley plains, consisting of thin to occasionally thick piles of assorted sediments.

### 7.6.1 AQUIFERS

Subsurface lithological data (Figure 7.6) indicate two broad groups of aquifers, e.g., (i) shallow water table and (ii) deeper water table or confined ones, separated by a system of aquicludes. The occurrence of groundwater in the hard rock inselberg areas is controlled by foliations, fractures/joints, and weathered zones. The physiographic features, types of drainage, depth and duration of precipitation also play an important role. There are shallow open wells and a few shallow tube wells yielding a limited supply of water. Many of the shallow open wells dry up by the month of March.

In the Tertiary sediments ground water occurs in the sandstones and along the fractures and solution cavities in the limestones. In the Cachar and the Mikir hills districts shallow tube wells tap the aquifers within a depth of 100m. Ground water occurs both under water table and confined conditions.

In the piedmont and alluvial fan deposits ground water occurs generally under water table conditions. The water table has, in general, a southerly slope corresponding roughly to the surface topography. The gradient of the surface topography is about 5m/km In addition to the general southward gradient, a centripetal hydraulic gradient is also discernible in the central part of the area of Kamrup district .

In the older and newer alluvium ground water occurs both under water table, and confined condition. The artersian belts are known to occur in a narrow strip between Dumni Chowki and Nalbari in Kamrup district, and within a small patch in Borpathar area of Golaghat subdivision in the Sibsagar district. Figure 7.6 shows a general hydrogeological map of Assam.

On the north bank, in the dissected flood plains between Abhayapuri and Beki-Chaulkhowa section, ground water generally occurs under unconfined or water table condition, at places it is perched where a thin veneer of sand or silt rests on the pockets of clay which in turn cause semiconfined condition in underlying aquifers. These are shallow aquifers lying within 15--20m below ground water level. However, deeper granular zones down to 90--100m at places are also recorded. They broadly belong to Hauli and Barpeta geomorphic units.

Within the inselberg zone, at places shallow unconfined aquifers occur right from ground level down to a depth range of about 30 m of Barpeta and Hauli geomorphic units. The sediments are mainly fine to coarse sand and silt. In these units the first saturated granular zone of medium to coarse sand and gravel with occasional

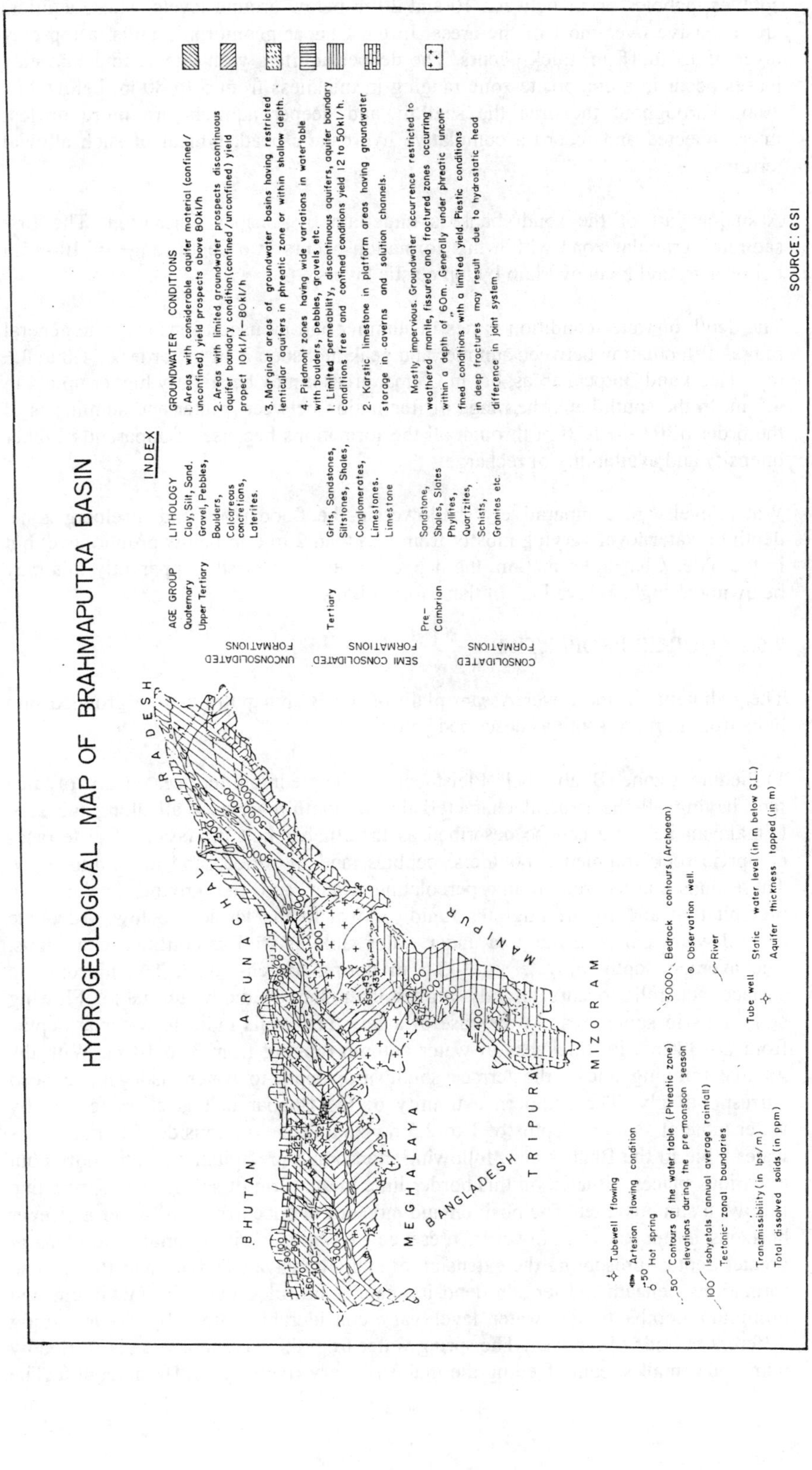


Fig. 7.6 Hydrogeological map of the Brahmaputra basin.

cobbles, pebbles occur between 10 and 20 m below ground levels. These aquifers are extensive over most of the areas .In the Chapar geomorphic units, a top clay layer of 15 to 18 m. thick occurs. The deeper aquifers with intervening thin clay lenses occur in a composite zone ranging in thickness from 5 to 80 m. below 25--30m. Throughout the area the shallow and deeper aquifers are more or less interconnected and record a cumulative hydrostatic head, typical of such alluvial terrains.

A major part of the south bank is underlain by Chapar Formation. The first saturated granular zone within this formation occurs in a depth range of 10--15m below a ground level overlain by top clastic layers.

The depth of water condition varies in summer and monsoon seasons. The general range of fluctuation between summer and peak monsoon is of the order of 1.8 to 3.6 m in Hauli and Barpeta areas, but in Chapar areas it may be slightly higher upto 4 to 4.5 m. In the south bank the range of fluctuation between winter and summer is of the order of 0.9 to 1.70 m through all the formations because of apparently higher intensity and availability of recharge.

Water levels are comparatively shallower in the flood plain and inselberg zone, depth to water level varying mostly from less than 2 to 6 m below ground level but in the older Chapar Formation, the deeper water level exists. Apparently this may be owing to higher elevations of these formations.

### 7.6.2 AQUIFER PROPERTIES

The sediments in the lower Assam plain of the Brahmaputra can be grouped into belts from north to south as described below:

1) Piedmont zone (Bhabar belt–Pleistocene): The piedmont terraces consist of talus fans having all the general characteristics of foothill terraces all along the sub-Himalayan belt and can be described as the Bhabar belt in Assam. The terraces comprise rock fragments, boulders, pebbles, and ill sorted sand and minor clay. These fans hardly retain any percolating water and the ground intake from precipitation and runoff migrates rapidly down to the terraces below. Adequate ground water exists in the area under water table as well as confined conditions. The average depth to water is 3--3.5 m during March--April. The piezometric surface generally occurs slightly higher than that of the water table. Flowing conditions in some areas in the case of a few tube wells indicate water at depths from 26–38 m with thickness of water column ranging from 3 to 14 m. With the gradual thinning out of the terrace southward, depth to water also gets reduced correspondingly. The southern extremity of the Bhabar belt is characterized by water table at shallow depths of 1 to 2.5 m. As a sequel to considerable rise in the water table in the Bhabar belt, following monsoonic precipitation, water oozes out at various places, situated on this border line, in the form of spring. The spring line follows a sinuous tract. The position and mode of the occurrence of water at deeper horizons is not known. However, presence of water bearing granular horizons at greater depths, presuming the extension of sub-Himalayan Tertiary and Pleistocene formations beneath the terrace deposits, cannot be ruled out. The maximum and minimum depths to the water level vary considerably, often by 50 m during different periods of the year. The spring water from the piedmont zone is carried by numerous small streams feeding the main tributary rivers of the Brahmaputra. The

stream water inflows into the highly permeable sediments of this zone and migrates down to emerge again as spring discharge at the junction of the Bhabar and its lower zone---the Tarai.

2) The Tarai zone (Pleistocene): The linear Tarai tract fringes the piedmont zone throughout, commencing from the spring line. It occurs below the piedmont zone comprising terraces of brown clay (so-called red bank terraces), coarse to medium sand with occasional pebbles and gravel beds.The depth to water table persists within depths of 1--2 m for about 4--5 km down south of the spring line. This 4--5 km width belt forms the main recharge zone for aquifers within relatively shallow depths in the sub-Tarai zone. Here the near surface aquifers are under water table condition but at greater depths in Pleistocene and Upper Tertiary sediments, the aquifers might occur under confined condition. This belt merges with the alluvial plain in the south almost imperceptibly characterized by alternating beds of sand, mixed with sand/gravel and subordinate clay. The depth to water increases from 1.5--2.5 m in the northern part to 2--4 m in the southern part. Thus the parallelism between the water table slope and the topographic gradient (from north to south) becomes clearly manisfested in this belt.

3) The present day flood plain deposits and their abandoned channels: They occur between the Tarai belt and the Brahmaputra River. The sediments are mainly gravel, sand, silt and clay. The near surface sand and gravel beds constitute the aquifers which contain ground water under water table conditions. Deeper aquifers are often under confined condition. The water table slopes more or less parallel to the topographical slope–about 2 m / km. The gradient flattens out where the flood plain is broader. In the central part of the plain, the water table stands at depth of 3--4 m below land surface during the months of March and April. Isolated pockets showing water table at shallower depth of 2--3 m and increased depth of 4--5 m are occasionally met within the central part. The bil areas, surrounding the central part, also show the water table at 2--3 m depth. Considering the position of the water table in the central part and the bil area together in an almost flattish topography, it appears that the water table has a centripetal gradient. On the southern periphery of the alluvial plain, in the vicinity of the Brahmaputra River, on its northern flank, a progressive increase in the depth to water from 2 to 5 m, is distinct. In the central part of the area where alternating sand and clay beds occur below the topmost clayey horizon up to 18 m thick, at least 3--4 significant confined aquifers have been tapped from 20--53 m depths. In the southern part of the alluvial plain, where sandy sediments predominate, small diameter tube wells tap water from the aquifers between 10--26 m depth. The aquifers are under semi-confined to free condition. Deeper aquifers down to 60 m depth are also under semi–confined condition as no impervious horizon of any importance exists within this depth. Piezometric level varies in the alluvial plain from 1.4 to 5.9 m below land surface. In most cases the level is 0.3 m above the prevalent level of the water table. The average piezometric level for the area is 3.2 m as against the average depth to the water table of 3.5 m. The water level rises up considerably in tube wells and at certain localities the piezometric surface may be above the land surface. A number of tube wells in the sedimentary plain have shown an artesian flow.

### 7.5.3 SHALLOW AQUIFERS AND THEIR HYDROLOGICAL PARAMETERS

On the southern bank of the Brahmaputra in the inselberg zone of the Kamrup district, Assam, ground water occurs in the weathered zone and in joints and fractures, whereas in the alluvial area, ground water occurs under water table condition. The aquifer materials in general, consist of white to grayish white, fine grained micaceous sand and the thickness ranges from 1.2 to 10.3 m. A top reddish brown clay zone varying in thickness from 6 to 10.6 m is followed by brownish grey silt and clay layer down to a depth of 12 to 32 m with a thin lense of sand of 1.2 to 6 m thick occurs in between. This silt and clay layer again is followed by bluish grey to brownish grey clay down to a depth of 73 m from land surface. The sand and clay ratio varies from 1:2.5 to as high as 1:26 in some boreholes. The bed rock occurs at a depth range of 30.4 to 39.5 m having a gentle slope from south to north as well as from east to west.

### 7.5.4 DEEPER AQUIFERS AND THEIR HYDROLOGICAL PARAMETERS

The deep borehole data from the northern part of the Kamrup district indicate the occurrence of granular zones varying in thickness from 6.8 to 18.6 m with a dark reddish brown to yellowish brown clay at a depth range of 3 to 11.8 m. The granular materials comprise grey to greysih white, fine to medium grained sand. The sand and clay ratio varies from 1:2 to 1:7. The effective size of the aquifer materials varies from 0.125 to 0.062 mm with uniformity co-efficient around 4.00. The aquifer materials show a porosity range from 38% to 42% with co-efficient of permeability varying from 304 to 390 galls per day/ft $^{2}$.

### 7.5.5 CHEMICAL QUALITY OF GROUND WATER

The ground water is generally mildly alkaline with pH value ranging from 6.5 to 8.5, total dissolved solids are low. Chloride (10--40 ppm ) and bi- carbonate (50--350 ppm ) values are considerably low. Iron content ranges from a fraction of a ppm to as high as 50 ppm having the higher values in the north eastern sector and varies both laterally and depth wise. Total hardness as $CaCO_3$ generally varies from 50 to 300 ppm, specific conductance at 25 °C generally varies from 150 to 650 mhos/cm. The chemical quality indicates a general flushing movement.

## 7.6 Ground Water Potential

The Brahmaputra basin has adequate surface and ground water resources. Besides, a few mineral water occurrences are also recorded. The Brahmaputra and its numerous tributaries like the Dhansiri, Diphlu, Deopani, etc., are the major sources of surface water in the area. They are being utilised effectively for agriculture, tea-plantation and domestic purposes.

Although ground water is available in plenty, its quality, because of high iron content, affects its potable nature. As for example, the sandy and gravelly zone of Bihora and Hatikuli formations serve as a high quality aquifer zone. But owing to oxidation of surficial sands, ground water available at shallow level is rich in iron content beyond this zone of oxidation. Moreover at a greater depth of 30 m, ground water is free from much of iron. In the Kaziranga unit, good quality ground water is available at a 14--30 m depth. But owing to the presence of clay horizons at some places, the ground water becomes brackish in nature.

The water level in the flood plains and inselberg zone are shallow (5--6 m) whereas in the older Chapar Formation, because of its higher elevation, the water levels exist at much deeper levels. The shallow water table is recharged by rainfall almost upto the ground level by mid monsoon period after which the rainfall is mostly rejected as run-off. The flood sediments have higher cumulative transmissibility than the alluvial sediments of Chapar Formation. In general, the yield of the tube wells of shallow aquifers varies from 1,440 litres to 33,750 litres /hour whereas the yield from the tube wells of deeper aquifers is 1,700 litres /per hour at a draw down of 13.41 m, specific capacity 21 litres per minute.

The temperature of the ground water is, in general, in the range of 23 °C-25 °C during winter, 24 °C-26 °C during pre monsoon period, and 27 °C- 28 °C during peak monsoon when the water table rises within 2 to 2.5 m of surface where ambient temperature operate. Thus the temperature data show an overall normal state of ground water motion. Owing to stagnation, the condition of the ground water of the back swamp lakes shows a higher temperature range of upto 31 °C-34 °C. The number of bils shows the temperature proximal to the ground water temperature, indicating a ground water feedback whereas the river water normally shows a lower temperature range of 18 °C-20 °C.

In the north bank the ground water has a general movement from north to south inspite of some local variation, i.e. towards the Brahmaputra across the main master slope of the country and also towards other tributaries which are all affluent in nature. The general hydraulic gradient is 1: 800.

In the south bank a general northerly movement of ground water towards the major affluent trunk channel is observed. The general hydraulic gradient is 1:300--400, i.e., it is comparatively steeper than that of the north bank. Moreover, the reversal of movement takes place during the peak monsoon period, causing aggradation of the basin by ground water movements from both north and south.

In Cachar and Mikir Hills districts a few shallow tubewells in Tertiary sediments yield a range of water from 200 to 300 l.p.m. The ground water occurs both under water table and confined condition and the aquifers occur within a depth of 100 m in the sandstone and along the fractured and solution cavities in limestone. The ferruginous sandstones of Tipum and Surma series give an excess of iron content as high as 12 ppm in the ground water.

The ground water yield from the older alluvium varied from 500--700 l.p.m. for a drawdown of about 3 m. The estimated transmissibility and coefficient of storage are of the order of 793 l.p.m./m. and $8.2 \times 10^{-3}$, respectively. The depth to water levels in the wells ranges from 5.3 m to 10 m below the land surface. In the Goalpara district the depth of tubewells ranges from 87 to 118 m with depth to water level varying from 0. 90 to 11 m below the land surface.

In the younger or newer alluvium ground water occurs both under water table and confined conditions. Depths to water levels vary from ground level to 10m below land surface. Depth to the tube wells ranges from 30 to 200 m and depth to water ranges from 6m below land surface to 2 m above land surface. The free flows of the following wells vary from trickles to 32 liters/minute in Sibsagar and Mikir Hills district. The yield of the deep tubewells ranges from 2 to 4 kl/minute for a draw down of 3m to 6m. The transmissibility of the aquifers varies from 69 to 1,600 l.p.s/m. and the co-efficient of storage co-efficient is of the order of 3.52 to $\times 10^{-2}$ (Raju, 1981).

7.6.1 MINERAL WATER

1. The occurrences of mineral water have been reported from three localities:

i) On the right bank of Kopili River, three sulphurous springs discharge hot water ranging in temperature from 53 °C to 58 °C. They occur on the faulted zone between the Therria Sandstone and Nummulitic (Sylhet) Limestone of the Jaintia Group.

ii) A sulphurous hotspring is located about 18 km south of Golaghat town with temperature about 35 °C and discharge about 35 liters/m.

iii) A warm spring with salt and sulphur has been reported by Evans within the Kopili shales and sandstones along the Samkhujan Stream.

## 7.7 Concluding Remarks

The Brahmaputra valley has a physical structure comparable to a winnowing fan. The geological history of the river shows that its is a very young river and its present configuration took shape only during the Pleistocene and recent times. The subsurface lithological data indicate the existence of both shallow water table and confined aquifers separated by aquicludes. There is therefore plenty of groundwater available in the basin. At several locations there is mineral water.

## Bibliography and References

Bristow, C.S. (1987). Brahmaputra River: Channel migration and deposition. In: Recent Developments in Fluvial Sedimentology, edited by F.G. Ethridge, R.M. Flores and D. Harvey, SEPM Spec. Publ. No.39, pp. 63--74.

Coleman,J.M. (1969). Brahmaputra River: Channel processes and sedimentation. Sedimentary Geology, Vol. 3, pp.129-239.

Gansser,A. (1964). Geology of the Himalayas. John Wiley, New York, pp.222--231.

Goswami,D.C. (1998). Fluvial regime and flood hydrology of the Brahmaputra River. Assam. Mem. Geol. Soc. Ind .No 41, p. 53--75.

Mathur,L.P. and Evans, P. (1964). Oil in India. I. G. C. 22nd Session, India.

Niyogi, B.N. and Narasimhan,Y.S.R.L. (1969). The problems of exploration and utilization of ground water in the Brahmaputra valley in lower Assam. G. S. I. Misc. Publ. No.14, pp.159--169.

Proceedings of the seminar on fluvial process and geomorphology of the Brahmaputra Basin.(1981). G. S. I. Misc. Publ. No. 46

Raju, K.C.B. (1981). Status of Ground water in Assam. In: G.S.I. Misc. Publ. No. 46. pp.133-140.

Sengupta, S. (1966). Geological and geophysical studies in western part of Bengal Basin, India. Bull. Am. Ass. Petrol. Geologists, Vol.50., No.5, pp.1001--1017.

# CHAPTER 8

# HYDROLOGY

B. DATTA & V.P. SINGH

Hydrology provides information fundamental to the design of hydraulic works: water resources planning, development and management; irrigation water management; environmental conservation; and so on. Despite the enormous importance of the Brahmaputra River basin to India, its hydrology has not received as much attention in the country as it deserves. The result is that the hydrological information about the basin is crude and primarily empirical. This chapter presents a review of the basin geomorphology, climate and weather, precipitation, discharges and floods, sediment loads, and erosion.

## 8.1 The Brahmaputra River Basin

### 8.1.1 THE BRAHMAPUTRA RIVER

The Brahmaputra River is one of the biggest rivers of the world. Originating at an elevation of 5,300 m from a glacier near the Kailash range of Himalayas located in the southwestern part of the Tibetan plateau of China, it traverses a distance of 2,880 km through three countries, namely, China, India, and Bangladesh, before joining the Bay of Bengal. It has a catchment area of 580,000 sq. km, average annual discharge of 19,820 cubic meters per second (cumecs), average annual sediment load of 735 million metric tonnes, and specific flood discharge of 0.149 cumecs/sq. km. The Brahmaputra River is highly braided in its course through the plains in India from Pasighat to the Indo-Bangladesh border for a length of 692 km and also in the reach within Bangladesh. The Brahmaputra River basin with the river system from its source to outfall is shown in the Figure 8.1.

Among the top ranking large rivers of the world, the ranking of the Brahmaputra River is fifth with respect to discharge, fourteenth with respect to the drainage area, twenty fifth with respect to length, second with respect to sediment load, and first with respect to specific discharge. The ranking of the Brahmaputra with respect to discharge, drainage area and length among the large rivers of the world is shown in the Table 8.1. In the table the top fifteen rivers are arranged in descending order of the average annual discharge.



*V.P. Singh et al. (eds.), The Brahmaputra Basin Water Resources*, 139-195.

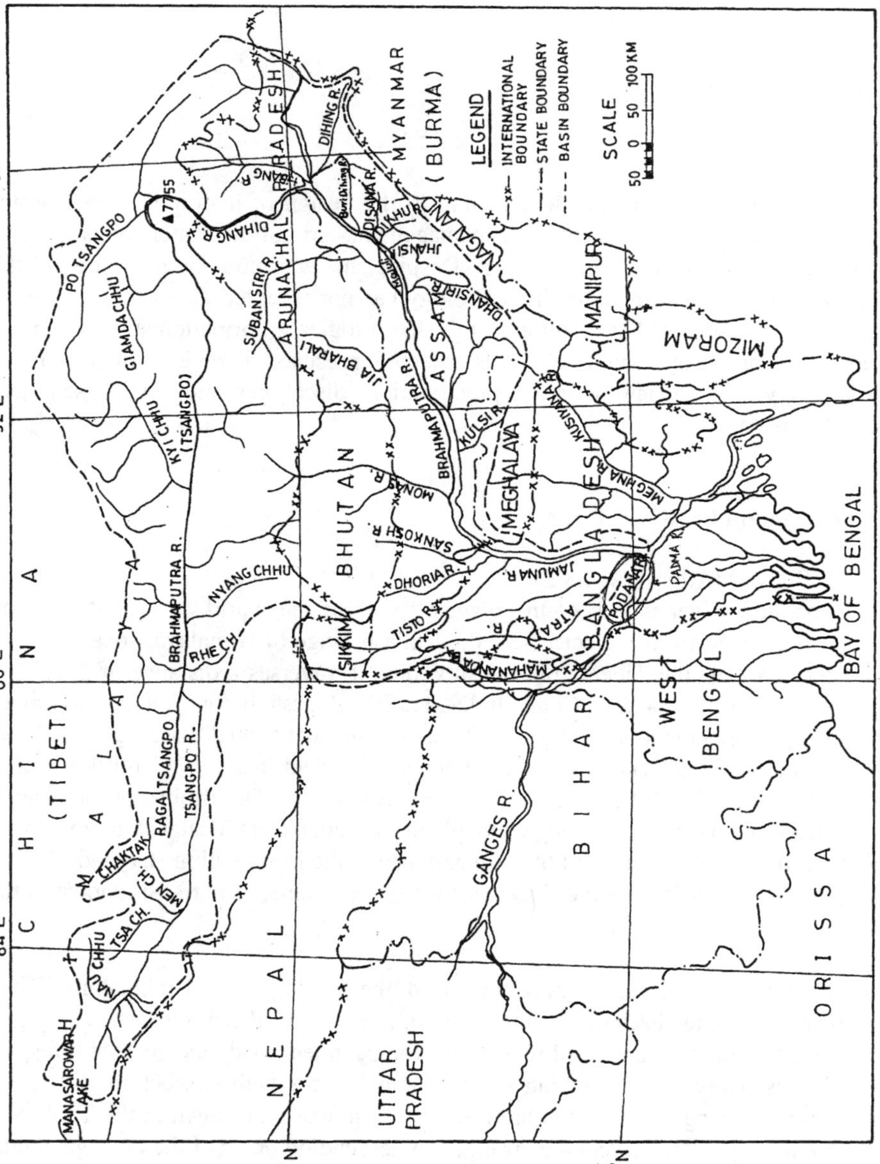


Fig. 8.1 The Brahamaputra River basin.

The total length of 2,880 km of the Brahmaputra River is divided into three reaches: upper, middle, and lower. In its upper reach the Brahmaputra River flows 1,625 km from the source to the Indo-China border through high Tibetan plateau with an elevation from 3,000 m to 5,000 m, mainly in the east direction almost parallel to the great Himalayan mountain ranges and north thereof. The river in the upper reach is known as Tsangpo, which means 'the purifier'. In the eastern part of the plateau the river takes a hairpin bend around the Namcha Barwa (7,755 m) mountain ranges and enters India flowing south as the Siang River. The river in this reach traverses through the southern part of the high Tibetan plateau of China.

In the middle reach, from the Indo-China border to the Indo-Bangladesh border, it flows 918 km through India, of which 278 km is through the mountainous state of Arunachal Pradesh where it crosses the great Himalayan mountain ranges through narrow valleys and deep gorges flowing mainly in the southern direction. The next 640 km length is through the valley in the state of Assam known as the Brahmaputra valley, where the river flows mainly in the west direction with some southern trend. In Arunachal Pradesh the river is known as the Siang in the upper reach and Dihang in the lower reach and the Siang and Dihang rivers cross the great Himalayan mountain ranges through narrow valleys and deep gorges traversing 226 km from the Indo-China border up to the Pasighat town flowing mainly in the southern direction till it enters into the plains of Assam at Kobo located in the northeast corner of the Brahmaputra valley. At Kobo the river meets two major trans-Himalayan tributaries, Dibong and Lohit, coming from northeast and east, respectively and the combined river, known as Brahmaputra, meaning 'son of the creator, Lord Brahma' as per Hindu belief, flows initially in the southwest direction and then in the west direction with a little southern trend through the valley in the state of Assam. The river in this reach is highly braided.

In the lower reach the river traverses 337 km through the plains of Bangladesh from the Indo-Bangladesh border to its outfall flowing in the southern direction. The river flows 225 km from the Indo-Bangladesh border up to Goalundo, initially as Brahmaputra and then as Jamuna. At Goalundo the river joins another major river, Ganges, coming from west and the combined river flows as the Ganges-Padma River mainly in the south direction for about 80 km. The river here meets with another major tributary, Meghna, coming from the northeast and the combined river flows as the Meghna River for another 32 km mainly in the southern direction and outfalls into the Bay of Bangal forming broad estuaries.

Although the river as a whole is known as the Brahmaputra, it has local names in some of its reaches. The river is known as Tsangpo in the Tibetan plateau. In India within the state of Arunachal Pradesh the river is known as Siang in the upper reach and Dihang in the lower reach. The river is known as Brahmaputra within the state of Assam (India) and in the upper part of the lower reach in Bangladesh. The lower part of the 225 km reach of the river in Bangladesh from the Indo-Bangladesh border to Goalundo is known as Jamuna.

A large number of tributaries join the river during its course through Tibetan plateau and India whereas only a few tributaries join the river in Bangladesh. Some of the important tributaries that join the river within the Tibetan plateau reach are Nam Chhu, Tsa Chhu, Man-Chhu, Rage-Tsangpo, Kyichhu, Po-Tsangpo, etc.. Lhasa, the capital city of Tibet, is located within the basin area and situated by the bank of a major tributary, called Kyichhu. Some of the important tributaries that join the Siang River and Dihang in Arunachal Pradesh are Yangsangchu, Yamme, Siku, Sibis, Ringong, Singong, Siyon, Shimong, etc. During its course through the valley in the state of Assam, 19 major tributaries originaing from the Himalayan mountain ranges join the river from the northern (right bank) and 13 major tributaries originating from southern (Assam) mountain ranges join the river from the south. Some of the major north bank tributaries are the Subansiri, Ranganadi, Jia-Bharali, Manas, Sankosh, etc., and some of the major south bank tributaries are Buri-dihing, Desang, Dikhow, Kopili, Krishnoi, Jinjiram, etc. Major tributaries in the lower reach within Bangladesh are Raidak, Dharla, Tista, Karatorya and Atrai; all these are right bank tributaries whereas Meghna is the only major left bank tributary.

The gradient of the Brahmaputra River varies from very steep near the source at the Tibetan plateau to very flat in the lower part of Bangladesh. The average river gradient in the Tibetan plateau is 1:385 whereas in Bangladesh it varies from 1:11,340 to 1:37,700 (Figure 8.2). The Brahmaputra River in the valley reach in Assam and in Bangladesh is highly braided and unstable in nature. The river flows into a number of interlacing channels separated by sandy islands. During high stages the braided channels get submerged to form a vast sheet of moving water. The average river width in the braided portion within valley reach is 8.08 km and varies from 6 km to a maximum of 18 km except in 9 nodal points where the width is restricted within stable banks and flows through deep and narrow channels followed by broad aggrading ones. The instability of the river is owing to high sediment load, steep slopes, transverse gradient and the occurrence of a number of major earthquakes whose epicentres were inside or very close to the basin area and which disturbed the topography and the river system to a great extent; the severest of earthquakes occurred in 1762, 1822, 1865, 1869, 1897, 1908, 1937, and 1950 as per the available record. The river is still under the process of attaining stability which was disturbed last by the major earthquake of 1950.

### 8.1.2 RIVER BASIN

The part of the Tibetan plateau bounded by high Himalayan mountain ranges in the south and Kailash (Himalayan) and Nyenchentanglha mountain ranges in the north falls under the basin area, the elevation of which varies from 3,000 to 5,000 m. This plateau area is dotted with numerous glaciers. The great Himalayan mountain ranges, with a width varying from 200 km to 300 km, lie in the eastwest direction. The Himalayan mountain region, falling within the basin area, covers 96% of the Himalayan Kingdom of Bhutan and parts of three states of India, namely, Arunachal Pradesh (97.6% of the state area), Sikkim (100%) and West Bengal (3.6%). The Brahmaputra basin is of irregular shape: the maximum eastwest length is 1,540 km and the maximum north-south width is 682 km. The basin lies between 23°N to 32°N latitude and 82°E longitude.

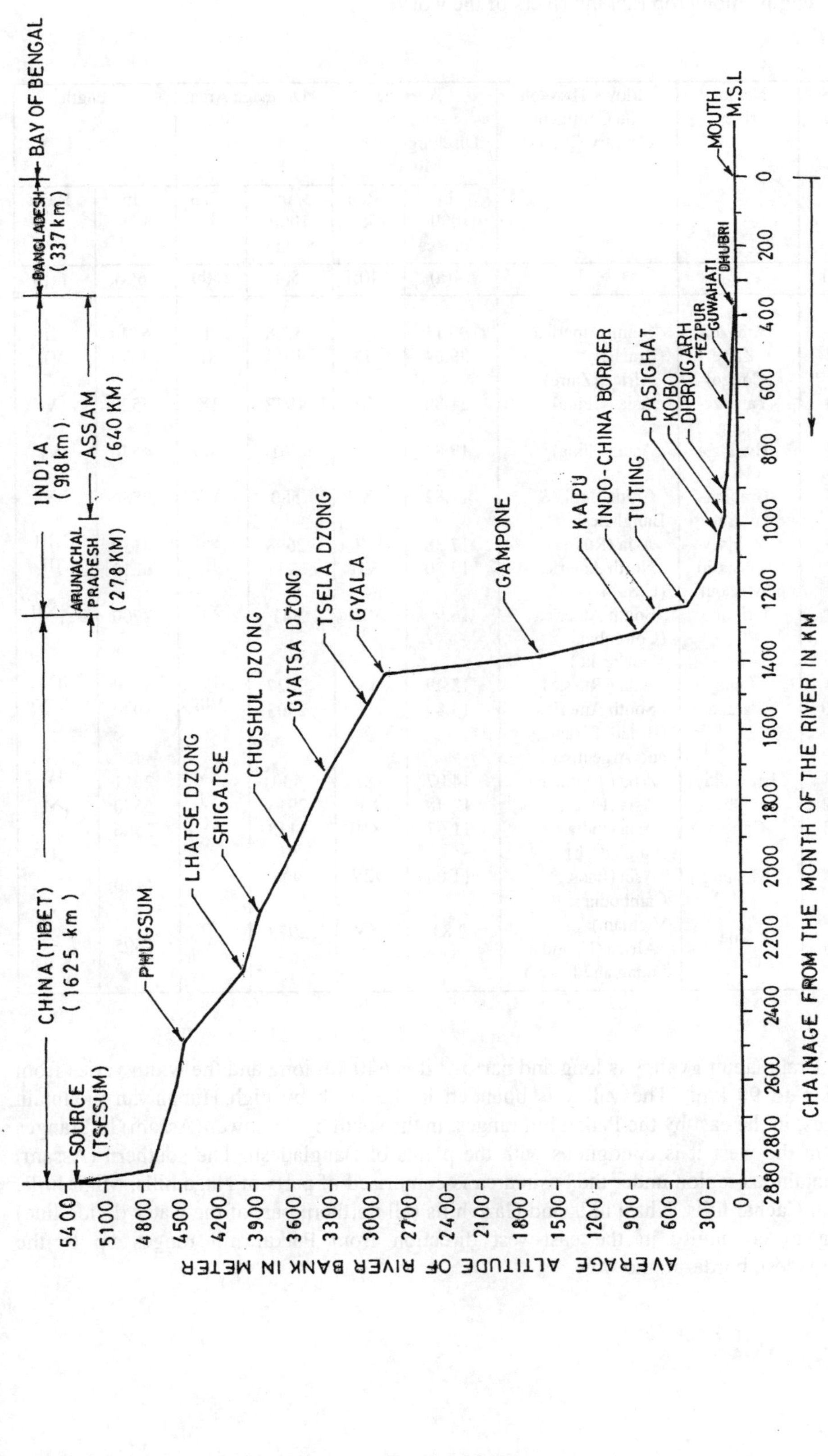


Fig. 8.2 The average bank profile of the Brahamaputra River

Table 8.1 Ranking of the Brahmaputra River with respect to discharge, drainage area and length among top ranking rivers of the world.

| S. N. | Name of river | Flows Through the Continent (Country/Countries) | Average Annual Discharge Near Mouth | | Drainage Area | | Length | |
|---|---|---|---|---|---|---|---|---|
| | | | In 1000 Cumecs | Rank | In 1000 sq. km. | Rank | In km | Rank |
| 1 | 2 | 3 | 4(a) | 4(b) | 5(a) | 5(b) | 6(a) | 6(b) |
| 1 | Amazon | South America (Brazil) | 99.11 | I | 5778 | I | 6440 | II |
| 2 | Zaire (Congo) | Africa (Zaire) | 39.64 | II | 4014 | II | 4670 | VII |
| 3 | Yangtze-kiang | Asia (China) | 21.80 | III | 1942 | IX | 5520 | V |
| 4 | Hwang - Ho | Asia (China) | 19.83 | IV | 761 | XIII | 4670 | VI |
| 5 | Brahma-putra | China, India & Bangladesh | 19.82 | V | 580 | XIV | 2880 | XXV |
| 6 | Yenisey | Asia (Russia) | 17.38 | VI | 2698 | VII | 4130 | XIII |
| 7 | Mississippi Missouri | North America (U.S.A) | 17.30 | VII | 3211 | III | 6210 | III |
| 8 | Orinoco | South America (Colombia, Venezeula ) | 16.99 | VIII | 841 | XII | 2900 | XXII |
| 9 | Lena | Asia ( Russia) | 15.49 | IX | 3027 | IV | 4270 | IX |
| 10 | Parana | South America (Brazil, Paraguay and Argentina) | 14.89 | X | 2305 | VIII | 3940 | XIV |
| 11 | Irrawaddy | Asia ( Mynmar) | 14.07 | XI | 430 | XV | 2011 | IV |
| 12 | Ob | Asia (Russia) | 12.48 | XII | 2913 | VI | 5570 | XI |
| 13 | Ganges | Asia (India, Bangladesh) | 11.67 | XIII | 1113 | X | 2506 | I |
| 14 | Mekong | Asia (Laos, Cambodia and Vietnam) | 11.04 | XIV | 906 | XI | 4180 | |
| 15 | Nile | Africa (Uganda, Sudan and Egypt) | 2.83 | XV | 2978 | V | 6695 | |

The Brahmaputra valley is long and narrow; it is 640 km long and the width varies from 64 km to 90 km. The valley is bounded in the north by high Himalayan mountain ranges, in the east by the Patkai hill ranges, in the south by the lower (Assam) hill ranges and in the west it is contiguous with the plains of Bangladesh. The southern (Assam) mountainous region under the basin area is comprised of parts of Naga hills, Mikir hills, North Cachar hills, Khasi hills and Garo hills (all northern part of the water divide line) lying in continuity in the east-west direction from Patkai hill ranges up to the Bangladesh border in the west.

The catchment area of the river falls in four countries; the portions of the catchment area within Tibet, India, Bhutan and Bangladesh are 293,000; 195,000 45,000 and 47,000 sq. km., respectively. Although the main river does not flow through the Himalayan kingdom of Bhutan, but 96% of the total area of the country falls under this basin.

The Brahmaputra River characteristics, including the local name, length, catchment area and gradient of the river from source to its outfall and nature of topography at different parts through which the river flows, are furnished in Table 8.2. In India the Brahmaputra basin covers part of the north eastern states, such as Arunachal Pradesh, Assam, Nagaland, Meghalaya and West Bengal and the whole of Sikkim. The statewise distribution of the Indian part of the catchment area is furnished in Table 8.3.

The Brahmaputra basin covers six different topographic regions extending over to three countries. These regions come from different parts of the basin and they are:

(i) high Tibetan Plateau (which covers 50.5% of total basin area),
(ii) high Himalayan mountainous region (which covers 23.6%),
(iii) Brahmaputra valley in Assam (which covers 9.7%),
(iv) southern (Assam) mountainous region of the total basin area (which covers 6.4%),
(v) plains covering part of West Bengal (India) and part of Bangladesh, and
(vi) coastal region of the Bay of Bengal that falls within the basin.

The plains and coastal region together cover the balance of 9.8% of the basin area. The topographical breakdown of the catchment area is furnished in Table 8.4.

### 8.1.3 BASIN CLIMATE

The great Himalayan mountain ranges with a width varying from 200 km to 300 km lie in the east--west direction just south of the Tibetan plateau and divides the basin into two distinct climatic zones (Figure 8.3). The northern part of the basin under the high Tibetan plateau, which covers 50.5% of the basin area, falls under climatic Zone III and is classified as 'Mountain Climate', which is cold, dry and arid. The southern part of the basin falls under climatic Zone I and is classified as 'Tropical Monsoon Climate'; the climate of this zone is relatively warm and humid, and experiences high amounts of widespread rainfall, especially during four months, June to September, under the influence of the southwest monsoon. These four monsoon months experience 65% of the average annual rainfall depending on the location and topography of the place. The northern part of the basin is cold, dry, and arid whereas the southern part is tropical, humid and experiences a high amount of rainfall mainly under the influence of the southwest monsoon. The general weather of the basin can distinctly be divided into four seasons: winter from December to February, Summer (or Pre-monsoon) from March to May, rainy season (or monsoon) from June to September and autumn in October and November.

The basin experiences four distinct seasons in a year: winter (December to February), summer or pre-monsoon (March to May), rainy season or monsoon (June to September),

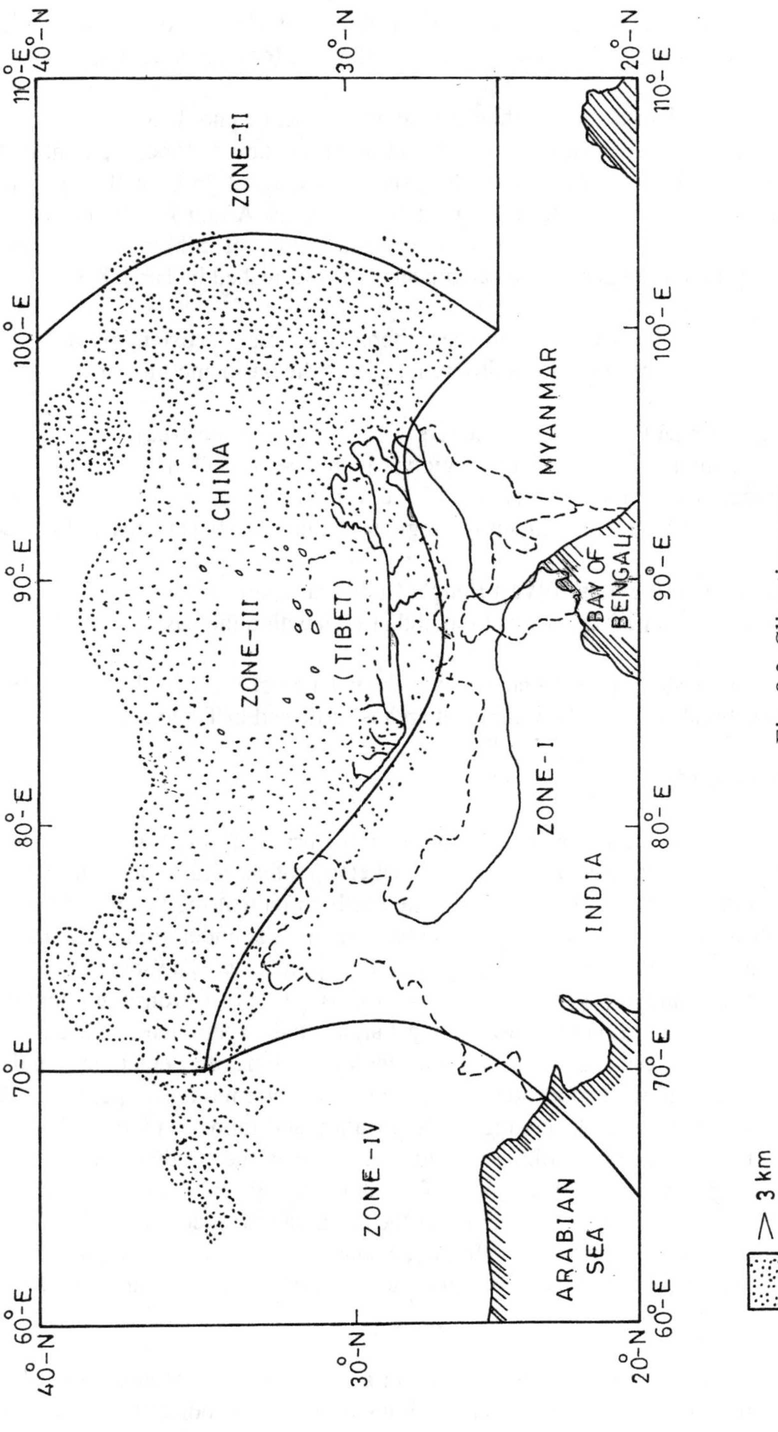


Fig. 8.3 Climatic zones.

Table 8.2 Reachwise distribution of length, catchment area, gradient of the Brahmaputra River and nature of topography through which the river flows.

| S. No. | Location (the Country the River Traverses through ) | Name and Length (in km) of the River in that Reach | Catchment Area in that part (sq. km) and % of total area. | Gradient | | Topography of the Area through which the River flows |
|---|---|---|---|---|---|---|
| | | | | Reach with Elevation (M) | Average Gradient | |
| 1 | 2 | 3 | 4 | 5(a) | 5(b) | 6 |
| 1. | Tibet (China)-(Upper Reach) | Tsangpo (1,25) | 293,000 (50.52%) | Source EL= 5300 m) to Indo-China Border (EL 660 m) | 1:385 | High Tibetan Plateau |
| 2. | India (Middle reach) | Siang in Upper reach & Dihang in Lower reach (278) | 195,000 (33.62%) | i) Indo-China Border to Kobo (EL 120 m) | 1:515 | Himalayan Mountain nous region |
| | i. Arunachal Pradesh<br>ii. Assam | Brahmaputra (640) | 45,000 (7.76%) | Kobo to Indo-Bangla-desh border (EL 28m) | 1:6,90 | Plains (Brahmaputr a Valley) |
| 3. | Bhutan | Main river does not flow through but 96% of the Country's area falls under the basin. | 47,000 (upto confluence with the river Ganges (8.10%) | - | - | Himalayan Mountainou s region |
| 4. | Bangladesh | Brahmaputra, Jamuna, Padma and then, Meghna (337km. Upto mouth) | | i) First 60 km. from India Border<br>ii)Next 100 km<br>iii) Next 90 km<br>iv) Rest portion upto outfall. | 1:11,40<br>1:12,60<br>1:27,00<br>1:37,00 | Plains including Coastal belt |

Table 8.3 Distribution of Indian part of the catchment area statewise.

| Sl. No. | Name of the State | Total Area of the State (sq. km) | Catch-ment Area under the state (sq. km) | Percentage of state Area under the basin | Percentage W.P. to total catchment area in India | Percentage W.R. to total catchment Area of the river |
|---|---|---|---|---|---|---|
| 1 | 2 | 3 | 4 | 5 | 6 | 7 |
| 1. | Arunachal Pradesh. | 83,578 | 81,600 | 97.6 | 41.85 | 14.07 |
| 2. | Assam | 78,523 | 70,700 | 90.0 | 36.26 | 12.19 |
| 3. | Meghalaya | 22,489 | 11,800 | 52.5 | 6.05 | 2.03 |
| 4. | Nagaland | 16,523 | 10,900 | 65.0 | 5.59 | 1.88 |
| 5. | Sikkim | 7,300 | 7,300 | 100 | 3.74 | 1.26 |
| 6. | West Bengal | 87,853 | 12,700 | 14.5 | 6.51 | 2.19 |
| | Total | | 195,000 | | 100% | 33.62% |

and autumn or post-monsoon (October and November). The basin experiences a wide seasonal variation of temperature under the influence of different prevailing winds. The temperature varies from place to place, depending on its latitude, altitude and distance from the sea (Bay of Bengal). The Tibetan plateau and higher reaches of Himalayan mountainous region are much colder owing to high altitude and in winter months Tibetan plateau experiences severe cold owing to the movement of very cold continental air moving outward from the Siberian anticyclones. These areas also experience light drizzle accompanied by snowfall during winter months. About 53,000 sq. km. of the basin area remains under permanent snow cover. The great Himalayan ranges shield the southern part of the basin from the Siberian cold air. The month of January is the peak winter month and minimum recorded temperature at Lhasa in -16°C. The minimum temperature in the plains and valley areas of the basin varies from 9°C in the western part to 4° C in the north eastern part.

The western part of the Brahmaputra valley and the northern part of the plains of Bangladesh experience hot summer in April and May under the influence of hot wind that originates from west Asia and blows over the states of north India from west to east. During summer the average maximum temperature in this part of the basin remains above 35 °C where as the recorded maximum temperature is 40 °C.

The basin experiences a wide variation of the average annual rainfall from less than 400 mm to than 6,000 mm. Orography plays an important role in the areal distribution of rainfall. A place, called Cherrapunji located in the Meghalaya state of India on the windward side of the southwest monsoon, just south of the basin boundary, experiences an average annual rainfall of 10,800 mm and is known to be the wettest place in the world.

Table 8.4 Topographic distribution of Brahmaputra basin area.

| Sl. No. | Nature of Topography | Basin Area under the Topography (sq. km) | Percentage of total Basin Area | Geographical Location. |
|---|---|---|---|---|
| 1 | 2 | 3 | 4 | 5 |
| 1. | High Tibetan Plateau | 293,000 | 50.5 | Southern Part of the Tibet province of China. |
| 2. | Himalayan | 137,050 | 23.6 | Part of Himalayan kingdom of Bhutan and part of three states of India namely Arunachal Pradesh, West Bengal and Sikkim (whole) |
| 3. | Brahmaputra Valley | 56,200 | 9.7 | Part of Assam State of India. |
| 4. | Lower (Assam) Mountainous Region. | 37,200 | 6.4 | Part of the three states of India namely Nagaland, Assam and Meghalaya. |
| 5. | Plains | 56,550 | 9.8 | Part of the two plains districts of west Bengal state (India) and part of Bangladesh. |
| 6. | Coastal Region | | | Coastal region of Bangladesh. |

The basin area under Zone I experiences 20% to 35% of the average annual rainfall during this summer period owing to north westerly storms that generally occur in a few days mostly in the afternoons. Sometimes this part of the basin also experiences widespread heavy rainfall for two to three days resulting from deep depressions that develop over the Bay of Bengal and pass over the basin. In general, the humidity and rainfall in the Brahmaputra basin area under Zone I are relatively high.

The four months from June to September constitute the rainy season. About 65% to 80% of the average annual rainfall takes place during this period under the influence of south-west monsoon. During the four monsoon months the Zone I area of the basin receives widespread heavy rainfall. The southeast part of the Tibetan plateau also receives some amount of monsoon rainfall during July and August.

The months of October and November, after the retreat of the southwest monsoon, constitute the autumn. This season is also called post-monsoon period. The weather during autumn remains very comfortable with relatively clear sky and moderate temperatures. On average these two months experience 3% to 6% of the average annual rainfall.

The humidity in the southern part of the basin under climatic Zone I is relatively high; the mean relative humidity at 8 hours in the forenoon is 80% and at 5 hours in the afternoon is 65%. The Tibetan part of the basin is relatively dry and arid.

The number of cloudy days annually in the Tibetan part of the basin is comparatively less; it is 98 days in Lhasa. The northeastern part of the Brahmaputra valley and lower mountainous regions of the eastern part of Arunachal Pradesh and Assam remain under cloud cover for more than 60% of the days in a year. The average number of cloudy days in the eastern part of the valley at Dibrugarh is 241 days, whereas it is 191 days at Guwahati located at the west central part of the valley.

The number of rainy days in the Tibetan plateau is less, although it experiences light drizzle accompanied by snowfall in some of the days during winter months. The number of rainy days at Lhasa is 48 annually. In the Brahmaputra valley the number of rainy days increases from west to east. The number of rainy days at Guwahati (chainage 542 km from the mouth of the river), the capital city of the Assam state is 115 days whereas it is 172 days at the Dibrugarh city (chainage 917 km) located on the eastern side of the valley.

The low altitude rain bearing clouds brought in by the southwest monsoon from the Indian Ocean and the Bay of Bengal during late May move in the northeast direction and get intercepted by the southern (Assam) hill ranges and causes extremely heavy rainfall along the Cherrapunji and Mawphalang Pynurala belt of the Meghalaya state. The wettest town of Cherrapunji is located in the Khasi hills, just south of the basin boundary and located towards the windward side of the southwest monsoon winds. The clouds that pass over the 1,800 m high mountain ridges of Garo hills and Khasi hills of Meghalaya, which means the abode of clouds, enter into the Brahmaputra basin and widespread rainfall takes place in the valley and the lower Himalayan mountain reaches of Bhutan and Arunachal Pradesh. The intensity and duration of rainfall increases towards the foothills region of the Himalayas and more towards northeast. The southwest monsoon wind crosses the great Himalayan mountain ranges only through the deep trenched valleys of the upper courses of some of the rivers like Brahmaputra, Salween, Mekong, etc. into the southeast part of the Tibetan plateau and rainfall occurs there in the months of July and August; rainfall in the plateau decreases as the monsoon winds advance towards the northwest direction. The central and western parts of the plateau are cold, dry and arid.

The average annual rainfall in the Brahmaputra valley in Assam varies from 1,500 mm in the western side to more than 4,000 mm in the east. There exist closed cells of maxima and rainshadow areas in the basin. There are three closed cells of maxima of more than 4,000 mm of the average annual rainfall; one is around Pasighat between latitude 27° N to 28°-45' N and longitude 93° E to 95° E; the second on the western side of the valley including the foothills region of the Bhutan between latitude 26° 15' N to 27°15' N and longitude 88° 30' E to 90° 0' E; and the third around Cherrapunji in the Meghalaya state. The northern half of the Cherrapunji cell falls under the basin area. In between the first and third cells of maxima, there lies a rainshadow area in the Mikir Hills region of the southern (Assam) hill ranges where the average annual rainfall is less than 2,000 mm. There is another rainshadow area in the northeastern part of Arunchal Pradesh (Lohit District) adjoining China where the average annual rainfall is less than 1,000 mm. Rainfall in general is less than 1,500 mm in the major part of the high altitude regions of Himalayas within Bhutan and Arunachal Pradesh. The variability of annual rainfall over the whole basin is 15 to 20%.

### 8.1.4. FLOODING

It is observed that some portions of the Brahmaputra valley in Assam and the adjoining mountainous regions or the whole of the valley and surrounding mountain regions receive heavy to very heavy rainfall in a spell of 2 to 3 days or sometimes even rainfall continues for 7 to 8 days during monsoon months under synoptic meteorological conditions. Such a spell of rainfall generally takes place under four meteorological situations, namely: (i) break monsoon situation; (ii) northwesterly to northern movement of monsoon depressions or storms from the Bay of Bengal; (iii) the formation and the movement of land depressions over northeast; and (iv) the upper air cyclonic circulation over the basin. Under the influence of such situations, one hour rainfall of 70 mm is common in areas favourable for heavy rainfall. The highest recorded 1-hour, 3 hour and 24 hour rainfalls within the basin area are 97.5 mm, 200 mm, 300 mm and 805 mm, respectively. Such a spell of heavy rainfall gives rise to flooding in the concerned tributaries or the main river along with tributaries and the Brahmaputra Valley in Assam and the Bangladesh part experiences severe flooding. So far the recorded maximum discharge at Guwahati, the capital city of Assam, a state of India (chainage 542 km from the month of the river located at latitude 26°11'N and longitude 91° 45'E ) is 72,794 cumecs where the catchment area is 424,100 sq. km (i.e., 73.12% of the whole). The specific flood discharge of the river at Bahadurabad in Bangladesh is 0.149 cumecs/sq. km and is the highest among large rivers of the world.

The Brahmaputra River in the plains generally experiences, on the average, 4 to 5 major flood waves annually during monsoon months from June to September; the probability of occurrence of high flood in July and August is high. The specific flood discharge of the river at Bahadurabad in Bangladesh is 0.149 cumecs/sq. km. and is the highest amongst large rivers of the world. In some years the flood wave remains above the danger level for a long period and thereby creates acute flood congestion in the valley reach in Assam and also in Bangladesh. On average the flood level remains above the danger level for 39 days in a year at Guwahati whereas the maximum recorded period is 70 days. In some years severe flooding takes place and results in huge losses of property, homesteads, crops, livestocks, loss of agricultural land, dislocation of communication networks, etc.

The extent of flood lift varies from reach to reach, depending on the width and depth of flow. The average lift at a braided reach near Dibrugarh city is 3.65 m where the river width is 8.03 km and the average flood lift is 9.91 m at a nodal point near Guwahati where the river width is 1.20 km which is minimum within the valley in Assam from Kobo to the Indo-Bangladesh border. On average a flood wave takes 5 days and 3 hours to pass through the valley from Pasighat to Dhubri, a distance of 672 km.

After cessation of monsoon by mid October, the water level and discharge of the river starts receding and attains the lowest annual water level (LWL) sometime in the late February or early March. The lean period discharge takes place generally from December to March and the discharge during this period is mainly contributed by snow melt and base flow. The recorded minimum discharge of the river at Guwahati during this period is 1,757 cumecs.

### 8.1.5 SEDIMENT LOAD

The sediment load of the Brahmaputra River is very high. The average annual sediment load of the river at Bahadurabad in Bangladesh is 735 million metric tonnes. The percentage breakdown of the surface sediment load is 12.5% coarse, 14.2% medium, and 73.3% fine. The tributaries that originate from the high Himalayan mountain ranges carry high sediment loads as these rivers flow through geologically younger Himalayan mountain ranges of unconsolidated sedimentary rocks with steep slopes and the area falls in one of the most earthquake prone belts of the world. In comparison, the tributaries originating from the southern (Assam) hill ranges carry less sediment load as these hill ranges are geologically much older and stable and of lesser height and have flatter bed slopes. Besides the natural causes, some human activities also contribute to the high rate of sediment load; most prominent among them is 'Jhuming', the shifting cultivation, which is done on the land obtained by burning down parts of the forest. This is extensively practised by the tribal communities living in the mountainous regions of the basin area. About 1.35 million hectares of the forest land remains under Jhum cultivation at any time. Heavy bank erosion by the Brahmaputra River takes place at different reaches owing to excessive sediment load, the erodible nature of the bank material, formation of char island, and consequent development of side channels.

### 8.1.6 THE BRAHMAPUTRA RIVER SYSTEM

The Brahmaputra River originates at an altitude of 5,300 m from Kanglung Kang glacier near Kailash range of Himalayas little southeast of Manasarowar Lake located in the south-west part of the Tibet province of China at $30^{\circ}$ $30'$ N latitude and $82^{\circ}$ $10'$ E longitude. The Tamchuk Khambab, a stream which emerges from the glacier, is considered as the head stream of the Brahmaputra River. A large number of tributaries, big and small, join the river during its course through the three countries. The basin falls in one of the highly seismic zones of the world and the river carries high amounts of sediment load.

*8.1.6.1 Upper Reach*

The Brahmaputra River originates from the Kanglung Kang glacier which is located very close to the Manasarowar Lake district of Tibet. From this district two other great rivers, Indus and Sutlej, also originate. The Tsangpo River traverses in the east direction whereas Indus and Sutlej rivers traverse in the northwest and west directions, respectively. Many tributaries join the infant river, Tsangpo, near its source from the pass of Mayum-La to Marnyak-La. The mountain ranges on which these passes are located separate the Brahmaputra basin from the Manasarowar Lake district. The catchment area of the Brahmaputra River at the Indo-China border is 293,000 sq. km; this part of the catchment falls under high Tibetan plateau and covers 50.5% of the total catchment area of the river. The shape of the basin in Tibetan part is long and narrow; the maximum length in eastwest direction is 1,540 km and the maximum width in northsouth direction is 310 km. The Tibetan part of the basin is bounded in the south by great Himalayan mountain ranges and in the north from west side initially by the Kailash ranges of Himalayas and then by the Nyenchentanglha mountain ranges.

Important glaciers that form the northern boundary of the basin from west to east are Kailash 6,100 m (Longitude 82°E to 84°E), Lunkar 6,000 m (Longitude 84° to 84° 30' E), Kangchung Kangri and Lapchung 6,000 m (Longitude 84°30' E to 86° 25'E) and Nyenchentanglha 7,088 m (Longitude 86° 25'E to 95° E). The width of the drainage area in the northern side of the river is comparatively more than that of the southern side and extends to a maximum of 250 km near Gyatsa Dzong (chainage 1,721 km from the mouth). The drainage width on the southern side varies from 10 km to 60 km as the river flows close to the Himalayan mountain ranges and almost parallel to it. In addition to contribution from rainfall, the river in this reach is also fed by snowmelt from numerous glaciers and snow covered high mountain ranges.

In this reach a number of tributaries join the Tsangpo River both on the north (left) bank and south (right) bank. Some of the important tributaries that join the river from west to east are:

(i) On the left bank are Nau Chhu, Tsa Chhu, Men Chhu, Chaktak Tsangpo, Raga Tsangpo, Tong Chhu, Shaug Chhu, Kyi Chhu, Giamda Chhu, Po-Tsangpo, Chindru Chhu, etc.;

(ii) On the right bank are Kubi, Kyang, Sakya Trom Chhu, Rhe Chhu, Nyang Chhu, Yearlang Chhu, and Truhing Chhu.

Some important places that are situated by the bank of river Tsangpo from the upstream to downstream are Lhatse Dwong, Shigatse (Rikaze), Chushul Dzong (Qushni Gougge), Gyatsa Dzong, Tsela Dzong (Ziang Zhong), Gyala (Linzhi), etc. The location, chainage (from the mouth) and altitude of the places are furnished in Table 8.5. The names of some of the places have been changed and the new names are shown within bracket.

Table 8.5 Location and altitude of some important towns/sites situated by the bank of river Brahmaputra (Tsangpo) within the upper reach (Tibetan plateau).

| SI. No . | Station (Town/Site) | Chainage (From mouth) km. | Location | | Altitude (m) |
|---|---|---|---|---|---|
| | | | Latitude | Longitude | |
| 1 | 2 | 3 | 4(a) | 4(b) | 5 |
| 1. | Source | 2,880 | 30° 30'N | 82°-10'E | 5,300 |
| 2. | Lhatse Dzong | 2,271 | | | |
| 3. | Shigatse(Rikaze) | 2,150 | 29° 14'N | 88° 54'E | 3,816 |
| 4. | Chushul Dz. (Qushui Gongge) | 1,933 | 29° 19'N | 90° 43'E | 3,565 |
| 5. | Gyatsa Dz. | 1,721 | 29° 8'N | 92° 46'E | 3,300 |
| 6. | Tscla Dz.(Ziang Zhong) | 1,515 | 29° 10'N | 93° 29'E | 2,957 |
| 7. | Gyala (Linzhi) | 1,433 | 29° 25'N | 94° 22'E | 2,835 |
| 8. | Indo-China Border. | 1,255 | 29° 07'N | 95° 06'E | 660 |
| N.B.: New name of the towns are shown within bracket. | | | | | |

Lhasa, the capital city of Tibet, lies on the bank of Kyi-chhu River, one of the major tributaries of the Tsangpo River. The town Chushul Dzong is situated at the confluence of Kkyi-chhu with the Tsangpo River. The city Shigatse the great trade centre and second largest city of Tibet with the great Monastery of Tashi Lhumpo, lies at the confluence of the tributary Nyang chhu with the main Tsangpo River.

The 650 km long reach downstream from Lhatse is one of the most remarkable inland navigation systems of the world where boats ply at an average altitude of 3659 m in the high Tibetan plateau. After 92°E longitude, the river enters into a difficult reach; it flows through deep gorges till it comes near 93° E longitude where a major town, Gyatse Dzong, is situated on the left bank of the Tsangpo River. The next important town on its course is Tsela Dzong located on the left bank at the confluence of the tributary river Giamda Chhu with Tsangpo. The river still continues to flow eastwardly up to Gyala (Linzhi), then it takes an abrupt turn towards northeast and flows through stupendous gorges between the huge mountain masses of Gyala Peri (7,150 m) and Namcha Bawrs (7,755 m). Here the river flows through a series of cascades and rapids taking numerous hairpin bends. Another major tributary, Po-Tsangpo, which is swift and torrentential, joins the river from north. From its confluence point the Tsangpo River, takes a southern turn and enters in the Indian territory of Arunachal Pradesh at an elevation of 660 m at a chainage of 1,255 km. (from the mouth of the river) as the Siang River.

A number of tributaries of the Ganges River, such as Arun Kosi, Kali Gandak and tributaries of the Brahmaputra River joining its middle reach like Subansiri, Manas, etc., originate from places very close to the catchment area of the Tsangpo River, i.e., the upper reach of the Brahmaputra River and find their way through the Himalayas (Trans-Himalayan rivers). The average gradient of the river in this upper reach is 1:385.

*8.1.6.2 Middle Reach*

From Pasighat up to the Indo-Bangladesh border for a length of 692 km, the river passes through alluvial plains; all along its course within this region, the river flows in a braided form and the main course oscillates from one bank to the other and the river forms many islands and sand chars. Most of these sand chars are not habitable as they get submerged during high stage. During monsoon months almost all the braided channels get submerged and form one vast sheet of moving water. The average gradient of the river from the Indo-China border to Kobo within the state of Arunachal Pradesh is 1:515.

One of the many islands of the Brahmaputra River, Majuli island (Reference location: Latitude 26° 53' N and Longitude 94°E) is the largest river island of the world. The island is inhabited by 150,000 people. But the island is under constant threat of bank erosion especially during the monsoon season when a number of flood waves flow down the Brahmaputra River. Owing to active bank erosion the land of the island has reduced from 1,300 sq. km as at 1,950 to 900 sq. km at present. Some important towns located by the bank of the main river some important tributary confluencing sites and discharge observation sites located within the middle reach of the river are furnished in Table 8.6.

*8.1.6.3 Tributaries*

A number of tributaries join the main river course. Siang and Dihang while traversing through the mountainous region of Arunachal Pradesh. Some of the important tributaries in this portion are Yangsangchu, Yemme, Sikk, Ringong, Sigong, Siyon, Shimong, etc.; the drainage characteristics of these tributaries are furnished in Table 8.7. Within the valley reach from Kobo to the Indo-Bangladesh border within the state of Assam 19 of the major tributaries originating from Himalayan mountain ranges join the Brahmaputra River on its north bank (right bank) and 13 of the major tributaries originating from the southern (Assam) hill ranges join the Brahmaputra River on its south bank (left bank).

Some of the important north bank tributaries confluencing with the Brahmaputra River within the valley reach from Kobo in the east to the Indo-Bangladesh border in the west are Simen, Jiadhol, Subansiri, Ranganadik, Borgong, Jia-Bharali, Gabharu, Belsiri, Dhansiri (north), Noa-nadi, Nonai, Barnadi, Puthimari, Pagladiya, Manas, Champamati, Gaurang, Tipkari, and Sankosh. All these rivers originate from Himalayan mountain ranges and traverse the upper portion through the narrow valleys and deep gorges of the high Himalayan mountainous terrain in steep gradient taking numerous hairpin bends.

These tributaries are also fed by numerous sub tributaries of different orders and carry a high amount of sediment load, most of which gets deposited at the foothill region where the river gradient changes suddenly to a much flatter one. These tributaries then traverse through the plains of Assam in braided form and outfall into the Brahmaputra River at its north bank (right bank).

Table 8.6 Location and altitude of some important towns/sites within the middle reach (from the Indo-China border to Indo-Bangladesh border) situated by the bank of the Brahmaputra River.

| Sl. No. | Station (Towns/Sites) | Chainage (from the Mouth) in Km. | Location | | | Altitude (m) |
|---|---|---|---|---|---|---|
| | | | State of India | Latitude | Longitude | |
| 1 | 2 | 3 | 4 | 5 | 6 | 7 |
| 1. | Indo-China Border. | 1,255 | International Border | 29° 07'N | 95°06'E | 660 |
| 2. | Tuting | 1,233 | Arunachal Pradesh. | 29° 06'N | 95°03'E | 440 |
| 3. | Pasighat | 1,029 | -do- | 28° 02'N | 95°18'E | 155 |
| 4. | Kobo | 977 | Assam | 27° 48'N | 95° 23'E | 120 |
| 5. | Dibrugarh | 917 | -do- | 27° 28'N | 94° 55'E | 106 |
| 6. | Dihingmukh | 877 | -do- | 27° 4'N | 94° 40'E | - |
| 7. | Bechamara | 832 | -do- | 26° 54'N | 94° 19'E | - |
| 8. | Dhansirimukh | 747 | -do- | 26° 38'N | 93° 40'E | - |
| 9. | Silghat | 689 | -do- | 26° 35'N | 92° 58'E | 67 |
| 10. | Tezpur | 672 | -do- | 26° 37'N | 92° 47'E | 64 |
| 11 | Guwahati | 542 | -do- | 26° 11'N | 91° 45'E | 55 |
| 12. | Jogighopa | 422 | -do- | 26° 14'N | 90° 35'E | - |
| 13 | Goalpara | 432 | -do- | 26° 10'N | 90° 30'E | 38 |
| 14. | Dhubri | 357 | -do- | 26° 02'N | 90° 0'E | 28 |
| 15. | Indo-Bangladesh Border | 337 | International Border. | 25° 50'N | 89° 50'E | 26.3 |

Average gradient of the river from Kobo to Indo-Bangladesh border is 1: 6990.

Table 8.7 Drainage characteristics of some major tributaries of river Siang and Dihang within the state of Arunachal Pradesh, India.

| Sl. No. | Name of Tributary | River Length (km/Basin Area(sq. km) | Drainage Density | Number of Streams in each stream order/Total length in Km. in each stream order. | | | | |
|---|---|---|---|---|---|---|---|---|
| | | | | 1st | 2nd | 3rd | 4th | 5th |
| 1. | Yangsang-chu | 60/1,292 | 0.72 | 286/640 | 50/205 | 14/56 | 2/50 | 1/10 |
| 2. | Yamme | 71/1,257 | 0.83 | 410/689 | 81/174 | 21/84 | 6/39 | 1/54 |
| 3. | Sikku | 28/256 | 0.90 | 100/166 | 19/41 | 3/27 | 1/13 | |
| 4. | Sibis | 32/256 | 0.67 | 24/113 | 7/38 | 3/19 | 1/2 | |
| 5. | Ringong | 50/883 | 0.34 | 150/251 | 30/97 | 4/34 | 1/21 | |
| 6. | Sigong | 61/966 | 0.37 | 66/257 | 12/61 | 4/17 | 1/22 | |
| 7. | Siyon | 160/5,110 | 0.44 | 581/1490 | 116/402 | 23/ | 5/99 | 1/77 |
| 8. | Shimong | 40/570 | 0.66 | 141/256 | 22/59 | 1826/47 | 1/26 | |

The two north bank tributaries Subansiri and Manas are trans-Himalayan rivers; besides these two, the major part of the catchment area of the other north bank tributaries like Ranganadi, Borgong, Jia-Bharali and Sankosh lie deep inside the Himalayan mountainous terrain of Bhutan and Arunachal Pradesh. Details about the river and the basins of the north bank tributaries are furnished in Table 8.8.

Some of the important south bank tributaries that originate from the southern (Assam) hill ranges and confluence with the Brahmaputra River on its left bank within the valley reach of Assam from Kobo to Indo-Bangladesh border are Dibong, Lohit, Dibru, Buri-Dihing, Desang, Dikhow, Jhanji, Dhansiri (South), Kopili, Kulsi, Krishnai, Jinari and Jinjiram. The southern (Assam) hill ranges are geologically much older and stable than the Himalayan mountain ranges, so the southern tributaries carry comparatively less sediment load than do the north bank tributaries. Details about the rivers and the basins of the south bank tributaries are furnished in Table 8.9.

The characteristics of the north bank tributaries are quite different than from those of the south bank tributaries owing to the following reasons:

(i) The north bank tributaries originate from higher Himalayan mountain ranges, and the river gradients are much steeper than those of the south bank tributaries.
(ii) Rainfall in the north bank tributary basins is much higher.
(iii) Catchment areas of north bank tributaries are much larger.
(iv) The north bank tributaries travel through geologically younger Himalayan mountain ranges which also contain unconsolidated sedimentary rocks and thereby carry higher amount of sediment load.
(v) Most of the north bank tributaries are flashy in nature.

Table 8.8 Some parameters of the north bank tributaries of the Brahmaputra River confluencing in the middle reach within the state of Assam (India).

| So No. | Name of Tributary | Chainage of confluence with Brahmaputra River (In km from the mouth) | Location of River Basin | |
|---|---|---|---|---|
| | | | Latitude (North) | Longitude (East) |
| 1 | 2 | 3 | 4(a) | 4(b) |
| 1. | Simen | 917 | | |
| 2. | Jiadhol | 877 | 27° 8' to 27° 45' | 94° 15' to 94° 38' |
| 3. | Subansiri | 820 | 27° 0' to 29° 0' | 91° 45' to 94° 45' |
| 4. | Ranganadi | 811 | 27° 0' to 27° 45' | 93° 15' to 94° 10' |
| 5. | Borgong | 719 | 26° 45' to 27° 0' | 93° 08' to 93° 20' |
| 6. | Jia-Bharali | 675 | 26° 30' to 28° 15' | 92° 0' to 93° 30' |
| 7. | Gabharu | 635 | 26° 55' to 27° 0' | 92° 5' to 93° 40' |
| 8. | Belsiri | 617 | 26° 36' to 27° 38' | 92°18' to 92° 36' |
| 9. | Dhansiri (N) | 607 | 26° 30 to 27° 03' | 92° 0' to 92° 17' |
| 10. | Noa-nadi | 567 | 26° 21' to 26° 58' | 91° 55' to 92° 0' |
| 11. | Nonoi | 552 | 26° 15' to 27° 45' | 9o° 45' to 92° 0' |
| 12. | Barnadi | 542 | 26°12' to 26° 52' | 91° 43' to 91° 54' |
| 13. | Puthimari | 509 | 26°14' to 27° 18' | 90° 31' to 91° 49' |
| 14. | Pagladiya | 507 | --- | --- |
| 15. | Manas | 422 | 26° 13' to 28° 35' | 90° 27' to 92° 27' |
| 16. | Champamati | 400 | 26° 15' to 27° 0' | 90° 17' to 90° 33' |
| 17. | Gaurang | 380 | 26° 12' to 27° 0' | 90° 15' to 90° 22' |
| 18. | Tipkai | 377 | 26° 03' to 26° 52' | 89° 30' to 90°14' |
| 19. | Sankosh | 337 | 26° 43' to 28° 18' | 89° 24' to 90°30' |

### 8.1.7 BRAIDING OF RIVERS

The Brahmaputra River is highly braided in nature in the valley reach from Pasighat to the Indo-Bangladesh border and also within Bangladesh. Width of the Brahmaputra River (at bank full level) from Kobo to the Indo-Bangladesh border within the state of Assam varies from 6 km to 18 km except in nine reaches (nodal points) where it traverses through deep and narrow throats. The width of the river at the nodal points varies from 1.2 km to 5.1 km. The reachwise average width, maximum width and width at nodal points of the Brahmaputra River are furnished in Table 8.10.

The river scour depth does not maintain a constant value along the entire length of the braided reach owing to variations in the river width specially near nodal points. In narrow or constricted reaches, scour occurs with increasing flow and fill with diminishing flow. In wide reaches fill generally occurs during flood and scour may not occur during low flows.

#### *8.1.7.1 Lower Reach*

It covers the lower portion of the river from Indo-Bangladesh border to its outfall into the Bay of Bengal for a length of 337 km. The whole portion of the lower reach of the river falls within Bangladesh. Near the Indo-Bangladesh border the Brahmaputra River takes a turn towards south and flows mostly in the southern direction with a little eastward trend.

Table 8.8 (Contd.)

| So No. | Altitude at source (m) | Length (km) | | | Catchment Area | | |
|---|---|---|---|---|---|---|---|
| | | In Hills | In Plains | Total | Total Area (sq. km) | Percentage In Hills | Percentage In Plains |
| | 5 | 6(a) | 6(b) | 6(c) | 7(a) | 7(b) | 7(c) |
| 1. | | | - | | | | |
| 2. | 1,247 | - | 130 | 99 | 1,346 | 22.7 | 77.3 |
| 3. | 5,389 | 312 | - | 442 | 37,000 | 95.7 | 4.3 |
| 4. | 3,440 | - | 21 | - | 2,940 | 76.2 | 23.8 |
| 5. | 700 | 21 | 66 | 42 | 550 | 63.6 | 36.4 |
| 6. | - | 198 | - | 264 | 11,843 | 71.9 | 28.1 |
| 7. | - | - | - | - | 295 | 19.3 | 80.7 |
| 8. | - | - | - | - | 751 | 24.6 | 75.4 |
| 9. | - | - | 75 | - | 956 | 34.8 | 65.2 |
| 10. | - | - | - | - | 366 | 18.6 | 81.4 |
| 11. | - | - | - | 120 | 860 | 23.8 | 76.2 |
| 12. | - | 10 | 102 | 112 | 739 | 17.2 | 82.8 |
| 13. | 3,750 | 74 | 116 | 190 | 1,787 | 33.4 | 66.6 |
| 14. | 1,300 | 19 | 180 | 197 | 1,820 | 24.2 | 75.8 |
| 15. | 4,900 | - | 105 | - | 41,350 | 85.9 | 14.1 |
| 16. | - | - | 60 | - | 1,038 | 13.2 | 86.8 |
| 17. | - | - | 57 | - | 1,023 | 18.5 | 81.5 |
| 18. | - | - | - | - | 1,744 | 9.8 | 90.2 |
| 19. | 7,300 | 214 | 107 | 321 | 10,345 | 92.4 | 7.6 |

The Brahmaputra River is known as Jamuna downstream from the confluence point of the Tista River near Bahadurabad (chainage 260 km) upto Goalundo (chainage 112 km). The river traverses 225 km from Indo-Bangladesh border upto Goalundo where the Brahmaputra River (Jamuna) joins another major river, Ganges-Padma, coming from west and the combined river flows as the Ganges--Padma for 80 km. Here the river joins again another major tributary Meghna coming from north-east near Rajabari, a little south of the tropic of cancer, and the combined river flows 32 km as Meghna River. The Meghna River in the downstream reach trifurcates into three channels forming delta. The three channels are Sandwip---the east channel, Shahbaz---the central, and Tutulia --- the western channel. These three channels ultimately outfall into the Bay of Bengal forming broad estuaries.

Table 8.9 Some parameters of the south bank tributaries of the Brahmaputra River confluencing in the middle reach within the State of Assam (India).

| Sl. No. | Name of Tributary | Chainage of confluence with the Brahmaputra (In Km from the mouth) | Location of River Basin | |
|---|---|---|---|---|
| | | | Latitude (North) | Longitude (East) |
| 1 | 2 | 3 | 4(a) | 4(b) |
| 1. | Dibong | 982 | 27° 0' to 29° 24' | 95° 15' to 96° 38' |
| 2. | Lohit | 977 | 27° 07' to 29° 37' | 95° 41' to 97° 46' |
| 3. | Dibru | 929 | 27° 23' to 29° 45' | 94° 56' to 95° 58' |
| 4. | Buri-Dihing | 877 | 26° 50' to 27° 45' | 94° 36' to 97° 09' |
| 5. | Desang | 852 | 26° 40' to 27° 0' | 94° 33' to 95° 35' |
| 6. | Dikhow | 842 | 26° 9' to 27° 05' | 94° 27' to 95° 16' |
| 7. | Jhanji | 832 | 26° 22' to 27° 00' | 94° 20' to 94° 44' |
| 8. | Dhansiri (S) | 757 | 25°22' to 26° 49' | 93° 10' to 94° 32' |
| 9. | Kopili | 557 | 24° 57' to 26° 38' | 91° 38' to 93° 40' |
| 10. | Kulsi | 477 | 25° 32' to 26° 0' | 90° 50' to 91° 48' |
| 11. | Krishnai | 445 | 25° 30' to 26° 08' | 90° 21' to 90° 52' |
| 12. | Jinari | 437 | 25° 40' to 26°10' | 90° 19' to 90° 44' |
| 13. | Jinjiram | 337 | 25°18' to 26° 09' | 89° 50' to 90° 37' |

Table 8.9 (Contd.).

| Altitude at source (m) | Length (km) | | | Catchment Area | | |
|---|---|---|---|---|---|---|
| | In Hills (km) | In Plains (km) | Total (km) | Total Area (sq. km) | Percentage in Hills | Percentage in Plains |
| 5 | 6(a) | 6(b) | 6(c) | 7(a) | 7(b) | 7(c) |
| 4,267 | 136 | 57 | 193 | 12,270 | 96.5 | 3.5 |
| 4,876 | 310 | 84 | 394 | 23,400 | 79.6 | 20.4 |
| - | 0 | 149 | 149 | 1,852 | 0 | 100.0 |
| 2,300 | - | - | 363 | 8,730 | 56.8 | 43.2 |
| 2,594 | 80 | 150 | 230 | 3,950 | 45.7 | 54.3 |
| 1,823 | 104 | 96 | 200 | 4,370 | 78.4 | 21.6 |
| 1,416 | 53 | 55 | 108 | 1,349 | 64.7 | 35.3 |
| - | 252 | 102 | 354 | 12,580 | 51.3 | 48.7 |
| 1,630 | 127 | 129 | 256 | 20,068 | 79.1 | 20.0 |
| - | 99 | 48 | 147 | 4,005 | 77.0 | 23.0 |
| - | 39 | 26 | 65 | 1,615 | 80.0 | 20.0 |
| - | 32 | 28 | 60 | 594 | 69.0 | 31.0 |
| - | 0 | 160 | 160 | 3,467 | 70.4 | 29.6 |

Table 8.10 Width of the Brahmaputra River at different reaches within the valley reach (from Kobo to Indo-Bangladesh border).

| Sl. No. | Reach (Chainage in Km from the mouth) | Length of The Reach (km) | Width of the River in Km. | | | |
|---|---|---|---|---|---|---|
| | | | Average width (km) | Max.width and (Chainage) km | Width at Nodal points (Chainage in km) | Place |
| 1. | Kobo (977) to Dihingmukh (877) | 100 | 8.19 | 14(967) | 4.8(977) | AT Kobo |
| 2. | Dihingmukh to Dhansirimukh (747) | 130 | 8.18 | 13(870) | 5.1(864)<br>3.7(832) | Near Desangmukh<br>At Bechamara |
| 3. | Dhansirimukh to Pandu (535), | 212 | 8.12 | 13.1(730) | 4.4(727)<br>3.6(674)<br>4.0(607)<br>1.2(535) | At Gamiri<br>Near Tezpur<br>-<br>At Paudu |
| 4. | Pandu (535) to Dhubi (357) | 178 | 7.93 | 18.5(503) | 2.4(517)<br>2.4(422) | -<br>At Jogighopa. |

Practically, there is no notable tributary on its east (left) bank within this reach except Meghna River. Important tributaries on its west (right) bank are Raidak, Dharla, Tista, and Atrai. The river system is shown in Figure 8.1. The river gradient in this reach varies from 1:11340 near the Indo-Bangladesh border to 1:37,700 near to mouth.

*8.1.7.2 Tributaries*

Some important tributaries of the lower reach are discussed below.

**(i) Raidak River.** The Raidak River is one of the main right bank tributaries of the Brahmaputra River in the lower reach. The river is known as Wong in its upper reach in Bhutan. The river traverses through the mountainous terrain in Bhutan and debouches into the plains in India (Jalpaiguri district of the West Bengal state) and after traversing through the Indian portion the river then enters Bangladesh. The Raidak River confluences with the Brahmaputra River (at chainage 327 km) within Bangladesh. The Torso River is an important tributary of Raidak River. The Torso River originates from the Chambi valley of Tibet where it is known as Machu, the river then traverses through Bhutan as Amo and after traversing through India it enters into Bangladesh and confluences with Raidak River.

**(ii) Dharla River.** The Dharla River is another important right bank tributary of the Brahmaputra River in this reach. In the upper reach the river is known as Jaldhaka and originates from Himalayan mountain ranges of the Sikkim state of India. The river traverses 186 km through Sikkim (India), Bhutan and then three districts of West Bengal (India) namely Darjeeling, Jalpaiguri and Cooch Behar and then enters into Bangladesh. The river confluences with the Brahmaputra River (at chainage 307 km) as Dharla a little downstream of Alipur in the Rangpur district of Bangladesh.

**(iii) Tista River.** The Tista River is the largest right bank tributary of the Brahmaputra River in the lower reach. The river originates from the Himalayan mountain ranges of the north east corner of the state of Sikkim (India) and traverses 138 km through the narrow valley and deep gorges of the Himalayan mountain ranges travelling mostly in the southern direction and debouches into the plains of the Jalpaiguri district of the West Bengal state (India) at Sevak and traverses through the district. The river then enters into Bangladesh and confluences with the Brahmaputra River near Bahadurabad (chainage 260 km) in Bangladesh. The Tista River has a number of tributaries and out of that worth mentioning are Rangni, Rangpa, Ranjit, Roll, Liah, Gish, Chal, etc.

**(iv) Meghna River.** The Meghna River is the only major left bank tributary of the Brahmputra River in its lower reach. The Meghna originates as Borak River from japvo mountain of Manipur hills (India) bordering the Nagaland state from an altitude of 3,015 m and flows south with a little westerly trend through the mountainous terrain of Manipur upto Tipaimukh near the trijunction of the three states of India, namely, Assam, Manipur and Mizoram. After this the river takes a hairpin bend and debouches into plains of Cacher District of Assam and flows north forming border of Assam and Mainpur states upto Jirimat little upstream of Lakhipur. The river then takes a turn and flows west in a highly meandering from through the three southern plain districts of Assam known as Borak valley and reaches to a place near Haritikar where the river bifurcates into two beaches, namely, Surma and Kushiyara. The southern branch, Kushiyara, flows in a western direction, forming an international boundary between India and Bangladesh for a length of about 32 km in the Karimganj district before entering into Bangladesh territory near Lafashail. The northern branch, Surma, initially flows towards north and then northwest, forming an international boundary between India and Bangladesh for a length of 22 km in the Cachar district before entering into the Bangladesh territory near Natanpur. The length of the river Borak from the source to the Indo-Bangladesh border along Kushiyara is 564 km and out of this length within the state of Assam is 143 km. The Borak River and its branches, Kushiyara and Surma, are fed by a number of tributaries. The tributaries joining within Indian territory are Jiri, Chiri, Badri, Madhura, Jatinaga, Banaimullaz, Gumrah, and Baleswar on the north bank and Sonai, Ghagra, Katakhal, Dhaleswari, Longai, and Kachna on the left lank. The flood gradient of the Borak River is very flat and varies from 1:10,000 in the upper reach to 1:20,000 in the lower reach within the Borak valley portion.

A good number of tributaries, Kushiyara and Surmalso, join the river during its course through Bangladesh. The Kushiyara River flows 338 km from the Indo-Bangladesh border up to Bhairab Bazar where the river again combines with the Surma River and the combined river flows as Meghna River up to a place near Rajabari-Mohanpur, a little south of the Tropic of Cancer and confluences with the Ganges-Padma River. The Borak River and its two branches, Kushiyara and Surma, are highly meandering in nature.

## 8.2 Hydrometeorology of the Basin

Hydrometeorology of the Brahmaputra basin varies greatly owing to the following reasons:

(I) The basin covers a vast spreading in three countries with wide variations in latitude, longitude and altitude. About 50.5% of the basin area lies on magnificent high Tibetan plateau just north of high Himalayan mountain ranges.

(II) The great Himalayan mountain ranges, the highest ranges of the world, pass through the basin dividing it into two distinct climatic regions.

(III) Orography plays a vital role in the hydrometeorology of the basin. The basin covers six distinct topographical regions including two distinct orographic regions. Besides these, small and medium mountain ranges are also spread out in other topographic regions which also influence hydrometeorology of the basin.

(IV) The location of the Bay of Bengal on the south of the basin influences greatly the hydrometeorology of the basin.

(V) Seasonal variations of weather.

(VI) Different parts of the basin in different seasons come under the influence of different winds, such as continental air moving under Siberian anti-cyclones, trade wind, monsoons, westerly disturbances, hot wind from south west Asia, cyclones developing owing to nonuniform surface heating and deep depressions that develop over the Bay of Bengal and many a times pass over the basins.

(VII) Location of numerous glaciers and permanent snow-covered regions.

(VIII) Influence of strong winds in high altitude regions.

(IX) Wide variations in temperature owing to non-uniform surface heating.

(X) Wide variations in vegetative cover and retention of soil moisture.

(XI) Wide variations in transparency and insolation depending on the altitude of the place.

### 8.2.1 CLIMATIC VARIATION

The basin can be divided climatically mainly into two zones: one that lies north to the high Himalayan mountain ranges in the high Tibetan plateau and falls under climatic Zone III and is classified as 'Mountain Climate'; and the remaining (southern) part of the basin covering parts of India, Bhutan and Bangladesh falls under climatic Zone I and is classified as 'Tropical monsoon climate'. The northern part under mountain climate, is in general, cold, dry and arid whereas the southern part under tropical monsoon climate is in general tropical, humid and experiences a high amount of rainfall mainly under the influence of the south-west monsoon.

*8.2.1.1 Climate of The Basin Area Under Zone III: Mountain Climate*

This zone emcompasses an area of 293,000 sq. km. which is about 50.5% of the total area of 580,000 sq. km. and is situated in the high Tibetan plateau. About 1,625 km long portion of the Brahmaputra River from the source to the Indo-China border flows through this zone. The basin area under this zone is located in the southern part of the high Tibetan plateau comprising the plateau of Chamdo, Changhai and southern parts of Singkiang.

The great plateau of Tibet is the most extensive of the world and lies 3,000 m to 5,000 m above the mean sea level and is known as the roof of the world. The plateau is wide and undulating with snow covered mountain ranges and spotted with numerous glaciers and lakes. Some of the glaciers stretch over several kilometres and extend into deep valleys. The physical features of the Tibetan plateau and the location of high Himalayan mountain ranges completely covering the southern part greatly influence the weather and climate of this part of the basin.

In winter months this part of the basin comes under the influence of continental air under Siberian anticyclone and experiences severe cold. The great Himalayan mountain ranges shield the southern part of the basin under Zone I from this Siberian cold air.

In summer months from June to September the southwest monsoon moves north east over the southern part of the basin and gets obstructed by the great Himalayan mountain ranges. The monsoon wind penetrates into south eastern part of this climatic zone through deeply trenched valleys of the upper course of the rivers like Brahmaputra, Salween, Mekong, and Yangtze Kiang and precipitates over these area. The amount and rate of precipitation decreases rapidly as the moisture-laden air advances in the north west direction towards the interior of the plateau. The central part of the basin under this zone is extremely dry and arid.

*8.2.1.2 Altitude Effect*

As the northern part of the basin under this mountain climate is situated in the high Tibetan plateau, it experiences different altitude effects, some of them are mentioned below:

(i) The atmospheric pressure is nearly half of that at sea level.

(ii) Correspondingly, the average atmospheric temperature is also less.

(iii) The transparency of the atmosphere is more as much of the haze and dust remain confined to lower altitude.

(iv) The insolation is stronger because transparency is more. This leads to a great contrast of temperature between the ground and air and the places under sun and shade. For this reason the mountain slopes exposed to sun experience burning heat while slopes under shadow experience cold.

(v) About 53,000 sq. km. area of the basin remains under permanent snow cover; the snow cover area increases as winter advances and in December and January snow covers almost the entire basin under this zone.

(vi) The water vapour content reduces with the increase in altitude. In the afternoons the humidity remains at its maximum and the chance of rain and thunderstorm increases.
(vii) High altitude areas come under the influence of strong winds. Low pressure and strong winds increase physiographic dryness.
(viii) The type of vegetation and plants and also the area of coverage varies with altitude depending on the temperature, rainfall and the period for which the area remains free from snow cover.

*8.2.1.3 Climate of the Basin Area Under Zone I: Tropical Monsoon Climate*
The southern part of the basin covering an area of 287,000 sq. km. falls under this climatic zone. This part of the basin area is comprised of an area of 195,000 sq. km in India, area of 45,000 sq. km in Bhutan and 47,000 sq. km in Bangladesh. A detailed countrywise breakdown of the catchment area is furnished in Table 8.2 and distribution of the basin area falling within India is furnished in Table 8.3. The lower reach of the river for a length of 1,255 km from the Indo-China border to its outfall into the Bay of Bengal falls under this zone. The topographical distribution of the basin area is furnished in Table 8.4. Out of the total area of 287,000 sq. km under this zone, an area of 137,050 sq. km., i.e., 47.75% falls under high Himalayan mountainous region. The rest of the area falls under the Brahmaputra valley and the plains of Bangladesh, including the coastal region.

The general climate of this zone is tropical, humid and experiences high amount of rainfall mostly for four months from June to September under the influence of southwest monsoon. About 65% to 80% of the average annual rainfall occurs during these four monsoon months. The three winter months from December to February are relatively dry and experience about 2% to 5% of the average annual rainfall. The three months, March to May, are generally known as pre-monsoon months or season and experience 20% to 35% of the average annual rainfall. The two months, October and November, are generally known as post-monsoon period and experience 3% to 6% of the average annual rainfall. Climate, weather, and rainfall distribution are greatly influenced by orography with minor mountain ranges, spurs and valleys of different widths and gaps.

### 8.2.2 SEASONAL WEATHER VARIATION

The weather of the basin varies greatly from place to place depending on its latitude, longitude, altitude, its topographic location, orographic effect and distance from the sea (the Bay of Bengal). The weather of a place is also greatly influenced by prevailing winds. But one thing is common, i.e., the whole of catchment area lies in the northern hemisphere, so general seasonal variation is almost similar though the duration of summer, rainy period and winter vary greatly from place to place, depending on the controlling factor already stated above.

Monsoon is the most important factor responsible for seasonal weather variation. An important characteristic feature of the monsoon wind is the complete reversal of winds which leads to the alteration of seasons. On the basis of monsoon variations, the year is divided into four seasons which are:

(i) Winter or cold season: Three months from December to February.
(ii) Summer, the hot weather: Three months from March to May; season.
(iii) Rainy Season or the South: Four months from June to September; west monsoon
(iv) Autumn or retreating: Two months October and November; south west monsoon.

*8.2.2.1 Winter Season*

Usually the three months from December to February are winter months. Starting in December, the cold weather season becomes fully established in January. In the Tibetan plateau and high altitude places of Arunachal Pradesh and Bhutan the duration of winter is much longer and covers five months from November to March and also the intensity of winter is much more severe. In this season the days are bright and sunny but nights are cold. Usually the fine weather of this season gets disturbed occasionally by the western disturbances which bring light rainfall and severe cold in the southern part of the basin, south of the Himalayas. The Tibetan part of the basin experiences severe cold and snowfall under the influence of very cold continental air that moves by the action of the Siberian anticyclone.

*8.2.2.2 Summer Season*

The period of three months from March to May is known as summer. This is the period of rising temperature and decreasing air pressure. The month of March is the spring time but the western part of the Zone I of the basin also experiences hot summer in April and May. Dry hot wind blows over the northern and western part of the basin under Zone I from west to east and sometimes it is associated with the dust storms. The Zone I also experiences northwesterly storms associated with short duration high intensity rainfall mostly in the afternoon part of the day. Sometimes, Zone I of the basin also experiences heavy rainfall for two to three days resulting from deep depression that develops in the Bay of Bengal and ultimately forms into cyclones and traverses a path of destruction over the basin mostly from the southwest to the northeast. This period is also called as pre-monsoon and experiences 20% to 35% of the average rainfall varying from place to place. Summer in the Zone III area of the basin is moderate and experiences very small amounts of rainfall.

*8.2.2.3 Rainy Season (Or Southwest Monsoon Season)*

The period of four months from June to September is the rainy season when the basin experiences 65% to 80% of the average annual rainfall under the influence of southwest monsoon. The southwest monsoon winds arising from the Indian Ocean and the Bay of Bengal moves from the southwest to the northeast direction and enters into the Brahmaputra valley by the first week of June. These monsoon winds get obstructed by the Patakii and Arakan mountain ranges in the east, Himalayas in the north and southern (Assam) mountain ranges in the south. After being obstructed the monsoon winds gradually start moving towards the west. Being obstructed by the mountain ranges, the Brahmaputra valley, the Lower (Assam) mountain ranges and the foothill regions of the Himalayan mountain ranges experience heavy monsoon rainfall during the period from June to September; the rainfall gradually decreases towards higher mountain ranges. Orography plays an important role and under its influence different parts of the basin come under the influence of different rainfall distribution patterns, some under heavy rainfall area and some under rainshadow area. Sometimes during this monsoon season cyclones develop over the Bay of Bengal and move over the main land of the Brahmaputra basin. Under their influence part of the basin under Zone I experiences heavy rainfall. The intensity and frequency of the cyclonic depressions originating from the Bay of Bengal and their crossing over to the main land as well as the path followed through the basin accounts for the variations in geographical distribution of rainfall in the basin.

*8.2.2.4 Autumn (Or the Retreating Southwest Monsoon)*

The period of two months, October and November, just after the monsoon and before the onset of winter, is the autumn. In this period the monsoon winds start retreating from the mainland towards the Indian ocean. From the middle of October the temperature begins to decline and the low pressure condition of northern India gradually disappears and shifts to the centre of the Bay of Bengal. The weather during this season is characterised by high day temperatures, clear sky and pleasant nights. The rain bearing clouds gradually retreat and are replaced by patches of bright and shining cirrus and cirrostratus clouds which give an indication of the arrival of Autumn. This period is also called as post monsoon period and experiences 3% to 6% of the annual average rainfall.

### 8.2.3 SOME ANNUAL WEATHER PHENOMENON

Different parts of the basin also experience some weather phenomenon like thunderstorm, duststorm, fog, etc.; the number and duration of occurrence varies from place to place. The number of cloudy days varies from place to place. The northeastern part of the Brahmaputra valley and the lower reaches of Arunachal Pradesh experience more rainfall and about 67% of the days in a year, the area remains under cloud cover. The number of cloudy days gradually reduces towards the western side of the basin. The Tibetan plateau generally remains clear and bright; only 10% to 25% of the days, area remains under cloud cover. This also declines from the southeast side towards the central part.

The number of occurrences of thunderstorms also varies from place to place. The western side of the Brahmaputra valley experiences more thunderstorms than does the eastern side. The number of thunderstorm days experienced at Dibrugarh, a city located at the eastern side of the valley, is only 31 days annually, whereas it is 62 days at Guwahati located towards the western side. The Tibetan plateau also experiences thunderstorms in about 15% of the days annually, mostly in the afternoons.

Some areas near the Brahmaputra River experience dust storms on some days in the months of February and March under the influence of strong winds. The dried-up fine silts on the bank of the Brahmaputra River gets lifted up and engulf the adjacent areas. The number of annual dust storm days is relatively more at Guwathti and Dhubri. The average number of annual dust storm days at Guwahati is 5.

The Brahmaputra basin experiences foggy nights during winter months. The northeastern parts of the valley and lower mountain reaches experience more number of foggy days specially during nights and mornings. The average number of annual cloudy days, thunderstorm days, dust storm days and foggy days of four stations Lhasa, Dibrugarh, Tezpur and Guwahati are furnished in Table 8.11.

### 8.2.4 TEMPERATURE VARIATION

The basin experiences wide seasonal variations of temperature. The variation also takes place owing to variations in geographical location, latitude and altitude of places. The basin as a whole experiences hot summer during April and May, cold winter in December and January and rainy days during four months from June to September under the influence of south west monsoon. During summer the north west part of the basin under Zone I comes under the influence of hot wind that blows from Central Asia. The winter period varies from three to five months depending on the latitude and altitude of the place and also owing to the influence of different cold waves that flow over the regions of the basin. The part of the basin under Zone I experiences very cold winter days under the influence of westerly disturbances. The Tibetan part of the basin experiences severe cold under the influence of very cold continental air that blows as Siberian anticyclone. The great Himalayan mountain ranges play an important role in obstructing cold Siberian winds coming to the southern part of the basin during winter and also obstruct to a great extent the southwest monsoon to cross over to the north. The weather remains very pleasant during March and November.

Table 8.11 Statement showing Annual Weather Phenomenon at some stations (towns) within the Brahamaputra basin.

| Sl.No. | Station (Towns) | Average (Annual Number of ) | | | |
|---|---|---|---|---|---|
| | | Cloudy Days | Thunder Storm (Days) | Dust Storm (Days) | Foggy Days |
| 1 | 2 | 3 | 4 | 5 | 6 |
| 1 | Lhasa | 98 | 53 | 0 | 41 |
| 2 | Dibrugarh | 241 | 31 | 1 | 52 |
| 3 | Tezpur | 147 | 16 | 0 | 7 |
| 4 | Guwahati | 191 | 62 | 5 | 45 |

The recorded mean daily maximum and minimum temperatures and recorded maximum and minimum temperatures of four stations, namely, Lhasa located in the Tibetan plateau and Dibrugarh, Tezpur and Guwahati located in the Brahmaputra valley in Assam (India) are furnished in Table 8.12. The monthly means of daily maximum and minimum temperatures of the above four stations are furnished in Table 8.13. The basin can be divided into five regions on the basis of temperature and its variation, discussed below.

*8.2.4.1 Region - I Tibetan Plateau*

It covers the Tibetan part of the basin where the altitude of the place varies from 3,000 m to 5,000 m. Owing to this high altitude, the average temperature is far less. Moreover the plateau comes under the influence of cold continental air from Siberian anticyclones during winter months and experience severe cold. The average minimum temperature during December and January comes down to -12 °C to -13 °C at Lhasa as recorded minimum temperature is (-) 16 °C in the month of January. The recorded maximum temperature at Lhasa is 32 °C which occurs in the month of June. The average maximum temperature varies from 25 °C to 28 °C in the period from May to August. The high altitude areas always remain under snow cover. The insolation is also large in this plateau.

*8.2.4.2 Region-II: Himalayan Mountainous Region*

The 200 to 300 km wide belt of the Himalayan mountain ranges pass through the basin. The states of Arunachal Pradesh and Sikkim of India and Kingdom of Bhutan fall in this temperature belt. Here the altitude of the mountain ranges vary from 1,000 m to 5,000 m. So the temperature variation is also wide. The high velocity winds that pass through the gorges and valleys also influence the temperature. The windward side of the mountains also experiences heavy rainfall and thereby it creates temperature differences between windward side and Leeward side. Again, the insolation results in wide differences of temperature between sunny areas and shadow areas in high altitude regions above 3 km, where the atmosphere is free from clouds dust and air pollution and thereby insolation plays an important role. High altitude places experience snow fall in winter months.

Table 8.12 Statement showing temperature and mean relative humidity at some stations within the Brahmaputra basin.

| SN | Station | Temperature (in °C) | | | | Relative Humidity (Mean) in % | |
|---|---|---|---|---|---|---|---|
| | | Mean Daily Max. | Min | Recorded Max. | Min. | At 0800 Hrs. | At 1700Hrs. |
| 1 | 2 | 3(a) | 3(b) | 4(a) | 4(b) | 5(a) | 5(b) |
| 1. | Lhasa | 16 | 1 | 32 (June) | -16 (Jan) | 69 | |
| 2. | Dibrugarh | 27 | 18 | 39 (June) | 4 (Feb) | 88 | 76 |
| 3. | Tezpur | 29 | 19 | 38 (May) | 6 (Feb) | 85 | 70 |
| 4. | Guwahati | 29 | 19 | 40 (April) | 5 (April) | 84 | 69 |

*8.2.4.3 Region-III: Valley and Northern Plains*

The Brahmputra valley in Assam and the northern part of Bangladesh fall under this region. The western side of this temperate region, approximately the area west of 93°E longitude, experiences heat wave during summer months of April and May. The average maximum temperature remains above 35 °C and the recorded maximum temperature is 40 °C. The temperature in the eastern side of the valley is slightly moderate. The three stations, Dibrugarh, Tezpur, and Guwahati, are situated in the Brahmaputra valley, and the average and maximum observed temperatures of these stations are furnished in Tables 8.12 and 8.13. The average minimum temperature during December and January varies from 7 °C to 9 °C whereas the recorded minimum temperature at Dibrugarh is 4 °C.

*8.2.4.4 Region IV: Southern (Assam) Mountainous Region*

Parts of the southern (Assam) mountainous region located south of the Brahmaputra valley covered by Naga hills, Mikir hills, Khasi hills, Garo hills, etc., from where the southern tributaries of Brahmaputra originate and traverse through the hilly portion of the catchment fall under this temperate region. The altitude of this region varies from 500 m to 3,000 m and the areas are moderately cool in summer and very cold in winter but not so severe. In some days during winter months of December and January the temperature drops below the freezing point especially in the high altitude places. Most of the places above 1500 m experience ground frost in peak winter nights. The region does not experience any snow fall.

*8.2.4.5 Region V: Coastal Belt*

The coastal areas of the basin near the Bay Bengal, where the river confluences, is influenced by coastal climate. The area also experiences annually a few cyclones that develop from deep depressions over the Bay of Bengal and pass over the belt. Generally, cyclones hit the coast during pre-monsoon and post-monsoon periods. The area does not experience extreme of heats.

## 8.2.5 PRESSURE VARIATION

The atmospheric pressure of places within the basin area varies widely depending on its altitude and seasonal variation. In general, the atmospheric pressure at the Tibetan plateau is almost half of that at MSL. The atmospheric pressure at Himalayan mountainous areas of Arunachal Pradesh and Bhutan varies greatly, depending on the altitude of place. The seasonal variation of the atmospheric pressure at MSL of the basin area is as follows:

(i) In January it varies from 1,019 mb to 1,018 mb.

(ii) In April it remains above 1,010 mb in major part of the basin whereas it varies from 1,008 mb to 1,010 mb in the south-west part.

(iii) In July it varies from 1,002 mb to 1,004 mb in the northern and eastern part and varies from 1,000 mb to 1,002 mb in the southern part.

(iv) In the month of October it remains high and above 1,013 mb in Tibetan plateau and the northern part of Zone-I and varies from 1,011 to 1,013 mb in southern and southwestern parts. Low pressure generally forms over the Bay of Bengal.

Table 8.13 Statement showing monthly mean of daily maximum and minimum temperature (in °C ) of some stations located within the Brahmaputra basin.

| Station Month | Lhasa | | Dibrugarh | | Tezpur | | Guwahati | |
|---|---|---|---|---|---|---|---|---|
| | Max. | Min. | Max. | Min. | Max. | Min. | Max. | Min. |
| 1 | 2(a) | 2(b) | 3(a) | 3(b) | 4(a) | 4(b) | 5(a) | 5(b) |
| January | 14 | -12 | 25 | 7 | 25 | 9 | 26 | 7 |
| February | 16 | -12 | 27 | 9 | 28 | 10 | 29 | 8 |
| March | 18 | -8 | 31 | 12 | 33 | 13 | 34 | 11 |
| April | 22 | -4 | 32 | 15 | 34 | 17 | 36 | 17 |
| May | 25 | 1 | 35 | 18 | 35 | 19 | 35 | 19 |
| June | 28 | 4 | 35 | 21 | 35 | 22 | 35 | 22 |
| July | 27 | 6 | 35 | 22 | 35 | 23 | 35 | 24 |
| August | 26 | 4 | 35 | 23 | 35 | 24 | 35 | 24 |
| September | 24 | 4 | 34 | 21 | 35 | 23 | 34 | 23 |
| October | 22 | -5 | 33 | 17 | 33 | 18 | 33 | 18 |
| November | 17 | -9 | 29 | 12 | 30 | 13 | 30 | 13 |
| December | 14 | -13 | 26 | 8 | 26 | 9 | 27 | 8 |

## 8.2.6 HUMIDITY VARIATION

The average humidity differs from place to place within the basin area depending on the latitude, altitude, wind velocity, sunshine hour, influence of different prevailing winds, precipitation, ground moisture, etc. The average humidity of places under climatic Zone-I is relatively high but it is low under Zone-III. The central part of the Tibetan portion of the basin under Zone III is relatively dry owing to its high altitude, coldness and situated beyond the coverage of the monsoon wind; low pressure and strong wind increase physiographical dryness.

The mean relative humidity of four stations Lhasa, Dibrugarh, Tezpur and Guwahati are furnished in Table 8.12. It shows that the mean relative humidity at 8 hrs of the three stations Dibrugarh, Tezpur and Guwahati, located in the Brahmaputra valley under Zone-I are above 80% whereas it is 69% at Lhasa under Zone-III. The mean relative humidity of the places at 1700 hrs is observed to be about 15% less than at 0800 hrs.

### 8.2.7 VARIATION OF INSOLATION

The transparency of the atmosphere increases with increase in altitude as most of the haze, dust, and air pollution remains confined at the lower layers of the atmosphere. With the increase of transparency, insolation also increases. The insolation at Tibetan plateau and in higher reaches of Himalayas is much stronger. Higher reaches also receive higher amounts of ultraviolet radiation than that at the sea level.

## 8.3 Precipitation

The amount of annual precipitation in different locations of the basin varies widely owing to the variation of climate, altitude, temperature pressure, latitude, and orography; and also owing to the influence of different prevailing winds like monsoon, trade wind, westerly disturbances, continental cold air, etc. The basin receives mainly two types of precipitation, namely, snowfall and rainfall. Depending on the amount of rainfall and snowfall, the basin may be broadly divided into two regions.

(i) The tibetan plateau that falls under climatic zone-III.
(ii) The remaining southern part of the basin that falls under climatic zone-I.

The southwest monsoon wind is the main factor responsible for wide spread rainfall in the basin area south of Himalayan mountain ranges under climatic zone I during four months from June to September. About 65% to 80% of the average annual rainfall takes place in this period and the amount varies from place to place depending on its geographical location, orography, direction of the movement of monsoon wind, altitude and distance of place from the Bay of Bengal.

The Tibetan plateau and higher reaches of the Himalayan mountain ranges mainly places above 3000 m altitude, receive snowfall during winter months from December to February. These higher reaches also receive snowfall. The southeastern part of the Tibetan plateau receives some amount of monsoon rainfall in the month of July and August. The amount of rainfall decreases gradually towards western and central part of the plateau; the general climate of this part is cold, dry and arid.

The average annual rainfall within the basin area varies from less than 400 mm to more than 6,000 mm. The mean annual rainfall over the whole catchment excluding the Tibetan part is around 2300 mm. Within the valley reach in Assam, the average annual rainfall is more towards the northeast and gradually decreases towards the west. The mean monthly and average annual rainfall, and the average number of rainy days in a year of four stations, Lhaea, Dibrugarh, Tezpur and Guwahati, are furnished in Table 8.14. It is observed that the average annual rainfall of Lhasa is only 400 mm whereas it is 2811 mm at Dibrugarh.

Depending on the prevailing climate and influence of different winds mainly monsoon, the year may be divided into the following four periods of rainfall:

I: Winter rainfall from December of February
II: Pre-monsoon rainfall from March to May
III: Monsoon rainfall from June to September
IV: Post-monsoon rainfall from October to November

The basin experiences very small amounts of rainfall in the winter months generally under the influence of western disturbances. The basin area under Zone I experiences occasional winter rainfall amounting to 2% to 5% of the average annual rainfall. Higher reaches of Arunachal Pradesh and Bhutan receive greater amounts of rainfall during these winter months amounting to about 10% of the average annual rainfall. The Tibetan part of the basin under Zone III experiences snowfall under the influence of continental cold air that blows owing to siberian anticyclones. The Tibetan plateau also receives some winter rainfall in the form of light rainfall for a number of days though total amount is not substantial.

The three months from March to May generally termed the pre-monsoon period, experience rainfall mainly from isolated thunderstorms occurring over small areas for short duration. Scattered thunderstorm activities are also observed on some days over large areas and generally such activities do not persist for more than a day or two. Widespread heavy rainfalls are sometimes observed during late May owing to the passage of cyclones over the basin under Zone I. These cyclones result from deep depressions that sometime form over the Bay of Bengal owing to uneven heating of the land surface and sea. This pre-monsoon period, in general, experiences 20% to 35% of the average annual rainfall. Higher reaches of Arunachal Pradesh and Bhutan experience about 35% of the average annual rainfall whereas the rest of the southern portion of the catchment experiences 20% to 30%.

The southwest monsoon plays the most important role for widespread rainfall during four months from June to September and agriculture mainly depends on this monsoon rainfall. Portions of the basin, particularly areas under Zone I, experience widespread rainfall starting from later part of May and in some years it continues up to the middle of October. Rainfall increases gradually from the month of April but abruptly from the end of May. The average normal monthly rainfall in descending order is July, August, June, May, September, October, etc. The average rainfall during these four monsoon months, from June to September, amounts to 65% to 80% of the average annual rainfall. Rainfall during the two monsoon months of July and August is the highest, and amounts to more than 33% of the total annual rainfall.

The monsoon rainfall is generally widespread and at places it is heavy. In most places the monsoon rainfall is received in showers of one to three hours duration. At some favourable places of maximum rainfall, heavy showers of six hours or more duration are also observed. Such situations generally persist for 2 to 3 days but sometimes continue for 4 to 5 days also.

The two months October and November are termed the post-monsoon period and experience very low amounts of rainfall of the order of 3% to 6% of the average annual rainfall. The rainfall in this period occurs generally from thunderstorm of isolated nature or from the cyclones that pass over the basin, being developed from the depressions in the Bay of Bengal, and move in the northeast direction towards the basin.

There is wide variation of rainfall within the basin. The average annual rainfall of the tributary basin shows a wide variation. The average rainfall of the northern and southern tributaries of the Brahmaputra River in the middle reach for the four seasons, namely, winter, pre-monsoon, monsoon, and post-monsoon, are furnished in Table 8.15 and Table 8.16, respectively. The mean annual rainfall of some tributary basins is much higher than the average of the whole basin (2,300 mm) such as 3,264 mm for Tipkai, 3,500 mm for Jiadhol, 3,860mm for Dibong, 2,991 mm for Lohit, etc. whereas some tributary basins experience less amount of rainfall like 1,668 mm for Belsiri, 1,672 mm for Kulsi, 1,713 mm for Krishna, 1,735 mm for Kopili, 1,767 for Dhansiri (S), etc..

Orography plays an important role in areal distribution of rainfall over the basin. The average annual rainfall is maximum on the southern slopes and the foothill region of Himalayas in Bhutan and Arunachal Pradesh. An isohyetal map of the average annual rainfall of the Zone I part of the basin except Bangladesh region is given in Figure 8.4.

Table 8.14 Mean monthly rainfall and rainy days of four stations within the Brahmaputra Basin.

| Station | Lhasa | | Dibrugarh | | Tezpur | | Guwahati | |
|---|---|---|---|---|---|---|---|---|
| Month | Rainfall (mm) | Rainy Days (mm) | Rainfall (mm) | Rainy Days (mm) | Rainfall (mm) | Rainy Days (mm) | Rainfall (mm) | Rainy Days (mm) |
| 1 | 2 | 3 | 4 | 5 | 6 | 7 | 8 | 9 |
| January | 3 | 1 | 38 | 7 | 15 | 3 | 10 | 2 |
| February | 13 | 1 | 62 | 10 | 27 | 6 | 4 | 5 |
| March | 8 | 1 | 103 | 14 | 48 | 8 | 25 | 6 |
| April | 5 | 1 | 241 | 18 | 153 | 12 | 145 | 10 |
| May | 25 | 3 | 307 | 19 | 271 | 19 | 236 | 15 |
| June | 53 | 8 | 500 | 22 | 308 | 18 | 312 | 18 |
| July | 122 | 13 | 536 | 24 | 348 | 19 | 312 | 20 |
| August | 89 | 10 | 451 | 22 | 331 | 18 | 261 | 16 |
| September | 66 | 7 | 352 | 21 | 210 | 16 | 167 | 15 |
| October | 13 | 2 | 152 | 8 | 104 | 6 | 71 | 5 |
| November | 3 | 1 | 53 | 4 | 23 | 3 | 14 | 2 |
| December | 0 | 0 | 16 | 3 | 6 | 2 | 4 | 1 |
| Total (Annual) | 400 | 48 | 2811 | 172 | 1844 | 130 | 1561 | 115 |

* Rainfall above 0.3 mm in a day is considered as a rainy day

It is observed from the isohyetal map that closed cells of maxima of more than 4000 mm exist where the southern slopes of Himalayan mountain ranges take a crucible shape at the entrance of the major northern tributaries to the plains. One such closed cell of maxima (Cell-I) is noticed in the area around Pasighat between latitude 27° N to 28° 45'

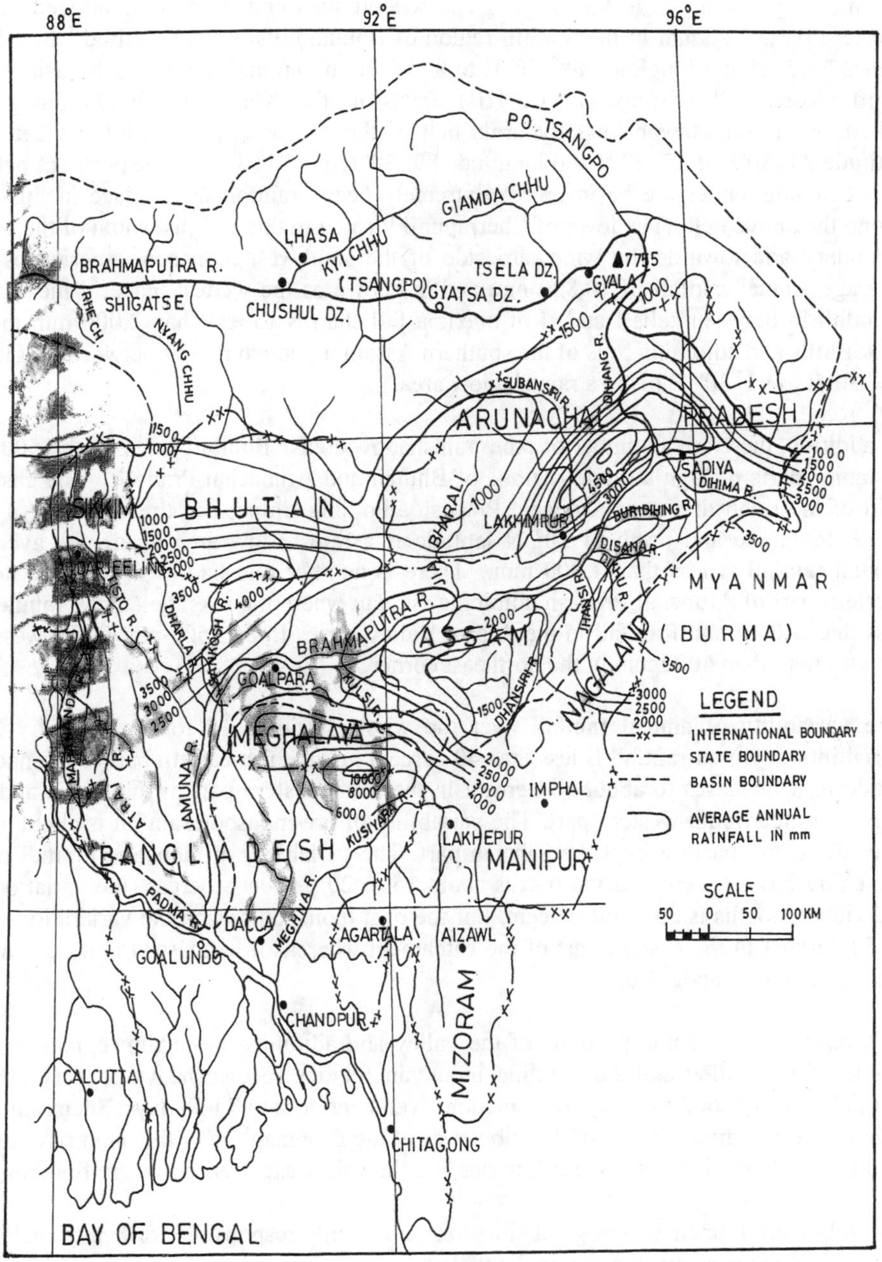


Fig. 8.4 Isohyetal map of the average annual rainfall.

N and longitude 93° E to 95° 30' E. The second such cell (Cell-II) is noticed in the western part of Assam in the foothill region of Bhutan hills between latitude 26° 15' N and 27° 15' N and longitude 88° 30' E to 90° E at and around the Sankosh basin. The third closed cell of maxima (Cell-III) exists in the Khasi hill region along the Cherrapunji and Mawphalang Pynurala belt of the Meghalaya state of India between latitude 24° 30' and 25° 30'N and longitude 89° 30' E and 91° 30' E. The northern half of this cell falls under the basin area. Extremely heavy rainfall takes place in this cell along the above belt. The town of Cherrapunji located in this cell just south of the basin boundary and towards the windward side of the southwest monsoon experiences the average annual rainfall of 10,800 mm and is known as the wettest place in the world. Rainfall in between cells I and II of maxima fall sharply to less than 2,000 mm in the Mikir hills and adjoining hills of the southern Assam mountain ranges between longitude 91° 30' E to 93° 30' E being a rain shadow area.

In addition, there are a number of other rainshadow areas. Rainfall is less than 1500mm in major parts of high altitude regions of Bhutan and Arunachal Pradesh. The eastern part of the Lohit disrict of Arunchal Pradesh adjoining China, i.e., the area east of 95° 30' E longitude and north of 28° N latitude is a rainshadow area where the average annual rainfall is less than 1,000 mm. There is another rainshadow area in the north western part of Arunchal Pradesh adjoining Bhutan where the average annual rainfall is less than 1,000 mm. Rainfall in the Assam Valley varies from 1,500 mm in the western side to more than 4000 mm in the northeast corner.

The variability of annual rainfall over the whole basin is 15 to 20 percent. The variability of winter rainfall is less than 40 percent in the north-eastern part of Arunachal Pradesh; it increases to about 65 percent in the south eastern part of the basin and 85 percent in the South western part. The variability of pre-monsoon rainfall is about 30% over the entire basin except the Tibetan part. The variability of monsoon rainfall over the entire basin except Tibetan part is about 15 to 20 percent whereas the variation in individual months is 30 to 40 percent. For the post monsoon season the variability is 40 to 50 percent in the eastern part of the catchment whereas it increases to 80 percent in the western part (under Zone I).

It is observed that some portions of the valley and adjoining mountain regions or the whole of the valley and surrounding mountain regions receive heavy to very heavy rainfall in a spell of 2 to 3 days or sometime even continues to 7 to 8 days. Such rainfalls give rise to flooding in some of the tributaries of the Brahmaputra or the main river itself or the main river along with the tributaries and the valley experiences severe flooding.

The following four meteorological situations are mainly responsible for heavy spells of rainfall and consequent flooding in the valley.

I. Break monsoon situation or when the axis of the seasonal trough shifts to the north from its normal position and lies close to the foothills of Himalayas.

II North-westerly to northern movement of monsoon depressions/storms from Bay of Bengal into the interior of the valley and the surrounding mountain ranges.

III. Formation and movement of land depressions over northeast India.

IV. Upper air cyclonic circulation over northeast India and its neighbourhood.

Table 8.15 The average rainfall of some north bank tributary basins of Brahmaputra River in the middle reach.

| Sl.No. | Name of River | Average Rainfall ( in mm) | | | | Annual |
|---|---|---|---|---|---|---|
| | | Dec.-Feb. (Winter) | Mar.-May. (Premonsoon) | June-Sept. (Monsoon) | Oct.-Nov. (Post-monsoon) | |
| 1 | 2 | 3 | 4 | 5 | 6 | 7 |
| 1. | Simen | - | - | - | - | 3,680 |
| 2. | Jiadhol | - | - | - | - | 3,500 |
| 3. | Subansiri | 108 | 461 | 1,450 | 131 | 2,150 |
| 4. | Ranganadi | 88 | 604 | 1,601 | 174 | 2,467 |
| 5. | Borgong | - | - | 1,455 | - | 2,216 |
| 6. | Jia-Bharali | 60 | 471 | 1,529 | 157 | 2,217 |
| 7. | Gabharu | - | - | - | - | 1,582 |
| 8. | Belsiri | 38 | 384 | 1,147 | 99 | 1,668 |
| 9. | Dhansiri(N) | 56 | 463 | 1,345 | 133 | 1,997 |
| 10. | Noa-Nadi | - | - | - | - | 1,722 |
| 11. | Nanoi | 50 | 539 | 1,359 | 126 | 2,074 |
| 12. | Barnadi | - | - | - | - | 2,037 |
| 13. | Puthimari | 46 | 608 | 1,407 | 93 | 2,154 |
| 14. | Pagladiya | 39 | 689 | 1,820 | 111 | 2,659 |
| 15. | Manas | 52 | 418 | 1,286 | 113 | 1,869 |
| 16. | Champamati | - | - | - | - | 2,667 |
| 17. | Gaurang | - | - | - | - | 3,123 |
| 18. | Tipkai | - | - | - | - | 3,264 |
| 19. | Sankosh | - | - | - | - | 2,863 |

It is observed that exceptionally heavy rainfall occurs over Assam and neighbouring hill states when westerly wave moving across Nepal--Assam Himalayas happens to synchronise with any of the above four meteorological situations. Such situations are generally called as synoptic situations. Under such situations heavy to very heavy rainfall of isolated nature takes place at different places of the valley or to the adjoining mountainous regions. The maximum rainfall of 1 hour, 3 hour, 6 hour, 12 hour, 24 hour, etc. are recorded at different places of maxima and are mentioned below.

Table 8.16 Average rainfall of some south bank tributary basins of Brahmaputra River in the middle reach.

| S.N. | Name of River | Average Rainfall ( in mm) | | | | Annual |
|---|---|---|---|---|---|---|
| | | Dec.-Feb. (Winter) | Mar.-May. (Premonsoon) | June-Sept. (Monsoon) | Oct.-Nov. (Post-monsoon) | |
| 1 | 2 | 3 | 4 | 5 | 6 | 7 |
| 1. | Dibong | 394 | 1,415 | 1,701 | 350 | 3,860 |
| 2. | Lohit | 216 | 840 | 1,776 | 159 | 2,991 |
| 3. | Dibru | 96 | 669 | 1,466 | 140 | 2,371 |
| 4. | Buri-Dihing | 113 | 631 | 1,567 | 175 | 2,486 |
| 5. | Desang | 110 | 613 | 1,535 | 174 | 2,432 |
| 6. | Dikhow | 113 | 594 | 1,553 | 178 | 2,438 |
| 7. | Jhanji | 80 | 654 | 1,277 | 186 | 2,197 |
| 8. | Dhansiri(S) | 58 | 449 | 1,116 | 144 | 1,767 |
| 9. | Kopili | 42 | 376 | 1,194 | 123 | 1,735 |
| 10. | Kulsi | 25 | 378 | 1,152 | 117 | 1,672 |
| 11. | Krishnai | 14 | 404 | 1,138 | 157 | 1,713 |
| 12. | Jinari | 28 | 835 | 1,842 | 345 | 3,050 |
| 13. | Jinjiram | 31 | 653 | 1,937 | 226 | 2,847 |

a) **1 Hour maximum recorded rainfall.** Rainfall of more than 40 mm in an hour is quite frequent in areas favourable for heavy rainfall. Rainfall of 70 mm per hour is also common. The highest recorded one hour rainfall is 97.5 mm at Saralpara under Tikpai (Tributary) River basin located at the already mentioned Cell-II of maxima where the average annual rainfall lies above 4,000 mm. One hour rainfall of above 60 mm is also seen in the Cell-I of maxima near Pasighat. The one-hour recorded maximum rainfall in areas of minima is observed to be around 30 mm.

b) **3 Hour maximum recorded rainfall.** Rainfall of 60 mm and above in 3 hour duration is quite frequent in areas favourable for heavy rainfall. Three hour rainfall of 200 mm is recorded in Cell-II of maxima at and around Saralpara. The maximum recorded 3 hour rainfall in Cell-I of maxima at and around Pasighat is 120 mm. The maximum recorded rainfall in areas of minima is around 40 mm.

c) **6 Hour maximum recorded rainfall.** Six hour maximum rainfall of 300 mm is recorded in Cell-II of maxima whereas it is recorded as 150 mm in Cell-I of maxima. The recorded 6-hour rainfall in areas of minima is around 60 mm. Closed cells of minima of 60 mm are seen in the South Central part of the catchment.

d) **24 Hour maximum recorded rainfall.** Rainfall of 200 mm and above in 24 hours is observed to be quite frequent in areas favourable for maximum rainfall like Cell-I at and around Pasighat and Cell-II at and around Sankosh basin; but rainfall of more than 300 mm occurs rarely. A cell of maxima of 500 mm in 24 hours is observed in Cell-II and a cell of 300 mm is observed in Cell-I near Pasighat and adjacent to this there is another cell of 250 mm rainfall near Likabali and Gerukamukh in the Himalayan foothills region. Very high amounts of rainfall in some selected localized places have occurred under some synoptic meteorological conditions in some exceptional cases; a few such recorded 24 hour rainfall amounts are mentioned below:

| | | |
|---|---|---|
| (i) | At Jamduar | 805.5mm on 7.9.76 |
| (ii) | At Jamduar | 603.2mm on 5.10.77 |
| (iii) | At Roing in A.P. | 745.0mm on 16.8.77 |

The 24 hour recorded rainfall for different months of the year for four stations, Lhasa, Dibrugarh, Tezpur and Guwahati, are furnished in Table 8.17.

### 8.3.1 RAINY DAYS

The number of rainy days monthwise of the aforementioned four stations are furnished in Table 8.14 (Rainfall of more than 0.3 mm in a day is considered as a rainy day). It is observed from the table that the station Lhasa experiences only 48 rainy days in a year compared to 172 days by the station, Dibrugarh. Rainshadow areas of the Mikir hills in the southcentral part of the basin experiences a low of 95 days. Almost 65% of the rainy days occur in the monsoon season and 20% in the pre-monsoon season.

Then Number of rainy days in the southern periphery of Arunahcal Pradesh is around 120 days and it increases to 170 days at the northern end. In the high altitude northern part of Arunachal Pradesh the number of rainy days in winter and pre-monsoon period is considerably high, although the total amount of rainfall is not much as the rainfall is mostly in the form of light rainfall.

## 8.4 Discharge

Among the large rivers of the world, the rank of the Brahmaputra River is fifth with respect to the average discharge whereas its rank is fourteenth with respect to the catchment area. The river, thus, has attained a distinction of having the highest specific yield with a of 0.149 cumecs/sq km. The average annual discharge of the river at its confluence with the Ganges River at Goalundo in Bangladesh is 19,820 cumecs. The recorded maximum and minimum discharges at Guwahati, where the catchment area is 424,100 sq km. (i.e. 73.12% of the catchment area) are 72,794 cumecs (recorded on 23-8-62) and 1,757 cumecs, respectively. The return period of this maximum discharge is 10 years. The three discharge measuring sites of the river at Tibetan plateau are Shigatse (the new name is Rikaze), Chushuldzong ( Qushui Gogge) and Tsela Dzong (Ziang Zhong ) located at an altitude of 3,816 m, 3,565 m and 2,957 m, respectively. The maximum recorded discharge at these three sites are 3,380 cumecs, 6,230 cumecs and 10,200 cumecs, respectively. The maximum recorded discharge at Tsela Dzong (chainage 1,515 km) is 14% of that of the maximum discharge at Guwahati whereas the catchment area at Tesladzong is 44% that of Guwahati. The discharge measuring site at Pasighat (chainage 1,029 km) is located in Arunachal Pradesh of India, just at the foothills, where the catchment area is 249,000 sq. km. (i.e., 42.93% of the total and 58.71% that of Guwahati) and the observed maximum discharge is 29,640 cumcs (40.72% that of Guwahati). There are three permanent discharge measuring sites within the valley reach of Assam, namely, at Bessamara (chainage 832 km), Pandu within Guwahati city (chainage 542 km) and Jogighopa ( chainage 422 km).

Typical discharge hydrographs of the Brahmaputra River as observed in a particular year for five stations, Chushuldzong, Tseladzong, Pasighat and Guwahati are represented in Figure 8.5. The discharge data at three sites of the Tibetan plateau are only available for the monsoon period. The discharge hydrographs of Pasighat and Guwahati clearly show that the river experienced number of floods from June to mid October. The annual peak discharge was observed at Guwahati on 18th September; this wave recorded peak discharge at Pasighat on 15th September, i.e, the flood wave took three days to travel from Pasighat to Guwahati. The flood peak discharge at Guwahati is little more than three times of the corresponding flood peak discharge at Pasighat. The minimum flow in the river is generally observed during late February or early March.

The percentage contributions to the annual yield of the river at Guwahati by the main river course at Pasighat and other main tributaries are as below:

| | | | |
|---|---|---|---|
| (i) | by the main river course Dehang at Pasighat | - | 34.00% |
| (ii) | by the river Dibong | - | 7.65% |
| (iii) | by the river Lohit | - | 9.50% |
| (iv) | by the river Subansiri | - | 10.30% |
| (v) | by the river tributaries joining between Pasighat and Guwahati. | - | 38.55% |
| | | Total | = 100.00% |

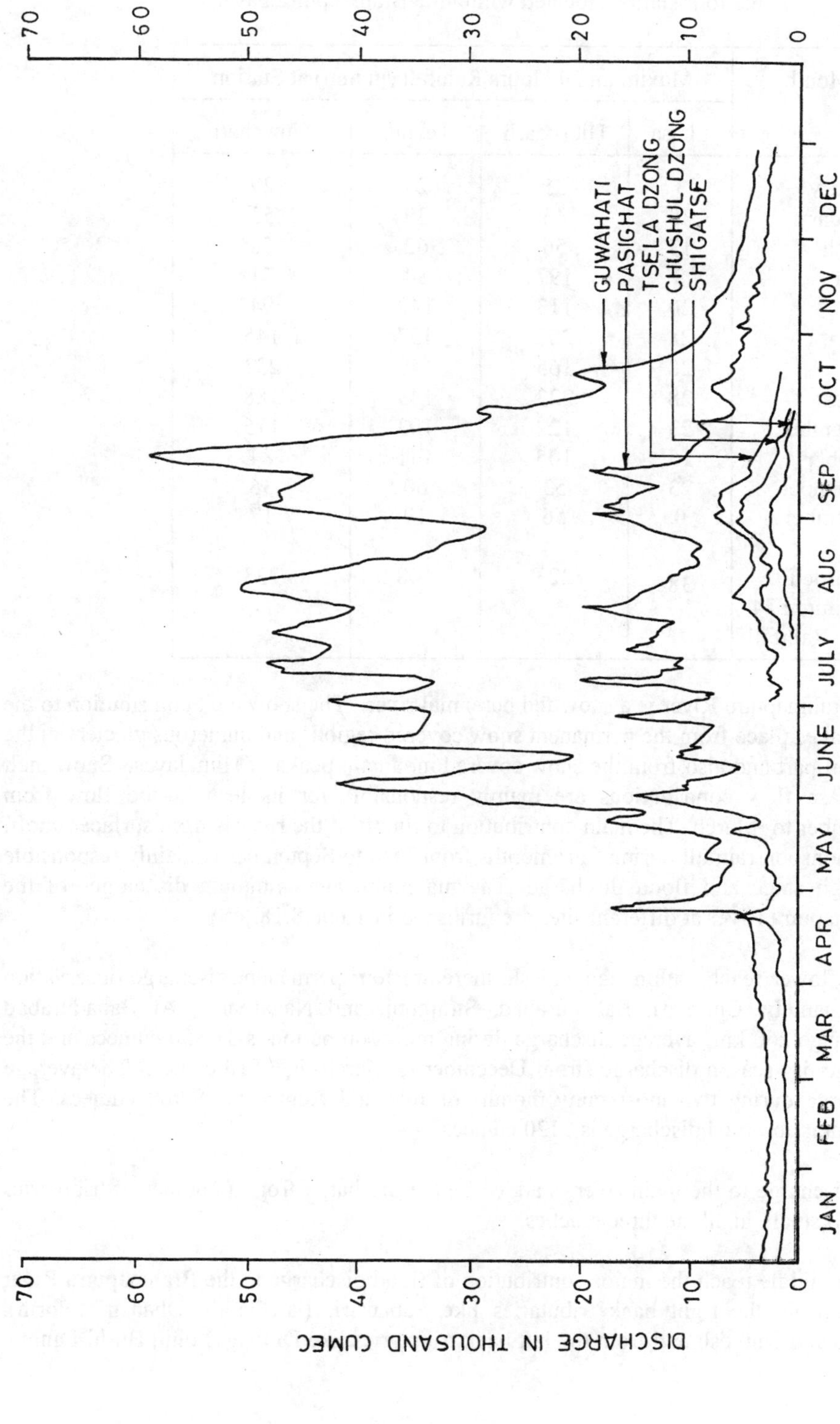


Fig. 8.5 Discharge hydrographs of the Brahamaputra River at different sites.

Table 8.17 Maximum recorded 24-hour rainfall in different months at the four stations located within the Brahmaputra basin.

| Month | Maximum 24 Hours Rainfall (in mm) at Station | | | |
|---|---|---|---|---|
| | Lhasa | Dibrugarh | Tezpur | Guwahati |
| January | 3 | 35 | 29 | 29 |
| February | 38 | 46 | 39 | 53 |
| March | 13 | 56 | 62 | 56 |
| April | 8 | 197 | 84 | 71 |
| May | 20 | 113 | 148 | 94 |
| June | 28 | 202 | 137 | 145 |
| July | 28 | 165 | 140 | 233 |
| August | 28 | 223 | 143 | 188 |
| September | 23 | 122 | 103 | 115 |
| October | 15 | 105 | 141 | 84 |
| November | 23 | 53 | 60 | 36 |
| December | 0 | 56 | 19 | 17 |
| Recorded Maximum 24 Hour Rainfall | 38 | 223 | 148 | 233 |

The Brahmaputra River is a snow fed perennial river. The snow melt contribution to the river takes place from the permanent snow covered regions and numerous glaciers in the Tibetan part and also from the snow covered mountain peaks of Himalayas. Snow melt and base flow contributions are mainly responsible for its lean period flow from December to March. The main contribution to runoff of the river is from surface runoff. The monsoon rainfall during four months from June to September is mainly responsible for high stage and flood discharge. The maximum and minimum discharges of the Brahmaputra River at different sites are furnished in Table 8.18.

At the lower reach within Bangladesh, there are four permanent discharge observation sites, namely, Chilmari, Bahadurabad, Sirajgonj, and Nagarbari. At Bahadurabad (Chainage 260 km) average discharge during monsoon period is 31,850 cumecs and the average dry season discharge (from December to March) is 6,510 cumes. The average discharge during two most rainy months of July and August is 39,900 cumecs. The minimum observed discharge is 3120 cumecs.

The discharge to the main river is added by the discharge from a number of tributaries big and small, in all the three reaches.

In this middle reach the major contribution of flood discharge to the Brahmaputra River comes from the right bank tributaries like Subansiri, Jia-Bharali, Dhansiri (North), Manas and Sankosh and from the left bank tributaries like Dibong, Lohit, Burhi-Dihing,

Desang, Dikhow, Dhansiri (South), Kopili, etc.. The recorded maximum and minimum discharge of some major right bank & left bank tributaries are furnished in Table 8.19.

Table 8.18 Recorded maximum and minimum discharge of the Brahmaputra River at different sites.

| Sl. No. | Name of the Brahmaputra River in that reach | Name of site (Location) State, Country | Chainage in Km from the mouth | Catchment area in sq. km. | Recorded Discharge in cumecs | |
|---|---|---|---|---|---|---|
| | | | | | Max. | Min |
| 1 | 2 | 3 | 4 | 5 | 6 | 7 |
| 1. | Tsangpo | Shigatse (Tibet, China) | 2,105 | 88,620 | 3,380 | --- |
| 2. | -do- | Chushuldzong (Tibet, China) | 1,933 | 113,500 | 6,230 | --- |
| 3. | -do- | Tseladzong (Tibet, China) | 1,515 | 186,730 | 10,200 | --- |
| 4. | Dihang | Pasighat (Arunachal Pradesh, India) | 1,029 | 249,000 | 29,640 | 1,076 |
| 5. | Brahmaputra | Guwahati (Assam, India) | 542 | 424,100 | 72,794 | 1,757 |

## 8.5 Floods

### 8.5.1 NORMAL AND HISTORIC FLOODS

On average the Brahmaputra River experiences 4 to 5 major flood waves annually during the monsoon months from June to September. The two most rainy months, July and August, experience most of the major flood waves. Sometimes, on rare occasions, the river also experiences major floods in the month of May, particularly in the latter part under typical meteorological situations.

The Brahmaputra valley in the State of Assam (India) and the plains of Bangladesh are historically subjected to frequent heavy floods from the Brahmaputra River and its tributaries resulting in extensive submergence of land, loss of property, dislocation of communication, loss of livestock and sometime even loss of human lives. Such historic unprecedented major floods that have occurred in recent times were in the years 1954, 1962, 1966, 1972, 1977, 1984 and 1988; floods of lesser magnitude occur every year. Such historical unprecedented floods generally occur as a result of heavy to very heavy rainfall in the catchment area during the monsoon months under the influence of any of the four typical metrological situations already discussed. A monthwise distribution in the percentage of occurrences of historic floods in the Brahmaputra River within the valley reach resulting from heavy rainfall under such four typical meteorological situations are furnished in Table 8.20.

Table 8.19 Observed maximum and minimum discharge of some major tributaries of the Brahmaputra River in its middle reach (in India).

| Sl. No. | Name of Tributary R. | Discharge Observation Site | Observed Discharge ( in Cumecs) | |
|---|---|---|---|---|
| | | | Maximum | Minimum |
| 1 | 2 | 3 | 4 | 5 |
| A. | Right Bank Tributaries | | | |
| 1. | Subansiri | Chawidhowghat | 18.799 | 138,00 |
| 2. | Ranganadi | Pahumaraghat | 1,911 | 13,50 |
| 3. | Jia-Bharali | N.T. Rd .Cr. | 9,939 | 53.00 |
| 4. | Dhansiri (N) | N.T. Rd .Cr. | 2,092 | 10.50 |
| 5. | Noa-nadi | N.T. Rd .Cr. | 176 | 0.14 |
| 6. | Nanoi | N.T. Rd .Cr. | 239 | 0.37 |
| 7. | Puthimari | N.T. Rd .Cr. | 861 | 1.01 |
| 8. | Pagladiya | N.T. Rd .Cr. | 1,737 | 1.70 |
| 9. | Manas | Mathanguri | 10,842 | 142.00 |
| 10 | Sankosh | Srirampur Rly. Bridge. | 6,391 | 87.00 |
| B. | Left Bank Tributaries | | | |
| 1. | Dibong | Jiagaon | 11,205 | 176.00 |
| 2. | Lohit | Digarughat | 12,350 | 209.00 |
| 3. | Burhi-Dehing | Khowang | 2,217 | 15.30 |
| 4. | Disang | N.h.Crossing | 1,093 | 3.40 |
| 5. | Dikhow | A.T. Rd. Cr. | 1,379 | 5.40 |
| 6. | Dhansiri (S) | Numuligarh | 2,296 | 9.30 |
| 7. | Kopili | Dharamtul | 1,942 | 17.80 |
| 8. | Kulsi | Kulurmara | 323 | 0.12 |
| 9. | Krishnai | A.T.Rd. Crossing | 834 | 1.60 |
| 10. | Janari | A.T.Rd. Crossing | 342 | 0.60 |

It is observed that the percentage occurrence of major floods owing to such four typical meteorological situations in the month of June, July and August are 25%, 35% and 37%, respectively.

The Brahmaputra River is turbulent and aggrading in nature; it is also braided in the valley reach and in Bangladesh. The slope of the river in the valley reach within the state of Assam is also much steeper than required for a poised river of its size. During flood, the river becomes burdened with a heavy sediment load, and a huge quantity of flood discharge flows down the river through its valley with a high velocity attacking its banks; it then looks as if the river is provoked.

Table 8.20 Monthwise distribution of occurrence of historic floods in the Brahmaputra River within the valley reach during monsoon months under four typical meteorological situations (in percentage).

| Month | Percentage Occurrence of Historic Flood | | | | |
|---|---|---|---|---|---|
| | Owing to Break Monsoon Situation | Owing to Movement of Depression / Storms From Bay of Bengal | Owing to Movement of Land Depression | Owing to Upper Air Cyclonic Circulation | Total Occurrence |
| 1 | 2 | 3 | 4 | 5 | 6 |
| June | 2 | 10 | 5 | 8 | 25 |
| July | 10 | 10 | 7 | 8 | 35 |
| August | 18 | 5 | 7 | 7 | 37 |
| Sept. | - | - | 3 | - | 3 |
| Total | 30 | 25 | 22 | 23 | 100% |

Generally flood waves remain above the danger level for long period in the valley reach and thereby create an acute flood congestion. The tributaries within the valley experience back flow from the main river when it is in high stage, extending up to 20 to 25 km from its confluence point with the Brahmaputra River.

Under such circumstances the tributary rivers cannot discharge their own flow. When tributaries together with the main river experience flood, it becomes a devastating one for the whole valley. It is observed that the normal duration for which the water level (W.L.) of the Brahmaputra River remains above danger level (D.L.) annually is from 40 to 50 days but in some years it extends even up to 70 days. Some of the flood characteristics like annual maximum and minimum number of flood waves and normal and maximum flood duration of some important tributaries are furnished in Table 8.21.

The main factors causing extensive floods in the valley are:

(i) adverse physiography of the valley;

(ii) heavy rainfall in the catchment of the tributaries joining in the valley reach;

(iii) excessive sediment load owing to frequent earthquakes and landslides and erosion of the top soil by heavy rainfall; and

(iv) inadequate waterway owing to encroachment of riverine areas.

Table 8.21 Number and duration of flood waves of some major tributaries of the Brahmaputra River at its middle reach.

| S.N. | Name of Tributary | Number of Flood Waves (Annually) | | Flood Duration - Annually (In Days) | |
|---|---|---|---|---|---|
| | | Max. | Min. | Normal | Max. |
| 1 | 2 | 3 (a) | 3 (b) | 4 (a) | 4 (b) |
| 1. | Burhi - Dihing | 5 | 2 | 19 | 41 |
| 2. | Desang | 9 | 2 | 12 | 36 |
| 3. | Dikhow | 5 | Nil | 3 | 13 |
| 4. | Jhanji | 3 | 1 | 2 | 4 |
| 5. | Dhansiri (S) | 10 | 1 | - | 55 |
| 6. | Kopili | 45 | 3 | 32 | 68 |
| 7. | Puthimari | 7 | 1 | 9 | 16 |

### 8.5.2 SPECIFIC FLOOD DISCHARGE

The specific flood discharge of the Brahmaputra River at Bahadurabad in Bangladesh is 0.149 cumecs/sq.km. and is the highest among the large rivers of the world.

### 8.5.3 GAUGE HYDROGRAPH

The gauge hydrographs of four stations, Dibrugarh, Guwahati, Goalpara and Dhubri, within the valley reach from upstream station Dibrugarh to downstream station Dhubri are shown in Figure 8.6. These are typical observed gauge hydrographs of four stations of a particular year. The hydrographs show that the gauge level started to rise gradually from the month of March but rapidly from May. From the month of June the river experienced flood. The danger level of the four stations is also shown in the figure. It clearly shows that the duration for which the flood remained above danger level (DL) gradually increased towards the downstream side.

### 8.5.4 HIGHEST FLOOD LEVEL (HFL)

Within the Brahmaputra valley reach in the state of Assam, the record shows that the observed HFL of all the stations are much higher than the D.L. of the stations. The observed HFL against D.L. of six stations, namely Dibrugarh, Neamati, Tezpur, Guwahati, Goalpara, of the Brahmaputra River are shown in Table 8.22. The return period of the observed H.F.L. is gradually reduced from 108 years for Dibrugarh to 21 years for Dhubri. The observed H.F.L. and D.L. of major tributaries of the Brahmaputra River within valley reach (middle reach) are shown in the Table 8.23.

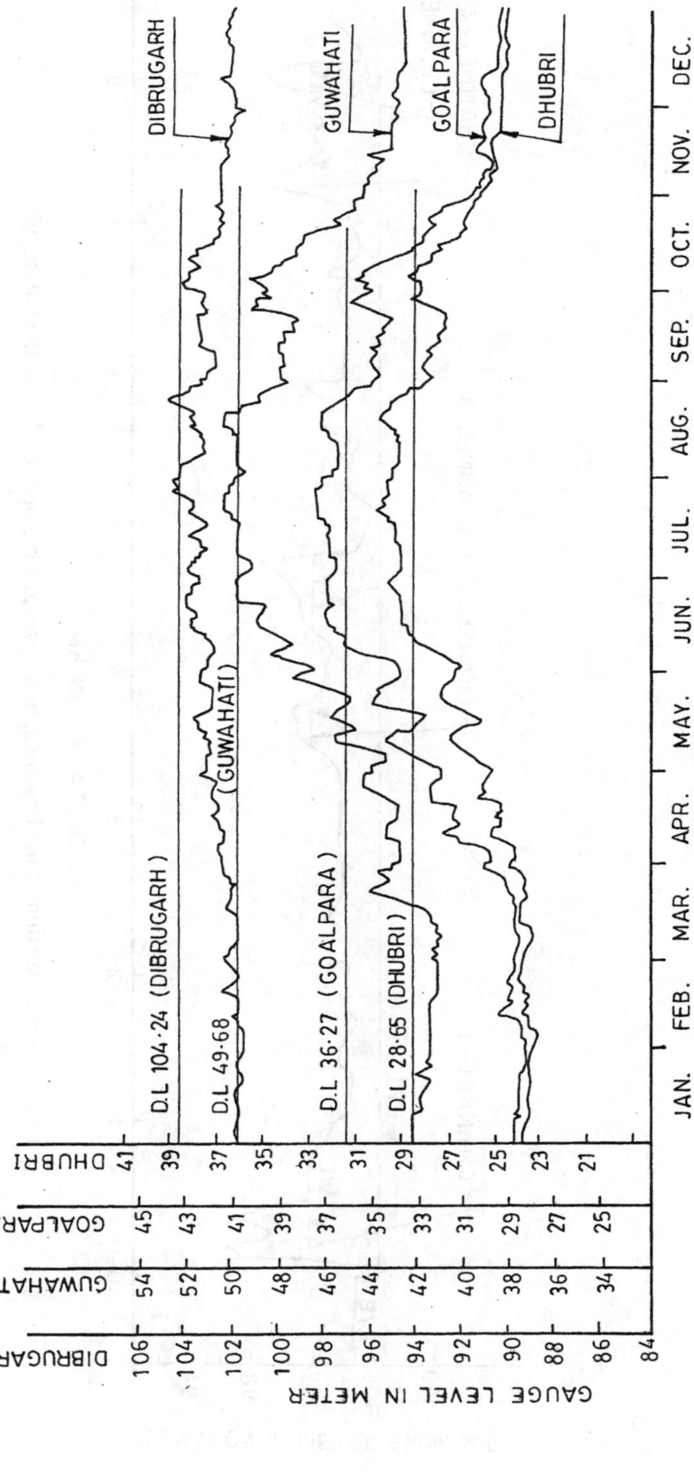


Fig. 8.6 Gage hydrographs of the Brahamaputra River at four sites within the middle reach.

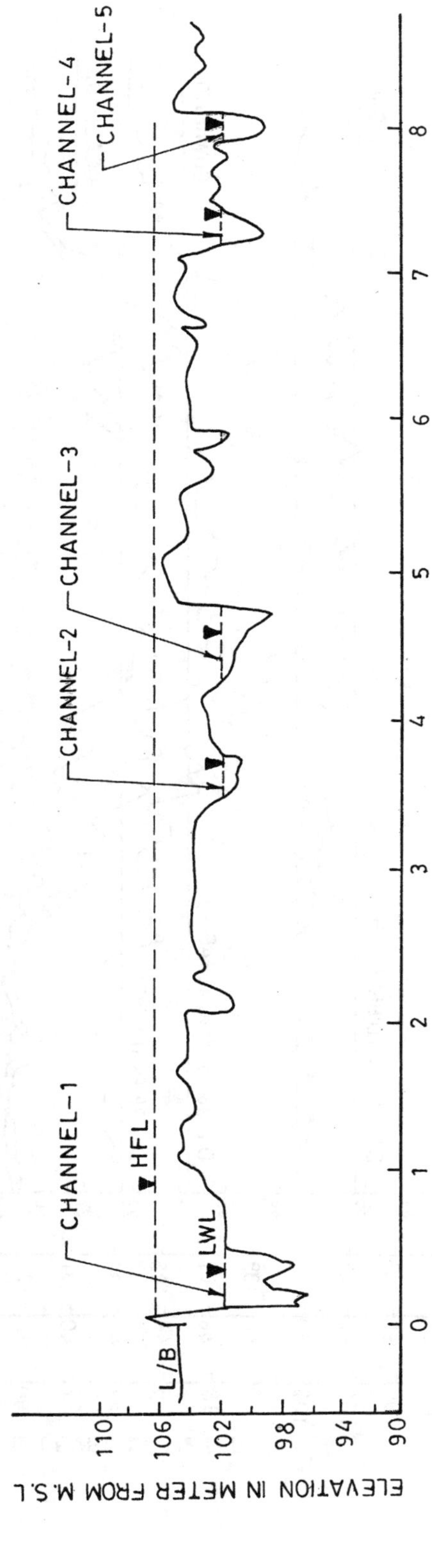


Fig. 8.7 Cross-section of the Brahamaputra River at a braided reach near the Dibrugarh.

The record shows that the observed H.F.L of the tributary rivers are all much above the D.L of the site. It clearly indicates that the tributary rivers are also experiencing high and devastating floods. There is a tendency of gradual increase of H.F.L. and L.W.L. particularly in the eastern part of the valley. The sudden rise of the bed level around Dibrugarh owing to the severe earthquake of 1950 is one of the factors which raised the water level. The rate of increase of L.W.L at Dibrugarh is the highest within the valley reach.

### 8.5.5 FLOOD LIFT

The extent of flood lift of the Brahmaputra River within the valley reach varies widely depending on the width and nature of river cross section. The flood lift of the Brahmaputra River at six stations within the valley reach (middle reach) is furnished in Table 8.24. The average flood lift at a braided reach near the Dibrugarh town is 3.65 m where the river width is 8.03 km. A typical cross section of the Brahmaputra River near Dibrugarh town is shown in Figure 8.7; the river is highly braided in this portion. The average flood lift at a nodal point near Guwahati is 9.91 m; at normal flood level, width at this nodal point is 1.20 km which is the minimum within the valley reach from Kobo to Dhubri. A typical cross section of the river is shown in the Figure 8.8. After the severe earthquake of 1950 a great variation in the flood lift has taken place almost throughout the valley except at Guwahati and Goalpara. The Flood lift of some of the tributaries of the Brahmaputra River within valley reach is furnished in the Table 8.25.

### 8.5.6 TIME LAG

On the average, a flood wave takes 5 days 3 hours to travel through the valley portion of the river from Pasighat (foothill) to Dhubri (near the Indo- Bangladesh border), a distance of 672 km. In this reach the average river gradient varies from 1:3700 near Pasighat to more than 1:8,875 near Dhubri. The average time taken by the flood wave to travel (time lag) reachwise the valley portion is furnished in Table 8.26.

Table 8.22 Observed highest flood level of the Brahmaputra River within the valley reach.

| Sl. No. | Name of Stations | Chainage from the Mouth (Km) | Danger Level (m) | H.F.L. (m) | Return Period of HFL (Yrs.) |
|---|---|---|---|---|---|
| 1 | 2 | 3 | 4 | 5 | 6 |
| 1. | Dibrugarh | 917 | 104.24 | 105.95 | 108 |
| 2. | Neamati | 809 | 85.04 | 86.84 | 32 |
| 3. | Tezpur | 672 | 65.38 | 65.66 | 30 |
| 4. | Guwahati | 542 | 49.68 | 51.04 | 25 |
| 5. | Goalpara | 432 | 36.27 | 36.88 | 23 |
| 6. | Dhubri | 357 | 28.65 | 29.97 | 21 |

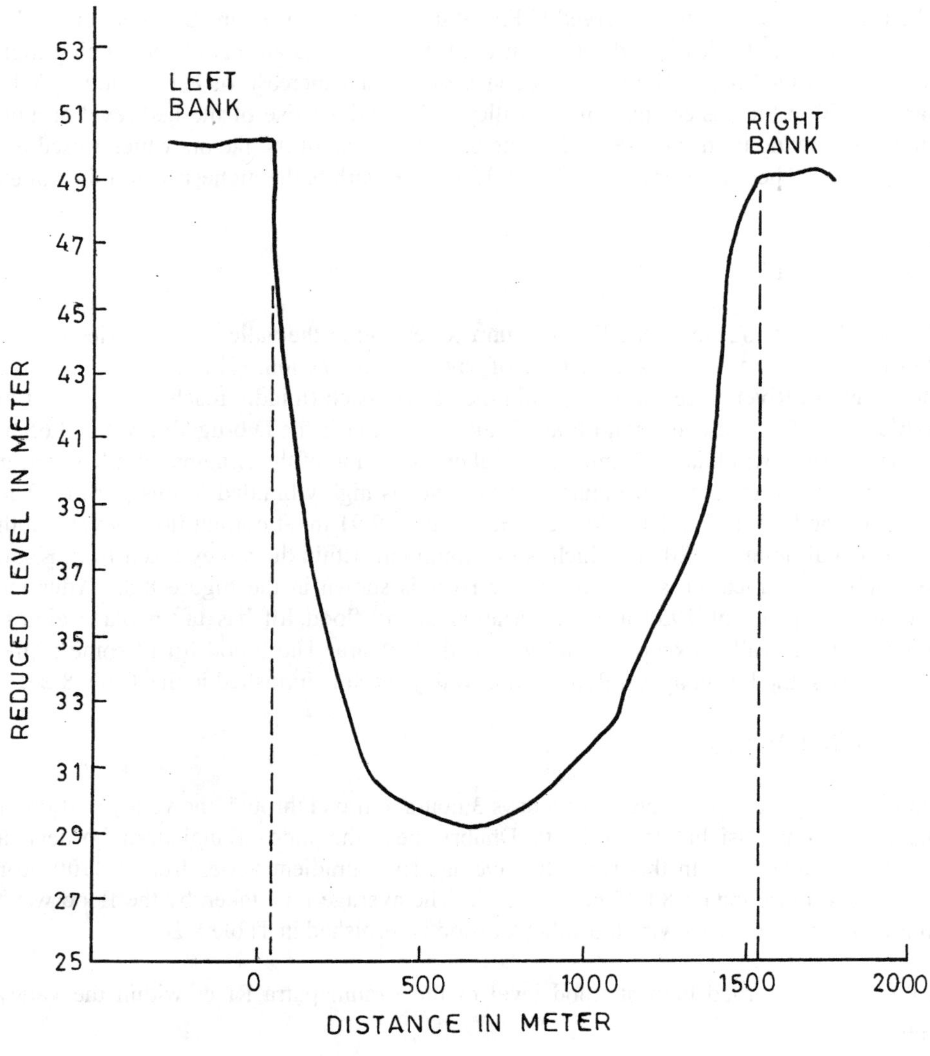


Fig. 8.8 Cross-section of the Brahamaputra River at a nodal point near Guwahati.

## 8.6 Sediment Load

The sediment yield of the Barhmaputra River is very high and its position is second amongst the large rivers of the world. The sediment yield of the river at Guwahati is 0.0755 ha.m/sq.km. of the catchment area; the maximum and minimum sediment concentration va;ues are 679 and 36 ppm, respectively. The average annual sediment load of the river at Bahadurabad in Bangladesh is 735 million metric tonne; the average maximum concentration of the river is 1,180 ppm. The percentage break up of the surface sediment load at Bahadurabad is 12.5% coarse, 14.2% medium and 73.3% fine.

Table 8.23 Observed highest flood level and danger level of some tributaries of the Brahmaputra River within the valley reach (middle reach).

| Sl. No. | Name of Tributary River | Name of Observation Site | Danger Level (m) | Observed HFL (m) | Year of Occurrence |
|---|---|---|---|---|---|
| 1 | 2 | 3 | 4 | 5 | 6 |
| 1. | Subansiri | Chowaldhowaghat | 100.43 | 101.31 | 1972 |
| 2. | Ranganadi | Pahumaraghat | - | 95.92 | 1979 |
| 3. | Jia-Bharali | N.T.Rod. Cr. | 77.00 | 80.89 | 1970 |
| 4. | Dhansiri (N) | N. T. Rd. Cr. | - | 84.14 | 1979 |
| 5. | Puthimari | N. T. Rd. Cr. | 51.82 | 54.20 | 1973 |
| 6. | Pagladiya | N. T. Rd. Cr. | 52.75 | 55.51 | 1970 |
| 7. | Manas | N. T. Rd. Cr. | 47.55 | 49.88 | 1974 |
| 8. | Sankosh | Srirampur | - | 46.71 | 1968 |
| 9. | Lohit | Dhola | 128.27 | 129.49 | - |
| 10. | Burhi-Dihing | Khowang | 102.11 | 103.51 | 1980 |
| 11. | Desang | N.H.Cr. | 92.00 | 93.88 | 1980 |
| 12. | Dikhow | A.T.Rd.Cr. | 93.30 | 95.60 | 1979 |
| 13. | Dhansiri (S) | Golaghat | 89.50 | 90.68 | 1971 |
| 14. | Kopili | Dharamtul | 56.31 | 57.85 | 1966 |

Within the valley reach in the state of Assam, the sediment yield of the north bank tributaries which originate and traverse through the Himalayan mountain ranges in their upper reach is approximately 5 to 6 times higher than that of the south bank tributaries. Over 70% of the annual average sediment load of the Brahmaputra River at Guwahati is contributed by the main river course Dihang and two tributaries Dibong and Lohit. These three rivers carry very high sediment loads.

Table 8.24 Flood lift of the Brahmaputra River within the valley reach from Kobo to Dhubri.

| S.N. | Name of Station | Chainage From The Month (km) | Width of The River at N.F.L. (km) | Flood lift | |
|---|---|---|---|---|---|
| | | | | Average (m) | Maximum(m) |
| 1 | 2 | 3 | 4 | 5 (a) | 5 (b) |
| 1. | Dibrugarh | 917 | 8.03 | 3.65 | 4.76 |
| 2. | Neamati | 809 | 7.64 | 4.53 | 6.33 |
| 3. | Tezpur | 672 | 3.60 | 5.76 | 7.16 |
| 4. | Guwahati | 542 | 1.20 | 9.91 | 10.39 |
| 5. | Goalpara | 432 | 2.01 | 7.96 | 9.33 |
| 6. | Dhubri | 357 | 8.72 | 5.63 | 6.60 |

Table 8.25 Flood lift of some tributary rivers of the Brahmaputra within valley reach.

| Sl. No. | Name of Tributary | Name of Site | Flood lift | |
|---|---|---|---|---|
| | | | Average (m) | Max.(m) |
| 1 | 2 | 3 | 4 (a) | 4 (b) |
| 1. | Burhi - Dihing | Khowang | 4.83 | 6.35 |
| 2. | Desang | N.H.Cr. | 6.21 | 6.60 |
| 3. | Dikhow | A.T.Rd. Cr. | 7.79 | 9.39 |
| 4. | Jhanji | A.T.Rd. Cr. | 4.59 | 5.39 |
| 5. | Dhansiri (S) | Golaghat | 6.70 | 10.18 |
| 6. | Kopili | Dharamtul | 8.06 | 9.87 |
| 7. | Kulsi | Kukurmara | 2.97 | 4.58 |
| 8. | Krishnai | A.T.Rd.Cr. | 4.67 | 5.15 |
| 9. | Jinari | A.T.Rd. Cr. | 3.39 | 4.49 |
| 10 | Jinjiram | Rangsai Bridge | 4.26 | 4.69 |
| 11. | Puthimari | N.T.Rd. Cr. | 3.50 | 5.31 |

Table 8.26 Reachwise travel time (time lag) of flood wave of the Brahmaputra River from Pasighat to Dhubri in the valley (middle reach) portion.

| Sl. No. | From Station (Chainage in km) | To Station (Chainage in km) | Distance Along the River (km) | Travel Time of Flood Wave (Hrs) |
|---|---|---|---|---|
| 1 | 2 | 3 | 4 | 5 |
| 1. | Pasighat (1029) | Dibrugarh (917) | 112 | 12 |
| 2. | Dibrugarh (917) | Neamati (809) | 108 | 24 |
| 3. | Neamati (809) | Tezpur (672) | 137 | 24 |
| 4. | Tezpur (672) | Guwahati (542) | 130 | 24 |
| 5. | Guwahati (542) | Goalpara (432) | 110 | 24 |
| 6. | Goapara (432) | Dhubri (357) | 75 | 15 |
| | Total | | 672 | 123 |

Among the north bank tributaries, the rivers like Subansiri, Jia-Bharali, Manas and Sankosh carry high sediment load. Among the south bank tributaries, the Burhi-Dihing River carries the highest sediment load; the other tributaries, like Kopili, Dhansiri (S) and Disang, carry significant amounts of sediment load. The average percentage breakup of the suspended sediment load of the north bank and south bank tributaries of the Brahmaputra River at its middle reach are furnished below:

| Tributaries of The Brahmaputra River At Middle Reach | Suspended Sediment Load ( in Percent ) | | |
|---|---|---|---|
| | Coarse | Medium | Fine |
| 1) North Bank Tributaries | 14.47 | 29.25 | 56.28 |
| 2) South Bank Tributaries | 8.24 | 17.00 | 74.76 |

The average annual suspended sediment load of some important tributaries of the Brahmaputra River within its middle reach in the valley is furnished in Table 8.27. The bed load of the Brahmaputra River is generally considered as 15% of the suspended sediment load.

Most of the tributaries of the Brahmaputra River form alluvial deltas just at the foothill region where the river gradient abruptly changes from a steeper one in the mountainous region to a flatter one in the plains. Most of the suspended sediment carried by the rivers from mountainous region gets deposited in the river bed. Sometimes these rivers are also found to change their course at this region when old channels get filled up with sediment. The north bank tributaries are more prone to such problems.

The main reasons of high sediment load in the Brahmaputra River are as below:

(1) Geologically the Brahmaputra River is a very young river and its present configuration took shape only during Pleistocene and recent times.

(2) The catchment area of the river falls in one of the well-known high seismic belts of the world. The basin experiences a number of earthquakes of moderate intensity every year and landslides and slips occur owing to such earthquakes resulting in high sediment load in rivers. The basin experienced several severe earthquakes in the past; as per the available record; the earthquakes of 1762, 1822, 1865, 1869, 1897 and 1950 were notable. The earthquakes of 1897 and 1950 were amongst the severest that have occurred in the world. These two major earthquakes disturbed the topography and the drainage system to a great extent, resulting in heavy sediment load and active bank erosion at different reaches. The river is still in the process of attaining stability.

(3) The river and its north bank tributaries traverse through unconsolidated sedimentary rocks of the Himalayas and high amount of sediment are washed down by the monsoon rains.

(4) High intensity rainfall erodes the top soil of the catchment area; landslides and slips also occur in many places which add high sediment load in the river.

(5) Steep slopes of river and its tributaries in the mountainous reaches bring down sediment.

(6) Forest fires also expose the soil to erosion.

(7) Gradual encroachment of forest areas.

Besides the above natural causes some human activities are also contributing to high rates of sediment load; one such human activity is 'Jhuming' of shifting cultivation which adds sediment load to a great extent. Jhuming is extensively practiced by tribal communities living in the mountainous regions of the catchment area specially in the Arunachal Pradesh and Southern Assam hill ranges and is done by burning the forest and exposing the soil for cultivation. About 1.35 million hectares of forest land remains under Jhum cultivation at one time or the other.

### 8.7 Bank Erosion

Heavy bank erosion by the Brahmaputra River at different reaches take place owing to excessive sediment load, erodible nature of bank material, formation of char islands and consequent development of side channels and varying bed width. Active bank erosion is generally observed to take place both upstream and downstream of the nodal points and also in the downstream reaches of the confluence of major tributaries. The instability of the Brahmaputra River coupled with silt and sand strata of its banks is also responsible for considerable bank erosion in its valley reach. There is a tendency of the Brahmaputra River to shift southward within the valley reach; this tendency has become more prominent after the great earthquake of 1950 which raised the whole land mass of the northeastern part of the valley, particularly north of the Brahmaputra River including the Himalayan foothills region by 3 to 4 metres. This southward thrust has initiated widespread erosion in the south bank near the Dibrugarh town which is still continuing at different reaches even after construction of different anti erosion schemes. A few kilometers downstream of the nodal point near Guwahati the river is observed to have a northern migration since 1920 and active bank erosion is observed to take place in the Nalbari and Barpeta districts of Assam. The situation is different further downstream of the next nodal point near Jogighopa where the river shows a tendency of migration towards the south. The south bank in this reach is facing active bank erosion.

Table 8.27 Average annual suspended sediment load of some important tributaries of the Brahmaputra River within its valley reach (middle reach ).

| Sl. No. | Name of Tributary River | Name of Obs. Site | Average Annual Suspended Sediment Load in Ha.m. |
|---|---|---|---|
| 1 | 2 | 3 | 4 |
| 1. | Dihang | Ranaghat | 7,694 |
| 2. | Dibong | Jiagaon | 2,620 |
| 3. | Luhit | Digarughat | 4,650 |
| 4. | Subansiri | Bhimpuraghat | 2,497 |
| 5. | Jia-Bharali | N.T.Rd. Cr. | 2,143 |
| 6. | Manas | Mathanguri | 2,038 |
| 7. | Sankosh | Srirampur | 594 |
| 8. | Burhi-dining | Khowang | 594 |
| 9. | Disang | N.H.Cr. | 70 |
| 10. | Dhansiri (S) | Numuligarh | 422 |
| 11. | Kopili | Dharamtul | 222 |
| 12. | Puthimari | N.T. Rd. Cr. | 371 |

In different reaches the soils forming the Brahmaputra bank are such that they cannot resist erosion. Erosion is more prominent during recession stages of the flood waters. Owing to the variation of the river width along its course, the velocity of flow also changes and as a result scouring and siltation also take place. This phenomenon also disturbs the channel configuration. In the last few years, owing to active bank erosion at some affected reaches, the river width has increased from 8 km to 10 km and in some other reaches from 15 km to more than 18 km.

### 8.8 Concluding Remarks

The Brahmaputra River is one of the largest rivers in the world. It is unique in the directions of its flow path. Its hydrology is also unique in that its basin enjoys a variety of seasons and climate ranging to very cold to tropical. It can serve as one of the most productive natural laboratories for hydrologic investigations. Being an international river, its study becomes even more challenging. In this millennium it is hoped that India, China, and Bangladesh will strive to foster a new spirit of cooperation for furthering scientific knowledge of the Brahmaputra basin and in turn establishing a sound basis for navigational, energy, and water resources development. This is the minimum the people of the basin deserve.

# CHAPTER 9

# CHANNEL PROCESSES

A. B. PALANIAPPAN

The Brahmaputra River has evolved in about two million years and rests on fractured pavements. Aggradation and degradation occur in alternate reaches of the river. The eroded sediment contributes to building the bars and islands within the river. This chapter presents a discussion of different channel processes that shape this huge river.

## 9.1 Introduction

The Brahmaputra River is one of the world's largest rivers. Hindus regard this river to be the son of Lord Brahma. It is ranked fourth in respect of the average discharge. The annual sediment load of 500 metric tons, that the river carries, ranks it as the second largest sediment laden river. The flow depth of the river varies from 4 m to 40 m. The difference in water levels between bankful to low flows in this river is of about 7 m. The width of the river varies from 2 km at a narrow section to 20 km. Some of the bed dunes of the river are of 15 m height (Coleman, 1969).

This Himalayan river, which was known as Sri Lauhitya in ancient times, rises in the Kailas range at an elevation of 5,150 m and drains an area of 580,000 km$^2$ out of which 187,110 km$^2$ lies in India. After flowing a distance of 1,700 km in the Himalayan range where it is known as the Tsongpo, the river enters India and is known as the Dihang. The Tsongpo flows through a mountainous course and has a steep slope which ranges from 4.3 to 16.8 m/km. In the Pasighat reach from which the Assam valley starts, the river slope is 0.27 m/km. Two rivers, the Dibang and Lohit, join the Dihang near Kobo about 52 km downstream of Pasighat. The river carrying the combined flow is known as the Brahmaputra. The world's largest single potential source of hydropower exits at the 'Great Bend' where the river drops 3,000m in about 500m. Numerous tributaries join the river from both north and south. A list of principle tributaries is given in the Table 9.1. The north bank tributaries differ from the south bank tributaries in many respects as given at the bottom of the table. The slope of the reach at Dibrugarh is 0.09 -- 0.17 m/km that reduces to 0.1 m/km within the Indian Territory. The high Himalayas in the north and the Shillong Plateau in the south form the Brahmaputra River valley in Assam. The valley is of 720 km long and about 70 to 80 km wide. The river width varies considerably.



*V.P. Singh et al. (eds.), The Brahmaputra Basin Water Resources*, 196-213.

The Brahmaputra alluvium has evolved in two million years and rests on fractured pavements. These blocks were subjected to differential vertical movement (Rao, 1975). This valley consists of 200 to 300 m of alluvium mainly of clay, silt, sand and pebbles (GSI, 1974). There is a disproportionate increase of sediment yield in respect of the drainage area in this region. The sediment yield increases by about 240% in between the sites Tsela D'zong and Ranaghat, India whereas the drainage area increases only by 28% in its drainage area. Considering the suspended load for the period 1971 --1979, Gosamy (1985) has concluded that the average denudation of the four Himalayan catchments (the Subansri, the Jia Bharali, the Manas and the Dihang) is 0.73 mm/yr and expected that the long time average would be twice this. He has pointed out that the rate of denudation is about 1 cm/yr (Menard, 1961) and the data of Khosla also indicated a value of 0.975 mm/yr. This fragile nature of slopes, monsoon downpours (500 cm/yr), and tectonic movements present a unique combination of natural events of importance to the river environment in the region.

Table 9.1 North and south bank tributaries of the Brahmaputra in India.

| Sl.no. | North Bank | Length (km) | South Bank | Length (km) |
|---|---|---|---|---|
| 1 | The Simen | 580 | The Dibru | 592 |
| 2 | The Jiya Dhol | 540 | The Burhi Dihing | 540 |
| 3 | The Subensri | 430 | The Disang | 515 |
| 4 | The Burai | 392 | The Dikhu | 505 |
| 5 | The Bargang | 382 | The Jhanzi | 495 |
| 6 | The Jia Bhareli | 338 | The Dhansiri(S) | 420 |
| 7 | The Gabbaru | 300 | The Kopili | 220 |
| 8 | The Belsiri | 280 | The Kulsi | 140 |
| 9 | The Dhansiri (N) | 270 | The Deosila | 130 |
| 10 | The Noa Nadi | 230 | The Dudhnai | 108 |
| 11 | The Nanai Nadi | 215 | The Krishnai | 107 |
| 12 | The Bar Nadi | 205 | The Jinari | 100 |
| 13 | The Puthimari | 172 | The Jiniram | 0 |
| 14 | The Pagladiya | 170 | | |
| 15 | The Beki | 115 | | |
| 16 | The Manas | 85 | | |
| 17 | The Champamati | 63 | | |
| 18 | The Gaurang | 43 | | |
| 19 | The Tipkai | 40 | | |
| 20 | The sankosh | 0 | | |

This region experiences floods owing to depressions in the Bay of Bengal in the period June to September. The monsoon rainfall is just enough to cause floods in this region. The infiltration capacity of the soil is very low and hence the major part of the rain runs into the rivers. Floods are so frequent that people have developed a tolerance to inundation in this region. The tributaries of the Brahmaputra remain at a high stage during floods in the main river. The drainage congestion exists throughout the river during this period of the year. The flood building process of the river is shown in Figure 9.1. There is a steep rise in the flood level because of the contribution from rivers Lohit and Dibang at about 55 km.

Table 9.2 Characteristic differences of tributaries.

| SN. | North bank tributaries | South bank tributaries |
|---|---|---|
| 1 | They have steep slope | Comparatively flatter slopes |
| 2 | Shallow and braided | Deep and meandering |
| 3 | Gravel bed | Sand bed |
| 4 | Heavy silt load | comparatively low silt load |
| 5 | Flood are often flashy | Very few |
| 6 | Angle of confluence differs | Flow parallel to the main river near the confluence |

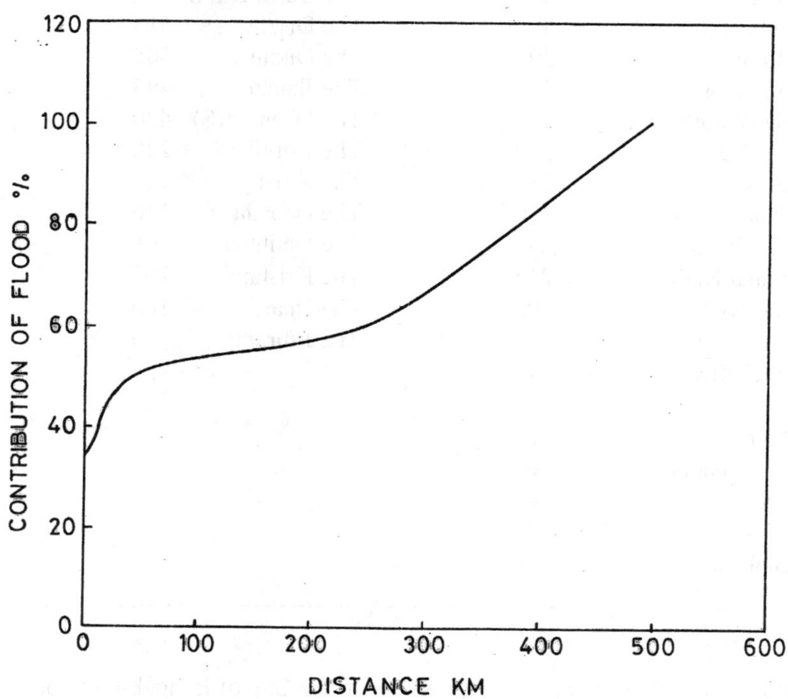


Fig. 9.1 Upstream building-up of flood in the Brhamaputra River at Pandu.

The average discharge of the river at Pandu (1955--1982) is about 15,675 $m^3/s$. The discharge per unit area of the drainage of the Brahmaputra at Pandu is 0.0306 $m^3/s$ and it is 0.0858 $m^3/s$ for the Jia Bharali River a north bank tributary of Brahmaputra. The river is used for navigation in Assam. Erosion processes cause a transverse movement of the river. Dibrugarh town, which was 6.5 km upstream of the confluence of the stream Dibru with the Brahmaputra as per the old maps, is now 13 km downstream of the confluence. Disi, a south bank tributary that flowed parallel to the main river has been engulfed by the main river and lost its 32 km of tail reach. Similar encroachments of the Brahmaputra can be seen in the case of Dhansri, and the Kulsi, indicating the erosional activity of the river. Figure 9.2 shows the variation of annual discharge with respect to drainage area in this region.

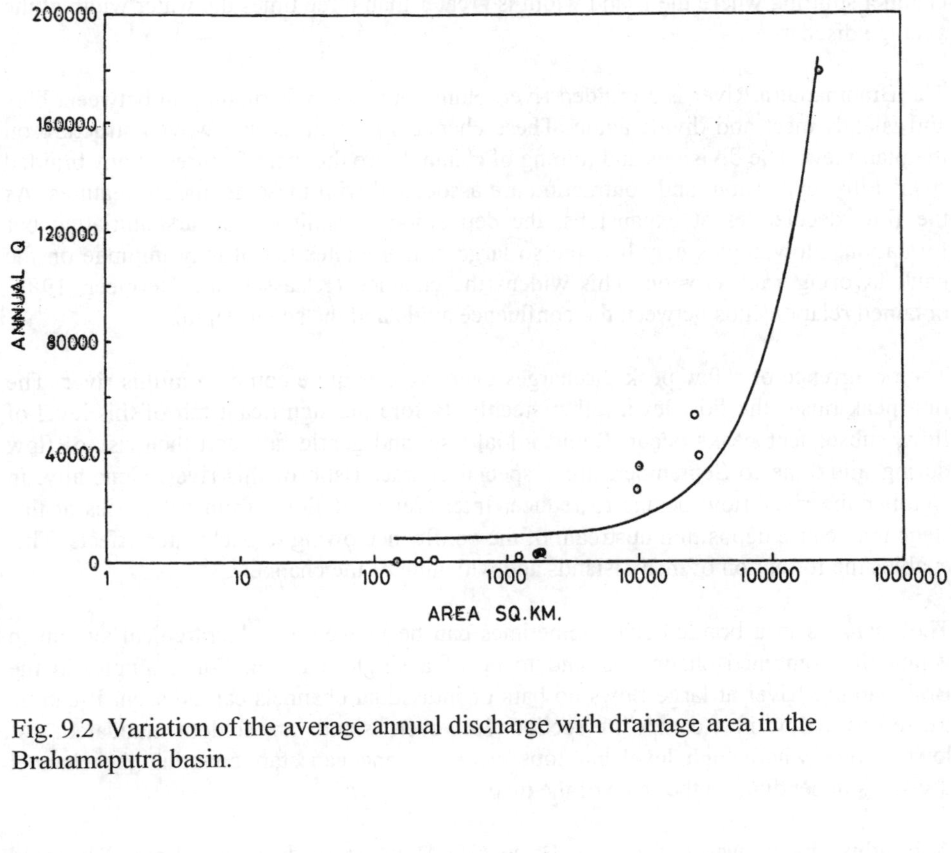


Fig. 9.2 Variation of the average annual discharge with drainage area in the Brahamaputra basin.

## 9.2 Braided River

Schumm (1977) defined braided channels as single channel bed load rivers, which at low water have islands of sediment, or relatively permanent vegetated islands, in contrast to multi-channel rivers in which each branch may have its individual pattern. A braided channel generally occurs on a relatively steep river with a heavy sediment load. Braiding starts when a submerged bar is formed at the center of a river. The initial cause of this development is a subject of conjecture at present. Some believe that it is owing to eddy processes. Once initiated, the bars can grow vertically and reach a height level with the flood plains. The cross-sectional area of the flow is affected by such a growth of a bar and the flow tends to erode either one or both banks of the river. The process may be repeated and thus multiple channel may be formed. The channel configuration is under constant change in a braided river.

The channel splitting is in some cases referred to as anastomosing or anabranching channel segments. These segments are longer than a curved channel segment around a single braid or point bar. The flow pattern in these cases is independent of adjacent channels, in contrast to the braided. The land areas between the anastomosed channel segments need not to be bars. An anabranching channel can be arbitrarily defined as channel splitting where the island width is greater than three times the water width at the average discharge.

The Brahmaputra River is a braided river. Numerous channels running in between, bars and islands meet and divide again. These channels present an interwoven structure on the plan view. The divisions and joining of channels are the main features of any braided river. Flow expansion, and contraction are associated with these geometric features. As the flow decelerates at expansions, the deposition of sediment occurs, initiating bar formation. Flow angles may become so large at many sites that it may impinge on the bank favoring bank erosion. This widens the channel. (Klaassen and Vermeer, 1988) obtained relationships between the confluence angle and the scour depth.

The occurrence of a few peak discharges every year is quite common in this river. The first peak raises the flow level rather steeply. Before the significant fall of this level of flow, subsequent peaks occur. Rapid initial rise, and gentle fall, and then rise of flow during mid June to September, are a special characteristic of this river. Generally, in October the river flow begins to reduce. Interference of flows from tributaries at this stage may cause deposition upstream of the confluence owing to backwater effects. This enables the formation of many islands and widening of the channel.

The channels in a braided river sometimes can be viewed as a hierarchical system in which the combined channel is said to be of a single channel. For example, in the Brahmaputra River at large flows no bars or individual channels can be seen. It can be treated as a flow in a composite section. However, islands are seldom inundated. At lower flows where high level bar tops are seen, one can start counting the channel divisions depending on the scale of the map.

A braiding index was defined by Brice (1964) while studying the Loup Rivers in Nebraska, USA as given below:

$$B_r = 2L_{IB} / L_R \qquad (9.1)$$

where $L_{IB}$ is the length of islands and bars in a reach and $L_R$ is the length of the reach along the centerline. Braiding indices of the river near Dibrugarh is 5.3 and it is 6.7 near Balas Bari (Goswami, 1985). Braiding index, as defined by (Brice, 1964) is worked out by the author for this river upstream of KamalaBari, a township in the Majuli Island, to be 6.95.

## 9.3 Bars

The American Society of Civil Engineers 'Task Force on bed forms (1966) defined '*bed forms having lengths of the same order as channel width or greater, and height comparable to the mean depth of the generating flows*' as bars. Bars are generally diamond-shaped in the plan view. Their longer axes are oriented parallel to the main flow direction. Since the flow direction changes upruptly within the braided width of the river the major axes of the bars can also be seen. Different shapes of bars in the Brahmaputra River are shown in Figure 9.3. The slope of the upstream end of a bar is steep. A scour pool can be spotted upstream of a bar. Bank erosion and consequent widening of the channel promote the growth of bars. The bed slope and the water surface slope will constantly change in these rivers. Brice (1964) defined bars as being unvegetated and emergent at the bankful stage. The degree of development of vegetation is related to the time that the bar surface is exposed out of water.

Channel splitting around bars (island) is generally recognised as braiding. However a distinction is necessary to differentiate it from *anabranching (anastomosing)*. One of the characteristic features of anastomosing (anabranching) channel segments is that they are larger than a channel that moves around a single braid or point bar. These segments exhibit independent flow pattern and behave independent of adjacent segments around bars. Many investigators have defined as a measure of the degree of braiding. The mean number of active channels or braid bars per transect across the channel belt is taken as a measure of the degree of braiding by some investigators. Some others defined it as the ratio of the sum of channel lengths in a reach to a measure of reach length --- referred as the total sinuosity. Many sedimentologists suggest that the river that transports a large proportion of the bed load relative to the suspended load tends to have a relatively low sinuosity and a high braiding index. Such streams are associated with relatively easily eroded banks of sand or gravel, and large channel slope. Vegetation helps in stabilization and may hinder lateral migration of the channel.

## 9.4 Islands

Islands are formed by the amalgamation of a cluster of braid bars and are relatively stable. They are generally diamond shaped in plan view and their long axes are in the direction of the flow. The height of the islands can not be greater than the maximum flood level. The surface of the island stabilises when a coat of silt covers (Thorne et al. 1993). The plant growth under favourable conditions would be rapid and islands would have matured thick vegetation. Islands do not migrate appreciably. Erosion of an island occurs when a high flow dissects at low elevations or when the banks are eroded. The dissections of the island in Brahmaputra River are shown in Figure 9.3.

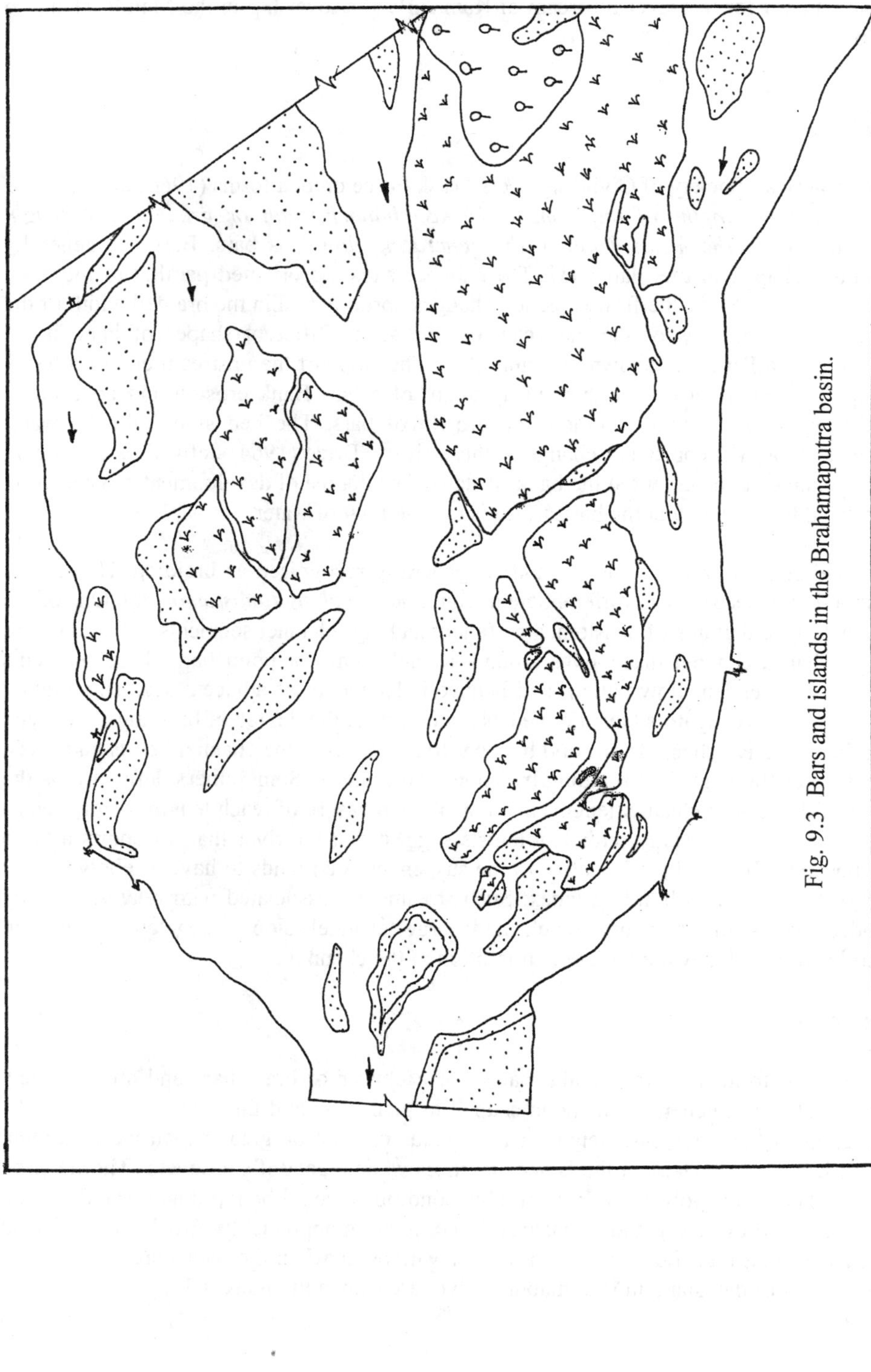

Fig. 9.3 Bars and islands in the Brahamaputra basin.

Cross-sections of most of the islands in this river are flat, but slightly saucer shaped top surface (Coleman, 1969; Bristow, 1987; Thorne et al., 1993). The maximum elevation change was 2 m in some islands, and agricultural activities and permanent settlements can also be seen. Some bars are also attached to these islands which are quite different in appearance to the island as can be seen in Figure 9.3.

### 9.5 Dunes

The echosounder profiles that were taken by the Inland Waterways Authority for navigational purposes indicate the presence of dunes on the riverbed (Thorne et al., 1993). Their studies indicate that dunes can migrate at speeds of up to 100 m/day. Dune movements cause the bed feature to undergo constant change. Dune movements in these rivers do not really contribute a change in the plan form of the river. However, they effect the flow velocity and sometime a change in the flow direction rapidly changing the banklines and islands.

Coleman (1969) suggested that the braided rivers can be divided into island reaches and nodes as the basis of the width and braiding intensity variation. Island reaches are much longer than the node reaches. However, considerable erosion and accretion are noticed in these reaches. Thorne et al. (1993) point out that the bank erosions in the island reaches are more than nodal reaches. Church and Jones (1982) observed that the island reaches are sediment storage zones and the nodal reaches sediment transmission zones. Yalin (1977) has brought out the concept that the length of storage zones (bars, etc.) is a simple function of the width of the channel as

$$L = a\,w \tag{9.2}$$

where L is the length of the feature, a is 3.142 and w is width of the channel. Working out the morphological feature of the Brahmaputra River, assuming the width of around 10km, the length of the bed form can be expected to be 30 km long. A typical cross-section of the Brahmaputra River will be seen to consist of an island or a large bar with a major anabranch on each side. Each one of the anabranches is divided by bar, which produces sub-channels. The morphological feature which divides the main channel will be larger than the feature that divides a sub channel owing to its width. A first order channel will have large features measuring tens of kilometers. Classifying the features pertaining to the order of the channel will be useful. It can be seen that the major left and right channels of the Brahmaputra will have braid bars of about 5 km width and. Islands of 30 to 60 km long. There is also a possibility of bars attaching the island and different features of elevation may appear in the river.

### 9.6 Bank Erosion

The bank of the Brahmaputra River is non-clayey. The composition varies in the proportion of silt and fine sand. The banks can be seen to have two distinct strata. The top one is fine grained and the bottom is coarser. The banks of the Brahmaputra River are made up of silt, sandy material. The clay portion is less than 5% .The sediment $D_{50}$ is of size 0.04 mm. Bank erosions are mainly owing to toe erosion and beam (overhanging) failure. Coleman (1969) observed an average of five slumps of bank per mile distance in this river. The bank line retreats under the combination of a favourable fluvial process clearing of the basal material derived out of the slips or other lump failures and the critical soil moisture of the bank. Throne et al. (1993) assessed that 40% of the eroded bank material in the Brahmaputra River goes as wash load and the rest goes into the bed material load. The river swings very often within these bank lines.

The bankline is also extremely unstable in many reaches of the river. A large scale slumping of banks is observed during the falling stage. It seems that soil moisture of the bank has a strong relation to its failure in addition to fluvial processes. Undercutting of banks and the development of overhanging cantilever lead to bank failure. The applied shear on the bank sediments or on natural levee deposits is quite large at higher stages. This can entrap voluminous material and cause fluvial failure of the bank. At this stage water is forced into the bank increasing the pore pressure. When the water level in the river falls quickly compared to the bank water condition, the water from the bank moves from the pores. The lateral flow takes away possible sediments into the channel. Caves appear on the banks. The overlaying bank portion fractures into number of blocks. These blocks eventually fail by tilt. The highly saturated bank will often liquefy and flow towards the channel.

Unlike the case of meandering rivers where one bank faces erosion and the other deposition, both banks are subjected to erosion in the case of braided rivers. It is only a matter of guesswork about the entrapped bank material. Field observations have established the occurrence of alternate river reaches of aggradation and degradation in the Brahmaputra River. Hence it may be expected that the growth of the bars and islands are owing to the erosion of the bank material locally or just upstream. An average of three slumps (shear failures) of the bank of the river Brahmaputra in Bangaladesh was found from aerial photographs taken in 1963 by Coleman(1969). The sizes of these slumps were found to vary from 20m to 300m.

The bank erosion in this river caused a number of retirement embankments to be constructed. Throne et al. (1993) hypothesized that it is the growth and evolution of the active sand bars that drives bank erosion. As erosion progresses, embayments develop on each side of the bar, providing space for consolidation and further growth of the original bar. This allows the formation of smaller, second generation bars, introducing curvature to the flow. The curvatures promote high velocities near bank regions producing deep scour. Scour depths of the order of 40 m were also seen (Bristow 1993) in this river. Flow erodes into the bar top exposing natural cut bank section up to a depth of 4 m. Bristow calculated the rate of deposition to be 19 cm per hour to 1 m per hour in this river, using a bulk density of 1.9 g/cm$^3$. The structure of these cut sections reveals the deposition pattern and plausible flow processes.

The suspended sediment lag distance in this river was found to be 20 m according to the theory given by Fredsoe (1981). Coleman (1969) observed bed forms of height 15 m in this river. The bed material in the Brahmaputra is in the range of fine sand-silt range.

The sediment size distribution varies alternately between finer to coarser gradation in the river reach between Dibrugarh and Golpara. The median size of the sediment, $D_{50}$, at Pandu and at Dibrugarh is 0.03 mm; at Golpara it is 0.07 mm. Goswami (1985) observed that the peak of the sediment load precedes the discharge peak by one to several days. He expected that the bank material might have contributed 19% of the sediment transport in 1971--79. In his calculation, he has assumed an average (spatial) recession of 100 m and an average bank height of 3 m.

## 9.7 Aggradation and Degradation

The left bank town, Dibrugarh, of the river is under constant attack by the river almost every year. The first known erosion was in 1880 when the river Dibru washed away the Strand road. In 1891 the bank of width 120m was eaten away by Dibru. In 1913 the Brahmaputra River engulfed a long stretch of the Dibru River and eroded the bazaar area of the town. Considerable aggradation took place in this reach after a series of earthquakes. The low water level on the river at this reach rose up by 3 m. Aggradation of the river bed caused much higher flood levels and inundated larger area than in the past. The highest flood level was 105.95 m that occurred in August, 1977.

The Brahmaputra River Training Studies carried out in 1990 by Bangladesh on bed levels indicated that the river is neither aggrading nor degrading over the past 40 years. (Halcrow, 1991). Combining the discharge frequency and the transport rate, it was found that the total sediment transport against discharge is bimodal. It is not necessary that only great floods transport major sediments. The above bimodal distribution indicated that in bank flows are responsible for large quantities of sediment movement. The suspended load (1971--1979) analysis carried out by Goswamy, showed the occurrence of aggradation in Bessamara-Bhurbandha and the Pandu-Jogighopa reaches, and the occurrence of degradation in Ranaghat-Bassamara and Bhurbandha-Pandu reaches (Figure 9.4).

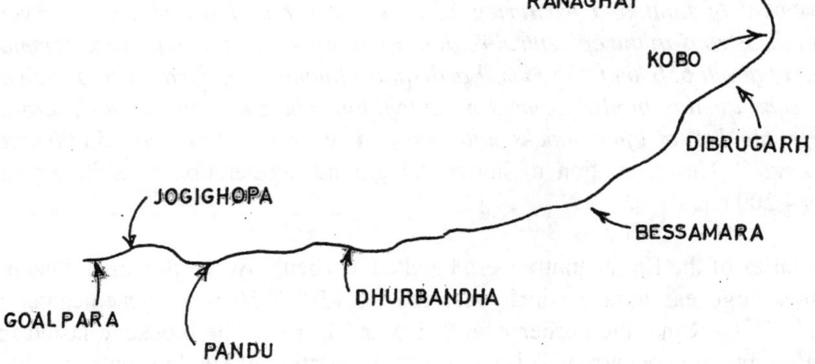


Fig. 9.4 Brahamaputra river showing relative locations of different sites.

## 9.8 Effect of Seismicity

The tectonic activities have a great bearing on the morphology of the Brahmaputra River. The valley is the result of uplift and subsidence of different blocks of geological formation. Tectonically it is a complex region. Major litho tectonic units of this regions are:

1) Shillong massif and its bounding faults;
2) Haflong - Disang- Naga thrust and other thrusts;
3) Arakan - Yoma fold belt along India-Burma plate boundary;
4) Mishmi and Lohit thrust;
5) Arunachal Himalayas north easterly orograhic trend; and
6) Brahmaputra valley.

Debris avalanches of considerable magnitude are common along the foothills of Arunachal Pradesh. The rapid toe erosion and consequent landslide are very common in Bhareli River in the Subansri area. The region had experienced a number of earthquakes of larger magnitude. The significant effect of the earthquake can be seen from the following description of the severe earthquake that occurred in the region.

One of the greatest earthquakes with epicenter 26° N and 91° E had occurred on 12th June 1897. Mr. R.D. Oldham of the Geological Survey of India gave a detailed account of it. In his own words: *"At about quarter past five in the afternoon of the 12 June 1897, there burst on the western portion of Assam an earth quake which in violence and extent has not been surpassed by any of which we have historic record. Lasting for about two and a half minutes, it had not ceased at Shillong before an area of 150, 000 square miles had been laid in ruins, all means of communication interrupted, the hill rent and cast down in land slips, and the plains fissured and riddled with vents, from which sand and water poured out in most astounding quantities; and ten minutes had not elapsed from the time Shillong was laid in ruins before about one and three quarter millions of square miles had felt a shock which was everywhere recognised as quite out of the common...... The vertical acceleration was so great that boulders were thrown upward, leaving cavity in the earth in which they had lain with slides almost unbroken. Visible movement of fault and fracturing of rocks occurred. Land slides were very common and disastrous in the epicentral region. Fissures in the alluvium were common over an area of about 650 km to 550 km. Earth quake fountains ejected sand to such an extent as to subsequently hinder farmers in cultivation. The earth quake was followed by very large number of after shocks upto the end of 1898. More than 5,000 after shocks occurred."* The estimation of horizontal ground acceleration at Shillong and Golpara was 4,200 cm/s$^2$.

Several tributaries of the Brahmaputra were blocked, particularly, the Subansri, Dihang. during another large earthquake which occurred on 15-8-1950 with the epicenter at (28.5° N and 97° E) Rima, the border with China and Burma. The blockage lasted for four days before bursting occurred. This caused the inundation of 800 km$^2$ with a sudden flood. Fissures and sand vents were extensive in the region. The river was silted up by 2.5 to 3 m at Kobo. Excessive sediment poured into the river for years after the earthquake. Although studies on the low water level could not point out the earthquake as a single cause of the rise in low water level, they strongly indicate its significant role.

### 9.9 Majuli Island

Majuli is the largest River Island in the world. It is situated in the Brahmaputra River where the tributary Subansri meets the river. The island is about 80 km long (eastwest direction) and about 10 km wide (northsouth direction). The Brahmaputra River and the Subansri River bound the island on three sides (Figure 9.5) and the Kherkatia Suti, a tributary of Subansri on the other side. It lies between latitudes 26° 52' N & 27° 10' N and longitudes 93° 30' E & 93° 35' E. Old maps indicate that Majuli land area was part of the main land in between the two rivers, the Dihing and the Lohit which were flowing almost parallel. The island would have been formed owing to the confluence of the Dihing River that was flowing parallel with the other Lohit River flowing in the southwesterly direction, as shown in Figure 9.5.

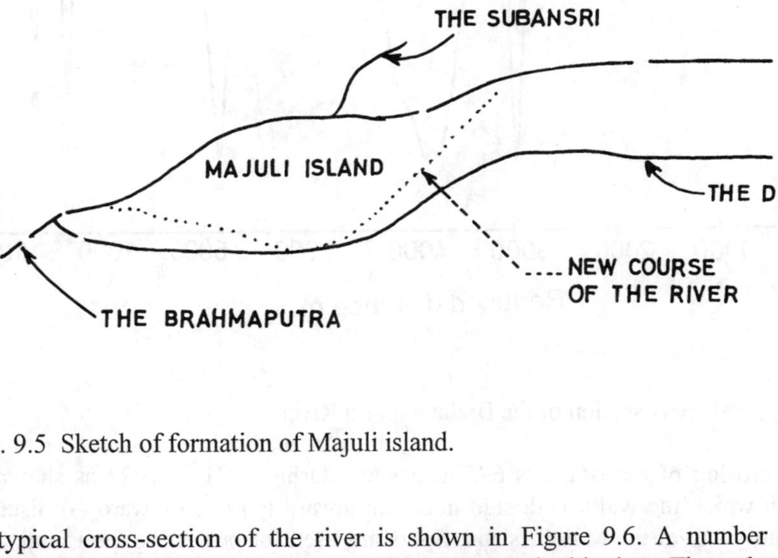


Fig. 9.5 Sketch of formation of Majuli island.

A typical cross-section of the river is shown in Figure 9.6. A number of channels carrying the flow can be seen in most of the transects in this river. The order of the depth of flows and the widths involved can be seen in such sections. The flow depth in the month of December would generally be at 12 m. At this stage six channels are visible at this location.

The area of the island in the year 1950 was 1,246 km$^2$ and in the year 1971 it was 924.6 km$^2$. A further reduction of about 45 km$^2$ of area of the island occurred up to 1993. The areal reduction indicates the extent of bank erosion in the vicinity of the island. The population of the island as per 1991 census is 135,378. Paddy is extensively cultivated in the island. Other principle crops grown are cereals, jute, sugarcane, etc.. There is about 360 km of road network. The Tuni River drains the island. Thirteen small islands joined to form the present island. The river has a deep gradient of 1 in 5595 in this reach. Flood inundation and bank erosion are two major problems. The land is submerged by 2--4 m of water for 10 to 15 days during flood. Erosion removes a land strip of 30 to 40 m size within a few hours.

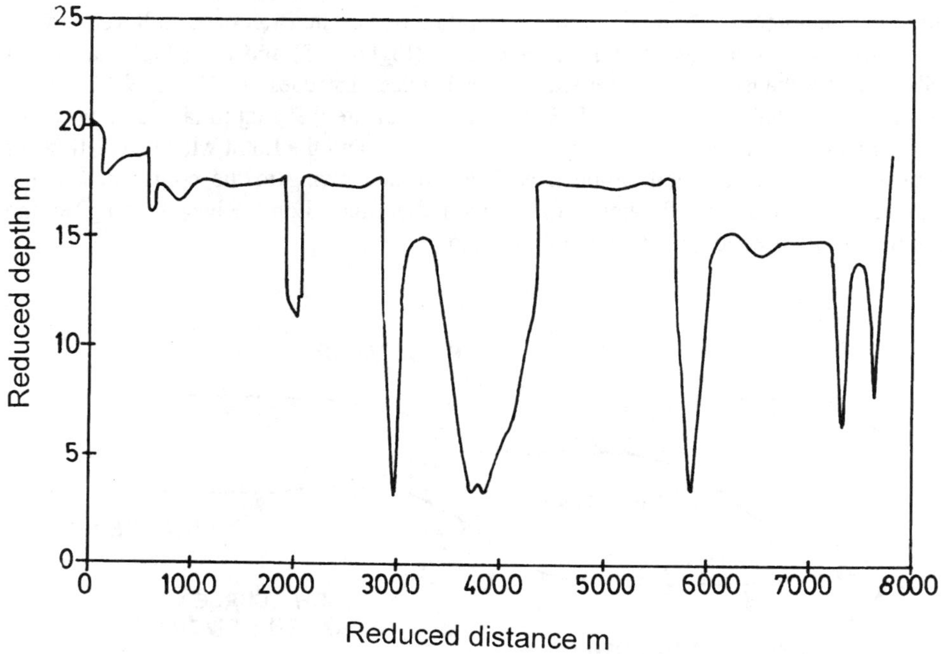


Fig. 9.6 A typical cross-section of the Brahamaputra River.

An average erosion of the island is 675 m per km during 1971 to 1982 as shown in Figure 9.7 in which the width with sign indicates inward (+) and outward (-) distance from the south bank of the Majuli Island. Spurs are used for protecting the island. The problem of bank line movements needs consideration of the fluvial processes of erosion, soil moisture management in flood plains and geotechnical processes leading to bank failures.

## 9.10 Mathematical Model

An understanding of the behavior of an alluvial river to man made changes is one of the challenging tasks of engineers. Although there is considerable research in establishing empirical relationships on various components like development of bed forms, sediment transport, etc., the mathematical modeling of the processes has not improved satisfactorily. The following presents a general structure of the model generally used. A few problems that need to be addressed are also given.

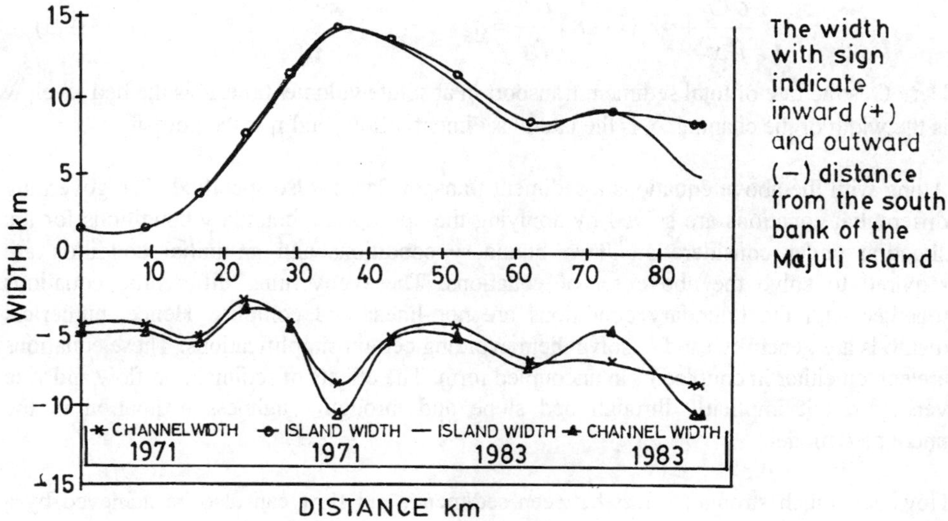


Fig. 9.7 Erosion in the vicinity of Majuli island.

### 9.10.1 BASIC EQUATIONS

Physical processes known to govern the flow are translated into mathematical or empirical equations for the purpose of modelling. The governing differential equations for a one-dimensional model are given below:

a) Continuity Equation:

$$\frac{\partial A}{\partial t}+\frac{\partial Q}{\partial x}=q \tag{9.3}$$

Here A is area of cross-section of the flow, Q is the water discharge, q is the lateral flow, and x, t are the space and time coordinates, respectively.

b) Momentum Equation:

$$\frac{\partial Q}{\partial t}+\frac{Q}{A}\frac{\partial Q}{\partial x}+Ag\,\frac{\partial d}{\partial x}+S_f-S_0=0 \tag{9.4}$$

where g is the acceleration owing to gravity, d is the depth of flow, $S_f$ is the energy slope, generally evaluated using uniform flow equations and $S_b$ is the bed slope.

c) Continuity Equation for sediment:

$$\frac{\partial G}{\partial x} + W(1-\eta)\frac{\partial z}{\partial t} = q_s \tag{9.5}$$

Here G is the rate of total sediment transport in absolute volume/ time, Z is the bed level, w is the width of the channel, $q_s$ is the lateral sediment inflow, and $\eta$ is the porosity.

Along with the above equations a sediment transport law is also specified. The governing differential equations are solved by applying the appropriate boundary conditions for the situation under consideration. Two boundary conditions and an initial condition are required to solve the above set of equations. These governing differential equations together with the boundary conditions are non-linear and complex. Hence, numerical methods are generally used to solve them avoiding certain simplifications. These equations are solved either in coupled or in uncoupled form. The effects of sediment on flow and vice versa are felt implicitly through bed slope and through roughness estimation in the uncoupled model.

However, much stronger effect between sediments and flow can also be achieved by a coupled formulation. A quasi steady flow condition where the cross-section is assumed to be rigid in computing the flow condition during a time step and then adjusting the cross-section is common in many models. The finite difference approach is a very common solution procedure. The usage of Finite Element method can also be seen in some of these models. The HEC-6, IALLUVIAL, MOBED, HEC2SR, CARICHAR are some of the commonly known models. These models differ mostly in the transport law used and in formulations of some of the processes, such as armouring. Unsatisfactory or inadequate formulations of some of the well-known models used in U.S.A., in respect of energy slope, sediment transport and armouring were discussed in a report of the National Research Council (1983). Some improvements in modelling armouring processes can be seen in Palaniappan (1991).

There are only a very few studies considering simulations for a long time and application of a model to a natural stream. Belleudy (1992) reported that SEDICOUP, an implicit coupled model, has been applied to a 16 year study of the Danube River. Loy et al. (1992) have also applied the same model to an erosion problem of Isar River downstream of a planned barrage at Pielweichs simulated fifty years of erosion.

The bed armouring aspect of degradation is considered in HEC-6, HEC2SR, IALLUVIAL and in SEDICOUP models. A layer of bed material of thickness of one or two grains is considered by HEC-2 and HEC2SR, respectively, and involves the armouring process for the case of a flat bed. Armouring of streambed with bed forms is considered in the IALLUVIAL model. But this formulation includes a number of correction factors.

The mathematical model developed by SOGREAH for application to sediment-laden water is CHAR2, which solves non-linear partial differential equations of the flow by Preissmann's semi- implicit finite difference scheme. Jehanne and Ulrich (1992) used this model to study the effect of different measures to limit the degradation of the riverbed and applied this model to a river with braided reach in it by 1.5 m.

FLUVIAL model has three major features: (1) computation of sediment transport capacity and corrections for availability, sorting, and diffusion; (2) upstream condition for sediment inflow; and (3) numerical solution of the basic equations. The difference in the sediment flow rate into and out of a reach is accounted by adjustments in the bed and the bank. The width adjustment is made using a stream power approach. Widening is expected during channel-bed aggradation and reduction in the width at the time of degradation. These adjustments are made based on the principle that the river will achieve an equal power expenditure along its course.

The TELEMAC system was developed within the EDF-DER's Laboratorie National d'Hydraulique. The suspended sediment transport and bed load transport are simulated within this systems using SUNIEF and TSEF respectively. SUBIEF uses finite element methods to vertically solve the averaged advection-dispersion equation. The deposition and erosion are related to the shear at the bottom. The consolidation of deposition is also modelled as well (Hervouet, 1996).

The demand of predictive models can be seen from their growth. They are mostly developed and tested for a particular river reach. Most of the complex processes like armouring are modelled approximately. These models are subject to continuous modification as and when new data or understanding of the process is available. Models accounting for sediment transport of different sizes present in the river distinctly are more realistic as the streambed naturally contains different sediment sizes. There are few formulations that can account for such transport.

## 9.11 Conclusions

The Brahmaputra River possesses a number of characteristics, which are not yet fully understood. An idea of its huge nature is provided in this chapter along with the general structure of a mathematical model. Aggradation and degradation occur in alternate reaches of the river. This suggests that the eroded bank material finds a way in building the bars or islands. It may be pointed out that mathematical models are inadequate to solve complex problems. The problem of bank line movements needs consideration of the fluvial processes of erosion, soil moisture management in floodplains and geotechnical processes leading to bank failures.

## References

American Society of Civil Engineers, Task Force on Bed Forms in Alluvial Channels, ASCE, Hydraulics Division, 92, pp 51-64, 1966.

Belleudy, P., Modelling of Danube and Isar Rivers Morphological evolution Pt. II: Symp. of River Sedimentation, Karlsruhe, 1992.

Brice, J.C., Channel patterns and terraces of the Loup Rivers in Nebraska, U.S. GS. Prof. Paper., 422-D, 27,1964.

Bristow, C.S., Brahmaputra River: Channel migration and deposition., In Recent Development in Fluvial Sedimentology. Edtr.., Ethridge, F.G. et al., Pub. Society of Economic Palaeontologists and Mineralogists Special Publications, 39, pp 63-74., 1987.

__________, Sedimentary Structure Exposed in Bar Tops in Brahmaputra River, Bangladesh, In Braided Rivers, Edts. J.L. Best and C.S. Bristow, Geological Society Spl. Pub. No. 75, pp277-289, 1993

Church, M. & Jones, D, Channel bars in gravel bed rivers, In Gravel Bed Rivers, Edr. Hey et al., Pub. Wiley, Chichester, pp 291-338.

Coleman, J.M., Brahmaputra River: Channel processes and sedimentation, Sediment. Geol., 3(2/3), pp 129-239, 1969

Fredsoe, 1981, Unsteady flow in straight alluvial streams, Pt.2, Transition from dune to plane stream bed. Journal of Fluid Mechanics, 102, 431-453

Garde, R.J. and K.G. Ranga Raju, Mechanics of Sediment Transportation and Alluvial Stream Problem. Second Edition, Wiley Eastern Ltd., New Delhi, 1985.

Goswami, D.C. (1985) 'Brahmaputra River, Assam, India: Physiography, Basin Denundation, and Channel Aggradation, Water Resources Research, Vol. 21, No. 7 pp. 929-978

GSI (1974), Geology and Minerals resources of the States of India, pp. 29-59, Misc. Pub. 30.

Halcrow, 1991, River training Studies of Brahmaputra River Second Interim Report to the Bangladesh Water Development Board Dhaka Office, Dec. 1991

Hervouet J. M. Introduction to the TELEMAC system, HE-43/96/073/A, EDF Electricite de France.

Jehanne and T.Ulrich (1992) Hydropower 92, Broch & Lysne (eds.) Balkema, Rotterdam ISBN 9054100540.

Kalaassen, G.L. & Vermeer, K. (1988) Confluence Scour in large braided rivers with fine bed material, International Conference on Fluvial Hydraulics, Budapest.

Karim, M.F. and J.F.Kennedy, IAlUVIAL: A Computer Based Flow and Sediment Routing Model For Alluvial Stream and Its Application to the Missouri River, IOWA Inst. of Hydr. Res., The Univ. of IOWA, IOWA city, (USA) Report no. 250, 1982.

Leopold, L.B. & Wolman, M.G. 1957. River Channel Pattern: braided, meandering and straight. US Geological Survey Professional Papers, 282B, 39-85.

Loy, G. Philippe Belleudy, Jurgen Kellermann, Bernhard Sohngen. Modelling Danube and Isar Rivers Morphological Evolution Part III: Range of Uncertainty caused by Scattered Field Data with special Respect to Grain Sorting, 5th Int. Symp. on River Sedimentation, Karlsruhe, 1992.

Menard, H.W., Some rates of regional erosion, J. Geol., 69(2), pp 154-161, 1961

National Research Council Advisory Board, On the built environment. An evaluation of Flood-Level Prediction Using Alluvial- River Models, Tech. Report, Committee on Hydrodyn. Comput. Models for Insurance Studies. Washington, D.C., 1983

Palaniappan, A.B., Numerical Modelling of Aggradation And Degradtion in Alluvial Stream. Ph.D., Thesis., Dept. of Civil Engg. Univ. of Roorkee, Roorkee, (India), 1991.

Rao, K.L. (1975) 'India's Water Wealth Its assessment, Uses and Projections', Orient Longman (p71)

Thorne, C.R., Russell,A.P.G. & Alam, M.K., Planform pattern and channel evolution of Brahmaputra River, Bangladesh, In Braided Rivers Edited by J.L.Best et al. Pp-276-257, 1993.

Yallin, M.S. Mechanics of sediment Transport. Pergamon, Oxford, UK. 1977

## Notation

| | |
|---|---|
| $A$ | area of cross-section of the flow |
| $B_r$ | Braiding Index |
| $D_{50}$ | Median size of the sediment |
| $d$ | depth of flow |
| $G$ | rate of total sediment transport in absolute volume/ time, |
| $g$ | acceleration owing to gravity |
| $L$ | Length of the morphological feature |
| $L_{IB}$ | length of islands and bars in a reach |
| $L_R$ | length of the reach |

| | |
|---|---|
| $Q$ | water discharge |
| $q$ | lateral flow |
| $q_s$ | lateral sediment inflow |
| $S_f$ | energy slope |
| $S_b$ | bed slope. |
| $w$ | width of the channel |
| $x$, $t$ | are the space and time coordinates respectively. |
| $Z$ | bed level |
| $\eta$ | porosity. |

# CHAPTER 10

# SPATIO-TEMPORAL MORPHOLOGICAL FEATURES

N. SHARMA, R. CHAKRABARTY & C. S. P. OJHA

The Brahmaputra River is a highly braided river, with meta-stable islands, a few nodal points, mobile sand bars, and shifting anabranches. The morphology of the river has undergone significant changes. Spatial and temporal changes occurring in the river regime are fundamental to the design of hydraulic works. This chapter presents a short discussion of river morphological changes.

## 10.1 Introduction

The Brahmaputra is one of the largest braided rivers in the world. The width of the river ranges from around 2k m to 20 km and most of the places it is highly braided. The Brahmaputra is one of the world's greatest rivers, ranking fifth in terms of discharge and eleventh in terms of drainage area. It also has a very high sediment discharge, ranking third in the world. The river is highly braided with meta-stable islands, a few nodal points, mobile sand bars and shifting anabranches.

The Brahmaputra River is a dynamic system and its complex geometry always poses difficulties in flow simulation. The river adjusts its channel geometry, pattern, and profile with changing climate, geological, hydraulic, sediment, and man-made constraints. The Brahmaputra River may be envisaged as a series of channel segments which divide and rejoin around bars and the division and joining of channels are essential features of the river. The behaviour of this large river is also quite erratic and it shifts its course abruptly causing erosion and deposition at various locations leading to immense loss of property. The Brahmaputra, having its origin in the Himalayan region, is a snow fed perennial river in which floods occur during the months of May to September in general. Over 80% of the annual runoff flows during this period. The river is known for the destruction caused by its periodic changes in its course from time to time. The erratic and dynamic nature of the river can easily be understood from its behavior at 'Majuli', the world's largest river island, where almost one third of its total area has been eroded by the Brahmaputra. The occurrence of flooding during monsoon has caused severe economic loss to the region.

The plan form of the river Brahmaputra reflects the hydrodynamics of flow within the channel and the associated processes of sediment transfer and energy dissipation. The



*V.P. Singh et al. (eds.), The Brahmaputra Basin Water Resources*, 214-228.

braiding of the Brahmaputra River occurs in a high energy environment with large and variable discharge carrying heavy sediment loads on steeper gradients and the river is characterised by a random pattern of multithread channel networks owing to the appearance of braid bars within the overall waterway of the river. The presence of braid bars increases the total flow resistance and the energy losses along the boundary, thereby promoting the development of hydraulically inefficient channel network. The chapter presents some of the details pertaining to the morphology of the Brahmaputra River, which has undergone significant changes in the past.

## 10.2 Spatio-Temporal Morphological Changes

Changes in river morphology are one of the most studied aspects of changing river channels, partly because of the relative ease of getting evidence through maps and aerial photographs and the impact of fluvial landforms and development, and the practical problems created by bank erosion and the channel mobility. The study on spatio-temporal morphological changes of the river attempt to describe and explain typical features of the river with respect to space and time. These features are formed as a result of complex dynamics of flow over a moving bed. In almost all the cases the river bed profile is irregular in shape and size. The complexity in representing this information mathematically increases when the sections are moderately or heavily braided. In such situations the fluvial land forms comprising the river cross-section undergo changes with respect to time and space. The analytical process requires extensive inputs from the field with regard to space and time. Hence such an analytical procedure which involve space and time both is known as spatio-temporal analysis. Spatio-temporal analysis allows for idealization of the changing morphological character of a braided river like Brahmaputra with respect to space and time.

A study of flow propagation in a braided river like Brahmaputra is fraught with difficulties on account of its complex channel plan form. The concept of the hydraulic and topological features of the Brahmaputra River poses problems because of the highly non-uniform and random pattern of variations of the cannel conveyance of the sub-channel across the cross-sections. Figure 10.1 shows the locations of six cross sections along the river for this study. Figure 10.2 shows the changes in two different cross sections for three different years. The channel pattern of a reach of the Brahmaputra River reflects the flow dynamics within the channel and the associated channel processes of sediment transport and energy expenditure. Adjustments of the equilibrium channel pattern may occur over a widely varying time scale, depending on the mobility of the channel forming sediments and may reflect the effects of any intervention of man for controlling water and sediment discharge that determine the fluvial processes. The channel pattern forms a continuum in response to varying energy conditions, ranging from straight or meandering channel to multi-thread braided channel as in Brahmaputra.

Most of the sections of stream bed profiles of the Brahmaputra River are irregular in shape and size, making it difficult to represent them mathematically by means of simple functions. The complexity of the mathematical representation increases in the case where the sections are moderately or heavily braided.

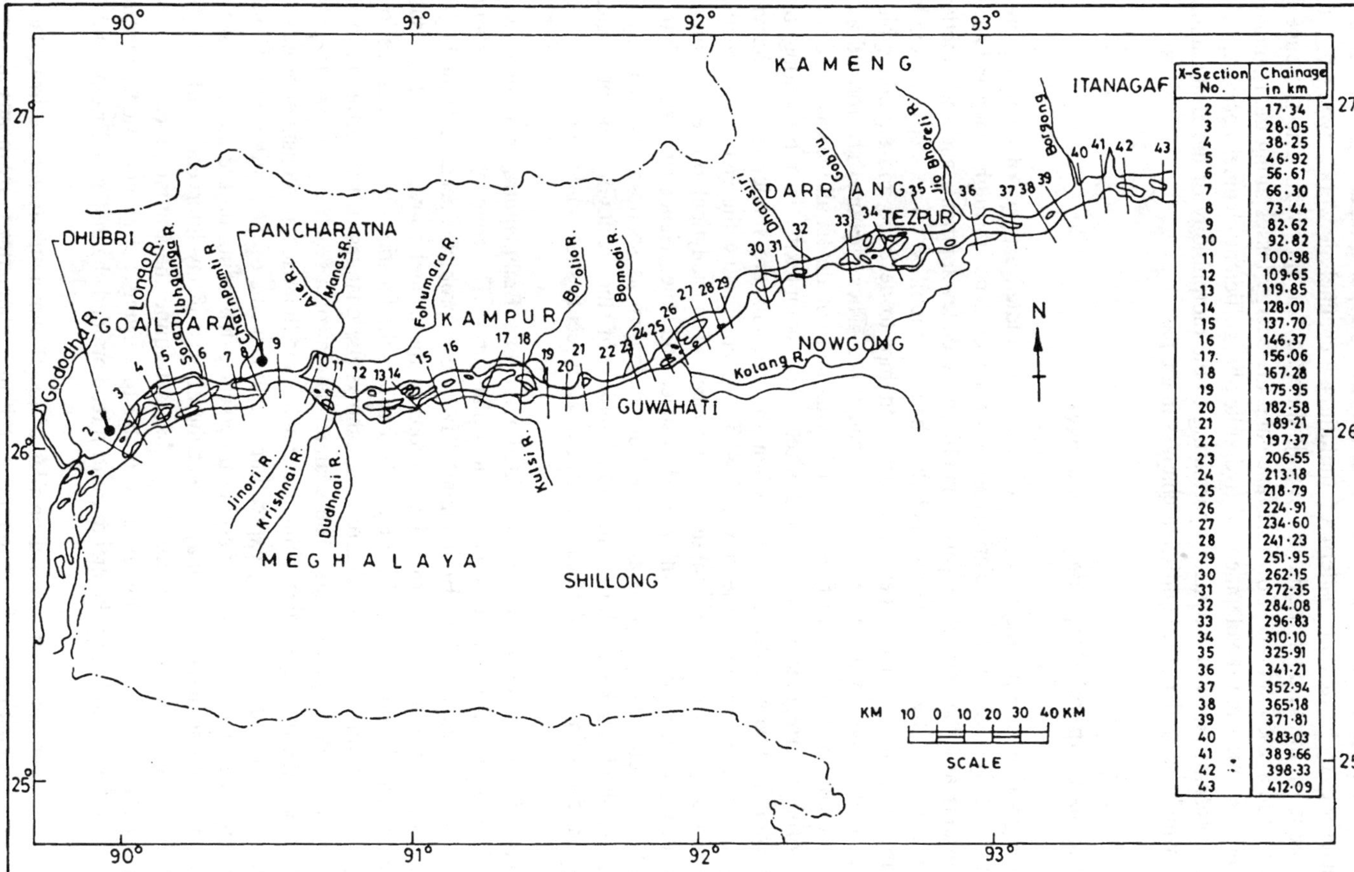


| X-Section No. | Chainage in km |
|---|---|
| 2 | 17·34 |
| 3 | 28·05 |
| 4 | 38·25 |
| 5 | 46·92 |
| 6 | 56·61 |
| 7 | 66·30 |
| 8 | 73·44 |
| 9 | 82·62 |
| 10 | 92·82 |
| 11 | 100·98 |
| 12 | 109·65 |
| 13 | 119·85 |
| 14 | 128·01 |
| 15 | 137·70 |
| 16 | 146·37 |
| 17 | 156·06 |
| 18 | 167·28 |
| 19 | 175·95 |
| 20 | 182·58 |
| 21 | 189·21 |
| 22 | 197·37 |
| 23 | 206·55 |
| 24 | 213·18 |
| 25 | 218·79 |
| 26 | 224·91 |
| 27 | 234·60 |
| 28 | 241·23 |
| 29 | 251·95 |
| 30 | 262·15 |
| 31 | 272·35 |
| 32 | 284·08 |
| 33 | 296·83 |
| 34 | 310·10 |
| 35 | 325·91 |
| 36 | 341·21 |
| 37 | 352·94 |
| 38 | 365·18 |
| 39 | 371·81 |
| 40 | 383·03 |
| 41 | 389·66 |
| 42 | 398·33 |
| 43 | 412·09 |

Fig. 10.1a Map showing CS2 to CS43.

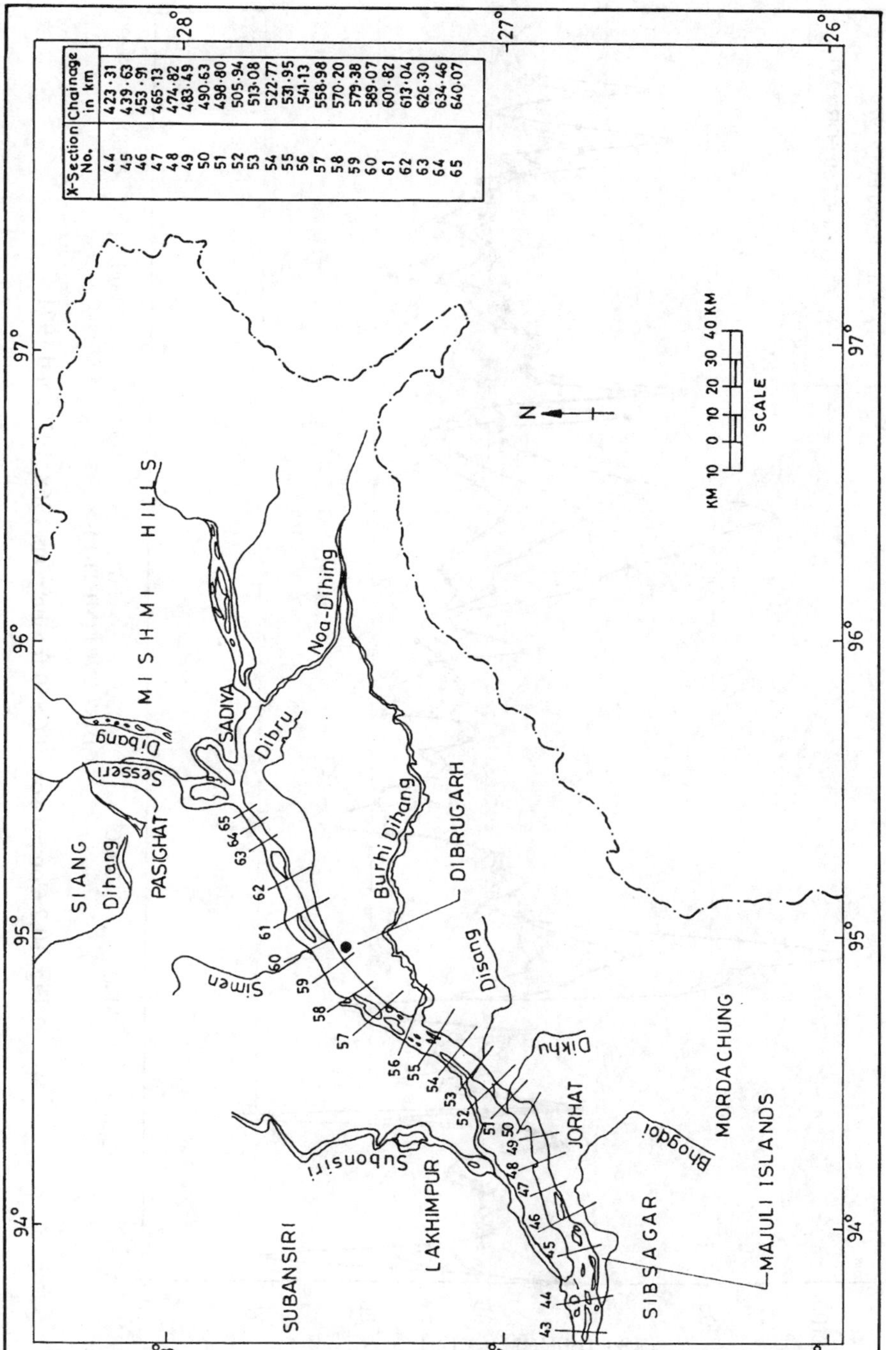


| X-Section No. | Chainage in km |
|---|---|
| 44 | 423·31 |
| 45 | 439·63 |
| 46 | 453·91 |
| 47 | 465·13 |
| 48 | 474·82 |
| 49 | 483·49 |
| 50 | 490·63 |
| 51 | 498·80 |
| 52 | 505·94 |
| 53 | 513·08 |
| 54 | 522·77 |
| 55 | 531·95 |
| 56 | 541·13 |
| 57 | 558·98 |
| 58 | 570·20 |
| 59 | 579·38 |
| 60 | 589·07 |
| 61 | 601·82 |
| 62 | 613·04 |
| 63 | 626·30 |
| 64 | 634·46 |
| 65 | 640·07 |

Fig. 10.1b Map showing CS44 to CS65.

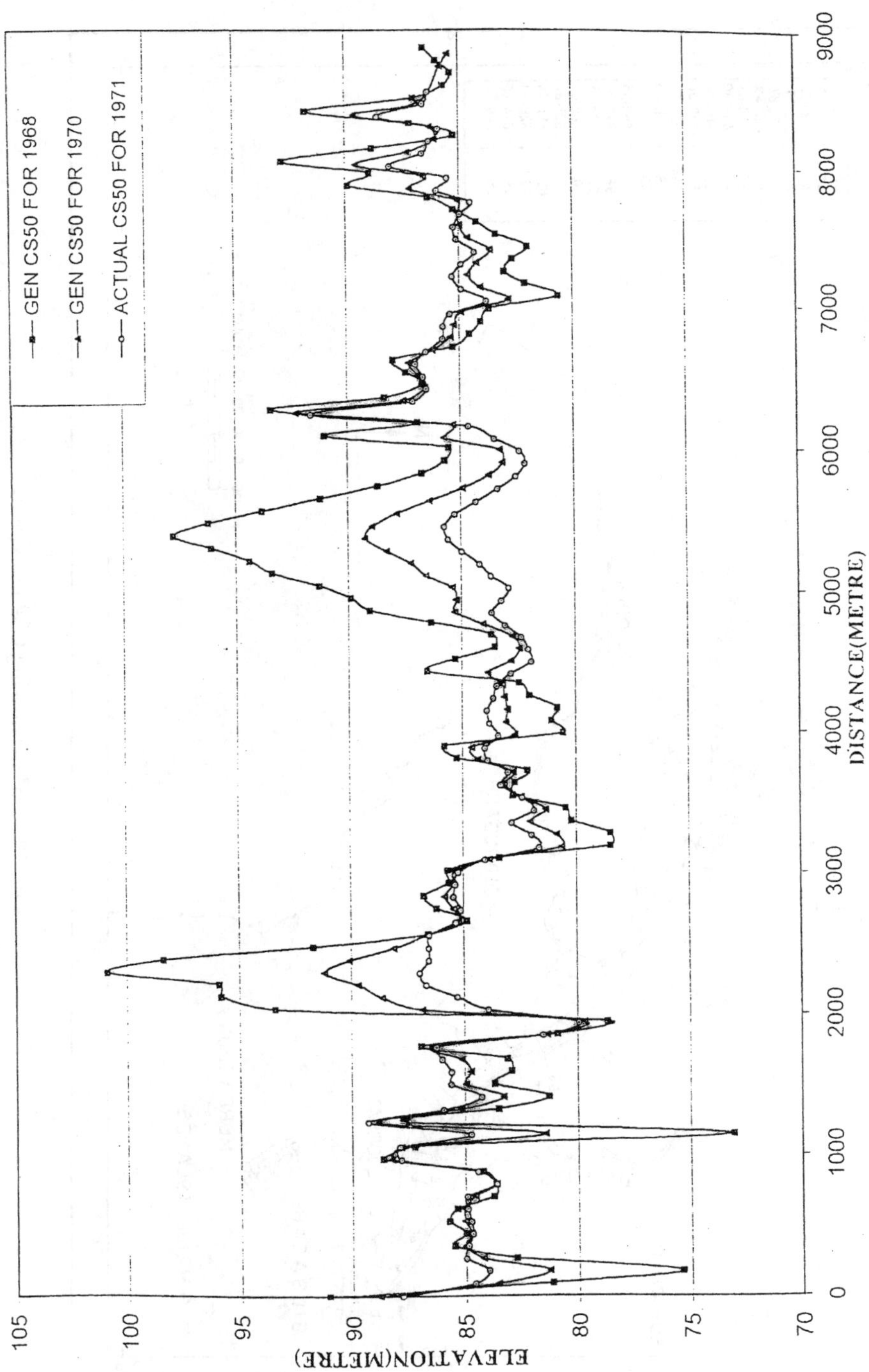


Fig. 10.2 Comparison of CS50 for the years 1968, 1970, and 1971.

The spatial discontinuity associated with environmental controls have been identified in a number of studies for different braided rivers. Sharma and Basumallic (1986) examined the impact of the knock-points on the downstream channel geometry along the Buri Dihing River, a tributary of the Brahmaputra River in India, which appears to have influenced the cross-sectional geometry more than the meander geometry. Garde et al. (1995) carried out a study to identify basic morphological parameters of the Brahmaputra River in Assam. Their study was pertaining to shoal formation in the river bed, river meandering, bed movement, river behaviour under special conditions and other hydrological aspects of the river.

## 10.3 Spatio-Temporal Analysis

Despite the aforementioned challenges, some progress has been achieved in spatio-temporal analysis of the Brahmaputra River. Basically this progress has been made on three fronts. First, attempts have been made to represent bed profiles as well as individual cross-sections using polynomials of different degrees. Second, the use of shape functions has been documented in relating various cross-sections and thus interpolating any intermediate cross section. Shape functions have been used for spatial as well as temporal interpolation. Third, Artificial Neural Networks have been used to train the data available for different cross sections. Such trained networks have been used to predict the changing morphology of the river.

For the convenience of a hydrographic survey the Brahmaputra River has been divided into different cross-sections which are numbered from 1 to 65, as shown in Figures 10.1a and 10.1b; cross section 1 falling in Bangladesh near the mouth of the river. The cross section (C/S) in India starts from 2 (Dhubri) near the international Border between India and Bangladesh and extends upto C/S No. 65 near Dibrugarh. These cross sections were decided during the British era, depending on the topography and accessibility of the river.

'Majuli Islands' fall within the reach from cross-section 44 (km 423.31) to cross-section 54 (km 522.27). With time, bed profiles as well as cross sections of the river have undergone changes with space and time. Modelling of the Brahmaputra River stretch includes the modelling of these transient bed levels as well as cross sections.

### 10.3.1 POLYNOMIALS

The use of polynomials is not new to the alluvial stream modelling. Chen (1973) used polynomials to model the flow cross-section of the Lower Mississippi for computation of the flow depth. A power series was adopted for representation of the river bed profile. Similarly, the need for improving the approximation of the river bed profile by the use of orthogonal polynomials has been suggested by Jansen et al. (1973).

A study was carried out by Pivadi (1992) and Singh (1995) at Rookee University, India, to provide a means for computing the cross sectional profile of a braided river like Brahmaputra at any location within the reach for any year within the period in which past cross sectional data are available. Two types of polynomials were attempted for approximating the selected cross sectional profile at the typical sections selected for the study, viz., ordinary polynomial and Legendre polynomial. To decide the degree of the polynomial most suited to the given cross sectional data, the criterion of Root Mean Square (RMS) Error was adopted and on the basis of error analysis, the RMS error had been computed. It was observed that a 10th degree polynomial yielded the minimum error and the error increased with higher order polynomials. Hence, the 10th order polynomial as well as Legendre polynomials were chosen for fitting the cross sectional data of typical cross sections. It was also observed that the Legendre polynomial tries to follow the actual cross-sectional profile more closely than an ordinary polynomial.

It is not intended to present the results of polynomial fits to any cross sectional data or bed profile as it is a well established mathematical result that any n-th degree polynomial will pass through exactly n data points. Furthermore, the use of polynomials poses additional difficulties while extrapolating or interpolating the temporal variations of the cross sectional details. This difficulty mainly occurs because of the widely varying nature of the coefficients associated with different order terms of the polynomials. Thus the development of a generalized polynomial which is a function of spatial as well as temporal coordinates has not been documented so far in the existing literature.

### 10.3.2 SHAPE FUNCTIONS

In preference to the polynomials, preserving the actual profile of the measured cross-section in the form of normalized data and carrying out spatio-temporal idealization with the help of shape functions is considered to be a better alternative for interpolating the intermediate profile of a braided stream. Shape functions are also widely used in the Finite Element Method, and hence any description about these is out of place here. Accordingly, the concept of a shape function has been applied for interpolating the cross sectional profile. The model uses the normalized data of measured cross sections and utilizes shape functions in the spatial as well as temporal direction for simulating the intermediate cross-sectional profile.

Considering the cross-sectional details for cross-sections between 44 to 56, Gupta (2001) has used shape functions to interpolate cross-sectional details at section 50 (Figure 10.2). It is noted that the section 50 is not utilized in the shape function representation. The agreement between the interpolated section 50 details with the actual ones are shown in Figure 10.2. This indicates the feasibility of using shape functions for interpolating the intermediate cross sections.

### 10.3.3 ARTIFICIAL NEURAL NETWORKS

Artificial Neural Networks (ANNs) take their name from the networks of nerve cells in the brain. The problems which are normally solved using ANNs are intractable or cumbersome with traditional methods. In particular, the neural network extracts relevant features from the input data and performs a pattern recognition task by learning from examples without explicitly stating the rules for performing the task. The use of ANNs has been found useful in several areas of science and engineering. Details about ANNs and their working and application to certain hydrological problems can be found in Ojha and Singh (2001).

Considering the potential of ANNs in other disciplines, Patil (2001) tried to develop an ANN architecture to model the morphology of the Brahmaputra River. The input variables used in the model include: (a) chainage along the Brahmaputra River starting from CS2 (17.34 km) up to CS65 (640.07 km); (b) year; (c) month; and (d) day. The cross sections available in the raw data are in the form of reduced distance (RD) from the left bank, and elevation (RL) of the river bed at different RDs. Each cross section has a varying number of such surveyed points depending upon the top width of the river at that cross section. However, as the neural network must have a fixed number of nodes in the output layer of an ANN, each cross-section was normalized for 25 equidistant points. At these points the bed level elevation was interpolated and used as an output. The activation function used in the model is sigmoidal. A total of 338 sets of data with cross sections from 2 to 65 have been processed for a period spanning forty years from 1957 to 1997. Amongst 338 data sets, 318 data sets have been used for training and the rest 20 cases have been used for validation. Figure 10.3 represents the validation potential of the trained ANN for two different years 1993 and 1998 for cross section no. 15.

Once ANN is developed it can be used to generate spatial as well as temporal cross section profiles, as shown in Figures 10.4 and 10.5. A comparison of two temporal profiles pertaining to two different years may also help in making comparisons at corresponding locations. For example, from the temporal series shown in Figure 10.6 it is possible to appreciate some of the channel processes including channel deepening, channel formation, change in bank profile, deposition in a certain portion of the channel, etc.. As the data on the Brahmaputra River is restricted in nature, it is not possible to elaborate more on the Figures or the data used in training as well as validation of ANN. However, that ANNs can prove to be a useful tool in morphological studies definitely finds support from the study by Patil (2001). There is a possibilty for further improving the ANN architecture to describe the morphology of this great river, and research at the University of Roorkee is on the way to exploring this aspect.

## 10.4 Remote Sensing

Application of remote sensing techniques is leading to important developments in morphological studies of braided river in both space and time. Hussain et al. (1993) identified areas of bank erosion at the Majuli island, the world's largest river island in the Brahmaputra River by interpreting IRS-IA date of 1992--93 collated with the base maps of the survey period between 1969--72. Majuli island is situated at the confluence of the Subansiri River with the Brahmapuatra River in the Jorhat district of Assam. The island has an extension of about 900 sq. km. and is facing severe erosion by both these rivers during floods. The erosion became a common prominent feature after the great earthquake of 1957. Conventional hydrographic surveys are time consuming and inadequate to assess and monitor the nature of erosion of this river island. Hussain et. al (1993) attempted to use IRS-1A data to assess the prevailing statue of bank erosion at the Majuli island. The study revealed that the Brahmaputra River eroded away a large tract of the eastern side of the Majuli island to increase its width from 6.75 km to 8.95 km. It was also observed that the entire north bank of the Brahmaputra River all along the southern side of the Majuli island is causing severe erosion to the island. Though the erosion is prominent on the northern side of the island, the rate of erosion is comparatively slow than it is on the southern side. At about 5 km south of Kamalabari in the Majuli island, a loop of about 3km length has been washed away together with a number of villages on the bank by the Brahmaputra River. The erratic nature of this mighty river causing severe erosion has threatened to cause a great disaster to the main township of the Kamalabari in the Majuli island. The analyses of multitude satellite data by Hussain et al. can be used as a tool for identifying and monitoring bank erosion taking place in the flood prone areas of the Brahmaputra basin.

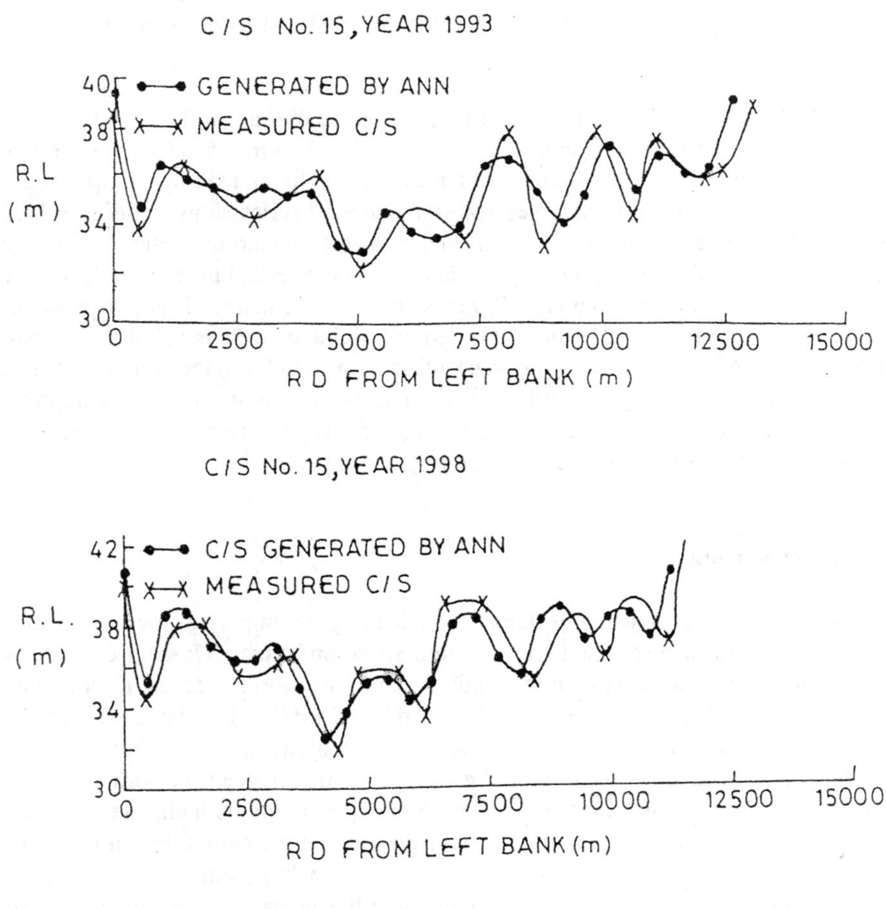


Fig. 10.3 Validation of an ANN model.

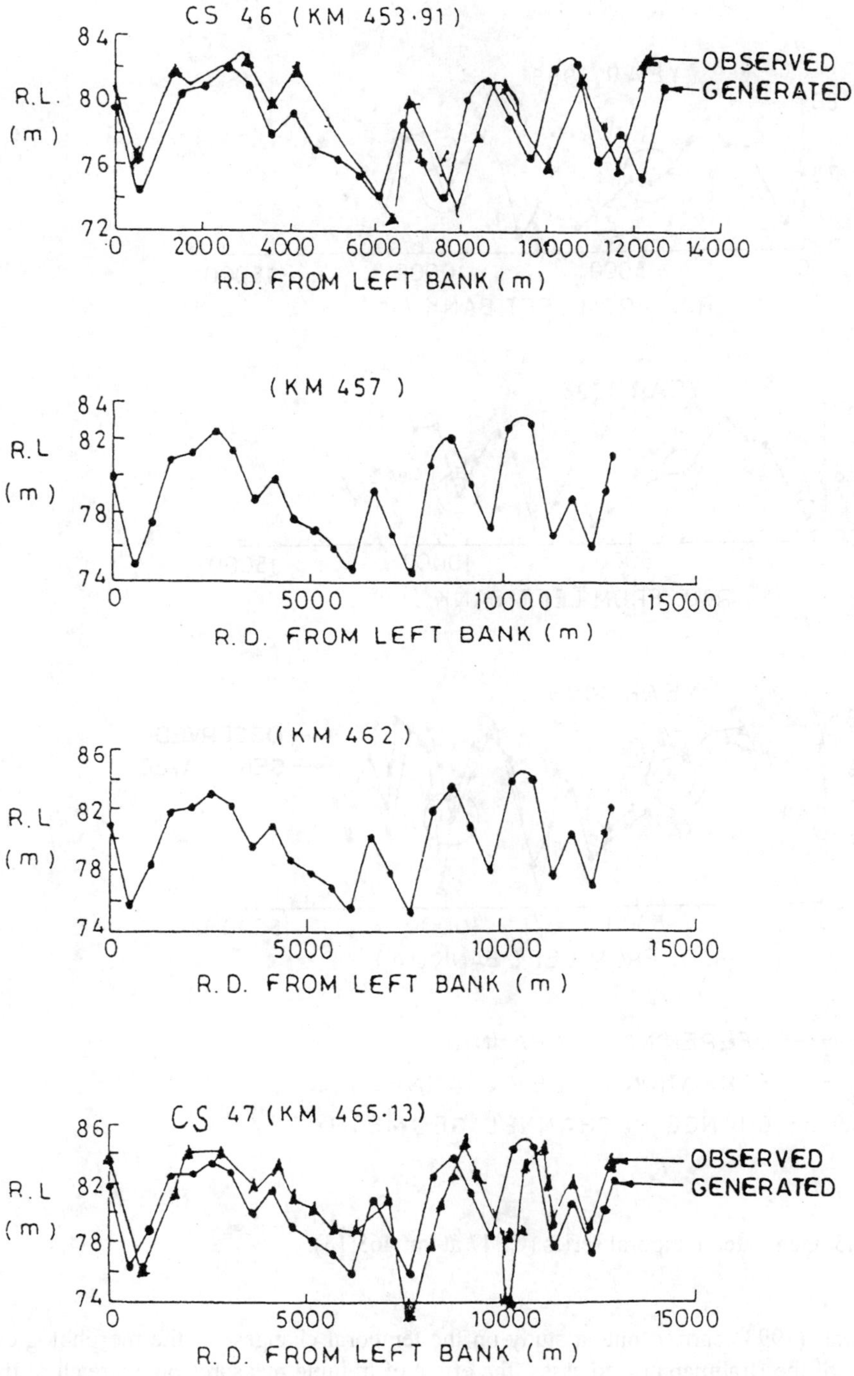


Fig. 10.4 Spatial series (year 1998, CS46 to CS47).

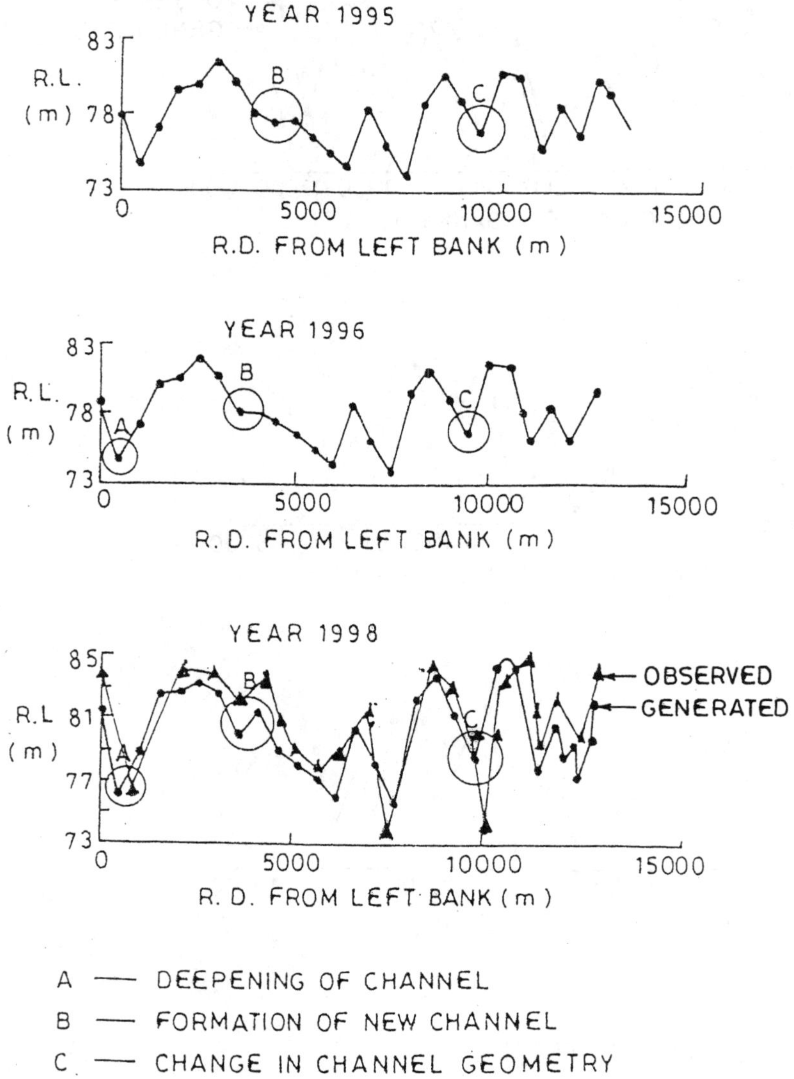


Fig. 10.5 Generated temporal series (CS47 at km 465.13).

Bhagawati (1993) carried out a study on the temporal changes in the morphological features of the Brahmaputra to asses the effect of training measures on a stretch of the river with the help of visual interpretation of multi-dated Landsat-5 data products and toposheets prepared by the Geological Survey of India. Relevant information in the form of satellite imagery were extracted by visual interpretation of FCC hard copies at two different scale, viz., 1:50,000 and 1:250,000 as available on different dates. Information regarding various morphological features like flood plain were extracted from imagery of the smaller scale which were enlarged to prepare overlays at a scale of 1:5,000. Some important topological features like roads, railway tracks, embankments, etc., were also included as reference object for the purpose of comparing multi-dated data by superimposing one over another. With the help of these permanent features a base line of 20 km length was adopted to measure the location of the bandline on different dates to enhance visual appreciation of the changes of the morphological feature. The study reveals that the braided Brahmaputra River follows a sinuous course in certain stretch with the convex bends appearing to be truncated by the enveloping lines of sinuous belt.

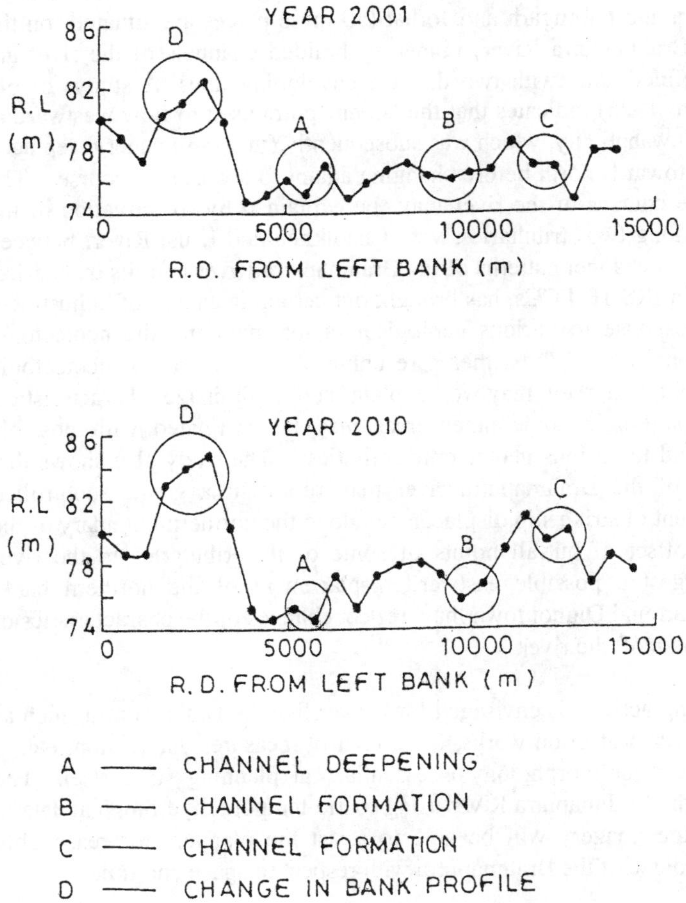


Fig. 10.6 Temporal series for CS at 457 km (between CS46 and CS47).

Lahiri (1996) carried out a case study in the Brahmaputra valley to establish that the channel pattern of the river as signature of the neotectonic movement. The stretch of the Brahmaputra river chosen for the study extends from Dibrugarh (latitude 27 ° 30'E, longitude. 94° 55'E) in the east to Dhuburi (latitude 26° 01'E, longitude 90° 00'E ) in the west, covering about 800 km length of the river. The area transgressed by the river with the adjoining sub Himalayan tract is susceptible to various neotectonic on satellite imagery which aims at synthesizing the available observed drainage characteristics in terms of possible neotectonic activities. To carry out the investigation IRS 1B B/W imagers and FCCs (1:250,000) together with topograhpical maps (1:250,000) have been used for comparing the drainage line. The investigation shows that the Brahmautra River follows a sinuous course in the stretch lying between 92° 00 E and 93° 00 E with its convex bends appear to be truncated by the enveloping lines of the sinuous belt and the width of the belt appearing to be truncated by the enveloping lines of the sinous belt and the width of the belt remains constant. It is interesting to observe that between Guwahati and Goalpara and Dibrugarh and Jorhat (all these places are situated on the southern bank of the Brahmaputra River) numerous braided channels of the river are restricted to a well defined zone with two discrete enveloping lines. A study of old toposheets (surveyed in 1926) indicates that the Brahmaputra used to flow westward at Palabari, west of the Guwahati city, which was subsequently (in 1986 imagery) replaced by a flow that swings towards north before changing again to a westerly course. The observed change in the courses of the river may be explained by reactivation of the northsouth lineament along two tributaries, viz., Chaolkhoa and Kulsi River, between 1926 and 1986. Study of channel patterns of the Brahmaputra River and its tributaries, as observed by Lahiri in IRS 1B FCCs, has brought out certain instances of adjustment of drainage lines in response to various geological factor including the neotectonic movement in the present area. Whilst there are unmistakable records of neotectonic movement in the present area, they may well explain certain drainage characteristics, changes in drainage pattern in some other areas may be correlated with possible adjustment of the ground to various neotectonic activities. The study also shows that the selected stretches of the Brahmaputra river may represent a greater controlled channel with a component of strike slip displacement along the northern boundary of the river. The observed offset of outfall points of some of the tributaries in the river Brahmaputra may suggest a possible eastward displacement of the northern bank. Local tilting of ground around Digboi town may reflect changes of the characteristics of drainage lines on the south of the river.

Whenever an engineering activity is envisaged for a river like the Brahmaputra, such as provision of bridges, river protection works, flood control measures, navigation, etc., a systematic study of the river morphology is essential at planning stage itself. The models formulated for the Brahmaputra River using spatio-temporal hydrological data in conjunction with satellite imagery will be able to assist the planners for reasonably predicting the further course of the Brahmaputra with respect to space and time.

## 10.5 Concluding Discussion

With the latest developments in data procurement and processing it appears that spatio-temporal modelling of the river has a bright future. Unfortunately, as the river data is classified its easy accessibility is not feasible. Unless the data from 1957 till the present for all the cross sections are used, the developed model may always have scope for further improvement. With the available graphics software such as EXCEL, it is feasible to describe a polynomial description of bed profiles as well as individual cross sections. However, the generalization of such polynomials in spatial and temporal domains is difficult and is not recommended. Also sometimes the simple polynomial fits fail to reproduce the physically plausible profiles. The use of Legendre polynomials has shown promise for application in modelling cross sections in the Brahmaputra River. Similar is the case for the use of shape functions. Their use is found to yield a better performance than polynomials, and shape functions are recommended to be used for interpolation of the intermediate cross section profiles. Out of the spatial and temporal interpolations it has been observed that spatial interpolations are surprisingly better.

Nevertheless, nothing seems to replace the potential of ANNs in the spatio-temporal morphological modelling of the Brahmaputra River. Despite limited data are used in training, the ability of ANN to learn the pattern of morphological changes is found to be worth appreciation. It not only helps to ascertain the changing bank profiles but also the channel processes, including the formation of new channels, deepening of channels, etc.. With rapid progress in the remote sensing arena, the future appears to be bright for forecasting the river morphology to a better accuracy. This is certainly expected to benefit the planning and execution of hydraulic structures with enhanced reliability.

It can be concluded from various studies that braided rivers like the Brahmaputra form an important topic of study of many research works because frequent channel migration and rapid bank erosion pose considerable problems. Most of the studies pertain to the parametric analysis of hydrographic data and are fraught with ambiguity owing to spatial and temporal variability, measurement error and many other parameters. Though a number of approaches have been explored for analyzing the behaviour of braided rivers including the mighty Brahmaputra, no study seems to exist which has addressed the behaviour of braided rivers including the random nature of the flow. It could be appreciated that there is a need to develop more versatile techniques for better simulation of the complex channel geometry of a highly braided river. River cross-sections are one of the important footprints which provide information regarding the river characteristics. This information can be used effectively in planning and designing river training measures or hydraulic structures. Also this information can be used for river navigation along any stretch of the river. Changes in stream bed profile over a reasonably long period of time can be an indicator of the future trend of the river and this information has an important bearing on the stability of any engineering activities in this mighty river.

Till now hardly any large scale river valley projects have been taken up in the Brahmaputra basin for control of flood, stabilisation of river morphology and harnessing the large water resources for the economic development of the North-East region of

India. As and when planning and implementation of any such large projects is considered, there will be an imperative need to use the findings of investigations related to spatio-temporal morphological changes of the Bhrahmaputra River for generating the probable future trend of the river.

## 10.6 Acknowledgments

The facts reported in this chapter are based on the work of several postgraduate students, including U.P. Gupta and T.S. Patil. These students have not only helped in obtaining processing the data but have also contributed immensely to the understanding of the subject. Authors would like to express their gratitude to these students.

## 10.7 References

Bhagawati, D. (1993). "Performance evaluation of river training measures using remotely sensed data : A case study on the Brahmaputra at Gumi." Proc. National Sympos. of Remote Sensing Applications for Resource Management with Special emphasis on NE region, Guwahati, 486-489.

Bristow, C.S. (1987). "Brahmaputra river: channel migration and deposition." Recent Development in Fluvial Sedimentology, 63-74.

Chen, Y.H. (1973) "Mathematical Modelling of Water and Sediment Routing in Natural Channel", Ph.D. Dissertation, Colorado State University, Fort Collins, Colorado, U.S.A.

Garde, R.J. et al. (1995). "A report on Brahmaputra", WAPCOS.

Hussain, I., Chaudhury, J.N., and Ghani, M.V. (1993). "River bank erosion of Majuli island as deciphered from IRS imagery." Proc. National Sympos. of Remote Sensing Applications for Resource Management with Special emphasis on NE region, Guwahati, 31-35.

Jansen et al. (1979) "Principles of River Engineering: The Non Tidal River", Pittman Publishing Limited , U.S.A.

Kar, M., Mishra, M., and Goswami, D.C. (1993). Runoff modelling using satellite remote sensing data : A case study of selected sub-basin of the Brahmaputra river, Assam." Proc. National Sympos. of Remote Sensing Applications for Resource Management with Special emphasis on NE region, Guwahati, 511-517.

Lahiri, S. (1993). "Channel pattern as signature of neotectonic movement : A case study from Brahmaputra valley in Assam." Photonirvachak, Vol. 24(4), 265-272.

Ojha, C.S.P. and Singh, V.P. (2001). "ANN Modelling in Watershed Hydrology", Chapter 2 in Mathematical Models of Water Shed Hydrology, published by Water Resources Publishers, Colorado, USA (Editors: Singh et al.), in press, 2001.

Pivadi, T. (1992). Stream profile modelling using Legendre polynomial. A special problem submitted in WRDTC, Univ. of Roorkee.

Patil, T.S. (2001). Artificial Neural Network (ANN) Based Spatio Temporal Studies For A Typical Stretch of The Brahmaputra, M.E. dissertation, WRDTC, Univ. of Roorkee.

Ramasamy, S.M., and Bakliwal, P.C. (1991). "Remote sensing and river migration in western India." Int. J. Remote Sensing, 12, 2597-2609.

Sarma, J.N., and Basumallic, S. (1986). Channel form and processes of the Buri Dihing River, Assam (India). Geografiska Annaler, Series A 68A(4), 372-382.

Sharma, N. (1995). Modelling of braided alluvial channel. Ph.D. Thesis , WRDTC, Univ. of Roorkee.

Singh, G.P. (1995). Spatio-temporal idealisation of typical cross section of a large braided alluvial river, M.E. Dissertation, WRDTC, Univ. of Roorkee.

# CHAPTER 11

# MATHEMATICAL MODELLING AND BRAID INDICATORS

N. SHARMA

Characteristics of braided rivers and braid indicators are vital for the design of levees and flood control works. These also provide useful information on the formation of islands. This chapter presents a review of braid indicators and develops a hydrodynamic model for the Brahmaputra River, which can be employed to derive a variety of river morphological features.

## 11.1 Introduction

The plan form of a reach of an alluvial river reflects the hydrodynamics of flow within the channel and the associated processes of sediment transport and energy dissipation. Braided streams occur in high energy environments with large and variable discharges carrying heavy sediment loads on steeper gradients. These streams are characterized by a random pattern of multi-thread channel networks owing to the appearance of braid bars within the overall waterway of the river. The presence of braid bars increases the total flow resistance and the energy losses along the boundary, thereby promoting the development of networks of hydraulically inefficient channels.

Simulation of flow propagation in a braided stream is fraught with difficulties on account of its complex channel plan form. These difficulties pertain primarily to the representation of the braided waterway of the stream for solution of the governing equations of a full dynamic wave model. The conceptualization of the hydraulic and topological features of a braided channel poses difficulties because of the highly non-uniform and random pattern of variation of the channel conveyance of the sub-channels across the cross section. The most ideal answer to modeling such a physical situation would be to consider the actual braided geometry for solving the three basic equations of water and sediment continuity and the momentum equations. In that case unresolved problems pertaining to correct estimation of flow division in multiple braid channels and their sub-channels with sharply varying conveyances will be encountered.

Whilst a considerable amount of work has been done on modelling of alluvial stream in recent decades, the difficulties of modelling braided alluvial channels briefly indicated above await comprehensive solutions from the modellers. Much work has not been done so far specifically for modelling of braided alluvial streams. The formulation of



*V.P. Singh et al. (eds.), The Brahmaputra Basin Water Resources*, 229-260.

branched and looped network models in a way implicitly accounts for the flow characteristics of a divided flow paths to some extent. Conceptually, in a branched network there exists only one flow path to another point, while there are, in general, several possible paths in a looped network. The drawback of the network models is that the inertial forces owing to temporal and spatial accelerations are not considered. Inclusion of the inertial terms in the looped models is likely to raise computational problems in a physical situation with a great number of loops and branches.

In view of the intricacies associated with modelling of braided streams, a simplifying concept is required to be adopted in such a manner as to obtain reasonable simulation of the fluvial characteristics in a real life situation. The objectives of this work were to constitute a model by adopting a simplifying but a reasonable strategy with certain modifications in the present knowledge of Saint Venant equations for flow simulation in a braided alluvial stream, and to formulate suitable parameters for analysing the attributes of braiding quantitatively. In pursuit of the above goal a model was constituted by adopting an uncoupled solution procedure in which the irregular geometry of the braided stream is temporarily substituted by an equivalent rectangular channel section in the water routing phase, whilst the actual cross section is recalled for consideration in the sediment routing and channel adjustment phase. For stability and accuracy of the numerical solution of the model, iteration procedures were introduced for computation of the coefficients of the downstream boundary condition and also for the sediment routing phase of the uncoupled scheme.

Furthermore, the model presented herein considers deposition in a braided channel to be in inverse proportion to the conveyance across the section. The rationale for this consideration is based on the understanding that the rate of sediment transport varies in direct proportion to the magnitude of conveyance in a flow channel. The model has been applied to the Brahmaputra, which is a highly braided river, and fairly good flow simulations could be achieved in terms of measured water stages. Furthermore, for meaningful analyses of the attributes of the braided channel from the model results, new braid indicators have been formulated in this work by incorporating the hydraulic and the microscale characteristics of the channel morphology. The existing braid indicators are not specifically geared to reflect the above attributes. Accordingly, new braid indicators, namely, Plan Form Index, Flow Geometry Index, and Cross Slope, were evolved and their functional relationships were derived from the simulation runs on the Brahmaputra River. The threshold limits of these braid indicators were also determined for the purpose of classifying the degree of braiding occurrence. The model behaviour was further studied through assumption of purely hypothetical channel improvement measures for the Brahmaputra River and the results were found to be reasonable, thereby demonstrating the potential use of the model.

## 11.2 Braided Channels and Braid Indicators

### 11.2.1 GENERAL

The channel pattern of a reach of an alluvial river reflects the flow dynamics within the channel and the associated channel processes of sediment transport and energy expenditure. Adjustments of the equilibrium channel pattern may occur over a widely varying time scale depending on the mobility of the channel forming sediments and may reflect the effects of any intervention by man for controlling water and sediment discharges that determine the fluvial processes. Channel patterns form a continuum in response to varying energy conditions, ranging from straight or meandering channels to multithread braided forms.

### 11.2.2 CHARACTERISTICS OF BRAIDED CHANNELS

Braided streams such as the Brahmaputra occur in high energy environments of large and variable discharges, heavy sediment load, and steeper gradients with erodible banks. Braided streams are characterized by wide and shallow cross sections with random bar formations creating flow divisions. Flume experiments by Leopold and Wolman (1957) suggest that a bar of coarse sand diverts flow to cause channel widening and positive feedback then accentuates bar development resulting in a braided channel pattern. Sediment transport takes place over the bar surface while incision in the lateral channels lowers the water surface to expose the bar, which then becomes dissected. The complex of islands is stabilized by vegetation in natural streams and experiences further high stage sedimentation (Richards, 1982).

The distributary channels formed by braiding are less hydraulically efficient than parent single channels. Braided reaches are characterized by steeper slopes for maintenance of stream power necessary for sediment transport. Braided reaches show greater complexity, that is, more bars and distributaries in the highest energy environments (Howard, 1970). This observation confirms that the braided channel pattern is optimal for the dissipation of excess energy in high energy streams, since the enhanced total flow resistance of the multithread channel results in rapid energy loss (Richards, 1982)

### 11.2.3 ADJUSTMENTS TO STREAM POWER

Normally a natural channel accommodates flows up to approximately the mean annual flood before overbank flooding occurs. If the channel capacity is too small, overbank and bankfull flood events are more common. Thus the threshold velocity or stream power of the perimeter sediment may be more frequently exceeded and sediment entrainment and perimeter erosion occur. The mean velocity over a period will be high in an initially small cross section, and the excessive momentum of flow will be transmitted to the banks and bed by strong velocity shear at the perimeter, causing erosion and channel enlargement until the frequency of the exceeding of threshold velocities is lessened, the velocities associated with discharges of a given frequency are reduced, and the channel is stabilized (Church, 1967). Establishment of an equilibrium section therefore reflects sediment properties as well as discharge magnitudes, since adjustment of the perimeter presupposes sufficient energy for entrainment. Energy of stream flow reflects the rate of conversion of its potential energy to kinetic energy and work, which depends on stream gradient and discharge. From the flume experiments by Wolman and Brash (1961) it was evident that maximum channel changes occurred with the maximum rate of energy expenditure expressed in terms of $\rho gQS$ where $\rho$ is density, $g$ is acceleration owing to gravity, $Q$ is discharge and $S$ is energy gradient. This suggests that the ultimate control of a channel's morphology is governed by the rate of loss of potential energy of the volume of streamflow generated at the frequency of channel forming events.

### 11.2.4 EXISTING BRAID INDICATORS

A number of indicators were proposed in the past as a measure of classifying streams in braiding category or as a measure of the degree of braiding. Leopold and Wolman (1957) gave a functional relationship between bed slope and discharge to identify braided reaches. Lane (1957), Henderson (1961) and Antropovskiy (1972) developed similar functions relating slope with discharge to predict braids at higher slopes and discharges. Smith (1970) formulated bed relief index to account for bar type and bed relief to implicitly characterize braids. Schumm and Khan (1972) plotted sinuosity as a function of experimental flume slope to provide the threshold when braiding begins. Subsequent authors have recognized different degrees not only of sinuosity but also of channel division quantified variously as twice the total length of

Table 11.1 Comparison of models on important aspects.

| S.N. | Name of model | Unsteady water | Flow of sediment | Non-uniform sediment | Coupled/ uncoupled | Lateral migration |
|---|---|---|---|---|---|---|
| 1 | DELFT. HYDRAULIC LAB MODEL | N | Y | N | U | N |
| 2 | HEC-6 | N | Y | Y | - | N |
| 3 | HEC-2 SR | N | Y | Y | U | N |
| 4 | SEDIMENT 4H | Y | Y | N | U | N |
| 5 | FLUVIAL II | Y | N | N | U | Y |
| 6 | KUWASER | N | Y | N | U | N |
| 7 | MOBED | Y | Y | N | U | N |
| 8 | UUWSR | Y | Y | N | U | N |
| 9 | IALLUVIAL | N | Y | Y | U | N |
| 10 | FLDWAV | Y | N | N | - | N |
| 11 | CHEN'S MODEL | Y | Y | Y | U | N |
| 12 | SEDIMENT ROUTING MODEL OF BORAH | Y | Y | Y | U | N |
| 13 | MIKE II | Y | Y | Y | U | N |
| 14 | DASS MODEL | Y | Y | Y | U | N |
| 15 | FCM MODEL | Y | Y | N | C | N |

Y = Considered N = Not Considered
C = Coupled U = Uncoupled

c bars divided by the reach length (Brice, 1964), one less than the average number of channels per cross section (Howard, 1970), the percentage of channel length that is divided (Brice, 1978), or the number of braid bars per meander wavelength (Rust, 1978). These braid indicators are inadequate in so far as these are not related to hydraulic efficiency and submerged subchannel configurations of the stream which are seen to be closely related to braiding process.

The review of the existing mathematical river models (Table 11.1) indicates that only a limited number of models consider bed changes and channel widening process associated with a braided alluvial stream. Almost all the models adopt an uncoupled solution procedure. The models are primarily oriented towards the irregular geometry of an unbraided alluvial stream where the problem of representation of a braided channel interspersed with large braid bars is not encountered. As such, there is a need to address the problem through some suitable modification in the hydraulic representation approach for braided stream.

Further, the pattern of adjustments of the braided channels could be seen to have a close relationship with the stream power and the energy loss. In this context, it may be logical to correlate the channel adjustments with the energy expenditure process while constituting a model for braided stream like the Brahmaputra. Also, it could be seen that the existing braid indicators do not reflect the hydraulic efficiency and the underwater bars, both of which are seen to have a close relationship with the braiding process of the Brahmaputra and other braided rivers.

## 11.3 Formulation of the Simulation Model and the Braid Indicators

### 11.3.1 GOVERNING EQUATIONS OF THE MATHEMATICAL MODEL

The governing equations of the model for the water and sediment are:

$$\frac{\delta Q}{\delta x}+\frac{\delta A}{\delta t}-q_1=0 \tag{11.1}$$

$$\frac{\delta \rho Q}{\delta t}+\frac{\delta \rho Q V}{\delta X}+gA\frac{\delta \rho Y}{\delta X}=\rho g A(S_0-S_f+D_1) \tag{11.2}$$

$$\frac{\delta Qs}{\delta x}+p\frac{\delta A_c}{\delta t}+\frac{\delta A_s}{\delta t}-q_s=0 \tag{11.3}$$

where $x$ = horizontal distance along the channel, $t$ = time, $Q$ = flow discharge, $A$ = cross sectional area of the channel, $Ac$ = volume of deposition of sediment per unit length of channel, the value of $A_c$ will be negative when bed erosion occurs, $A_s$ = volume of sediment suspended in water over the cross section per unit length of channel, $Qs$ = sediment discharge, $q_l$ = lateral flow of water per unit length of channel, $q_s$ = lateral flow of sediment per unit length of channel, a positive value indicates inflow while negative figure denotes outflow, $\rho$ = density of sediment laden water which is given by $\rho_W + C_S(\rho_S - \rho_W)$, $C_S$ = sediment concentration = $Q_S / Q$, $\rho_W$ = density of water, $\rho_S$ = density of sediment, p = volume of sediment in a unit volume of bed layer, V = mean velocity of flow, $Y$ = depth of flow, $S_o$ = bed slope, $S_f$ = energy slope, $D_l$ = dynamic contribution of lateral discharge = $\dfrac{Q1.V1}{g.A}$.

The following assumptions were made in the governing equations: (i) The flow is one dimensional, i.e., the velocity is uniform over the cross-section; (ii) The streamline curvature is small and vertical accelerations are negligible, i.e., the pressure is hydrostatic only; (iii) The average bed slope is small; (iv) The water surface slope is small; (v) The unsteady flow resistance coefficient is assumed to be the same as steady flow in alluvial channels; and (vi) The water surface across the cross section is horizontal. Furthermore, the model assumes uniform sediment size for purposes of computations involved in sediment routing and channel adjustments.

In a short time period, the change in So is very small, and the value of $\delta Ac/\delta t$ is usually much smaller than that of $\delta A/\delta t$. Hence the solution yielded by solving the three equations simultaneously can be obtained using a simplified uncoupled procedure by first solving equations (11.1) and (11.2) and then refining the solution of equations (11.1) and (11.2) by using equation (11.3). To solve, equations (11.1) and (11.2) are rewritten, assuming fluid density '$\rho$' as constant as:

$$\frac{\delta Q}{\delta X} + \frac{\delta y}{\delta t} - q_1 = 0 \tag{11.1a}$$

$$\frac{\delta Q}{\delta t} + \frac{\delta\left(\dfrac{Q^2}{A}\right)}{\delta x} + gA.\frac{\delta y}{\delta x} + gA(Sf - So - D1) = 0 \tag{11.2a}$$

11. 3. 2 SUPPLEMENTARY RELATIONS

The model requires mathematical functions for the supplementary relations as described below.

*11.3.2.1 Energy Slope*

The energy slope is expressed in the form:

$$S_f = \frac{Q^2}{K^2} \tag{11.4}$$

where K = conveyance

$$K = \frac{AR^{2/3}}{n}$$

$n$ = Manning's roughness coefficient, and $Q$ = discharge.

The model provides the option to either adopt a changing roughness by using the Garde--Ranga Raju velocity--depth predictor or a constant value of Manning's '$n$' on the basis of calibration. The Garde-Ranga Raju predictor requires, depth, width and energy slope. The average value of $S_f$ between sections '$i$' and '$i+1$' has been used for better representation of the section reach.

*11.3.2.2 Sediment Discharge*

If measured data on sediment discharge values are not available with respect to time and space, mathematical functions are required to generate these data. Eight sediment discharge predictors are incorporated into the model and any one of these may be used for computation of sediment load. One may decide the suitable method from amongst the eight methods by comparing the computed results with the measured values of the available fluvial data.

*11.3.2.3 Boundary Conditions*

Normally, discharge versus time or stage versus time relationships are furnished as the upstream boundary condition, whilst the stage versus discharge rating curve is provided as downstream boundary condition. To represent the downstream boundary, the following power function is fitted to the observed data by multiple regression analysis separately for the rising and falling limbs of the hydrograph.

$$h = c.Q_m^a \, .Q^b \tag{11.5}$$

where $h$ = gauge level, $Q_m$= threshold discharge value after which the trend of the hydrograph has changed, and $a$, $b$, $c$ = regression coefficients depending upon rising/ falling limbs of hydrograph.

*11.3.2.4 Representation Of Braided Channel Geometry*

Simulation of highly braided large alluvial streams presents difficulties on account of their complicated flow geometry. These difficulties primarily pertain to the hydraulic representation of the braided waterway for numerical solution of the governing equations. A practical solution to the problem can be modifying the braided geometry as an equivalent rectangular channel for the purpose of the water routing phase and the sediment routing can be carried out considering the actual river cross section. The equivalent rectangular channel substituting for the braided river cross section can be hypothesized as a hydraulically similar rectangular channel with its depth and width determined based on the weightage of the subchannel conveyance to the total for the stream.

The channel geometry of an alluvial river is highly irregular and varies from section to section. In fact, most of the alluvial streams display braided channel characteristics with highly undulating cross sectional profile. The computer programme developed herein is geared for calculating the geometric properties of a braided river cross section for given bed levels subsection wise and given water levels. Adopting the trapezoidal rule, it computes wetted perimeter, channel top width, cross sectional area and deepest bed elevation of the stream.

*RD(i,l)* = Distance from left bank of sub section *l* and section *i*,

*ZO(i, l)* = Elevation of subsection l of section *i*.

The channel is divided into subsections as per data input and the actual flow area and other geometric properties are worked out taking due cognizance of braided island if any within the cross section. Channel width has a profound influence on wave propagation speed. Similarly, flow depth affects celerity, Manning's `n' and flow dynamics.

Therefore the active width and the representative for a braided alluvial stream need to be assigned with care. Adopting the top width of a braided stream as an active width may lead to oversimplification of highly irregular channel cross section. To account for the irregular cross section shape of a braided channel a weighted depth or equivalent depth is calculated.

$$\text{Equivalent depth, } EQD = \frac{\sum_{i=1}^{n} D_{av}.A_i.D_{av}^{2/3}}{\sum_{i=1}^{n} A_i.D_{av}^{2/3}} \quad (11.6)$$

$$\text{Equivalent width, } EQD = \frac{\sum_{j=1}^{n} A_i.D_{av}^{2/3}}{(EQD)^{5/3}}) \quad (11.7)$$

The equivalent width or effective width is calculated from equivalent effective depth to preserve the proper $A(D^{2/3})$ for the cross section.

$A_i$ = area of subsection i, and

$D_{av}$ = average water depth at subsection.

*11.3.2.5 Stream Bed and Width Changes*

The changes in stream bed width of an alluvial river are reflected in the change in cross sectional area, δAc, for any time increment. In an uncoupled solution the solution of equation (11.3) yields the amount of change in cross-sectional area of the channel bed. This amount of the cross-sectional area change may present a scenario in which it is possible for the channel bed to undergo deposition or erosion. The bed deposition is normally associated with channel widening. It can be reasonably adopted that these channel adjustments that one normally observes in a braided alluvial stream follow the path of minimum stream power in the direction of new equilibrium for the reach. It is apparent that the volume of sediment deposition and bank erosion explicitly affects the change in channel width. The application of the minimum stream power entails that the width changes at all the cross sections take place in such a manner that the total stream power of the reach is minimized. However, these changes should be limited by physical constraints like sediment rate, bank soil properties, etc.. These factors limit the maximum possible width changes for a time increment. In the present model, the maximum possible width increase within a time interval by lateral bank retreat is estimated by the method of Osman and Thorne (1988).

The critical shear stress, $\tau_c$, for cohesive bank material is determined by method adopted by Osman and Thorne from The U.S. Army Engineer Waterways Experiment Station, Vicksburg (Arulanandan et al., 1980), while $\tau_c$ for non cohesive bank material is computed using the non dimensional parameters of the analytical form of Shields' diagram given by Leo C.Van Rijn. The procedure for computing the limiting width is given below:

(i) Determine the average shear stress, $\tau$;

(ii) Determine the critical shear stress $\tau_C$ depending on soil properties;

(iii) Check $\tau \geq \tau_c$;

(iv) If $\tau > \tau_c$, determine the rate of soil erosion, which when multiplied by the time duration gives the lateral erosion for the time step. The simulation of variation of channel cross section profile is based upon the concept of minimum stream power. The channel width at time $t+\Delta t$ may be obtained using the width at time t and a search procedure such that the total stream power, $P = \gamma QS$ at time $t + \Delta t$ is a minimum subject to the given constraints . A simplified search procedure given by Chang (1984) satisfying the minimum stream power criterion is used in the present model. In this search procedure the minimum total stream power can be obtained when the channel cross section profiles are so adjusted that the variation in energy gradient for adjacent sections are reduced, using the following expression. For a section $i$, the relation for weighted average energy gradient of its adjacent section is:

$$\overline{S_i} = \frac{S_{i-1}\Delta X_1 + S_{i+1}\Delta X_{i-1})}{2(\Delta X_{i-1} + \Delta X_i)} \tag{11.8}$$

in which $\overline{S_i}$ = weighted average energy gradient. The actual energy gradient at section $i$ , $S_i$ is compared with the weighted average energy gradient $\overline{S_i}$ for the channel widening process. These changes in channel cross section are subject to the limitations for width adjustment and physical constraints.

For each channel cross-section, the change in channel bed deposition area, $\delta A_c$ is distributed subsection wise to obtain the new cross sectional profile. In the case of an aggrading channel section, firstly the width adjustments are made as per Chang's expression. As a result of the width increase process, the volume of bank material supplied by erosion is added to the bed deposition area $\delta A_c$ for spreading out in the channel bed as per the deposition process to follow subsequently. For the purpose of bed deposition, a simple but rational approach is formulated in this work to distribute the bed deposition area in inverse proportion to the conveyance as below:

$$\frac{\Delta A_{bk}}{\Delta A_b} = \left[ 1 - \frac{A_K D_K^{2/3}}{\sum_{i=1}^{K} A_K . D_K^{2/3}} \right] \tag{11.9}$$

where $\Delta A_{bk}$ and $\Delta A_b$ are bed deposition area of *Kth* subsection and the total river cross section respectively. $A_k$ and $D_K$ are area and depth of the *Kth* subsection. The rationale for this consideration is based on the understanding that the rate of sediment transport varies in direct proportion to the magnitude of conveyance. In the case of a degrading section, the changes in channel bed area, $\delta A_c$, is distributed in direct proportion to the effective tractive force as below:

$$Z_s = \frac{\tau_o - \tau_{ocr}}{\sum_W [\tau_o - \tau_{ocr}]\Delta y} \Delta AC \tag{11.10}$$

where $Zs$ = Local scour depth, $\tau_0$ = Local tractive force, $\tau_{0cr}$ = Critical tractive force, $\Delta Y$ = Horizontal distance between two points along the channel boundary, $W$ = Width of stream channel, and $\Delta AC$ = Change in channel bed.

*11.3.2.6 Computation Scheme For Channel Width Adjustments*

The scheme for channel increase along with bed elevation changes is developed as described below.

*ZO(i,l)* = Elevation at subsection *l* of section *i* before width increase,

*ZON(i,l)* = New Elevation at Subsection *l* of section *i* after width increase,

*WIMAX* = Maximum possible width increase of one bank within the time interval,

*WIL, WIR* = Step size of width increase in left and right banks at each section,

*NSUB1* = Subsection dividing the cross section, and

*ZMIN1* = Minimum bed elevation.

*WIMAX* is divided into step size *WIL* and *WIR* to obtain new widths at the cross section. At each iteration, the new bed elevation at each subsection is computed by the expression as below:

$$ZON(i,1) = ZO(i,1) - [WIL/(WIL + WSUB(L))] \times [ZO(I,l) - ZO(I,l+1)] \tag{11.11}$$

for the left hand side of the cross section from ZMIN. Similarly, for computation of width increase on the right hand side from ZMIN, the following expression is used,

$$ZON(i,1)=ZO-(i,1)[WIR/(WIR+WSUB(l)]x[ZO(i,1)-ZO(i,l-1)] \tag{11.12}$$

Iterations for adjustment in channel widths are continued till the actual energy gradient comes close to the value given by equation (11.8).

**11.3.2.7 Computation Scheme For Bed Degradation.** After the sediment routing phase, the bed degradation process is undertaken if $\delta Ac$, the bed deposition area, has a negative value. The scheme for channel bed degradation is formulated as described below:

*ZO (i,1)* = Elevation of subsection *1* of section *i*,

*WSUB (i,1)* = Width of subsection *l* of section *i,*

*WSUB1* = Width of that portion of the subsection *1* coming within water line,

*H* = Water level,

*DPRV* = Previous subsection depth = *H - ZO (i,1)*, and

*DSUB* = Subsection depth = *H - ZO (i,1)*.

The degraded bed profile is computed subsection wise using the expression:

$$ZO(i,1) = ZO(i,1) + \frac{\tau_0 - \tau_{er}}{\sum_1^W \tau_0 - \tau_{er}.WSUB1}\delta AC \tag{11.13}$$

where $\tau_0$-$\tau_{cr}$ is the net tractive force acting on the bed. The term *WSUB1* is computed separately for left bank side (Case 1) and right bank side (Case 2) using the expressions given below.

**Case I :**

$$\frac{WSUB1}{WSUB(i,1)} = \frac{DSUB}{DSUB + (-DPRV)} \tag{11.14}$$

or

$$WSUB1 = WSUB(i,l-1) \times \frac{DSUB}{(DSUB - DPRV)} \tag{11.14b}$$

**Case II:**

$$\frac{WSUB1}{WSUB(i,l-1)} = \frac{DPRV}{DPRV + (-DSUB)} \tag{11.14a}$$

or

$$WSUB1 = WSUB(i,l-1) \times \frac{DPRV}{(DPRV + (-DSUB))}$$

As can be seen from equation (11.10), the degraded bed levels at each subsection is worked out by distributing the bed degradation area δAC in direct proportion to the ratio of the net tractive force of the subsection to the total tractive force of the entire flow cross section as stated earlier.

### 11. 4 Numerical Solution of the Governing Equations

The numerical solution of the governing equations are obtained in an uncoupled manner in which the flow continuity equation (11.1a) and the flow momentum equation (11.2) are solved first. The solution is then refined by solving the sediment continuity equation (11.3). The partial derivatives in the governing equations are replaced by quotients of finite differences using the implicit scheme. In the present model the Linear Fully Implicit Scheme is adopted for solution of the flow continuity and momentum equations, which is briefly described below. For stability and accuracy of the numerical solution of the model an iteration procedure has been adopted for computation of the coefficients of downstream boundary condition C12 and C14 till the changes in discharge of successive iterations are within tolerance limit. Similarly, the sediment routing and its subsequent adjustment processes are subjected to an iteration procedure in the model till the changes in computed water levels of successive iterations satisfy the accuracy limit.

#### 11.4.1 LINEAR IMPLICIT METHOD

With reference to Figure 11.1, it is assumed that all the variables are known at all nodes of network on the time $t^j$. It is desired to compute the values of the variables at all nodes on the time $t^{j+1}$.For convenience, equations (11.1a) and (11.2b) are rewritten as below:

$$\frac{\delta Q}{\delta X}+T.\frac{\delta Y}{\delta T}-q_1=0 \tag{11.15}$$

$$\frac{\delta Q}{\delta T}+\frac{\delta}{\delta X}\left(\frac{Q^2}{A}\right)+gA.\frac{\delta Y}{\delta X}+gA(S_f-S_o-D_1)=0 \tag{11.16}$$

or

$$\frac{\delta Q}{\delta T}+\frac{2Q}{A}.\frac{\delta Q}{\delta X}+\frac{\delta Y}{\delta X}(gA-\frac{Q^2.T}{A^2})+gAS_f=\frac{Q^2}{A^2}\left(\frac{\delta A}{\delta X}\right)_y+gA(S_o+D_1) \tag{11.17}$$

$$T=\frac{\delta A}{\delta Y} \quad and \quad A_x^y=\left[\frac{\delta A}{\delta X}\right]_y=cons\tan t$$

which represent the departure from a prismatic channel.

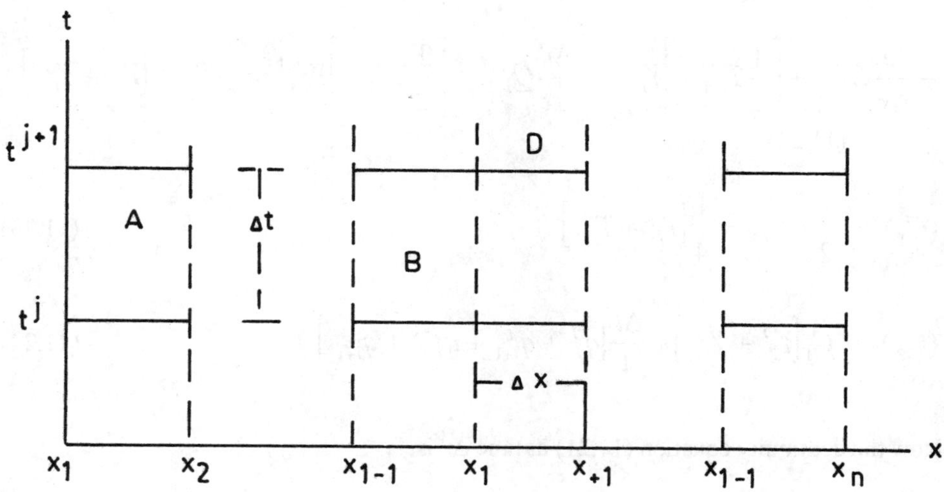


Fig. 11.1 Network fro implicit method.

Equations (11.15) and (11.17) are applied to a point $D$ within the four point grid, formed by the intersections of the lines $X_I$ and $X_{I+1}$ with the time lines $t^j$ and $t^{j+1}$. At the point $D$ the average values and the partial derivatives of a function $f$ are expressed as:

$$f_{i+1/2}^{j} = \frac{1}{2}(f_i^j + f_{i+1}^j) \tag{11.18}$$

$$\frac{\delta f}{\delta x} = \frac{1}{\Delta x}(f_{i+1}^{j+1} - f_i^{j+1}) \tag{11.19}$$

$$\frac{\delta f}{\delta t} = \frac{1}{2\Delta t}\left\{\left[f_i^{j+1} - f_i^j\right] + \left[f_{i+1}^{j+1} - f_{i+1}^j\right]\right\} \tag{11.20}$$

where $f$ may represent discharge $Q$, flow depth $d$, etc.. Using the above finite difference relations into equations (11.15) and (11.17) yield a system of nonlinear algebraic equations. In the linear fully implicit scheme which is used for better numerical stability, the nonlinear term in the momentum equation is linearized. This eliminates the difficulty of solving nonlinear algebraic equations by iteration. The continuity equation (11.15) in discretized form is:

$$\frac{1}{\Delta X}\left[Q_{i+1}^{j+1} - Q_i^{j+1}\right] + T_{i+1/2}^{j}\frac{1}{2\Delta t}\left[\left(Y_i^{j+1} - Y_i^j\right) + \left(Y_{i+1}^{j+1} - Y_{i+1}^j\right)\right] = q_{i+1/2}^{j+1/2} \tag{11.21}$$

Multiplying by $\Delta t$ and then rearranging, one obtains

$$-\frac{\Delta t}{\Delta x}Q_i^{j+1} + \left[\frac{1}{2}T_{i+1/2}^j\right]Y_i^{j+1} + \frac{\Delta t}{\Delta x}Q_{i+1}^{j+1} + \left[\frac{1}{2}T_{i+1/2}^j\right]Y_{i+1}^{j+1} = \frac{1}{2}T_{i+1/2}^j\left[Y_i^j + Y_{i+1}^j\right] + \Delta t \; q$$

(11.22)

$$(C_1)_i = \frac{1}{2}T_{i+1/2}^j = \frac{1}{4}\left[T_i^j + T_{i+1}^j\right] \tag{11.23}$$

$$(C_2)_i = C_1\left[Y_i^j + Y_{i+1}^j\right] + \frac{\Delta t}{4}\left[q_i^j + q_{i+1}^j + q_i^{j+1} + q_{i+1}^{j+1}\right] \tag{11.24}$$

Now the continuity equation (11.21) at time $\Delta t^j$ is

$$-\theta Q_i + (C_1)_i Y_i + \theta Q_{i+1} + (C_1)Y_{i+1} = (C_2)_i \tag{11.25}$$

The momentum equation in discretized form is:

$$\frac{1}{2\Delta t}\left[\left(Q_i^{j+1}-Q_i^j\right)+\left(Q_{i+1}^{j+1}-Q_{i+1}^j\right)\right]+\left[\frac{2Q}{A}\right]_{i+1/2}^j.\frac{1}{\Delta x}\left(Q_{i+1}^{j+1}-Q_i^{j+1}\right)+\left[\frac{2Q}{A}\right]_{i+1/2}^j.\frac{1}{\Delta x}\left[Q_{i+1}^{j+1}-Q_i^{j+1}\right]+\left[gA-\frac{Q^2T}{A^2}\right]_{i+1/2}^j$$
$$.\frac{1}{\Delta x}\left[y_{i+1}^{j+1}-y_i^{j+1}\right]+(gA)_{i+1/2}^j(SF)_{i+1/2}^{j+1/2}=\left[\frac{Q^2}{A^2}A_x^y\right]_{i+1/2}^j+(gA)_{i+1/2}^j\left[(So)_{i+1/2}^j+(D_1)_{i+1/2}^{j+1/2}\right] \tag{11.26}$$

Now the momentum equation (11.26) becomes

$$(C_3)_i Q_i^{j+1}+(C_5)_i Y_i^{j+1}+\left(C_4\right)_i Q_{i+1}^{j+1}+(C_6)_i Y_{i+1}^{j+1}=(C_7)_i \tag{11.27}$$

where

$$(sf)_{1+1/2}^{j+1/2}=\frac{1}{4}\left[(sf)_{i+1}^j+(sf)_i^j+(sf)_{i+1}^{j+1}+(sf)_i^{j+1}\right] \tag{11.28}$$

$$(sf)_i^{j+1}=(sf)_i^j+\left[\frac{\delta Sf}{\delta Q}\right]_i^j\left[Q_i^{j+1}-Q_i^j\right]+\left[\frac{\delta Sf}{\delta y}\right]_i^j\left[y_i^{j+1}-y_i^j\right] \tag{11.29}$$

$$\frac{\delta Sf}{\delta Q}=\frac{\delta\left[\dfrac{Q^2}{K^2}\right]}{\delta Q}=\frac{2Q}{K^2}-\frac{2Q^2}{K^3}+\frac{\delta K}{\delta Q}=\frac{2Q}{k^2}+\frac{2Q^2}{K^3}.\frac{k}{\eta}\frac{\delta n}{\delta Q} \tag{11.30}$$

$$\delta SQ=\frac{\delta Sf}{\delta Q}=2sf\left(\frac{1}{Q}+\frac{1}{n}.\frac{\delta n}{\delta Q}\right) \tag{11.31}$$

$$dsy=\frac{\delta Sf}{\delta Y}=-2Sf\left[\frac{1}{A}\left(\frac{5}{3}T-\frac{2}{3}R\frac{\delta P}{\delta y}-\frac{1}{n}\frac{\delta n}{\delta y}\right)\right] \tag{11.32}$$

Therefore,

$$(Sf)_{i+1/}^{j+1/2}=\frac{1}{4}[2(sf)_i^j+(dSQ)_i.(Q_i^{j+1}-Q_i^j)+2(Sf)_{i+1}^j+$$
$$(dSQ)_{i+1}(Q_{i+1}^{j+1}-Q_{i+1}^j+(dSY)_i(Y_i^{j+1}-Y_i^j)+(dSY)_{i+1}(Y_{i+1}^{j+1}-Y_{i+1}^{j)}] \tag{11.33}$$

$$(C_3)_i = \left[\frac{1}{2} - 2\theta\left(\frac{Q}{A}\right)^j_{i+1/2} + \Delta t(gA)^j_{i+1/2} + \frac{1}{4}(dSQ)^j_i\right] \tag{11.34}$$

$$(C_4)_i = \left[\frac{1}{2} + 2\theta\left(\frac{Q}{A}\right)^j_{i+1/2} + \Delta t(gA)^j_{i+1/2} + \frac{1}{4}(dSQ)^j_{i+1}\right] \tag{11.35}$$

$$(C_5)_i = \left[-\theta\left[gA - \frac{Q^2T}{A^2}\right]^j_{i+1/2} + \Delta t(gA)^j_{i+1/2} + \frac{1}{4}(dSY)^j_i\right] \tag{11.36}$$

$$(C_6)_i = \left[\theta\left[gA - \frac{Q^2T}{A^2}\right]^j_{i+1/2} + \Delta t(gA)^j_{i+1/2} + \frac{1}{4}(dSY)^j_{i+1}\right]$$

$$(C_7)_i = \Delta t\left[\frac{Q^2}{A^2}.A^y_x\right]^j_{i+1/2} + \theta(gA)^i_{i+1/2}(Z^j_i - Z^j_{i+1}) + \Delta t(gA)^j_{i+1/2}$$

$$(D1)^{j+1/2}_{i+1/2} + \frac{1}{2}\left(Q^j_i + Q^j_{i+1}\right) - \frac{\Delta t(gA)^j_{i+1/2}}{4}$$

$$\left[2(Sf)^j_i + 2(Sf)^j_{i+1} - (dSQ)_i Q^j_i - dSY_i.Y^j_i - (dSQ)_{i+1}Q^i_{i+1} - dSY_{i+1}.Y^j_{i+1}\right] \tag{11.37}$$

In the above equations all the values are evaluated at time $t^j$ and therefore are known. There are four unknowns, viz., $Q_i$, $Y_i$, $Q_{i+1}$ and $Y_{i+1}$ in equations (11.25) and (11.27) at time step $t^{j+1}$ which cause the system to be indeterminate. However, two unknowns are common for two neighboring rectangular grids. Consequently, *(N-1)* pairs of equations contain *2N* unknowns. Two additional equations supplied by the boundary conditions make this system of equations mathematically determinate.

At the upstream section, it is assumed that the gauging station flood hydrograph supplies the flow depth or discharge as a function of time. The relation for upstream boundary condition can be written as

$$C_9\,Q_1 + C_{10}Y_1 = C_{11} \tag{11.37}$$

But usually one out of $C_9$ and $C_{10}$ is zero, and the other is unity. At the downstream section, the gauging station rating curve is expressed in segment form as

$$C_{12}Q_N + C_{13}Y_N = C_{14} \tag{11.38}$$

Equations (11.25), (11.27), (11.38), and (11.39) constitute a system of *2N* linear algebraic equations in *2N* unknowns. Any of the standard methods such as the Gaussian elimination method or the Matrix Inversion method can be used for its solution. In the present case the Gauss--Jordan method is used for solution of the equations. The above numerical scheme is called linear fully implicit method. As demonstrated by Chen (1973), this method is found to be the one of the best numerical schemes for solving gradually varied unsteady flow equations and also has good numerical stability.

The flow conditions at each node section at the time step $t^{j+1}$ are computed by solving equations (11.1) and (11.2) numerically. For better accuracy of the model results, an iteration procedure has been adopted for refinement in the computation of downstream boundary coefficients $C_{12}$ and $C_{14}$. In this iteration procedure the coefficients $C_{12}$ and $C_{14}$ are refined by repeating the water routing till the change in discharge values from successive iterations at the downstream boundary comes within the tolerance limit.

## 11.5 Sediment Routing

The phenomena of erosion and deposition in alluvial water courses are basically three dimensional in nature, with the assumption that the net deposition or erosion of sediment between two neighboring sections or computational points in general uniformly distributed over the channel bed in both longitudinal and transverse directions and its nature could be simulated by a one dimensional approach. In the present case the equation (11.3) is solved for sediment routing by adopting the one dimensional approach and subsequently an attempt is made to introduce a lateral distribution of deposits or erosion related to the tractive force as already described earlier. Equation (11.3) is rewritten as:

$$\frac{\delta Qs}{\delta x} + p\frac{\delta Ac}{\delta t} + \frac{\delta As}{\delta t} - qs = 0 \qquad (11.40)$$

where $As$ = volume of suspended sediment in water over the cross section per unit length of channel.

This equation is expressed in finite difference form and ultimately can be written as follows to compute the value of $\delta Ac$, i.e., the change in the volume of deposition of the sediment on a unit length of the channel bed.

i) For the interior points:

$$\frac{\Delta Qs}{\Delta x} = \left[Qs_i^j - Qs_{i+1}^j + Qs_i^{j+1} - Qs_{i+1}^{j+1}\right] \div \left(4.\Delta x_i\right) \qquad (11.41)$$

$$\frac{\Delta As}{\Delta t} = \left[2A\rho_i^{j+1} - 2A\rho_i^j + A\rho_{i-1}^{j+1} - A\rho_{i-1}^j + A\rho_{i+1}^{j+1} - A\rho_{i+1}^j\right] \div (4.\Delta t) \quad (11.42)$$

$$qs = \frac{1}{4}\left[qs_i^j + qs_i^{j+1} + qs_{i+1}^j + qs_{i+1}^{j+1}\right] \quad (11.43)$$

ii) At the upstream boundary:

$$\frac{\Delta QS}{\Delta x} = \left[Qs_2^j - Qs_1^j + Qs_2^{j+1} - Qs_1^{j+1}\right] \div (2.\Delta x) \quad (11.44)$$

$$\frac{\Delta Qs}{\Delta t} = \left[A\rho_1^{j+1} - A\rho_1^j + A\rho_2^{j+1} - A\rho_2^j\right] \div (2.\Delta t) \quad (11.45)$$

$$qs = \left[qs_1^j + qs_2^{j+1}\right] \quad (11.46)$$

iii) At the downstream boundary:

$$\frac{\Delta Qs}{\Delta t} = \left[Qs_{n-1}^j - Qs_n^j + Qs_{n-1}^{j+1} - Qs_n^{j+1}\right] \div (2.\Delta x) \quad (11.47)$$

$$\frac{\Delta As}{\Delta t} = \left[A\rho_n^{j+1} - A\rho_n^j + A\rho_{j+1}^{j+1} - A\rho_{n-1}^j\right] \div (2.\Delta t) \quad (11.48)$$

$$qs = \left[qs_n^j + qs_n^{j+1}\right] \div (2.0) \quad (11.49)$$

$$\Delta Ac = \frac{\Delta t}{p}\left[qs - \frac{\Delta Qs}{\Delta x} - \frac{\Delta As}{\Delta t}\right] \quad (11.50)$$

The sediment routing and the associated channel adjustment processes are carried out iteratively till the change in the water levels between two successive iterations comes within the limit of tolerance. If the desired accuracy is not satisfied, then the latest value of water level and area are used for a fresh calculation of sediment discharge and energy slope for repeating the sediment routing and the channel adjustment processes.

## 11.6 Formulation of New Braid Indicators

### 11.6.1 BRIEF BACKGROUND

The channel pattern or plan form of a reach of an alluvial river reflects the hydrodynamics of flow within the channel and the associated processes of sediment transfer and energy dissipation. Channel patterns form a continuum in response to varying energy conditions ranging from straight and meandering to braided forms. Usually braiding is favoured by high energy fluvial environments with steeper gradients, large and variable discharges, dominant bed load transport and non-cohesive banks lacking stabilization by vegetation.

There is an imperative need to formulate appropriate indicators to describe braiding phenomenon and fluvial landform pattern in quantitative terms on a more logical basis. It may be mentioned here that the braided channels are hydraulically less efficient and the formation of braid bars plays an important role in the modification of the energy losses owing to friction. With a view to incorporate the effect of the above hydraulic variables, indices are formulated in this work as given in the following section.

### 11.6.2 FORMULATION OF NEW INDICES

For a more logical and quantitative description of the braiding phenomenon, the following indices are proposed hereunder:

Plan Form Index (PFI):

$$PFI = \frac{\frac{T}{B} \times 100}{N} \tag{11.51a}$$

where with reference to Figure 11.2, $T$ = Flow top width, $B$ = Overall river width, and $N$ = Number of braid channels.

Flow Geometry Index (FGI):

$$FGI = \frac{\sum di.xi}{R.T.} \times N \tag{11.51b}$$

where with reference to Figure 11.3, $d_i$ and $X_i$ are depth and width of submerged sub-channel, $T$ = Flow top width of the stream, $R$ = Hydraulics mean depth of the stream, $N$ = Number of braid channels.

The plan form Index represents the percentage of the actual flow width over the overall river width per braid channel. Obviously, this index reflects the fluvial landform disposition with respect to a given water level and its lower value is indicative of higher degree of braiding.

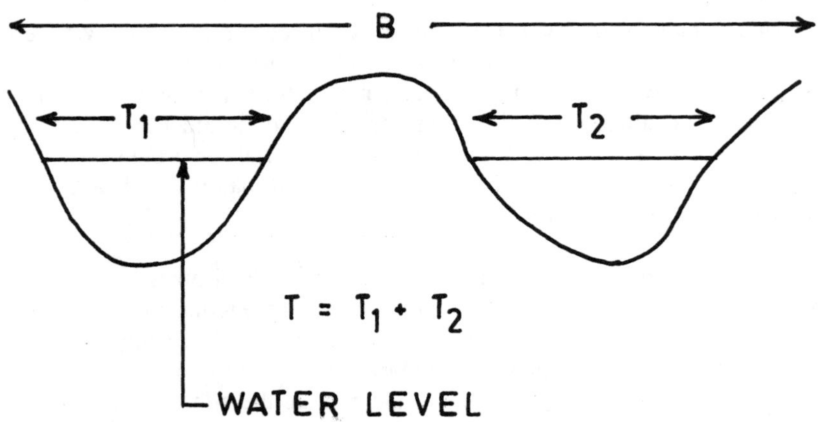


Fig. 11.2 A typical flow cross-section.

The Flow Geometry Index reflects the underwater sub-channel disposition and the hydraulic efficiency of a braided stream. Its higher value signifies the occurrence of higher degree of braiding. The cross slope is basically a form ratio indicator and its higher value indicates higher braiding intensity.

## 11. 7 Model Application

### 11.7.1 GENERAL

The simulation model formulated herein is applied to a reach of the Brahmaputra River between Guwahati and Jogighopa falling within the Assam state of India. The Brahmaputra is one of the largest alluvial rivers in the world both in terms of its size and discharge and exhibits highly braided channel characteristics throughout its course in India. These attributes make the Brahmaputra River an ideal example for undertaking the application of the model.

The Brahmaputra is the largest river in the Indian subcontinent. This river travels 918 km in India and flows into Bangladesh. The average bed gradient of the river within India (Assam) is 1 in 6,990. The maximum observed discharge of the river is 72,974 $m^3$ /s at Pandu on 23.8.1962, the minimum observed discharge is 1,757 $m^3$ /s at Pandu on 20-2-1968 and the average dry season discharge (October to March) is 4,420 $m^3$ /s. The Brahmaputra, having its origin in the Himalayan region, is a snow fed perennial river in which floods occur during the months of May to September in general. Over 80% of the annual runoff flows during this period. The average annual yield of the river at Pandu amounts to 49.43 million ha.m. Tectonic upheavals have resulted in a number of disastrous earthquakes, which are responsible for major geomorphological changes within the basin.

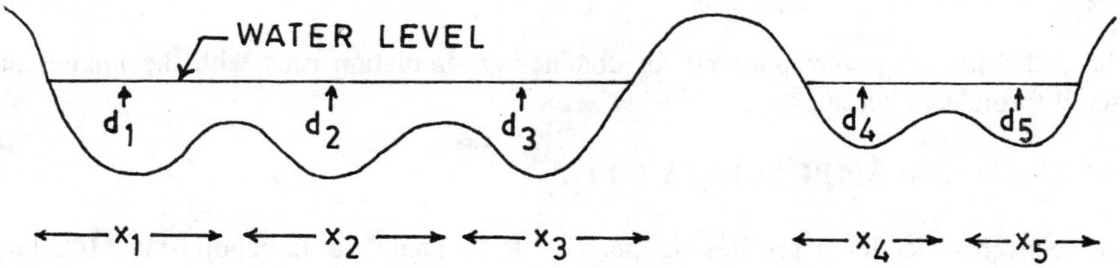


$$\text{Cross Slope} = \frac{\frac{L}{2}}{\text{Av. Bank level - Av. bed level}}$$

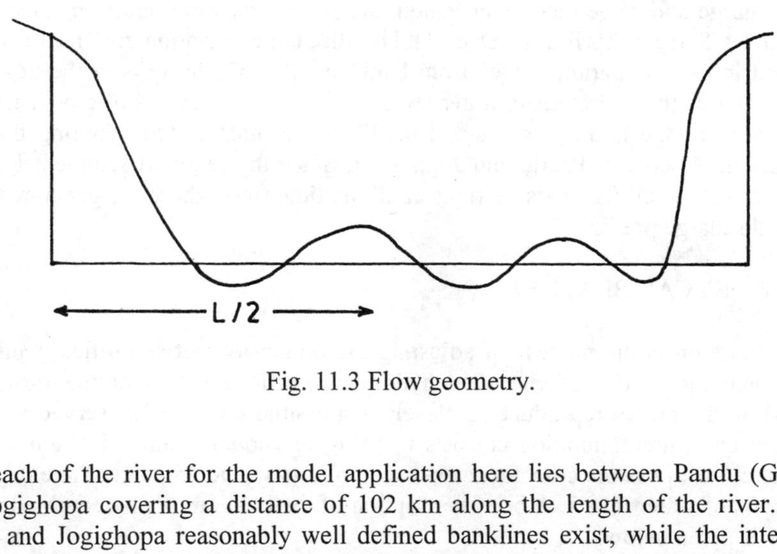


Fig. 11.3 Flow geometry.

The reach of the river for the model application here lies between Pandu (Guwahati) and Jogighopa covering a distance of 102 km along the length of the river. At both Pandu and Jogighopa reasonably well defined banklines exist, while the intermediate portion of the reach exhibits a wide and shallow braided cross sectional shape. At both ends of the reach under study, two important towns are situated namely Guwahati at the upstream end and Goalpara at the downstream end, where gauge and discharge measurements are made daily by the Flood Control Department of Assam. The contribution from the tributaries in the reach forms hardly 0.5 to 1 percent of the main stem peak flow at Pandu. The river in this reach has constricted width at some points with rock outcrops confining the river. The minimum width of the river in the reach is 3.5 km. Mostly the river flows in braided channels between alluvial banks. During floods, the river becomes one sheet of water from bank to bank and occupies a maximum width of about 19.5 km. The average bed gradient of the river in the reach is about 1 in 8,875.

The following data were adopted for conducting simulation runs with the numerical model formulated herein.

### 11.7.2 HYDROGRAPHIC SURVEY DATA

Fourteen cross-sectional profiles of the river from Pandu to Jogighopa in a 102 km reach length pertaining to year 1977 were obtained from the records of Flood Control Department, Assam. These cross sections were used in the simulation runs. The cross-section No. 22 is located in the upstream end at Pandu near Guwahati, while the cross section No. 9 is situated at the downstream end of the study reach, i.e., Jogighopa. C/S 2 and C/S 9 have relatively rigid constricted bank lines due to presence of rock outcrops at both the sites. While a road-cum-rail bridge is existing at C/S No. 22, a bridge has recently been constructed at C/S No. 9 making these two sites nodal points in the reach. The remaining 12 of the intermediate cross-sections of the reach depict the braided character of the river with a shallow and wide channel configuration.

### 11.7.3 HYDROLOGICAL DATA

Daily discharge and stage records at Pandu and Jogighopa were obtained for the years 1977 and 1978 from April 1 to Dec. 31.The discharge variation for the hydrograph considered in the simulation ranges from 8,884 $m^3/s$ to 53,389 $m^3/s$.In the absence of observed records the intermediate water levels at the intermediate 12 of cross sections at the beginning of the hydrograph, i.e., 1.4.1977 were interpolated adopting the water surface gradient between Pandu and Jogighopa. Since the required sediment load data are not available at all the cross-sections at all the time steps, these are generated by the sediment discharge predictor.

### 11.7.4 MODEL CALIBRATION

Model calibration is the process of adjusting the dimensions of simplified geometrical elements and the values of empirical hydraulic coefficients so that the flow events simulated on the model reproduce as closely as possible the actual observed values. In the present case, the calibration consists in adjusting model features in such a way to obtain the closest possible coincidence between observed and computed time dependent hydrographs of water levels. A brief description of the calibration procedure adopted in this study is given below:

i) The space increment of the model $\Delta x$ is kept the same as per the actual available cross- sectional data. As such the space increment varies from reach to reach as per actual survey.

ii) The size of time increment Δt is kept the same as per the actual interval of observations which is 24 hours.

iii) The lateral flow, ql in the absence of available data, is estimated by computing the difference in volume between the inflows at the upstream boundary at Pandu (Guwahati) and at Jogighopa for a period of time with the assumption that the change in storage for this period will be negligible. Here the lateral flows are calculated on monthly basis.

$$q_1 = \frac{-\sum q_{u/s} - \sum Q_{d/s}}{L} \tag{11.52}$$

where $Q_{u/s}$ and $Q_{d/s}$ are the upstream and downstream discharges respectively and L is the total length of the reach. The lateral flow variation within the month period is approximated by observing the rising and falling limbs of the hydrograph, ensuring that the mean of the distributed lateral flow over the period of time remains the same as given by equation (11.52). The inflows from two tributaries namely the Pagladiya between section 19 and 18, and the Manas between sections 10 and 9 are incorporated at the above specified locations for the total time period.

iv) Downstream boundary condition: The available stage discharge records for the period 1977 at downstream section (Jogighopa) were used for the purpose of developing the relationship for downstream boundary condition. To represent the downstream boundary, the following power function is fitted to the observed data by multiple regression analysis for falling and rising limbs of the hydrograph separately.

$$h = CQ_m^a Q^b \tag{11.53}$$

where $Q_m$= threshold discharge after which the trend of the hydrograph has changed, $Q$ = Discharge at the section, $h$ = Stage corresponding to discharge $Q$, and $a$, $b$, $c$ = Regression coefficients depending upon rising/falling limb of hydrograph.

After carrying out multiple regression analysis of the stage discharge data, the computed values of the regression coefficients for both rising and falling discharges of the hydrograph are as follows:

$a = -0.01$
$b = 0.087$ for rising discharge
$c = 15.136$

and

$a = 0.007$
$b = 0.072$ for falling discharge
$c = 14.958$

The boundary condition is represented in a linear form as stated earlier with the following equation:

$$C12Q + C13Y = C14$$

For $C13 = 1$, the equation becomes:

$$C14 = C12.Q + (h - z),$$

where $h$ is stage and $Z$ is bed elevation. It may be noted that $C12$ and $C14$ as calculated in the above equation is a function of discharge $Q$ itself. Therefore, it is revised iteratively till the change in the value of discharge due to the revision of $C12$ and $C14$ is within the accuracy limit.

v) **Manning's Roughness Coefficient '*N*'**

The roughness coefficient '*N*' is calibrated by varying its value till the cumulative error between the computed and observed water levels at Jogighopa comes to a minimum possible value. On the basis of the above exercise the minimum cumulative error was obtained at $N = 0.038$. This calibrated value was used as Manning's N in the computations.

vi) **Initial Flow Conditions**

In the absence of observed data the initial stage and discharge at each section (other than the first and the last sections) are estimated by linear interpolation between upstream and downstream discharges at $t = 0$.

vii) **Sediment Discharge**

In the absence of measured sediment discharge data for every time step at the 14 of cross-sectional points, the same were generated by using the sediment transport relationship given by Swamee and Ojha. To ascertain the suitable method for the particular case, computer runs were made with the different sediment discharge predictors available in the model and the resulting stages were compared with the corresponding observed ones. The methods given by Swamee and Ojha as well as that of Karim and Kennedy indicated minimum error compared to the other sediment discharge predictors in this case. Hence, the method of Swamee and Ojha was adopted for generation of sediment discharge data.

### 11.7.5 SIMULATION STUDIES

With the available data of the Brahmaputra River, simulation studies were conducted for the purpose of studying the model behaviour on a large alluvial river. These studies were planned with a view to (i) verify the model with the help of measured stage data available for Jogighopa site on the Brahmaputra, (ii) derive functional relationships for the braid indicators proposed herein on the basis of the simulation studies, (iii) determine threshold values for the braid indicators for classifying the degree of braiding, and (iv) study the model response on the face of different hypothetical river improvement measures.

### 11.7.6 MODEL VERIFICATION

Since the cross sectional geometries were available only for the beginning of the simulation period, model calibration and verification was done with the help of measured water stages at Jogighopa site for the years 1977 and 1978, respectively. Figs. 11.4 and 11.5 show the computed and observed water stages at Jogighopa for the years 1977 and 1978 respectively. From the figures, it can be seen that the simulated stages are in fair agreement with the measured values. Amongst the hydrological data, water level can be considered to be the most reliable primary data with minimum error in comparison to other data. Water stages get affected by overall channel changes such as aggradation and degradation which are part and parcel of the channel dynamics of an alluvial stream. In the above context, the good reproduction of stages can be viewed as quite encouraging for the simplified modelling approach considered for highly braided alluvial streams like the Brahmaputra River.

## 11.8 Evaluation of Functional Relationships for Newly Defined Braid Indicators for the Brahmaputra

On the basis of the computer-based simulation studies undertaken in this work functional relationships have been evolved for the braid indicators newly defined to study their variation as well as to make them useful for practical applications subsequently. Towards this end the output of the model runs were processed, and plots were developed and best fit lines were drawn for obtaining the functional relationships as given below:

$$PFI = 92038\ (B/D)^{-1.40} \quad (11.55)$$
$$FGI = 0.025\ (B/D)^{+0.84} \quad (11.56)$$
$$PFI = 27087\ (C_r\ S)^{-1.35} \quad (11.57)$$
$$FGI = 0.05\ (C_r\,.S)+0.50 \quad (11.58)$$
$$FGI = 130.07\ (PFI)^{-0.98} \quad (11.59)$$
$$FGI = 0.8\ S\ w^{-0.447} \quad (11.60)$$
$$FGI = 1413\ S_w + 7.8 \quad (11.61)$$
$$(S_F/F_R) = 2.74 \times 10^{-6}\ (C_r\,.S)^{0.53} \quad (11.62)$$
$$(S_F/F_R) = 1.235 \times 10^{-6}\ (B/D)^{0.83} \quad (11.63)$$

where *PFI* = Plan Form Index, *FGI* = Flow Geometry Index, *B/D* = Overall Width/Av. Depth, $C_r\ S$ = Cross Slope, $S_w$ = Water Surface Slope, and $S_F/F_R$ = Energy Slope/Froude number.

### 11.8.1 THRESHOLDS FOR NEW BRAID INDICATORS

The basic components of the channel form are the mean dimensions of width and depth of the cross section. Form ratios considered in this work namely B/D ratio and Cross Slope are the major controlling variables wielding a profound influence on the intensity of braiding in an alluvial stream. There is reason to believe that the form ratios like B/D and Cross Slope closely reflect the integrated effects of discharge, sediment supply, local bank erodibility along with various associated channel processes like aggradation and degradation. Furthermore, in practical cases these form ratios can very easily be worked out without requiring much detailed informations. In this backdrop it was intended to classify the braiding phenomenon by correlating the numerical values of the form ratio with the indices proposed in this work. Towards this end the trend of the available surveyed cross sections of the Brahmaputra River along with the computed values of the braid indicators were analysed together and the following thresholds were identified to provide a broad range of classification of the braiding phenomenon:

| Parameter | Range for Moderately braided | Range for Highly braided |
|---|---|---|
| B/D | $350 \leq B/D \leq 1000$ | $B/D > 1000$ |
| Cross-Slope | $170 \leq C_r.S \leq 800$ | $C_r . S > 800$ |
| Plan Form Index | $19 \geq PFI \geq 4$ | $PFI < 4$ |
| Flow Geometry Index | $7 \leq FGI \leq 35$ | $FGI \geq 35$ |

### 11.8.2 MODEL RESPONSE WITH RIVER IMPROVEMENT WORKS

To study the model behaviour with different river improvement works, simulation runs were made for the study reach of the Brahmaputra River. The simulation runs were organized for the following three cases :

i) With natural river condition without any intervention;
ii) By jacketting the river with dykes from the upstream end to section 17; and
iii) By jacketting the river with dykes along with dredging up to section 17.

The second and third cases were conceived as purely hypothetical situations for purposes of studying the model behaviour with interventions in the natural river environment.

## 11.9 Discussion of Results

As stated earlier, the reasonably good correspondence between computed and measured water stages, vide Figures 11.4 and 11.5, at the Jogighopa site is a clear indication that the simplified strategy adopted in this work is a viable solution for modelling of braided water courses involving large widths and sustained high discharges. The evolved relationships of Plan Form Index with *B/D* ratio and Cross slope for the Brahmaputra River show an exponentially decaying function for the braid indicator. With increasing width to depth ratio the Plan Form Index displays a decreasing trend thereby registering an increasing level of braiding as can be seen from the threshold limits. Similar findings are observed in the relationships between Flow Geometry Index and with the *B/D* ratio and cross slope. The Flow Geometry Index increases with increase in the values of the two form ratios which is again indicative of the increasing intensity of braiding as evident from the threshold ranges. The relationship between the Plan Form Index and

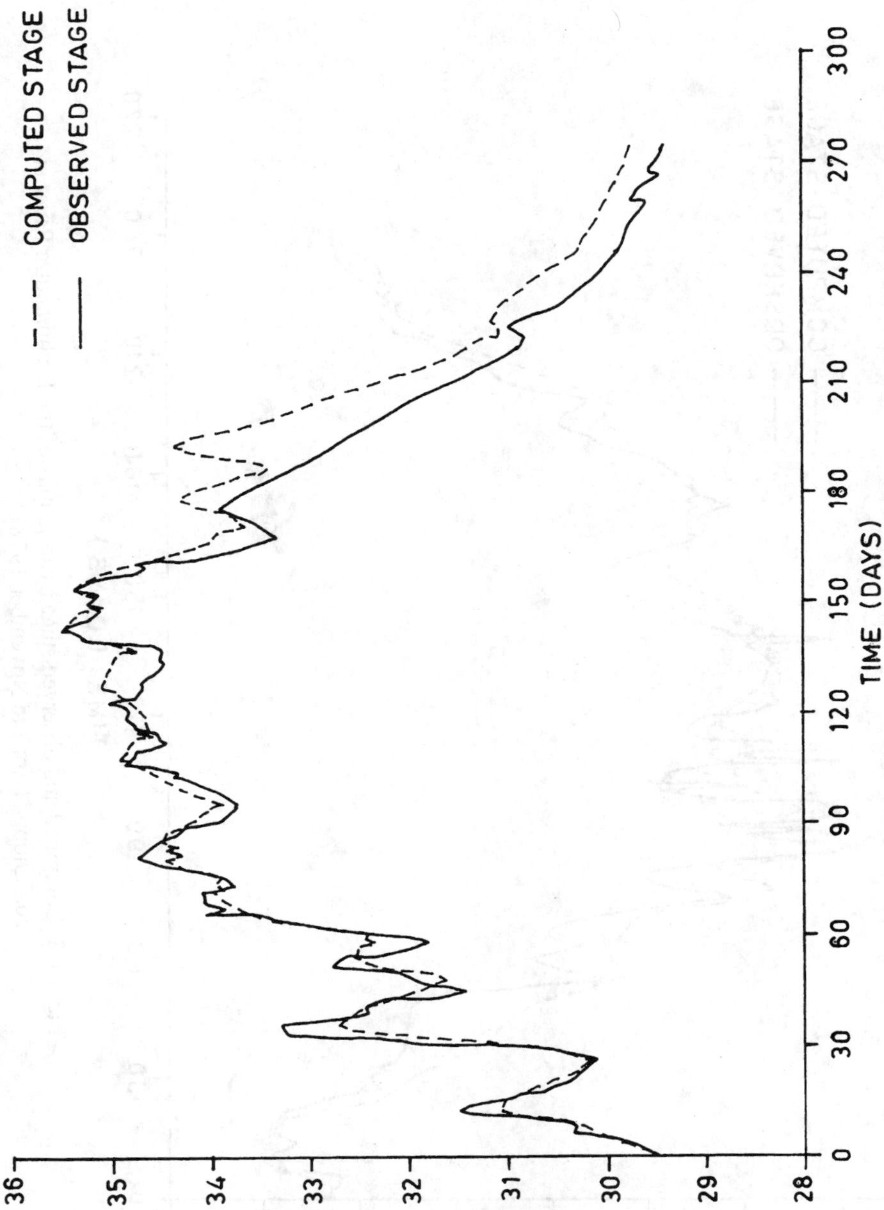


Fig. 11.4 Computed and observed stage hydrographs of the Brahmaputra River at Jogighoppa.

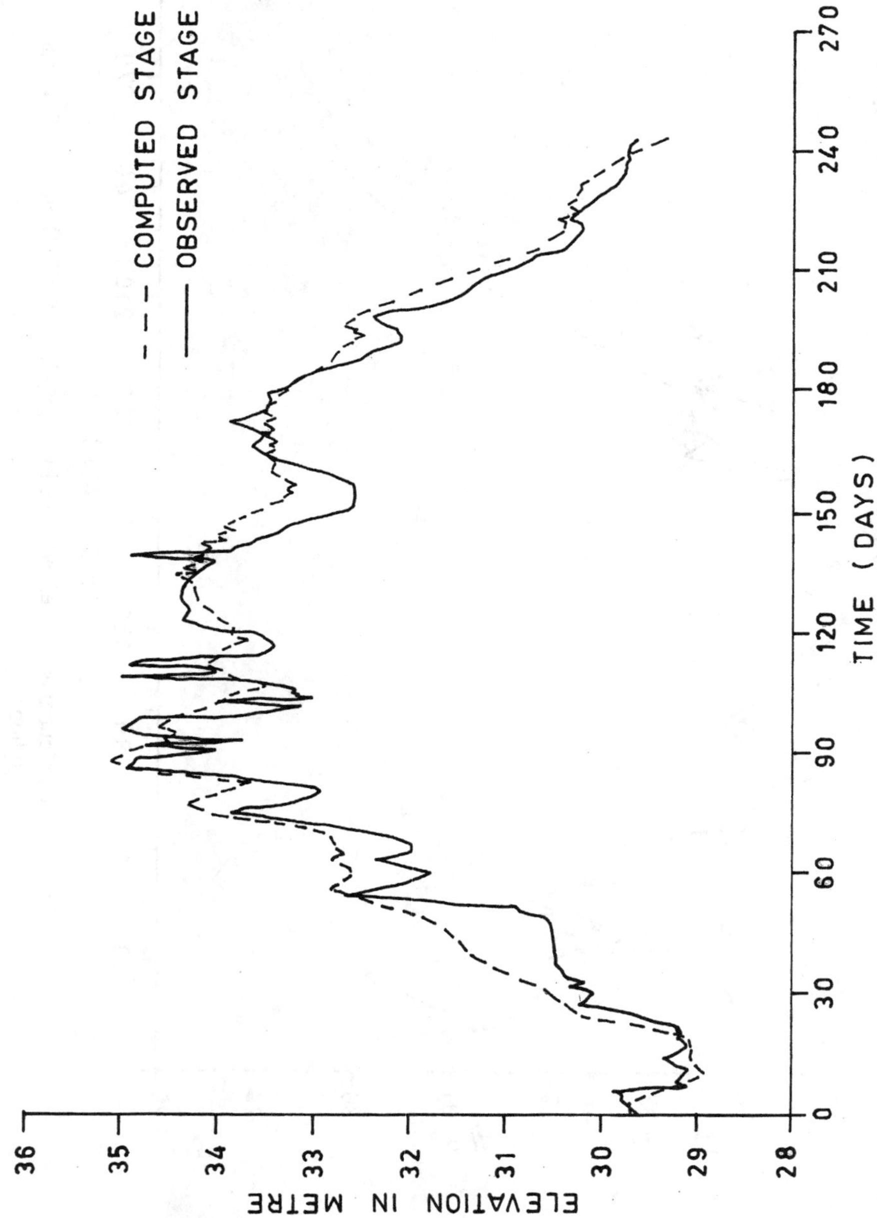


Fig. 11.5 Computed and observed stage hydrographs of the Brahmaputra River at Jogighoppa (April to November 1978).

the Flow Geometry Index conforms to their pattern of variation as per threshold limits. Their relationships clearly demonstrate the rationality of the braid indicators formulated in this work and conform to the actual physical situation of the phenomenon of braiding. The basic aim of these relationships was to further verify the rationality of the above two indices with regard to their general trend with longitudinal water surface slope. It is now an established fact that relatively steeper gradients are indicated in braided reaches and their magnitude registers an increase with rise in the intensity of braiding. The general trend displayed by the relationships conforms to the above stated pattern of variation.

On a line similar to that of Parker (1976), graphical plots made to study the variation of form ratios with the energy loss parameter $S_f / F_r$, where $S_f$ is energy gradient and $F_r$ is the Froude number, has confirmed that braided channels are less hydraulically efficient than single channels and so braid represents a major modification of energy losses. The braided pattern is conducive for the dissipation of the excess energy in high energy streams, since the enhanced total flow resistance of the multithread braided channel results in a rapid energy loss. In the light of the above, it can be seen that the energy loss parameter $S_f/F_r$ registers an increment with increase in *B/D* and cross slope. This signifies an occurrence of increased energy loss through increased boundary friction and results in lower velocities in the braided channel.

The results of the simulation runs to study the model behaviour with different hypothetical interventions through dyking and dredging were processed and have exhibited an acceptable river response to the simulated channel improvement interventions. Expectedly, dredging with dyking results in reduction of braiding for obvious reasons. Also the plot indicates that cross sections 18 and 13 are the highest braided sections in the study reach. These graphical plots highlight that the *B/D* ratio and cross slope convey a fairly good description of the extent of the overall braiding quantitatively at the macro scale level, but the Plan Form Index and the Flow Geometry Index provide a better indication of the braiding by accounting for the channel bed topography at the micro scale level.

While the results of the above simulation studies appears to be reasonable, the sensitivity of the model could not be verified as no observations on such modifications exist. The outcome of the simulation runs for the Brahmaputra River can be generalized as below:

i) Dyking caused deposition in the thalweg as well as raising of the average bed elevation in the leveed reach.

ii) The backwater effect of the water surface profile owing to the presence of the levees gets eliminated through supplementary channel improvement measures in the form of dredging operations.

iii) On the downstream portion of the leveed reach, dyking did not indicate any significant effect, while dyking and dredging in the upstream reach caused lowering of the thalweg in the downstream reach.

iv) The general tendency of the Brahmaputra River in the reach under study is found to be raising of its stream power in the post flood period with a corresponding decrease in the actual flow area owing to bed deposition. This trend can be arrested by resorting to dyking and dredging which obviously results in the corresponding reduction in braiding, as characterized by the decrease in the form ratio and the Flow Geometry Index and increase in the value of the Plan Form Index.

## 11.10 Summary and Conclusions

In this work a model has been developed for simulating water and sediment routing along with aggradation, degradation and width adjustment processes in braided alluvial channels on the basis of some modification in the existing technique and also evolved suitable braid indicators and their thresholds from its application to the Brahmaputra River. The new braid indicators provide a quantitative measure of the braiding phenomenon and may be utilized in river morphology studies as well as for assessment of channel improvement programmes.

The model is based on unsteady flow equations of water and sediment flow. Braided streams such as the Brahmaputra River are interspersed with the formation of braid bars on a highly random pattern with flow channels of variable conveyances. In the case of large alluvial streams like the Brahmaputra with high discharge and heavy sediment load, the braid bars quite often assume sizes which may run into a few kilometeres on both longitudinal and transverse directions of the river flow. This kind of channel morphology poses difficulties at the time of numerical modelling for flow simulation, especially for hydraulic representation of the complex channel geometry. The main thrust is in numerical modelling of braided alluvial streams such as the Brahmaputra River.

The model herein adopts an uncoupled solution procedure in which the highly irregular braided stream section is substituted by an equivalent rectangular section in the water routing phase. The equivalent depth and width of the above rectangular section is worked out in direct proportion to subchannel conveyances. After completion of the water routing phase the actual river cross section is considered for the purpose of sediment routing and channel adjustment processes. In addition, the bed deposition is considered to be in inverse proportion to conveyances of the subsections, whilst the degradation is effected in direct proportion to the net tractive force across the cross-section. The above apparently simplified strategy has been applied to a 102 km long reach of the Brahmaputra River in India, which is a highly braided large alluvial river. The model was verified with the help of observed water stages as available at the outflow section of Jogighopa.

Furthermore, in this work an attempt has been made to formulate rational new braid indicators which duly reflect and account for the presence of submerged bars in addition to fluvial landform. Two new indices, namely the Plan Form Index and Flow Geometry Index, along with a form ratio are proposed in this work and their functional relationships are evolved from the simulation model runs made on the Brahmaputra River. In addition, thresholds for these braid indicators are determined after analysing the trend of the available hydrographic data along with the simulated information from the model runs. The model behaviour was studied by conducting simulation runs for the Brahmaputra River for hypothetical cases of dyking and dredging. This also facilitated testing of the newly formulated braid indicators with real life data of the Brahmaputra River.

On the basis of the above study, the conclusions are summarized below:

1. The model application on the Brahmaputra River has amply brought out that numerical modelling in large alluvial rivers with highly braided channels can be carried out by adopting an uncoupled solution procedure wherein the braided channel is substituted by an equivalent channel in the water routing phase, while the actual braided cross section can be considered for sediment routing processes.

2. The braid indicators evolved in this work, namely, Plan Form Index, Flow Geometry Index and Cross Slope were found to be rational and meaningful for purposes of a quantitative description of the braiding phenomenon. The relationships developed for the braid indicators suggest that the degree of braiding is directly proportional to the overall form ratio of the braided channel.
3. The simulation results indicate the occurrence of increased energy loss through enhanced boundary friction over a larger wetted perimeter and resulting lower velocities in braided sections. Predictably this makes the braided channels hydraulically less efficient than the unbraided water courses and as such braid represents a major modification of the energy loss process which profoundly affects the stream power and channel morphology.

4. The present study has indicated that in high energy fluvial environments of a large alluvial stream, such as the Brahmaputra characterized by large and variable discharges, dominant sediment transport and erodible banks, the thresholds of form ratio like width/depth for indicating braiding lie in the higher range than those of the smaller alluvial streams which is of the order of about 50 as indicated in literature.

5. The results of the simulation runs with the data of about 102 km long reach of the Brahmaputra River appear to be reasonable, demonstrating the potential application of the model. The general observation emerging from the simulation studies with hypothetical interventions indicate the efficacy of dyking with dredging as a channel improvement measure which however needs economic analysis before adoption. However, the sensitivity of the model could not be verified as no observations on such modifications exist.

## References

Chang, H.H. (1982). "Mathematical Model for Erodible Channels", Journal of Hydraulic Division, ASCE, VOL. 108, No. HYS, May.

Chen, Y.H. (1973). "Mathematical Modelling of Water and Sediment Routing in Natural Channels", Dissertation presented to the department of Civil Engineering, Colorado State University, Fort Collins, Colorado, in partial fulfillment of the requirements for the degree of Doctor of philosophy.

Cunge, J.A., Holly, F.M., and Verway, A. (1980). "Practical Aspects of Computational River Hydraulics", Pitman Publishing Limited, London.

Dass, P. (1975). "Water and sediment routing in non-uniform channels" Doctoral Dissertation submitted to Colorado State University, Fort Collins, Colorado, 1975.

Grijsen, J.G. (1986). "River Flow Simulation", Chapter 9 of River Flow Modelling and Forecasting", Edited by D.A. Kraijenhoff and J.R. Moll, D. Reidel Publishing Company.

Karim, M.F. and Kennedy, J.F. (1990). "Menu of Coupled Velocity and Sediment-Discharge Relations for Rivers", Journal of Hydraulics Division, ASCE, Vol. 116, No.8, August. .

Mahmood, K., and Yevjevich, V. (1975). "Unsteady Flow in Open Channels", Vol.I, Water Resources Publications, Colorado.

Garde, R.J. and Ranga Raju, K.G. (1988). " Mechanics of Sediment Transportation and Alluvial Stream Problems", Wiley Eastern Ltd., New Delhi.

Osman, A.M., and Thorne, C.R. (1988). "River bank stability analysis I: Theory", Journal of Hydraulics Division, ASCE, Vol. 114, No.2, February.

Sharma, N. (1995). " Modelling of braided alluvial channels", Dissertation presented to University of Roorkee in fulfillment of the requirement for the award of the degree of Doctor of Philosophy.

Swamee, P.K. and Ohja C.S.P. (1991). "Bed Load and Suspended Load Transport of Non-Uniform Sediment", Journal of Hydraulics Division, ASCE, Vol.117, No.6, June.

# CHAPTER 12

# HYDRAULIC STRUCTURES

A. K. SARMA

The water resources potential of the Brahmaputra basin is the highest in India, yet its utilization is the lowest. The basin is yet to have a major hydraulic structure for harnessing its tremendous water resource potential. Of all the several major dams which have been suggested for construction, none have been completed. This chapter reviews the various types of hydraulic structures that have been completed or are under construction.

## 12.1 Introduction

Brahmaputra, the gift of nature to the northeastern part of India, is now becoming a source of misery for the people residing in this region. The water resource potential of this region is the highest in the country while the present utilization is the lowest. The basin carries 585 billion cubic meters (BCM) of annual surface flow, 1,869 BCM being the total annual flow of all the basins of India. Under-utilization of fabulous water resources potential of this region, which amounts to 38% of the country, has manifested itself in the form of flood--- a disaster for this economically backward region. The basin is yet to have a major hydraulic structure for harnessing the tremendous water resources potential. Although the Brahmaputra Board has suggested construction of dams at Pagladia, Tipaimukh, Dihing, and Subansiri, the project implementation is still suffering severe setback owing to various specific problems. Significant progress has been made only in the construction of a Concrete Dam on the Ranganadi River, a major tributary of the Brahmaputra. However, the Irrigation Department of Assam has completed construction of some barrages under various irrigation schemes. On the other hand, the length of the embankments constructed in this region to provide temporary protection against floods amounts to 28% of those in the country. Therefore the hydraulic structures constructed so far in the Brahmaputra basin can be discussed under the following two headings:

1. River dykes or embankments constructed for flood protection
2. Hydraulic structures constructed under irrigation schemes.

## 12.2 River Dykes

The devastating effect of floods is mitigated all over the world by adopting several temporary and permanent flood control measures. Out of all such measures the construction of embankments and a few bank protection works have been adopted so far



*V.P. Singh et al. (eds.), The Brahmaputra Basin Water Resources*, 261-273.

in the Brahmaputa and the Borak basin. The northeastern part of India contains 12% of the total flood prone area of the country as against 7.7% of its geographical area. The approach of controlling flood by constructing embankments is not new in this region. Embankments were constructed along the Brahmaputra, Dihing, Disang, Dikhow, and Dorika rivers of Sibsagar district during the Ahom period (1570--1642). The problem of flood and erosion in the Brahmaputra valley was examined by a number of committees and experts in the past. In 1947 Sri S.C. Mazumdar, member of the Central Water Irrigation and Navigation Commission, opposed the construction of embankments. In fact he suggested the construction of reservoirs and gradual removal of embankments. However, taking all other constraints into consideration the High Level Committee on floods, appointed by the Government of India in 1958, found this proposal impracticable and advocated in favor of embankment construction for flood protection.

At present, as per records of 1995, the state of Assam contains about 4,721 km of embankments, which constitute about 85% of those in Northeast India. These embankments serve the purpose of protecting 1.745 Mha of flood prone areas of Assam. But the performance of these embankments is not at all satisfactory. The specification of the embankments had to be changed from time to time. Up to 1960, embankments having a crest width 8 to15 ft. and H.G. line gradient of 1:4 to 1:6 were adopted. But after the devastating flood of 1962 this specification had to be modified to convert all the 1:4 H.G line coverage to 1:6 and a provision for wider crest width (>15ft) had to be made. In spite of such modifications breach of embankments continued to occur. In 1970 the Brahmaputra Flood Control Commission was constituted. After analyzing the whole situation the Commission finalized the following specification for the Brahmaputra dyke:

| | |
|---|---|
| Crest width | = 4.5m |
| Free board | = 1.5m |
| H.G. line | = 1:8 |
| River site slope | = 1:3 |
| County side slope | = 1:2 & 1:8 (H.G. line) |
| H.G. coverage | = 0.6m |

It was proposed to follow these specifications both for construction of new embankments as well as to raise the existing embankments. But experience has shown that sudden and frequent failure of embankments is continuously causing great damage to nearby areas. Some investigators are continuously putting their effort towards the development of a new technology of dyke construction so that the sudden failure of dykes can be controlled. Recently Sarma and Katoki (2000) carried out an experimental study to investigate the possibility of controlling dyke failure by using geotextile and encouraging results were obtained.

### 12.2.1 CAUSES AND TYPE OF FAILURE

An earthen embankment may fail as a result of various causes. The causes and types of river dyke failure can be grouped into the following categories:

(a) Hydraulic failure

i) Overtopping
ii) Erosion of upstream
iii) Erosion of downstream owing to heavy rain
iv) Erosion of downstream owing to tail water

(b) Seepage failure
i) Piping through embankment
ii) Sloughing of downstream owing to seepage

(c) Structural failure
i) Sliding of upstream face owing to sudden draw down
ii) Sliding of downstream slope owing to continuous saturation
iii) Inadequate cross-section
iv) Insufficient H.G. line coverage

(d) Others
i) Owing to afflux of bridge
ii) Structural failure owing to sudden thrust of floodwater
iii) High pressure of floodwater from county side

(e) Cutting of embankment
i) Cut by miscreant
ii) Public cut

347 breach records collected from the river dykes of Assam from the year 1990 to 1994 have revealed that most of the failures occur from hydraulic causes. The percentage of different types of failure are given in Table 12.1. However, failure can occur from the combined effect of various causes as well. For example, an insufficient H. G. line coverage may lead to sloughing of the downstream toe owing to seepage, which in turn may result in sudden failure as the cross-section becomes inadequate to maintain structural stability. Again both seepage and over topping failure may occur simultaneously. Therefore the causes of failure mentioned in Table 12.1 should be considered as the prime cause only and not as the lone ones.

Table 12.1 Percentage of different types of breaches

| Sl. No. | Causes | Percentage | |
|---|---|---|---|
| 1 | Hydraulic failure | | 86% |
| | (i) Overtopping | 64% | |
| | (ii) Erosion | 22% | |
| 2 | Seepage | | 7% |
| 3 | Overtopping & Seepage | | 2.5% |
| 4 | Others | | 4.5% |

## 12.2.2 BREACH CHARACTERISTICS

### *12.2.2.1 Shape of Breach*

The data of 326 breach records have revealed that the length of breach is quite large as compared to the depth of breach. In most cases breach length becomes more than ten times the depth of breach. Therefore in case of dyke failure the shape of breach can reasonably be assumed as rectangular.

*12.2.2.2 Extent of Breach*

In the case of failure owing to overtopping, the extent of breach depends upon the duration of overtopping, material of construction and some other minor factors. Owing to overtopping an initial breach is formed, which gradually enlarges if overtopping continues. This enlargement continues till the inflow velocity drops to such a level that the shear force exerted by the flowing water becomes just insufficient to cause further erosion, i.e., until equilibrium is reached. In the case of pipe failure the initial breach formation is, however, quite different. Once the pipe is formed, the initial breach progresses quite rapidly until the pipe becomes unstable and collapses. But after this the process of breach enlargement is the same as that of the overtopping failure. From the above discussion regarding the mechanics of breach formation, it appears that the final breach width is independent of the type of failure. Therefore for predicting final breach width, no special consideration as to whether the failure is a piping failure or an overtopping failure is needed.

Different models with various degrees of sophistication are available for computing the final breach width in case of failure of earthen dam. But for implementation of these models a wide range of data about soil characteristics, breach development time, etc., are necessary. Unfortunately for the river dyke of the Brahmaputra basin no records about breach development time is available. Again unlike a dam, dykes of the Brahmaputra basin are distributed over a large area. Physico-chemical properties of soil will obviously differ significantly from place to place. Therefore, development of a generalized deterministic model for computing breach width for river dyke of Brahmaputra basin is not at all justified. Sarma and Das (1999) proposed a probability based approach to predict a reasonable breach width for the purpose of simulating flood propagation from an opening in the river dyke. From the breach record of 326 cases it was observed that the final breach width ranges from as small as 5 m to as large as 800 m. However, for most of the failures breach width lies between 10 m to 60 m. Figure 12.1 shows frequency of failure against breach width. From the available data the curve showing the probability of breach width being less than a particular value can be developed, as shown in the Figure 12.2. One should refer to the breach width corresponding to higher probability if the failure would result in great damage to the nearby area.

## 12.3 Hydraulic Structure Constructed Under Irrigation Project

The need of harnessing water resources potential of the mighty Brahmaputra cannot be over- emphasized. However, taking up a project in the Brahmaputra River is actually a difficult task. Difficulties arise not only in the technical aspect but in the political aspect as well. In most cases a project on the Brahmaputra River requires the consent of two or more neighboring states. Although the North Eastern Council of India (NEC) has sponsored some proposed projects of such a nature, these are yet to be implemented. Therefore to utilize the water resources of Brahmaputra basin for the purpose of irrigation, the Irrigation Department of Assam has taken up several projects in the tributaries of the Brahmaputra River. Depending upon the value of the Culturable Command Area (CCA), the schemes are classified as major, medium, and minor. Minor schemes basically aim at exploring ground water for irrigation purposes by installing a deep tube well. The major and medium schemes aim at tapping the water from the rivers and tributaries and therefore involve construction of hydraulic structures. Hydraulic structures such as barrage, canal systems, aqueduct, canal drop, cofferdam, etc., have been constructed under these irrigation projects. A list of completed or ongoing important projects taken up by Irrigation Department of Assam is given in Table 12.2.

12.3.1 CONSTRUCTED/PROPOSED HYDRAULIC STRUCTURES

A brief description of the various hydraulic structures constructed or proposed for construction under different Irrigation schemes is given below.

*12.3.1.1 Sukla Irrigation Project*

This major project has the capacity of irrigating 808,000 acres of land annually. Under this scheme a weir has been constructed across the Sukla River, a tributary of Puthimari River in the Brahmaputra basin, in the Kamrup district of Assam. The salient features of hydraulic structures are given bellow.

**The weir:**

| | | | |
|---|---|---|---|
| Overall length: | 156 m | River sluice: | 2 Bays, 6.00 m each |
| Maximu discharge: | 50,000 cumecs | Piers for Sluices: | 2, 1.50 m each |
| Spillway: | 5 Bays, 24.40 m each | Fish pass bay: | 1 Bay, 2.13 m each |
| Piers for spillway: | 6 , 1.50 m each | Piers of Fish ladder: | 1 , 1.50 m each |
| Scouring sluice: | 2 Bays, 6.00 m each | Divide wall: | 2 , 1.50 m each |

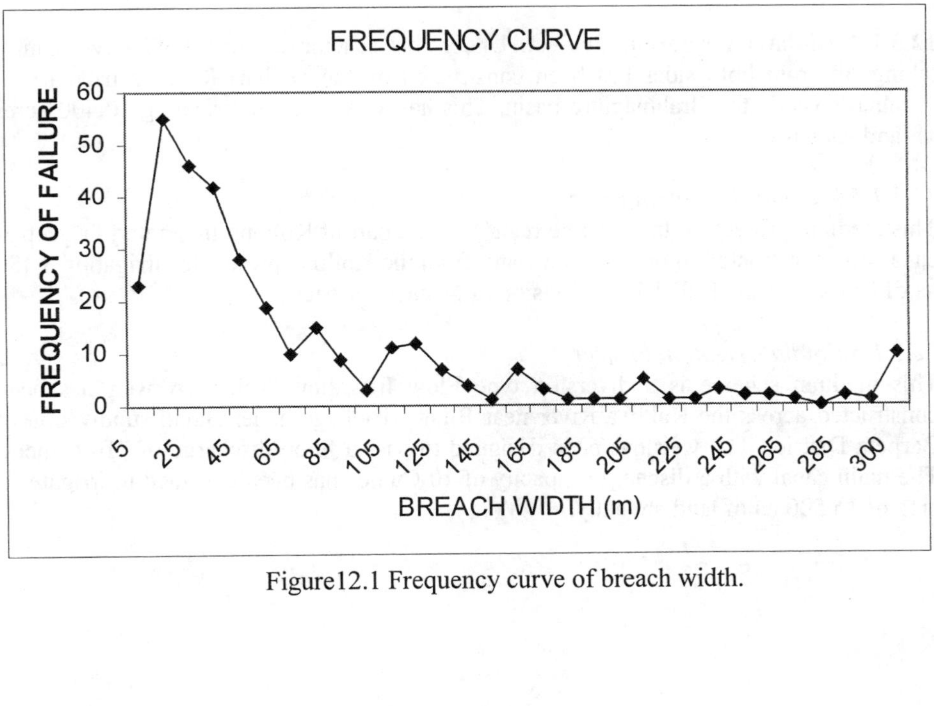


Figure12.1 Frequency curve of breach width.

*12.3.1.2 Jamuna Irrigation Project*

This major scheme has a capacity of irrigating 95,740 acres of land annually. A concrete weir has been constructed across the Jamuna River. Weir and distribution canals are the hydraulic structures constructed under this scheme.

**The weir:**

| | | | |
|---|---|---|---|
| Overall length: 302.4 m | | Width of Divide wall: | 1.53 m |
| Spillway: 18 Bays, 12 m each | | Divide walls(D/S): | 1 of 75.3 m |
| Piers of Spillway: 17,1.53 m each | | Divide walls(U/S): | 2 of 32.5 m |
| crest level of weir: R.L. | 91.60 m | Width of Fish ladder: | 2.13 m |
| Under Sluice: 3 Bays of 6 m each | | Crest level of sluice: | R.L. 89.50 m |
| River Sluice: 4 Bays of 6 m each | | Pond level: | R.L. 92.80 m |

**Canal System:**

| | |
|---|---|
| Length of the main canal : | 4725.5 m |
| Length of branch and distributaries canal: | 263705.5 m |

*12.3.1.3 Longa Irrigation Project*

This medium scheme on Longa River in the Kokrajhar district in the Brahmaputra basin has the capacity of irrigating 19,620 acres of land annually. Details of the barrage constructed under this scheme are given bellow.

**The Barrage:**

| | | |
|---|---|---|
| Design discharge:18,000 cumecs | Pond Level: 60.30 m | |
| Floor length: 36.5 m | H.F.L. | 60.92 m |
| Floor width: 75.6 m | Crest level: | 57.90 m |
| Bays: 12 of 5.8 m | U/S Floor level: | 56.55 m |
| Piers: 11, diameter1.53 m | D/S Floor level: | 55.95 m |

**12.3.1.4 Dikhari Irrigation Project.** Under this scheme a barrage with two canals taking off from both sides has been constructed on the Dikhari River, a tributary of Jamuna River in the Brahmaputra basin. This has a capacity of Irrigating 10,400 acres of land annually.

*12.3.1.5 Kaliabar Irrigation Project*

This medium project, in fact, can be regarded as a part of Kollong Integrated Project. A canal was constructed to divert some water from the Kollong project for irrigating 8,480 ha of land under the Kaliabar subdivision of Nagaon District.

*12.3.1.6 Kaldia Irrigation Project*

This medium scheme is a diversion type Flow Irrigation Project. A weir has been constructed across the Kaldiya River near Rihabari village under Bajali subdivision of Barpeta District. The weir has been designed to pass a flood discharge of 710 cumecs. The main canal with a discharge capacity of 10 Cumec has been designed to irrigate an area of 16,500 ha of land annually.

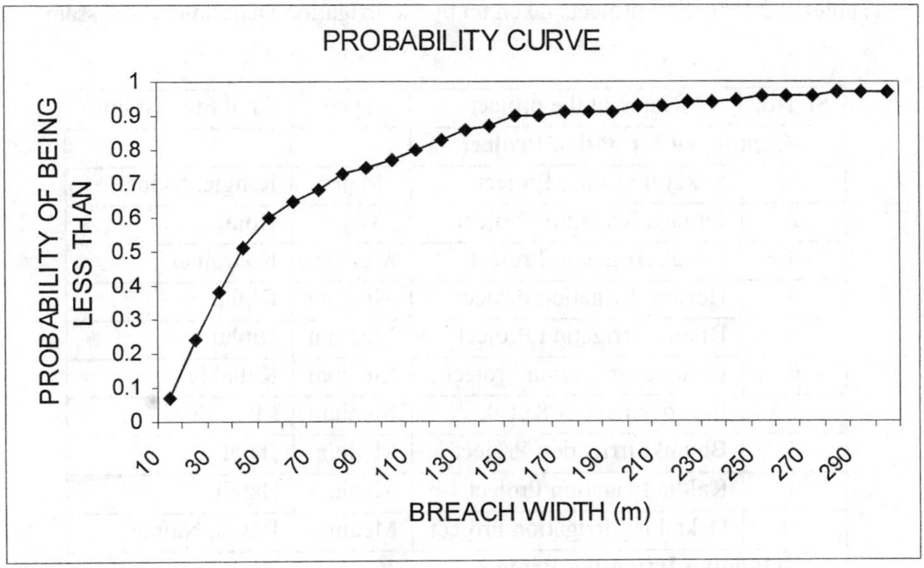


Figure 12.2 Probability of being less than breach width.

*12.3.1.7 Dekadong Irrigation Project*

This Medium Irrigation Project is situated on Dekadong River, which originate from sub-soil water about 3 km upstream of the village Dihira. Under this project a weir has been constructed to pass a design discharge of 425 cumecs.

*12.3.1.8 Bardikarai Irrigation Project*

This major project is located on the Bardikarai River in Sunitpur District of Brahmaputra basin. It has the capacity of irrigating 34,042 ha of land annually. The details of head works are given bellow.

**The Head works:**

Length of the barrage: 260.50 m
Design discharge: 2250.00 cumecs

Head regulator: 1
Design discharge: 36.00 cumecs

**The Canal System:**

| | |
|---|---|
| Length of the main canal: | 8.495 km |
| Design discharge of the main canal: | 30.1 cumecs |
| Total length of branch canal: | 165.477 km |

*12.3.1.9 Dhansiri Irrigation Project*

Dhansiri Irrigation project is one of the biggest project in Assam. Apart from irrigating 83,400 ha of land annually, it will generate 16.5 MW of electricity also. The barrage is located at Bhirabkunda across the Dhansiri River and in the trijunction of Bhutan Arunachal, Pradesh and Assam. The salient features of the hydraulic structures are as given below.

Table 12.2 Irrigation projects taken up by the Irrigation Department of Assam.

| Sl. No. | Name of the project | Type | Civil Subdivision |
|---|---|---|---|
| **Completed Irrigation Project** | | | |
| 1 | Sukla Irrigation Project | Major | Rangia, Nalbari |
| 2 | Jamuna Irrigation Project | Major | Hojai |
| 3 | Longa Irrigation Project | Medium | Kokrajhar |
| 4 | Horguti Irrigation Project | Medium | Diphu |
| 5 | Dikhari Irrigation Project | Medium | Diphu |
| 6 | Kaliabor Irrigation Project | Medium | Kaliabor |
| 7 | Improvement of Kulsik | Medium | Mangaldoi |
| 8 | Bhumki Irrigation Project | Medium | Bajali |
| 9 | Kaldia Irrigation Project | Medium | Bajali |
| 10 | Dekadong Irrigation Project | Medium | Bajali, Nalbari |
| **Ongoing Irrigation Project** | | | |
| 11 | Bardikarai Irrigation Project | Major | Sonitpur |
| 12 | Dhansiri Irrigation Project | Major | Udalguri |
| 13 | Champamati Irrigation Proj. | Major | Kokrajhar |
| 14 | Integrated Kollong Irr. Proj. | Major | Nagaon |
| 15 | Borolia Irrigation Project | Medium | Nalbari, Rangia |
| 16 | Pahumara Irrigation Project | Medium | Bajali |
| 17 | Rupahi Irrigation Project | Medium | Bajali |
| 18 | Buridihing Lift Irr. Scheme | Medium | Dibrugarh |
| 19 | Hawrapur Lift Irr. Scheme | Medium | Karbi-Anglong |
| 20 | Kolonga Lift Irr. Scheme | Medium | Hamreng |
| **Proposed Irrigation Project** | | | |
| 21 | Burisuti Irrigation Project | Medium | Barpeta |
| 22 | Garufella Irrigation Project | Medium | Gosaigaon |
| 23 | Rengma Irrigation Project | Medium | Golaghat |
| **Proposed NEC sponsored Project** | | | **States Involved** |
| 24 | Jiri Irrigation Project | - | Assam, Manipur |
| 25 | Kundil Balhan Irr. Project | - | Assam, Arunachal |
| 26 | Doyang Irrigation Project | - | Assam, Nagaland |
| 27 | Jhanji Irrigation Project | - | Assam, Nagaland |
| 28 | Katakhal Reservoir Project | - | Assam, Mizoram |
| 29 | Sonai Reservoir Project | - | Assam, Mizoram |

**The barrage:**

| | |
|---|---|
| Pond level: R.L. 210.80 m | River sluice: 4 bays, 16 m each = 64 m |
| Average bed level: R.L. 207.24 m | 3 piers, 2 m each = 6 m |
| Crest level: R.L. 208.00 m | Under sluice: 2 bays, 16 m each = 32 m |
| High flood level: R.L. 213.41 m | 1 piers, 2 m each = 2 m |
| Max. flood level :R.L. 215.35 m | Spillway: 3 bays, 16 m each = 48 m |
| Design flood: 4250 cumecs | 2 piers, 2 m each = 4m |
| Length of Barrage: 160 m | Divide wall: 2, 2 m each = 4 m |

**Head Regulator on right flank:**

| | | | |
|---|---|---|---|
| Length: = 37.50 m | | Pond level: | R.L. 210.80 m |
| No. of bays: 6, 5 m each. | = 30.00 m | Crest level: | R.L. 209.50 m |
| Piers: 5 , 1.5 m each | = 7.50 m | Design capacity: | 68.0 cumecs |

**Canal System:**
Main canal: 21.20 KM
Branch canal & Distributaries: 414.41 KM
Chute Fall in the main canal: 5 Nos. of 30 ft drop enabling micro power generation.

*12.3.1.10 Champamati Irrigation Project*
This major irrigation scheme envisages irrigation of 24.994 ha of land annually by using the water of Champamati River. The hydraulic structures proposed to construct under this scheme are given bellow.

**The Barrage with 4.5 M R.C.C. bridge:**

| | |
|---|---|
| Length of barrage: | 258.5 m |
| No. of bays: | 18 |
| Design Flood: | 2.266 cumecs |
| Pond level: | 70.00 m |

**Head Regulator:**

| | |
|---|---|
| Design discharge on left side: | 18.90 cumecs |
| Design discharge on right side: | 14.40 cumecs |
| Full supply level: | 69.70 m |

**Canal System:**

| | |
|---|---|
| Length of main canal (left side) : | 14.36 km |
| Length of main canal (right side): | 8.77 km |
| Branch canal and distributaries (left side): | 127.33 km |
| Branch canal and distributaries (right side): | 80.33 km |

*12.3.1.11 Integrated Irrigation Scheme for Kollong*
The back flow from the Brahmaputra River to its sub-channel (Island channel) Kollong induced regular flooding in the Kollong basin. To avoid this, a barrier was constructed in the Kollong River at the confluence on public demand. But as a result of such a step the agricultural field of the Kollong basin had to suffer a lot owing to the scarcity of water. As a remedy to this, the Integrated Kollong project was formulated. This is a Lift Irrigation Scheme, where water lifted from Brahmaputra River is carried to the

command area through the Kollong River and Sonai River. Four lift points have been constructed with small weirs erected on downstream of each lift point to raise the flow depth to facilitate lifting. The length of Feeder Canal is 2.65 km, design discharge being 25.68 cumecs.

*12.3.1.12 Borolia Irrigation Project*

This ongoing project proposes to Irrigate 37,050 ha of land annually on the right bank of Borolia River. The salient features of the hydraulic structures are as given bellow.

**The weir:**

Barrage: 10 bays @ 10 m
Under Sluice: 2 bays @ 6 m
Maximum design flood: 1050 cumecs
Water way between abutment: 92.00 m
H.F.L. 71.26 m
Pond level: 69.33 m
Permissible afflux: 1.20 m
Crest level of Barrage: 68.33 m
Crest level of Under Sluice: 66.83 m
Water way of head regulator: 14.40 m

**Canal System:**

Main canal (1 No.): 10.48 km
Distributaries (21): 108.75 km
Sub Minors (3): 8.81 km

*12.3.1.13 Pahumara Irrigation Project*

This ongoing medium project envisages irrigation of 12,995 ha of land annually by using water from Pahumara River, a tributary of the river Brahmaputra. The salient features of the Hydraulic structures are:

**The Barrage:**

Pond level: 49.60 m
Canal F.S.L.: 49.30 m
Crest level of Under Sluice: 44.75 m
Crest level of barrage: 45.00 m
No. of gates: 26

**Canal System:**

Main canal: 4.243 km
Branch canal: 75.00 km
Main canal discharge at head regulator: 9.92 cumecs
Canal type: Earthen/Lined canal

**12.3.1.14 Rupahi Irrigation Project.** This medium irrigation project is taken up on the Rupahi River of the Brahmaputra basin. The barrage is located in a village Bhuyapara. The location of the headwork site is about 24 km from Sarupeta. This scheme propose to irrigate 5688 ha of land annually. Important features of hydraulic structures are as given bellow.

**The Barrage:**

Pond level: 55.00 m
Canal FSL: 45.88 m
Crest level: 50.55 m
Overall Length: 81.00 m
Nos. of gate: 11 @ 6 m
Design discharge: 56,657 cumecs
Top level of guide bund: 58.00 m
Head regulator: On the right flank

**Canal System:**

Length of the main canal:3.75 km
Length of the branch canal: 15.064 km
FSL of main canal: 54.88 m
Discharge of main canal: 4.75 cumecs

Minor and Sub-minors: 27.67 km Type of canal: Earthen/Lined

*12.3.1.15 Buridihing Irrigation Project*

This is a lift irrigation scheme, where required water is pumped from the Buridihing River, a major tributary of the Brahmaputra River, by 7 pumps installed on a floating barge. The canal system consists of 18.30 km of Main Canal and 12.17 km of the Branch Canal.

*12.3.1.16 Hawrapur Irrigation Project*

This is a Lift Irrigation Scheme situated in the Karbianglong district of Assam. The Intake Well of 9.70 m diameter was constructed on the Kapili River just on the downstream of the confluence of Kapili and Doyang rivers. The canal system consists of 6.00 km of main canal, design discharge being 2.26 cumecs. The total length of branch canal is 19.5 km.

### 12.3.2 HYDRAULIC STRUCTURES IN THE BRAHMAPUTRA BASIN AND ITS RELATED PROBLEMS

In spite of having tremendous water resources potential, the Brahmaputra basin is yet to have any major hydraulic structure for harnessing its potential. Although various projects have been taken up by different Government organizations, significant progress is yet to be made. While a majority of the projects are facing problems other than technical, a few are suffering badly owing to genuine technical problems. The Dhansiri Irrigation Project is one such example. After construction of the barrage the main stream deviated from the original course of flow inducing severe setbacks to the progress of the project. Heavy siltation in the barrage site and rolling down of elephant size boulders from the upstream added another dimension to this problematic site. Much additional work had to be undertaken for overcoming these unanticipated problems. The Dhansiri Project, once claimed to be a glory of the Irrigation Department of Assam, has now become a major challenge for the Department. To avoid such circumstances in future, the effect of any hydraulic structures on the fluvial process and environmental change as a whole need to be studied elaborately with due weightage and utmost care.

However, many other problems still persist in outnumbering the technical problem related to the construction of a hydraulic structure. Paucity of funds, environmental hue and cry, insurgency problem in the northeastern part of India, lack of lateral coordination amongst the neighboring states are now becoming obstacles in the way of desirable water resources development in the Brahmaputra basin.

Environmentalists of this region are continuously raising their voice against construction of large dams in the region. But the quantification of the environmental damage always depends upon the weightage given to various factors. The anticipated area of inundation owing to construction of dams under major water resources projects in the North East India were estimated as given in the Table 12.3 (Rajib et al., 2000). But the impact on the environment of this inundation is yet to be quantified. On the other hand construction of the reservoir will definitely solve the flood problem of this region substantially, which can be regarded as a major environmental hazard faced by the people of this economically backward region. Therefore before saying a NO or YES to a dam project, proper study, without any bias, involving experts from different disciplines is mandatory.

Progress in the construction of hydraulic structures is also suffering badly from the insurgency problem in the northeastern part of India. Major Hydraulic Projects are always located in the interior places. Therefore incidents such as kidnapping of the technical experts involved in such projects or extortion of huge amounts of money by such extremist groups from the contractors handling such projects are now becoming a regular phenomenon in this region, creating a great impediment in timely execution of projects.

Table 12.3 Anticipated Inundation by various water resources project.

| Name of the Project | Inundation(sq. km) | Name of the Project | Inundation (sq. km) |
|---|---|---|---|
| Subansiri | 154.00 | Siang | 179.00 |
| Lohit | 39.00 | Someswari | 9.20 |
| Dibang | 30.00 | Pagladia | 51.50 |
| Noadihing | 15.00 | Kulsi | 68.60 |
| Tipaimukh | 82.72 | Um N Got | 11.59 |
| Bhairabi | 106.00 | Kameng | 27.00 |

The overall development of North east India depends, to a great extent, on its ability to control the Brahmaputra River. Therefore construction of suitable hydraulic structures in the Brahmaputra basin is quite essential. The Brahmaputra, being an international river passing through the three most densely inhabited countries of the world, China, India, and Bangladesh, calls for management of its vast water resources for diverse use and adequate regional as well as international cooperation. Apart from the need for technical expertise, what is most required at this hour is a strong political will both at the state and national level. Let us hope that humanity and technology will join hands together to face the challenges posed by this mighty river and turn the Brahmaputra River into a river of prosperity.

## 12.4 Concluding Remarks

It is reasoned that without construction of major hydraulic works the Brahmaputra basin, and hence the northeast of India, cannot be developed fully. Indeed, the development of India is closely linked with the development of her northeast region.

## 12.5 Acknowledgment

The author expresses his sincere thanks to the Flood Control Department and the Irrigation Department of Assam for extending their help in collecting the required data. Er. Mukut Goswami, assistant executive engineer and Er. Abdul Baser, Director of Planning, Irrigation Department, Assam, deserve special thanks in this regard.

## References

Flood Control Department of Assam (1977). Comprehensive Plan for Brahmaputra River, Assam, India.

Flood Control Department of Assam (1990). Annual Flood Report of Brahmaputra and Barak valley, Hydrology Div. of Flood Control Department, Assam, India.

Flood Control Department of Assam (1991). Annual Flood Report of Brahmaputra and Barak valley, Hydrology Div. of Flood Control Department, Assam, India

Flood Control Department of Assam (1992). Annual Flood Report of Brahmaputra and Barak valley, Hydrology Div. of Flood Control Department, Assam, India.

Flood Control Department of Assam (1993). Annual Flood Report of Brahmaputra and Barak valley, Hydrology Div. of Flood Control Department, Assam, India

Flood Control Department of Assam (1994). Annual Flood Report of Brahmaputra and Barak valley, Hydrology Div. of Flood Control Department, Assam, India.

Goswami, D. C(1998). Fluvial Regime of Flood Hydrology of the Brahmaputra River, Assam Memoir Geological Society of India, No.41, pp. 53-75.

Goswami R.K. Mahanta C. and Singh B (2000). River System, Drainage Basins and anthropogenic Change: the Brahmaputra River Basin, Proc. of XVI National Convevtion of Environmental Engineering, Guwahati, Assam, India, PP. I-7—I-15.

Singh, V.P. (1996). Dam Breach Modeling Technology, Kluwer Academic Publisher, Dordrecht, Boston, London.

Sarma, A.K. (1999). A Study of Two- Dimensional Flow Propagating from an Opening in the River Dike, Unpublished Ph.D. thesis submitted to Gauhati University, Assam, India.

Sarma, A.K. and Kakati, K. (2000). Mitigation of Flood Disaster by Controlling Dyke Failure Using Geotextile, Proc. of XVI National Convention of Environmental Engineering, Guwahati, Assam, India, pp. 47-50.

# CHAPTER 13

# SCOUR AROUND SPURS AT GUMI SITE

C. S.P. OJHA, N. SHARMA AND V.K. TALWAR

The Brahmaputra Riber basin is one of the most erodible basins in India. Erosion of the Brahmaputra River banks is severe and spurs have been employed to control erosion. This chapter describes scour around spurs at the Gumi site of the river. The experience gained will be valuable in design of similar measures for river bank protection.

## 13.1 Introduction

Rivers have always played an important role in the history of Indian civilization. However, their role has been more devastating during periods of floods. Erosion of river banks is a very critical issue, and unless suitable measures are taken to safeguard river banks a significant loss of human life and property may take place. In the context of the Brahmaputra River certain useful information exists regarding the breaching of river banks in a river reach length of nearly twenty kilometres from Palasbari to Gumi, a site close to Guwahati, the capital city of Assam. The severity of the bank erosion in this reach resulted into construction of seven spurs at Gumi site of the reach. Since the introduction of spurs in 1986 there have been attempts to monitor the river cross-sections at the site of spurs, in addition to an appraisal of river bed erosion at the nose of spurs. One obvious question which needs an elaborate modelling effort is whether the spurs will be able to sustain themselves in the future. To be more specific, will the noses of spurs be large enough to endanger the stability of spurs? In this context it becomes crucial to examine the performance of spurs with the kind of data available since their operation. An analysis of such a type is not only helpful in assessing the role of spurs at the Gumi site, but may also prove valuable for undertaking similar measures of river bank protection in different stretches of the river.

In this chapter the details of the river reach followed by a review of past anti-erosion measures are presented along with the data collected on scour around spurs. The main emphasis has been given to the modelling of scour around spurs, as it helps to assess the stability of spurs at the site under consideration.

## 13.2 Gumi Site

A schematic layout of a certain reach of the Brahmaputra River along with the location of Gumi site is shown in Figure 13.1. Up to a distance of nearly 25 km upstream of Agiathuri the river reach is a constricted section bounded by scattered rockyhills on both



*V.P. Singh et al. (eds.), The Brahmaputra Basin Water Resources*, 274-298.

sides. The width of this reach is about 2 km only and the flow velocity is large enough to carry the sediment load. At Agiathuri, one km downstream of the Brahmaputra bridge, the river, on entering the alluvial plains, fans out into channels, depositing its sediment load in between. The bulk of the river flow is diverted to the northern channel. Owing to the spur action of the Ghatia Hill, the flow is further deflected to a south westerly direction in which it further experiences the spur action of Dakhala Hill. In the downstream of Dhakhala Hills, a deep embayment exists for a length of seven kilometres. At the exit of the embayment the river bifurcates into two channels, of which one moves in the northwesterly direction, and the other, known as Lakhitary channel, finds its way in the Westerly direction.

It has been observed that during floods the river throws huge quantities of its suspended load in the Lakhitary channel. This results in the creation of many sand bars, which have further pushed the Lakhitary channel to the south. As a consequence every year the southern bank experienced severe erosion problems. The Gumi site, which is located on this southern bank, therefore needed certain measures to stop the migration of river towards it. To achieve this, a project to protect the river bank from Palasbari to Gumi, a stretch of 19 km, was undertaken. The importance of the project can be realised because the area protected as a result of this project is nearly 100 sq. kms. Also the implementation of this project was necessary in view of the Kulsi River, which originates from Meghalaya hills, and flows almost parallel to the Brahmaputra River bank, as shown in Figure 13.1. It is interesting to observe the short distance between the two rivers. One can also imagine at this juncture the possible consequence had the two rivers joined and submerged the area from Plasbari to Gumi, which consists of thickly populated villages, mini industrial places, forests, valuable agricultural land, and many government and semi-government institutions.

### 13.3 A Review of Anti-Erosion Measures from 1955 Onwards

In 1955 the southern bank of the Brahmaputra River from Palsbri to Gumi was embanked for a length of 21 km. From its inception the embankment came under the attack of the Brahmaputra River and had to be retired several times for providing protection to the affected areas. Figure 13.2 shows the different works taken up in this reach from time to time. It can be seen from the figure that except the five to six year period of 1955--1961, there has been retirement of the bank in the following years. The breach of the bank is reported in years 1969--70, 1974--1975, 1980--81, 1981--82, 1982--83, and 1983--84. Attention must be given to the fact that breaching had occurred more frequently in the period from 1980--84. In the year 1978--79 certain dredging operations were also performed. Certain anti-erosion measures were also taken from 1974. Although the provision of anti-erosion measures slowed the erosion process and provided a short term relief, it failed to achieve a permanent solution to the problem of continued migration of the river bank line. The need for a permanent solution could only be realised when the distance between the Brahmputra River and the Kulshi reduced to the very short distance of only couple of hundred meters. It was strongly felt that the confluence of the Kulsi with the Brahmaputra could affect as large as 400 sq. km of the adjoining area. Thus the construction of spurs received top priority

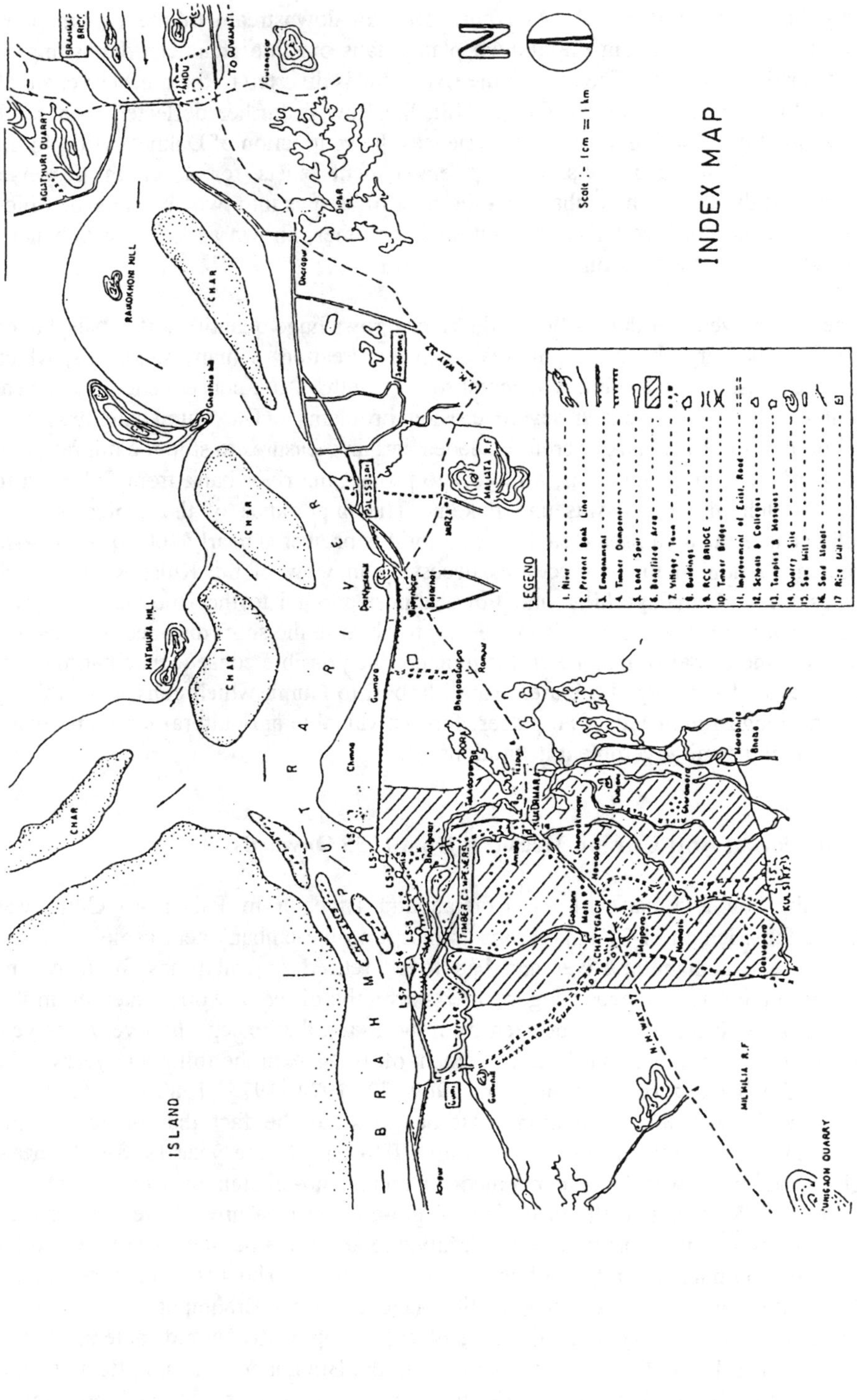


Fig. 13.1 A schematic layout of a Brahamaputra reach.

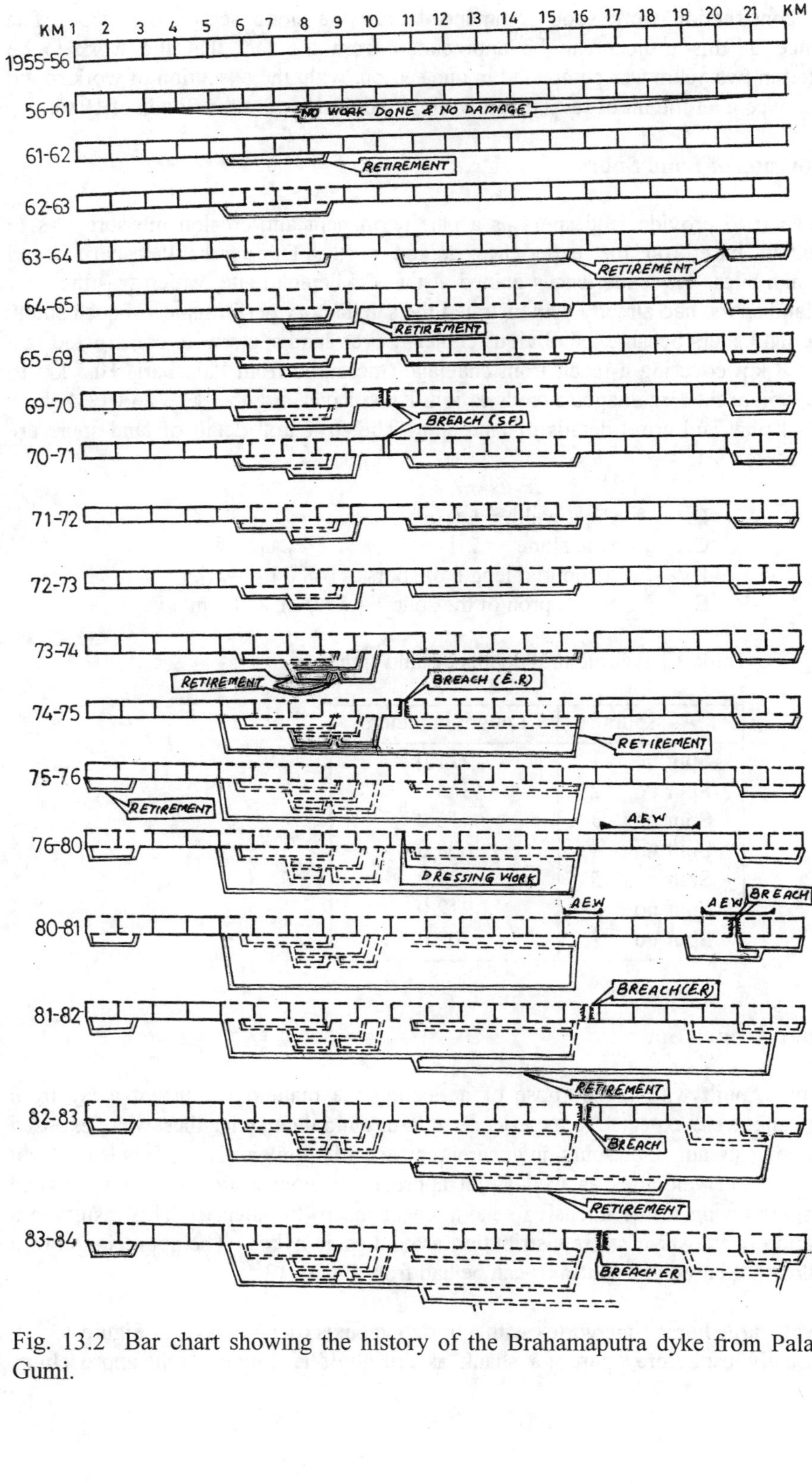


Fig. 13.2 Bar chart showing the history of the Brahamaputra dyke from Palasbari to Gumi.

and the construction work soon commenced after the flood season of 1985. The importance of this project can be appreciated from the fact that the work to be completed in two years was completed in one season, with the execution of work round the clock except a little bit of revetement work in the days before the floods of 1986.

### 13.4 Provision of Land Spurs

The decision to provide land spurs as a part permanent anti-erosion measure was to some extent based on the experience gained at the Dibrugarh, Palasbarim, and Kakilamukh sites. The experiences gained at the Kakilamukh site, which was the only user of land spurs, had already demonstrated the satisfactory performance of such spurs. Thus the land spurs became the obvious choice. Seven land spurs were constructed in a length of 8 km covering a reach from chainage (measured from Palasbari) 10.5 km to 18.5 km, where the Brahmaputra tends to join Kulsi River. Figures 13.3 and 13.4 show the typical plan and cross details of land spur. The structural detail of land spurs are given in Table 13.1.

B. Crest width = 6.0 m
C. Side slope = 2:1
D. Armoured length of spurs = 148.0 m
E. Nose apron of the spur = 54.0 m × 2.7 m

Table 13.1 Structural details of land spurs.

| A. Spurs | Length in meters |
|---|---|
| Spur no. 1 | 721.0 |
| Spur no. 2 | 343.0 |
| Spur no. 3 | 415.0 |
| Spur no. 4 | 608.0 |
| Spur no. 5 | 327.0 |
| Spur no. 6 | 212.0 |
| Spur no. 7 | 202.0 |

### 13.5 Scour Around Spur

Seven spurs, built with the purpose of deflecting the main river current away from erodible bank, were expected to experience scour around them, once these were exposed to the flowing stream. The spurs, in general, constrict the flow in river. This leads to an increase in local velocities as well as an appreciable modification of the flow field around spurs owing to the intensive generation of macro-turbulence. This results in a modification of the shear stress distribution around spurs/dykes. The appreciation of a typical flow field around a spur-axis can be had from Figure 13.5.

A relatively large horse shoe vortex with a horizontal axis (denoted as 1 in Figure 13. 5), occurs usually just before a part of a shank, as a result of heading up of the approaching

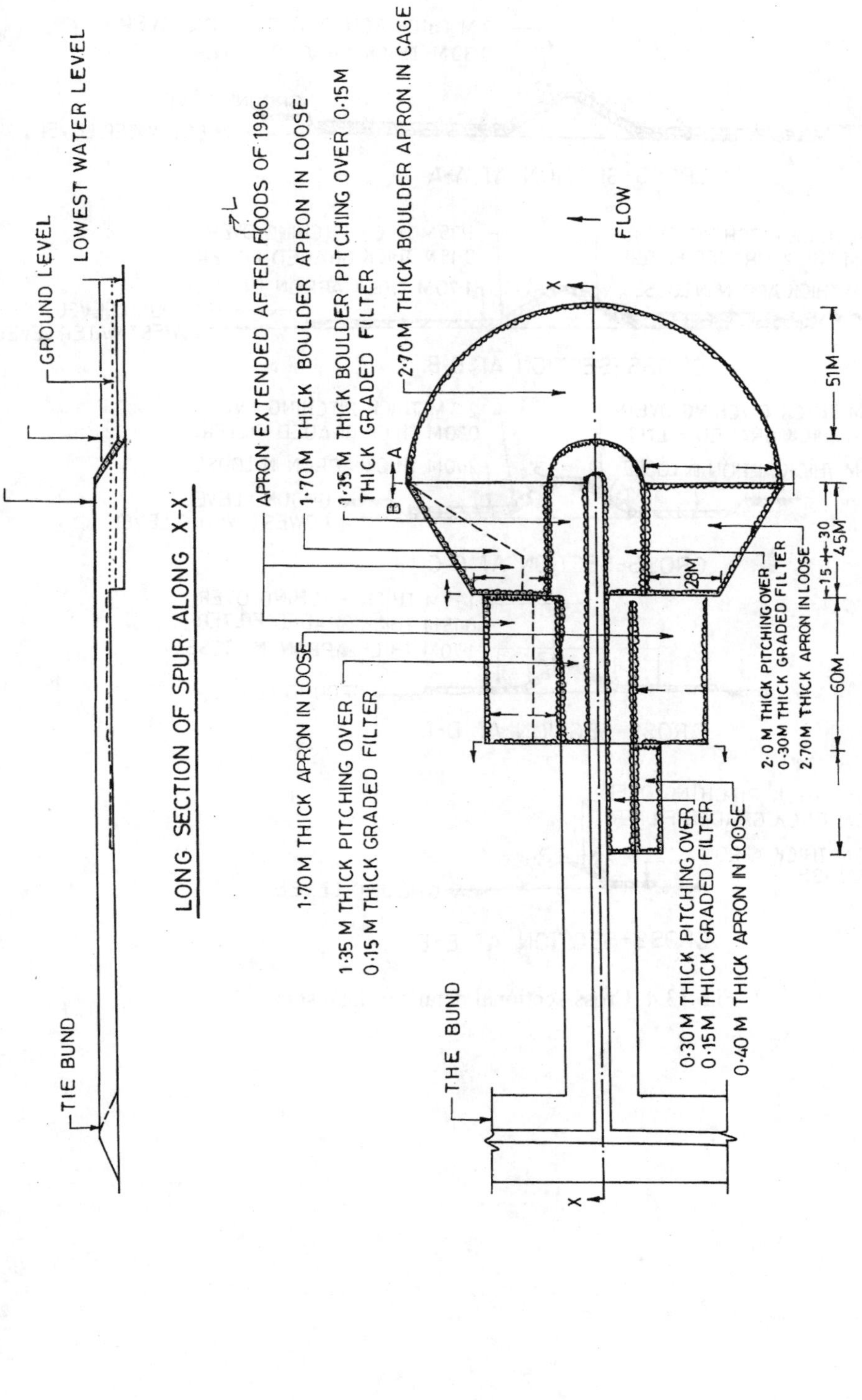


Fig. 13.3 Typical plan of land spur.

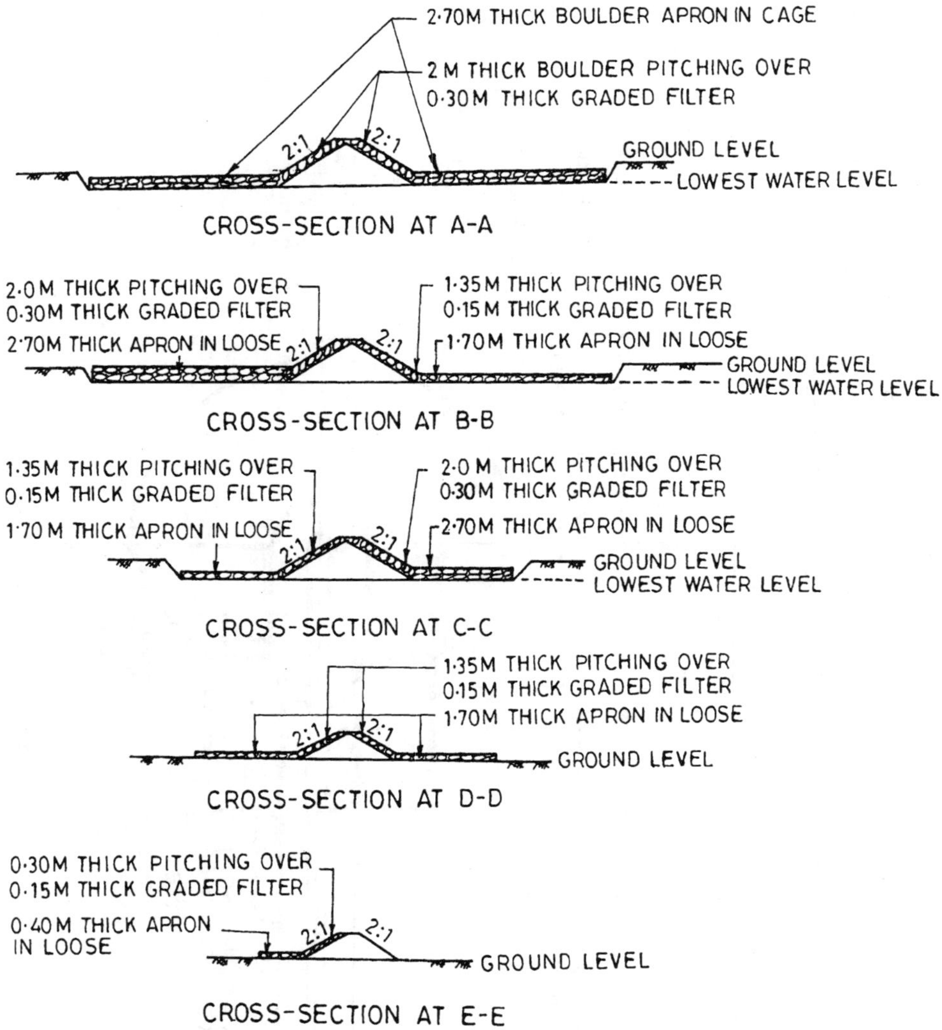


Fig. 13.4 Cross-sectional details of land spur.

Fig. 13.5 Flow around groyne: (a) surface current and (b) near bed current.

flow. Along the separation line downstream of the groyne head, vortices with vertical axes (see 2 in Figure 13.5), also known as von Karman vortex street, occurs. This vortex system is formed up of shear layers between main stream and a stationary whirlpool with vertical axes (3 in Figure 13.5) in the wake of spur. The von Karma vortices act a little like tornadoes by sucking sediment into a low pressure core. The effect is regarded as one of the major causes of erosion below spurs. The structure of flow around spurs is not only affected by their length but also by their shape, side, head slopes, location (in series or isolation), the spacing among the spurs, and the type of river bank (straight, concave, or convex). Also the scour around spurs is sensitive to the incoming sediment load.

## 13.6 Field Data

It is possible to perform controlled experiments in the laboratory. However, owing to scaling effects it is less likely that the models based on laboratory data will yield a better performance in the field. Also the collection of field data is more challenging, especially during the period of floods. Thus the variable which may appear in the lab-based models, because of the ease of their monitoring, may be unknown while using the lab-based models for the field conditions. Thus it is very important to know the kind of data which has been obtained in the field. For the present site the data were not available in a format suited to a quick analysis. Much of the information was available in the form of contours and figures. Certain details pertaining to the data are as follows.

## 13.7 Stage-Discharge Curve

The project site experienced two major floods in 1987 and 1988. Based on the data from 1986--1989, the stage versus discharge data at spur no. 2 is given in Table 13.2. Using the stage versus discharge data, the following relationship is obtained between stage, Y, and discharge, Q:

$$Y = 26.03\ Q^{0.057} \tag{13.1}$$

The plot of equation (13.1) is shown in Figure 13.6 along with the observed data points.

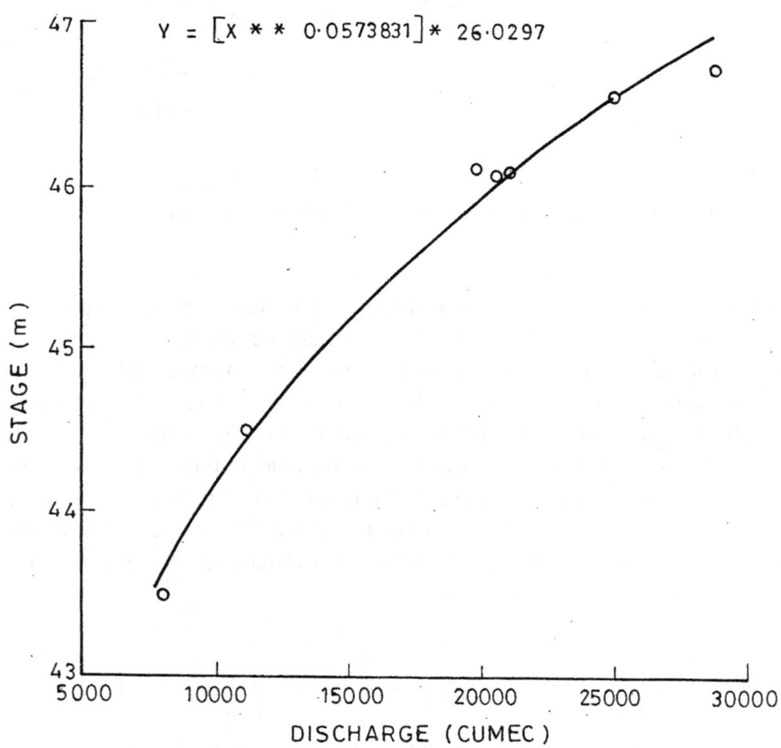


Fig. 13.6 Stage-discharge curve at spur no. 2.

Table 13.2 Stage versus discharge data at spur no. 2

| Stage (m) | Discharge (cumecs) |
|---|---|
| 43.49 | 8,060.00 |
| 44.50 | 11,100.00 |
| 46.11 | 19,800.00 |
| 46.07 | 20,531.00 |
| 46.55 | 24,983.00 |
| 46.71 | 28,762.00 |
| 46.09 | 21,064.00 |

## 13.8 Particle Size Distribution

As information pertaining to the particle size distribution of bed and bar material is used in many scour predictive formulae, the same is available in the form of Figure 13.7. It can be seen from this figure that the bed material is sandy in nature. The average size of the bed particles (50% finer) is 0.15 mm.

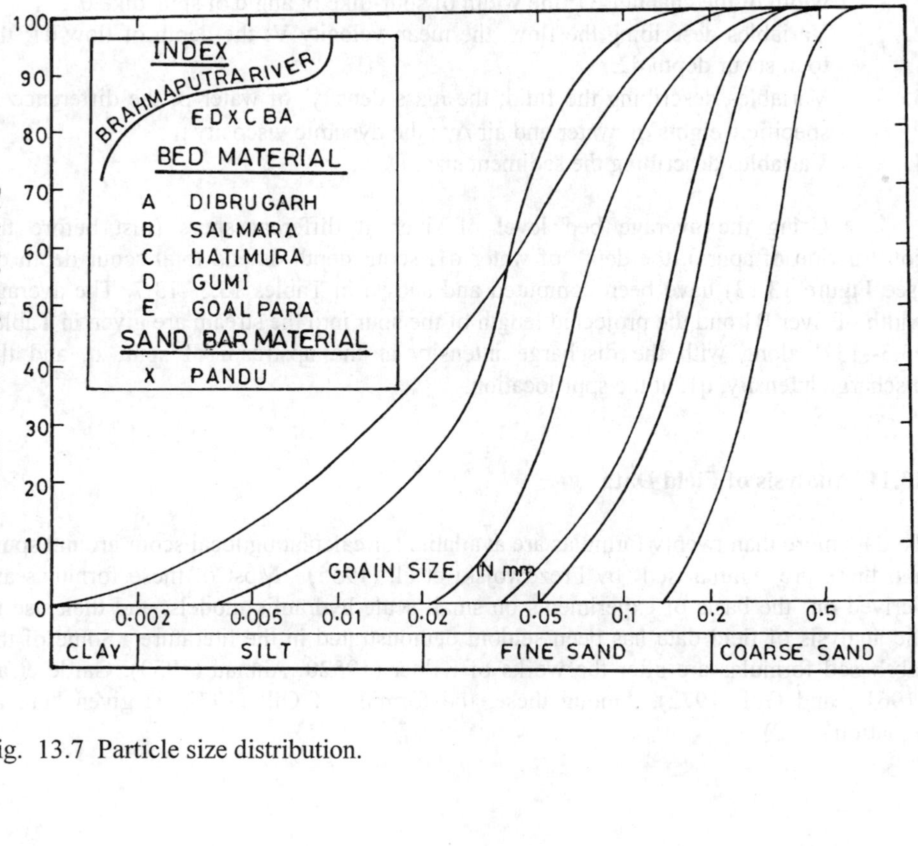


Fig. 13.7 Particle size distribution.

## 13.9 River X-Section Along Spur Axes

Figures 13.8--13.12 provide an information regarding the bed level variation along the axes of spur nos. 2, 3, 5, 6 and 7. The information pertaining to spur nos. 1 and 4 is not available as these were not exposed to the river flow. In each case the bed levels are gives for six different times. The observations of March 1986 represent the situation at the time of construction of spurs. The observations of September--October 1986 represent the bed level conditions after the flood of 1986. From Figure 13.8, one can note the severity of scour near spur no. 2. Figure 13.8 also shows that after flood of 1987 and 1988 there has been siltation around spur no. 2. This is why the scour around spur no. 2 has been reduced. Similarly, one can have a better understanding of the relative contribution of different spurs from perusal of Figures 13.9--13.12. The point to be noted from these figures is that there is not only variation in terms of the magnitude of scour depth but also in terms of its occurrence from the nose of spurs.

## 13.10 Observed Scour Data and Related Variables

To appreciate different notations used to define scour and other flow variables, one can refer to Figure 13.13. The various variables can be summarized as follows:

1. Variables describing the geometry of the channel and of the spur-dike; the width of the channel B1; the width of spur-dike b; angle of spur-dike $\theta$.
2. Variables describing the flow: the mean velocity V; the depth of flow d1; the total scour depth d2.
3. Variables describing the fluid; the mass density of water $\rho$; the difference in specific weights of water and air $\Delta\gamma$ ; the dynamic viscosity $\mu$ .
4. Variables describing the sediment size, D.

Using the average bed level of river at different spurs (just before the construction of spurs), the depth of water d1, scour depth ds and total scour depth d2 (see Figure 13.13) have been computed and shown in Tables 13.3--13.7. The average width of river B1 and the projected length of the spur into the stream are given in Tables 13.3--13.7 along with the discharge intensity at the upstream of spur, q, and the discharge intensity, q1, at the spur location.

## 13.11 Analysis of Field Data

To date more than twenty formulas are available for calculating local scour around spurs and these are summarised by Prezdwojski et al. (1955). Most of these formulas are derived on the basis of experiments on small scale hydraulic models, and their use to the analysis of field data has been seldom demonstrated in the literature. Some of the lab-based formulas are after the works of Kilner (19520. Ahmad (1953), Garde et al. (1961), and Gill (1972). Among these, the formula of Gill (1972) is given here as equation (13.2):

Table 13.3 Field data four spur no.2
The average bed level = 38.95 m, B1 = 3211.0, B2 = 2868.0 m

| Date | water level (m) | bed level around (m) | $Q(m^3/s)$ | $q(m^3/m/s)$ | $q1(m^3/m/s)$ | d1 (m) | ds (m) | d2 (m) |
|---|---|---|---|---|---|---|---|---|
| 25-03-86 | 40.67 | 37.00 | 2598.99 | 0.81 | 0.91 | 1.72 | 1.96 | 3.67 |
| 7-07-86 | 44.95 | 36.00 | 14596.61 | 4.54 | 5.09 | 6.00 | 2.96 | 8.96 |
| 25-09-86 | 44.74 | 26.30 | 13455.22 | 4.19 | 4.69 | 5.79 | 12.65 | 18.44 |
| 25-10-86 | 43.62 | 29.46 | 8690.85 | 2.71 | 3.03 | 4.67 | 9.49 | 14.16 |
| 6-10-87 | 45.41 | 31.82 | 17395.64 | 5.41 | 6.06 | 6.48 | 7.13 | 13.59 |
| 25-09-88 | 43.97 | 33.40 | 9974.75 | 3.11 | 3.48 | 5.02 | 5.55 | 10.57 |
| 26-10-89 | 43.75 | 39.20 | 9148.38 | 2.85 | 3.19 | 4.80 | -0.25 | 4.55 |
| 30-10-90 | 42.79 | 41.79 | 6240.42 | 1.94 | 2.18 | 3.84 | -2.84 | 1.00 |
| 12-11-91 | 41.93 | 39.33 | 4220.01 | 1.31 | 1.47 | 2.89 | -0.38 | 2.50 |
| 30-12-92 | 42.88 | 40.88 | 6470.62 | 2,20 | 2.26 | 3.93 | -1.93 | 2.00 |
| 21-11-93 | 42.68 | 41.18 | 5969.54 | 1.86 | 2.08 | 3.73 | -2.23 | 1.50 |
| 20-11-94 | 42.86 | 41.86 | 6418.79 | 2.00 | 2.24 | 3.91 | -2.91 | 1.00 |

Table 13.4 Field data for spur no.3
The average bed level = 38.92 m, B1 = 3211.0 m, B2 = 2796.0 m

| data | water level (m) | bed level around (m) | Q(cu m/s) | q(cu m/m/s) | q1(cu m/m/s) | d1(m) | ds(m) | d2(m) |
|---|---|---|---|---|---|---|---|---|
| 15.09.86 | 44.74 | 42.20 | 14254.53 | 4.44 | 5.10 | 5.82 | -3.28 | 2.54 |
| 31.07.87 | 44.95 | 33.50 | 15448.96 | 4.81 | 5.53 | 6.03 | 5.42 | 11.45 |
| 26.08.87 | 45.95 | 32.00 | 22548.96 | 7.02 | 8.08 | 7.03 | 6.92 | 13.95 |
| 8.10.87 | 45.26 | 33.00 | 17385.64 | 5.41 | 6.22 | 6.34 | 5.92 | 12.26 |
| 25.09.88 | 43.80 | 34.20 | 9896.82 | 3.08 | 3.54 | 4.88 | 4.72 | 9.60 |
| 3.11.89 | 43.50 | 41.30 | 8794.48 | 2.74 | 3.15 | 4.58 | -2.38 | 2.20 |

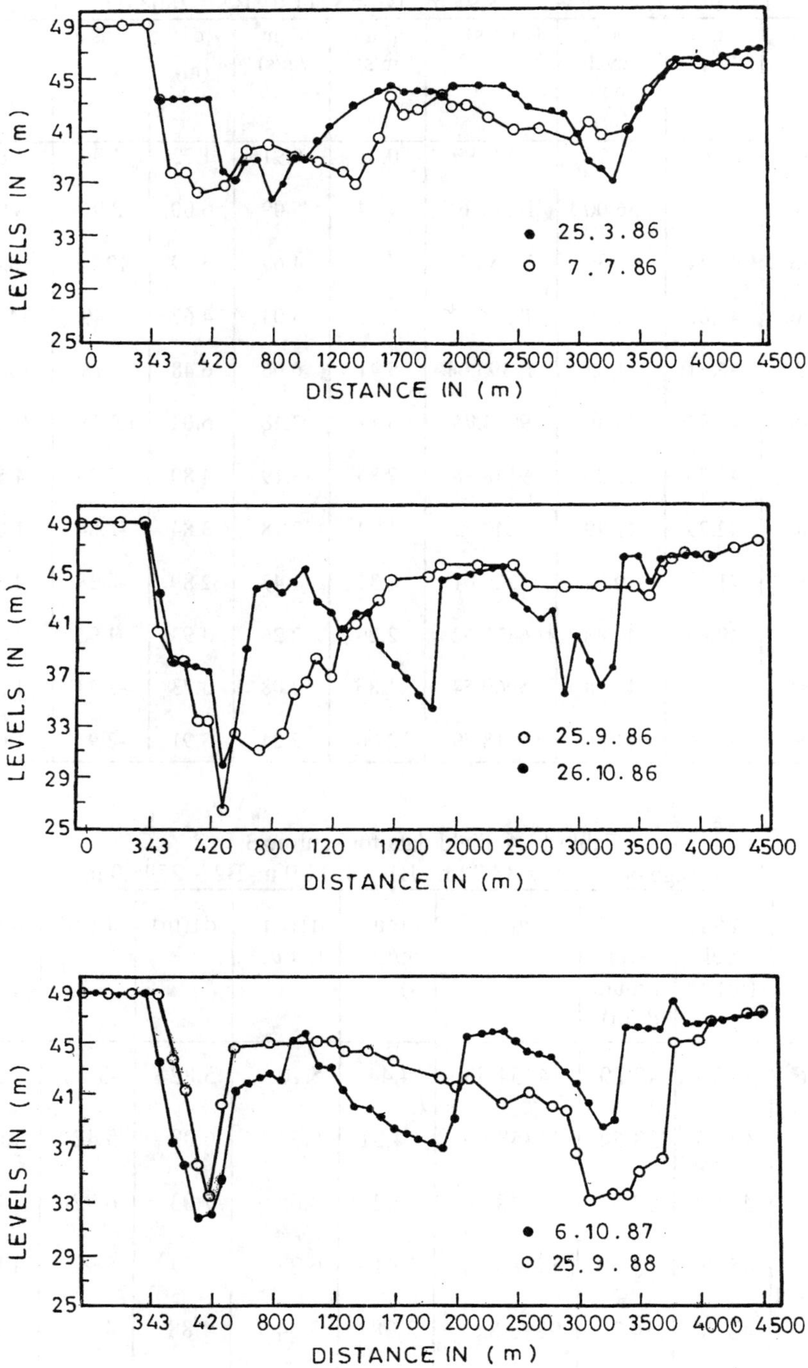


Fig. 13.8 River section along the axis of spur no. 2.

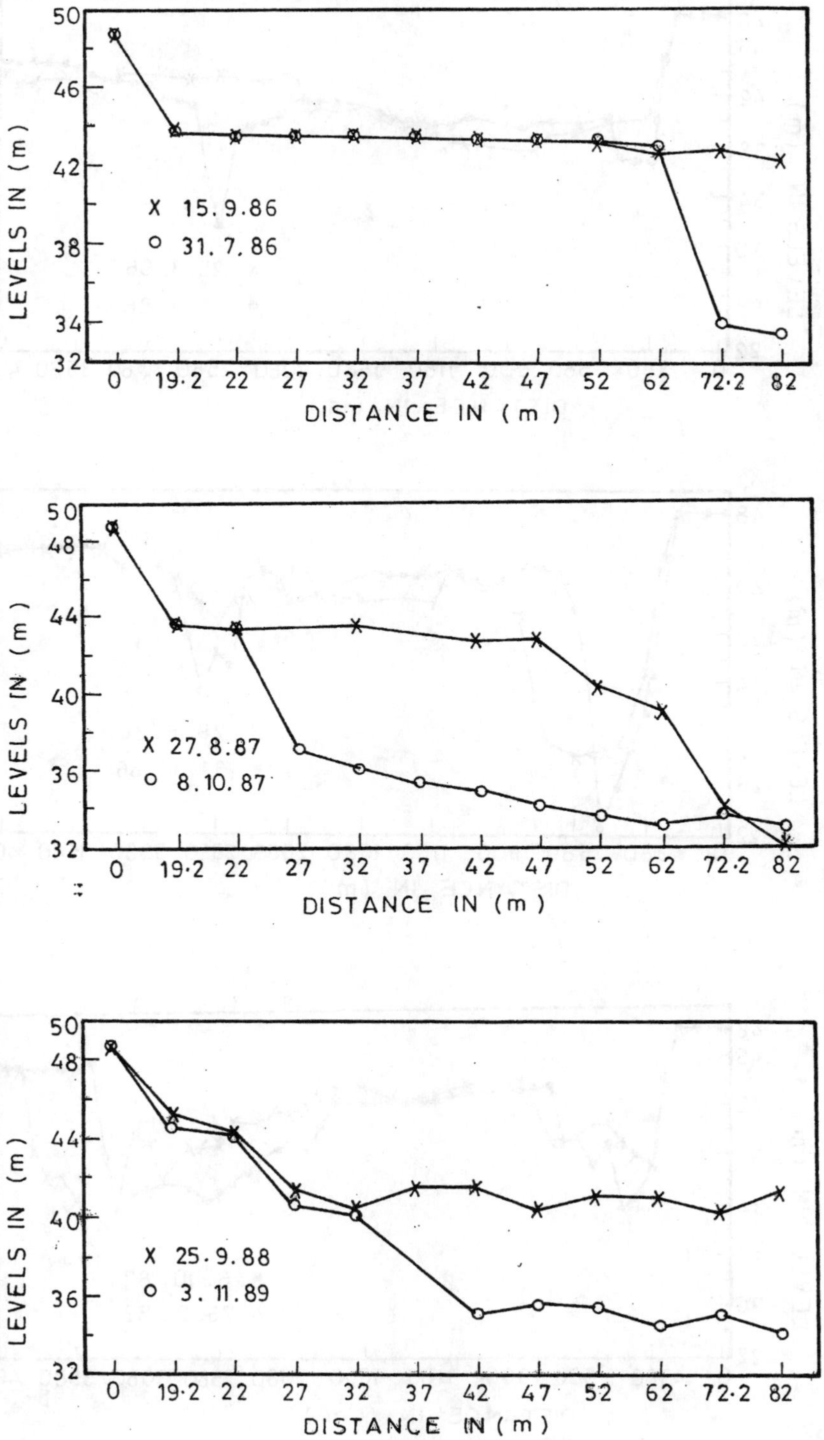


Fig. 13.9 River section along the axis of spur no. 3.

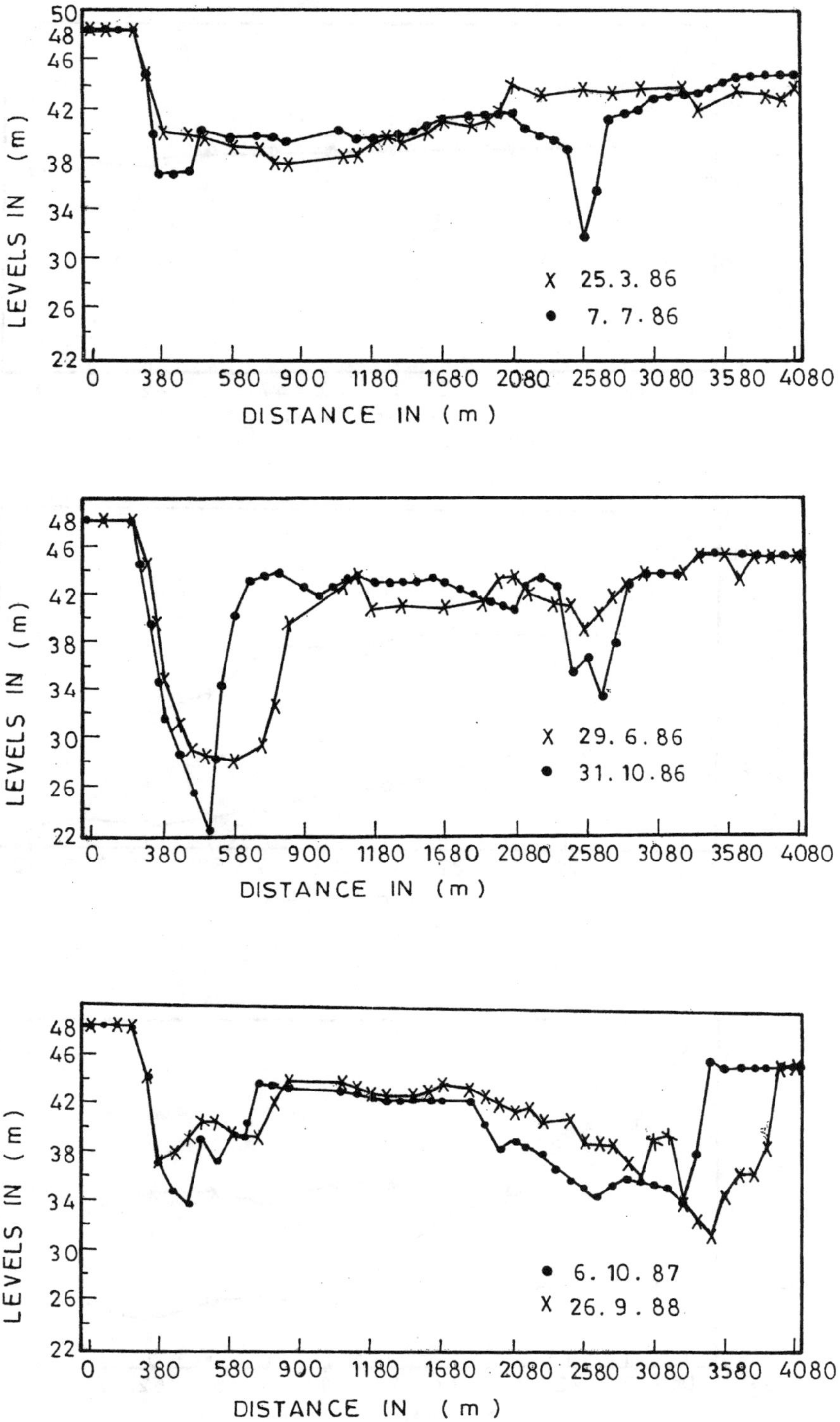


Fig. 13.10 River section along the axis of spur no. 5.

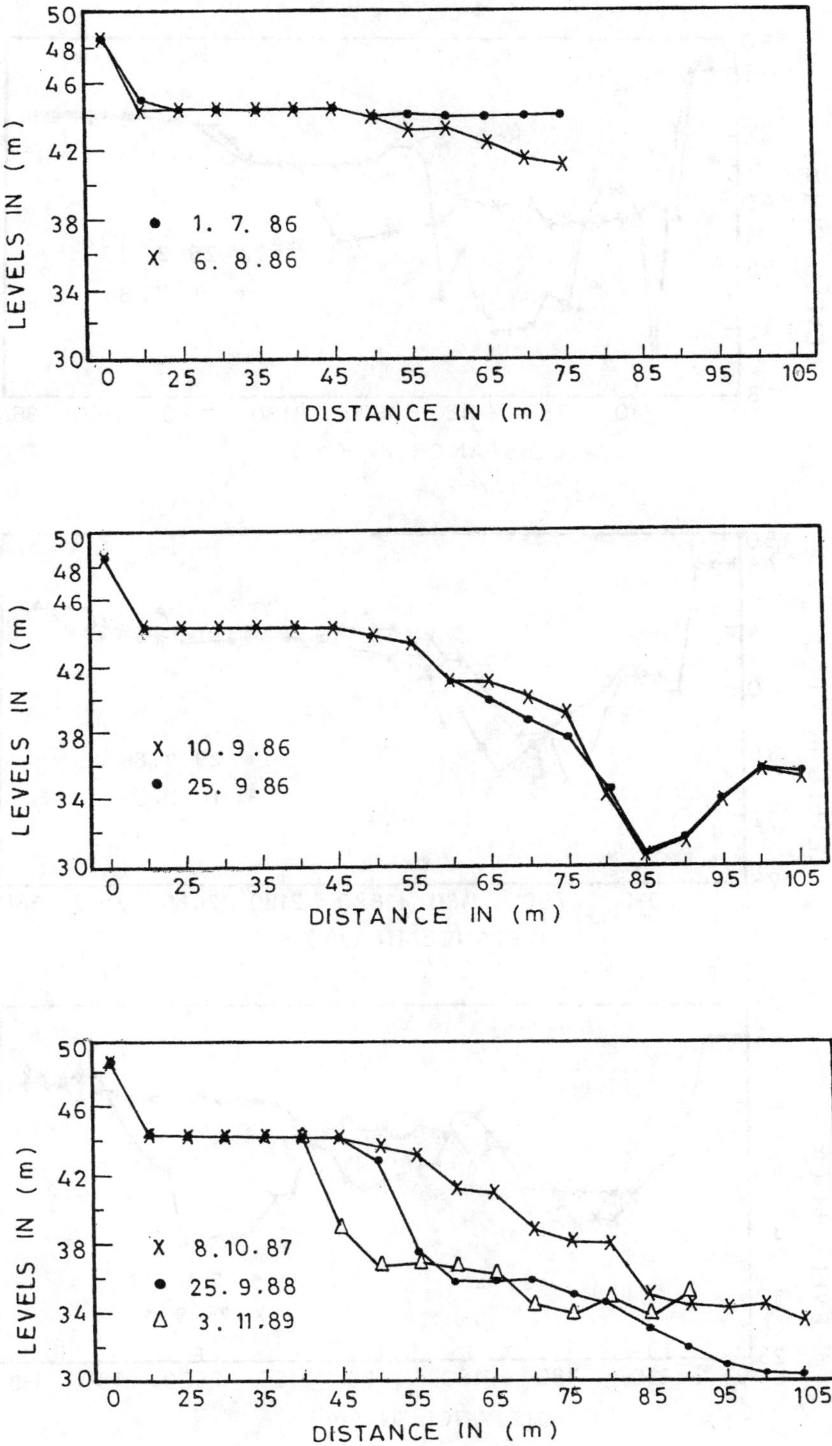


Fig. 13.11 River section along the axis of spur no. 6.

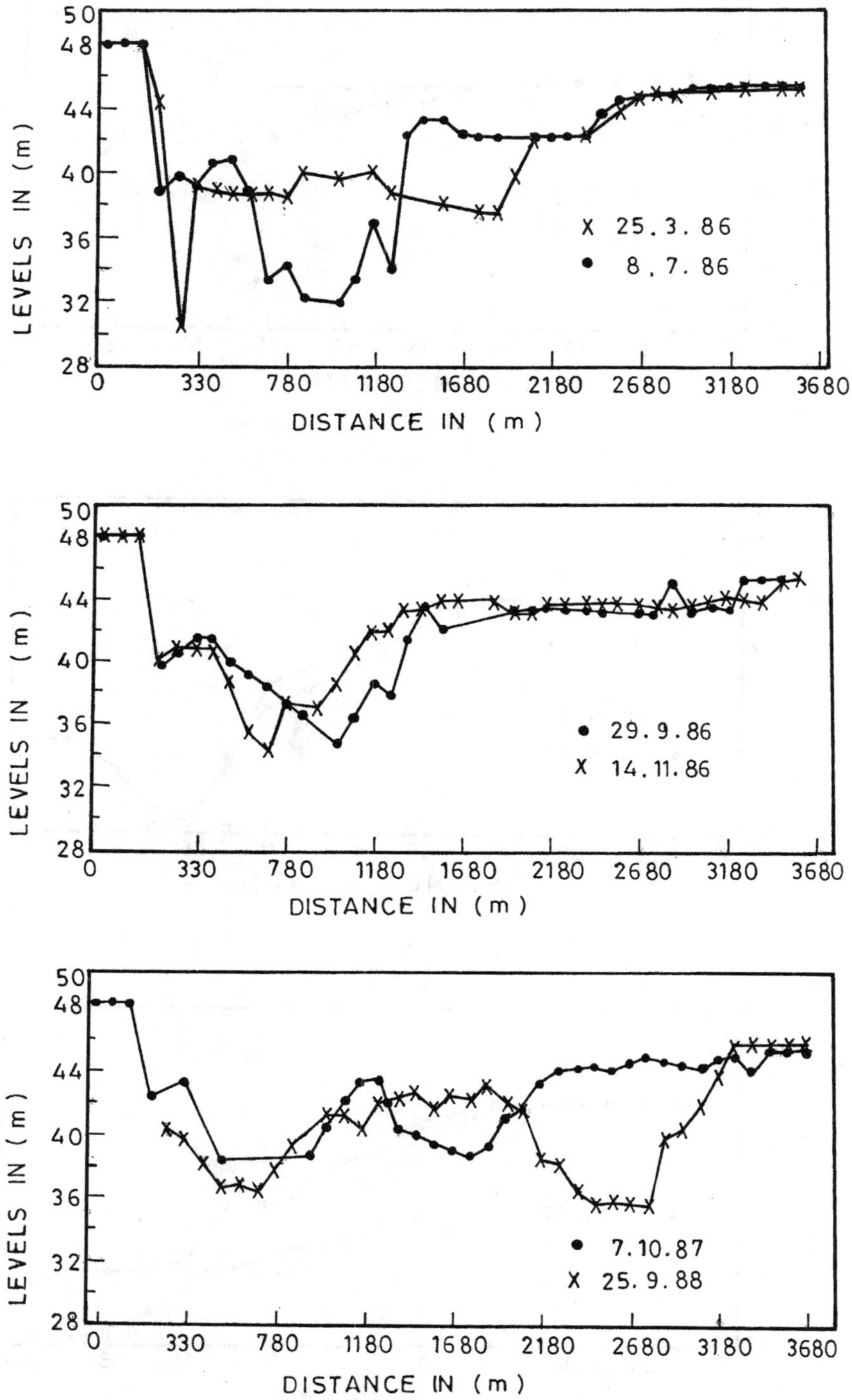


Fig. 13.12 River section along the axis of spur no. 7.

Table 13.5 Field data for spur no. 5.
The average bed level = 38.9 m, B1 = 3653.0 m, B2 = 3326.0 m

| data | water level (m) | bed level around (m) | Q(cu m/s) | q(cu m/m /s) | q1(cu m/m/s) | d1(m) | ds(m) | d2(m) |
|---|---|---|---|---|---|---|---|---|
| 25.3.86 | 44.40 | 38.90 | 2804.17 | 0.77 | 0.84 | 1.50 | 0.00 | 1.50 |
| 7.7.86 | 44.59 | 39.73 | 15098.41 | 4.13 | 4.54 | 5.69 | -0.83 | 4.86 |
| 6.8.86 | 44.61 | 39.50 | 15214.42 | 4.16 | 4.57 | 5.71 | -0.60 | 5.11 |
| 10.9.86 | 44.40 | 33.10 | 14037.12 | 3.84 | 4.22 | 5.50 | 5.80 | 11.30 |
| 26.9.86 | 44.35 | 28.00 | 13769.74 | 3.77 | 4.14 | 5.45 | 10.90 | 16.35 |
| 31.10.86 | 44.11 | 22.20 | 26757.97 | 7.32 | 8.05 | 7.21 | 16.70 | 23.91 |
| 8.10.87 | 45.00 | 33.25 | 17651.57 | 4.83 | 5.31 | 6.10 | 5.65 | 11.75 |
| 31.10.87 | 44.69 | 33.90 | 15686.91 | 4.29 | 4.72 | 5.79 | 5.00 | 10.79 |
| 25.9.88 | 43.55 | 36.00 | 10092.74 | 2.76 | 3.03 | 4.65 | 2.90 | 7.55 |
| 26.10.89 | 43.20 | 38.20 | 8794.48 | 2.41 | 2.64 | 4.30 | 0.70 | 5.00 |
| 30.10.90 | 42.52 | 39.02 | 6708.81 | 1.84 | 2.02 | 3.62 | -0.12 | 3.50 |
| 12.11.91 | 41.50 | 40.50 | 4433.67 | 1.21 | 1.33 | 2.60 | 1.60 | 1.00 |
| 30.10.92 | 42.58 | 41.08 | 6872.16 | 1.88 | 2.07 | 3.68 | -2.18 | 1.50 |
| 17.11.93 | 42.55 | 41.31 | 6790.02 | 1.86 | 2.04 | 3.65 | -2.41 | 1.24 |
| 20.11.94 | 42.53 | 41.51 | 6735.78 | 1.84 | 2.03 | 3.63 | -2.61 | 1.02 |

$$d2/d1 = \alpha(B1/B2)^{6/7}\{(B1/B2)^{1/n}(1-\tau_c/\tau_1) + \tau_c/\tau_1\}^{-3/7} \qquad (13.2)$$

In equation (13.2) $1 - \tau_c/\tau_1 = 0$ for $\tau_c/\tau_1$ less than 1. The notations *B1, B2, d1*, and *d2* are same as shown in Figure 13.3; *n* = numerical exponent; and $\alpha$ = a coefficient which is dependent on flow depth and the size of bed material. Gill (1972) has suggested the following expression of $\alpha$:

$$\alpha = 8.375\,(D/d1)^{0.25} \qquad (13.3)$$

The value of $\tau_c/\tau_1$ in the case of Gumi site is << 1. Also, at any spur site, the ratio *B1/B2* is a constant. Thus for a given bed material *d2* becomes a function of d1 only. With this in view, Figure 13.14 was developed which shows the variation of *d2* with *d1* in case of five spurs. The excellent agreement in the case of spur no.3 is remarkable in view of the complexity of the flow field around spurs and the simplicity of the relationship. The resulting equations of the best fit line in the case of spur nos. 2,3,5,6, and 7 were obtained as follows:

$d2 = 1.41 + 2.10\, d1$; spur no. 2 (13.4a)
$d1 = -0.31 + 2.00\, d1$; spur no. 3 (13.4b)
$d2 = -19.69 + 5.76\, d1$; spur no.5 (13.4c)
$d2 = 6.90 + 0.58\, d1$; spur no. 6 (13.4d)
$d2 = 7.19 + 0.33\, d1$; spur no. 7 (13.4e)

The limitation of best-fit line equations is these do not satisfy the boundary conditions of *d2* being zero when *d1* approaches zero. Also no unique relationship seems to exist for all the spurs. This is possibly owing to the differences in the flow conditions encountered around spurs because of the varying angles of incoming flow.

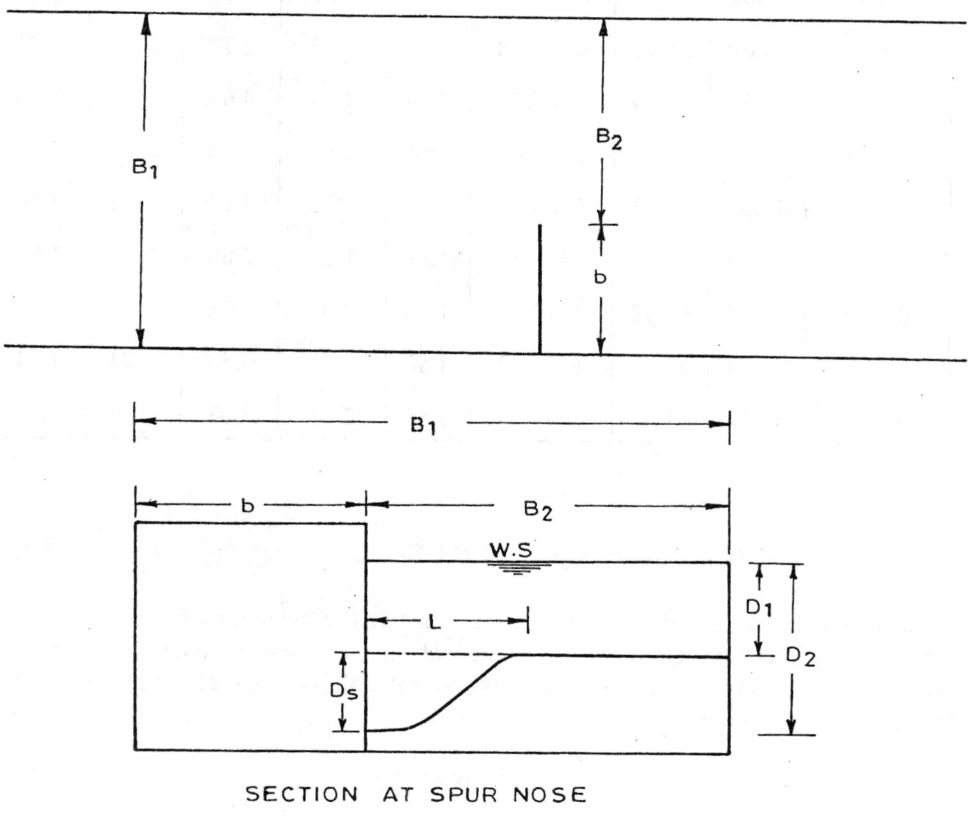


Fig. 13.13 Definition sketch.

Table 13.6 Field data for spur no.6.
Average bed level = 38.9 m, B1 = 2652.0 m, B2 = 2440.0 m

| date | water level (m) | bed level around (m) | Q($m^3$/s) | q ($m^3$/m/s) | q1 ($m^3$/m/s) | d1 (m) | ds (m) | d2 (m) |
|---|---|---|---|---|---|---|---|---|
| 1.7.86 | 44.34 | 43.91 | 14530.76 | 5.48 | 5.96 | 5.42 | -4.99 | 0.43 |
| 6.8.86 | 44.40 | 41.10 | 14868.89 | 5.61 | 6.09 | 5.48 | -2.18 | 3.30 |
| 10.9.86 | 44.20 | 35.21 | 13769.74 | 5.19 | 5.64 | 5.28 | 3.71 | 8.99 |
| 25.9.86 | 44.15 | 35.42 | 13507.17 | 5.09 | 5.54 | 5.23 | 3.50 | 8.73 |
| 8.10.87 | 43.50 | 33.70 | 10495.60 | 3.96 | 4.30 | 4.58 | 5.22 | 9.80 |
| 25.9.88 | 43.35 | 30.40 | 9896.82 | 3.73 | 4.06 | 4.43 | 8.52 | 12.95 |
| 3.11.89 | 43.25 | 35.00 | 6393.02 | 2.41 | 2.62 | 3.33 | 3.92 | 7.25 |

Table 13.7 Field data for spur no.7.

The average bed level = 38.93 m, B1= 794.0 m, B2=1592.0 m

| date | water level (m) | bed level ar-ound, m | Q ($m^3$/s) | q ($m^3$/m/s) | q1 ($m^3$/m/s) | d1 (m) | ds (m) | d2 (m) |
|---|---|---|---|---|---|---|---|---|
| 25.3.86 | 40.00 | 38.70 | 2688.14 | 1.50 | 1.69 | 1.07 | 0.23 | 1.30 |
| 8.7.86 | 44019 | 32.03 | 14530.76 | 8.10 | 9.13 | 5.26 | 6.90 | 12.16 |
| 29.9.86 | 44.00 | 36.64 | 13507.17 | 7.53 | 8.48 | 5.07 | 2.29 | 7.36 |
| 14.11.86 | 43.70 | 34.20 | 12028.14 | 6.70 | 7.56 | 4.77 | 4.73 | 9.50 |
| 7.10.87 | 44.20 | 37.05 | 14596.61 | 8.13 | 9.16 | 5.27 | 1.88 | 7.15 |
| 25.9.88 | 43.35 | 36.55 | 10495.60 | 5.85 | 6.59 | 4.42 | 2.38 | 6.80 |
| 25.10.89 | 43.27 | 35.90 | 10172.13 | 5.67 | 6.39 | 4.34 | 3.03 | 7.37 |

| | | | | | | | | |
|---|---|---|---|---|---|---|---|---|
| 30.10.90 | 42.31 | 33.81 | 6955.24 | 3.88 | 4.37 | 3.38 | 5.12 | 8.50 |
| 12.11.91 | 41.32 | 32.31 | 4657.49 | 2.60 | 2.93 | 2.39 | 6.62 | 9.01 |
| 30.10.92 | 42.40 | 34.90 | 7210.17 | 4.02 | 4.53 | 3.47 | 4.03 | 7.50 |
| 17.11.93 | 42.14 | 34.65 | 6496.69 | 3.62 | 4.08 | 3.21 | 4.28 | 7.49 |
| 20.11.94 | 43.35 | 31.80 | 10495.60 | 5.85 | 6.59 | 4.42 | 7.13 | 11.55 |

Attempts were also made to test the efficacy of some of the field-based models of scour prediction. Among the performances of one of the very popular models which relates *d2* to Lacey's regime depth (Inglis, 1949) is worth mentioning. With the use of Lacey's model, the percentage error between observed and computed *d2*, defined with respect to the observed *d2*, did indicate the occurrence of at least 85% values within an error band of ± 35% However, the major criticism with the use of discharge intensity based equations, which is the case with all the available field data-based equations of scour around spurs, is that the scour around spurs is a local phenomenon and thus the appropriate parameter to be used in the analysis should be that value of the discharge intensity which really exists near the spur nose. In a stream of a few kilometres of length, it is not rational to divide the flow by such a large width of the river and use it in developing some sort of a predictive relationship for the site under consideration.

Truly speaking, the situation, encountered at the Gumi site, is a multiple spur system. With the exception of Suzuki's (1987) model in which the ratio of scour depth at downstream spur to the upstream spur is a linear function of the ratio of the distance between the two spurs to the length of the spur, none of the existing models deal with the multiple spur system of the type encountered at the Gumi site. The main limitation of Suzuki's model is that it is based on the experiments with spurs having an equal length and spacing. Thus there is a need to explore the modelling of a multiple spur system having unequal spur spacing as well as varying spur lengths.

It is also interesting to find that spur nos. 1 and 4 never came into action. It is quite possible that at least the construction of spur no. 4 could have been avoided by conducting certain flow visualisation experiments. In future, trial experiments at the laboratory scale may offer better insight into the number of spurs as well as their deflection angles.

The problem of siltation at the spur nose was also observed. In the literature on scour, it is well recognised that the sediment load plays an important role in the modelling of the scour phenomenon. Thus, it will be useful to monitor the sediment load also along with scour observations. This may further help the economical design of spurs as the presence of sediment load is expected to reduce the extent of scour around spurs.

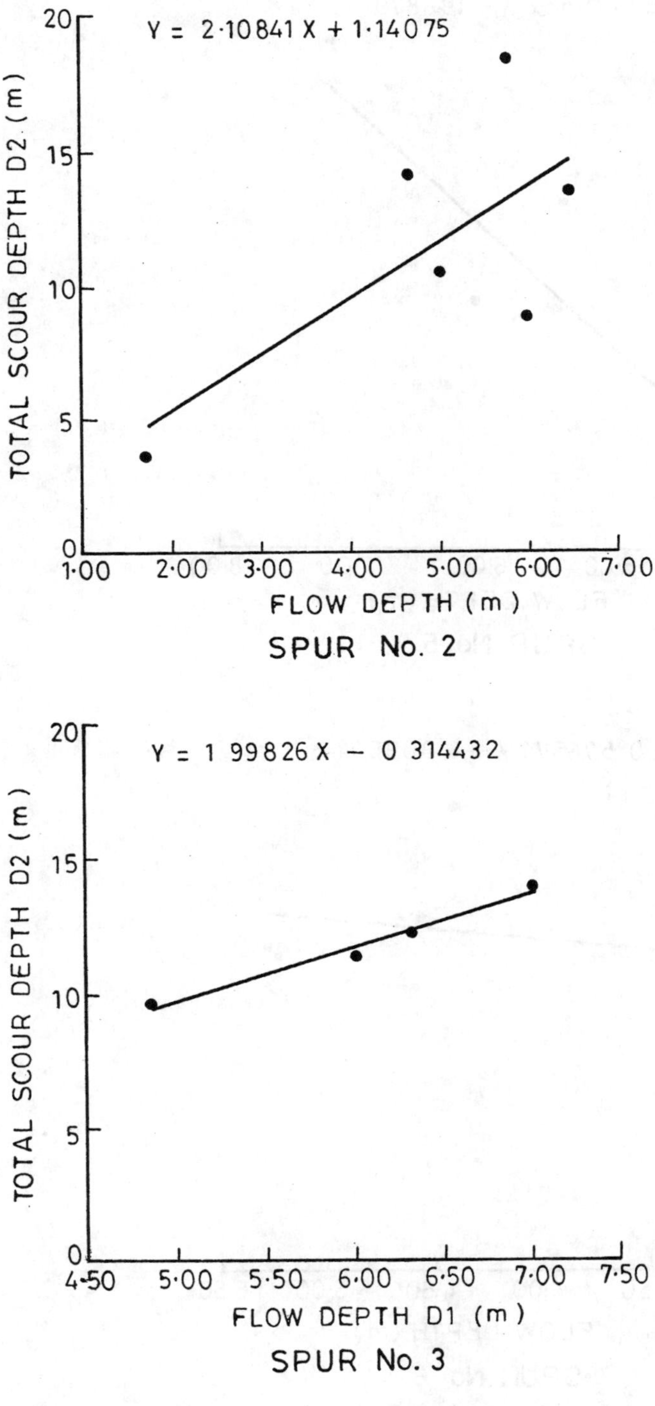


Fig. 13.14a Relation between D1 and D2 for spur no. 2 and 3.

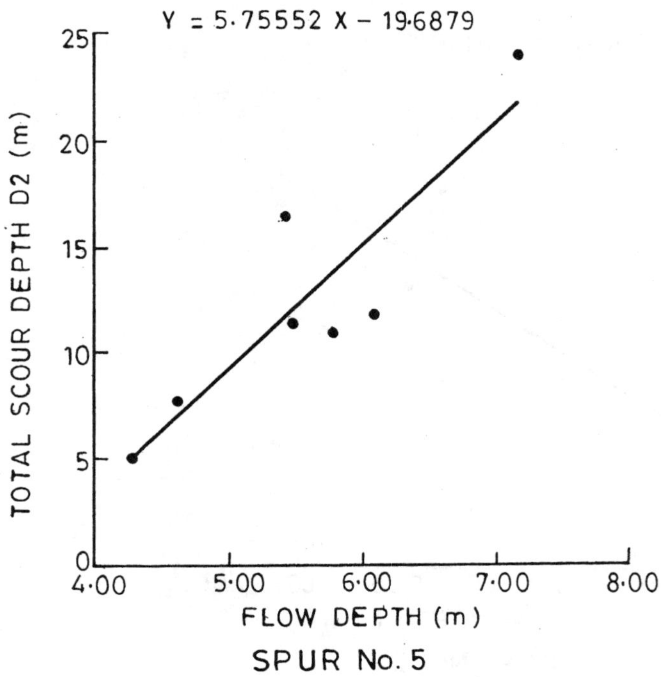


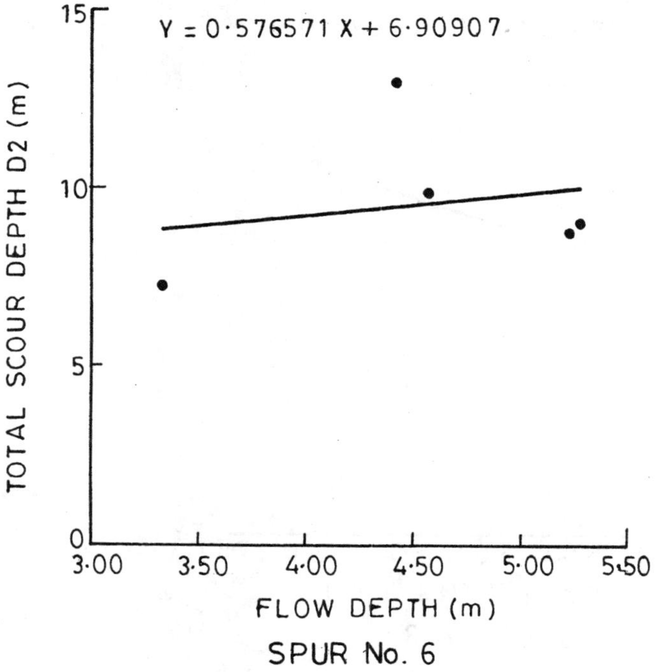


Fig. 13.14b Relation between D1 and D2 for spur no. 5 and 6.

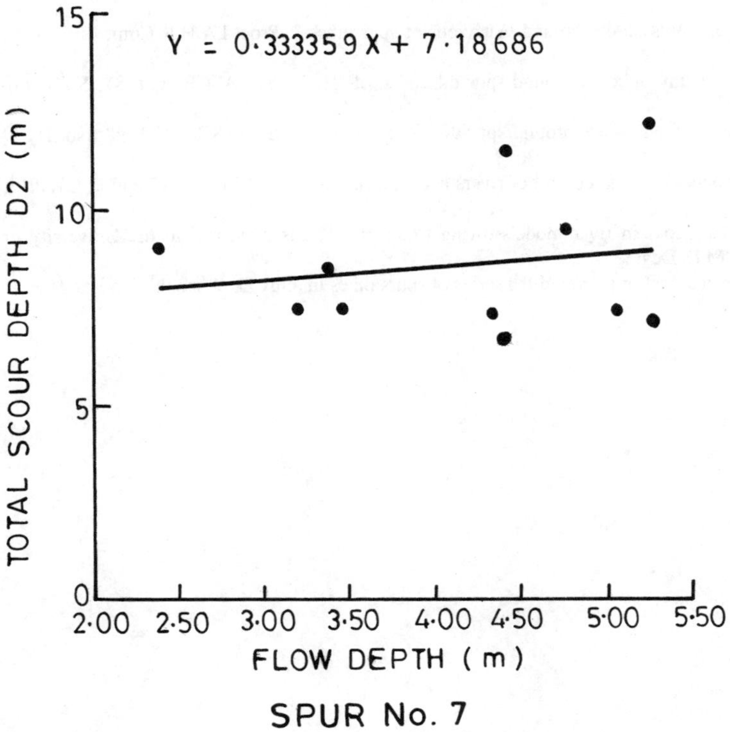


Fig. 13.14c Relation between D1 and D2 for spur no. 7.

### 13.12 Concluding Remarks

The main emphasis in this study has been to test the potential and limitations of the available spur scour predictors. However, it is found that none of these are suited to the system of multiple spurs, as encountered at the Gumi site. Also, owing to the lack of a comprehensive data base of all the variables affecting flow around spur, no ambitious attempt has been made to develop a sophisticated model of scour around spurs. Rather, using Gill's (1972) model it has been shown that for the present flow conditions, a relationship between scour depth and flow depth is feasible for different spurs. Although an acceptable agreement between the scour and flow depth could not be obtained for all the spurs, considering the complexity of the problem, inadequacy of the existing models, and simplicity of the relationship between scour depth and flow depth, the present analysis gains importance. Also the areas in which further work is needed for developing proper knowledge and data base are indicated. The provision of land spurs has proved to be a successful strategy. However, attempts are needed to minimize its cost. To be specific to the Gumi site, it may also be interesting to test the validity of the predictors developed in this study with the available data base in the future.

## Reference


Ahmad, M. (1953). "Experiments on design and behaviour of spur dikes." Proc. I.A.H.R Conf. Minnesota, 149-159.

Garde, R.J. et al. (1961). "Study of scour around spur dikes." J. of Hyd. Div., ASCE, Vol. 87, No. Hy 6, Nov. 23-27.

Gill, M.A. (1972). "Erosion of sand beds around spur-dikes." J. of Hyd. Div., ASCE, Vol. 98, No. Hy 9, Sept.,

Inglis, C.C. (1949). "The behaviour and control of rivers and canals." Res. Pub. No. 13, pt I and II, CWPRS, Pune.

Kilner, F.A. (1952). "Local scour in river models owing to spurs." Thesis presented to the University of London, England for M.E. Degree.

Suzuki, Koichi (1987). "Local bedform around the series of spurs dikes in alluvial channel." J. of I.A.H.R.

# CHAPTER 14

# WATER RESOURCES PLANNING

S.N. PHUKAN

Water resources planning in the northeastern region of India can be summed up as the water resources planning in the Brahmaputra River basin. The development of this region depends on the utilization of the resources of the basin. Likewise, many of the problems faced by the region are caused by the river. This chapter makes a few preliminary remarks about water resources planning in the Brahmaputra basin.

## 14.1 Introduction

While the Brahmaputra and its tributaries offer a unique scope for development of its resources for irrigation, hydropower generation, navigation, pisciculture, etc., they also cause large scale damages in the form of floods and erosion. In order to achieve the benefits of its development potential and to alleviate its damaging attributes, careful planning of river valley projects on the main branch called the Dihang and the 35 odd tributaries at intervals along the 650 kilometers long Assam valley is a prime necessity. Before the planning of the river valley projects is discussed, a brief description of the river is called for to understand its characteristics.

## 14.2 Brahmaputra Basin

The Brahmaputra River is the most dominant natural feature of India's north-eastern region. It has the distinction of being the seventh largest river in the world considering either its length of 2,906 kms or its annual average flow of nearly 54 million hectare meters, measured a short distance before it leaves the Indian territory to enter Bangladesh. The Brahmaputra River originates in western Tibet (China) about 63 km south east of the lake Mansarowar at an approximate altitude of 5,300 meters above the mean sea level and is known as the Zangbo (Tsangpo) there. The river journeys along the Tibetan plateau for a distance of 1,625 kilometers before it enters the Indian territory in the state of Arunachal Pradesh, where it is called the Siang of Dihang. The average gradient of the river in Tibet is 1 in 385 although for a reach of nearly 650 kilometers at an elevation of about 3,600 meters the river is gentle enough to allow navigation. Towards the end of its Tibetan journey the river changes direction in quick succession, firstly towards the north east, then north, and then sweeping with a hairpin bend cascades down around the Gyalaperi and Namcha Barwa massifs towards the south and south west entering the Indian territory. In the last 260 kilometers of its travel before debauching into the Indian territory, the Tsangpo drops through a height of 2,200



*V.P. Singh et al. (eds.), The Brahmaputra Basin Water Resources*, 299-308.

meters. This is perhaps one of the largest potentials for hydropower generation in the world, should China consider developing the drop! The Siang (Dihang), as the river is known in India, flows in the southerly direction through the fairly steep hills of Arunachal Pradesh for a distance of 226 kilometers up to Pasighat in the foothills. Thereafter the Dihang flows more gently for 52 kilometers up to Kobo, where it is joined by the Dibang and the Lohit; from the confluence of these three rivers the river is called the Brahmaputra. The river gradient between the Indo-China border and Kobo is 1 in 515. The elevation at Kobo is 197 meters above the mean sea level.

The Brahmaputra cuts through the valley of Assam from east to west for a distance of 650 kilometres before it turns south into Bangladesh, where it is called the Jamuna. Between Kobo and Dibrugarh, a distance of 70 kilometers, the river has a fairly steep gradient of 1 in 770 and thereafter it flattens out. From an elevation of 106 meters above MSL at Dibrugarh the river drops to 30 metres above MSL near the Bangladesh border in a distance of 580 kilometres. The river was navigable throughout from the sea upto Dibrugarh until 1950.

## 14.3 Brahmaputra Basin Characteristics

The extent of the Brahmaputra basin lies between latitude 25° and 32° north and longitude 82° and 97° east. The total basin area is 580,000 sq. km. of which 293,000 sq. km. is in China (Tibet), 240,000 sq. km. in India and Bhutan and 47,000 sq. km. in Bangladesh.

The catchment falls in different climatic zones. In the Tibetan catchment, situated 4 to 5 kilometers above sea level, snow melt and rainfall combine to feed the river. The catchments within Indo—Bhutan--Bangladesh territory contribute to the flow entirely from rainfall. The precipitation varies from area to area. In some areas the monsoon is active, being in line with the path of the south west monsoon, whereas in some areas the monsoon path is obstructed by the hills making these rainshadow areas. Orography plays a major role in distribution of monsoon rainfall.

The south west monsoon sets in early June and withdraws in early October. Thunderstorms and pre-monsoon showers in mid-April and May are usual. The mean annual rainfall over the catchment excluding that of Tibet and Bhutan is around 2,300 mm. The mean annual rainfall varies from a low of 1,750 mm in the Kopili sub-catchment on the south bank to 2,600 mm in the Dihang sub-catchment on the north bank. Heavy rainfalls up to 4,000 mm in a year have been recorded in some pockets in some years. Nearly 80 percent of the rainfall occurs during the monsoon months of June to October.

The monsoon rains and snow melt combine to contribute to the flow of tributaries and the Brahmaputra. Quite often all the tributaries are in spate simultaneously, causing the Brahmaputra to flood the entire valley. The maximum flood flow recorded at Pandu, 10 km downstream of Guwahati, is 72,794 cusecs. The recorded minimum is 1,757 cusecs. The variation in water level is 9 metres.

After the confluence of the Dihang and the Lohit the Brahmaputra River receives 20 tributaries from the north and 13 from the south within the Assam valley. The tributaries from the north have steep slopes and shallow braided channels and carry fairly heavy sediment load. The south bank tributaries have gentler slopes and deep meandering channels with much lower sediment load. The Brahmaputra is a wide and braided river in most of its length, occupying 6 to 19 kilometres of the narrow Assam valley, which measures about 80 kilometres from foothill to foothill.

Geologically the hills of Arunachal Pradesh lying north of the main river, are formed of the relatively young Himalayas of the tertiary group of rocks. The east west ranging Assam and Meghalaya hills on the south bank are of the older Archaean group of rocks.

The north eastern region is seismically active and is covered under the highest seismic classification. Two major earthquakes within the last 100 years, one in 1897 and the other in 1950, had exceeded magnitude 8 in the Richter scale. The 1950 earthquake (Richter 8.7) caused extensive damage in the valley, particularly in the upper Assam. The morphology of the Brahmaputra River and some of the tributaries underwent major changes as a result of the earthquake. The bed levels of the river at Kobo and Dibrugarh had risen by as much as 3 meters. In the subsequent years, the sediment load of the river increased enormously.

### 14.4 Flooding in the Basin

Floods of the Brahmaputra and some of the tributaries cause serious damage to property, crops, roads, railways and habitations. An area of up to 2.9 million hactares are observed to be flood prone. This is 37 percent of the land area of Assam. The other serious problem caused by the river is the erosion of the banks. A map study of the bank lines of the Brahmaputra River as it existed in 1930 and as it was in 1967 reveal that over 200,000 hectares of land were lost to the river. Towns like Sadiya and Palasbari no longer exist. Part of the town of Dibrugarh was eaten away by the river and the remaining part was saved at great cost.

The recurring floods and erosion kept everyone concerned with the river works busy in planning to mitigate the adverse impacts arising out of these twin problems. There was really neither any time nor resource in the hands of the planners to consider planning for development of the water resources for beneficial use. Plans were made for construction of a network of embankments for flood protection and some erosion control measures. The man power and financial resources were tied down between 1954 and 1985, 934 kilometres of embankments along the banks of the Brahmaputra and about 2500 kilometres of embankments along the tributaries have been completed. These have provided a reasonable degree of protection to about 1.3 million hectares of flood-prone area.

## 14.5 River Valley Development in India: Brahmaputra Board

In India all river valley development works are planned and executed by the state or the Central Government or government agencies. The flood control works in the Assam valley including that of the Brahmaputra were carried out by the Public Works Department of the state government. The need for an exclusive organisation to take charge of flood control and river training works was felt when the unusually heavy floods of 1954 caused extensive damage. In 1954--55 a government department in the state was created to take over flood control and river training works. This department was re-organised in 1970 as the Brahmaputra Flood Control Commission with wider powers. During the late seventies some studies and field investigations for storage dams on the Dihang and Subansiri rivers were carried out by the commission under the advice and guidance of the Joint Rivers Commission. The JRC was a technical body formed by the governments of India and Bangladesh to study the augmentation of waters of the Ganga downstream of the Farakka barrage.

In 1982 the government of India established a new organisation, called the Brahmaputra Board, to undertake the work of planning of the entire water resources of the river within the north eastern region. While the main objectives of the plan would be control of floods and erosion, the plan would require to include development of the potential of the water resource for all beneficial purposes. The board was also authorised to undertake implementation of major projects under the plan, subject to approval of the concerned state government and the central government. The planning of storage reservoirs was actively taken up by the Board.

### 14.5.1 PLANNING FOR FLOOD MITIGATION

An effective solution to the problem of recurrent floods during the monsoon months as also utilisation of the water resources lies in providing for adequate storage of the seasonal monsoon flows in the reservoirs. As nearly 80 percent of the total flow occurs during the four or five months of the monsoon, there is no alternative to its utilisation other than by providing for storage. Difficult and inaccessible terrain, large volumes of flow, lack of infrastructure in the north eastern region, seismic activities, etc., did not initially allow for consideration of storage dams in the Brahmaputra valley. With the advanced technology and expertise now available and examples of construction of large dams in similar areas within the country and all over the world now make it possible to consider suitable structures to provide for storage in the major tributaries of the Brahmaputra.

Careful and detailed studies were carried out for fairly large storage dams over the Dihang and Subansiri (S) rivers and preliminary studies for storage in other tributaries and these have shown that reservoirs can be constructed with great benefit. One major obstacle is the submerging of some important towns and villages in Arunachal Pradesh for which reason alternatives are now being studied. The storage possibilities of some of the major tributaries and the Dihang are as given below:

| River | Average runoff (Million meters) | Possible live storage (Million meteres) | Total storage (Million meters) |
|---|---|---|---|
| Dihang | 17.90 | 3.55 | 4.70 |
| Subansiri | 5.27 | 1.00 | 1.40 |
| Dibang | 3.39 | 0.47 | 0.62 |
| Lohit | 3.70 | 0.33 | 0.52 |
| Total | 30.26 | 5.35 | 7.24 |

It is observed that the contribution of these four rivers is over 60 percent of the total annual runoff of the river as measured at Guwahati. Mathematical model studies of the effect of storage on the river downstream have shown considerable reduction in flood levels as well as flood frequencies. The reservoirs will also reduce the transportation of bed load sediments considerably. The planning provides for a life expectancy of 100 years for the reservoirs and the dead storage capacity has been based accordingly.

Studies for providing storages have been carried out for the following tributaries also, each of which is now considered feasible.

| North Bank tributaries | | South Bank tributaries | |
|---|---|---|---|
| 1. | Jia Bharali | 1. | Noa Dihing |
| 2. | Pagladiya | 2. | Disang |
| 3. | Puthimari | 3. | Dikhu |
| 4. | Manas | 4. | Jhanzi |
| | | 5. | Doyang/Dhansiri (S) |
| | | 6. | Kopili |
| | | 7. | Kulsi/Khri |

Two reservoirs have already been built over the Kopili River, but these are mainly for hydropower generation. The storage dam over the Pagladiya River is in a fairly advanced stage of planning. The total expected storage on the proposed south bank tributaries comes to about 0.61 million hectare metre. The total storage on the four north bank tributaries is expected to be around 1.65 million hectare metres. The overall effect of the smaller storages will be to reduce the floods that occur in individual river basins. In addition, the storages will enable utilisation of the water resource for hydro-power generation and irrigation.

## 14.5.2 HYDROPWER GENERATION

Along with planning of the storages on the tributaries, a detailed study of the hydroelectric power generation potential has been made. The river basin hydropower potential is appended. It also lists the hydro projects already developed, purely as power projects. The total hydropower potential works out to 40,594 MW on the basis of providing the installed capacity at a 30 percent load factor.

**Hydropower potential of the Brahmaputra valley (as of June 1997):**

| | River | Potential | Status |
|---|---|---|---|
| | Umiam-Umtru-Khri-Basin: | | |
| 1. | Umtru | 11.2 MW | in operation |
| 2. | Umain stage-I | 36.0MW | `` |
| 3. | Umiam stage-II | 18.0MW | `` |
| 4. | Umiam-Umtru Stage - III | 60.0MW | `` |
| 5. | Umiam-Umtru Stage - IV | 60.0MW | `` |
| 6. | Umiam-Umtru Stage - V | 36.0MW | Planned & investigated |
| 7. | Umiam-Umtru Stage - VI | 120.0MW | |
| | Total of basin = | 341.0 MW | |

| | | | |
|---|---|---|---|
| Lower Umiam Basin: | | | |
| 1. | Umswai (Lower Umiam) | 120.0 MW | Planned & Investigated |
| Borpani Basin: | | | |
| 1. | Upper Borpani | 60.0 MW | Partially Investigated |
| 2. | Intermediate Borpani | 60.0 MW | `` |
| 3. | Lower Borpani (Langpi) | 100.0 MW | Under construction, work stopped for non-technical reasons. |
| | Total = | 220.0 MW | |
| Kopili Basin: | | | |
| 1. | Kopili Stage -I | 150.0 MW | In operation |
| 2. | Kopili Extension Phase -II | 100.0 MW | Work completed, under test run |
| 3. | Kopili Extension Phase -II | 50.0 MW | Planned |
| 4. | Lower Kopili | 100.0 MW | Investigated & report approved. |
| 5. | Amring | 33.0 MW | Partially Investigated |
| | Total Basin = | 433.0 MW | |
| Dhansiri South Basin (Doyang) : | | | |
| 1. | Doyang Hydro | 75.0 MW | Under construction. |
| Lohit Basin: | | | |
| 1. | Lohit Storage | 3000.0 MW | Planned |

| | | | |
|---|---|---|---|
| 2. | Lohit run of the river | 1000.0 MW | Partially investigated |
| | Total of Basin = | 4000.0 MW | |

Dibang Basin :

| | | | |
|---|---|---|---|
| 1. | Dibang Storage : | 2500.0 MW | Planned |

Subansiri Basin :

| | | | |
|---|---|---|---|
| 1. | Subansiri storage | 4800.0 MW | Investigated and DPR prepared. Alternative sites being examined to to reduce submergence |

Ranganadi Basin :

| | | | |
|---|---|---|---|
| 1. | Ranganadi runoff scheme | 405.0 MW | Under construction. |
| 2. | Ranganadi storage | 100.0 MW | DPR prepared. |
| | Total Basin = | 505.0 MW | |

Jiabhroli Basin (Kameng):

| | | | |
|---|---|---|---|
| 1. | Kameng runoff the river scheme | 600.00 MW | DPR prepared. |
| 2. | Jiabhoroli storage | 2000.0 MW | Planned. |
| | Total Basin = | 2600.0 MW | |

Manas Basin :

| | | | |
|---|---|---|---|
| 1. | Manas storage | 5000.0 MW | Partially Investigated |

Siang Basin :

| | | | |
|---|---|---|---|
| 1. | Dihang Storage | 20,000.0 MW | Investigated & feasibility Report prepared. Project being revised to reduce submergence. |
| | Total Potential | 40,549.0 MW | |

Installed Capacity at 30 percent Plant Capacity Factor.

### 14.5.3 IRRIGATION POTENTIAL

The geographical area of Assam, Meghalaya, Nagaland, and Arunachal Pradesh, the four states adjoining the Brahmaputra Valley, is about 20 million hectares. The area under all crops in these four states has been reported as 4.5 million hectares. The figure for Assam is 3.6 million hectares which includes 2,37,000 hectares under tea. Out of the 4.5 million hectares of the cropped area, only 0.7 million hectares are irrigated.

The main crops in the region consist of rice, wheat, maize, pulses, oilseeds, sugarcane, certain vegetables, jute, and mesta. The total food grain production for all crops excluding tea but including sugarcane is around 6.2 million tonnes. The production of rice in Assam in 1994--95 is about 3.4 million tonnes. The average yield of about 1,300 kg/hectares is much below the figure of 3000 kg/hectare in other parts of the country.

The population of the north eastern states has gone up from 10.36 million in 1951 to 31.55 million in 1991. The staple diet for people in the region is rice. In order to provide foodgrains to the increasing population, increase in productivity has to be achieved. One of the major inputs for increased production is to provide for adequate irrigation.

Unlike other states of India, irrigation was hardly practiced for agricultural production in the north eastern region. Most of the major crops depended on rainfall. In the mid sixties a few flow irrigation projects were developed in Assam. The cultivatable area in the region has been estimated to be about 7 million hectares. The irrigation potential has been estimated to be 3.6 million hectares.

There is no shortage of water for irrigation. Apart from the huge surface water resources in the Brahmaputra valley, the ground water which can be utilized has been estimated to be over 2 million hectare meters. Even with the optimum provision for irrigation, the annual requirement of water will not exceed a figure of $6*10^6$ hectare meters.

Surface water for irrigation can be drawn from the perennial tributaries both in the north bank and the south bank of the Brahmaputra. There are tributaries at distances of 7 to 10 kilometers except in some reaches where the gap goes up to 40--50 kilometers. These surface water resources can be tapped either by diversion structures or by pumping for providing irrigation. Release from storages will also help in basins where storage projects are completed. The planning for irrigation has to be under taken subbasin wise to utilise the full irrigation potential.

### 14.5.4 NAVIGATION

The Brahmaputra River served as the principal route for transportation of goods between Calcutta and places in the Assam valley. Regular steamer services were operated by the River Steam Navigation (RSN) company from 1863. The RSN company was merged with the India General Steam Navigation Company in 1899 and formed what was popularly known as Joint Steamer Companies. They provided an excellent freight service between Calcultta and as many as 16 riverside towns from Dhubri in the west to Dibrugarh in the east along the 650 kilometres long river route in Assam after traversing about 370 kilometres from Calcutta Port. Outgoing traffic from the valley to Calcutta was also heavy.

Following the 1950 earthquake steamer services beyond Neamatighat became difficult. The 1965 Indo--Pak war disrupted the steamer services for quite sometime. The central government took over the steamer Company and formed a Public sector company owned by the Government of India called the Central Inland Water Transport Corporation in 1957 (CIWTC). A protocol was signed between the Governments of Bangladesh and India in 1972 for operating the freight service via Bangladesh. The CIWTC is continuing to operate the service now.

However, with the opening of the national highway via Siliguri and the conversion of the meter gauge rail link to broadgauge, the traffic on the river route got diverted to road and rail. Besides, industrial goods produced in different parts of the country found it more convenient to take the direct fail or road route without touching Calcutta. In spite of the loss of traffic to road and rail, the traffic potential on the river route is estimated to be about 1.1 million tonnes by the year 2004--2005. Efficient and quicker turnaround is likely to improve the traffic potential between Calcutta and the Assam valley.

The planned release of water from major storages will improve the draught in the river during the lean season by 1.5 to 2 meters. This will help in providing more efficient and economical freight traffic on the river route. The navigation needs should be kept in view while planning the operation of the reservoirs.

### 14.5.5 FISH PRODUCTION

Fish is an important component of diet for the people of the north eastern region. The production of fish in the region is 170,000 tonnes annually. However, the requirement for the region is about 280,000 tonnes a year. The major fish production/cultivation is done in small wetlands and depressions called beels and derelict water. The total area of such wetlands, lakes and beels is about 230,000 hectares, the majority of which is located in the Assam valley. The construction of the smaller reservoirs in the tributaries will offer an opportunity to increase cultivation and production of fish to a large extent. The planning of reservoirs has therefore to keep this aspect in view and integrate the requirement of fish culture.

### 14.5.6 ENVIRONMENTAL IMPACT

The most serious thought will, however, be required to be given to the danger of any adverse environmental impact which is bound to occur in the building of storage reservoir projects. The hilly areas still have a low density of population and therefore resettlement of people may not be very difficult but would need very careful planning and advance action. Treatment of the catchment areas, alternative afforestation for loss of forests, alternatives for loss of cultivatable land would need meticulous planning for each storage project. These are however usual steps for planning of any reservoir project. The adverse impact of breeding of mosquitoes, snails etc. need to be taken care of. However, these should not pose as insurmountable problems if these are tackled in a proper manner. Many reservoirs have been built all over the world where these problems have been solved. The main issue is to weigh the benefits against the losses.

## 14.6 Concluding Remarks

It will be obvious from the foregoing analysis that the water resources of the Brahmaputra offer a huge potential for the benefit of the people. Only planning and construction of storage reservoirs on the tributaries would allow people to obtain the major benefits like flood and erosion control, irrigation, and hydropower generation, and lesser benefits like navigation and pisciculture. Even trans-basin transfer of surplus water to deficit areas of the country may perhaps be considered. However, all this would need the will and determination of the highest authorities of the government and the capability of the leadership to get the co-operation of the people for the difficult tasks involved in building the reservoirs.

# CHAPTER 15

# BRAHMAPUTRA FLOODING: PROBLEMS FOR ADMINISTRATION AND PROSPECTS

K. S. KRISHNA

Flooding in the Brahmaputra basin is an annual occurrence. Every year flooding occurs and causes havoc. Despite all the advances in hydraulic engineering, the flood menace has hardly diminished. This chapter examines the flooding problem from an administrative perspective.

## 15.1 Introduction

Disaster of any kind is a curse for human society. A disaster is also a challenge for administrators. It is the time when an administrator can prove his calibre by taking quick and efficient steps. The earlier the crisis is met, lesser will be human suffering and losses. It is a time when all departments concerned are to coordinate and work hand in hand so that all related hazards are dealt with by proper care and attention.

In Assam floods are a recurring phenomenon. Every year one or the other part of the state is engulfed by floods, resulting in much loss of life and property. Floods have caused incalculable damage to crops, livestock, bridges, communication and transport networks, irrigation and flood control structures and various other private assets and public property.

Apart from floods, other disasters such as earthquakes, epidemics, drought, and famine, fire breakouts, landslides, etc. also occur in Assam, though on a relatively smaller scale. Floods occur in Assam owing to heavy and incessant rainfall during Monsoon. This leads to excessive flow of water in Brahmaputra, resulting in floods. Prevention of floods has been attempted through the construction of embankments, dikes, etc.. Measures of rescue, relief, and rehabilitation have been devised and put in use on a systematic basis.

The danger level of the river at each strategic point has been determined after elaborate assessment. The travel times of flood from one point to another have also been worked out. The flood control department has also prepared district maps showing the very vulnerable areas, the vulnerable areas, and the protected areas. Maps are revised every year. An elaborate flood warning system has also been worked out. The warning system is quite elaborate and only human failure or breakdown of the communication network might cause disasters without notice.



*V.P. Singh et al. (eds.), The Brahmaputra Basin Water Resources*, 309-320.

Certain guiding principles were communicated by the Government of India to State Governments in 1983 in the shape of a provisional code for the rest of the phases of rescue, relief and rehabilitation. In Assam the Revenue Department is the nodal department for relief operations. Other concerned departments, namely, Agriculture, Education, Forest, Health, Inland Water Transport, Irrigation, Public Works, Public Health Engineering, Police, Information and Public Relations, Civil Supplies, Social Welfare, Animal Husbandry and Veterinary, have been assigned particular jobs to perform at different stages.

The most important role is that of the Deputy Commissioner. Next is the Sub-Divisional Officer, who works under the guidance and supervision of the Deputy Commissioner. They are asked to minimise damage to life and property by a predetermined, planned, and timely approach to the problem. The factors which cause such damage are to be identified and responsible officers are required to take a timely action in respect of each of these factors. The Deputy Commissioner is required to be acquainted with the history of floods in his district, the causes of damage owing to floods in the past, and steps devised to overcome them.

A Contingency plan is kept ready before a flood, to deal effectively with the situation arising from the flood. The plan contains details regarding identification of relief camps, formation of relief parties, association of voluntary agencies, plans for test relief work, facilities to be provided in the voluntary areas, estimate of the requirement of essential commodities, facilities to stock them etc.

The Deputy Commissioner, along with engineers of concerned departments, e.g., Flood Control, Irrigation, etc., has to undertake tours to see that embankments and other measures taken for protection are in good condition and ensure that no breaches or fissures have taken place. On receipt of a flood warning the Deputy Commissioner has to take all precautionary measures, including evacuation to places earmarked for it earlier.

On occurrence of a flood the Deputy Commissioner has to visit the places, ascertain the nature and extent of flooding, make prompt operational decisions, arrange for co-ordinated aid by the concerned departments including voluntary organisations and rescue assistance from the army when the situation so demands. The Deputy Commissioner also deputes relief parties to the affected areas. Different relief teams carry different items of necessity like personnel, boats, vehicles, equipments medicines, food, fodder, etc.. Immediate relief is to be given to people as well as animals. Once the flood subsides rehabilitation grants for reconstruction of houses, taking up of agricultural activities are to be distributed. Loan and subsidy for purchase of plough, etc., are also given.

The Brahmaputra is such a huge river that its control by the traditional method of damming is a gigantic task. It will require enormous capital investment. International cooperation between China, India, and Bangladesh will be a pre-requisite for such a venture. Recently the Brahmaputra River Commission has been established to dam or otherwise contain some of the tributaries, instead of attempting to control the main river.

One way to minimise the damage from flooding is to change the cropping pattern in such a way as to avoid the flood season, which has been recommended in a report of the National Commission on Agriculture in 1976. The humanitarian aspect of the problems and needs of people in misery should not be lost sight of. In certain areas people have to live in riverine tracts in order to take out a living based on fishery and crops such as oilseeds and pulses.

The experience of years of flood relief came in handy when riots and arson in the wake of the general elections in February 1983 created a holocaust. On time, the administration veered around to meet the challenge of this unsuspected problem. Immediately 250 relief camps were set up in eight districts to accommodate 310,732 affected people. They were provided with gratuitous relief and also test relief. Later they were properly rehabilitated in their own villages. This operation cost the exchequer Rs. 49.71 crores [one crore = 10 million]. This massive operation has proved that the Assam administration has the ability to respond to disasters in a positive and adequate manner.

There is a necessity for an integrated and systematic approach to disaster management. It has been universally accepted that governments and concerned organisations should make preparations in advance for management of disasters. A considerable sum of money is spent every year by the central and state governments on relief and rehabilitation of people affected by disasters, and on reconstruction and restoration of roads, buildings, bridges and other public properties. It has been estimated that the total annual expenditure on natural calamities in India is in the vicinity of Rs. 2000 crores.

To manage disasters many state governments have drawn up manuals and evacuation schemes beside laying down guidelines relating to principles, objectives, priorities and thrusts in respect of different phases of disaster management, such as warning, rescue, relief and rehabilitation.

The advance preparation presupposes that the people who would implement the disaster management plan are fully posted about the subtle nuances of the plan and be made aware about the tasks each would be expected to perform. This involves training. Not much importance has been laid so far on training in this field in India. However, as recommended by the United Nations Disaster Relief Commission, "high priority should be given to education, training and preparedness both by individual governments and international organisations" (Water: Resources and Hazards, Geneva, 1977).

Training is important because it will condition the reflexes of trainees to act in the expected manner in response to emerging situations of different nature and magnitude. This prevents panic and inculcates disciplined thinking, mental equipoise and coherent action in face of a disaster. For training, the first and foremost is to identify and categorise the different groups of people involved in disaster management, because different types of training are to be imparted to each group, depending upon the different types of duties and responsibilities assigned to each of them. For example, when people are likely to be affected by a flood they should be trained to be ready for all eventualities, including evacuation to previously identified safer places. Here the required information may be disseminated through proper and extensive publicity.

The information should include assessment about the probable magnitude of the apprehended disaster, the simplified guidelines to be followed and that suffering and misery caused by disaster can be overcome only through orderly, disciplined adherence to guidelines issued for that purpose. The information should include details about the areas affected with different gravity, persons or institutions to be contacted for help, location of transport and medical units, where and how to evacuate and arrangements made for food and fodder supply.

The information may be disseminated through the mass media, pamphlets and handouts, house to house contact, announcements over microphones, meetings in public places and educational institutions and various other means. People for these purposes should be selected from among the people who actually participate in the field. There should be an in-built mechanism for periodical feedback from the public. Based on this the feedback modifications must be made in the plan and fresh publicity be made to keep people informed. The same format may be followed with minor variations to deal with other disasters affecting a mass of people, such as cyclones, earthquakes, fires, epidemics, etc..

The second category to be trained are the community leaders, including leading members of the Non-Governmental Organisations, Teachers, Local Panchayat Functionaries, Municipal Administrators, Revenue officials and others who can be expected to work as agents, commissioner motivators, and mobilisers. In their case the training will have to be more elaborate and intensive. Besides, they will have to be very thoroughly trained in the methods of first aid, the strategy of communication with a mass of people, the organisation of rescue and relief and in assisting the authorities in the event of evacuation, disciplined and orderly response to situation in hand.

For this category of people training may be broadly similar for almost all types of disasters. However, if disasters of special nature are apprehended to occur, such as gas leaks (e.g. Bhopal) or nuclear catastrophe (e.g., Chernobyl) particular attention will have to be given to train them for meeting such eventualities in areas where such disasters might take place, such as in the vicinities of fertiliser plants, oil refineries, nuclear power stations, etc..

The third category of persons to be trained are the members of specialised groups who are already in existence, such as Homeguards, Civil Defence Force, Para Medical, N.C.C., A.C.C. Territorial Army, Scouts and Guides, etc. Each group will have to be trained to deal with situations arising in their particular field of activity. The homeguard, for example, will have to be trained to assist the police to deal with law and order situations and to maintain discipline during rescue and relief operations. They should also get training to maintain essential services in case of a breakdown of these services. This aspect of training should be given priority in case training of Civil Defence Forces. In the case of para-medical personnel training should include treatment of both physical and psychological shocks. They should be trained to identify major cases requiring specialised medical attention and refer such cases to the appropriate hospitals or medical units. As far as N.C.C., A.C.C. scouts and guides, etc. are concerned, they should receive general training in first aid, traffic, management, rescue operations and distribution of relief.

General guidance to the specialised groups will have to be quite vigorous. During the training simulation of actual conditions should be attempted to the extent possible. The specialised groups are quite important as they will play a critical role in the implementation of the disaster management planning. Therefore the maximum attention should be paid to their training.

The fourth and the last category of people to be trained are those who will draw up the whole planning and manage the entire operation. This category will include the policy makers in Secretariat and Heads of Departments, Deputy Commissioners, Sub-Divisional Officers and top functionaries of the local governments, such as Panchayat Presidents, Mayors, and Municipal Commissioners. While imparting training to this category, the basic principles of psychology, human management and methods of dealing with stresses and strains of different types of disasters should be kept in mind. The design of the training programme should be tailored to the needs of each category and the needs of the situation that each will be expected to face.

It is important to realise that the Central, State and Local Governments must combine their resources, and manpower facilities to impart training so that the affected people are kept well aware about the way they should meet the disaster and the rescue and relief workers are properly equipped to meet the challenge of the disasters with minimum cost and maximum relief possible.

Disasters evoke a variety of behavioural response in different individuals, communities and countries. In fact, the behavioural pattern reflects the innate character of the affected people and also their capacity to absorb shock as a result of training and advance preparation (Das, 1987). In the case of unpredictable disasters, however, the behavioural pattern is completely different. Fire, industrial accidents, volcanoes, civil conflicts, mass exodus, etc. are examples of such disasters. In such cases the first reaction is panic. People suddenly find themselves at a loss to think coherently. When a disaster is known for its periodicity in particular regions and in identifiable localities, a disaster subculture begins to develop (Raghavulu, 1982).

As a result of disasters, both natural and man made, millions become refugees mainly in the third world countries. In a number of cases it has been noticed that they remain refugees long after the disaster. Their pitiable condition resembles beggars (Church, 1991). Meanwhile, the curious phenomenon of 'donor fatigue' develops which reflect the international community's frustration in coping with the increasing number of disaster victims (Franket, 1991).

Many such instances can be cited. In the third world countries people living below the poverty line suffer the most. Whether it is the slum dwellers of shanty towns of South America or the famine - stricken people from Africa or the landless farmers of the Indian sub-continent it is the people at the margin of existence who are affected the most by such disasters. Moreover, it is the children, women, and the other weaker sections amongst the victims who are the worst sufferers. It is especially the backward areas which suffer most from disasters (Planning Commission, 1981).

The response pattern of the defected people, N.G.O.s and Government, have been studied and commented upon by many. (Thaha, and Brauham, 1986; Ahmed and Kumar, 1981; Ragharulu, 1982). However, the conclusions arrived at are at best tentative because the human behaviour itself does not follow any definite course and is not always amenable to theoretical generalisations. Impressions gathered from experience and case studies are the only pointers to probable human behaviour.

The main exogenous response to disasters must be from the Government. The expectations of victims being high, the Government assistance becomes a critical variable in the behavioural pattern before, during, and after a disaster. There is, of course, a belief that "non-governmental efforts have enjoyed a greater acceptability, respectability and credibility than government sectors" (Secha, 1991). However, experience of actual work in the field suggests that a co-ordinated effort is more effective. The leadership in such co-ordinated effort has to be taken by the governmental agencies (Das, 1971).

Any kind of disaster is primarily to be managed by the administration. An administrator's work in a disaster is of great importance, because only he can manage a quick and efficient plan to meet with the crisis, with help and co-operation from other organisations and all concerned with the tragedy.

## 15.2 A Case Study of Floods in the Morigaon Sub-Division in 1988

### 15.2.1 INTRODUCTION

After completion of his Civil Service Training in Mussoori, the author joined as Sub-Divisional Officer (Civil) of Morigaon Sub-Division under the erstwhile undivided Naogaon District (Morigaon Sub-Division became a District from 14th October, 1989) in May 1987 and continued his service there up to July 1989. During his tenure as Sub-Divisional Officer (Civil), he faced the challenging job of combating flood menace in the Sub-Division in 1988. There were five successive waves of floods from the month of May to October, all the waves were equally devastating, displaying aggravation of overall tendencies. However, the devastation and havoc came to an end after the onset of autumn, to the great relief of the suffering people. Although the breached portions of embankments were not plugged till that time owing to the paucity of funds and inclement weather.

Morigaon became a Sub-Division in 1972. The area of the Sub-Division was 1,559 sq. km. and of the total 423,901 population. Its total population was 70,113 (16.54%), the S.C. population was 52,952 (12.49%) and the rest of the population was 300,836 as per census, 1971. The population was a mixed one; immigrant Muslim and Bengali Hindu formed a sizeable portion of the total population of the tribal population. Tiwas

(Lalungs) were the main components.

The Sub-Division consisted of five Revenue Circles, viz., Morigaon Sadar, Mikirbheta, Lahorighat, Bhuragaon and Mayang and three Development Blocks, viz., Lahorighat, Bhurbandha and Mayang. The Sub-Division was very backward at that time and lagged behind in many fields, although the road distance from Guwahati to Morigaon Town, the Sub-Division Headquarters is only 78 km. The situation has not improved much till date. The roads and wooden bridges over the rivers were in dilapidated condition. All the gravelled P.W.D. roads leading to Circle headquarters were in a horrible condition and bus services to those places were almost on the verge of breaking down. The main arterial P.W.D. road leading to Morigaon Town was half pitched and very narrow for passing two heavy vehicles at a time.

### 15.2.2 FEATURES

Flood is almost an annual feature in the Morigaon Sub-Division. Floods have been causing large human sufferings and incalculable damage to crops, livestock, bridges, communication and transport networks, irrigation and flood control structures, and various other private assets and public property.

It is beyond doubt that after the earthquake in 1950, floods in the Brahmaputra valley have been aggravated by the change of course of rivers, change in bed topography, heavy landslides, and silting. The river beds of the Brahmaputra and all other tributaries have been raised considerably by the impact of the 1950 earthquake owing to heavy deposit of silts, leaving very little scope for good drainage during peak monsoon period. The floods in the Brahmaputra valley have also been accentuated by minimisation of catchment areas, indiscriminate earth cutting, abrupt decline of forest, perceptible use of land for Jhum cultivation, and other factors. Floods basically occur when the river carries a flow in excess of its normal transporting capacity. The major cause of floods is, of course, the incidence of heavy rainfall along with melting of snow in the Himalayas and other allied mountain ranges. Again, although embankment construction has got a fillip since independence, impractical methods of construction of embankments along with the two sides of the rivers lead to erosion and inundation of new areas causing perennial floods in Assam.

The Morigaon Sub-Division witnessed devastating floods in 1988. In that year widespread damage was caused to agriculture, cattle population, and residential property and several human lives were lost. Virtually all the roads went under water for days in every wave of the floods, communication between Morigaon and Guwahati was totally disrupted and Morigaon and Naogaon are somehow maintained. Many wooden bridges over the Public Works Department (P.W.D.)-built roads were washed away by the turbulent currents of floods. Embankments along the Brahmaputra River at Lahorighat and Morigaon and along the river Kollong at Jagibhakatgaon (near Jagiroad) were breached. All the Circle officer, Block Development Officer, S.D.O. (Civil) office including S.D.O. (Civil)'s bungalows were in knee deep water. Not only the riverine areas of the Sub-Division but about 90% of the total villages were inundated by flood waters.

Added to flooding were the associated problems of erosion which assumed serious proportion in the Morigaon Sub-Division. The Brahmaputra River had slightly changed its course at places bordering the river in the Bhuragaon, Mayang and Lahorighat Circles following the natural inundation causing constant erosion of embankments and other agricultural and homestead land and silting on the opposite side. More than thirty villages had been engulfed by the mighty Brahmaputra from 1970 when erosion first started. Many hundreds of hectares of valuable tracts had been eroded by the Brahmaputra in the Sub-Division and the embankments at Bhuragaon, Lahorighat and Mairabari had been threatened. Although the E & D Department had taken up massive anti-erosion schemes for protection of Dhing-Morigaon-Lahorighat-Bhuragaon embankments from the clutches of erosion, no remedy to the problem was witnessed in actual terms.

### 15.2.3 ACTION PLAN FOR COMBATTING FLOODING

The authors had seriously taken up the nagging and critical flood and erosion problem of Morigaon Sub-Division as an unprecedented flood was witnessed at the unusual time of the month of October 1986 in the Sub-Division prior to his joining. A meeting of the Sub-Divisional Relief Committee was convened in the first week of April and all the Sub-Deputy Collectors (S.D.Cs.), Senior Block Development Officer/Block Development officer (Sr. B. D.Os/B.D.Os ), Executive Engineer of Irrigation and Public Health Engineering, Assistant Executive Engineer, P.W.D. (Roads), Morigaon (East & West) Divisions, P. W. D. (Building) and E & D, Sub-Divisional Medical and Health Officer (S. D.M. & H.O.), Sub-Divisional Animal Husbandry and Veterinary officer (S.D.A.H. & V.O.), Sub-Divisional Agriculture Officer (S.D.A.O.), Sub-Divisional Police Officer (S.D.P.O.), local M.L.A. and selective representatives of Non-Governmental organisations (N.G.O.s), including local unit of Indian Red Cross Society took part in the meeting. All officers present in the meeting in respect of specific flood and erosion prone areas and proceedings of that meeting were circulated among all the officers with a copy to the Government.

Contingency plans were received from all concerned and a consolidated Contingency plan of the Sub-Division was prepared much ahead of the onset of the monsoon. Circle maps identifying flood prone areas, probable relief camps and vulnerable P.W.D. and village roads along with week bridges were also obtained from concerned S. D. Cs. Meetings of Circle Level Relief Committees had also been held. All officers and staff of all Govt. offices of the Sub-Division were asked not to leave headquarters except with prior permission from S.D.O. (Civil).

### 15.2.4 RELIEF MEASURES: RELIEF CAMPS AND RESCUE OPERATION

For smooth functioning of relief operation, each revenue circle was divided into Zones. The S.D.C. was the Relief officer in his circle and he was assisted by Sr. B.D.O/B.D.O. and other Government officers/Teachers employed in the circle. Two or three officers from Sub-Divisional Headquarters were sent to each circle to assess the situation and make liaison in the matter. During floods in 1988 almost all the villages of this Sub-Division went under water and people were given shelter in relief camps established in school buildings, embankments and high portion of P.W.D. roads. Tin sheds were provided to flood affected people obtaining C.I. sheets from Engineering Departments. Makeshift latrines and urinals were also constructed in the relief camps. As road communication in flood affected areas got disrupted, country boats (both manually and engine driven) were pressed into service for rescue of the marooned people. List of such boats had been obtained beforehand by the S.D.Cs. through concerned Lat Mandals and Gaonburas.

### 15.2.5 GRATUITOUS RELIEF (G.R.)

Gratuitous Relief was provided as per Assam Relief Manual to the people stationed in the Relief Camps. The scales were - Adult: Rice - 600 gms, Dal -100 gms, Salt and Mustard oil - 30 gms and Minor: Rice - 400 gms, Dal - 100 gms, Salt and M.oil - 30 gms. Sufficient stocks of relief commodities were maintained in the statefed godown at Morigaon Town. As there was no F.C.I. godown at Morigaon, relief materials had to be carried from the F.C.I. godown of Naogaon. This transhipment caused much difficulties as the wooden bridges over the P.W.D. roads between Morigaon and Naogaon were so weak that only loaded Mini Trucks could pass over the bridges and at two approach points of a bridge thelas and labourers were engaged to carry relief materials from one truck at one approach to another truck at the approach of the bridge on the other side. Rolling stocks of petroleum products in sufficient quantities were maintained in the only depot at Morigaon town and the other depot at Jagiroad.

### 15.2.6 MEDICAL CARE

The Sub-Divisional Medical and Health Officer had made out a contingency plan beforehand for maintaining adequate stock of essential medicines and also to keep Doctors and para-medical staff ready in all Public Health Centres/State dispensaries. The plan was executed smoothly during the floods. Mobile medical teams moved to far away places to render medical assistance to distressed people. One medical team comprising junior doctors from Guwahati Medical College had offered free and voluntary medical service to the people camped in relief centres and flood ravaged villages. Local unit of Indian Red Cross Society distributed some valuable medicines, baby food, and other related items among flood affected people. There was no report of outbreak of epidemics and starvation death during and after the floods.

### 15.2.7 VETERINARY ASSISTANCE

Animal care was properly taken care of by the Veterinary Department. The Sub-Divisional Animal Husbandry and Veterinary Officer maintained stock of veterinary medicines and cattle food like rice bran, lab bran, kept as rolling stock in rice mills, at Naogaon town in liason with District Administration of Naogaon and District, A.H & V.O., Naogaon. Green fodder was also available locally although not sufficiently. These articles were distributed by Veterinary staff with the help of the S.D.C. Concerned. Veterinary Doctors and field staff were detailed for duty zone wise by Sub-Divisional Animal Husbandry and Veterinary Officer as per contingency plan under the overall supervision of S.D.Cs.

### 15.2.8 DRINKING WATER FACILITIES

The Executive Engineer, Public Health Engineering (P.H.E.) chalked out an elaborate plan for catering to the need of the people in flood affected areas in respect of providing drinking water. Special mobile teams comprising engineers and other technical staff of P.H.E. Department. were also formed to find out the inundated places. Although most of the villages of the Sub-Division were provided drinking water facilities by P.H.E. Department in the form of tube wells and pipe water schemes. Many tube-wells had also been installed by the Block Development officer, Mohkuma Parishad and were provided under different development schemes, all the Tube wells had gone under water as soon as the flooding occurred. Special Drive was taken by P.H.E. Deptt. to repair defective tube wells and to disinfect all water sources before the onset of the monsoon. Moreover, the P.H.E. Department installed tube wells in flood affected villages as per necessity on the basis of report of concerning D.D.C. These tube-wells were used temporarily and taken back by the P.H.E. Department after the flood season.

### 15.2.9 COMMUNICATION RESTORATION WORKS

The National Highway 37 passing through the southern tract of the Sub-Division had been frequently submerged at places between Amsoi gate to Thekraguri and the submerged portion were extensively damaged for which plying of heavy vehicular traffic was suspended for days at a stretch. Besides these, almost all the P.W.D. and village roads of flood affected areas including culverts and wooden bridges were either washed away or badly damaged. Assistant Executive Engineer, P.W.D. (Roads) East and West Divisions took up necessary steps day and night to repair the roads including culverts and bridges in order to restore at least the skeleton road communication network as inadequacy of fund badly hampered the works.

### 15.2.10 E & D WORKS

Assistant Executive Engineer, E&D, detailed his staff for keeping a close watch on the embankments during the flood time day and night to prevent possible cutting of weak embankments by miscreants and give prior warning to the adjacent villagers residing near the embankments to shift their families and other belongings to safer places if any mishap happened, in case of any breach of embankments. The daily water level report of the Brahmaputra and other rivers of the Sub-Division was submitted by the Department to the Relief Branch of the author's office regularly from 15th May, 1988.

### 15.2.11 ARMY ASSISTANCE

Considering the gravity of the situation, the services of Army were requisitioned. Army authorities at Missa, Naogaon were sent S.O.S. to help the Sub-Divisional Administration in rescuing marooned people and the Army tried their best. Several Army boats were pressed into service by Army personnel for rescue operation.

### 15.2.12 POLICE FORCES

Police forces helped to a great extent in the rescue operation to save lives of distress people from difficult places braving the turbulent current of floods. They also co-operated in relief distribution works and for the agile mobility of the S.D.P.O. and other officers no untoward incidents reported during the relief works of five waves of floods in Morigaon Sub-Division.

### 15.2.13 AGRICULTURE AND IRRIGATION DEPARTMENT WORKS

Sub-Division Officers (Agriculture) were asked to assess the extent of damage to total crop area of the Sub-Division and submit daily report in matter to the author beginning from the onset of flood. The Sub-Divisional Agriculture Officer was also directed to prepare a detailed plan and estimate for raising Community Nursery to supply seedlings to flood affected farmers immediately after the flood season was over. The Executive Engineer, Irrigation, was informed to submit exhaustive plan for repairing irrigation structures damaged in the flood and offer irrigation facilities to cultivators of flood prone areas in their fields of Rabi crops.

### 15.2.14 REHABILITATION GRANT (G.R.) AND EX-GRATIA GRANT

All the S.D.Cs were asked to submit Rehabilitation proposal for allowing homestead and other land to flood affected people and sanction of Rehabilitation Grant for fully damaged and partially damaged households. Ex-gratia grants were paid to the next of kin of the all flood victims.

### 15.2.15 CONTROL ROOM AND FLOOD REPORT

A Control Room with Telephone at Morigaon Circuit House started functioning round the clock as soon as the water level approached the danger level during the flood season. Officers and staff from different offices of the Sub-Divisional headquarters were detailed for control room duty. Flood Situation Report in 'Appendix 4' of Assam Relief Manual was sent regularly from the date of occurrence of flood to the Government in Revenue (Relief and Rehabilitation) Department, Dispur by a wireless telegraph (W.T.) message.

## 15.3 Concluding Remarks

Disasters are of two types: periodic and unpredictable. The behavioural pattern of people is different in different cases. Since flood in Assam is a periodic phenomenon the behaviour of the people is different than it will be in unpredictable disasters. Here surprise element and unpreparedness sometimes play a vital role in behavioural pattern. The Brahmaputra River is a very vast river-- -one of the biggest in the world--- and it runs through a relatively less developed area. Around its very fertile valleys, people are dependent mostly on agriculture.

A long term programme is necessary for taking preventive steps. The Government and its functionaries take all steps to provide relief to affected people. Hundreds of thousands of people stay on embankments for months which is a very sad part of life. They fight for their existence. Lots of people loose their property and become refugees. After the flood season is over, the Government and its agencies need to take up proper planning by which affected people can be (1) provided adequate rehabilitation grants to construct and repair their damaged houses, (2) provided proper plots of land for those who lost their landed properties, (3) provided proper drinking water facilities to escape from gastro-enteritis which is very common disease after flood in these areas, (4) provided a job under rural development schemes to earn a livelihood, etc.

The administration has a very vital role to play during a flood such as rescue, distribution of relief materials, prevention of breaches in embankments, etc., and after the flood season such as rehabilitation of all flood-affected people. This is a test for administrators to prove their ability and management skills in such a disaster.

# CHAPTER 16

# THE BRAHMAPUTRA IN ASSAM'S ECONOMY

B.K. BHATTACHARYA

The Brahmaputra River is the lifeline of Assam and its economy. It can be argued that, if properly developed, the Brahmaputra basin can be one of the most prosperous regions of India. The key is to harness the Brahmaputra's water resources and other resources that accompany it. This chapter makes a few preliminary remarks about the role of the Brahmaputra in Assam's economy, especially from a navigational viewpoint.

## 16.1 Introduction

From the dawn of civilization the Brahmaputra in Assam had been providing livelihood and sustenance to the people through agriculture on its alluvial soil. It has also served as a means of transport and communication of merchandise and people from and to Assam. Long before the advent of the railway system into Assam, the river maintained the link with other parts of the country through the network of its waterways and hundreds of boatmen were employed in its services. It was the lifeline of Assam. In this connection it may be added that the modern steamship service between Assam and Calcutta was first introduced by a Calcutta-based British-owned river transport company, viz, the India General Navigation and Railway Company Ltd. It was founded around 1844.

Another British-owned company which started operation was the River Steamer Company. It was founded in 1867 and was subsequently renamed as River Steam Navigation Company. Later on both these companies came to be known as the Joint Steamer Companies. In 1967 the Central Inland Water Transport Corporation---a Government of India undertaking--- became the legal successor of the joint steamer companies, Calcutta. The river service was well organised, an infra-structure developed on the riverine route and it was one of the world's most flourishing internal trade routes. Above all, the movement from Assam through the Brahmaputra was maintained up to Allahabad in Upper India via Calcutta and Patna through the Ganges. It was the bulk carrier of Assam's tea and jute to Calcutta---the country's valuable foreign exchange earners. Even after India's partition in August 1947, Assam continued to be well connected with Calcutta by the riverine route through the former East Pakistan particularly when Assam lost her direct link with the rest of India for a few years. Its role was pivotal in the economic of life of Assam and its hinterland and was in operation for more then hundred years. However, after independence cracks surfaced, troubles started for these and other Calcutta based companies as they were plying through East Pakistan (now Bangladesh) waters. On a number of occasions they faced severe stress



*V.P. Singh et al. (eds.), The Brahmaputra Basin Water Resources*, 321-325.

and strain in their operations. The steamer services from and to Calcutta were finally closed down during the Indo--Pakistan conflict of September, 1965 and thus came to an end of the century old checkered history of the inter-state trade and commerce.

Discussing the role of the inland water ways in the economic life of Assam, the National Council of Applied Economic Research observed as follows: "The internal waterways of Assam are said to be over 6,000 miles. Approximately 1,000 miles are navigable by steamers and large country boats and larger mileage by small country boats. The Brahmaputra is the main artery of the state and its important tributaries, the Subansiri, the Dihang, the Burhi-Dihang and the Desang are also navigable for varying distance in their lower reaches". The earthquake of 1950 changed the course of the Brahmaputra owing to heavy silting and consequently the river bed was raised and "affected steamer navigation in the upper reaches of the Brahmaputra for 46 miles downstream of Dibrugarh, and the main line steamer services had to be terminated at Disangmukh." In 1958 "the inter-state traffic, i.e., the traffic from Assam to Calcutta and vice versa is 84.8 per cent of the total traffic and local traffic amounts to only 5.12 percent. The main exports of Assam are tea, leather, jute grown in Assam have to find their way to Calcutta for internal distribution in India and export overseas. These have necessarily to be carried by the inland waterways. Tea has to be carried by the river route not only because the railways cannot carry the entire traffic, but because the waterway is quicker than by rail and the tea warehouses in Calcutta are so located that they can be used in conjunction with water transport." Imports into Assam from Calcutta were food grains, foodstuffs, salt mustard oil, iron and steel materials, cement, cloth, textiles, etc., Table 16.1 below illustrates the position of movement.

Furthermore, in an article on inland water transport its writer while emphasizing the importance of the waterway in the economy of the North-East India, added: "Even in a cursory glance at the topography of the Northeast India would convince any impartial observer that the rivers of this rather remote region have from time immemorial provided that the natural highway along whose course civilization spread and trade and commerce progressed. Since then there have been many changes both in the pattern of trading and the modes of transport used, but the importance of island water transport in the economic life of Northeast India has not waned. By maintaining a steady flow of traffic to and from this region, not only the economic stability of the area is ensured but industrialization is greatly helped. What is perhaps its greatest contribution is its carriage of two of the largest foreign exchange earners of the India tea and jute from North-East India to Calcutta. Few people appreciate that the approximately 93.2% of Assam's tea and 91.2% of her jute crops are annually imported to Calcutta by river from internal distribution or export to overseas markets." In terms of the quality, movement of jute from Assam to Calcutta through the Brahmaputra during 1956--57 and 1957--58, were 175,000 tonnes and 174,000 tonnes, respectively. Mr. Man Singh, the former Director of Waterways of the Central Waterways, Irrigation and Navigation Commission in a note dated the 3$^{rd}$ September 1947 discussed the possibilities of developing the river potentials in Assam on a multi-purpose basis said: "...The water resources of the Province .... are tremendous. They have not, however, been harnessed to serve man. A fairly heavy crop of paddy is reaped from the valley areas, but most of the water after causing devastating floods here, erosion there, passes wastefully into the sea through the

Table 16.1 Summary traffic dealt within 1958 at each of the ghats on the Brahmaputra.
(Figures in mounds)

| S.N | Name of Ghats | Outward to Calcutta | Inward from Calcutta | Outward | inward | Total |
|---|---|---|---|---|---|---|
| 1 | Dibrugarh | 1,331 | 43,249 | 384 | 203 | 45,167 |
| 2. | Rajbari | 97,127 | 19,433 | nil | 154 | 116,714 |
| 3. | Disangmukh | 276,407 | 233,783 | 3,296 | 45,283 | 558,769 |
| 4. | Neamati | 667,182 | 738,173 | 24,643 | 318,966 | 1748,964 |
| 5. | Via Neamati | 2288,890 | 1916,871 | 157,643 | 156,071 | 4439,475 |
| 6. | Subansirimukh | 4,688 | 4,680 | 45,005 | 150 | 54,531 |
| 7. | Borduttaghat | 101,701 | 206,689 | 139,903 | 32,181 | 480,474 |
| 8. | Shilikhaguri | 15,509 | 47,477 | 3,956 | 39,517 | 106,459 |
| 9. | Dhansirimukh | 175,811 | 68,881 | 229 | 730 | 245,651 |
| 10. | Gamirighat | 89,596 | 92,903 | 145,001 | 41,112 | 368,612 |
| 11. | Behalimukh | 43,279 | 39,563 | 31,777 | 23,513 | 158,132 |
| 12. | Kathanibari | 74,323 | 53,466 | 99,046 | 3,419 | 130,254 |
| 13. | Bishnath | 128,628 | 205,769 | 162,752 | 17,903 | 515,052 |
| 14. | Silghat | 124,058 | 61,086 | 1,080 | 654 | 186,878 |
| 15. | via Silghat | 2,764 | 50993 | Nil | 2,375 | 56,132 |
| 16. | Tezpur | 327,090 | 491,102 | 306,095 | 44,023 | 1168,310 |
| 17. | Singrighat | 17,765 | 10,338 | 934 | 416 | 29,444 |
| 18. | Singri depot | 33,167 | 13,077 | 133 | Nil | 46,377 |
| 19. | Dhekiajuli depot | 32,127 | 42,881 | 1,563 | 615 | 77,186 |
| 20. | Kharupatia | 58,844 | 124,036 | 879 | 174 | 183,933 |
| 21. | Bharangaon | 203,836 | 8,358 | Nil | Nil | 212,194 |
| 22. | Guwahati Bazar Ghat | 626,258 | 2069,279 | 45,258 | 1035,755 | 3776,577 |
| 23. | Via Guwahati a/c SGHO | 15,673 | 67,530 | 3,764 | Nil | 86,967 |
| 24. | Shillong out-agency | 21,941 | 72,836 | 1,667 | 89 | 96,533 |
| 25. | Via Pandu | 1303,944 | 1987,362 | Nil | 9,951 | 3301,257 |
| 26. | Kholbanda | 170,881 | 69,717 | Nil | 457 | 241,055 |
| 27. | Goalpara | 100,233 | 136,698 | 27,364 | 325 | 264,620 |
| 28. | Dhubri | 2144,371 | 469,834 | 553,619 | 27,896 | 3195,720 |
| 29. | Via Dhubri | 444,186 | 1281,392 | 20,136 | 612 | 1746,325 |
| 30. | Palasbari | 204 | 8 | Nil | Nil | 212 |
| 31. | Fanshi bazar depot | 352,290 | 393,871 | 66,253 | 66,761 | 879,355 |
| | Total (Ton) | 9511,832 | 10632,083 | 1796,127 | 1802,552 | 23742,594 |

(Source: Techno-Economic Survey of Assam by National Council of Applied Economic Research, New Delhi, 1962)

mighty Brahmaputra on one side and Meghna on the other..". Mr. Man Singh further observed that a "study of the topography of the Province from the published maps and rainfall data reveals that a very large potential for hydropower exists. A very rough estimate it should be possible to generate as much as a 1.5 million kilowatt of continuous power in province by developing eleven sites on different rivers." However, the huge hydel potential of the Brahmaputra is yet to be harnessed.

The Brahmaputra becomes furious and turbulent and uncontrollable during the monsoon season. It inundates vast tracts of land, destroys standing crops and brings untold misery to the people of Assam. It is an annual ordeal. Table 16.2 below indicates the colossal havoc brought by the river to Assam during the period 1954-55.

Table 16.2 Colossal havoc to Assam by the Brahmaputra River during period 1954--55.

| Sl. No. | Damages | Figures |
|---|---|---|
| 1. | Area affected in million hectares | 40.52 |
| 2. | Population in million numbers | 92.50 |
| 3. | Damage to crops in million hectares | 4.53 |
| | Value of which in terms of crores of rupees | 2,845.83 |
| 4. | Damage to houses | |
| | (a) Total number | 3132,000 |
| | (b) Value in terms of crores of rupees | 0<br>265.45 |
| 5. | Cattle lost in numbers | 425,450 |
| 6. | Human lives lost in numbers | 1,650 |
| 7. | Damage to public utilities in crores of rupees | 832.42 |

(Source: Memorandum to the Hon'ble Prime Minister of India by All Party Delegation of Assam Legislative Assembly, 28th, August)

Owing to the devastation caused by floods every year, the State could not achieve self-sufficiency in food production and is required to import huge quantities of food grains regularly from other states of India, viz., Punjab, Haryana, and Uttar Pradesh, resulting in a continuous outflow of a substantial amount of resources and it is one of the reasons for which capital formation in Assam in meagre. Many a time it is felt that had the Brahmaputra been controlled and its water harnessed for the peoples benefit of Assam, it would have given a tremendous boost to production of food grains and other agricultural products. Apart from meeting her own requirements, Assam could have met the requirements of other rice-eating Northeastern States which are heavily deficient, controlling floods in Assam may also reduce the severity of floods in Bangladesh and will help increase the production of food grains and thereto meet the hunger of her teeming millions. After the creation of Bangladesh in December 1971 the old river from Calcutta to Assam through Bangladesh waters was revived. It was expected that the Indo-Bangladesh Treaty of Friendship, Cooperation, and Peace of March 1972 would usher in a new era "of harnessing the waters of the Brahmaputra, Meghna and Ganga" to the benefit of the peoples. However it did not occur, although the Central Inland Water

Transport Corporation maintained a small service between Calcutta to Assam for some time, which has since been discontinued to the disadvantage of the peoples of the two countries. In the interest of Assam, her neighboring Northeastern States, and Bangladesh, it is essential to revive the century-old river services and run the same between Calcutta and Assam and vice versa. Another aspect of the river service is that a regular plying of steamers automatically does some dredging of the rivers, which, in turn, helps reduce the impact of floods. The Brahmaputra basin consisting of Assam and some parts of Bangladesh has the full potential for development of agriculture. The main question is to tame the mighty Brahmaputra and utilize its waters for the people's wellbeing. The Brahmaputra basin may, therefore, become one of the most prosperous zones of the Indian subcontinent.

## References

Nehru, J. (1956). The Discovery of India (p. 38). Meridian Books Limited, London.

Children's Britannica (1960, 1961). Encyclopedia Britannic Limited, Volume 3, page 268, London. Brahma himself is the father god of the trinity, the supreme God who created universe. The other two members of the trinity are Vishnu, the preserver and Shiva the destroyer,.... Brahma, Vishnu and Shiva therefore stand for the three process of creation, preservation and destruction that are continually going on in the Universe".

National Council of Applied Economic Research, (1962). Techno-economic survey of Assam, pp. 127--128, New Delhi,

The Statesman (1959). Inland water transport vital for jute, by Brahmaputra, Calcutta, September 30.

Guha, K.D. (1948). "Industrial Planning and Development of Assam-first approximation of a five year plan."

Bhattacharya, B.K. (1972). "Task before NEC : Transport Needs of North-East India", The Economic Times, Bombay, June.

# CHAPTER 17

# FINANCIAL CONSTRAINT IN ASSAM

A. SHARMA

There is a multitude of factors that have contributed to the low level of development of the northeastern region of India in general and Assam in particular. Although the lack of finance is an acute problem, there are other equally potent impediments to the development of the northeastern region. Failure of state and local governments, poor infrastructure, natural calamities, deplorable law and order situation, and most importantly the lack of political will at practically all levels are some of the reasons, besides finances, for the present condition of Assam and the northeast in general. These issues are discussed using statistics made available by the government offices.

## 17.1 Introduction

The creation of many a nation state may be accidents of history. The North East (NE) as an integral part of India is an example of such accidents of history. This region that was never occupied by any ruler from the Indian mainland came under the British because one of its leaders sought the help of the British to overthrow the Burmese invaders. What if Badan Barphukan, a patriotic general himself, had not gone to seek the help of the British! Would the NE then have become a country belonging to the South East (SE) Asia --- as another poor country like Laos, Kampuchia or as rich as Thailand, Indonesia, Malaysia, or even Singapore? Anyway, the fact of the matter is that it is not a South East Asian country, but one of the poorest regions of the most powerful South Asian country, a country known more for its population pressure and poverty than for the muslin, silk, handloom, and handicraft, tea, or spices. The NE region and Assam in particular have been much in the news because of the problems created by insurgency. The lack of development is perhaps the main cause of insurgency which in turn has been blamed for strangulation of the development process. In this chapter an attempt is made to look closely into other factors like finance which may also contribute to the vicious circle.

Had India been a rich country, most probably the NE would have been happy with the accident of history. Had the NE been a rich part of the country, then too the accident of history may have been overlooked. But what is noticed is that while on the eve of independence Assam, which practically meant today's NE, was one of the richest parts of the country, which is fifth in terms of per capita income, today it is one of the poorest region (sixth from the bottom). There is now a tremendous sense of alienation of the people of the area. While there is some agreement that the state of affairs is the result of



*V.P. Singh et al. (eds.), The Brahmaputra Basin Water Resources*, 326-335.

the people themselves, nevertheless the role of the Centre has not been perceived as sympathetic until recently.

### 17.2 Growth Rates

The downward trend in growth rates of Assam, in relative terms, seems to have started soon after the planning era. In 1951--52, the state's per capita income, at 1980--81 prices, was Rs 1,173 compared to India's Rs 1127. By 1955--56, the per capita income of Assam was Rs1227 compared to India's Rs 1,229. It is seen that the ratio of NSDP to GDP itself has not changed much (North Eastern Institute of Bank Management (NEIBM) study), but because of the higher population growth in the state compared to the national growth rate, the growth of per capita income has been slower. By 1997--98, the per capita income of the state reached Rs 1,673 whereas the all India figure was Rs 2,848. Table 17.1 below shows the growth rates of Assam compared with that of the whole of India. It is seen that while the rates have been increasing for India, those for Assam have been going down.

Table 17.1 Growth rates of NSDP/NNP at 1980--81 prices (in %).

| Year | Assam | India |
|---|---|---|
| 1981--82 | 11.31 | 5.83 |
| 1982--83 | 4.61 | 2.19 |
| 1983--84 | 4.37 | 8.09 |
| 1984--85 | 0.50 | 3.41 |
| 1985--86 | 6.59 | 3.90 |
| 1986--67 | -2.84 | 3.75 |
| 1987--88 | 4.47 | 3.84 |
| 1988--89 | 0.76 | 10.66 |
| 1989--90 | 7.35 | 6.98 |
| 1990--91 | 4.29 | 5.15 |
| 1991--92 | 4.41 | -0.14 |
| 1992--93 | 1.01 | 5.05 |
| 1993--94 | 3.90 | 5.96 |
| 1994--95 | 3.44 | 6.80 |
| 1995--96 | 4.25 | 7.30 |
| 1996--97 | 3.51 | 7.85 |
| 1997--98 | 2.67 | 4.85 |
| 1998--99 | 1.8 | 6.8 |
| Trend Growth Rate | 3.33 | 5.16 |

Source: Directorate of Economics and Statistics, Govt. of Assam

Since the growth rates in the NSDP alone would not give the whole picture, the sectoral growth rates are also given below. *Agri and allied sector* which had a share of 54% of NSDP in 1980--81 contributed only 38.48% in 1998-99 this decrease being expected in a developing country. *Mining and Manufacturing* came up from 11.6% to 13.5% although manufacturing alone came down from 12.46% in 1990--91 to 8.2% in 1998--99. It is the *tertiary sector* which really grew consistently from 25.78% in 1980--81 to 42.49% in 1998--99. These trends give a contrasting picture when the Indian sectoral growth is considered.

Table17.2 Sectoral Contribution to SDP (at 70--71 prices).

| | 1970--71 | 1980--81 | 1990--91 | 1998--99 (at 80--81 prices) |
|---|---|---|---|---|
| *Primary (%)* Agri & allied, Mining & Quarrying | 63.96 | 63.05 | 50.40 | 43.77 |
| *Secondary (%)* Manufacturing, Construction etc. | 14.00 | 11.17 | 17.48 | 13.73 |
| *Tertiary Sector (%)* Transport, Communication, Trade & Hotel, Banking, Real Estate, Public Administration, etc. | 22.04 | 25.78 | 32.12 | 42.49 |

**Source**: *i) "An Enquiry into the Institutional credit Flow in Assam" by North Eastern Institute of Bank Management, Assam*
*ii) Directorate of Economics and Statistics, Govt. of Assam*

### 17.3 Factors Contributing to Downward Growth

In case of the national economy the fall in contribution of agricultural and allied sectors have been much faster while the growth in the secondary sector much higher compared to Assam's. In 1998--99 agriculture, forestry and mining sectors in India contributed 29.16 %, while manufacturing and construction contributed 24.71%. There is thus an indication that India as a whole is developing faster than Assam. The failure of the government, poor infrastructure, floods, unavailability of timely and adequate finance, poor law and order situation and the lack of local leadership can be reasoned as the main factors causing the fall in growth rates in the state. Among the three sectors, industry or the secondary sector depends mostly on government's initiative and patronage compared to agriculture and tertiary sector. There is also evidence that there has been a downward slide in the quality of infrastructure in the state and this has affected the growth of the industrial sector. The infrastructure development index was 84 in 1990--91 to 78.9 in 1993--94. Mere policies, guidelines declared from time to time, are not good enough unless there is a drastic change in the benefits to the private sector compared to the risk associated with starting an industrial venture in this area. *Bandhs*, strikes in the state, have not affected the agricultural sector as much as it has the industrial sector. The service sector too has not been affected so much. In fact, the prices in the NE are higher than those in the rest of the country only because of the poor infrastructure and the unrest here and this way the profit margins are taken care of.

## 17.4 Financial Constraint

Amongst all the constraints, the one which affect all the sectors equally and which can be removed without major overhaul is the problem of availability of finance in a timely fashion and in adequate amounts. Both term loan and working capital loans are in question here. Factors like infrastructure, insurgency, and governance, need a much larger effort and depend on a host of other factors for their solution. In the case of finance even a directive from the Government for nationalized banks or even the All India Financial Institutions (AIFIs) to increase their lending may do the job. If there could be a directive regarding priority sector lending, so there could be one regarding spatial or regional lending. Banks in the NE must have their C/D ratio at least 60% and similarly the NPA norms could be relaxed for these branches. Such suggestions are often seen in the various reports on industrializations of the NE.

### 17.4.1 A MACRO VIEW

A macro view of things is important in certain cases of policy making. But often micro views have to be taken into account for them to become more useful. In fact, a micro-view of things based on the macro and consistent with it have become the most recent approach in the theory of economic policy known as the micro foundation of macroeconomics. In India the economic policies are normally made for the country as a whole. This may be one reason why these policies have usually failed in their implementation. No part of the country can be ignored when formulating a policy. Heterogeneity cannot be wished away. This has been realized and there is an effort now to take this into the ambit while formulating the plans.

### 17.4.2 LACK OF INFRASTRUCTURE

The NE is a region that stands out because of its difference with the rest of India. The narrow chicken neck of 22 km (98% of the total boundary) connecting the NE with the Indian mainland underlines the difference. The main problem is of connectivity – both physically and psychologically. People in the mainland think of the NE as a remote area. Likewise people of the NE too have a feeling of living in an isolated place. Perhaps this was not true before independence when the borders were open to reach the ports of Chittagong, Rangoon, or Calcutta directly. The presence of hostile neighbours is one strategic reason why many infrastructural projects especially large Hydel projects may not have come up despite having 40% of the country's hydro power potential. With a potential of about 38,000 MW, the shortage of power for the 33 million people in the NE is an inexplicable situation. A large number of airports in the state that were in use during the British rule and beyond are now lying abandoned. The Shukla Commission (March 1997), appointed by the Centre to study the backlog of infrastructure and basic necessities, found that about Rs 94,000 crores would be needed to bring the NE to the level of the rest of India in terms of infrastructure.

### 17.4.3 BUDGETARY ALLOCATION

One area where a macro outlook may not have helped the NE states is that the NE states basically follow the Planning Commission's directive while making their plans. This may mean that their own plans are getting influenced by the aggregate plan for the whole of India. And this may be the reason why despite agricultural sector contributing almost 40% of the NSDP of Assam as against 26% for India's GDP, budgetary allocation for agricultural and allied services is less than 5%. Even when differences are recognized as being industrially backward, the backward areas of the developed regions or state may garner all the benefits. Of late there has been some realization of the above factors. In 1995 there was a proposal in the Central budget for a development bank for the North East which gave rise to the birth of the North Eastern Development Finance Corporation Ltd. Since 1995 the last three Prime Ministers have all made trips to the NE to announce a special package for the NE. In January 2000 Prime Minister, A.B. Vajpayee announced a financial package of Rs 10,271 crores for the development of the NE. There is now a non-lapsable pool at the Centre, earmarking unspent budgeted funds of the Central Ministries for the NE which also ensures availability of funds for the region.

According to a report in 'India Today' February 14 2000, the North Eastern States, along with Himachal Pradesh and Jammu and Kashmir receive 20% of the total central Government assistance, although they have less than 4% of the country's population. As in Table 17.3, per family distribution of central assistance is very high for the North Eastern states including Assam. The funds are also cheap as 90% of it is grant and only 10% is loan. The other states on the other hand receive 70% as loan assistance and only 30% as grant.

Table 17.3 Per family per month Central Assistance to NE States (1999--2000).

| States | Per family Per Month (in Rs) | Central assistance (1999--2000 Rs crore) |
|---|---|---|
| Assam | 867 | 4,666 |
| Arunachal Pradesh | 5,714 | 1,186 |
| Manipur | 2,767 | 1,220 |
| Meghalaya | 2,417 | 1,030 |
| Mizoram | 5,569 | 922 |
| Nagaland | 4,169 | 1,211 |
| Tripura | 2,367 | 1,566 |

Source: *India Today* magazine (14th February 2000)

According to Gulshan Sachdeva Assam received the following amounts from the Centre (Table 17.4):

Table 17.4 Devolution and transfer of resources from Centre to Assam.

(Rs in crores)

| | 1990--91 | 1994--95 | 1997--98 | 1998--99 |
|---|---|---|---|---|
| Gross Devolution | 1,955.85 | 2,512.06 | 3,905.96 | 4,068.71 |
| Net Devolution | 1,399.40 | 1,730.46 | 3,182.56 | 3,297.83 |
| Total revenue receipts | 1,776.65 | 2,961.41 | 4,325.66 | 5,254.97 |

The above amounts are important, given the state's size because the average contribution from the Centre to Assam's budget has been about 69% for almost each of the last 14 years whereas the all India average for the states is around 42%. Assam contributes around 20% from its tax revenue collection and less than 10% from non-tax revenue to its revenue expenditure. The All India average is about 43% and 14%, respectively. More efforts are needed to raise internal revenue and to unproductive expenditure. This will help the state govt. to spend more in welfare activities. It is seen that the Centre also allows the state 20% diversion of plan funds to meet revenue requirements. This and other concessions for the NE have created a sense of complacency and it is now becoming hard for these states to come up with even the 10% matching funds.

### 17.4.4 ABSORPTIVE CAPACITY

Another problem plaguing the state is its the lack of absorptive capacity. It is one thing to come up with matching sums for central schemes, but quite another thing to be able to spend the money in a timely fashion for completion of the projects. Time and cost overrunning in the NE has become a problem unlike in any other region. Some technical problems could be there but mostly it is inefficiency, corruption, and the lack of commitment of the leaders, which have led to failure of many projects. As recommended by the S.K. Shukla Commission, a close monitoring of the projects need to be ensured. Participation of stakeholders in the projects can ensure some improvement in implementation. May be the Central Govt. should implement the project itself through Central government organizations like CPWD or BRO or its own consultants.

### 17.4.5 LACK OF FINANCES

Given the market friendly environment since 1991, the private sector in Assam too would like to take up new ventures. But for this the entrepreneurs would need financial assistance from banks and financial institutions (FIs). While in the advanced states, getting a loan for projects with reasonable feasibility outlook is not a problem and venture capital funds have mushroomed, this is not true in Assam. If a term loan is difficult then working capital loan is almost impossible to get. Banks and FIs have pointed to low recovery rate as the main obstacle.

### 17.4.6 FINANCIAL INSTITUTIONS

Assam is served by many financial institutions most of them are under the public sector. The Assam Industrial Development Corporation (AIDC) and Assam Financial

Corporation (AFC) are the two main ones for financing industries. Both of them are in the doldrums because of bad repayment and new financing has almost come to halt. Even refinancing has been kept on hold by the refinancing institutions like SIDBI and IDBI. Therefore, entrepreneurs have to look upto the all India Financial Institutions (AIFIs). But they too have not done much. Table 17.5 below reflects the disparities in per capita sanctions and disbursements experienced by Assam vis-à-vis India.

Table 17.5 Per capita assistance sanctioned and disbursed (Assam versus India).

| Financial Institute | Per capita sanctions (Cumulative up to end of March 1999) in Rupees | | Per capita disbursement (Cumulative up to end of March 1999) in Rupees | |
|---|---|---|---|---|
| | Assam | India | Assam | India |
| AFIs | 784.5 | 5,091.5 | 511.9 | 3,516.5 |
| AIFIs | 785.6 | 4,928.8 | 509.3 | 3,369.5 |
| IDBI | 250.8 | 1,750.4 | 207.4 | 1,173.3 |
| IFCI | 113.7 | 488.1 | 106.5 | 366.9 |
| ICICI | 302.1 | 1,515.4 | 119.1 | 909.9 |
| SIDBI | 41.6 | 352.7 | 34.0 | 275.1 |
| IIBI | 18.4 | 84.6 | 16.6 | 59.8 |
| RCTC | 0.4 | 1.6 | 0.4 | 1.4 |
| LIC | 22.4 | 229.4 | 12.1 | 211.2 |
| UTI | 36.3 | 484.3 | 13.3 | 358.0 |
| SFC | 42.8 | 313.8 | 39.4 | 256.9 |
| SIDC | 22.8 | 177.0 | 21.6 | 133.3 |

Source: *IDBI Report on Development Banking in India, 1998--99*

Note: AFIs (All Financial Institutions) comprise all-India Development Banks viz., Industrial Development Bank of India (IDBI), IFCI (Industrial Finance Corporation of India), ICICI (Industrial Credit and Investment Bank of India), SIDBI (Small Industries Development Bank of India) and IIBI (Industrial Investment Bank of India); Specialized Financial Institutes (SFIs) viz., Risk Capital and Technology Finance Corporation Ltd. (RCTC), ICICI Venture and Tourism Finance Corporation of India (TFCI) and Investment Institutions viz. Life Insurance Corporation of India (LIC), Unit Trust of India (UTI) and General Insurance Corporation (GIC) and State Level Institutions viz. State Financial Corporations(SFCs) and State Industrial Development Corporations (SIDCs).

AIFIs (All India Financial Institutions) comprise all the above except the state level institutions.

It is noted that AIDC and AFC have been disbursing less than Rs 2 crores and Rs 5 crores, respectively, since 1995--96. The total sanctions by AIFIs in Assam is only 0.44% of all India. This works out to Rs 785 per capita, while the same for Punjab is Rs 4470.9/-. In the case of major FIs like the IDBI, ICICI, IFCI, SIDBI and IIBI the percentage is 0.47. These amounts should be compared with those for other states like

Punjab (2.34 %), Andhra Pradesh (7.62%) to show that Assam has been a poor beneficiary. (Figures are cumulative up to March end 1999).

In the case of banks the disparities are even more marked. The number of bank branches in the country increased between 1994 and 1999 by 4.35 %. In the NE the same is 2.31%. In Assam the growth rate is 0.12% from 1293 in 1994 to 1301 in 1999.

In terms of the deposit mobilization, the total in Assam was Rs.3,154 crores The per capita bank deposit of the state is Rs 2,868/ in 1999 as against Rs.1,358 in the year 1994. This compares poorly with the corresponding All India figures of Rs.13,912 and Rs.7,059.

In terms of credit the total amount deployed in Assam was Rs 2,214 crores of which rural semi-urban and urban break up were Rs 713 crores, Rs 601 crores and Rs 899 crores respectively. The above figures point to a low C/D ratio of 31.5% in 1999. This is as against the all India average of 55.5%. Is this low credit deployment made up by investment in government securities by the banks? It is seen that the total investment in government and other approved securities stood at Rs 979 crores, giving a (C+I)/D ratio of 44% in 1999 as against 83% for all India.

After the introduction of market friendly policies in India, FDI has increased to significant levels (over $4 billion in some years). Here too, the developed states have attracted most of the investment while only an insignificant amount has filtered into the North East. Out of the total number of 37,402 IEMs (Industrial Entrepreneur Memorandum) applications filed under delicensed sector and LOIs (Letters of intent) issued in respect of items under the licensed sector between August 1991 and September 1998, Assam's share is only 86 with proposed investment of Rs 2580 crores (For the nation, it is Rs 6,36,340 crores).

From the above it is seen that while the state government is getting a sizeable amount of fund from the Centre, the financial institutions have neglected their development role somewhat. Is this going to be the trend given that liberalization hurts the weak and helps the strong? If one looks at the banking sector, at least this seems to be the case since before 1991 the credit to deposit ratio was comparably high 53.09% in 1990.

### 17.4.7 PRIORITIZATION OF LANDING

The priority sector lending does not present a satisfactory picture in the state although it is amongst the most important goals of social banking. The per capita deployment of priority sector lending is Rs 350 in 1999 as against more than Rs.750 in the country as a whole. In case of agriculture, the priority sector advances constituted 20 % only in 1999 and outstanding loans per hectare of gross cropped area was amongst the lowest in the country. As mentioned above, agriculture is the main contributor to the state economy and this sector has been almost completely ignored by the Financial Institutions. The NEIBM study shows that proportion of bank credit to agricultural sector was 62 % in 1958 and in 1995 the same was only 16 %. The same study also mentions that in the 1990s almost 50 % of the total credit has been cornered by five main towns having 4% of the population while the rest is shared by 96 % of the people, mostly rural.

It is not that there have been no efforts by the government to rectify this. In the SLBC meetings and other forums the banks have been repeatedly requested and even warned. But even then there does not seem to be any perceptible change. There was a Central government directive for the banks to open 100 more branches in the NE in 1999-2000. But only three branches have come up.

17.4.8 RECOVERY

Recovery has been a problem in the North East. A sample survey by NEIBM mentions that in 1995, the rate was slightly more than 10% for bank loans. Experience at NEDFi, IDBI, and SIDBI has also been very bad regarding repayment of loans. But there is scope to believe that the problem is not wilful default but because of projects going bad owing to various technical, market as well as institutional problems. Financial institutions need to appreciate these problems. Repayment has been found to be better for small loans in rural areas, particularly for agriculture. In urban areas the recovery rate is better for bigger loans. Thus, most of the loans have been going to this sector although agri and trade sectors have been contributing more to NSDP. The credit flow must improve in the state just as deposit has been increasing. For a poor region, the deposit mobilized should at least remain within, if not in credit then in investment in designated securities.

There are complaints that the system of keeping the officers in the NE for two years has not helped the area. The way the officers have been pulled up for bad loans even after ten years, make them very cautious because of the bad recovery record in Assam. This has resulted in dilatory handling of credit applications resulting in turning away of good borrowers. Earlier defaults, mostly arising out of factors beyond the control of the borrowers, have given rise to a problem of severe credit restriction and corresponding adverse selection and moral hazard problem. The Government too has not given the help required to expedite the *bakijai* cases accentuating the problem of adverse selection further. These problems, of course, are not analysed by the people from outside who merely view the NE as a problem area. That their views themselves have worsened the problems need to be stressed.

## 17.5 Concluding Remarks

To conclude, the above analysis shows that while the government has been getting substantial amounts of funds from the Centre, the intention, andefficiency are in question. A large amount of funds have also been tapped from the World Bank and other international funding institutions. For instance, under the Assam Rural Infrastructure and Agricultural Services Project (ARIASP), the World Bank has given an assistance of Rs 567.65 crores in 1995. However, the revenue expenditure has been increasing everyday; wages, salaries and pension taking a major chunk. Diversion of funds from plan to non-plans heads, inability to provide matching amounts for central schemes have become very common. At this juncture the private sector should have been able to fill the vacuum. But there is severe funds crunch – the AIFIs and banks, as well as the state level organizations, have been disbursing increasingly small amounts in the region. Compared to other states, there is hardly any Foreign Direct Investment in the state. Under these circumstances this poor state will have become poorer and insurgency may prosper, in spite of the initiatives taken by the Central Government.

## 17.6 Acknowledgment

The author is grateful to Ms Ajanta Phatowali for the assistance rendered in preparation of this chapter.

## References

Gulshan Sachdeva: Economy of the North East: Policy, Present Conditions and Future Possibilities"
India Today (2000). India Today magazine, 4th February 2000 Issue.
V S Kaveri (2000). Banking in the north Eastern Region – an interstate Analysis.
IDBI (1998-99). Report on Development Banking, 1998-99.
NEIBM (1998). An Enquiry into the Institutional Credit Flow in Assam.
Observer Research Foundation (2000). India: 2000 Observer Statistical Handbook.
Government of Assam (1999). State Finances–The Factual Position : Finance (Economic Affairs) Department.
Government of Assam (2000). Assam 2000 --- A Hand Book, Directorate of Information & Public Relations.

# CHAPTER 18

# SOCIETY IN THE BRAHMAPUTRA VALLEY

M. HUSSAIN

Assamese society is a diverse, multi-religious, and plural society. Indeed Assam can be divided into three regions which are not only distinct geographically but also culturally, ethnically, nationally, economically, politically, and psychologically. This chapter traces the history of the evolution of the texture and structure of Assamese society. This historical perspective is vital to planning development schemes.

## 18.1 Introduction

This chapter seeks to describe the contemporary society in the Brahmaputra Valley located in the North Eastern part of India in the state of Assam. Needless to say, Assam as a political entity emerged during the colonial period (since 1826 till 1947) and its territory has changed several times during the postcolonial period in the wake of the creation of several smaller states for the hill tribes of North East India. Three new small tribal states were created from the erstwhile state of Assam, i.e. Nagaland for the Nagas, Meghalaya for the Khasis, Jaiantias and Garos and Mizoram for the Mizos. Assam may face still more territorial reorganization in the wake of two important ongoing tribal movements i.e. the Bodo movement and the Karbi-Dimasa movement for a separate tribal state. If Karbi Anglong and North Cachar districts are separated from Assam, the Barak/Surma valley comprised of Cachar, Hailakandi and Karimganj districts will automatically get severed/separated from the present state of Assam and the Brahmaputra Valley. It is pertinent to point out that the Asamiyas are the numerically major and politically dominant national group in the Brahmaputra Valley. They have been historically sharing this valley as their homeland with innumerable autochthon tribals, like Bodos, Sonowals, Rabhas, Tiwas, Mishings, Deuries, etc.. The Karbis and the Dimasas are the major ethnic groups living in the hills of Assam. Both Hindu and Muslim Bengalis live in the Barak/Surma Valley. These three regions of Assam are not only very distinct geographically but also culturally, ethnically/ nationally, economically, politically, and psychologically.

During the colonial and the postcolonial period the geographical boundary of Assam has remained somewhat fluid. Therefore it is crucial to point out the time and space /territory of Assam specifically whenever one writes about it. Here we intend to describe sociologically the society in the Brahmaputra Valley. The society that has emerged over a long period of history is fundamentally a multi-racial, multi-ethnic, multi-religious,



*V.P. Singh et al. (eds.), The Brahmaputra Basin Water Resources*, 336-350.

multi-caste, multi-cultural, and multi-lingual one in which Asamiya is the majority language. It is the mother tongue of more than two thirds of the population of the Valley.

## 18.2 The Pre-Colonial Society

By the way of preliminary remarks it should be pointed out that the pre-colonial society in the Brahmaputra Valley was fundamentally semi-tribal and semi-feudal in nature, which ensured the mixture of more than one classical mode of production. It generated a very limited surplus and obviously had a very limited market. Though it was an oppressive system with its resultant backward economy even as compared to neighboring Moughal India, it must be admitted that it was largely a self-sufficient economy and therefore could maintain its significant distance from the rest of India. This, together with its peculiar geographical factors, helped the perpetuation of the Ahom dynasty for 600 years till the British colonization in 1826. The geographical location of the Brahmaputra Valley at the easternmost part of the Indian sub-continent occupies a very peculiar position. Being surrounded by neighboring hill tribes, Tibet and Bhutan on the northern side and India (Bengal) and Nepal on the western side, and Myanmar on the southern side, the Brahmaputra Valley had all the characteristics of a frontier region. It has been the meeting place of two great world civilizations: Indian and Chinese. It has been receiving various groups belonging to different races, languages, culture and religions. It is estimated that at the time of immigration of the Aryans into India various groups belonging to Mongoloid groups entered the Brahmaputra Valley from different directions: China, Tibet, and Burma. Later the Aryans, too, came in, obviously through Bengal. Interaction, assimilation and integration of various cultures, religions, races, and civilizations produced a distinctive synthesis of the Valley's culture, economy, and polity.

Though the Brahmaputra Valley/Ahom Assam was largely successful in maintaining her isolation from mainland India, it would be wrong to ignore its connection with India. About the ancient relations Suniti Kumar Chatterjee (1970:16) observed "The *Kirata* or Mongoloid people were certainly very well known to the Vedic Aryans and we find mention of them in *Yajur* and *Artharva Vedas*. Even the epic *Mahabharata* mentions about Assam's Indo-Mongoloid king Bhagadatta who together with his *Kirata and Cino* or Chinese soldiers fought the battle at Kurukhshettra as an ally of the Kauravas. He further observed that by the end of the Vedic age around 1,000 B.C. the Mongoloid ruling figure in Assam had come to the fold of the Brahminical world as developed in the mid-land India and paved the way for the entire non-Aryan people to merge into the Brahminical fold in the Brahmaputra Valley (Ibid, 16,17). We, however, feel that this Brahminical Hinduization process, which sociologists would prefer to call the *Sanskritization* process, was not complete. It remained incomplete even in colonial and post-colonial India. The Mongoloid groups in the Brahmaputra Valley who accepted the Hinduism paradoxically continued some of their pre-Hindu rituals and culture together with the *Sanskritized* rituals and practices. Hence one can locate a very distinctive kind of Hinduism being practiced in the Brahmaputra Valley.

The society that emerged in the valley was composed of various migrant groups. However, it did not take place at one time. Since ancient times many migrant groups from different directions settled down in the Brahmaputra Valley. A very significant group that entered the Valley from Burma in the 13th century through the Patkai hills was the Ahoms, a Tai-Shan tribe who gave it its ruling dynasty from 1228 AD till the British annexation in 1826 AD. Before the arrival of the Ahoms, many Muslims entered Assam and settled down permanently. Though very limited in number, the Ahom rulers too encouraged migration of priests, learned men and technicians from outside, i.e., India. The society that emerged in the pre-colonial Brahmaputra Valley was composed of many distinct social groups. It was obviously coalitional in nature and highly plural in composition. Historically speaking, the Brahmaputra valley emerged as a shared homeland of many groups/communities.

It would be pertinent to broadly specify the broad social groups that emerged during the pre-colonial period. Figure 18.1 presents a broad idea about the various groups/communities lived in the pre-colonial Brahmaputra Valley. Though we find the presence of Aryan groups composed of various Hindu castes and Muslims, it must be pointed out that an overwhelming majority of the social groups racially belonged to the non-Aryan Mongoloid groups. Despite its Mongoloid predominance, Hinduism too played a very important role. Although the caste system was relatively less rigid in the Brahmaputra Valley compared to the rest of India, the Brahmins and other high castes played a very significant elite role in the society. During the later stage of the Ahom rule the Brahmin priests could successfully reduce the importance of the traditional Ahom priests, the Deodhai Bailungs. The Ahom rulers together with the Brahmin and other castes played a decisive role in the society and polity of pre-colonial Brahmaputra Valley.

As a result of the pan-Indian *Bhakti* movement, *Vaisnavism* attained popularity in the medieval Brahmaputra Valley. Prior to the emergence of *Vaisnavism*, the *Shaktaism/Shakti* cult, i.e., the worshiping of a Mother Goddess in various manifestations, was dominant. Saint Shankardeva (1449-1568) popularized *Vaisnavism* among the masses of the Brahmapuitra Valley. Gradually the *Satras* (monasteries) of *Vaisnavism* became popular and powerful social institutions. This later led to the conflict between the clerics of dominant *Shaktaism* and the emerging *Vaisnavaism.* The Ahom kings at times treated both the sects equally and patronized them accordingly with liberal grants of revenue-free land. However, at a later stage they favoured the clerics of *Shaktaism* more than *Vaisnavism*. The clerics of *Vaisnavism* faced state persecution because of their popularity, which threatened the clerics of *Shaktaism* on the one hand and the state on the other. The growing economic power of the *Satras* as a parallel semi-feudal institution within the state made the clerics of *Shaktaism* insecure. Obviously, they and the Ahom rulers did not like the growth of the *Satras* as a powerful religious-feudal institution and they together persecuted the clerics of the *Vaisnavism,* particularly during the 18th and early 19th century, which resulted in the rebellion of the *Moamoria Vaisnavas* (for details see Bhuyan, 1974; Neog, 1965; Dewan, 1838; Bora, 1983; and Gait 1905.) The rebellion was virtually transformed into a long drawn out civil war in a semi-feudal society wherein the Ahom state failed to resolve the internal conflict and paved the way for external intervention, which finally led to the Brahmaputra Valley's

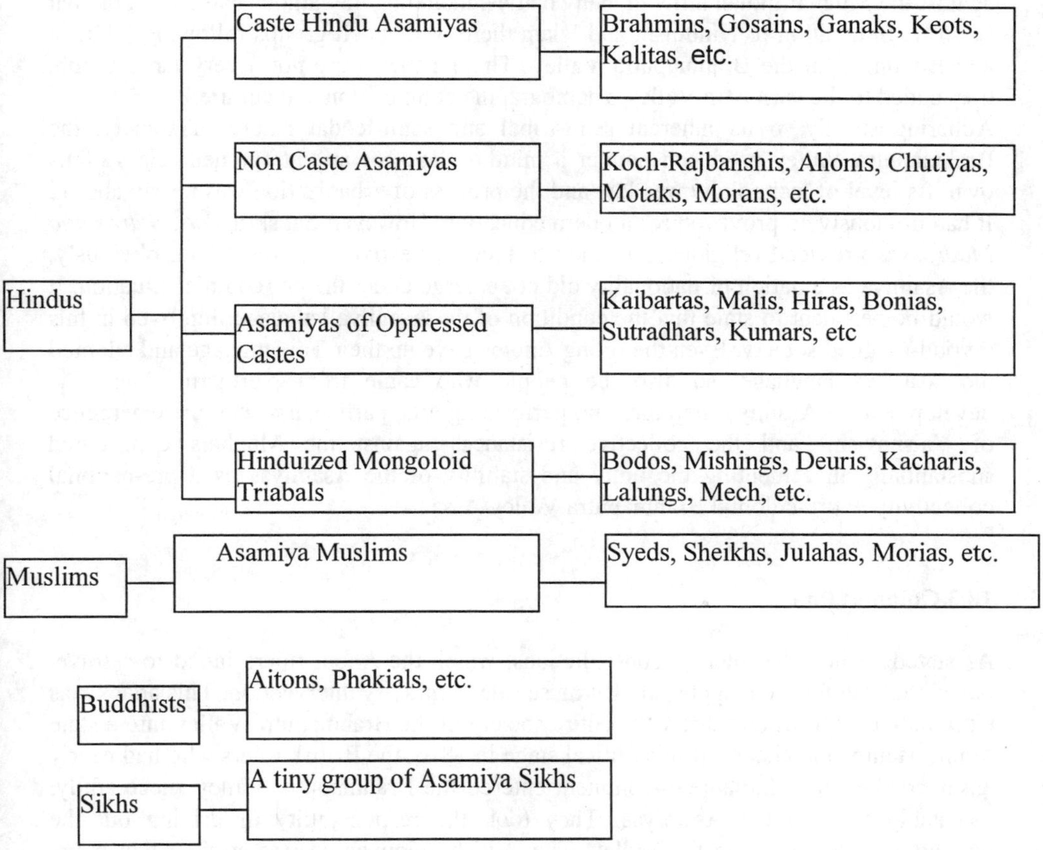


Figure 18.1 Social groups in the Brahmaputra Valley.

incorporation into the British colony. However, it would be relevant to record here that with the emergence of *Vaisnavism*, the Asamiya language, literature and performing arts such as dance and drama developed very significantly. The popularity of *Vaisnavism* helped in reducing the gap of the Brahmaputra Valley's isolation from India. It also helped in detribalizing the society in the Valley. Besides these two major sects of Hinduism we find a significant number of followers of Islam in the Brahmaputra Valley (for details see Hussain 1987a, 1987B, 1988; Bhuyan, 1975; Barua, 1985; Choudhuri, 1982; Sattar, 1972; Gohain, 1988; Ali, 1986; Saikia, 1986).

According to a report published in 1841, a British colonial administrator estimated the total Muslim population at one sixth of the state (Robinson 1841:252). Significantly, the Muslims had syncretized Islam in the Brahmaputra Valley and obviously there was no Hindu--Muslim conflict in the pre–colonial Assam. It is a historical fact that the

Muslims of the Valley together with others in the Ahom militia successfully fought against the Mughal Indian army in many battles, including the famous battle of Saraighat in 1671. In addition to Hinduism and Islam there were small groups following Sikhism and Buddhism in the Brahmaputra Valley. Though they were not a very large group, they added to the variety in Valley's demographic composition and culture.
Adhering strictly to its inherent semi-tribal and semi-feudal nature of society, the Brahmaputra Valley produced neither a middle class nor an entrepreneur class of its own. Its level of technology was low and the process of urbanization was nearly absent. It had obviously no provision for modern education. However, Sanskrit *Tols, Satras and Madrassas* provided religious education to their respective followers. And, obviously, the Asamiya as a nation or nationality did not emerge under the pre-colonial situation. It would be pertinent to state that the condition of the Asamiya language improved in this myriad-tongued society. Even the ruling Ahoms gave up their Tai language and adopted the Asamiya language, so also the people who came from Northern India. The development of Asamiya literature and performing arts, particularly with the emergence of Vaisnavism, and the collective resistances against the Mughals contributed substantially in cementing the unity and stability of the Asamiyas as a pre-national collectivity in pre-colonial Brahmaputra Valley/Assam.

## 18.3 Colonial Phase

As stated earlier, the internal contradictions, which the Ahom rulers failed to resolve, paved the way for the neighboring Burmese ruler's military intervention. This attack was extremely barbarous and threw the entire society of the Brahmaputra Valley into a state of uncertainty and chaos. At this critical stage in 1826, the British rulers who had nearly enslaved the entire Indian sub-continent entered the Brahmaputra Valley successfully, ostensibly to save the Asamiyas. They took the responsibility of driving out the marauding Burmese from the Valley. The British colonization was so tacit that many Asamiyas could not even understand that they were being made a dependent subject of a colony. The Asamiyas being a highly disorganized and a frustrated community at the time of British colonization felt relieved and rescued from a long drawn internal turmoil.

The colonization gradually broke the isolation of the Brahmaputra Valley by making it a part of British India and thereby linking it with the colonial capitalist world economy. The penetration of colonialism added new complexities with serious socio-economic and political ramification into the existing semi-feudal and semi-tribal society. Initially the colonial rulers made the Brahmaputra Valley a part of larger Assam. It was made a division of Bengal. However, Assam was made a new province of British India in1874 and the same arrangement continued till the end of colonial rule in 1947. The new province of Assam included the Sylhet region of Surma Valley/East Bengal now in Bangladesh, the entire hill areas of North East India and Cachar district, in addition to the traditional Asamiya homeland, i.e., the Brahmaputra Valley, which they have been sharing with many tribal groups like the Bodos, Mishings, Tiwas, and Rabhas, etc.. Though the province was named Assam, it was in fact " an amalgam of Asamiya speaking, Bengali speaking and myriad tongued hill and tribal areas in which Asamiya was claimed as the mother tongue of less than a quarter and Bengali more than 40 percent of the population"(Guha1980: 1701). The size of pre-colonial Assam swelled significantly in colonial Assam wherein the Asamiyas became a minority and the second largest group after the Bengalis. Sylhet was never a part of Assam, historically and culturally belonged to Bengal. Obviously it was a colonial arrangement to weaken both the Asamiyas and the Bengalis, and to pave the way for Asamiya--Bengali competition and conflict in colonial Assam. This had a decisive impact on the society and politics of colonial as well as post-colonial Brahmaputra Valley/Assam.

The Brahmaputra Valley's incorporation into the colonial system led to the abandonment of the Ahom way of administration, which was replaced by a new British--Mughal system of administration. As the pre-colonial aristocracy did not fit into the new system based on modern bureaucratic principles of maintaining formal and written records, moderate western education became a pre-condition for entry into jobs in the new colonial administration. The British colonizers had no intention of educating the Asamiyas immediately in order to ensure their continuity in the new colonial set-up. This led to a decline of the pre-colonial Asamiya aristocracy immediately, and more significantly it delayed the process of middle class formation among the Asamiyas. This also pertinently explains the reason behind the import of the *baboos* from neighboring Bengal to man the junior level positions in the colonial administration in Assam. According to Guha, without any investment in western education in Assam, the colonizers profitably availed themselves of the services of already surplus educated unemployed persons from the Bengal Presidency (Guha, 1977:58). It would be worthwhile to recollect that the British colonized Bengal much earlier, and a new middle class emerged with the expansion of western education in the Bengal Presidency from the Bengali Hindus, mainly from the *Bhadraloks/*high castes like Brahmins, Kayasthas, and Vaidyas (see Misra 1961, Seal 1971, Broomfield 1986, etc). In the absence of an Asamiya middle class in the early colonial Assam, the Bengalis monopolized nearly all jobs meant for Indians in the colonial administration. However, later on, with the emergence of an Asamiya middle class, the competition and conflict between the Bengalis and the Asamiyas sharpened.

The colonial state introduced a qualitatively different land revenue system as opposed to the pre-colonial *Paik* syatem, known as *Ryotwari* system in all the districts of the Brahmaputra Valley with the exception of Goalpara district, which retained the *Zamindari* system like Bengal (for details see Hussain 1993:39-40). In contrast to the earlier system, wherein the *Paik* had to offer periodic service to the state for land, the colonial system monetized the land revenue system wherein the landowner had to pay the land revenue in the form of money. Again, in contrast to earlier pre-colonial system, land became a saleable, purchasable and inheritable commodity in colonial Assam. The colonial state used the land as a major source to mobilize its revenue.

Because of a thin population the colonial state had to accept the reality of limited revenues. Hence, it had to look for additional sources of revenue for the colonial coffers. In the absence of any significant industrial and commercial development in the province, the colonial state engaged itself in the sale of opium through government agencies at a very exorbitant price. Next to land, Guha (1977:55) observed "opium was the most important source of revenue in colonial Assam." This deeply affected the health, productivity, morality, and economic conditions of the Asamiya peasants and the migrant labors in the newly emerging tea plantations. It also produced a group of Asamiya middlemen in the opium business. A sizeable section of rich Asamiya families in Upper Assam today owe their wealth and status to the boom in the opium business in colonial Assam.

One of the important gifts that Assam received from the British colonial state was the tea plantation. As an inherent rule of colonialism, the British rulers tried to exploit Assam economically to the maximum. Thy found the Brahmaputra Valley, particularly its foothills, physically most suitable for growing tea plants on a large scale. The powerful British capital backed by their own colonial state apparatus succeeded in their venture and soon the Brahmaputra Valley/Assam became one of the important tea producers in the world. It still continues to be so. However, it must be pointed out that in the absence of a strong local labour force the most important impact the society had to face was that of mass-migration of labour, mostly black tribal from the Jharkhand region to the Brahmaputra Valley. Their massive migration transformed the demographic structure very significantly. Davis (1951:115) observed that such massive migration made Assam demographically the fastest growing province in colonial India. A new and distinct social group of 'Tea laborers' belonging to various tribes like the Santhals, Mundas, Orans, etc., emerged in the Brahmaputra Valley. According to the colonial historian Gait (1905:362) the population of tea labor in 1921 was 1.3 million, or one sixth of the total population of Assam. Obviously they were the most oppressed social group in colonial Assam.

In order to dispatch tea to the market through the Calcutta port, the colonial rulers had to improve the communication network. As a consequence communication improved very significantly in the colonial Brahmaputra Valley. Besides the steamer service to Calcutta through the Brahmaputra River and road transport, a new railway network was established by the end of the 19$^{th}$ century. By 1903 Assam had 712 miles of railway lines, an impressive development indeed in a colonial situation. In addition to the tea plantations, the coal and petroleum industries were set up under the colonial aegis, which forced the colonial state to build up an effective communication network as an integral part of their infrastructure for profitable exploitation of the region.

It would be worthwhile to point out that by the time the Brahmaputra Valley was colonized its population declined substantially owing to earlier civil war and the Burmese intrusion. In 1826 the population of the Brahmaputra Valley was estimated at 830,000 (Barpujari, 1977:59). According to the Mills Report the population of the Brahmaputra Valley was 12,01,151 (Mills, 1854:56). According to Barpujari (ibid: 59), the population of the Valley rose to about 15 hundred thousand in 1872. However, according to the Census Report of 1872 the population of the entire Brahmaputra Valley stood above 18 hundred thousand. All this points to a low density of population in the Brahmaputra Valley in the early colonial period. However, the consolidation of colonial economy opened the floodgates of migration to Assam---mainly to the Brahmaputra Valley. The East Bengal peasants, mainly from the neighboring Mymensing district, migrated to the Valley's wastelands. Their migration helped to increase the quantum of land revenue and agricultural production. An eminent economist observed: '*One welcome result of the influx of these farm settlers is the improvement of farming practices developed the whole agricultural system in Assam. It helped to improve the health of the countryside by clearing the jungles and marshes, and converted the wild areas into a populous agricultural region. The introduction of jute, vegetables, etc., as a commercial crop in Assam, has largely been owing to the immigrants' (Goswami, 1988:29).* As most of the peasant migrants were Muslims, it increased the Muslim population significantly in the Brahmaputra Valley

As a result of the massive migration, distinct and new social groups emerged in the Brahmaputra Valley in addition to existing pre-colonial social groups. The British, though very small in numbers, were obviously the most powerful social group in the colonial society and the Brahmaputra Valley. Most of them were officers in the colonial administration, both civil and military, tea planters or Christian missionaries. The Hindu Bengalis who came from Bengal to assist the colonial rulers as their subordinates formed another group. Besides, some of the Bengalis were engaged in various professions and small businesses. The Nepalis too came to Assam as a part of the colonial army. In the absence of an indigenous business caste or class the Marwaris from Rajputana filled up the vacuum in business. Many Biharis came in as laborers. The migrant Muslim peasants from East Bengal occupied most of the waste and low-lying land in the Brahmaputra Valley. The tribal plantation laborers came in mainly from the Jharkhand region. All these migrations were propelled by the distorted process of development under the colonial rule. Migration altered the social composition of the society very drastically. The colonial society became more heterogeneous and more complex, and came to be linked with the capitalist world.

In order to understand the colonial society better, it would be necessary to briefly comprehend the emerging class structure and social dynamics in colonial Brahmaputra Valley in addition to social groups. After colonization the pre-colonial aristocracy started gradually declining. However, significantly, the high caste Asamiya Hindus succeeded in maintaining their position because of similar and selective land revenue policy adopted by the colonial rulers towards them. Like the Ahom rulers, the colonial rulers, too, patronized the Hindu clerics by maintaining a liberal policy and state patronage. Their ownership of reasonably large tracts of land together with their positive response to modern education helped them to avert their decline in colonial situation. However, the other groups falling into the category of pre-colonial aristocracy like Ahoms experienced a decline.

The colonial land revenue system produced a new class of *Mauzadars,* the revenue collecting agents for a cluster of villages, throughout the Brahmaputra Valley except the Goalpara district. They became important collaborators of the British colonialism and mediators between the rulers and the ruled. Most of the *Mauzadars* were appointed from the high castes and the Asamiya Muslims. They joined and supported the colonial policy of oppression and exploitation of peasants.

The colonial rulers and the planters backed by the colonial state and capital were at the top of the class structure in colonial Brahmaputra Valley. Next to them, but much below, stood the Asamiya planters and the Marwari traders. Next to them stood a relatively large Bengali middle class and a small Asamiya middle class. The *Zamindars* of Goalpara district and the Mauzadars of the rest of the Valley formed the rural gentry. Most members of the emerging Asamiya middle class largely belonged to the high castes and had their rural link as absentee landlords. Below them stood the fragmented peasantry: medium and small. The migrant peasants of the East Bengal origin and the tea plantation laborers, most of whom were the tribal from the Jharkhand region, stood virtually at the bottom of the class structure of colonial Brahmaputra Valley.

Table 18.1 New Major Social Groups in the Colonial Brahmaputra Valley.*

| Nationality/ Community | Major occupation | Size and social standing |
|---|---|---|
| 1. British | Colonial administration: civil and military | Small but most powerful |
| | Tea planters | Small and powerful |
| | Members of Christian Missions | Small and influential |
| 2. Americans | Member of Christian Missions | Small and influential |
| 3. Marwaris | Trade, commerce, and banking | Small and rich |
| 4. Hindu Bengalis | Middle class jobs in colonial administration | Larger than group 1, 2, & 3, worked as collaborator |
| 5. Nepalis | Army, dairy, and construction | Small, hardworking, oppressed |
| 6. Biharis | Laborers | Small, poor and oppressed |
| 7. Muslim Bengalis | Peasants | Large, poor and oppressed |
| 8. Jharkhandi Tribal | Labor in tea-plantations | Large, poor and oppressed |

*In addition to the social groups included in the Figure 18.1.
Source: Hussain (1993:47)

At this stage, it would be relevant to make a few sociologically pertinent comments on education during the colonial period. By and large, education was neglected in the Valley, particularly more so during the 19$^{th}$ century. The response to western education among the Asamiyas was crippled by the colonial suppression of the Asamiya language from 1837 to 1874. Asamiya was replaced by Bengali language in Assam/Brahmaputra Valley (for details see Hussain 1993). The suppression of the Asamiya language to a large extent suppressed the popular urge for western education in Assam. Besides, the popular enthusiasm for the development of education was nearly absent unlike the northern and other parts of India where non-government associations took initiative to expand the western education in their areas and communities. Needless to say, western education was the major institution in the colonial period for upward social mobility.

The first and second high schools in the Brahmaputra Valley were started in 1835 and 1841 respectively, which is fairly late. Neighboring Bengal got its first university (of Calcutta) in 1857 and Assam had to wait for the establishment of a University (of Gauhati) till the end of the colonial period, i.e., in 1948. Even to start a college Assam had to wait till the end of the 19$^{th}$ century. The first college in the Brahmaputra Valley, i.e., the Cotton College was established in Guwahati in 1901. All this is an indication of the belated development of education in the Brahmaputra Valley and its educational backwardness as compared to Bengal. Even under such colonial constraints, a new Calcutta oriented Asamiya middle class gradually emerged in the late 19$^{th}$ century. Obviously it was a weak and very small middle class located in the colonial hinterland. The incipient middle class took a special interest in the development of the Asamiya language and literature. Gradually language became an important and perhaps the most sensitive, symbol of the Asamiya middle class and its nationality.

The colonial situation imposed on the Asamiya middle class stiff competition from the migrant Bengali middle class in the Brahmaputra Valley. In such a situation the Asamiya middle class had to play a collaborative role with the colonial rulers to keep pace. However, with the emergence of the national movement for the independence, the growing popularity of the Congress, and the consolidation of the Asamiya middle in the 20$^{th}$ century, a large section of this class could gradually overcome this collaborative role and step into newer roles. The anti-colonial national movement for independence of India helped the emerging Asamiya nationality to identify itself with the great Indian nationalism. It broke the political isolation and marked its integration with Indian nationalism. However, the horizon of the Asamiya nationality did not expand much despite massive migration, and was largely confined to the social groups of pre-colonial Assam.

## 18.4 Post Colonial Development

India entered into the post-colonial phase in 1947 with the joy of independence and pain of partition at the same time. As a result of independence and partition Assam lost Bengali Muslims and Bengali Hindus dominated and highly populated district of Sylhet to East Pakistan. This substantially reduced the number and percentage of both the Muslims and the Bengalis in Assam. However, the predominantly Bengali speaking Cachar district in Surma Valley remained within the truncated province of Assam. Except the North East Frontier Agency (NEFA), now Arunachal Pradesh, Manipur, and Tripura, the entire hill region of the North East with numerous tribes standing at uneven levels of socio-economic development remained with Assam in addition to the Assam proper, i.e., the Brahmaputra Valley.

Independence and partition made the Asamiyas numerically and politically stronger in post-colonial Assam. The Asamiyas for the first time became the single largest and majority group in Assam. The Bengalis in Assam lost their size in number and the partition affected them perniciously. On the other hand, the Muslims who were involved in the Muslim League's divisive politics lost their political relevance and suffered frustration and shame. The tribal groups both in the hills and the Brahmaputra Valley remained backward, and consequently failed to assert their position in the state in the early years of the post-colonial phase. Besides, the large non-caste Mongoloid groups, like the Ahoms and the Koch-Rajbonshis too, remained largely backward. In such a situation, the Asamiya middle class mainly composed of high castes Hindus of the Brahmaputra Valley consolidated their position and increased their power in Assam in the changed post-colonial situation. They began dominating the state politics and exercising their hegemony over the Asamiyas, the Na(neo)-Asamiyas and the non-Asamiya groups in Assam. The exertion of this hegemony and the response of various groups towards such hegemony determined the politics of the hill and the Valleys during the post-colonial period.

Besides other historical reasons, the tribal fear of Asamiya dominance ultimately led to several reorganizations of the state of Assam. In the process the Naga Hills became Nagaland in 1963, the Khasi Hills, Jaiantia Hills, and Garo Hills together became Meghalaya in 1969 and the Mizo Hills became Mizoram in 1972. However, the Karbi Hills and the North Cachar Hills remained with Assam. In these two hill districts, too, there is a popular movement for an autonomous state within the state of Assam. The size of Assam reduced drastically during the post-colonial period. Ultimately the state of Assam may be confined to the Brahmaputra Valley alone.

We have noted earlier that the plain's tribals have been sharing their traditional homeland, i.e., the Brahmaputra Valley, with the Asamiyas. It is, therefore, almost impossible to isolate the colorful tribal elements from the composite Asamiya society, culture, and nationality. According to the 1991 Census the scheduled tribes constituted more than 10% of the total population in the Valley. The Bodos, Mishings, Sonowals, Ravas, and Tiwas demographically occupy first, second, third, fourth, and fifth positions, respectively. Needless to say, the tribals are the first natives of the Brahmaputra Valley. Unlike the tribals of the hills, the tribals of the Valley own land individually, not communally. They could not make any significant progress during the colonial period, but they did make some progress during the post-colonial period; other groups particularly the high caste groups progressed even further, thereby increasing the social distance between them and the tribals of the same Valley. The politically dominant groups in the Brahmaputra Valley have successfully kept the tribals from acquiring even the limited benefits that accrued to the hill tribals. The Indian Constitution provided the hill tribals some autonomy in managing their society through the provisions of the sixth Schedule. This was not, however, extended to the tribals of the Valley. It would be worthwhile to state that the tribals are perpetually experiencing the problem of land alienation, poverty, indebtedness, severe unemployment, economic exploitation, and political oppression. In colonial Assam the tribal languages remained neglected. However, during the post-colonial period, with the growing consciousness of their identity the tribals have been receiving support from the progressive and democratic sections of the Asamiyas. As a result some attention has recently been paid to the development of tribal languages. For instance, the Bodos have developed their language in the Devnagari script and it is now the medium of instruction up to the secondary level in the Bodo-dominated areas.

In the absence of a well developed language of their own, the tribals of the Brahmaputra Valley accepted the Asamiya language voluntarily as the medium of education and inter-tribal communication. As a consequence they were regarded as sub-nationalities within the composite Asamiya nationality. However, the official imposition of the Asamiya language in 1960 had resulted in a severe backlash from the tribals both in the hills and the Valley. Besides, from the late sixties onwards the tribals of the Valley have become more conscious and articulate about their ethnic identity, using it to gain political power to overcome their backwardness and oppression. The problems regarding their rights to land, language, and script, culture, identity, and, more significantly, demands for political autonomy are yet to be resolved democratically.

## 18.5 Demography: Transition and Plurality

Historically speaking, Assam and its Brahmaputra Valley have been highly plural in demographic composition. It has been a shared homeland of various racial, religious, ethnic, linguistic and cultural groups. Demographic plurality of the pre-colonial society increased further as an inseparable part of the colonial transformation during the British rule. Obviously it had grown gradually from a land abundant to a land scarce valley during the later part of colonial period and more so during the post-colonial period.

As per the 1991 Census, the population of Assam stood at 22414,322 persons in which the share of the Brahmaputra Valley was 1,91,09,300 persons. However, as per the provisional Census Report of 2001, the population of Assam has reached 26638,407 persons (as reported in The Assam Tribune 27 March 2001). During the post-colonial period the Brahmaputra Valley had shown a rapid growth of population up to 1991 Census. Assam's population has grown from 7970,998 persons in 1951 to 22414,322 persons in 1991, i.e., within a period of 40 years during the post-colonial period. However, very significantly, the growth rate of population declined substantially from 24.24 percent during the decade 1981--91 to only 13.85 percent during the decade 1991--2001.

The literacy rate, too, has increased from 56.08 percent in 1991 to 66.08 percent in 2001. However, the gender gap continues to be an existential reality in the state, the literacy among the women is only 56.03 percent compared to 77.58 percent among men.

The density of population in Assam has increased from 286 per square kilometer in 1991 to 340 per square kilometer in 2001. However, if we deduct the remaining hill areas of Assam, the density of population of the Brahmaputra Valley is definitely much higher than the average of the state. Notwithstanding the low growth of population in the last decade, the Brahmaputra Valley is obviously a thickly populated plain wherein the man to land ratio is not favorable, given the low economic and technological development in a largely agrarian society located on India's periphery.

## 18.6 Linguistic Plurality

As noted earlier, the Brahmaputra Valley continues to be a multi-racial, multi-lingual, and multi-religious society. According to the Census of 1991 there are more than 26 linguistic communities living in the Brahmaputra Valley. Asamiya is the most popular language of the state of Assam and its Brahmaputra Valley. In the Valley Asamiya is the recorded mother tongue of more than two thirds of the population, i.e., 67.40 %. Only 14.50 % recorded Bengali as their mother tongue, and the remaining 18.10 % recorded other languages like Bodo, Hindi, Mishing, Nepali, Santhali, etc., as their mother tongue.

One must, however, note, despite that, the number of people who can speak Asamiya is much higher than that has been shown in the 'dead' statistics of Census. Most of the non-Asamiya population of the Brahmaputra Valley can speak and communicate well in the Asamiya language. The prevailing social and economic situation constantly forced not only the Bengalis but also other non-Asamiyas living particularly in the Brahmaputra Valley to learn the Asamiya language---the way many non-Hindi speakers are learning Hindi in their own interest. The tribals of the Valley have historically been bi-lingual. Besides their respective language/dialect, known as Asamiya, some of them are now struggling to establish their distinct ethno-linguistic identity vis-à-vis the Asamiyas. Asamiya has been one of the fastest growing languages during the post-colonial period (see Hussain 1993:226-253). Despite the high rate of migration and massive growth of population during the colonial as well as in the post-colonial period, the volume of the Asamiya-speaking population is also increasing at a higher rate than that of others, ensuring the dominant demographic status of the Asamiya speakers in Assam in general and the Brahmaputra Valley in particular. Historically this demonstrates the positive capability of the Asamiya nationality to absorb many permanent migrants into its fold. Now the Asamiya language is irreducibly well entrenched in its historical homeland in a situation of linguistic plurality.

## 18.7 Religious and Racial Plurality

The pre-colonial religious plurality of the Brahmaputra Valley continued during the post-colonial period. Even the partition of India in 1947 failed to affect the demographic composition of the valley in any significant manner. As a result of rioting in the wake of partition, more than one hundred thousand Muslims living in lower Assam/western part of the valley, involuntarily migrated to East Pakistan leaving behind their land and property in the Brahmaputra Valley. However, following the Nehru--Liaquat Ali Pact of 1950 between India and Pakistan, which assured their safe return and rehabilitation, many returned home within two years. Hence it did not affect the religious composition of the valley unlike the Panjab and West Bengal. Not only the Brahmaputra Valley as a geographical entity, but the Asamiyas as a community or as a national entity continued to be a multi-religious one. The Asamiya nationality includes the Hindus, Muslims, Christians, Buddhists and a tiny community of Sikhs within its fold.

Out of the Brahmaputra Valley's 19109,300 population, 12993,429; 5287,571 and 591,148 are Hindus, Muslims, and Christians respectively (Government of India 1991). The Hindus, Muslims and Christians constitute 68 %, 27 %, and 3 % of the valley's population, respectively. However, the Muslims, if we look at their demographic status within the Asamiya nationality, would be no less than two thirds of the total Asamiya population. Hence, while looking into the nationality question in Assam, one should not ignore the multi-religious character of the Asamiya nationality.

As noted earlier, historically the Brahmaputra Valley continues to be a multi-racial society and so does the Asamiya nationality. Hence we find that demographically the Brahmaputra Valley is a highly plural geographic entity. It has been a shared homeland of various racial, religious, linguistic and cultural groups wherein the Asamiyas are the numerically major and dominant nationality. They have been living side by the side with many distinct smaller ethnic/tribal groups like the Bodos, Mishings, Tiwas, Rabhas, and Sonowals, etc.. Ignoring this historical plurality and its resultant multi-culturalism, and the reality of a shared homeland, will be both ahistorical and immoral. Admittedly, this shared homeland/living space has shrunk substantially in the distorted process of the socio-economic transformation of Assam/Brahmaputra Valley during the colonial and post-colonial period. This has contributed towards the sharpening of competition and conflict for 'space'. In the absence of an effective intervention from the state and the civil society such a conflict has not been resolved and this has affected the society in the Brahmaputra Valley for more than two decades now.

## References


Ali, A.N.M. Irshad (1986). Asamat Islam Drarmar Vistar: Asamiya Janasankhyar Gathan Aru Sanskritir Sangmisran. Asamor Manoh (in Asamiya), edited by B.M. Das, Asam Sahitya Sabha, Jorhat.

Barpujari, H.K. (1977). Political History:1826-1919. Govt. of Assam, Guwahati .

Bhuyan, S.K. (1975). Swargadeo Rajeswar Simha ( Asamiya). Publication Board, Guwahati

Broomfield, J.H. (1968). Elite Conflict In A Plural Society. University of California, California,

Chatterji, S.K. (1951). Kirata Janakriti. Asiatic Society of Bengal, Calcutta.

Chatterji, S.K. (1970). Place of Assam in the History and Civilization of India. University of Gauhati, Gauhati

Choudhuri, M. (1982). Luit, Barak Aru Isam.

Davis, K. (1951). The Population of India and Pakistan. Princeton University Press, Princeton, N.J.

Dewan, M. (1836). Buranji Vivek Ratna (in Asamiya). DHAS, Gauhati.

Gait, E. (1905). A History of Assam. Reprint edition, LBS Publications, Guwahati.

Gohain, H. (1988). Adrusta Aru Asam (Asamiya). Shanti Prakashan, Guwahati.

Government of India (1951). Census of India , Report on Assam

Government of India (1961). Census of India, Report on Assam.

Government of India (1971). Census of India, Report on Assam,

Government of India (1988). Census of India, Report on Assam.

Government of India (1988). Census of India. Provisional Report (as published in The Assam Tribune, 27th March 2001).

Goswami, P.C. (1988). The Economic Development of Assam. Kalyani Publishers, New Delhi.

Guha, A. (1977). Planter Raj To Swaraj: Freedom Struggle and Electoral Politics in Assam 1826-1947. Indian Council of Historical Research, New Delhi.

Guha, A. (1980). Little Nationalism Turned Chauvinist: Assam's Anti Foreigners Upsurge, Economic and Political Weekly, Vol.XV, N0. 41, 42 and 43.

Hussain, M. (1987a). Muslims of the Indian State of Assam. Journal: Institute of Muslim Minority Affairs, London.

Hussain, M. (1987b). Muslims of Contemporary Assam: Sociological Introduction, Journal of North East India Council of Social Science: Research, Vol. XI, N0. 2.

Hussain, M. (1987c). Tribal Movements for Autonomous State in Assam, Economic and Political Weekly, Vol.: 22, No 32, 8 August.

Hussain, M. (1988). Caste Among Non-Hindu Indians: An Exploratory Study of Assamese Muslims, The Eastern Anthropologist, vol:41: 4, October-December.

Hussain, M. (1993). The Assam Movement: Class, Ideology and Identity. Manak Publications, New Delhi.

Hussain, M. (2000). State, Identity Movements and Internal Displacement in the North East. Economic and Political Weekly, Vol: XXXV, No 51, December 16.

Mills, M. A. J. (1853). Report On The Province of Assam, Reprint Edition, Publication Board, Guwahati, Assam.

Neog, M. (1965). Sankardeva and his Times: Early History of Vaisnava Faith and Movement in Assam. Lawyers Book Stall, Gauhati.

Misra, B.B. (1961). The Indian Middle Class: Their Growth in Modern Times, Oxford University Press, New Delhi.

Robinson, W. (1841). A Descriptive Account of Assam, Reprint edition, Sanskaran Prakash, Delhi, 1975.

Saikia, M. K. (1986). Asomolo Islam Dharmar Agaman Aru Sangmisran. B.M.Das ed.Asamor Manoh, (in Asamiya), Asam Sahitya Sabha, Jorhat.

Sattar, A. (1972). Sanmisronat Asamiya Sanskriti,( Asamiya) Granthapeeth, Jorhat

Seal, A. (1971). The Emergence of Indian Nationalism: Competition and Collaboration in the Late Nineteenth Century, Cambridge University Press, Cambridge.

# CHAPTER 19

# ECOLOGY

B.K. TALUKDAR

Assam contains a wide diversity of flora and fauna and other natural ecosystems. There are 51 different forest types and subtypes, forming an amalgam of Indian, Malayan, and Tibetan elements. Assam is also famous for its unique wildlife in the world. This chapter describes forests, wildlife, and wetlands and emphasizes that their conservation is essential for preservation of Assamese culture and life and their promotion and growth are fundamental to development of the Brahmaputra valley.

### 19.1 Introduction

Assam is one of the most important areas in the Indian Subcontinent for developing a comprehensive network of biodiversity conservation. It covers an area of 78,438 sq km and lies between latitudes 24° 8' to 27° 55' and longitudes 89° 45' E to 96° 02' E. Assam comprises four distinct physiographic units, viz, the Brahmaputra valley – the Assam proper, the Barak valley, the Karbi plateau, and the North Cachar hills. The Brahmaputra valley is the eastern continuation of the Indo–Gangetic plains. It covers an area of 56,480 sq km with a length of about 800 km and the average width of 80 km enclosed by hills on all sides. The mighty Brahmaputra River flows through the heart of Assam's alluvial plains with tortuous courses forming a number of islands including the largest river islands–the Majuli. The valley as a whole slopes towards the south –west and west. In the northeastern end the plains have an altitude of about 120 MSL, while the western end is 30 MSL. The numerous outcrops of low hills scattered near Silghat, Tezpur, Guwahati, Goalpara, and Dhubri are, in fact, the outlaying portions of the Karbi-Meghalaya plateaus. The old and new alluvial materials brought by the Brahmaputra and its tributaries have formed the plains. It contains a series of swamps, lakes, and riverine sandy areas subjected to annual floods. With an average discharge of about 19,830 $m^3 s^{-1}$ (= 625,350 $m^3 yr^{-1}$), the Brahmaputra is the fifth largest river in the world in terms of the average discharge, and its average sediment load of about 402 million tonnes $yr^{-1}$ (Mahanta and Subramanian, 1993). One of the majestic river systems in the world, the Brahmaputra encompasses a $1.6 \times 10^6$ sq. kms. The drainage basin shared by Tibet, India, and Bangladesh. Rising at the glaciers of the Kailash range of the northern Himalayas at an elevation of 5,150 m at an latitude 30° 31' N and longitude 82°10' E, south of the lake called 'Konggya Tsho', it traverses a distance of about 2,900 kms before joining the Bay of Bengal. The river belongs to the family of east flowing rivers of Asia like that of Hwang Ho, Yangtze-kiang, and Mekong, and possesses most of the riverine characteristics (Barthakur, 1986). The Brahmaputra valley is in a state of



*V.P. Singh et al. (eds.), The Brahmaputra Basin Water Resources*, 351-365.

perpetual flux owing to rapid geo-morphological changes over the basin and the periodic earthquakes in the area aggravate this geologically unstable zone.

The state has a typical subtropical monsoon climate with an average annual rainfall varying from 1,000 mm to 3,750 mm. The temperature varies from 6 °C in winter to about 38 °C in summer. The relative humidity ranges in between 40--95% at different locations. As the valleys have an extremely low gradient, inundation and water logging are therefore common to the flood plains. The flood plains of the Brahmaputra valley have sandy soil with a thin silt cover and almost regular inundation could be observed during the monsoon.

## 19.2 Forest

The forests of Assam contain a great diversity of flora and fauna and other natural ecosystems. There are about fifty one different forest types and subtypes occurring in the region, forming an amalgam of Indian, Malayan and Tibetan elements. Five major groups of forests in Assam have been identified by Champion and Seth (1968) as revised forest types. These are:

Group-1: Tropical Wet Evergreen Forest;
Group-2: Tropical Semi-evergreen Forest;
Group-3: Tropical Moist Deciduous Forest;
Group-4: Littoral and Swamp Forest; and
Group-5: Tropical Dry Deciduous Forest.

## 19.3 History of Forest Conservation in Assam

Forestry has been playing a very intimate role in the social lives of the people living in the region from ages and the importance for judicious management of these natural resources was lost sight of even by the earliest rulers of the state. Historical data revealed that during the Ahom dynasty, a specific administrative branch was there to look after the management of timber, exploitation and timber depots. This is evident from the surnames being carried till today after the titles conferred by the Ahom kings on the professionals engaged in their respective area of activities, e.g., Hatibaruah (the person engaged to look after Elephants), Kath Bharali (the persons engaged to look after the timber depots), etc.. After that the British rulers also took keen steps to enhance conservation of forests and the Assam Forest Department came into being with the appointment of a Deputy Conservator of Forests in charge in 1860s. A total area of around 697 sq km was constituted as Reserve under the Act of 1865. In 1879 Mr Gustav Mann was appointed as Conservator of Forests. Since then there were increased activities towards promotion of Forests conservation in Assam and by the year 1918 the total area of Reserved Forest at 13,934 sq km. The compilation of flora of Assam was completed in the year 1913. By the end of the year 1939, the area of Reserve Forests increased up to 17,084 sq km of forests dedicated to wildlife conservation by constituting sanctuaries. As of 1992, the Reserved Forests in Assam stands at 17,580.54 sq km and around 2,334 sq km has been dedicated to wildlife Sanctuaries and National

Parks. However, by the year 2000 the area under wildlife sanctuary and national parks have been increased up to around 3,000 sq km.

The Assam Remote Sensing Application Centre (ARSAC), based on the satellite data of late 1980s, revealed that the state of Assam as a whole has an existing forest cover of 21.98 percent equalling around 17263,87 ha as against 25.2 percent during 1980--82. The percentage of forest cover in the flood plains stand at around 17.2 percent, while in the hills it is 42 percent. The district wise forest cover, based on satellite data, are summarised in Table 19.1.

Table 19.1 Districtwise Forest Cover of Assam based on Satellite Data. (Source: Assam Remote Sensing Application Centre).

| DISTRICT | FOREST LAND (in sq km) | | | | |
|---|---|---|---|---|---|
| | *Evergreen/ Semi-evergreen* | *Deciduous* | *Degraded/ Scrubland* | *Forest Plantation* | *Total* |
| BARPETA | ----- | 194.13 | ----- | 39.98 | 234.11 |
| BONGAIGAON | ----- | 20.62 | 32.92 | ----- | 53.54 |
| CACHAR | 1,286.51 | ----- | 143.02 | ----- | 1,429.53 |
| DARRANG | 121.49 | ----- | 103.68 | 6.98 | 232.15 |
| DHEMAJI | 450.96 | ----- | 18.06 | ----- | 469.02 |
| DHUBRI | 85.35 | 379.90 | ----- | ----- | 465.25 |
| DIBRUGARH | 290.36 | 124.52 | 37.87 | ----- | 452.75 |
| GOALPARA | ----- | 49.35 | 97.32 | 81.83 | 228.50 |
| GOLAGHAT | 170.84 | ----- | 135.21 | ----- | 306.05 |
| HAILAKANDI | 351.65 | ----- | 152.89 | ----- | 504.54 |
| JORHAT | 252.21 | ----- | 24.03 | ----- | 276.24 |
| KAMRUP | ----- | 642.81 | 378.39 | 151.40 | 1,172.60 |
| KARBI-ANGLONG | 2,476.40 | 1,696.61 | 68.44 | 267.48 | 4508.93 |
| KARIMGANJ | 186.97 | ----- | 274.94 | 1.37 | 463.28 |
| KOKRAJHAR | 688.62 | 809.30 | 46.04 | 14.83 | 1,558.79 |
| LAKHIMPUR | 45.43 | ----- | 10.33 | ----- | 55.76 |
| MORIGAON | ----- | 366.94 | 19.46 | 10.42 | 396.82 |
| NALBARI | 29.18 | 80.08 | 16.02 | ----- | 125.28 |
| N.C. HILLS | 1,791.32 | ----- | 79.38 | ----- | 1,870.70 |
| NAGAON | 56.71 | 34.91 | 100.71 | 151.36 | 343.69 |
| SIBSAGAR | 174.56 | 10.47 | 93.94 | 8.00 | 286.97 |
| SONITPUR | 814.01 | 3.33 | 236.57 | 24.70 | 1,078.61 |
| TINSUKIA | 525.12 | 124.36 | 68.62 | 32.66 | 750.76 |
| STATE TOTAL | 9,797.69 (12.47%) | 4,537.33 (5.78%) | 2,137.84 (2.72%) | 791.01 (1.01%) | 1,7263 (21.98%) |

The Reserved Forests occupy around 1503,018 hectares accounting for 19.14 percent of the State's total geographical area. The forested area outside the notified Reserved

Forests boundaries cover an area of 567,747 hectares accounting for 7.2 percent of the state. Most of these forest areas are under extreme stress owing tomassive deforestation, large-scale encroachment and continued intense degradation. The encroachment position in various forest divisions of Assam has been summarised in Table 19.2 and the forest areas under encroachment in the inter-state border of Assam has been shown in Table 19.3.

Table 19.2 Encroachment Position in Forest Areas in Assam as on 1st January 1992. (Source: Department of Forest, Assam)

| No. | NAME OF FOREST DIVISION | ENCROACHMENT AREA (in sq km) | NUMBER OF ENCROACHERS |
|---|---|---|---|
| 1. | Kachugaon | 11.99 | 577 |
| 2. | Aie-valley | 63.78 | 5,821 |
| 3. | Haltugaon | 29.51 | 2,697 |
| 4. | Dhubri | 18.21 | 1,219 |
| 5. | Darrang East | 108.25 | 2,616 |
| 6. | Darrang West | 145.02 | 12,991 |
| 7. | Lakhimpur | 190.89 | 6,275 |
| 8. | Nagaon | 5.13 | 1,139 |
| 9. | Nagaon South | 2.50 | 584 |
| 10. | Kamrup East | 2.01 | 622 |
| 11. | Kamrup West | 6.86 | 605 |
| 12. | Kamrup North | 35.11 | 2,533 |
| 13. | Goalpara | 7.46 | 1,059 |
| 14. | Doomdooma | 7.54 | 628 |
| 15. | Digboi | 8.03 | 386 |
| 16. | Dibrugarh | 0.38 | 56 |
| 17. | Sibsagar | 243.50 | 912 |
| 18. | Golaghat | 799.40 | 18,803 |
| 19. | Silchar | 3.45 | 1,162 |
| 20. | Karimganj | 35.44 | 2,717 |
| 21. | Karbi-Anglong East | 352.19 | 2,143 |
| 22. | Karbi-Anglong West | 51.06 | 8,200 |
| 23. | Hamren | 41.49 | 2,413 |
| 24. | N.C. Hills | 16.88 | 334 |
| 25. | Western Assam Wildlife | 6.00 | 2,375 |
| 26. | Eastern Assam Wildlife | Nil | Nil |
| TOTAL | | 2,192.08 | 78,849 |

A glimpse of the worse scenario is provided by the fact that the Gohpur RF has no trace of forest left; the Nambor (South) RF has merely 3.6 percent forest cover, Doyang RF only 2.5 percent and Diphu RF hardly 8.2 percent. The reserved forests have highly depleted forest cover in the districts of Golaghat (12.13 percent), Goalpara (16 percent), Barpeta (41.2 percent), Darrang (40.4 percent), Karimganj (37.3 percent), Lakhimpur (44.5 percent), Sibsagar (56.3 percent) and Sonitpur (58.9 percent). The status of reserve forests is most deplorable in the Golaghat district with only 12.13 percent good forest left. Increased population pressure culminated with influx from neighbouring Bangladesh has sharply increased the human population from around 14625000 in the year 1971 up to 22414000 in the year 1991. The density of humans per sq km has sharply increased from 186 in the year 197 up to 286 in the year 1991. This population increase has resulted in an increased encroachment in the forestlands.

Table 19.3 Forest areas encroached in the Inter-state Border of Assam [RF= Reserved Forest; PRF = Proposed Reserved Forest]. Source: Department of Forest, Assam.

| NAME OF THE AREA | No. of RF and PRF | AREA ENCROACHED (in sq km) |
|---|---|---|
| Assam--Nagaland Border | 10 | 1017.50 |
| Assam--Meghalaya | 31 | 7.62 |
| Assam--Arunachal Pradesh | 30 | 250 |
| Assam--Mizoram Border | 3 | 2.40 |

**19.4 Wildlife of Assam**

The state of Assam is one of the most legendary sites for in the world for unique wildlife. The conservation of various species of endangered wildlife is always a challenging task. Assam harbouring the big five mammals---Rhino, Tiger, Wild Buffalo, Gangetic Dolphin and Elephant--- has been in limelight for its role in the conservation of the big five. Owing to the immense human population explosion, the areas for the wildlife get reduced and encroached, resulting in increased stress on these animals. Assam has a century long conservation history related to Wildlife conservation. Indigenous mammals such as the Golden Langur and Pigmy Hog are found only in the Assam and adjoining Bhutan areas in the world. Most of the Indian wildlife species, either originated from Assam or ends their distribution in Assam. The spotted deer, which is found in India, has its last distribution range in western Assam up to the Kachugaon forest, while the Hoolock Gibbon starts its distribution range from Assam to other southeast Asian countries. The Malayan Sun Bear is found in the southern part of Brahmaputra valley as the Hoolock Gibbon. The Brahmaputra River acted as a natural barrier for the Hoolock Gibbon and hence it is only found in the southern Brahmaputra valley. Similarly the River Sankhose has restricted the distribution of the Golden Langur and hence it is found only in a few pockets in northern Assam between the Sankhose River and Manas River.

The long term commitment of Assam towards conservation of wildlife reverberates at the number of Protected Areas for Wildlife declared by the State Government. At the beginning of the year 2000 Assam has five national parks, and over 10 wildlife sanctuaries. The National Parks are: Kaziranga (430 sq. km), Manas (500 sq km), Nameri (200 sq. km), Dibru-saikhowa (340 sq. km) and Orang (78.80 sq. km). The salient wildlife sanctuaries are: Pabitora, Barnadi, Burachapori, Panidihing, Hollangapar Gibbon Sanctuary, Chakrasila, Bordoibam-beelmukh, Garampani, Padumoni-Bherjan-Borajan, and Laokhowa. All the national parks and sanctuaries are situated in the Brahmaputra valley of Assam. The Manas National park that was declared as a World Heritage Site is also one of the important Tiger Reserves in India. In March 2000 the Ministry for Environment and Forests, the Government of India has also declared Nameri National Park as a Tiger Reserve. Dibru-saikhowa National Park has been declared as a Biosphere Reserve with a total area of around 740 sq. km of which 340 sq. km remained as core area and the rest 400 sq. km as buffer zone.

The Kaziranga National Park (KNP) of Assam is a place of not only regional and national importance, but also of international importance because of its mega-diversity in flora and fauna along with a diverse ecosystem. The Kaziranga was first declared as a reserved forest in the year 1908 after realising the potentiality of its habitat towards conservation of the Great Indian One Horned Rhino *Rhinoceros unicornis*. At that time it was expected that there were only a dozen rhinos left in the area. In the year 1916 the area was declared a Game Reserve to promote conservation of rhinos and other animals. During the year 1950 the Kaziranga was declared as a Sanctuary, and in the year 1974. It was upgraded to Kaziranga National Park with an area of around 430 sq. km. Since then Kaziranga has contributed immensely towards conservation of the rhinos along with other fauna, such as Elephant, Tiger, Swamp Deer, Wild Buffalo, etc.. Owing to the constant efforts from the forest staffs of KNP, the IUCN/UNESCO, declared KNP as a World Heritage Site in the year 1985.

The KNP lies within the geographical limit of two civil districts, viz., Nagaon and Golaghat at present. However, there is a plan to add around 400 sq. km areas with the KNP that fall under the Sonitpur district of Assam. If the proposed area is added to the KNP then the total area of the KNP will exceed 800 sq. km. The KNP is situated in the flood plain of the Brahmaputra River. The park spreads from the Brahmaputra River in the north. The terrain of the park is flat (55-80 m above MSL) with an inclination from east to west. Being in the flood plain of the Brahmaputra River, the soil of the park is rich in alluvial deposits. Flood is an annual phenomenon and excess water submerges the area, covering 50--75 percent of the total landmass. The flood waters generally start receding after 8--12 days. Areas along the base of the Karbi Anglong hills are at higher elevation (80--100 m above MSL) and form the natural highlands, which do not get flooded. Erosion and shifting of the Brahmaputra induces a constant change in the total area. The vegetation of the park could be broadly divided into four categories:

1. Eastern Wet Alluvial Grassland;
2. Eastern Dillenia Swamp Forests;
3. Riparian Fringing Forest; and
4. Assam Alluvial Plains Semi-evergreen Forest.

In the KNP a total of 478 species of birds have been recorded out of which 25 species are globally threatened. Some of the salient species of Birds in the KNP are the Greater Adjutant Stork (*Leptoptilos dubius*), Lesser Adjutant Stork (*Leptoptilos javanicus*),

Black necked Stork (*Ephippiorhynchus asiaticus*), Spot billed Pelican (*Pelecanus philippensis*), Bengal Florican (*Eupodotis bengalensis*), Swamp Francolin (*Francolinus gularis*), etc.

The first scientific census of animals in the KNP was carried out in the year 1966 and thereafter the process was repeated every sixth year. The rhino census figures in the KNP since 1966 has been summarised in the following Table 19.4.

Table 19.4 Rhino census record in Kaziranga National Park of Assam (Source: Forest Department, Assam).

| Year | 1966 | 1972 | 1978 | 1984 | 1991 | 1993 | 1997 | 1999 |
|---|---|---|---|---|---|---|---|---|
| Rhino | 366 | 658 | 939 | 1080 | 1129 | 1164 | 1129 | 1552 |

The Great Indian One Horn Rhino (*Rhinoceros unicornis*) is one of the three Asian Rhino species found in Assam. The other two rhino species are the Sumatran Rhino (*Dicerorhinus sumatrensis*) and Javan Rhino (*Rhinoceros sondaicus*). The current status of these three Asian Rhinos are summarised in Table 19.5. See Figure 19.1 for the number of rhinos in three protected areas of Assam. The KNP holds almost two thirds of the Indian Rhino Population. However owing to poaching and the subsequent trade of rhino horns, the species is in danger of extermination. The rhino has been protected under different laws and International agreements. The rhinos are killed only for its horn, which is often used in medicines in oriental region to reduce fever and high pressure, treatment of paralysis. The horns are also mistaken as an aphrodisiac.

Table 19.5 Current Population of Asian Rhino In the wild (Estimated in the year 1999).

| COUNTRY | INDIAN RHINO *Rhinoceros unicornis* | JAVAN RHINO *Rhinoceros sondaicus* | SUMATRAN RHINO *Dicerorhinus umatrensis* | TOTAL |
|---|---|---|---|---|
| INDIA | 1820 | Not Found | Not Found | 1820 |
| NEPAL | 650 | Not Found | Not Found | 650 |
| INDONESIA | Not Found | 50 | 110-200 | 160-250 |
| MALAYSIA | Not Found | Not Found | 120-160 | 120-160 |
| VIETNAM | Not Found | 5-8 | Not Found | 5-8 |
| LAOS PDR | Not Found | Not Found | ? | ? |
| THAILAND | Not Found | Not Found | ? | ? |
| MYANMAR | Not Found | Not Found | 10? | 10? |
| TOTAL | ≅ 2470 | ≅ 60 | ≅ 300 | ≅ 2830 |

The main enemy of the rhino in the KNP seems the poachers. Although there are records of tiger predation of rhino calf, but poaching remains the greatest threat to the rhino. Efforts have been made by the manager of the KNP to reduce poaching by strengthening the anti-poaching network. The poaching of rhino could be reduced to only 9 in the year 1998. In 1999 till October the poachers could poach only three rhino. Aaranyak Nature Club, a local NGO is helping the Forest and Police staff in the KNP area to minimise poaching by establishing a good network and research in the area since 1997. There are other NGOs also coming forward to assist in the conservation of rhino and other threatened animals and plants in the KNP. Some NGOs even donated materials to the KNP to enhance protection measures.

The global trade in wildlife has been estimated by economists at more than 25 billion US dollars annually, of which more than 40 % is illegal. The increase in wildlife trade has been of serious concern at the regional, national and international level. In India the illegal trade in Ivory, Tiger skin and parts, rhino horns, musk, etc., has already raised the eyebrows of many conservationists, including the Government. In response to this threat the Government of India has promulgated many policies and laws to protect the wildlife and to halt wildlife crime. However only the efforts from the government alone will not yield the desired results if people in general do not come forward to join together to protect and conserve the diverse array of flora and fauna of the country. The state of Assam is particularly important as it harbours more than 70% of the Great Indian Rhino Horn population of the world. It also provides shelters to over 400 tigers and 5,600 wild elephants. Hence the state of Assam is ecologically an important site for the conservation of globally endangered species, and it could be termed the last stronghold of many endangered species, including rhinos.

Ironically over poachers in Assam have killed the last fifty years almost 700 rhinos, however there is only one instance where poachers have been convicted. There are lots of cases where poachers were arrested by the enforcing agencies, but they are able to come out of the punishment under the law. The lacunae in this matter have to be sorted out and international pressure must come on the Indian Government to strengthen the enforcement and conviction under the law. In fact there is no meaning in detecting cases and culprits with conviction. Conviction is a must to reduce and halt the assault on the wildlife. As wildlife can't stand in Court to represent themselves, we the superior animals in the world must come forward to assist the cause of wildlife conservation for the benefit of the present and future generations, so that the children in future could see the diverse array of the wildlife in Assam.

## 19.5 Wetlands

Wetland has been regarded as one of the most productive ecosystems in the world by International agencies like International Union for the Conservation of Nature and Natural Resources (IUCN), World Wildlife Fund (WWF), Wetland International, etc.. The IUCN has defined wetlands as "All the submerged or water saturated lands, natural or man made, inland or coastal, permanent or temporary, static or dynamic, vegetated or non-vegetated, which necessarily have a land--water interface, are defined as wetlands". As per the Ramsar Convention which was held in Ramsar, Iran in the year 1971, the wetlands are "areas of submerged or water saturated land, whether natural or artificial or both or permanent or temporary and whether the water is static or flowing or fresh, brackish, or saline including areas of marine waters, the depth of which at low tide does not exceed six meters". This includes marshes, ponds, lakes, rivers, streams, reservoirs, and marine waters of the shoreline.

The wetland ecosystem, though highly efficient, is very fragile and easily disturbed by human actions. The toxic substances produced as a result of industries in the catchments may reach wetlands, and in due course get accumulated in components of the ecosystem leading to the problem of biomagnifications. Avifauna is one of the most spectacular life form in the wetland ecosystem and they are indicators of the health of a wetland. Increased incident of wetland destruction and degradation owing to the biomagnifications has resulted into demand for an international convention for the safety of the water birds and wetland habitats. The Ramsar Convention has taken the initiatives to conserve wetlands mainly for water birds, declaring come of the crucial wetlands throughout the globe as "Wetlands of International Importance".

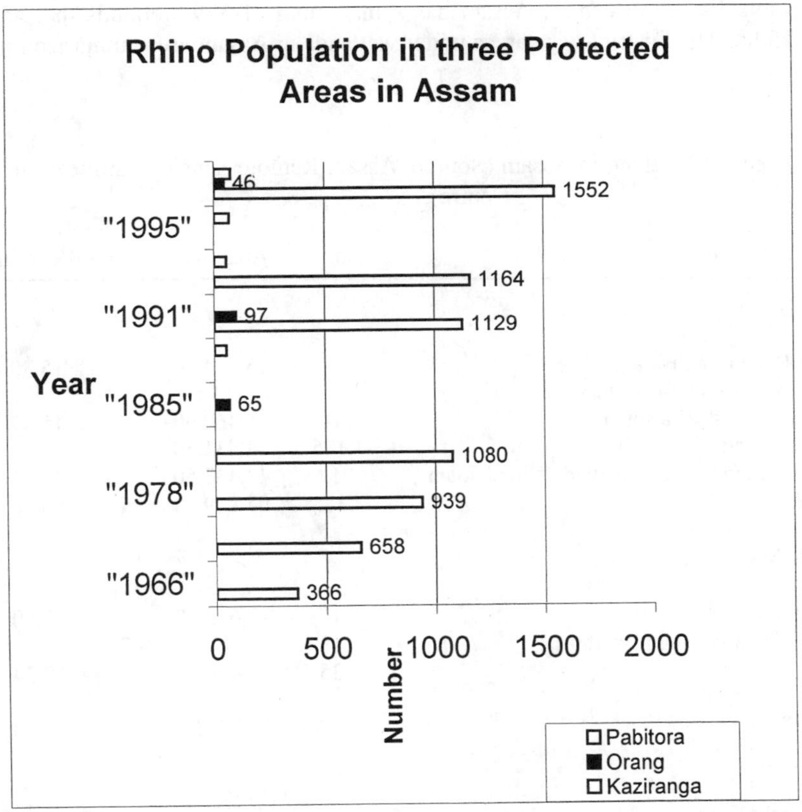


Figure 19.1 Number of Rhino recorded in Three Protected Areas in Assam.

The wetlands in Assam are primarily fresh water and riverine in nature. The state of Assam is one of the potential areas for the conservation of wetlands with more than 3500 wetlands covering an area of around 101,231.60 ha. The types of wetlands found in Assam is summarised in Table 19.6. A study carried out by the Assam Remote Sensing Application Centre (ARSAC) based on satellite map, revealed that out of the total 3,513 wetlands recorded, 2,278 wetlands are with low turbidity, 346 with moderate and 178 wetlands with high turbidity. High turbidity was found prevalent in waterlogged and swampy/marshy areas, while low turbidity was prevalent in waterlogged as well as lakes/ponds. Turbidity was high in most of the natural wetlands compared to that of man-made wetlands. The study further revealed that 282 wetlands are vegetated at fringes, 25 vegetated islands, and 1,145 wetlands with no vegetation, 999 wetlands with partially vegetated while 61 wetlands are completely vegetated.

## 19.6 District Wise Distribution of Wetlands

As mentioned earlier, the state of Assam comprises 23 districts. The Morigaon district in Central Assam has the maximum area under wetlands (ca. 1,658.00 ha) in Assam, while the Nagaon district in central Assam has a maximum of 379 wetlands that are larger than 2.25 ha. The District-wise areas under wetlands in Assam are summarized in Table 19.6.

Table 19.6 Types of Wetland in Assam (Source: Assam Remote Sensing Application Centre).

| Wetland Type | No. | Area (ha) | Per cent |
|---|---|---|---|
| NATURAL | | | |
| Lake/Pond (locally called 'Beel' | 690 | 15,494.00 | 15.30 |
| Ox-bow Lake/Cut-off Meander (locally called 'Era suti') | 861 | 15,460.60 | 15.27 |
| Waterlogged (seasonal) | 1,125 | 23,431.50 | 23.15 |
| Swamp/Marsh (locally called 'Pitoni/Jolah') | 712 | 43,433.50 | 42.91 |
| Total | 3,388 | 97,819.60 | 96.66 |
| MAN-MADE | | | |
| Reservoirs | 10 | 2,662.50 | 2.60 |
| Tanks (locally called 'Pukhuri') | 115 | 749.50 | 0.74 |
| Total | 125 | 3,412.00 | 3.34 |
| Grand Total | 3,513 | 101,231.60 | 100.00 |

Table 19.6 Number of wetlands in each district of Assam with areas. (Source: ARSAC).

| DISTRICT | NUMBER OF WETLANDS | AREA (ha) |
|---|---|---|
| Barpeta | 97 | 3,301.00 |
| Bongaigaon | 100 | 3,158.50 |
| Cachar | 340 | 7,188.00 |
| Darrang | 103 | 3,515.50 |
| Dhemaji | 139 | 3,960.00 |
| Dhubri | 233 | 6,459.70 |
| Dibrugarh | 86 | 2,752.50 |

| | | |
|---|---|---|
| Goalpara | 165 | 3,832.50 |
| Golaghat | 330 | 5,467.50 |
| Hailakandi | 47 | 840.00 |
| Jorhat | 109 | 2,108.50 |
| Kamrup | 352 | 11,407.00 |
| Karbi-Anglong | 77 | 897.00 |
| Karimganj | 70 | 5,719.50 |
| Kokrajhar | 85 | 1,578.40 |
| Lakhimpur | 151 | 3,033.50 |
| Morigaon | 183 | 11,658.00 |
| Nagaon | 379 | 11,295.50 |
| N.C. Hills | 10 | 2,552.50 |
| Nalbari | 68 | 1,988.00 |
| Sibsagar | 109 | 2,135.00 |
| Sonitpur | 206 | 3,651.00 |
| Tinsukia | 74 | 2,732.50 |
| TOTAL | 3,513 | 101,231.60 |

There are different kinds of wetlands found in Assam and district wise type of wetlands along with their areas are summarised in Table 19.7.

Table 19.7 District Wise Wetland Details in Assam.

| Name of the District | Wetland Type | Number | Area (in ha.) |
|---|---|---|---|
| BARPETA | Lake/Pond | 17 | 252.50 |
| | Ox-bow Lake | 30 | 390.00 |
| | Seasonal Waterlogged | 7 | 57.50 |
| | Swamp/Marsh | 42 | 2,583.50 |
| | Reservoir | 1 | 17.50 |
| | Tank | -- | -- |
| BONGAIGAON | Lake/Pond | 25 | 1,365.50 |
| | Ox-bow Lake | 36 | 670.50 |
| | Seasonal Waterlogged | 17 | 237.50 |
| | Swamp/Marsh | 21 | 880.00 |
| | Reservoir | -- | -- |
| | Tank | 1 | 5.00 |
| CACHAR | Lake/Pond | 57 | 1,151.50 |
| | Ox-bow Lake | 29 | 592.50 |
| | Seasonal Waterlogged | 231 | 4,869.50 |
| | Swamp/Marsh | 19 | 564.50 |
| | Reservoir | -- | -- |
| | Tank | 4 | 10.00 |
| DARRANG | Lake/Pond | 11 | 42.50 |
| | Ox-bow Lake | 45 | 655.00 |
| | Seasonal Waterlogged | 22 | 540.50 |
| | Swamp/Marsh | 18 | 2,213.00 |

| | | | |
|---|---|---|---|
| | Reservoir | -- | -- |
| | Tank | 7 | 65.00 |
| DHEMAJI | Lake/Pond | 36 | 460.00 |
| | Ox-bow Lake | 22 | 317.50 |
| | Seasonal Waterlogged | 61 | 1,212.00 |
| | Swamp/Marsh | 17 | 1,963.00 |
| | Reservoir | -- | -- |
| | Tank | 3 | 7.50 |
| DHUBRI | Lake/Pond | 73 | 1,816.50 |
| | Ox-bow Lake | 68 | 1,300.00 |
| | Seasonal Waterlogged | 25 | 425.50 |
| | Swamp/Marsh | 65 | 2,912.50 |
| | Reservoir | -- | -- |
| | Tank | 2 | 5.00 |
| DIBRUGARH | Lake/Pond | 13 | 115.00 |
| | Ox-bow Lake | 36 | 1,502.50 |
| | Seasonal Waterlogged | 23 | 320.00 |
| | Swamp/Marsh | 12 | 807.50 |
| | Reservoir | -- | -- |
| | Tank | 2 | 7.50 |
| GOALPARA | Lake/Pond | 31 | 822.50 |
| | Ox-bow Lake | 32 | 255.00 |
| | Seasonal Waterlogged | 31 | 232.50 |
| | Swamp/Marsh | 68 | 2,515.00 |
| | Reservoir | -- | -- |
| | Tank | 3 | 7.50 |
| GOLAGHAT | Lake/Pond | 113 | 1,707.50 |
| | Ox-bow Lake | 104 | 1,563.00 |
| | Seasonal Waterlogged | 80 | 1,392.00 |
| | Swamp/Marsh | 29 | 760.50 |
| | Reservoir | 2 | 37.50 |
| | Tank | 2 | 6.00 |
| HAILAKANDI | Lake/Pond | 7 | 322.50 |
| | Ox-bow Lake | 4 | 37.50 |
| | Seasonal Waterlogged | 36 | 480.00 |
| | Swamp/Marsh | -- | -- |
| | Reservoir | -- | -- |
| | Tank | -- | -- |
| JORHAT | Lake/Pond | 39 | 1,171.50 |
| | Ox-bow Lake | 24 | 347.50 |
| | Seasonal Waterlogged | 34 | 369.50 |
| | Swamp/Marsh | 9 | 207.50 |
| | Reservoir | -- | -- |
| | Tank | 3 | 12.5 |
| | Lake/Pond | 49 | 1,575.50 |
| | Ox-bow Lake | 65 | 600.00 |

| | | | |
|---|---|---|---|
| KAMRUP | Seasonal Waterlogged | 64 | 822.00 |
| | Swamp/Marsh | 155 | 8,109.50 |
| | Reservoir | 1 | 220.00 |
| | Tank | 18 | 80.00 |
| KARBI-ANGLONG | Lake/Pond | 6 | 100.00 |
| | Ox-bow Lake | 37 | 415.00 |
| | Seasonal Waterlogged | 21 | 259.50 |
| | Swamp/Marsh | 3 | 67.50 |
| | Reservoir | -- | -- |
| | Tank | 10 | 55.00 |
| KARIMGANJ | Lake/Pond | 6 | 95.00 |
| | Ox-bow Lake | 1 | 87.50 |
| | Seasonal Waterlogged | 53 | 4,667.00 |
| | Swamp/Marsh | 10 | 870.00 |
| | Reservoir | -- | -- |
| | Tank | -- | -- |
| KOKRAJHAR | Lake/Pond | 12 | 190.00 |
| | Ox-bow Lake | 38 | 763.40 |
| | Seasonal Waterlogged | 20 | 145.00 |
| | Swamp/Marsh | 14 | 472.50 |
| | Reservoir | -- | -- |
| | Tank | 1 | 7.50 |
| LAKHIMPUR | Lake/Pond | 38 | 382.50 |
| | Ox-bow Lake | 43 | 536.00 |
| | Seasonal Waterlogged | 56 | 660.00 |
| | Swamp/Marsh | 7 | 1,400.00 |
| | Reservoir | -- | -- |
| | Tank | 7 | 55.00 |
| MORIGAON | Lake/Pond | 37 | 935.50 |
| | Ox-bow Lake | 41 | 2,143.00 |
| | Seasonal Waterlogged | 4 | 1,521.00 |
| | Swamp/Marsh | 59 | 7,051.00 |
| | Reservoir | -- | -- |
| | Tank | 3 | 7.50 |
| NAGAON | Lake/Pond | 68 | 2,175.50 |
| | Ox-bow Lake | 71 | 1,746.00 |
| | Seasonal Waterlogged | 138 | 2,559.00 |
| | Swamp/Marsh | 92 | 4,764.50 |
| | Reservoir | -- | -- |
| | Tank | 10 | 50.00 |
| N.C. HILLS | Lake/Pond | -- | -- |
| | Ox-bow Lake | -- | -- |
| | Seasonal Waterlogged | 6 | 187.50 |
| | Swamp/Marsh | -- | -- |
| | Reservoir | 4 | 2,365.00 |
| | Tank | -- | -- |

| | | | |
|---|---|---|---|
| NALBARI | Lake/Pond | 10 | 52.50 |
| | Ox-bow Lake | 29 | 380.00 |
| | Seasonal Waterlogged | -- | -- |
| | Swamp/Marsh | 24 | 1,515.00 |
| | Reservoir | 2 | 22.50 |
| | Tank | 3 | 17.50 |
| SIBSAGAR | Lake/Pond | 16 | 505.00 |
| | Ox-bow Lake | 49 | 500.00 |
| | Seasonal Waterlogged | 16 | 412.50 |
| | Swamp/Marsh | 8 | 450.00 |
| | Reservoir | -- | -- |
| | Tank | 20 | 267.50 |
| SONITPUR | Lake/Pond | 23 | 220.00 |
| | Ox-bow Lake | 35 | 406.00 |
| | Seasonal Waterlogged | 110 | 1,509.00 |
| | Swamp/Marsh | 22 | 1,432.50 |
| | Reservoir | -- | -- |
| | Tank | 16 | 83.50 |
| TINSUKIA | Lake/Pond | 3 | 35.00 |
| | Ox-bow Lake | 22 | 252.50 |
| | Seasonal Waterlogged | 31 | 552.50 |
| | Swamp/Marsh | 18 | 1,892.50 |
| | Reservoir | -- | -- |
| | Tank | -- | -- |

The destruction of the Brahmaputra valley wetland system started with the arrival of the water hyacinth from Central America more than a century ago. Extensive growth of this fast growing weed can cut off sun light from the micro flora and also produces faster eutrophication by slowing down water current and depositing debris at the bottom. The second phase of enhanced eutrophication took place with the raising of earthen bunds along the banks of almost the entire length of the river and many of its tributaries after the 1950 earthquake. These artificial levees cut off, to a great extent, the periodic flushing out of the wetlands by the monsoon flood. The third and the final onslaught on the wetlands took place with the arrival of human settlers in the sand bars and the minor riverine islands, mostly in the lower parts of Assam. This has turned the wetlands into rich agricultural zones and vegetables but totally denuded of wild-life. With vanished wetlands, gone also are the rich supplies of fish, a compulsory item of the Assamese menu and a good source of protein for rural masses. In spite of the presence of the mighty Brahmaputra, its numerous tributaries, and the large number of wetlands. Assam today imports 20,000 (0.2 lakh) tonnes of fish annually to satisfy the domestic market. Out of this, 0.14 lakh tonnes is consumed in Assam. The total fish production from Assam's wetlands is 1.55 lakh tonnes per year. Thus, a total of 1.69 lakh tonnes of fish is supplied from imports as well as local wetlands. The total demand for fish in the state, on the other hand, is estimated at 2.21 lakh tonnes per year, 6.68 per cent of which is met by imports from other states source: Directorate of Fisheries, Government of Assam. As a result there is a deficit of 0.52 lakh tonnes of fish every year and consequently the price of fish has rocketed up to such an extent that the poor man simply cannot afford to buy it even once a week. The production potential of wetlands in the state is estimated at 400--500kg ha year after development. The progressively short supply of fish in Assam is a direct consequence of mismanagement and neglect of its wetland ecosystems.

In this gloomy picture a few examples of healthy wetland systems do exist. The Kaziranga National Park, the last refuge of the great Indian one–horned rhino, is such an example. Another good example is the Panidihing wetland situated on the southern bank of the river, between the tributaries of Disang and Buridihing, where the splendour of thousands of migratory waterfowl, the indicator of a healthy ecosystem, can be seen every winter between September and April. But unlike Kaziranga this wetland is humming with human activities. The rhinos and the other animals like Tigers and elephants have already gone and a good number of water birds are fighting a losing battle against the man in this wetland. There is an imperative need to conserve these wetlands and protect their unique biodiversity. If properly managed, the wetlands are going to be a source of immense wealth for the state of Assam leading also to enrichment of the quality of its environment.

Although Assam has several thousands of family-owned small size tanks, which is less than 2.5 ha, the satellite study done by ARSAC could not incorporate the number of small size tanks available in Assam owing to scale factor.

## 19.7 Concluding Remarks

The Brahmaputra basin is endowed with a rich variety of ecosystems and flora and fauna. Its wildlife is unique and legendary in the world. The wetlands are primarily freshwater and riverine. However, they are being destroyed indiscriminately and the governmental agencies are paying little attention. It will be a human tragedy to allow them to vanish. If the north eastern region of India has to prosper, its ecosystems must be preserved and fostered.

## Bibliography

Barthakur, K. (1986). Proceedings of the Symposium on Harnessing the River Brahmaputra, Assam Science Society, pp. 6-16.

Mahanta, C. and Subramanian, V. (1993). Particle Size Distribution, Mineralogy and heavy metal composition of the suspended sediments in the Brahmaputra River.

Champion and Sethe (1968).

# CHAPTER 20

# WETLANDS AND THEIR ROLE IN GEO-ECOLOGICAL ENVIRONMENT

P. SHARMA

For conservation of biodiversity, production of fish and other aquatic resources, flood control, and ecological balance, wetlands play a fundamental role. Unfortunately, wetlands have been neglected in the Brahmaputra River basin and as a result they are degrading. This chapter presents a discussion of wetlands in the basin and their role in geo-ecological environment.

## 20.1 Introduction

The term 'wetland' is used for diverse habitats in different climatic zones of the earth. Accordingly, wetlands have been defined variously from a simple working definition to highly technical ones. A few of such important definitions are given by the agencies like the U.S. Fish and Wildlife Service, Environmental Protection Agency (EPA), International Union for the Conservation of Nature (IUCN), International Biosphere Programme (IBP), etc.. Here in this study the modified IUCN definition has been taken into consideration in which wetlands mean 'all the submerged or water saturated lands, natural or man made, inland or coastal, permanent or temporary, static or dynamic, or vegetated or non vegetated, which necessarily have a land-water interface'.

A truly riverine state amidst the hilly areas of the north eastern region, Assam is drained mainly by two river systems, the Brahmaputra and the Barrak. The Brahmaputra besides being the major river of the valley is the only large drainage outlet of the entire north east India. About 254,000 sq. km of the catchment area is drained at the section where it enters Indian Territory and an area of 487,000 $km^2$ where it leaves India to enter Bangladesh. High seasonal rainfall and high humidity are the main characteristics of the climate of Brahmaputra valley. The prevailing patterns of weather and climate have a strong bearing on the occurrence and intensity of flood hazard in this region. Although the average annual rainfall in the Brahmaputra basin is 230 cm, there is a marked variation in its distribution. For example, the rainfall in the north eastern part is 410 cm while in Kopili basin of the south central part accounts only 173 cm. Rainfall at the foothills of the Himalayas is normally high around 500 cm. Again, monsoon rains from June to September account for about 70 percent of the annual total.

In the flood plains of this river are found a large number of natural lakes and swampy areas. The lake-like water bodies are locally known as beels, whilst the swampy areas are variously called as 'Jalah', Doloni', 'Pitoni', Doba', 'Hola', etc. In certain areas, the beels are often called as 'haor' or 'gadeng'. All these natural features may be included



*V.P. Singh et al. (eds.), The Brahmaputra Basin Water Resources*, 366-375.

under the comprehensive term wetlands. Normally beels represent a vast sheet of water with varying shape, size and depth. Some of them are partly or fully infested with floating mats of aquatic plants. Reeds, grasses and bushes comprise the common natural vegetation in the shore areas of beels. The swampy and marshy areas, such as jalah, pitoni, and doloni, are shallower than the beels. They are normally developed close to river banks and on ageing shallow beels. Some of them are very extensive, although most are smaller in size. The wetlands of the state are the home of a large variety of fauna and flora, some of which are of extremely rare and endangered type while others are of great ecological or economic value.

The beels of Assam are traditionally used as natural fisheries which are potentially rich in fish and other aquatic resources. The genesis and development of beels are closely related to the geomorphic and tectonic history of the region, hydrologic behaviour of rivers and the prolonged human use of the adjoining lands.

## 20.2 Distribution of Wetlands

As many as 3,513 number of wetlands have been identified in the state by the Assam Remote Sensing Application Centre (ARSAC) using satellite imageries with 1:50,000 scale and topographical maps. The present shape, size, and area of wetlands are delineated from satellite images which are supplemented by field observations. These wetlands have covered an area of 1,012.31 sq. km which accounts for 1.29 percent of the total geographical area of the state. Form the data presented here (Table 20.1) it is revealed that natural wetlands account the lion share, i.e., 96.66 percent of the total geographical area under wetland category. Of this around 43 percent of the area is occupied by swamps and marshes. The Ox-bow lakes which play a major role in production of fishes and other aquatic economic resources are 861 in number and cover around 15 percent of the total wetland area.

Table 20.1 Distribution of different types of wetlands in Assam.

| Wetland Type | No. | Area (ha) | Percent |
|---|---|---|---|
| **NATURAL** | | | |
| Lake/pond | 690 | 15,494.00 | 15.31 |
| Ox-bow Lake / Cut-off Meander | 861 | 15,460.60 | 15.28 |
| Waterlogged (Seasonal) | 1,125 | 23,431.50 | 23.15 |
| Swamp / Marsh | 712 | 43,433.50 | 42.91 |
| Total | 3,388 | 97,819.60 | 96.65 |
| **MAN-MADE** | | | |
| Reservoirs | 10 | 2,662.50 | 2.60 |
| Tanks | 115 | 749.50 | 0.74 |
| Total | 125 | 3,412.00 | 3.34 |
| **G. Total** | 3,513 | 101,231.60 | 99.99 |

(Source: ARSAC Report on wetlands of Assam, 1997)

The state has 23 districts of which two districts, North Cacher and Karbi Anglong, are full of hills and are uplands where wetlands are very few in number. The Nagaon district has highest number, 379 wetlands of sizes greater than 2.25 hectares. On the other hand, the Morigaon district has a maximum area under wetlands, i.e., 11,658 ha. As a whole, the districts falling in the Brahmaputra valley account for 2,969 numbers of wetlands and cover an area over 84,000 hectares.

Table 20.2 District –wise distribution of wetlands in Assam.

| District | Number | Area (ha) |
|---|---|---|
| Barpeta | 97 | 3,301.00 |
| Bongaigaon | 100 | 3,158.50 |
| Cacher | 340 | 7,188.00 |
| Darrang | 103 | 3,515.50 |
| Dhemaji | 139 | 3,960.00 |
| Dhubri | 233 | 6,459.70 |
| Dibrugarh | 86 | 2,752.50 |
| Goalpara | 165 | 3,832.50 |
| Golaghat | 330 | 5,467.50 |
| Hailakandi | 47 | 840.00 |
| Jorhat | 109 | 2,108.50 |
| Kamrup | 352 | 11,407.00 |
| Karbi Anglong | 77 | 397.00 |
| Karimaganj | 70 | 5,719.50 |
| Kokrajhar | 85 | 1,578.40 |
| Lakhimpur | 151 | 3,033.50 |
| Morigaon | 183 | 11,658.00 |
| Negaon | 379 | 11,295.50 |
| N.C.Hills | 10 | 2,552.50 |
| Nalbari | 68 | 1,988.00 |
| Sibsagar | 109 | 2,135.00 |
| Sonitpur | 206 | 3,651.00 |
| Tinsukia | 74 | 2,732.50 |
| Total | 3,513 | 101,231.60 |

Source: ARSAC Report on wetlands of Assam (1997).

## 20.3 Formation of Wetlands

Formation of beels and other wetlands in Assam depends both on geomorphological processes as well as environmental changes. Environmental changes are not necessarily natural, rather such a change may take place as a result of the human interference at various levels. Broadly, the formation processes of all category wetlands are divided into two (i) autogenic and (ii) allogenic.

The autogenic process is related to the river regime, i.e., channel migration, cutoff, loop formation, etc.. On the other hand, allogenic processes include various types of human interference. The wetlands in the state are formed mainly by the hydraulic action of rivers. However, a few wetlands, such as Sundubi of Kamrup district, were formed by the effect of an earthquake. Sundubi was formed by the subsidence of an extensive forest area during the great earthquake of 1897. Not only that, some reports of channel changes are recorded in connection with the last two major earthquakes of 1897 and 1950 for which the wetland scenario in the state changed to a great extent. The entire Brahmaputra valley in Assam particularly from Dibrugarh to Dhubri is a flat plain with a very gentle slope. In this part the average slope is less than 10 cm per km. As soon as the Brahmaputra River and its tributaries debouch into the plain after their sudden drop from the lofty mountains of Arunachal Pradesh, Nagaland, Manipur, and Meghalaya they become sluggish and develop a meandering pattern. Consequently they lose their stability. As a result many ox-bow beels and cut off meanders have formed in this part of the region (Sharma and Goswami, 1993).

The flood plans of the Brahmaputra, as well as other tributaries, in Assam possess levees or naturally raised banks. However, in some rivers these features are absent. These levees are the main cause of back swamp development. Most of the swamps and many shallow beels in the state are formed by this process. These swamps and beels generally obstruct entry of small tributaries to the main stream.

The Brahmaputra River is characterized by an interlacing arrangement of channels and bars. Here sandbars or chars are important geomorphic features. These are generally transient landforms exposed during low flows and formed mainly of sand, sometimes partly covered with vegetation. Some chars migrate downstream during high flows of the flood season. This change in location of channels and sandbars is responsible for formation of many beels and swamps. Most of the beels and swamps in Kaziranga, Laokhowa, and Pabitara reserved forests are formed either a result of such migration or as back swamps of rivers.

The allogenic changes that take place mainly as a result of human activities are responsible for formation of many beels and swamps in the state. Major changes are observed in the river regime after the commissioning of regulatory works on rivers. Construction of embankments, bunds, etc. for flood control and irrigation causes drastic changes in the river course, thereby altering the wetland environment in recent years. In addition, the decreasing forest cover in the entire northeast is also significantly affecting the flow regime and the denudation pattern of rivers, accelerating the siltation rate of the river as well as of beels and swamps.

The channel shortening owing to meander cut off or lengthening owing to intensification of meander loops causes a dramatic change in the slope and flow pattern of rivers. As a result the formation of new beels and degeneration of the existing ones are induced. The construction of bunds at the mouth of rivers, particularly at the mouth of Kalong an important river in central Assam, causes a complete transformation of river morphology. A large number of wetlands have been formed after the blockade of the mouth of the Kalong River. All these allogenic processes change the pattern of beels, swamps and other wetlands formed by the fluvial action of rivers.

A large number of channels criss-cross the flood plains to irrigate the agricultural land. The lifting of water from beels to agricultural fields is a common sight throughout the state. Innumerable roads, bridges, railways, irrigation canals, and embankments interrupt the natural flow of water in various parts. In the low flow stage, the bankful stage, and in extreme flood situations, the natural flow of waters is blocked by these structures. As a result the formation process of beels and swamps become more complicated.

From the paleo-channel study of some parts of the Brahmaputra, Kopili, and Kalong basins with the help of topographical maps and satellite images it is observed that the course of the river has changed in many places during recent years. Thus in certain places many new ox-bow lakes have been formed, whilst other old dead channels get rejuvenated owing tothe change of the channel. For example from 1905--06 the Kopili River changed its course almost throughout its entire length, especially in its lower course (Figure 20.1). From Kharikhana which is located at the state boundary, to Kampur located further downstream, as many as four important beels have developed in the recent past by the channel change of the Kopili River. The effect of channel change is observed to be more pronounced in the case of feeder channels which could be said as the lifeline of wetlands.

## 20.4 Degeneration of Wetlands

In general the beels and swamps provide an ideal habitat for feeding, resting and breeding of a large number of fishes, birds and various aquatic animals. But it has been observed that most of the wetlands of Assam are now being degenerated. The main causative factors responsible for degeneration of beels and swamps are (i) inorganic bottom deposits, (ii) organic bottom deposits, (iii) clockade of feeder channels, (iv) unscientific construction of engineering structures, (v) encroachment and cultivation in marginal areas, (vi) release of industrial and urban wastes, and (viii) fishing methods and gears.

Inorganic bottom deposits include the accumulation of sand, silt, clay, pebbles and other inorganic materials. High seasonal rainfall, active seismocity, rapid channel aggradation, massive deforestation, intense landuse and ad-hoc type temporary measures of flood control are the major factors of inorganic bottom deposits.

Most of the beels in Assam are infested with water hyacinths. The growth of this aquatic plant is normally very high owing tohigh rainfall, high humidity, moderately high temperature and low sunshine. Except a little amount grazed by some herbivorous animals most of these get deposited at the bottom after their death. This process of accumulation of the organic deposit over the years leads to degeneration of wetlands.

The blockade of feeder channels is another cause of degeneration of wetlands. Usually the young and brood fishes enter the wetlands from river through the feeder channels. But it has been observed that owing tothe accumulation of petrified substances over the years, these feeder channels have become shallow and hamper the autostocking process. The blockade of feeder channels owing to construction of roads, railways, embankments, etc. accelerates the degeneration of feeder channels as well as the wetlands proper.

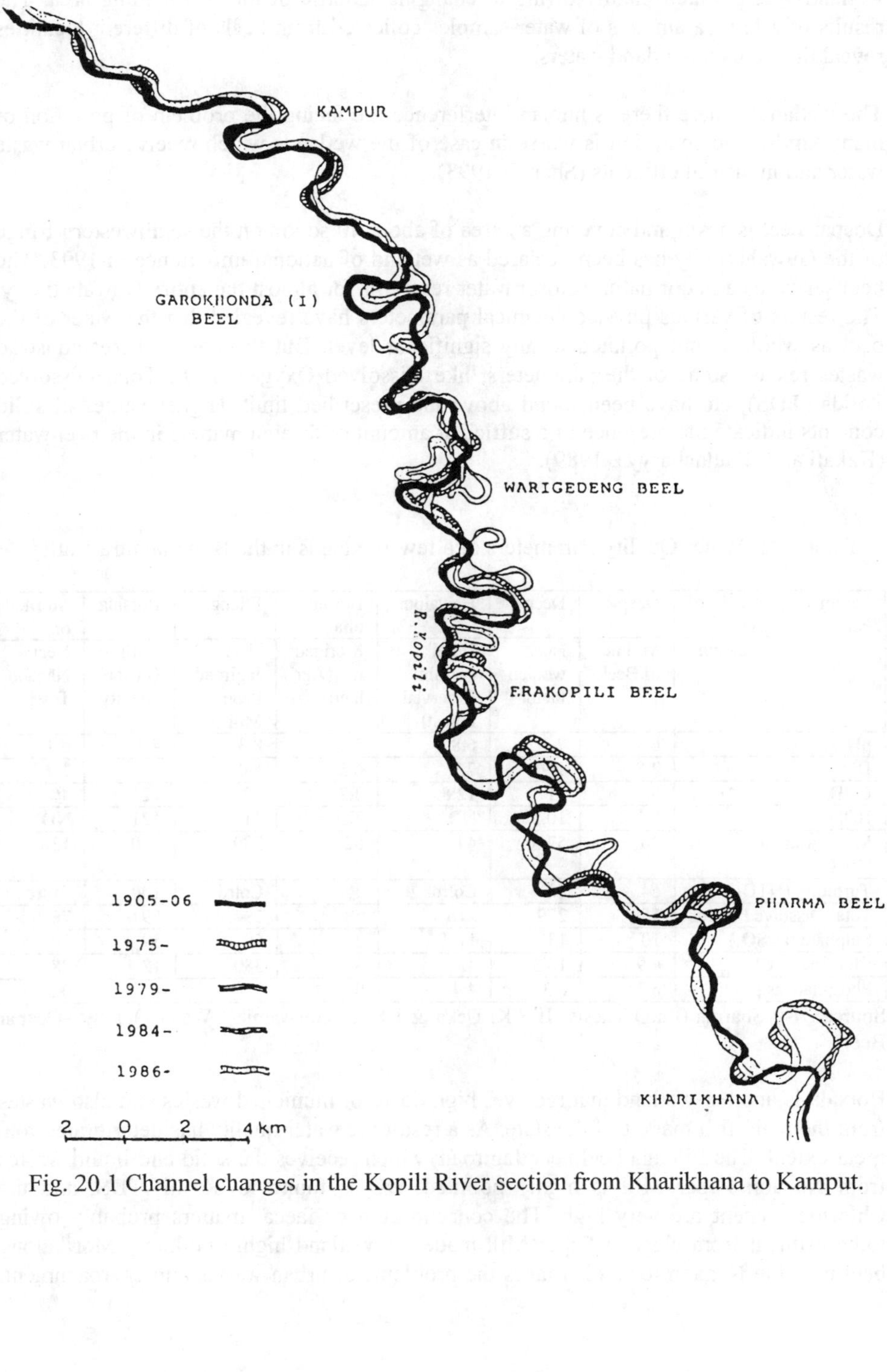


Fig. 20.1 Channel changes in the Kopili River section from Kharikhana to Kamput.

Owing to a tremendous increase of population in the recent years many of the wetlands are now being encroached for agricultural and other purposes. As a result it reduces not only the wetland areas but also make the wetlands less productive. Many of the wetlands get polluted mainly owing to changing landuse of the surrounding area. The results of a few parameters of water samples collected from beels of different localities reveal the status of wetland waters.

The wetlands where there is human interference are facing the problem of pollution of many kinds. The condition is worse in case of the wetlands which receive urban waste water and industrial effluents (Sharma 1993).

Deepar beel is a wetland covering an area of about 40 sq km on the southwestern fringe of the Guwahati city has been declared as wetland of national importance in 1993. The beel serves as a major natural storm water reservoir for almost the entire Guwahati city. The results of various physico chemical parameters have revealed that the water of the beel as whole is not polluted to any significant level. But the areas where industrial wastes release some of the parameters, like Dissolved Oxygen (DO), Total Dissolved Solids (TDS), etc have been found above the prescribed limit. Higher values of solid contents indicate the presence of a sufficient amount of foreign matters in the beel water (Kakati and Bhattacharyya, 1989).

Table 20.3 Water Quality Parameters of a few Wetlands in the Brahmaputra Valley.

| Parameters | Name of Beels | Deepar | Deepar | Hawaipur | Bormonoha | Ellenga | Borsala | Morikalong |
|---|---|---|---|---|---|---|---|---|
| | Location | Middle of Beel | Near woolen factory | Hojai (within Reserved Forest) | Morigaon (Agri land) | Near Jagiroad Paper Mill | Within Guwahati City | Near Nagaon Town |
| pH | | 6.9 | 7.3 | 6.8 | 7.2 | 8.3 | 4.2 | 6.1 |
| DO | | 2.8 | 1.9 | 7.3 | 6.5 | Nil | 3 | 1.2 |
| COD | | 58 | 71 | 12.8 | 3.2 | 496 | 121 | 40 |
| BOD | | 8.1 | 10.3 | ND | ND | 115 | 121 | ND |
| M Alkalinity as $CaCO_3$ | | 40 | 57.5 | 44 | 62 | 280 | 160 | 138 |
| Turbidity (NTU) | | 64 | 78 | Colrd | 40 | Colrd | 100 | Colrd |
| Total Dissolved Solids | | 159 | 279 | 216 | 58 | 194 | 59 | 76 |
| Sulphate (as $SO_4$) | | 10 | 13 | 11 | 2 | 50 | 39 | 10 |
| Chloride as Cl | | 4.9 | 10.3 | 4. | 4 | 380 | 12.4 | 28 |
| Phosphate as p | | 6.2 | 6.3 | 4.4 | 0.1 | 2.9 | - | 0.8 |

Source: I-P Sharma (Ph.D Thesis) II-S.K. Deka & DS. C. Goswami – 'Water Quality –Deepar Beel'.

Borsala is another wetland that receives high doses of municipal wastes and also wastes from the main fish market of the state. As a result the water quality has deteriorated to a great extent. The Ellenga beel near Jagiroad, which receives the solid and liquid wastes from Nagaon Paper Mill is highly polluted. DO is found to be nil; TDS, and the Chloride content are very high. The concentration of faecal matters probably owing tothe effluent from Nagaon Paper Mill made the wetland highly polluted. Morikalong beel near the Nagaon town also faces the problems of urban wastes and encroachment.

Morikalong beel, which once could supply half of the total fish requirement of Nagaon town, now producing a meager amount of fishes owing to deterioration in the quality of beel water and undesired human interferences.

During the pre-monsoon season the DO value was found to be only 1.2. But there are many wetlands in the Brahmaputra Valley, where the human interference is very little, having good quality water. The results of water quality parameters of two wetlands, Bormonoha and Howaipur, have been presented here, where the natural quality of water is still satisfactory. Hawaipur is one of these two located amidst a reserve forest while the other one, Bormonoha lies amidst the agricultural land. Some changes in water quality in the recent years have been observed, probably owing to the changing land use pattern of the surrounding areas of the Bormonoha beel.

Apart from deteriorating quality of water, in most of the wetlands over fishing, the use of nets with small mesh size, traditional gear and methods, poaching and hunting cause further deterioration of these precious resources of the valley.

## 20.5 Measures for Wetland Conservation and Management

In order to stop the present trend of degeneration of wetlands and improve their productive potential, the following few administrative, socio-economic, and ecological measures are suggested.

- It is necessary to remove the weeds from beels on a regular basis. Application of an insect, viz., Flea beetle (Agasiches Hydrophilla), some fungi like Chercospora rodamanii, Achromonium Zonatum, and Uredes eichhorria and to encourage compost manure from the water hyacinth may be taken as alternative measures in controlling the water hyacinth in many of the wetlands (Chakravarty, 2000).
- Indiscriminate construction of engineering structures across wetlands by various departments for different purposes should be stopped.
- Wetland boundaries should be clearly demarcated and encroachment should be stopped. The unregistered beels should be brought into the registered category so the some of the development measures could be taken.
- The use of the nets of undersized mesh for fishing should be totally banned, particularly during the months from May to August. The community poaching which is common in almost all the beels is found to be very harmful. Therefore this practice should be stopped without delay.
- The wetlands in Assam are at present controlled by four different departments of the state government. Overlapping in departmental management creates various problems. Therefore, all the wetlands should be brought under a single management organization.
- The beel water of Assam is comparatively more productive than the water of the reservoirs of other states of the country. Moreover, it has already been proved that beels of less than 10 hectares can easily be converted into culturable fisheries. As such, there is enough potentiality to develop the beels

and swamps below this size, which remain grossly underutilised at present, into productive fisheries.

- All sorts of effluents from industries and liquid and solid waste discharge from municipality areas should be restricted from entering wetlands.
- As the siltation of beel bottoms is becoming a serious problem leading to the degeneration of these water bodies, the sediments should be trapped in the catchment areas through afforestation and by constructing contour bunds parallel to the shoreline with the accumulated silt in the littoral zones.

## 20.6 Conclusions

The wetlands of Assam are extremely significant from the point of view of bio-diversity conservation, production of fish and other aquatic resources and controlling the intensity of flood during rainy season. Wetlands improve water quality through the processes of cleansing and detoxification of the polluted water. They help in groundwater discharge, recharge, and water storage The large scale import of fish in recent years from states like Andhra Pradesh has been a great drain on the state economy. If the beels and swamps in Assam are properly managed it may bring a revolutionary change in the agro-industrial sector in both the river valleys. Wetlands will also help in maintaining ecological balance as well as environmental quality in the entire region. Many of the wetlands can be easily put on the tourist map of the country, which has not been done so far. For overall development and management of these precious wetlands, co-ordination and co-operation amongst the various concerned departments is most vital without which deterioration of wetlands cannot be checked, nor their potential properly utilized.

## Bibliography

Baruah B.K and Das M. (1997), Effect of Paper Mill Effiuent on Plankton Population of Wetland. Environment & Ecology 15(4), pp. 707-777.

Barua P., Sharma P. and Goswami D.C. (1998), A Database for Wetlands of Assam; A study Using remote Sensing Technique: Proceedings NAGI(En Region) National Conference on Environment and sustainable Development, pp. 27-34.

Chakraborty D. N. (2000), Utilization and Management of Wetlands in Assam --A Challenge Ahead: A paper presented in seminar on wetland development and Management in Assam. Organized by State Landuse Board and NEOLAND Technologies, Guwahati.

Deka S. K. and Goswami D.C. (1993), Water Quality and sediment characteristics of wetlands: A Geo-environmental Study of Deepar Beel, Assam. Gauhati University J. Sci. XXXII pp. 93-105.

Goswami M (1985), Limnological Investigation of tectonic Lake of Assam, India and their Bearing of Fish Production, Unpublished Ph. D. Thesis Gauhati University pp. 305-95.

Jhingran A.G.V (1987), Ecology and Management of Beels in Assam : A case study of Dhir Beel, compendium, Workshop on development of Beel Fishery in Assam, Assam Agricultural University, Khanapara Campur, pp. 16-28.

Lahon B (1983), Limnology and fisheries of Some Commercial Beels of Assam, Unpub. Ph. D. thesis, Gauhati University, Guwahati.

Narayan, A.K. (1991); Overall Strategy Being adopted at the Government Level to Encourage Conservation of important Wetlands of India; Wetlands of India.

Sharma P. and Goswami D.C. (1993) Geo-environmental Study of Wetland (Beel) in the Nagaon and Morigaon Districts of Assam Using Satellite Data. In the Proceedings of National Symposium on remote Sensing Application for Resource Management with Special Emphasis on North-Eastern

Region, ASTE Council. Assam pp. 232-40.
Geo-Environmental Status of Selected Beels (wetlands) (1995). In Indian Geomorphology Vol. II (S.R. Jog, ed) Rawat Publication, New Delhi pp. 237-46.
Sharma P. (1993) Geo-Ecological study of Beels and Swamps in Nagaon and Morigaon Districts, Assam, Unpub. Ph. D. thesis Gauhati University.

# CHAPTER 21

# WATER QUALITY, MINERAL TRANSPORT AND SEDIMENT BIOGEOCHEMISTRY

C. MAHANTA AND V. SUBRAMANIAN

The chemistry of the Brahmaputra River water changes during its course but the spatial changes of several water quality parameters do not exhibit a consistent spatial pattern. During dry months chemical weathering is predominant whilst physical weathering is so during the monsoon. This chapter discusses water quality, mineral transport and sediment biogeochemistry of the Brahmaputra River water.

## 21.1 Introduction

The rivers of Asia and the Pacific Islands contribute 35% of the river flow and 50%--70% of the total sediment delivered by all rivers to the oceans (Milliman and Meade, 1983; Hu and others, 1998). However, very few Asian rivers have been investigated in detail so far, the Brahmaputra perhaps the least. Only brief reports have appeared about the part of the river falling in India, based on very few samples (often only one or two) of the water chemistry (Subramanian, 1979; Hu Ming-Hui and others, 1982; Sarin and Krishnaswami, 1984; Sarin and others, 1989), sediment characteristics (Subramanian, 1980; Goswami, 1985) and heavy metal concentrations in sediments (Subramanian and others, 1987). In recent years Datta and others (1997, 1998, 1999) have studied the texture, mineralogy and sediment biogeochemistry of the Ganges--Brahmaputra--Meghna system as a whole in Bangladesh.

The knowledge of elemental transport in a river like the Brahmaputra is crucial for quantifying the global nutrient and trace metal budgets. Hardly anything is known about the nutrient budget in the Brahmaputra River basin. Linkages between main stem, flood plain and tributary processes have not been established clearly. Large gaps exist in the understanding of how nutrients like C, N, P form, in what quantities, to what depths, and with what mobility and availability. Therefore having an understanding of the erosional flux and water--sediment exchange in the flood plains of the Brahmaputra and transformation between solute and the suspended load is of considerable importance. The findings presented in what follows are based on a preliminary study dealing with the entire part of the basin lying within the geographical boundary of India and depicts seasonal and spatial variations of several biogeochemical parameters.



*V.P. Singh et al. (eds.), The Brahmaputra Basin Water Resources*, 376-400.

## 21.2 The Brahmaputra River Basin

With a total length of over 2,900 km and a drainage basin covering an area of $1.6 \times 10^6$ sq. km., the Brahmaputra is considered to be the fifth largest river in terms of the average discharge at the mouth, with a flow of 19,830 $m^3s^{-1}$ (Goswami, 1985). The Brahmaputra receives extensive sediment inputs from at least forty major tributaries from both sides. As a result the main channel has undergone frequent diversions and migration. Chemically, too, the sediment and nutrient loads have been continuously changing. The average annual suspended load of the Brahmaputra at Pandu, Assam is more than 402 million tonnes $yr^{-1}$ (Goswami, 1985), and considering the increasing human impacts this is likely to rise in future. As increasing human settlement in the watershed is converting more and more soil to sediment by erosion, it is high time that at least an initial understanding of the direct and indirect human impacts on the biogeochemistry of this large river is developed.

The Brahmaputra is characterized by high rates of basin erosion, high sediment yield, and frequent channel changes (Goswami 1988). The geology of the Tibetan Himalayas where the river has originated is poorly known. Hu Ming-Hui and others (1982) have described the dominant lithology of this region to be of reduced shales, gneisses, carbonates, and perhaps some volcanic rocks at a few places. Inside India, whilst the upper reaches of the river basin predominantly drain the Siwalik sediments including limestones, dolomites and calcareous shales, the drainage lithology downstream is composed of gneisses, and schists of the Precambrian age (Figure 21.1a). The Brahmaputra valley in Assam is underlain by a 200--300 m thickness of recent alluvium. Oldham (1900), Evans (1964) and many other authors have suggested a tectonic origin of the Brahmaputra basin.

## 21.3 Methodology

Water samples were collected from the main channel during June--July 1992 and January-February '94. Sampling locations are shown in Figure 21.1b. Altogether twenty six samples of water were collected in 1 litre and 5 litres wide mouthed polyethylene bottles along the middle part of the river by holding the bottles against the flow at about mid-depth. Considering the great width but comparatively shallow depth of the river, it was, however, difficult to obtain a single well representative sample, because complete lateral mixing of tributary inputs (which is significant) might not have taken place at each sampling location. Sampling locations were selected at one site before a major tributary joins the main river, and after the confluence of a major tributary, taking both the banks of the main river into account. Besides, the locations of the sampling stations were guided by the presence of a major tributary, wherever possible, also by the location of major industry and pathway of major effluent channels.

Suspended sediments were derived from 5 litres composite water samples by settling and decanting. Water samples for analysis of major ion chemistry were obtained by filtering each 1 litre sample through a 0.45 μm Millipore membrane filter. 13 bed sediment samples were collected during June--July 1992 from the same locations by

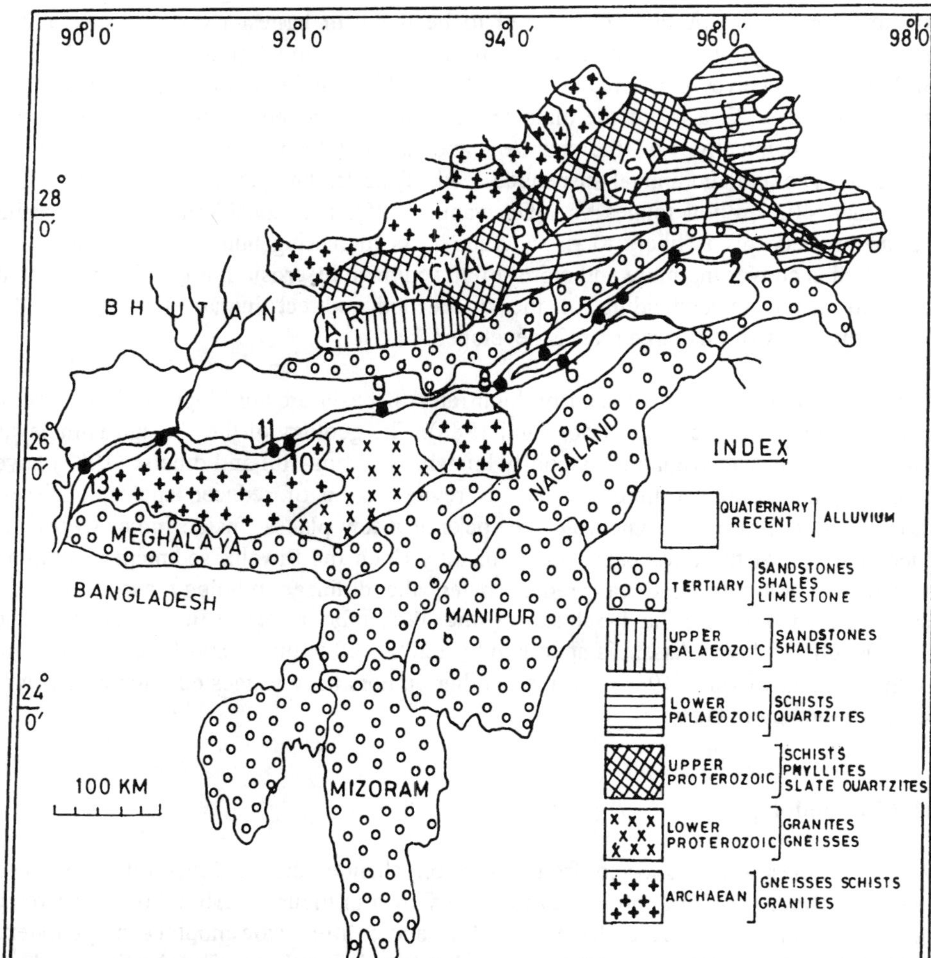


Fig. 21.1a Geological map of the Brahamaputra basin in India.

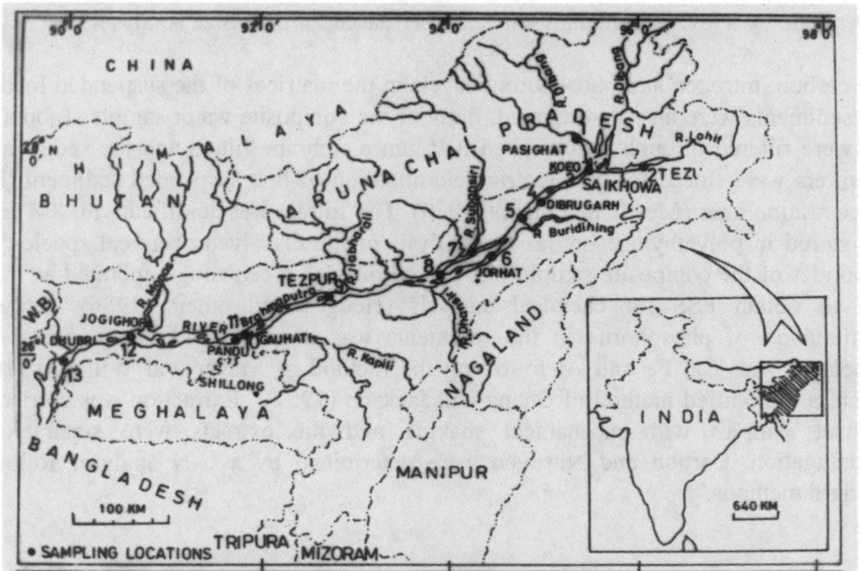

FIG.1A MAP OF THE STUDY AREA WITH SAMPLING LOCATIONS

Fig. 21.1b Map of the Brahamaputra basin in India with sampling locations.

scooping freshly deposited materials with a plastic spade. The materials were transferred to plastic bags taking care not to lose the fines. After measuring pH, EC and alkalinity of the water samples in the field, all the water, bed, and suspended sediment samples were kept refrigerated at 2° to 4°C until they were processed for laboratory analyses.

Standard analytical procedures were followed for water analyses ($Cl^-$ by Mohr's titration, $F^-$ by Corning pH ion meter, $SO_4^{2-}$ by Barium perchlorate titration and phosphate and silicate by calorimetric technique, using a Cecil double beam spectrophotometer). $Ca^{2+}$ and $Mg^{2+}$ were analysed in absorption mode and $Na^+$ and $K^+$ in emission mode, using an atomic absorption spectrophotometer (GBC 901). $SiO_2$ and $Al_2O_3$ in suspended and bed sediments were analysed by preparing 'solution A' (Shapiro and Brannock 1962). Other cations in the suspended and bed sediments were determined by AAS after their complete digestion (by adding 0.5 ml of $HNO_3$, 1.5 ml of HCL, and 5 ml HF to 0.1 g of sediment and keeping the mixtures in a teflon bomb inside an oven at 100°C for 90 minutes) to solution formation. Low concentrations (in ppb level) were analysed by Graphite Furnace. Accuracy and precision of the measurements were checked by comparing with values obtained for various USGS rock standards (MAG-1, SCO-1, SDC-1, G-2) (Flanagan, 1976). Slides of suspended sediments were prepared by the drop on slide technique (Gibbs, 1967). After being glycolated they were run on a

Philips X-ray diffractometer with Cu-K$\alpha$ radiation using an Ni filter. Mineral identification and estimation of their abundance were done by the methods of Biscaye (1965) and Carrol (1970). The grain size distribution of suspended sediments was determined by a FRITSCH Analysette-22 laser based Particle Size Analyser.

The carbon, nitrogen and phosphorous levels in the matrices of the suspended load and bed sediments were also investigated. Parts of the composite water samples (about 300 ml) were filtered through pre-weighed 0.45 um membrane filters and the sediments in the filters were stored for gravimetric determination of fine suspended sediment (FSS) concentration later (Meade and others, 1979). The filtrate was acidified with 8M $HNO_3$ and stored in polyethylene bottles for analysis of the dissolved chemical species. The remainder of the composite sample was passed through a centrifuge operated at 15,000 rpm to obtain FSS for chemical analysis (Hedges and others, 1986). Inorganic fractionation of phosphorus in the sediments was done to find the amounts of P associated with Ca, Fe and Al following the method of Mayer and Williams (1981) which is a modified method of Chang and Jackson (1957). Extraction was carried out on wet samples with mechanical shaking and the extracts were separated by centrifugation. Carbon and Nitrogen were determined by a C-N analyser following standard methods.

## 21.4 Water and Sediment Biogeochemistry

### 21.4.1 MAJOR ION CHEMISTRY

Table 21.1 shows the results of the water analyses. The water of the Brahmaputra is moderately alkaline (average pH 7.5). The pH values do not show pronounced seasonal or spatial change. Electrical Conductivity (EC) during monsoon ranges between 121 $\mu s.cm^{-1}$ and 139 $\mu s.cm^{-1}$ whilst it is 139 $\mu s.cm^{-1}$ to 172 $\mu s.cm^{-1}$ during the low flow period of winter season. Correspondingly, the ionic strength increases from about 0.002 to 0.003 during the dry season indicating an increased concentration of major ions owing toevaporation and influence of groundwater.

The $Cl^{-1}$ concentrations in winter (11-23 ppm) are consistently higher relative to those during monsoon (4-6 ppm) which suggests a dilution effect during high discharge period. Higher $Cl^{-1}$ values during non-monsoon may also represent enrichment from the reported halite deposits in the upper Tibetan reaches of the Brahmaputra (Hu Ming-Hui and others, 1982). This observation is supported by a similar trend shown by the Na values.

Bicarbonate in the Brahmaputra is variable between 48 ppm and 62 ppm. Carbonate dissolution as well as primary and secondary mineral weathering can be the sources of $HCO_3$. Following Raymahasay's (1987) method it was calculated that on average 40% bicarbonates in the water of the Brahmaputra comes from the carbonate minerals of the basin and the rest from the silicates. Based on the $HCO_3$ content and pH, the $P_{co2}$ values

Table 21.1 Chemical composition of water of the Brahmaputra River

(Units : $mg.l^{-1}$, except $p^H$ and cond; Conductivity in $\mu S.cm^{-1}$)

| Sample No | Sampling Period | $p^H$ | EC | $Cl^-$ | $HCO_3^{-1}$ | $SO^4$ | $PO_{43}$ | $F^-$ | $H_4SiO_4$ | $Na^+$ | $K^+$ | $Ca^{2+}$ | $Mg^{2}{}_{+}$ | TDS | dE- | dE+ | TSM |
|---|---|---|---|---|---|---|---|---|---|---|---|---|---|---|---|---|---|
| 1. | m | 7.6 | 124 | 4 | 54 | 2.3 | 0.09 | 0.2 | 8 | 2 | 1 | 12 | 5 | 89 | 1.06 | 1.12 | 135 |
| | d | 7.2 | 139 | 13 | 48 | 3.6 | 0.07 | 0.1 | 4 | 8 | 2 | 14 | 5 | 98 | 1.30 | 1.51 | 5 |
| 2 | m | 7.7 | 124 | 4 | 57 | 1.5 | 0.05 | 0.1 | 11 | 2 | 2 | 14 | 4 | 87 | 1.08 | 1.17 | 124 |
| | d | 7.6 | 156 | 19 | 52 | 3.9 | 0.06 | 0.1 | 5 | 7 | 3 | 16 | 5 | 111 | 1.47 | 1.59 | 60 |
| 3. | m | 7.7 | 121 | 6 | 53 | 1.7 | 0.05 | 0.1 | 10 | 2 | 2 | 12 | 5 | 92 | 1.08 | 1.15 | 185 |
| | d | 7. 5 | 152 | 19 | 49 | 3.7 | 0.08 | 0.1 | 6 | 8 | 3 | 15 | 5 | 109 | 1.43 | 1.60 | 31 |
| 4. | m | 7.3 | 138 | 5 | 61 | 2.0 | 0.06 | 0.1 | 9 | 3 | 2 | 16 | 5 | 103 | 1.19 | 1.39 | 178 |
| | d | 7.8 | 172 | 23 | 55 | 3.4 | 0.04 | 0.1 | 5 | 8 | 3 | 16 | 6 | 120 | 1.48 | 1.71 | 35 |
| 5. | m | 7.3 | 126 | 4 | 61 | 1.8 | 0.04 | 0.1 | 9 | 2 | 1 | 17 | 4 | 100 | 1.15 | 2.29 | 492 |
| | d | 7.6 | 168 | 19 | 54 | 4.9 | 0.04 | 0.1 | 6 | 10 | 3 | 18 | 6 | 121 | 1.52 | 1.60 | 66 |
| 6. | m | 7.8 | 130 | 5 | 59 | 2.4 | 0.07 | 0.1 | 12 | 4 | 3 | 13 | 5 | 104 | 1.16 | 1.31 | 427 |
| | d | 7.8 | 169 | 18 | 56 | 4.8 | 0.05 | 0.1 | 6 | 12 | 2 | 15 | 6 | 120 | 1.50 | 1.80 | 19 |
| 7. | m | 7.3 | 117 | 5 | 56 | 1.8 | 0.07 | 0.1 | 10 | 2 | 2 | 12 | 4 | 93 | 1.10 | 1.07 | 391 |
| | d | 7.3 | 164 | 17 | 54 | 5.2 | 0.05 | 0.1 | 6 | 14 | 4 | 14 | 4 | 118 | 1.35 | 1.75 | 29 |
| 8. | m | 7.5 | 139 | 4 | 61 | 2.6 | 0.07 | 0.1 | 7 | 2 | 2 | 15 | 4 | 98 | 1.12 | 1.22 | 468 |
| | d | 7.4 | 166 | 11 | 58 | 5.3 | 0.06 | 0.1 | 6 | 15 | 3 | 16 | 4 | 118 | 1.37 | 1.86 | 21 |
| 9. | m | 7.3 | 126 | 5 | 54 | 1.6 | 0.12 | 0.2 | 9 | 2 | 2 | 12 | 4 | 90 | 1.07 | 1.02 | 960 |
| | d | 7.4 | 160 | 13 | 52 | 7.0 | 0.06 | 0.1 | 6 | 14 | 4 | 14 | 4 | 114 | 1.38 | 1.74 | 90 |
| 10 | m | 7.5 | 119 | 5 | 58 | 2.0 | 0.10 | 0.2 | 8 | 2 | 3 | 14 | 4 | 96 | 1.14 | 1.19 | 820 |
| | d | 7.4 | 168 | 13 | 56 | 7.5 | 0.07 | 0.1 | 6 | 14 | 4 | 14 | 5 | 120 | 1.44 | 1.80 | 40 |

| | | | | | | | | | | | | | | | | | |
|---|---|---|---|---|---|---|---|---|---|---|---|---|---|---|---|---|---|
| 11. | m | 7.5 | 139 | 5 | 61 | 2.1 | 0.14 | 0.2 | 9 | 5 | 3 | 17 | 3 | 105 | 1.20 | 1.44 | 790 |
| | d | 7.8 | 170 | 15 | 59 | 7.2 | 0.06 | 0.1 | 7 | 11 | 2 | 17 | 4 | 122 | 1.54 | 1.71 | 30 |
| 12 | m | 7.4 | 124 | 5 | 53 | 1.9 | 0.16 | 0.2 | 11 | 2 | 3 | 12 | 4 | 92 | 1.11 | 1.09 | 660 |
| | d | 7.8 | 151 | 15 | 51 | 5.4 | 0.07 | 0.1 | 8 | 11 | 2 | 12 | 5 | 110 | 1.30 | 1.54 | 20 |
| 13 | m | 7.7 | 127 | 6 | 62 | 2.0 | 0.18 | 0.2 | 11 | 2 | 3 | 14 | 3 | 103 | 1.24 | 1.11 | 1500 |
| | d | 7.9 | 169 | 15 | 58 | 9.0 | 0.07 | 0.1 | 7 | 12 | 2 | 15 | 4 | 123 | 1.52 | 1.68 | 30 |

m: monsoon (June-July 92)
d : non monsoon (Jan-94-Feb-94)
$E^+$ and $E^-$ : lonic balance in milliequivalensts /ltr.

Table 21.2 Grain Size distribution of suspended sediments in the Brahmaputra River (in %).

| Size class (micron) | | | Frequency (%) | | | | | | | | | | |
|---|---|---|---|---|---|---|---|---|---|---|---|---|---|
| Low | Average | High | Loc.1 | Loc.3 | Loc.2 | Loc.4 | Loc.6 | Loc.8 | Loc.9 | Loc.10 | Loc.11 | Loc.12 | Loc.13 |
| 0.4 | 1.2 | 2 | 5 | 6 | 3 | 5 | 6 | 8 | 17 | 10 | 11 | 21 | 17 |
| 2 | 3 | 4 | 5 | 7 | 4 | 7 | 7 | 10 | 18 | 10 | 11 | 20 | 17 |
| 4 | 6.5 | 9 | 7 | 12 | 7 | 13 | 15 | 18 | 32 | 18 | 18 | 33 | 29 |
| 9 | 13 | 17 | 15 | 13 | 10 | 16 | 22 | 23 | 33 | 19 | 21 | 26 | 32 |
| 17 | 26 | 35 | 68 | 62 | 76 | 59 | 50 | 41 | - | 43 | 39 | - | 5 |

-: < 0.5 %

calculated for the Brahmaputra lie between $10^{-2}$ to $10^{-3.5}$ atm. which supports the global trend that rivers are not in equilibrium with the atmosphere.

The sulphate concentration (average 3.8 ppm) is lower than the average for Indian rivers (13 ppm: Subramanian, 1987). Oxidation of pyritic sediments (Gansser, 1964) and gypsum or anhydrites may have major contribution to sulphates. This is indicated by $SO_4$ having a good correlation with Calcium. The average dissolved silica concentration for Brahmaputra is 8 ppm, little higher than the Indian average (7 ppm, Subramanian and others, 1987) but less than the world average (10.4 ppm; Martin and Meybeck, 1979). Owing to extensive draining of the catchment area by heavy rains (of the order of 1650 mm during June to September against about 60 mm during December to January), dissolved silica shows relatively higher concentration (lowest 4 ppm, highest 12 ppm) during monsoon.

Weathering of carbonates, sulphates and silicates releases most of the Ca (12--18 ppm) in the Brahmaputra. Mg (3-6 ppm) is released from dissolution of dolomitic rocks and also likely from some volcanic rocks reported (Sarin and others, 1989) in the source area. Na shows greater mobility and pronounced seasonal variation than K in the Brahmaputra basin. The Na values also indicate an increasing trend downstream, particularly in case of lean period samples. The TDS values reflect a similar profile. Considering the variation of the weight ratio Na/(Na + Ca) as a function of the Total Dissolved Solids (Gibbs, 1970), it was indicated that the ion chemistry of the Brahmaputra is mainly controlled by the source rock composition and to some extent by precipitation (Figure 21.2).

The results of the analysis of the single water sample collected from Brahmaputra by Hu Ming-Hui and others (1982) are similar to the present values obtained from the same station (both collected during the monsoon) except for Mg. The values obtained by Sarin and others (1989) from three different stations fall outside the range of the present results with respect to $HCO_3^-$, $Cl^-$ (present values are higher) and $Mg^{++}$ (present values are lower). Samples by these workers were, however, collected during April '82 and December '82. Thus the sampling seasons were different from those of the present study.

### 21.4.2 SUSPENDED SEDIMENT SIZE AND MINERALOGY

Surface suspended sediments of the Brahmaputra range from fine sand to clay with a mean size of 9 μm (Table 21.2). The fraction greater than 12 μm constitutes the important size population of the suspended load. Dominance of detritals was found to increase with the increasing grain size of the surficial suspended sediments (Table 21.3a).

Whilst in the upstream, a relatively higher size fraction ranging from 17 μm to 34 μm are dominant, towards extreme downstream, most of the suspended sediments belong to the size of less than 17 μm. Within the silt fraction, distribution of different size populations vary spatially but does not show any trend. Along downstream, the percentage of finer materials increases in the suspended sediments compared to upstream (Table 21.2). Surface suspended sediments of the Brahmaputra are moderately to poorly sorted (S.D. = 0.83 to 1.43: Table 21.4). Positive to very positive skewness of the suspended sediments suggests that the amount of finer material is more in the distribution. However, since these suspended sediments were collected during low flow and from shallow depths, the possibility, that the present samples may not represent the total sand fraction in suspension, cannot be ruled out. The suspended load in a dynamic river like the Brahmaputra is expected to include particles of larger size than that observed. Samples collected from greater depth may provide a different particle size distribution pattern.

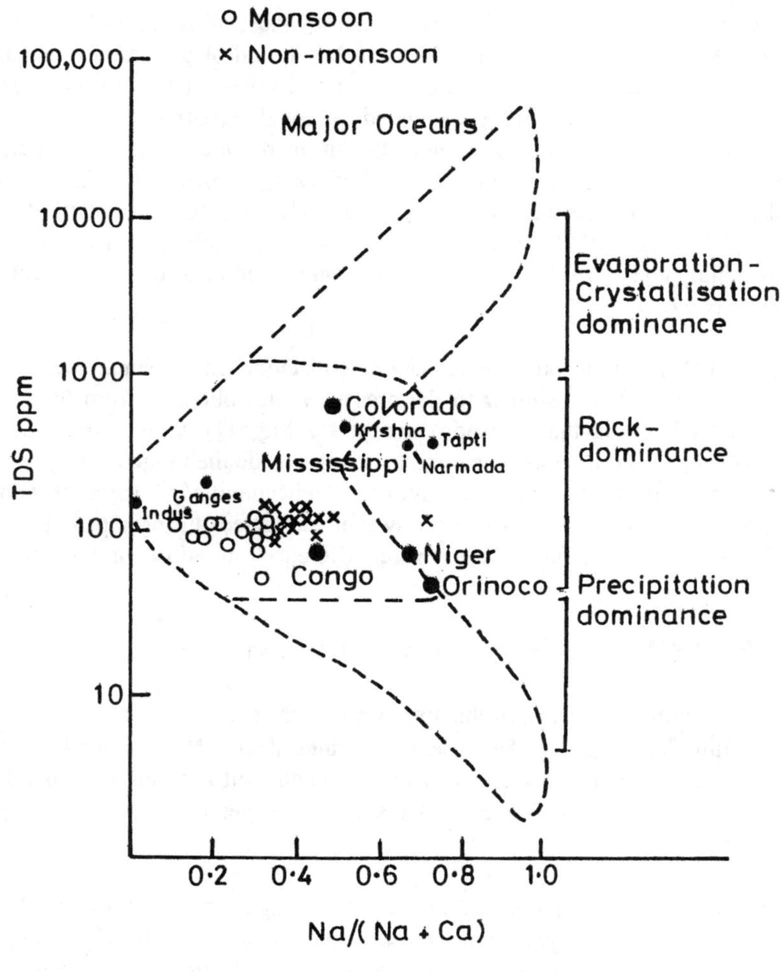


Fig. 21.2 Major factors controlling the water chemistry of the Brahamaputra River.

Table 21.3a summarises mineral abundance in the suspended load of the Brahmaputra. Suspended sediments are dominated by quartz, felspar, and clay minerals. The basin geology is the main controlling factor in the mineral distribution. However, in an actively degrading river system such as the Brahmaputra, the mineral composition of the sediments is controlled, to a considerable extent, by the composition of the local sediment sources as well as by continuous addition of detritus from bank cuttings. Quartz and Felspar are present as detritals, mainly derived from the pre-Cambrian granite and gneisses. The river is characterised by the presence of significant amount of illite followed by almost equal amounts of chlorite (measured at 4.72 a°; Biscaye, 1965) and Kaolinite (7 a°). Very little or no mixed layer clays or montmorillonites were observed in the suspended load. Unlike the peninsular rivers, chlorite is a dominant clay mineral in the Brahmaputra basin (Tables 21.3a and 21.3b). The composition of water plotted in the mineral stability diagrams for the predicted minerals in the Brahmaputra also indicate Chlorite and Kaolinite to be the expected clay minerals in the silicate systems and dolomite as a stable phase in the carbonate system (Figure 21.3a and 21.3b). However, this mineralogy characterises the suspended sediments quite upstream of the river mouth and may not represent exactly the final material transferred to the oceans after estuarine mixing. Mallick (1976) observed illite and kaolinite at the mouth of the Ganges-Brahmaputra with chlorite being abundant in individual samples, whereas Murthy and Srivastava (1979) found illite and quartz to be dominant in these samples.

Present studies show that in the Brahmaputra the detrital contribution in the form of quartz, felspar and mica make up more than 80% of the suspended sediments upstream. Then it gradually decreases downstream. Clay minerals, on the other hand, show an increasing trend (Tables 21.4) as the river flows downstream. Chlorite, illite, and kaolinite form more than 95% of the total clay minerals. Illite, being much more dominant in the upstream, suggests its primary origin to be from the granitic and metamorphic source rocks as a result of pre-dominant physical weathering in a cold and dry climate of the Tibetan Plateau.

Table 21.3a Semi-quantitative estimates of minerals in suspended sediments (bulk) (in %).

| Loc. No. | Quartz | Feldspar | Carbonates | Clay minerals |
|---|---|---|---|---|
| 1 | 41 | 39 | 4 | 16 |
| 2 | 43 | 29 | - | 28 |
| 3 | 42 | 27 | 9 | 32 |
| 4 | 53 | 37 | - | 10 |
| 5 | 63 | 14 | 4 | 19 |
| 6 | 53 | 13 | 2 | 32 |
| 7 | 61 | 10 | 15 | 14 |
| 8 | 36 | 20 | 14 | 30 |
| 9 | 44 | 19 | 6 | 31 |
| 10 | 33 | 15 | 3 | 49 |
| 11 | 42 | 18 | 2 | 38 |
| 12 | 45 | 14 | 8 | 43 |
| 13 | 32 | 10 | 5 | 53 |

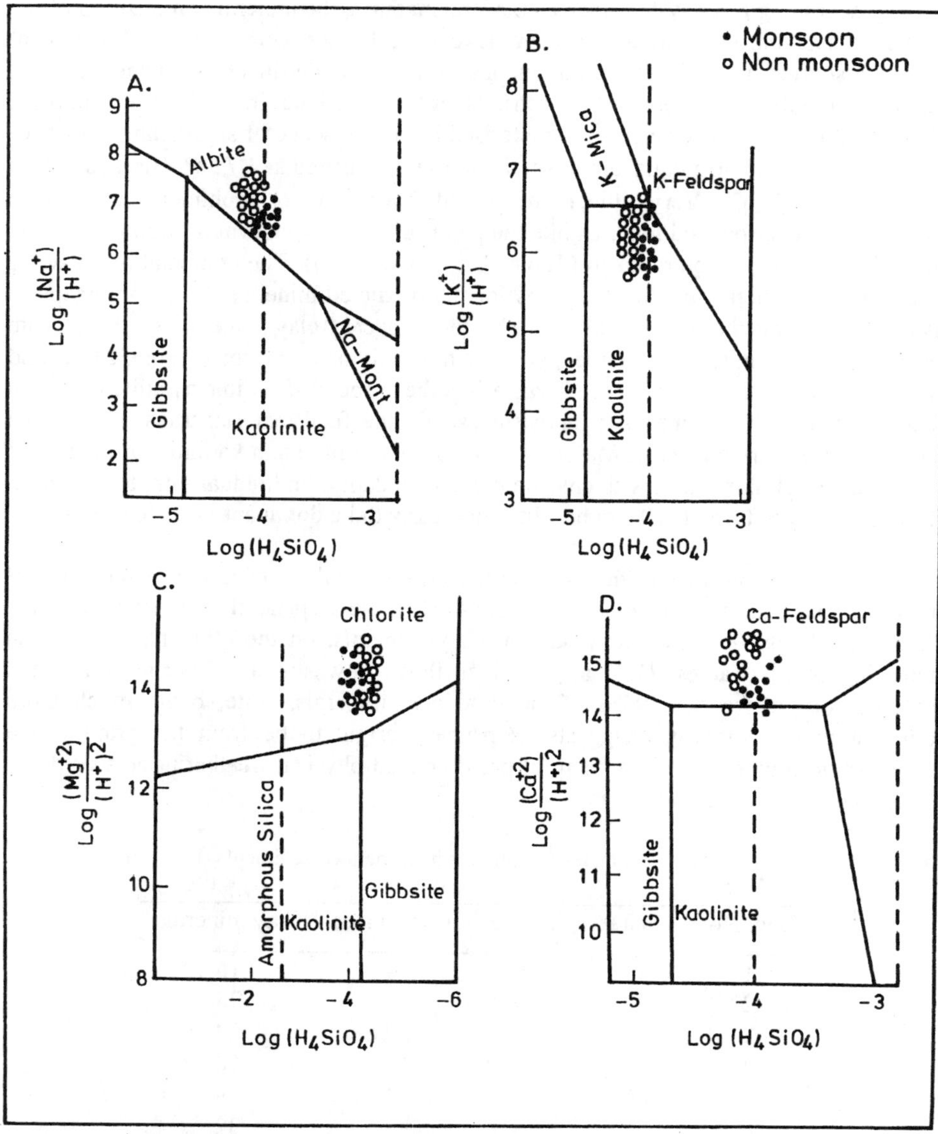


Fig. 21.3a Mineral stability diagrams for the Brahamaputra River in silicate system.

Table 21. 3b Percentage of clay minerals (weighted peak area % ) in suspended sediments (< 2u).

| Loc. No. | Chlorite | Lllite | Kaolinite | Montmorillonite |
|---|---|---|---|---|
| 1 | 22 | 59 | 19 | - |
| 2 | 36 | 25 | 39 | - |
| 3 | 25 | 47 | 28 | - |
| 4 | 21 | 61 | 18 | - |
| 5 | 17 | 61 | 13 | 9 |
| 6 | 20 | 64 | 16 | - |
| 7 | 34 | 39 | 27 | - |
| 8 | 32 | 38 | 28 | 2 |
| 9 | 35 | 40 | 25 | - |
| 10 | 36 | 27 | 35 | 2 |
| 11 | 42 | 25 | 33 | - |
| 12 | 45 | 18 | 37 | - |
| 13 | 36 | 21 | 31 | 2 |

- not detected,

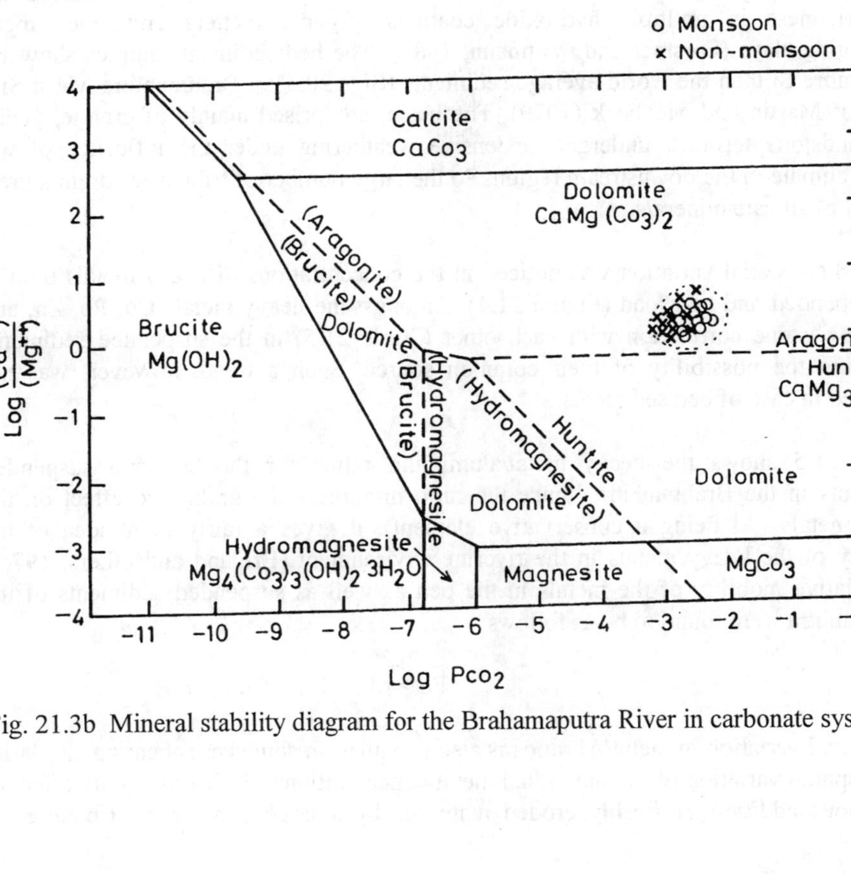


Fig. 21.3b Mineral stability diagram for the Brahamaputra River in carbonate system.

### 21.4.3 SEDIMENT CHEMISTRY

The results of the chemical analyses of bed and suspended sediments have been plotted in Figure 21.4. Since the chemical data on the source rocks were not available, the sediment chemistry was compared with the average composition of surficial rocks exposed to weathering. The abundance and the relative mobility of the analysed elements in Brahmaputra are of the following order:

Bed sediments:

Si > Al > Fe > Ca > Mg > Na > K > P> Mn > Zn > Cu > Cr > Pb > Cd

Suspended sediments:

Si >Al > Fe > Na > K > Ca > Mg > P > Mn > Zn > Cu > Cr > Pb > Cd

Except in the cases of Si, Al, Ca, Na and Mg, the suspended sediments are more enriched in the remaining elements relative to the bed sediments (considered in bulk, not size segregated). The enrichment factor in bulk suspended sediments relative to the bed sediments are 1.3 for Fe, 1.8 for Pb, 1.6 for Cu, 1.6 for Cr, 1.2 for Zn, 1.4 for Mn and 1.3 for Cd. This is generally attributed to the relative fineness of the suspensions and their richness in multiple hydroxide coatings, organic matters and trace metal scavenging clays (Forstner and Wittmann, 1981). The bed sediment samples show 1.3 times more Si than the world average sediments (Si = 28.5%), Brahmaputra Mean Si = 36%) of Martin and Meybeck (1979). The basin, comprised mainly of granite, gneiss and sandstone terrains, undergoes extensive weathering under the influence of wet humid climate in the downstream region. So the large number of tributaries drain a great amount of silicate minerals.

No marked spatial variation was noticed in the concentrations of heavy metals both in the suspended and bed load (Figure 21.4). Amongst the heavy metals Cu, Pb, Zn, and Mn show some correlation with each other (Table 21.5) in the suspended sediments indicating the possibility of their common source. Such a trend, however, was not observed in case of bed sediments.

Figure 21.5 shows the heavy metal/aluminium ratios for the bed and suspended sediments in the Brahmaputra. Since the ratio minimises the grain size effect on the heavy metals (Al being a conservative element), it gives a fairly good idea of the mobility of the heavy metals in the riverine environment (Bruland and others, 1974). The relative mobility of the metals in the bed as well as suspended sediments of the Brahmaputra were found to be as follows:

Fe > Mn > Zn > Cu > Cr  Pb

The spatial variation in metal/Al ratios is also irregular, in some cases bearing similarity to the spatial variation of the individual metal concentrations. Whilst the likely effect of continuous addition of freshly eroded materials by a large number of tributaries at different points is somewhat obvious by this erratic trend, location 5 shows a peaking of almost all the metal concentrations and also that of Ca and Mg (Figure 21.3) suggesting the possibility of enriched input by the tributary *Buridihing* just above this location. Since the *Buridihing* drains coal mining and oil field areas in the upstream, metal flux from anthropogenic sources in the sediment of this river cannot be ruled out. However, the absence of similarly higher concentrations of metals in the suspended sediment of the same location contradicts this observation. The overall metal concentration in the sediments of the Brahmaputra, however, appears to owe their origin more to natural sources than any anthropogenic source since industrialisation is very limited in the catchment area (Table 21.6). Although the large network of tributaries drain diverse mountainous and agricultural terrains, the present data are limited to distinguishing between natural and human sources of metal concentration in the sediments, since the inputs from domestic, agricultural and industrial sources may be diluted owing tothe enormous amount of the sediment load. This may also be the reason why the samples collected from the downstream of Guwahati (location 11) did not show any significant rise in elemental concentration despite being the biggest urban settlement along the river with population of more than 1.5 million.

Table 21.4 Correlation matrix for the suspended sediments of the Brahmaputra River basin.

| Element | Si | Fe | Al | Ca | Mg | Na | K | Cu | Zn | Pb | Cr | Mn |
|---|---|---|---|---|---|---|---|---|---|---|---|---|
| Si | -- | | | | | | | | | | | |
| Fe | -0.24 | -- | | | | | | | | | | |
| Al | 0.35 | 0.59 | -- | | | | | | | | | |
| Ca | -0.46 | -0.06 | -0.07 | -- | | | | | | | | |
| Mg | -0.50 | -0.12 | -0.12 | 0.75 | -- | | | | | | | |
| Na | 0.03 | -0.14 | -0.34 | 0.26 | -0.12 | -- | | | | | | |
| K | 0.29 | 0.15 | 0.42 | -0.02 | 0.31 | -0.96 | -- | | | | | |
| Cu | -0.80 | 0.44 | 0.11 | 0.72 | 0.54 | 0.14 | 0.07 | -- | | | | |
| Zn | -0.70 | 0.40 | 0.02 | 0.84 | 0.66 | 0.10 | -0.09 | 0.84 | -- | | | |
| Pb | -0.60 | 0.09 | -0.37 | 0.44 | 0.64 | -0.31 | 0.32 | 0.33 | 0.66 | -- | | |
| Cr | -0.08 | -0.35 | -0.03 | 0.56 | 0.82 | -0.40 | 0.52 | 0.90 | 0.38 | 0.60 | -- | |
| Mn | -0.74 | 0.57 | 0.10 | 0.70 | 0.51 | 0.24 | -0.07 | 0.94 | 0.92 | 0.43 | -0.47 | -- |
| Cd | -0.80 | -0.06 | -0.38 | 0.53 | 0.25 | 0.04 | 0.06 | 0.67 | 0.54 | 0.38 | 0.04 | 0.53 |

Table 21.5a Average chemical composition of Indian rivers (in mg/l).

| RIVER | $HCO_3$ | Cl | $SO_4$ | $SiO_2$ | Ca | Mg | Na | K |
|---|---|---|---|---|---|---|---|---|
| Krishna | 178 | 38 | 49 | 24 | 29 | 8 | 30 | 2 |
| Cauvery | 53 | 18 | 39 | 8 | 15 | 16 | 30 | 3 |
| Godavari | 105 | 17 | 9 | 10 | 22 | 5 | 12 | 3 |
| Ganga | 100 | 6 | 9 | 3 | 41 | 6 | 9 | 4 |
| Brahmaputra | 56 | 11 | 4 | 7 | 14 | 5 | 7 | 3 |
| Indus | 64 | 9 | 15 | 5 | 27 | 1 | 1 | 2 |
| Narmada | 225 | 20 | 5 | 9 | 14 | 20 | 27 | 2 |
| Tapti | 150 | 65 | 1 | 16 | 19 | 22 | 48 | 3 |

- Present study; Reset from Subramanian 1987.

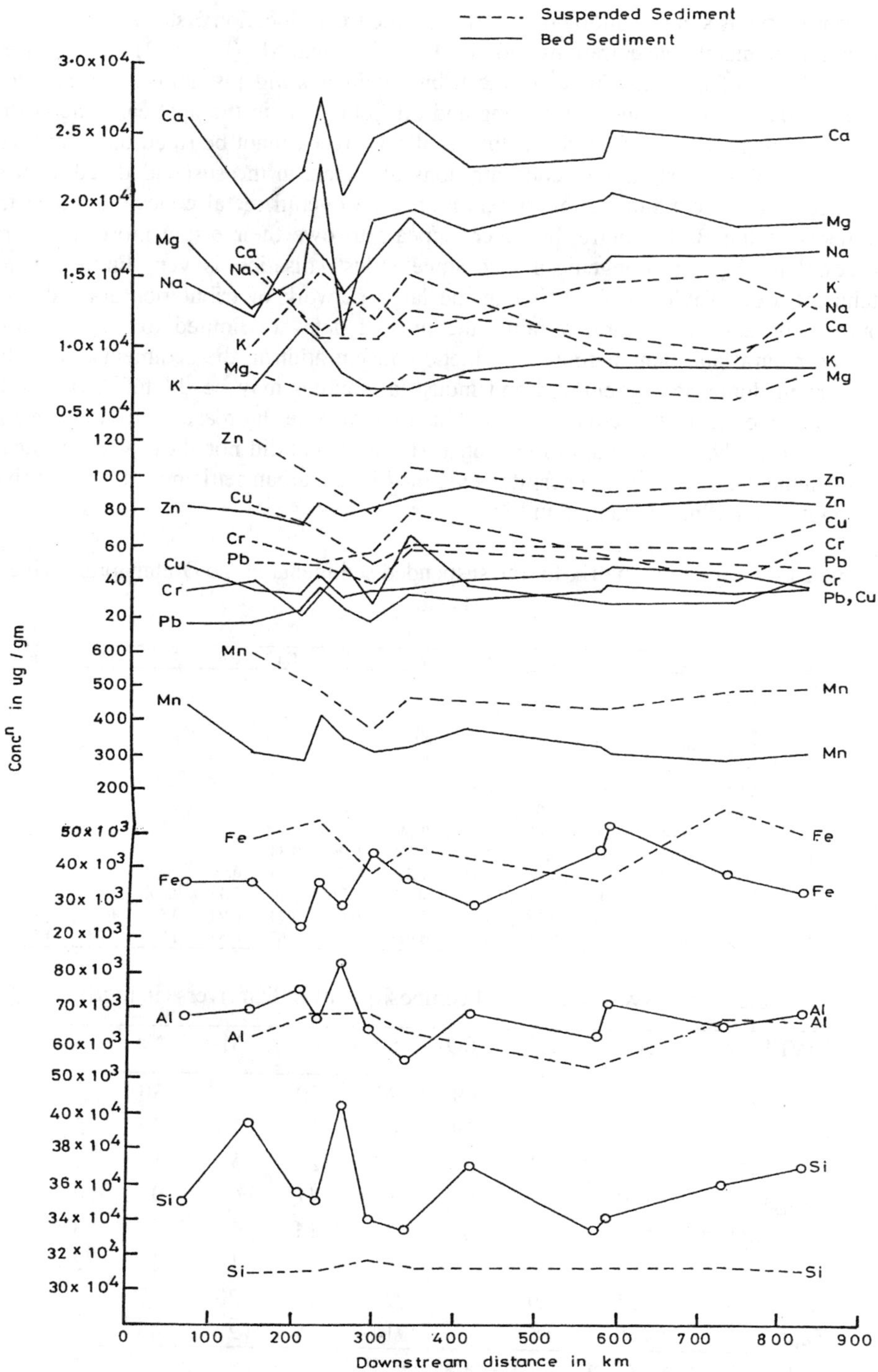


Fig. 21.4 Downstream variation in elemental concentrations including heavy metals, in bed sediments and suspended sediments of the Brahamaputra River.

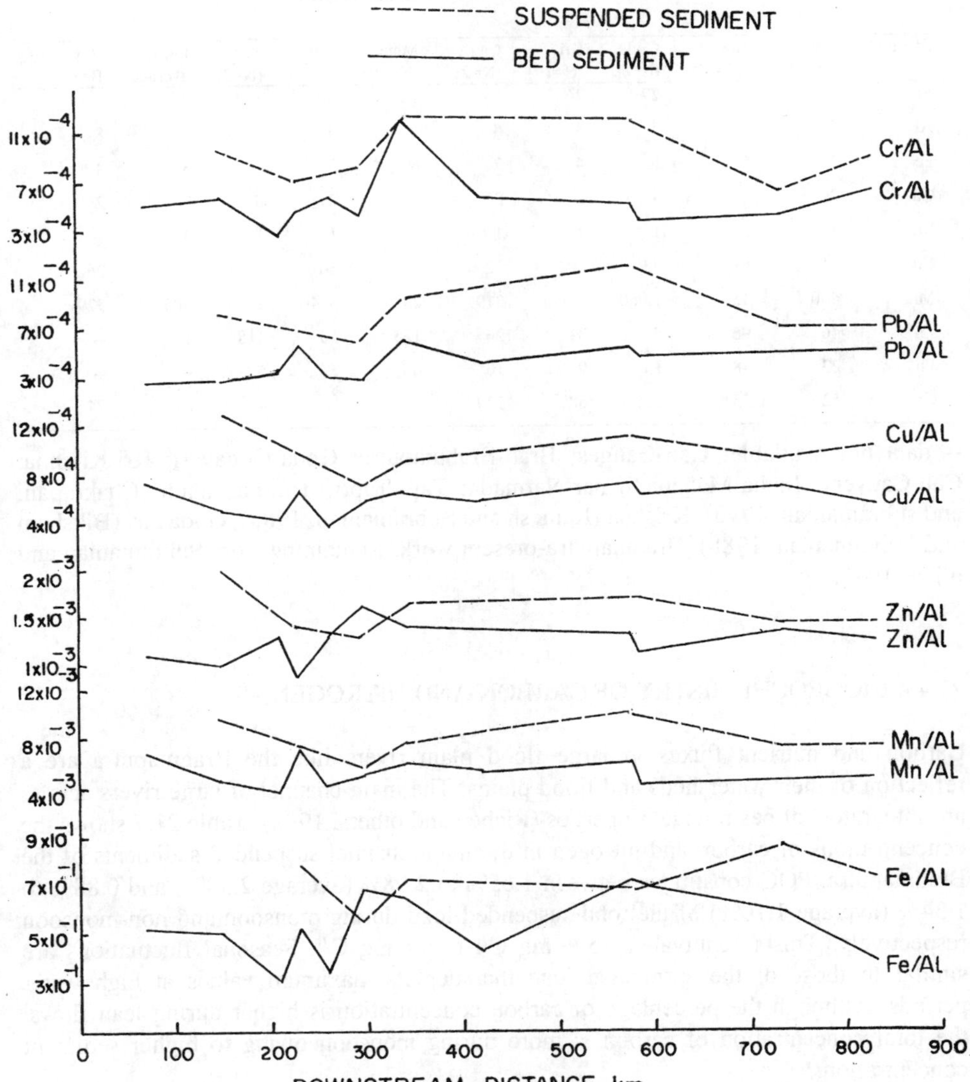


Fig. 21.5 Downstream variation in metal/A1 ratios in bed sediment and suspended sediment of the Brahamaputra River.

Table 21.5b The Brahmaputra River sediment chemistry (Si to Mg in %, other in μg/gm comparison with other Indian rivers.

| Element | Gan. (N=35) | Bra. (N=26) | Goda. (N=23) | Kri. (N=19) | Cau. (N=21) | Maha. (N=19) | Nar. (N=3) | Tap. (N=3) | Indian average | Bay of Bengal |
|---|---|---|---|---|---|---|---|---|---|---|
| Si | 31 | 32 | 27 | 26 | 35 | 12 | 30 | 32 | 25 | -- |
| Al | 5 | 6 | 5 | 3 | 5 | 6 | 3 | 4 | 5 | 8 |
| Fe | 2 | 5 | 6 | 4 | 2 | 6 | 3 | 1 | 3 | 4 |
| Ca | 2 | 2 | 4 | 5 | 2 | 1 | 2 | 8 | 2 | 2 |
| Mg | 1 | 2 | 1 | 1 | 1 | 1 | 1 | 1 | 1 | 1 |
| Cu | 21 | 61 | 73 | 49 | 12 | 57 | 56 | 126 | 28 | 26 |
| Mn | 400 | 457 | 1,060 | 1,040 | 319 | 2,020 | 5140 | 1,300 | 605 | 529 |
| Zn | 46 | 96 | 53 | 31 | 26 | 125 | 50 | 118 | 16 | -- |
| Pb | 23 | 48 | 13 | 9 | 10 | 60 | 5 | 5 | -- | -- |
| Cr | 52 | 55 | 126 | 68 | 129 | 15 | -- | -- | 87 | 84 |

-- data not available; Gan-Ganges; Brah-Brahmaputra; Goda-Godavari; Kri-Krishna; Cau-Cauvery; Maha-Mahanadi; nar-Narmada; Tap.-Tapti; data-mahanadi (Chakrapani and subramanian, 1990), Krishna (Ramesh and Subramanian, 1988), Godavari (Biksham and Subramanian, 1988); Brahmaputra-present work, Remaining from Subramanian and other, 1985.

### 21.4.4 BIOGEOCHEMISTRY OF CARBON AND NITROGEN

Carbon and nutrient fluxes in large flood plain rivers like the Brahmaputra are a reflection of their watersheds and flood plains. The main channel of large rivers acts as an 'integrator' of basin wide properties (Richey and others, 1991). Table 21.7 shows the concentrations of carbon and nitrogen in the main channel suspended sediments of the Brahmaputra. POC constitutes between 1.65% to 2.98% (average 2.38%) and 0.88% to 1.34% (average 1.16%) of the total suspended load during monsoon and non-monsoon respectively. This is equivalent to 1 mg C/l to 17 mg C/l. Seasonal fluctuations are similar to those of the suspended load that depicts maximum values at high water periods. Although the percentage of carbon concentration is higher during lean flows, the total concentration of carbon is more during monsoon owing to higher sediment concentration.

The particulate nitrogen in the Brahmaputra varies between 0.21% and 0.38% during lean flow and between 0.08% and 0.11% during peak flow. Based on an analysis of a limited number of samples, it was observed that particulate nitrogen concentration increases along flow downstream during lean flow. The C/N ratio is, however, variable over the basin including the tributaries (Mahanta, 1998). C/N ratio in the main channel varies from 10.2 in the lean flow period to 18.4 during monsoon rains. The tributary inputs of particulate N have been found to be relatively higher (0.09% to 0.29%) from the south bank tributaries than that from the north bank tributaries (0.05% to 0.17%) when only monsoon samples were considered (Mahanta, 1998).

Table 21.6 Particulate carbon and nitrogen concentration in the Brahmaputra mainstream.

| Loc. No. | C (total) % | C (inorg.) % | C(org.) % | N (total) % | C (total) N (total) | Dist. / Downstream |
|---|---|---|---|---|---|---|
| 1p | 2.69 | 1.04 | 1.65 | 0.21 | 12.8 | 70 |
| m | 1.12 | 0.24 | 0.88 | 0.07 | 16 | |
| 2p | 2.98 | 0.86 | 2.12 | | | 140 |
| m | 1.61 | 0.27 | 1.34 | | | |
| 3p | 2.83 | 0.82 | 2.01 | | | 160 |
| m | 1.63 | 0.37 | 1.26 | | | |
| 4p | 3.2 | 0.65 | 2.55 | 0.24 | 13.3 | 200 |
| m | 1.47 | 0.33 | 1.14 | 0.11 | 13.4 | |
| 5p | 2.4 | 0.46 | 1.94 | | | 250 |
| m | 1.58 | 0.4 | 1.18 | | | |
| 6p | 2.67 | 0.7 | 1.97 | 0.21 | 12.7 | 300 |
| m | 1.41 | 0.35 | 1.06 | 0.09 | 15.7 | |
| 7p | 2.81 | 0.56 | 2.25 | | | 350 |
| m | 1.49 | 0.35 | 1.14 | | | |
| 8p | 3.12 | 0.81 | 2.31 | | | 400 |
| m | 1.73 | 0.42 | 1.31 | | | |
| 9p | 3.04 | 0.73 | 2.31 | 0.26 | 11.7 | 450 |
| m | 1.66 | 0.44 | 1.22 | 0.1 | 16.6 | |
| 10p | 3.58 | 0.69 | 2.89 | | | 550 |
| m | 1.58 | 0.4 | 1.18 | | | |
| 11p | 3.87 | 0.92 | 2.95 | 0.28 | 13.8 | 600 |
| m | 1.47 | 0.35 | 1.12 | 0.08 | 18.4 | |
| 12p | 3.26 | 0.79 | 2.47 | 0.32 | 10.2 | 700 |
| m | 1.74 | 0.44 | 1.3 | 0.11 | 15.8 | |
| 13p | 3.66 | 0.68 | 2.98 | | | 830 |
| m | 1.36 | 0.39 | 0.97 | | | |
| Mean | 2.31 | 0.56 | 1.75 | | | |

:- Not analysed; p: Pre-monsoon samples; m : Monsoon samples

### 21.4.5 BIOGEOCHEMISTRY OF PHOSPHOROUS

The phosphate concentrations in the Brahmaputra are low at flood level and high during lean flow which indicates a simple dilution pattern. The suspended sediments analysed in this study were representative of both the high flow (monsoon) and lean flow (pre-monsoon) periods. Whilst the high flow discharge is dominated by rain water, the lean flow discharge results from snowmelt. Most of the suspended sediment samples show similar C, N, P concentrations and composition for a particular season. For example, the

P composition of most samples was indicative of a low P availability, because there was no significant change downstream. The NAIP content ranged between 5% and 24% (Mahanta, 1998). The bio-availability of a particular fraction of particulate P is determined and characterised from interpretations of the results of sequential extractions. The time a certain particle is accessible to algae before sedimentation or washout affects the actual bio-availability of sediment-bound P. The availability of P after re-suspension of particles or after P mobilisation in the sediments and a subsequent upward transport of dissolved P can be significant. The OP content which could add marginally to the bio-availability was slightly higher ranging between 20% and 25 %. 22% to 27 % of the P was not bio-available because of the low solubility of the primary phosphate mineral Apatite. There is no clear relationship between relative composition and land use. For example, in the sample collected from station no. 11 in the mainstream before which lies the largest urban settlements of the area under study (Guwahati: population about 1.5 million), only OP was somewhat elevated compared to the average. The composition was very close to the average at some other locations (e.g., no. 3, 8) in which the river flows through some extensively cultivated region. The fertiliser consumption in the basin is however, very low and may not have a perceptible effect on the P distribution over the basin.

The phosphorous is mineral grains of Apatite (AP) decreases downstream and is considered biologically unavailable because the grains of Apatite rapidly and the pore water of sediments is usually super-saturated with respect to apatite (Emmerson, 1976). Although both NAIP and OP is considered potentially bio available to phytoplanktons, in a dynamic river like the Brahmaputra, there may not be sufficient time to fully utilise the NAIP and OP, specially if the biochemical processes in bottom sediments are not suitable for the P release.

The rate of suspended sediment transport by Brahmaputra at the downstream location of Pandu is $1.2 \times 10^{15}$ gms per year (Goswami, 1985). Multiplying by the concentration of 700 ug/gm of the total P we obtain the transport of P to be $84 \times 10^{10}$ gm P/year. This is distinctly larger than the flux of dissolved P by the river, $12 \times 10^{10}$ gm P/year. However, the enhancement of the dissolved P flux by solubilized sediment P is quite possible at the oceanic shelf interface, once it reaches the mouth. The total flux of P from the Brahmaputra at Dhubri is about three times greater than that carried in dissolved form. The present global flux of the total sediment to the oceans is $14.24 \times 10^{15}$ gm P/year (Milliman and Syvitski, 1992). If the pre-human sediment flux to the ocean was half of that at present (Berner and Berner, 1987) and the flux of solubilized sediment P (from the study of Berner and Rao, 1994) of 5 micromole/gm. (155ug/gm) for the Amazon is taken to be applicable to the entire (natural) world, one obtains a global solubilized sediment P flux of $110\text{-}190 \times 10^{10}$ gm P/year. In contrast, the natural dissolved P flux is estimated to be $40\text{-}80 \times 10^{10}$ gm P/year. (Froelich et. al., 1982). Thus the resulting pre human total flux of $150\text{-}270 \times 10^{10}$ gm P/year is more than three times higher than the dissolved flux (Mahanta, 1998).

## 21.4.6 CHEMICAL AND PHYSICAL EROSION IN THE BRAHMAPUTRA

The Brahmaputra together with the Ganges account for nearly 90% of the total sediment transfer from the Indian subcontinent (Subramanian and others, 1987). The basin area of the Ganges--Brahmaputra system (60% area lying in the mountains) is less than that of the Amazon, but sediment contributions to the ocean by both are nearly the same. The control of relief on the weathering environment in the drainage basin of the Ganges--Brahmaputra system is thus obvious.

Owing to the tectonic instability in the Himalayas and the relatively younger lithology of the lesser Himalayas as well as of the outer Himalayas, extensive physical weathering takes place in the catchment area, resulting in a very high sediment load of the river, particularly during the rainy season. The basin geology and the runoff also have a major control over the sediment yield in the basin. Despite being a river of such magnitude, the Brahmaputra still has a natural flow regime without having many dams in its course. This has resulted in the predominance of physical over chemical erosion. Raymahasay (1970) had suggested that the ratio of mechanical to chemical erosion in case of the Brahmaputra is around 9 to 1. Local factors, such as deforestation, shifting channels, earthquakes and landslides, are of significant importance in enhancing the rate of mechanical erosion in the Brahmaputra basin.

Coleman (1969) had reported that the water discharge and suspended sediment transport by the Brahmaputra used to be in excess of 2.5 million cubic metre per second and 13 million tonnes per day respectively during flood. More than thirty years afterwards, today the suspended load of the Brahmaputra appears to have further increased, apparently owing to extensive deforestation and other human induced activities such as faulty land use practice in the catchment area. The annual average water discharge and sediment load data of the Brahmaputra and its major tributaries for the period 1972 to 1988 have been plotted in Figure 21.6. The sediment loads for both Brahmaputra and its tributaries are closely related to the discharge. The ratio of sediment load to discharge is generally found to increase with increasing discharge and sediment load. This can also be observed from a plot of sediment load against discharge (Figure 21.7). According to Goswami (1985), the peak of the sediment load normally precedes the discharge peak by one to several days owing to the high concentration of relatively finer sediments.

The average sediment composition of the Brahmaputra has been calculated taking 90% of the suspended sediment composition and 10% of the bed sediment composition. A comparison of the average ion chemistry and sediment composition of the Brahmaputra with those of other major Indian rivers shows considerable difference in their geochemical behaviour (Tables 21.6a, 21.6b). All these rivers vary in their size, precipitation, topography, geology, vegetation, land use and anthropogenic exploitation. The ways and the extent to which these factors influence the individual behaviour of each of these rivers is not within the scope of this chapter. However, it is clear that the Ganges, which also originates in the Himalayas and flows through both hills and plains, has almost two times the $HCO_3$ and $SO_4$ and three times the Ca in the dissolved load than the Brahmaputra, whilst Cl and $SiO_2$ concentrations are comparatively higher in

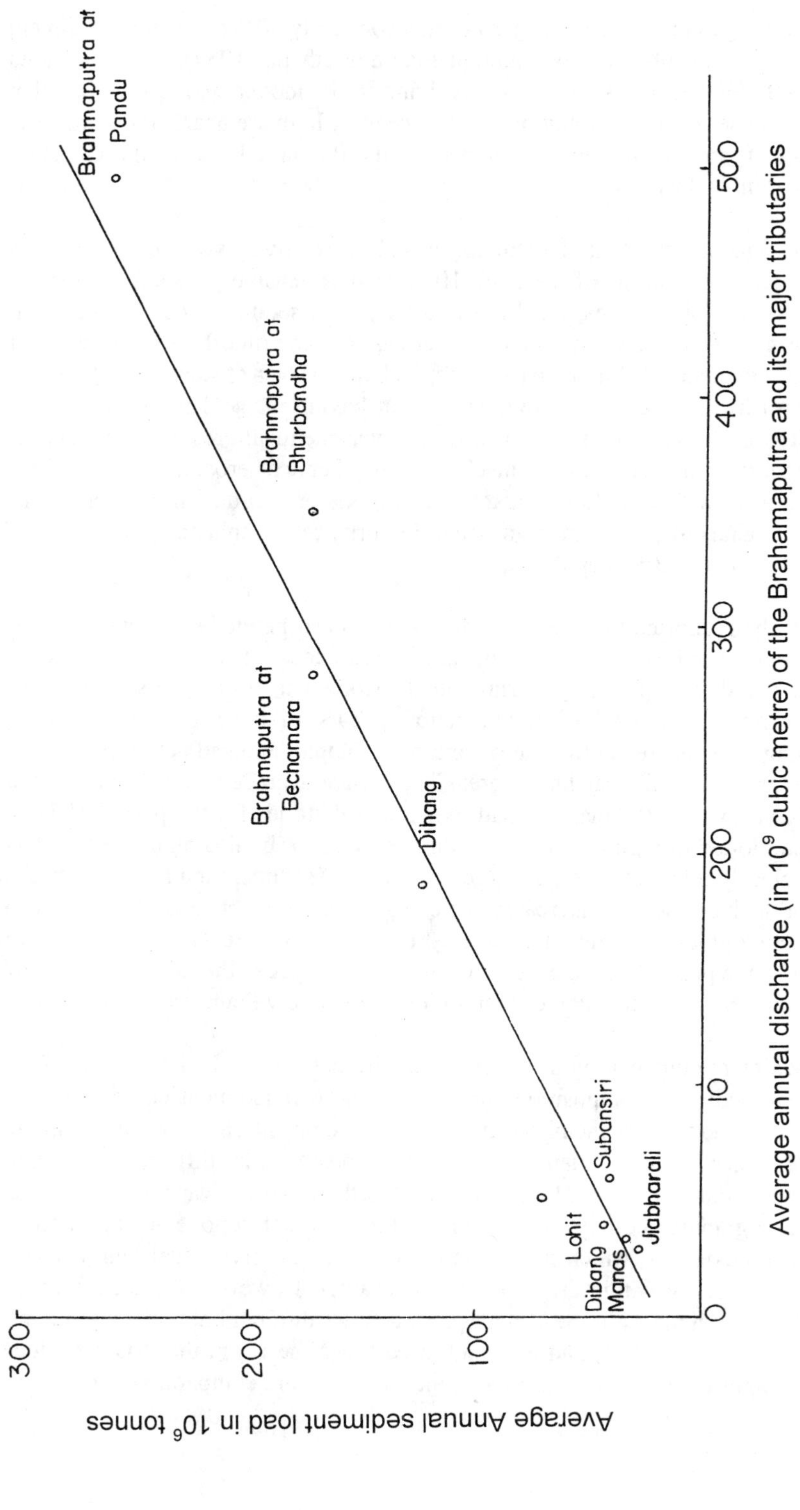


Fig. 21.6 Plot of average annual sediment load and average annual discharge of the Brahamaputra based on the data obtained from a few selected locations.

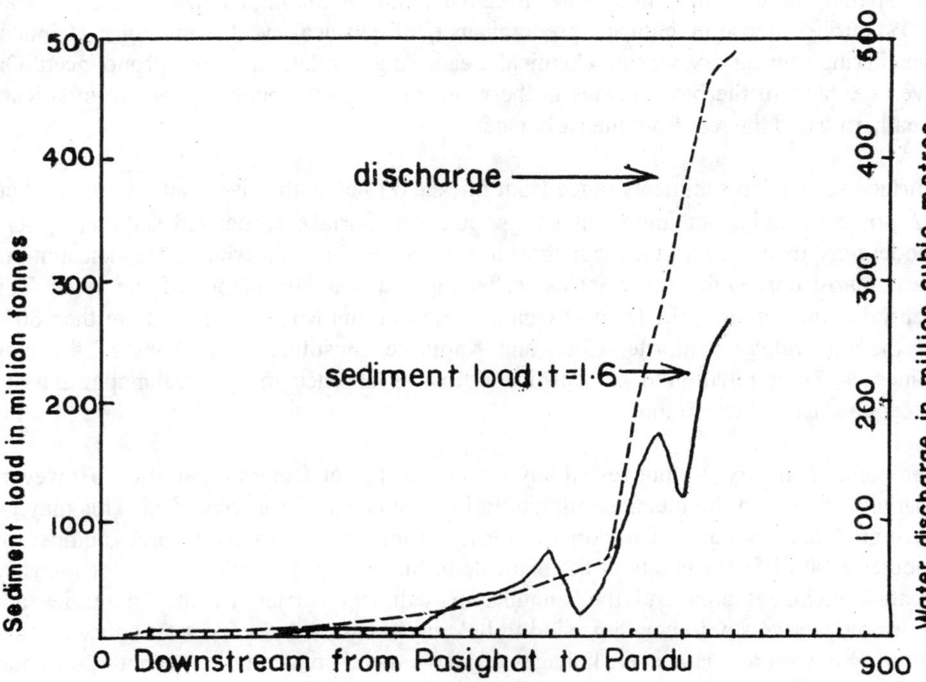


Fig. 21.7 Average annual discharge and sediment load of the Brahamaputra River for the period 1972 to 1988, based on the data collected from the Assam Flood Control Department.

case of the latter. Sediment chemistry also reveals relatively higher amounts of Al, Fe, Cu, and Zn in the sediment of the Brahmaputra than in the case of Ganges.

According to Goswami (1985), as a whole the Assam section, i.e., the mid-stream part of the Brahmaputra experiences an aggradation of about 16 cm/yr if the bulk density of the sediment is considered to be 2.65 t/m$^3$ . In fact, this computes as high as 27 cm/yr if the bulk density is considered to be 1.6 t/m$^3$ , which seems more realistic considering the grain size of the sediments. Subramanian and others (1989) computed the rate of sedimentation in the Brahmaputra by Pb$^{210}$ method from five core samples collected over the basin. The range given by them was of the order of 1.7 mm/yr to 15 mm/yr.

## 21.5 Conclusions

The major ion chemistry of the Brahmaputra is characterised by high bi-carbonate content and source rock influence. As the river flows down from the mountainous terrain of the Himalayas to the Assam plains, changes occur in case of several parameters but the spatial variation present a rather irregular trend. While higher values of TSM than TDS during monsoon indicate predominance of physical weathering over chemical weathering, during dry season, chemical weathering is relatively more pronounced. On average, 60% of the bicarbonates in the water of the Brahmaputra comes from silicate weathering and the rest from the carbonates.

Surface suspended sediments range from fine sand to clay, the size fraction greater than 12 μm constituting an important size population. Surface suspended sediments were moderately to poorly sorted with greater amount of finer material in the distribution, particularly during the rainy season reflecting the dynamic nature of the river. The detrital contribution in the form of Quartz, Feldspar and Mica make up more than 80% of the mineralogy. Chlorite, Illite, and Kaolinite constitute about 95% of the clay minerals. Montmorrillonite is conspicuous by its absence in the Brahmaputra basin except occasional exceptions.

Sediment chemistry did not reveal any marked spatial or temporal variation. However, spatial variation in the metal/aluminium ratio in some cases was observed. This may be attributed to continuous addition of freshly eroded materials by a large number of tributaries at different points of the main stem. Since there is hardly any major industry in the catchment area and the amounts of sediment carried by the tributaries are enormous, there is not enough ground to link any occasional rise in metal concentration to any point source. However, linkages of heavy metal concentration to both point and non-point sources of anthropogenic heavy metal concentration would require wider investigation.

The major sources of carbon in the Brahmaputra basin, other than the natural source, are deforestation and fossil fuel consumption. In the case of nitrogen, precipitation, soil erosion, drainage and fertilizer application constitute the main sources. The current data are not adequate for establishing possible linkages, however. Phosphorous carried by the suspended load makes up a substantial part of the P loading in the Brahmaputra. A comparison of different forms of P between suspended and bed sediments indicates a decrease in organic-P, Fe-P and total-P after deposition. The low P bioavailability in suspended sediments may have a considerable effect on management plans for controlling the P flux through the Brahmaputra. Planners should consider both the total P loading and compositional information of other nutrients to determine which areas are contributing large amounts of biologically available nutrients. Since much of the suspended sediment-transported P is not available biologically, land management practices to minimise sediment input to the river should produce no significant reduction in biological availability of P. Thus the effect of curtailing suspended sediment inputs in the tributaries as well as the main stream of the Brahmaputra may be more beneficial compared to many other rivers, since NAIP and OP inputs from domestic sewage, fertilizer, etc., to the river are still not high. The total flux of P from the Brahmaputra ($150 -- 270 \times 10^{10}$ gm P/year) constitutes nearly 5% of the global flux and is about three times greater than that carried in dissolved form. Although data are limited, considering the mobilization of P by fertilizer use, it is reasonable to consider that the P concentrations will be further increasing.

The composition of the sediment load of the Brahamaputra River also appears to be controlled to a considerable extent by source rock influence and the local sediment sources by continuous addition of detritus from bank cuttings and resuspension of materials from the bed load. Suspended sediments play a significant role on the overall solute and sediment biogeochemistry of the river. Biogeochemistry of the Ganges and that of the Brahmaputra were found to be markedly different with respect to $HCO_3$, $SO_4$, Cl, $SiO_2$, Ca and a few heavy metals such as Cu, Fe, and Zn. It may be possible that apart from the natural factors, such as geology, topography, vegetation, and precipitation, this is also owing to the difference in the intensity of utilisation of the two rivers and anthropogenic impacts on their watersheds which may be different, having a direct reflection on their geochemical behaviour. Obviously, to clearly establish this interelationship within the respective watersheds, it is essential to understand the overall biogeochemistry of this hitherto scientifically unexplored river, and to do so, investigations at a greater scale are mandatory.

## References


Biksham, G and Subramanian, V. (1988). Sediment Chemistry of the Godavari River basin.Chemical Geology, 70: 225--236.

Biscaye, P.E. (1965). Mineralogy and sedimentation of recent deep sea clay in the Atlantic Ocean and adjacent seas and oceans. Geol. Soc. Am. Bull., 76: 803--832.

Bruland, K.W., Bertine, K., Koide, M. and Goldberg, E.G. (1974). History of metal pollution in southern California coastal zone. Environmental Science & Technology, 8, 425--433.

Carrol, D. (1970). Clay minerals - a guide to their identification. The Geol. Soc. Ame., Special Paper 126, 80 pp.

Chakrapani, G.J.and Subramanian, V. (1990). Preliminary Studies on the Geochemistry of Mahanadi river basin. Chemical Geology, 81: 241--253

Chang and Jackson, M.C. (1957). Fractionation of soil phosphorus. Soil Sc., 84: 133--144.

Coleman, J.M. (1969). Brahmaputra river channel processes and sedimentation. Sediment. Geol., 3; 129-239.

Datta, D and Subramanian, V. (1997). Texture and Mineralogy of sedimentsfrom the Ganges-Brahmaputra-Meghna river system in the Bengal Basin, Bangladesh and their environmentalimplications. Env. Geology, 30(3/4): 181--188

Datta, D and Subramanian, V. (1998). Distribution and Fractionation of heavy metals in the surface sediments of the Ganges-Brahmaputra-Meghna river system in the Bengal basin. Env. Geology, 36(1--2): 93--100

Datta, D, Gupta, L.P. and Subramanian, V. (1999). Distribution of C, N and P in the sediments of the Ganges-Brahmaputra-Meghna river system in the Bengal basin. Organic Geochemistry, 30: 75--82

Emmerson, S. (1976). Early diagenesis in anaerobic lake sediments: Chemical equilibria in interstitial water. Geochim Cosmochim Acta, 40: 925--934

Evans, P. (1964). The Tectonic frame work of Assam. Jour. Geol. Soc. India, 5: 80 -- 96.

Flanagan, F.J. (1976). Dispersion and analysis of eight new USGS rock standards. U.S. Geol. Surv. Prof. Pap., pp. 840.

Forstner, U. and Wittmann, G.T.W. (1981). Metal Pollution in the aquatic environment. 2nd ed. Springer - Verlag, Berlin - Heidelberg - New York., 486 pp.

Froelich, P. N., Bender, M. L., Luedtke, N. A., Heath, G. R. and DeVries, T. (1982). The marine phosphorous cycle. Amer. J. Sci., 282: 474--511.

Gansser, A. (1964). Geology of the Himalayas. Interscience, New York. 290 pp.

Gibbs, R. J. (1970). Mechanisms controlling world water chemistry. Science, 170: 1088--1090.

Gibbs, R.J. (1967). The Geochemistry of the Amazon river system. Part I : The factors that control the salinity and the composition and concentration of the suspended solids. Geol. Soc. of Am. Bull., 78: 1203--1232.

Goswami, D.C. (1985). Brahmaputra river, Assam, India: Physiography, basin denudation and channel aggradation. Water Resour. Res., 21 (7) : 959--978.

Goswami, D. C. (1988). Estimation of bed load transport in the Brahmaputra River, Assam. Indian Journal of Earth Science, 15(1): 14--26.

Hedges, J. I., Clark, W. A., Quay, P. D. Richey, J. E. Devol, A. H. and. Santos, U.de M (1986). Composition and fluxes of particulate organic matter in the Amazon river. Limnol. Oceanogr., 31: 717--738.

Hu, D., Satio, Y. and Kempe, S (1998). Sediment and Nutrient transport to the coastal zone. Asian Change in the Context of Global Climate Change. Ed. J. N. Galloway and J. M. Melillo. Cambridge University Press, pp: 245--270

Hu, Ming - Hui., Stallard, R.F. and Edmond, J. M. (1982). Major ion chemistry of large Chinese rivers, Nature, 298 (5874): 550--553.

Mahanta, C. (1998). Carbon, Nitrogen and Phosphorous distribution in the sediments of the Brahmaputra river basin of North East India. In Proceedings of the Indo-German Workshop on Biogeochemistry, New Delhi, Dec. 13--18, 1998. pp. 121--142.

Mallick, T. K. (1976). Shelf sediments of the Ganges delta with a special emphasis on the mineralogy of the western part, Bay of Bengal, Indian Ocean. Mar. Geol., 22: 1--32.

Martin, J. M. and M. Meybeck. (1979). Elemental mass - balance of material carried by major world rivers. Marine Chemistry, 7: 173--206.

Meade, R.H., Nordin Jr., C.F.. Curtis, W.F., Mahoney, A.H. and Delaney, B.M. (1979). Suspended sediment and velocity data, Amazon river and its tributaries June -July 1976 and May-June 1977. US Geol. Surv. Open -File Rep. 79--515, 42 pp.

Milliman, J. D. and Syvitsky, J. P. M. (1992). Geomorphic/tectonic control of sediment discharge to the ocean: the importance of small mountainous rivers. Jour. of Geol., 100: 525--544.

Milliman, J. D. and Meade, R. H. (1983). World - wide delivery of river sediment to the oceans. Jour. of Geol., 91: 1--21.

Murthy, M. R. and Srivastava, P. C. (1979). Clay mineralogy of shelf sediments of north-west Bay of Bengal. Mar. Geol., 31: M21--M32.

Oldham, R.D. (1900). Assam Earthquake of 12th June, 1897. Mem 29, G.S.I.

Ramesh, R. and Subramanian, V. (1988). Temporal, spatial and size variation in the sediment transport in the Krishna river basin. J. Hydrol., Vol. 98: 245--256.

Richey, J. E., Victoria, R.L., Salati, E. and Forsberg, B.R. (1991). The biogeochemistry of a major river system: The Amazon case study; Biogeochemistry of major world rivers, Ed. E. T. Degens, S. Kempe and J. E. Richey. John Wiley and Sons, pp. 57-73.

Raymahasay, B.C. (1987). A comparative study of clay minerals for pollution controls. J. Geol. Soc. Ind., 3: 408-413.

Raymahasay, B. C. (1970). Characteristics of stream erosion in the Himalayan region of India. In Proc. Symp. on Hydrogeochemistry and Biogeo- chemistry, 1: 82-89.

Sarin, M.M., and Krishnaswamy, S. (1984). Major ion chemistry of the Ganga - Brahmaputra river system, India. Nature, 312 (5994): 538 -- 541.

Sarin, M.M., S. Krishnaswamy., K. Dilli., B.L.K. Somayajulu. and W.S. Moore., 1989. Major ion chemistry of the Ganges- Brahmaputra river system - weathering processes and fluxes to the Bay of Bengal. Geochim. Cosmochim. Acta, 53 : 997--1009.

Shapiro, L. and W. W. Brannock., 1962. Rapid analysis of silicate, carbonate and phosphate rocks, U.S. Geol. Surv. Bull. No. 1143, 56 pp.

Subramanian, V., Van Grieken and L. Van't Dack 1987. Heavy metal distribution in the sediments of Ganges and Brahmaputra Rivers. Environ. Geol. Water Sci., Vol. 9, No. 2: 93--103.

Subramanian, V., 1980. Mineralogical input of suspended matter by Indian rivers into the adjacent areas of the Indian Ocean. Mar. Geol., 36: M29--M34.

Subramanian, V., 1979. Chemical and suspended sediment characteristics of rivers of India. Jour. Hydr., 44: 37--55.

Subramanian, V., AL. Ramanathan and P. Vaithiyanathan., 1989. Distribution and fractionation of heavy metals in the Cauvery estuary, India. Marine Pollution Bulletin, 20: 286--290.

# CHAPTER 22

# CROPPING SYSTEMS

D. N. BORTHAKUR

The agriculture of the northeast region of India has lagged far behind the developed states of the country. There is a multitude of factors responsible for this deplorable state. Reflecting on the present status, this chapter discusses crops and cropping patterns, strategy for cropping patterns, cropping patterns for flood prone areas, and intensive cropping systems.

## 22.1 Introduction

The North Eastern Region of India, comprising the seven states, namely, Arunachal Pradesh, Assam, Manipur, Meghalya, Mizoram, Nagaland and Tripura, presents a specialised geographic area because of (a) the variation in altitude ranging from near sea level to over 5,000 meters above the mean sea level (m.s.l.), (b) high rainfall with high humidity, (c) undulating topography with varying agro-climatic situations (d) geographic isolation, being connected with the rest of the country by a 56 km long narrow corridor, the width of the corridor being 38 km. The entire boundary of the region, except the 38 km wide corridor, is with foreign countries, namely, China (1,350 km), Bangladesh (900 km), Burma (1,350 km) and Bhutan. Nearly 90 percent of the population is rural, and agriculture is the most dominant vocation of the people.

There are vast potentialities for development of agriculture in the northeastern region. The wide variability in the agroclimate permits cultivation of all types of agricultural and horticultural crops, plantation crops, medicinal and aromatic plants, etc.. The region is well known for its biodiversity and it is also the primary or secondary source of origin of many crops and plants. The other assets for agriculture include the abundant sources of water, fertile soil and potentialities for development of a wide range of agro-based enterprises.

In spite of the enormous potential, the agriculture of the northeastern region has remained rather under-developed because of a number of constraints, which can be broadly classified under (a) infrastructural constraints, (b) climatic constraints, (c) biophysical constraints, (d) socio-economic constraints and (e) constraints of management. In addition to these, there are two special problems, namely, shifting cultivation or jhumming in the hills and the chronic floods in the plains. Assam can be divided into two physiographic divisions---the hills and the plains. The hills division separates the two valleys of the Brahmaputra and the Barak rivers. The Brahmaputra



*V.P. Singh et al. (eds.), The Brahmaputra Basin Water Resources*, 401-410.

River enters the north eastern region through Arunachal Pradesh before entering Assam. It traverses right across the state of Assam and then flows into Bangladesh. The Brahmaputra valley covers 56,000 sq. kms of alluvial plains interspersed by small hillocks and swampy low lands. The Brahmaputra in its course in Assam receives over 35 tributaries, all of which emanate from the hills of the neighbouring states surrounding Assam.

## 22.2 Present Status

While discussing the present status of performance in the agriculture sector, including cropping patterns, one has to consider the prevailing environment for growth. Because of the various constraints under the different categories, mentioned earlier, the agriculture of the region as a whole has remained far behind that in the developed states in the country. The average size of a farm holding in the state is only 1.31 ha, and 82.54% of the total holdings are marginal and small holdings. While the investment capability of the farmer is very poor, the availability of institutional finance is also very meagre. The power in the farm sector is practically not available. The communication network is very poor. Agri-business does not belong to the farmers and the middlemen take advantage of the lack of marketing infrastructure and storage. Above all, the overall rural environment is not at all conducive to agricultural growth. Road transport, and communication, storage, market, literacy, good schools, medical facilities, sanitation and hygiene, nutrition, post harvest facilities, etc., all make up for a favourable rural environment for agricultural growth. The rural areas and farmers are practically deprived of all these amenities.

## 22.3 Crops and Cropping Patterns

The major field crops raised in Assam include rice, wheat, maize & other cereals, pulses, oilseeds, jute, sugar cane, and potato. The horticultural crops grown in the state are fruit crops, plantation crops, tuber crops, spices, and vegetables. Cultivation of ornamentals for commercial purposes has been initiated only recently. Rice is by far the most important crop of Assam, occupying an area of 25.80 lakh hectares or about 80 percent of the annual cropped area. Of this area, winter or kharif rice (known as 'Sali') occupies 70 % of the total area under rice. Sali is cultivated from June--July to November--December. Next in importance, as far as the area is concerned, is the 'ahu' or autumn rice which occupies 25% of the total area under rice. The 'ahu' crop season ranges from March--April to July. The third crop of rice occupies only 5% of rice area in the state. 'Boro' and early ahu crop are raised in this season spreading from December-- January to April -- May. Area, production and productivity of the three rice crops during 1990-91 to 1992--93 are presented in Table 22.1. The area covered under high yielding varieties is little over 40% of the total cropped area.

The area under wheat in the state increased from 5,000 ha in 1965--66 to 93,000 ha in 1985--86. However, the area under wheat indicates a declining trend since 1990-91 as evident from the following table (Table 22.2). The area under maize (corn) and other cereals is rather low being only 21,000 and 15,000 hectares, respectively.

Table 22.1. Area (in ha), production (in tonnes), and average yield (in kg/ha ) of 3 classes of rice.

| | Sali | | | Ahu | | | Boro/Early ahu | | |
|---|---|---|---|---|---|---|---|---|---|
| year | Area(A) | Production (P) | Yield /ha (Y) | A | P | Y | A | P | Y |
| 1990--91 | 1801,220 | 2565,423 | 1,446 | 608,150 | 522,189 | 873 | 117,374 | 182,581 | 1,556 |
| 1991--92 | 1804,541 | 2486,646 | 1,399 | 638,952 | 494,223 | 785 | 128,468 | 216,295 | 1,684 |
| 1992--93 | 1776,189 | 2442,472 | 1,396 | 635,131 | 613,696 | 982 | 147,957 | 243,454 | 1,648 |

Source: Department of Agriculture, Assam.

Table 22.2 Area, production, and productivity of wheat (area in lakh ha, production in lakh tonnes, yield in kg/ha).

| Year | Area | Production | Average Yield |
|---|---|---|---|
| 1990--91 | 0.84 | 1.05 | 1,248 |
| 1991--92 | 0.76 | 1.11 | 1,455 |
| 1992--93 | 0.74 | 0.79 | 1,066 |

Source: Department of Agriculture, Assam.

Pulses are cultivated primarily during the rabi season, the important pulses raised during the season being black gram, green gram, pea, lentil, and rajmah. About 10--15 thousand hectares, out of the total area of 1.2 lakh ha under pulses, are cultivated during the kharif season, the crops raised being arhar (pegion pea), summer greengram and summer blackgram. The productivity of pulses is rather low being around 450--550 kg per hectare. Pulses are grown in the rain-fed areas only. Edible oilseed crops grown in the state include rape and mustard, niger, soyabean, sesamum and ground nut, while linseed and castor are the non-edible crops. However, rape and mustard alone occupy 92% of the total area under oilseeds which is around 3.5 lakh hectares. Productivity of oilseeds is very low since the crops are grown in the rain-fed areas during the rabi season, which is a dry season. Jute is the primary fibre crop cultivated in around one lakh hectares although mesta is also grown in about 7 thousand hectares. The production of jute ranges from 8.5 to 10.5 lakh bales annually. Sugar cane is another important crop in the state grown in about 40 thousand hectares, the total production being around 14.5 to 15.5 lakh tonnes annually. The area under potato is around 63.5 thousand hectares, the average yield being about 6 tonnes per hectares.

A large number of horticultural crops are raised in the state. The important horticultural and plantation crops include orange banana, pineapple, papaya, limes and lemons, ginger, turmeric, chillies, tapioca, sweet potatos, arecanut, coconut, betel vine, black paper, an array of summer and winter vegetables, and onion, etc.. The area under fruits is around 0.78 lakh hectares the total production being about 9.2 lakh tonnes. Similarly, the total production of vegetables is around 19 lakh tonnes, the crops occupying an area of about 1.55 lakh hectares.

Mixed farming is usually practised by farmers because rearing livestock and raising of crops are practised commonly. Growing of horticultural crops is primarily confined to homesteads, except pineapple and oranges which are raised commercially. Shifting cultivation and the 'bun' system of farming (raising crops in bunds along the slope) are practised in the hills only. As indicated earlier, over 90% of the total cropped area is rain fed. Hence the rain-fed cropping system is the primary system of farming. The Brahmaputra River has made the valley fertile and has contributed to the favourable moisture regime in the area. The high rainfall has also contributed to adequate water supply so that the rainfed farming has become possible.

The cropping intensity in the state of Assam is around 125 percent only. The lack of irrigation facility is the primary cause for such a situation. There are heavy rains and floods during monsoon season, but from October to May, when irrigation is required for intensive cropping, there is practically no irrigation at all. Production and productivity of the major crops during 1991 to 1993 are shown in Table 22.3.

Table 22.3 Production and productivity of four major crops of Assam.

| | 1990--91 | | 1991--92 | | 1992--93 | |
|---|---|---|---|---|---|---|
| Crops | Production | Productivity | Production | Productivity | Production | Productivity |
| Rice | 32.70 | 1,313 | 31.97 | 1261 | 33.0 | 1,308 |
| Rape & Mustard | 2.95 | 535 | 3.03 | 586 | 2.90 | 476 |
| Pulses | 1.21 | 433 | 1.26 | 472 | 1.19 | 475 |
| Jute | 866 | 1,632 | 8.67 | 1,537 | 10.34 | 2,011 |

Production: in lakh tonnes (Jute in lakh bales )
Productivity: in kg/ha.
Source: Department of Agriculture, Assam

As for the cropping pattern, Ahu rice (summer paddy) followed by Kharif rice (Sali or winter paddy) and jute followed by Kharif rice are the two primary cropping patterns in the low lying areas. In the rainfed uplands Ahu/olitorins jute followed by mustard/wheat is the common cropping pattern where double cropping is practised. Comparatively there are more areas under double cropping in the districts of Goalpara, Kamrup, Nowgong and Barpeta. The sandy loam soil of these areas is a major contributing factor followed by the hard working nature of the farmers for such a practice of double cropping. The availability of irrigation water, however, remains the primary factor in determining the multiple cropping practice.

There are specific problems in raising the ahu or summer rice. The dependence on rains forces the farmers to raise ahu paddy as a broadcast crop since the land cannot be puddled. Further tillage and broadcasting has to be done around the end of March and April, depending on the onset of rains. Weeds crop up profusely in the season and their control poses great problems, particularly in the clayey soils. The crop is ready for harvest during June or even July when there are usually heavy rains. As a result it is difficult to harvest and dry the crop. Threshing and storage is not possible unless the grains can be dried. The seeds may sometimes germinate or may be spoiled by fungal growth if kept in wet conditions for long. There is another social problem related to cultivation of ahu rice. The farmers usually let the cattle loose for grazing during this period. Unless the community as a whole cultivate during this period, fencing as well as keeping cattle off the field become costly for individual farmers. Furthermore, the damage of small plots by birds also creates problems. Thus the provision of irrigation and motivation of the farmers to raise an early ahu crop will go a long way to increase the cropping intensity.

## 22.4 Strategy for Cropping Pattern

There are a number of factors that influence the adoption of cropping patterns. These are:

(i) Land and soil: Suitability of the land for specific crops and the type of soil will have to be taken into consideration for deciding the crops to be raised. Low lying areas and upland sandy and clayey soils, etc., are some such considerations;

(ii) Irrigation: The availability of irrigation or otherwise will influence the crop as well as the cropping system;

(iii) Environmental and climatic factors: Precipitation, temperature, soil moisture status, etc., also determine the type and number of crops that can be raised;

(iv) Market: The dearth of proper marketing facilities acts as a disincentive for raising 2 or more crops, particularly if the product is perishable;

(v) Time gap between two crops: For raising a crop after harvesting, a minimum time gap is required for processing the first crop and to prepare land for the second. The time required will depend on the horsepower available with the farmers in terms of tools and equipment and labour force;

(vi) Capability of farmers: The small and marginal farmers in most cases do not have the resources, including the capability of investment, to engage in for intensive cropping.

Some of the other limitations upon intensive cropping are inadequacy of management techniques, lack of post-harvest facilities, and the dearth of grazing lands for cattle.

## 22.5 Scope for Intensive Cropping

As indicated earlier, the primary patterns of intensive cropping are (a) double cropping of ahu/jute and Sali paddy practiced on a limited area, (b) mixed cropping of ahu and bao (deepwater) rice (or floating rice) in pockets of lowlying inundated areas, (c) double cropping of boro (spring rice) and Sali rice, (d) double cropping of ahu/summer pulse and wheat/mustard and double cropping of winter and summer vegetables. The intensity of cropping is still around 125 to 130 percent only.

There is adequate scope for inter-cropping, which is now rarely practised. For example, green and black gram can be raised as an intercrop with maize, ladies finger, or even sugar cane in the first year of planting. Ginger, turmeric, and chillies can be raised as an intercrop with arecanut and coconut etc. Seasonal leguminous crops, such as pulses, soybean, etc., can also be raised as an intercrop with many horticultural trees/plants. The practice of relay cropping was in vogue in the past in some areas, where farmers sowed by broadcasting the ' Khesaridal ' (Lathyrus) in the paddy field prior to the harvesting of rice, taking advantage of the residual moisture in the fields. The practice has died out after prohibition of cultivation of Khesari dal. It is, however, possible to cultivate lentil and pea as relay cropping in the rice field. Various forms of multiple cropping can be practised, depending on the irrigated land, rainfed lowland and upland, etc.. Some such feasible cropping patterns are indicated in what follows.

## 22.6 Cropping Pattern for Flood Prone Areas

The cropping pattern for flood-prone areas should be such that crops can be raised during the flood-free period. The seasons for raising crops in such areas are pre-flood and post-flood seasons. This will require restructuring of the cropping pattern. Such restructuring of cropping patterns will also help to increase productivity of the major crop, i.e., rice. The scope for restructuring of the cropping pattern and feasibility for growing various crops are indicated in Figure 22.1. Similarly the scope for growing rice in flood prone areas is shown in Figure 22.2.

It may be mentioned in this connection that the productivity of rice during the main season, namely, kharif, from July--November--December is much lower compared to boro or early auh season owing to a variety of reasons. The high rainfall, cloudy weather, lesser sun-shine hours, low intensity of sunlight, waterlogged condition, and the downward trend of temperature towards the ripening stage of grains, are all unfavourable factors during the kharif season, leading to an abundance of pests and diseases, low response to fertilisers, and finally to lower productivity. On the other hand, these unfavourable factors are not present during the early ahu season from February to May. There are more sunshine hours, no cloudy sky, better intensity of sunshine hours, as well as favourable temperatures, so that with less incidence of pests and diseases, and much better response to fertiliser, the productivity of rice is much higher. Thus for raising productivity, and hence production, of rice it is essential that the early ahu season be converted to the major rice growing season for the state. Such a change will also help cultivation of rice, mitigating floods and also the high rainfall during harvest. However, irrigation needs to be ensured for this purpose. It is possible to develop irrigation during this season through various means of minor irrigation systems, particularly utilising the abundance of ground water in the state.

Table 22.4 Feasible cropping pattern.

| Situation | Cropping pattern | | |
|---|---|---|---|
| | Feb.-- Mar.-- May | July -- Oct. | Oct. -- Nov. -- Feb. |
| (1) | (a) | (b) | (c) |
| Flood free Irrigated Area. | Early ahu | Sali | Mustard/Pea/Lentil/Rajma/ vegetables/Potato /Garlic |
| | Early ahu | Green gram/ Blackgram (Sept. Oct.) | Pea/Lentil/Mustard/Potato/ vegetables wheat |
| | Early ahu | Soyabean | Pea/Lentil/vegetables. |
| Flood-free Rainfed areas (Low & Medium land) | Jute (March to July) | Greengram/ Blackgram. | Wheat/Mustard/Potato Vegetables /Pea/Lentil |
| | Jute Direct Sown Ahu (April - July) | Sali/Soyabean Green/Black-gram | Lentil/Pea/Vegetables Wheat/Mustard/ Vegetables/Pea/ Lentil/ Onion/ Potato |
| Rainfed Upland | Blackgram/ Greengram | Summer Vegetables. | Mustard/Winter Vegetables |
| | Olitorius Jute | Greengram/ Back gram | Mustard/buck wheat vegetables/Pea/Onion. |
| | Sugar cane intercropped with green/black gram | | |
| | Maize inter-cropped with green/ Black gram | Summer vegetables /Black/ green gram. | Mustard/Vegetables/Pea/ Onion |
| | NOTE: Ground nut (Peanut) can also be raised successful in sandy loam soil. | | |

## 22.7 Intensive Farming Systems

Since agricultural land is shrinking with growth of population and development, it has become essential to engage in for intensive farming systems. Various forms of agro-forestry-based farming system are feasible in the state. Such a farming systems may include syli-pastoral, horti-pastoral, agri-horti-pastoral cum fisheryland livestock based farming systems, etc.. An integrated approach to farming systems through scientific management of land and water on the principles of watershed management is also feasible.

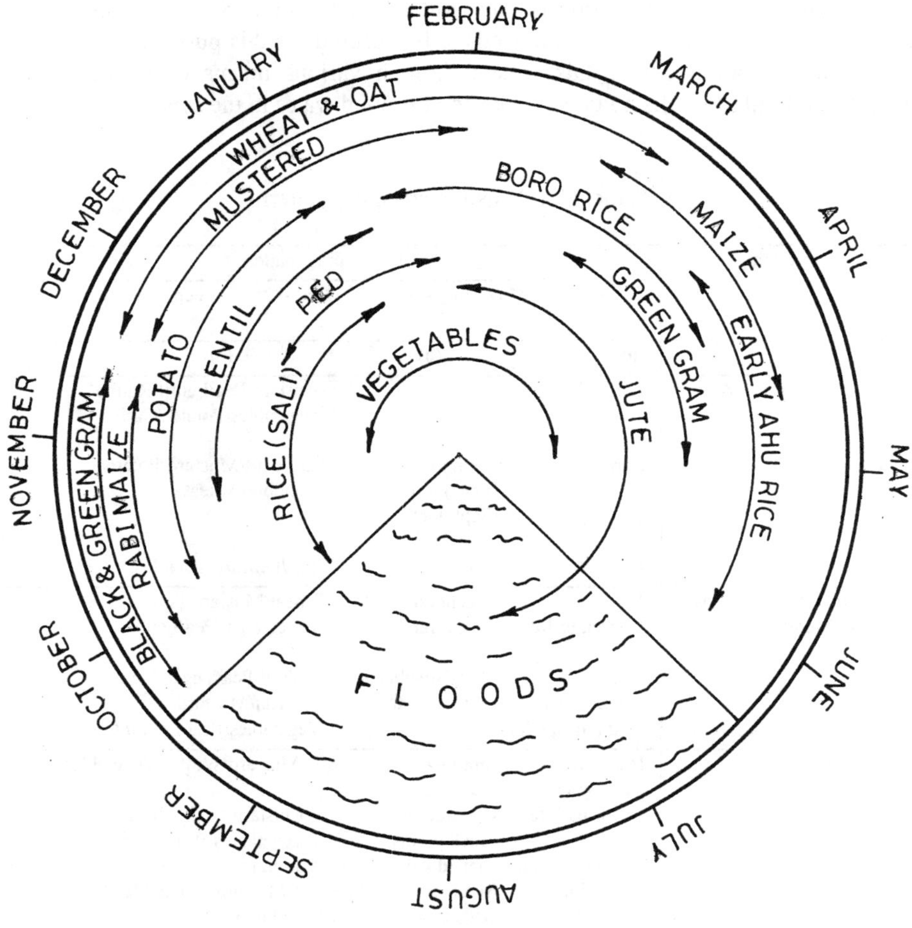


Fig. 22.1 Restructuring of cropping patterns.

## 22.8 Multilayer Cropping

Various forms of multi-layer cropping, particularly in respect of horticultural and plantation crops, can be developed and practised as a measure of intensive farming. A few examples of such feasible multi-layer cropping are:

(i) Arecanut with betelvine/black pepper as a companion crop, and ginger, turmeric, and chillies as intercrop.

(ii) Coconut - Papaya/lime - pineapple/ginger.

(iii) Banana - pineapple - cowpea.

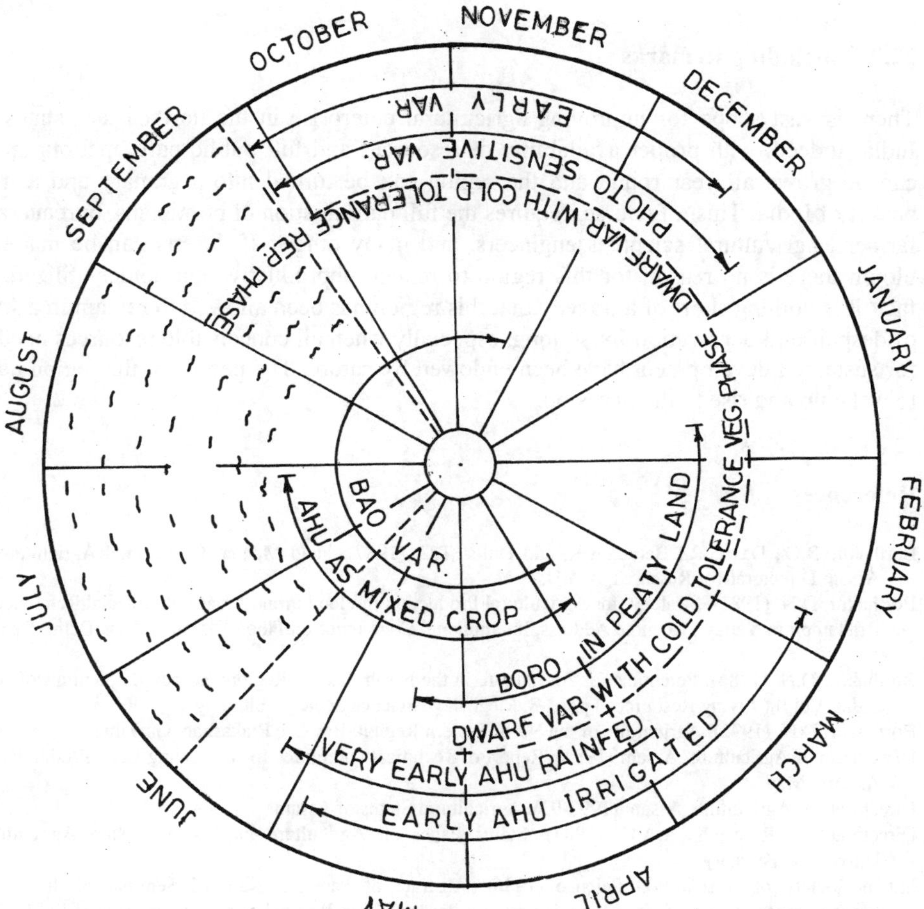


Fig. 22.2 Rice cultivation during flood free period.

Since agriculture in Assam is still primarily rain fed, appropriate technology for rainfed farming, including cultivation, preparation of land, appropriate crops and varieties, technology for harvesting and preservation of rain water, moisture retention, etc., need to be followed for ensuring productivity. Principles of watershed management are important in this regard. It would be evident from the preceding discussions, that there is ample scope for diverse cropping systems and patterns with immense potentialities to increase agricultural production. However, need based research, effective extension programmes and appropriate public policies must continuously back up the agricultural production systems. Similarly, certain infra-structural facilities will have to be catered for as intensive cropping patterns are adopted. These include rural storage structures, regulated market, effective institutional credit, timely supply of other inputs. In the case of horticultural crops, even concentrated plants in production centres and proper link up with processing plants and markets, etc. may be required.

## 22.9 Concluding Remarks

There is vast scope for improving agricultural enterprise in the northeastern states of India. Indeed, with proper availability of resources and full public participation, crops can be grown all year round and the region can be turned into a granary and a fruit basket of India. This, of course, requires the full participation of politicians, bureaucrats, farmers, agricultural scientists engineers, and many others. If deserts can be made to bloom there is no reason for this region to remain unproductive and under-utilized. In fact, it is nothing short of a tragedy that this region has been allowed to remain in a state of despair and deprivation for so long, especially when all conceivable resources needed for sustained development have been endowed by nature. The people of the region need to wake up and rise to the occasion.

## References

Barthakur B.C., Das G.R., Barua, A.R, and Dutta, T.C. (1987). Field Manual for Rainfed Agriculture in Assam Directorate of Research, A.A.U.

Borthakur, D.N. (1981). Strategy for Agricultural Production in Flood Prone Areas with Special Reference to Brahamputra Valley : Keynote Address, International Conference on Flood Disasters, New Delhi, Dec. 3-5.

Borthakur, D.N. (1984). Potentials for Agriculture in the North Eastern Region: Resource Potentials of N.E. India, Vol II ( Living Resources ) pp. 1-8, Meghalaya Science Society, Shillong.

Borthakur, D.N. (1992). Agriculture of the North Eastern Region: Bee Cee Prakashan, Guwahati.

Directorate of Agriculture, Assam (1984). Report of Technical committee for Improving Crop Production in Assam, May.

Directorate of Agriculture, Assam (1992-93). Agricultural Status of Assam.

Directorate of Research, A.A.U. (1984). Status Paper on Agriculture in Assam: Assam Agricultural University, February.

Indian Society of Agricultural Science (1986). Abstract of Papers: National Seminar on Integrated Management: Approach for Maximising Crop Production in Rainfed and Problem Areas: IARI, New Delhi, February.

International Council for Research in Agroforestry (1983). Guidelines for Agroforestry, Diagnosis and Designs, Working Paper No. 6, Nairobi.

# CHAPTER 23

# GROUND WATER DEVELOPMENT

B. PURKAIT

The surface water resources of the Brahmaputra basin are not adequate, under present condition to meet water demands during the rabi season. There is, however, ample ground water potential in the basin but it has not yet been fully exploited. Therefore groundwater needs to be developed without further delay. This chapter discusses ground water prospecting, development and management, as well as the ensuing environmental impact.

### 23.1 Introduction

Ground water is an integral part of the hydrological cycle, and is therefore inseparably linked with the waters of the rivers, lakes, reservoirs, etc. and also to rainfall. It must be acknowledged that there cannot be one standardized set of laws to control and regulate groundwater in all basins as each is likely to involve varying hydraulic, economic and legal problems. Ground water occurs both under free and confined conditions. The ground water reservoir receives recharge through the direct precipitation over the area and also through influent seepage from the surface runoff and return of the applied irrigation water.

The piedmont plains and alluvial cones contain coarser sediments and have high intake capacities and deep water table, and generally form recharge areas. The quaternary sediments of the Brahmaputra basin contain some of the most prolific aquifers. The heavy rainfalls of about 2500 mm, a good recharge zone fringing the hill front, and the presence of thick granular zones at shallow depths are all definite indications that the potential of ground water resources of the basin should be rated as high.

The surface water resources of the Brahmaputra basin are not adequate during the Rabi season as rainfall is low and erratic. Therefore it is essential to develop the dependable ground water resources expeditiously through tube wells.

### 23.2 Ground Water Prospecting and Exploration

Direct recharge by rainfall percolation is the most prominent source for any ground water reservoir anywhere. It not only operates for narrow tracts of the Brahmaputra basin but a large amount of subsurface inflow also takes place from the vast area to the north up to the foothills in the north bank where the rate of infiltration is quite high. The



*V.P. Singh et al. (eds.), The Brahmaputra Basin Water Resources*, 411-418.

upper part of the piedmont belt, comprising mainly unconsolidated boulder-cobble-pebble-sand and silt/clay beds, serves essentially as a recharge zone for ground water. As the water table is moderate to very deep (7 to 29 m below ground level), special drilling techniques employing percussion rigs are required and the tube wells have to be necessarily deep (100 to 150 m) to register a discharge of about 90,000 to 100,000 liters per hour.

The type of drilling equipment required depends on the geological formations to be pierced through. In the zone of about 15--20 km in width along the foothills of the Himalayas the boulder formation is likely to encounter a very shallow depths, and cable-tool or percussion rigs capable of drilling 60 cm to 25 cm diameter are more suitable for the construction of telescopic holes down to 300 m. In the area south of it up to the Brahmaputra River, except for the Pagladya basin, boulder gravel beds occur at various levels within a depth of 200 m. In this area Rotary-cum-percussion rigs and Reverse Circulation Rotary rigs capable of negotiating 15 cm diameter boulders, and capable of drilling 60 cm through 30 cm diameter telescopic holes down to 300 m, are suitable. In the southern bank boulder gravel beds are not frequently encountered, and both Direct and Reverse Circulation Rigs capable of constructing 200 m deep heavy duty tube wells will be suitable. The proposed field should be in a circular or square pattern.

Sand samples for each borehole of the aquifer zones should be collected at an interval of 3 m depth for size analysis while selecting the slot opening. When there is a marked change of strata, it should be preserved for detailed sedimentological studies. For accurate subsurface correlation to depict the subsurface geology, the elevation of the borehole sites with reference to the mean sea level (m.s.l.) should be recorded. The accurate lithologic boundaries can be obtained by electric logging. After selecting the proper slot size the length and diameter of the strainer required depends on the total open area required to allow an entrance velocity of 3 cm/sec for the anticipated discharge. An extra 50% to 60% of open area is provided for the anticipated blockage of the open area owing to clogging and encrustation. The tube wells should be gravel packed after lowering the assembly. The gravel pack depends on the grain size characteristics of the aquifer zones as well as the slot size derived from the size analysis data. The gravel pack should neither be too thin nor too thick. In a very thin pack, some portions of the screen may not be covered, especially when the borehole is out of plumb or the tube well assembly is not exactly in the center of the bore. On the other hand a very thick pack hinders the development of the aquifer in a borehole drilled by a Direct Rotary Rig because it is difficult to remove the invaded drilling mud from the aquifer. It is observed that a gravel pack of 10 to 20 cm thick is the most suitable. In the case of boreholes whose diameter is restricted to the method of drilling adopted, pre-packed gravel filters can be used. The gravel suitable for shrouding tube wells is easily available from the river beds of these Himalayan streams in piedmont and alluvial fan areas.

After completion of tube wells, these should be developed thoroughly by using different development methods, e.g., surging, back-washing, over-pumping, etc. so that the entire aquifer zone is clear of the invaded drilling and all the fines are removed and the water is free of sand for the optimum rate of pumping. Many of the well failures are attributed to the improper development and the inadequate open area provided. After completion of the development a pumping test is conducted for determining the aquifer characteristics and the well capacity for pump selection and its setting.

## 23.3 Ground Water Development

The river systems of the Himalayas have a vast potential for water resource development. The objective of water resource development is to equally distribute the supply and transfer of water from surplus to deficit areas. Hence it requires the intervention of natural flow of water with proper planning minimizing the likely adverse effects like submergence of forests, cultivable land and human habitats along with such effects on flora and fauna. The planning for water resource development is to design safe facilities like dams and diversion structures and to take care of the seismo-tectonic set up of the Himalayas. A dam which is designed and constructed to account for the destabilizing forces with sufficient factor of safety is safe against failure. The destabilizing forces for which the dam is designed, one must accommodate forces owing to the occurrence of a most likely event during the life time of the structure.

Three types of dams, including gravity, arch, and embankment dams, are constructed. Gravity and arch dams are founded on rocks. The properties of the foundation materials are studied to withstand the transient vibrations caused by the design earthquake. The cross-section of a dam takes care of the probable maximum earthquake forces through additional defensive measures which can prevent any damage of the dam from an incipient state of stress.

On the other hand the embankment dams are founded on any foundation materials. They have a better energy capacity for healing because they can admirably absorb strain through compaction and adjustment of particle without getting distressed beyond a reparable state. They are easily amenable to the introduction of defensive measures in the construction. Therefore in the Himalayan region they may be a natural choice.

The Bureau of Indian Standards (BIS) deals with standardization of design of dams through River Valley Divisional Council and the RVD 9 Committee. At present the BDC 39 Committee takes care of the remaining structures. Of late the RVD 9 has undertaken revision of IS 1893 laying down 'criterion' for Earthquake Resistant Design of Structures entitled as 'Guidelines for A Seismic Design of Dams' in 3 different parts to deal with different components of the safety measures.

All the Himalayan South Asian countries depend on the vast river system of the Indus-Ganga and the Brahmaputra. They are mostly fed by the south west monsoons and to some extent by the summer melt of the winter snow. The availability of waters from these rivers is extremely variable in time and space because of higher intensity of precipitation and limited span of rainy days in a year. During the monsoon period, the river flow is about 80% of the total flow of the year. Therefore it cannot be utilized evenly or as per requirements during the year, unless stored in reservoirs created behind dams. The water resources drained by the Ganga-Brahmaputra-Meghana (Barak) basins annually are of the order of 1,110 billion cubic metres (BCM). The break-up of Ganga,

Brahmaputra and the Meghana runoff is roughly 525,537 and 48 billion cubic meters, respectively.

Being perennial in nature, the water resources of these rivers have been easily harnessed by constructing diversion structures and inundation canals since historic times. Since the rate of growth of population is gradually increasing in this region, the need for storages to meet the ever-increasing needs of the dependent population is obvious. The surface waters recharge the ground waters in the plains which are being increasingly utilized as they are found more reliable in meeting critical requirements. As a result the recharge owing to river flow gets artificially augmented, causing a reduction in the surface flow itself making the rivers gradually dry up in the post-monsoon months. The storage also reduces the post-monsoon flow. Hence, proper planning is required to regulate the release for maintaining the carrying capacity of the rivers. The water utilisation from these rivers has so far relied on diversion structures rather than storages.

The water resources development of the Brahmaputra and Barak basins are as follows (Thatte, 1996):

| Sl. No. | Particulars | | Brahmaputra | Barak |
|---|---|---|---|---|
| 1 | Annual average run-off | (BCM) | 537 | 48 |
| 2 | Utilisable run-off | (BCM) | 24 | 10 |
| 3 | Present use | (BCM) | 3 | 1 |
| 4 | Planned, 2000 AD | (BCM) | 6 | 3 |
| 5 | Storages (10 MCM & more) | | | |
| | (a) completed | Nos. | 4 | nil |
| | (b) under construction(UC) | Nos. | 2 | 2 |
| 6 | Major projects completed/UC | Nos. | 0/3 | 0/0 |
| 7 | Multi-purpose projects | | | |
| | (a) completed/UC | Nos. | 7/7 | 1/0 |
| | (b) identified | Nos. | 166 | 32 |

The water resource development from shared resources between Bhutan and India is being planned in mutual consultation and agreement. In the Ganga-Brahmaputra basin the scope for transfer of water is from east to west because of the availability per capita for the basin of Ganga, Meghana, and Brahmaputra in the proportions of 1:2:7. The proportion of the availability per unit of cultivable area is also similarly higher in eastern rivers

The zonal planning based on the concept of watershed management appears to hold much promise for optimal development of water resources and for maximization of regional welfare (Goswami,1992b). Watershed management practices would include the various measures adopted to protect the watershed from erosion and degradation such as afforestation, soil conservation, etc.. The change in the existing crop calendar and cropping pattern is suggested by many as a viable method of avoiding the impact of floods and ensuring sustained productivity especially in respect of the Sali crop which coincides with the flood season (Goswami,1990a; Barthakur,1981).

## 23.4 Ground Water Management

The entire hilly catchment of the Brahmaputra needs to be studied in detail in order to delineate areas which are more susceptible to erosion. Rigorous afforestation of such areas which are more susceptible to erosion is required to be carried out in order to check erosion and overcharging of the streams with debris which eventually contribute to flooding. Forests are known to lessen the rate of runoff, which will thus help in preventing floods. Check dams on the tributary streams may also be considered to control silt, but these dams may only provide short term flood protection because they are likely to be filled up soon. Straightening of the river channel, man made levees, and higher bridges are other measures which can be considered for the flood of the mighty Brahmaputra.

The ground water possibility in the area covered by inselbergs is very limited and can be exploited through dug wells constructed in the weathered mantle. The presence of inselbergs or shallow buried pediments may act as barriers to ground water movements and locally the subsurface flow paths may be sinuous. All the run off systems are effluent and ultimately along with all the tributary base flows, ground water moves slowly to the Brahmaputra itself to augment its run off. During monsoons this process is likely to be reversed temporarily. It has to be noted that the direct recharge by rainfall percolation is the most important source of the ground water reservoirs. In delineating the ground water potentiality of the area, basement slopes are to be recorded. In the Brahmaputra basin the basement, in general, has a gentle slope from south to north and from east to west.

The nature, composition and water yielding capacities of aquifers in the vicinity of the inselberg zone up to a distance of 0.5 km suggest that the ground water can be developed with the help of shallow tube wells. On the other hand, the nature, composition and hydrological characteristics of aquifers in the alluvial area 0.5 km away from the inselbergs indicate that the large-scale development of ground water by deep tube wells may not be feasible. However, a few medium duty tube wells down to a depth of 105 m may be possible.

The quality of ground water is good except for a high concentration of iron, which can be removed by aeration and subsequent filtration. Thus a systematic qualitative and quantitative assessment of ground water potential should be carried out to boost rural economy by adopting multi-cropping practices. It is essential to search for suitable aquifers at depths where the permissible amount of iron is present. A systematic deep drilling programme may be taken up to achieve this goal.

Tapping of ground water for irrigation and domestic purposes is a serious problem in the piedmont zone. Here the water goes down very deep owing to thick terrace deposits. As such the construction of dug wells becomes a difficult and risky venture. A programme of drilling using percussion drills and development is necessary to tackle this problem.

Subsurface geology is also vital for a comprehensive knowledge of the ground water condition. Stress should be given to preserve the lithologs in respect of borings conducted by private or Government agencies. To know the geohydrological features of the area a systematic recording of the seasonal fluctuation of the water table and piezometric level by periodic measurements over a protracted period in suitably selected observation wells is to be carried out. Basement configurations are to be obtained based on the results of geophysical studies for a clear picture of the subsurface geology.

## 23.5 Environmental Impact

The Brahmaputra basin is underlain, for the most part, by very young and unweathered sedimentary formations. The riverscape is dynamic. The trend and tempo of its morphological evolution are determined by tectonic and fluvial processes. Consequently, the geological, seismic and fluvial hazards have great impacts on the environmental set up of this area.

### 23.5.1 GEOLOGICAL HAZARDS

The causative processes of geological hazards can be grouped as (i) endogenic and (ii) exogenic. The endogenic process includes earthquakes owing to the movement of the earth's crust and the failure of crustal blocks along discrete fractures in response to deep seated crustal processes. The morphotectonic history of the Himalayas and the Shillong plateau are greatly affected by this type of movement. The exogenic process includes the migration of rivers, land erosion, and floods. The instability of the fluvial regime is caused by the fluctuation of water discharge, high incidence of rainfall, vulnerable erosion of the youthful catchment areas of the Himalayan rivers causing high sediment discharge, occasional landslides, and spasmodic dislocation of equilibrium by earthquakes and tremors.

### 23.5.2 SEISMIC HAZARDS

The valley particularly its upper part and its adjoining hill ranges are seismically very active, causing major earthquakes of 1897 and 1950. The intensity of the earthquake was 8.7 on the Ritcher scale and caused extensive landslips on the Himalayan hill slopes, subsidence and fissuring of the ground in the valley, and changes in the course and geomorphic configuration of several tributary rivers as well as the main Brahmaputra (Oldham, 1899; Poddar,1952).

The magnitude--frequency analysis of earthquakes in NE India during 1920 to1969 is given below:

| Magnitude (Ritcher scale) | Frequency |
|---|---|
| 5 to 6 | 237 |
| 6 to 7 | 161 |
| 7 to 8 | 15 |
| 8 above | 3 |

However, the destructive effects are likely to vary in different geomorphic units. In the metastable slopes of the inselberg zones the landslides are expected to be triggered. The flood plains are mostly affected by changes of river courses, subsidence and flowage of sand by liquefaction. Moreover, engineering structures in the areas where a thick gravel bed exists in the subsurface may be least affected owing to the damping effect of the gravel on propagating shock waves.

The seismic activity has a great impact on the fluvio-sedimentary regime of the Brahmaputra River and its tributaries. The effect of such an earthquake of 1950 on the highest and the lowest flood levels of the Brahmaputra is that the lowest water level rose as high as 3 m owing to this earthquake (Dulal, 1998). This earthquake caused aggradation and excessive bank erosion in the upper Brahmaputra, but its impact appears to have petered off upstream of Gauhati. There was a 1.5 to 10 m rise in the bed levels in some cases. Consequent on the reduction of channel capacities, the flood problem in Brahmaputra has been greatly aggravated. Hence the construction of storage dams across its main tributaries is necessary as an effective flood moderation measure. Besides, the earthquakes of 1889 and 1988 are known to have affected considerable changes on the Brahmaputra River morphology.

### 23.5.3 FLUVIAL HAZARDS

The Brahmaputra and its tributaries cause extensive floods and land erosion, which constitute a perpetual hazard in the lower Assam. Discharge and hydraulic geometry of the river, nature of bank materials, and the behavior of groundwater during the rising and falling stages of the rivers play a great role to the rate of land loss. Of these factors, discharge plays the dominant role. Therefore the land destroyed annually by the main Brahmaputra River is of several orders higher in intensity than the loss inflicted by its tributaries.

The instantaneous economic loss caused by the erosional activity is slowly recovered. But the time gap between the destruction and the slow recovery keeps the people of the area in a state of perpetual economic stress. The long term consequences of the erosional process, on the other hand, are beneficial to man for the development of new fertile lands. The neo-flood plains become more fertile than the old terraces because of more mineral nutrients and water.

Thus the tectonic instability, highly potent monsoonal rainfall, and snowmelt from Himalayan glaciers, steep slopes, and easily erodible soil of the catchment area combined provide the Brahmaputra its unique flow regime. The high sediment yield owing to high rates of unit discharge causes drainage congestion in the valley. However, the quantum of morphological transformation owing to the cumulative activity of erosion and deposition is not balanced on both banks. The contrast may be attributed to the geological setting. Resistant crystalline rocks (of inselbergs) and plastic clay (of Chappar Formation) are more abundant along the south bank than the north bank in some reaches where such contrasts are observed. The Kopili and Kalong rivers cause a lot of damage every year. It is advisable to launch a river action plan to control these rivers by various treatments such as diversion of streams, construction of bank line, etc.. If part of the flood waters of the tributaries of the Brahmaputra in the foothills are diverted to spread over the area by heavy intake through canals or any other means at the time of peak floods, the area susceptible to floods in the lower reaches will be reduced. To select suitable areas for diverting the flood waters, accurate stream gauging stations will have to be set up at shorter intervals on these tributaries and also before the confluences of minor tributaries with major ones.

## 23.6 Concluding Remarks

The Brahmaputra River basin possesses some of the best aquifers in the country but they have not yet been exploited. Direct recharge by rainfall is one of the most prominent source of groundwater recharge in the basin. The quality of groundwater is good except for a high concentration of iron, which can be removed without much difficulty. While developing the groundwater resources care has to be exercised in order not to damage the environment.

## Further Readings

Barthakur, D.N. (1981). Strategy of agricultural production in flood prone areas with special reference to the Brahmaputra Valley. In: Proceedings of International Conference on Flood Disasters, December, 3--5, 1981, New Delhi, Mimeo.

Dulal, D.C. (1998). Fluvial regime and flood hydrology of the Brahmaputra river, Assam. Memoir Geological Society of India. No. 41, pp. 53--75.

Goswami, D.C. (1990a). Floods and their impact on agriculture of Assam. In: Agriculture in Assam. P.C.Goswami (ed), pp. 191--207.

Goswami, D.C. (1992b). Zonal planning in water. In: Planning for the Northeastern and Eastern Zone. Adisheshiah, M. S .(ed.), Lancer Publishers, New Delhi, pp. 9--23.

Oldham, R. D. (1899). The great earthquake of 1897. Geological Survey of India, Memoir. 29.

Poddar, M.C. (1952). Preliminary report of the Assam earthquake of 15th August, 1950. Bulletin Geological Survey of India, V.2, pp.11--13.

Thatte, C.D. (1996). Water resources development and seismotectonic set up in Himalayas- constraints and prospects. Geological Survey of India Special Publication, 21(2), pp.1--9.

# CHAPTER 24

# IRRIGATION DEVELOPMENT

P.S. GOSWAMI, N. SHARMA, AND C.S.P. OJHA

A lack of consolidation of fragmented and scattered land holdings can be attributed as a major impediment to development of extensive irrigation needed for modernization of agriculture in Assam and in other north eastern states. Providing a brief discussion of closely related issues of rainfall, flooding, land holding, and cropping systems, this chapter discusses irrigation development

## 24.1 Introduction

Physiographically, the state of Assam is divided into three divisions: (1) Brahmaputra valley; (2) the Barak Valley; and (3) the two hills districts of Assam sandwiched between the Brahmaputra and Barak valleys. Among these the share of the Brahmaputra valley is nearly two thirds. This magnifies the role of Brahmaputra River in the development of state's agricultural sector whose contribution to the overall economic growth of the state has been fifty percent or more since independence. From a survey based in 1981--82, out of the total geographical area of 78.52 lakh hectare of Assam, the total cropped area was 34.60 lakh hectares. To date the holdings of agricultural land of various sizes exist in the state. A lack of consolidation of fragmented and scattered holdings remains a major impediment to the propagation of extensive irrigation for modernisation of agriculture in Assam. Despite these constraints, irrigation has to play its role in the development of the agriculture sector of the state which is often subjected to an erratic, uneven and inconsistent distribution of rainfall. This chapter discusses the development of irrigation in Assam along with a brief description of rainfall, flooding, landholding, etc., which are very closely related with irrigation development.

## 24.2 Rainfall, Flooding and Land Holding

Most of the crops in Assam are grown under rain fed conditions. A major constraint of rain fed agriculture stems from the vagaries associated with the behaviour of rainfall. Figure 24.1 shows a typical distribution of annual and monthly rainfall over Assam. From this figure, it can be inferred that in the middle region of the Brahmaputra valley comprising Guwahati sub-division and Naogaon and Sonitpur districts of Assam rainfall is comparatively less than at the extreme parts of the valley. An important fact to note is that Lanka and Lumding areas of Naogaon district receive rainfall of less than 50 inches per year because of their unfavourable physical disposition---both are placed on the



*V.P. Singh et al. (eds.), The Brahmaputra Basin Water Resources*, 419-435.

leeward side of the Garo---Khasi table land A measure of the wide variation in annual total rainfall in the state can be had from a comparison of the annual total precipitation (in mm) in North Lakhimpur-a high rainfall centre and Lumbding-a low rainfall centre. In 1980, the annual total rainfall in North Lakhimpur was 3071, while in Lumbding in the same year, the figure was 591. In 1982 the corresponding figures for the two centres were 2449 and 971, respectively.

In 1985, while North Lakhimpur recorded a higher figure of 3481, Lumbding recorded only 288. The latest figures available for 1988 are 3449 for North Lakhimpur and 1963 for Lumbding. The significant rise in the rainfall figure for Lumbding, a low rainfall area, in 1988 (1963 mm) over the 1985 figure (288 mm) is explained by the fact that during the monsoon period of 1988 the state received heavy rainfall which resulted in the occurrence of several waves of widespread floods.

The rainfall conditions in the Lower Brahmaputra valley are ideal for rice and jute, while those in the Upper Brahmaputra valley are ideal for tea. On average the lower Brahmaputra valley records a rainfall of between 25 cm and 38 cm in the pre-monsoon months and of between 125 cm and 175 cm in the monsoon months, both within optimum conditions for rice and jute production. The Upper Brahmaputra valley which records an average rainfall of 250 cm presents ideal conditions for tea cultivation. The Cachar valley, the second important region of Assam, presents the broad pattern of the Brahmaputra valley in this regard.

Table 24.1 indicates the flood damage is Assam. The data available till 1988 indicates that in the year 1986-1988, the damage has been significant. Assam witnessed two unprecedented natural calamities during 1986. From mid-July to mid-August in 1986 the state was hit first by a prolonged dry spell, a phenomenon which had been quite unknown to the state till then. Immediately there was a devastating flood that spread over several districts of the state. The occurrence of these two natural calamities in quick succession led to a severe setback in the state's economy, particularly in the agricultural sector.

Table 24.2 shows the land utilisation statistics of Assam for a typical year 1981--82. By restoring the fallow lands it is certainly possible to increase the net area sown in Assam. The available statistics, however, do not indicate much efforts by Government to utilise the fallow land directly in the path of rainbearing monsoon clouds. Rainfall is heavy in the eastern part of the Assam Himalayas.

## 24.3 Cropping Pattern

Rice is the main staple of the state. However, during the long 19 year period-1968-69 to 1987-88 the area under rice registered an insignificant annual increase of 1 percent, from 19.53 lakh hectares in 1968-69 to 24.36 lakh hectares in 1987-88. This once again brings to focus the crucial fact that the scope for extensive agriculture, especially in rice cultivation, is practically exhausted. It may be noted that winter rice (Sali and Boro) is the principal crop of the state, accounting for about 57 percent of the total cropped area

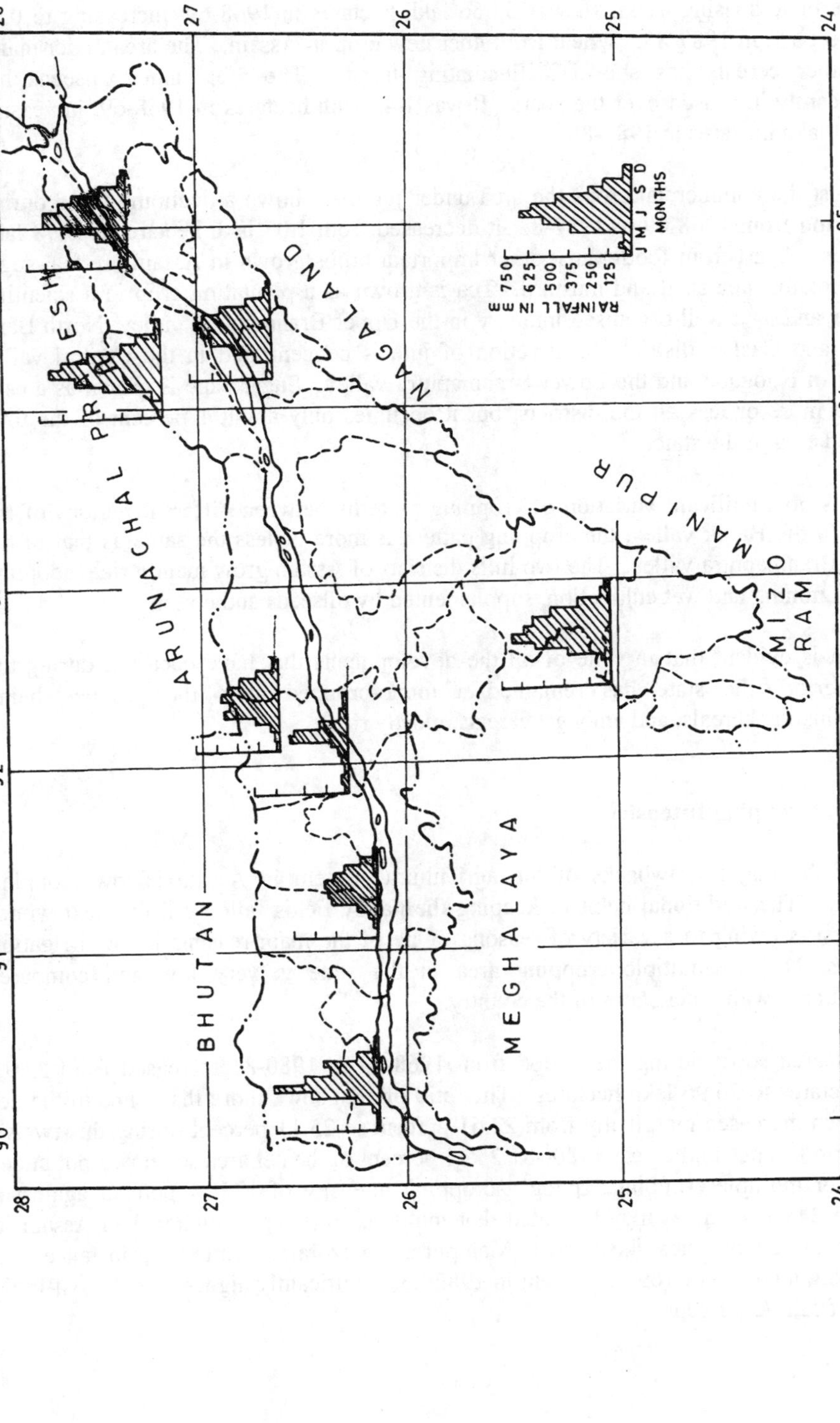


Fig. 24.1 Typical distributions of monthly and annual rainfall in Assam, India.

and 72 percent of the total area under rice in the state. The area under wheat has shown a more or less rising trend. It was 0.066 lakh hectares in 1968-69, increasing to 0.98 lakh hectares in 1987-88. Wheat is a rather new crop in Assam. The area under maize and other cereals has shown a fluctuating trend. The area under oilseeds has significantly increased over the years. It was 1.43 lakh hectares in 1968-69, increasing to 3.64 lakh hectares in 1987-88.

Amongst the commercial crops the area under jute has shown a declining trend during the period from 1968-69 to 1987-88; it decreased from 1.05 lakh hectares to 0.98 lakh hectares. Apart from foodgrains, other important crops grown in Assam are tea, sugar cane, potato, rape seed and mustard. Tea is grown as a plantation crop in a scientific manner and as a well-organised industry in the Upper Brahmaputra valley, North Bank region and Cachar district. Cultivation of jute is concentrated in the Central valley district of Naogaon and the Lower Brahmaputra valley. Sugar cane is grown as a cash crop in more or less all the districts, but it occupies only about 1 percent of the total cropped area of the state.

There is no significant variation of cropping patterns between different regions of the state. In the Barak valley, the cropping pattern is more or less the same as that of the Upper Brahmaputra valley. The two hills districts of Assam grow mainly rice, adopting both jhumming and wet cultivation, supplemented by oilseeds and cotton.

Thus, it is evident that in spite of all the developments that have occurred during the plan period, the state has remained a mono-cropped area, the produce being predominantly cereals, and amongst cereals, mostly rice.

## 24.4 Cropping Intensity

One of the major drawbacks of the agricultural system in Assam is low cropping intensity. The traditional habit of keeping the paddy fields fallow till the next winter still persists owing to a variety of reasons. One of the main reasons is low irrigation facilities. Double multiple cropping area in the state is very low and compares unfavourably with other states of the country.

The net area sown during the period from 1968-69 to 1980-81 increased from 22.91 lakh hectares to 26.96 lakh hectares. The ratio of area sown more than once to the net area sown increased marginally from 22.61 percent to 25.11 percent during this twelve year period. Thus in the year 1980--81 25.11 percent of the net area sown was put under double or multiple cropping, giving a cropping intensity of 125.11 percent against a possible 244 percent. It may be noted that multiple cropping is higher than Assam in other north eastern states like Tripura, Manipur and Mizoram. Tripura, for instance, had a cropping intensity of 169.41 percent in 1982-83, significantly higher than the All-India figure of 121.72 percent.

Table 24.1 Flood damage in Assam [1 lakh = 100,000]

| Year | Area affected (lakh ha) | Cropped area damaged (lakh ha) | Value of crops damaged (lakh Rs.) | Population affected lakh) |
|---|---|---|---|---|
| 1 | 2 | 3 | 4 | 5 |
| 1954 | 29.00 | 3.05 | 1175.14 | 13.00 |
| 1955 | 13.50 | 0.73 | 238.43 | 1.77 |
| 1956 | 5.13 | 0.69 | 254.55 | 2.71 |
| 1957 | 3.95 | 0.25 | 100.71 | 3.16 |
| 1958 | 12.29 | 0.59 | 144.33 | 4.04 |
| 1959 | 7.58 | 1.44 | 486.59 | 11.72 |
| 1960 | 4.68 | 2.21 | 762.92 | 13.22 |
| 1961 | 1.89 | 0.13 | 29.90 | 2.21 |
| 1962 | 15.95 | 6.61 | 1,848.57 | 39.08 |
| 1963 | 5.67 | 0.74 | 197.40 | 8.80 |
| 1964 | 6.02 | 1.23 | 238.07 | 7.65 |
| 1965 | 3.22 | 0.24 | 88.56 | 2.58 |
| 1966 | 15.11 | 3.69 | 2,149.04 | 36.24 |
| 1967 | 2.45 | 0.88 | 133.44 | 4.50 |
| 1968 | 3.76 | 1.25 | 801.40 | 8.35 |
| 1969 | 10.63 | 0.69 | 335.73 | 8.90 |
| 1970 | 7.58 | 2.26 | 1,042.52 | 18.91 |
| 1971 | 4.48 | 1.12 | 469.88 | 6.59 |
| 1972 | 9.97 | 3.59 | 2,221.41 | 29.52 |
| 1973 | 24.09 | 1.64 | 1,440.04 | 18.47 |
| 1974 | N.A | N.A | 1,366.11 | N.A. |
| 1975 | 1.24 | 0.17 | 124.55 | 2.32 |
| 1976 | 2.52 | N.A | 865.13 | 4.40 |
| 1977 | 10.24 | N.A | 2,654.00 | 45.49 |
| 1978 | 3.06 | N.A | 393.00 | 9.17 |
| 1979 | 6.73 | N.A | 2,614.00 | 24.51 |
| 1980 | 10.60 | N.A | 3,237.00 | 33.59 |
| 1981 | 4.57 | N.A | 701.00 | 13.58 |
| 1982 | 68.85 | N.A | 469.000 | 14.24 |
| 1983 | 6.95 | 1.25 | 1,032.00 | 21.21 |
| 1984 | 9.36 | 3.57 | 4,899.00 | 38.79 |
| 1985 | 6.46 | 0.82 | 8,290.00 | 24.66 |
| 1986 | 4.26 | 3.22 | 33,867.00 | 24.45 |
| 1987 | 25.73 | 10.70 | 36,859.00 | 94.60 |
| 1988 | 46.50 | 13.35 | 33,410.00 | 126.77 |

Source: i) Draft comprehensive Plan for Brahmaputra Basin, Vol. I.B.F.C.C., Govt. of Assam, 1977.

ii) Directorate of Economic and Statistics, Govt. of Assam.

iii) Darth 7th. Plan Assam, Govt. of Assam

iv) Revenue Department, Govt, of Assam.

Table 24.2 Land utilisation in Assam during 1981-82 (in thousand hectares).

| Category of use | Area in thousand hectares | Percentage |
|---|---|---|
| 1.Total Reporting Area | 7,852 | |
| 2. Forests | 1,984 | 25.27 |
| 3. Land not available for cultivation | | |
| i) Land put to non-agricultural uses. | 914 | 11.64 |
| ii) Barren and uncultivable | 1,541 | 19.63 |
| 4. Other uncultivated land excluding fallow land | | |
| i) Permanent pastures and other grazing land | 184 | 2.34 |
| ii) Land under misc. tree crops & groves not included in net area | 247 | 3.15 |
| iii) Cultivable waste land | 104 | 1.32 |
| 5. Fallow land | | |
| i) Current fallow | 88 | 1.12 |
| ii) Fallow land other than current fallow | 84 | 1.07 |
| 6. Net area sown | 2,706 | 34.46 |
| 7. Area sown more than once | 754 | 9.60 |
| 8. Total cropped area | 3,460 | 44.60 |

Source: Statistical Hand Book, Assam, 1988, Pg. 58-59. Directorate of Economic and Statistics, Govt. of Assam, Guwahati.

"Given that the gross cropped area in Assam is more than 2.5 times of that in other six states of the region, Assam can easily increase its crop production by raising its intensity of cropping to the levels of Manipur and Tripura", as described in the report of the Assam Agriculture Commission (1975), there is very little rotation of crops in Assam. As Assam lies in the high rainfall zone of the country, the general impression is that there is no necessity for irrigation in the state. Although the state usually receives good rainfall almost every year spreading over eight months (March to October), it is not evenly distributed over time and space. Sometimes, monsoon is delayed or terminates early. The erratic nature of rainfall results in floods and droughts, which adversely affect agricultural productivity. Further there are certain specific areas which fall under the rain shadow belt. Agriculture in Assam is thus a gamble in the monsoon.

Furthermore the vulnerability of the Kharif crops, the principal crops of Assam to the twin natural calamities: flood and drought makes it all the more important that the area under Rabi crops should be expanded rapidly to compensate for the crop losses that may result during the Kharif season. The Rabi season, which times with the dry months of the year, is dependent for its proper development on adequate and assured irrigation water. Unfortunately, during winter season not more than 5 percent of the net sown area (in Assam) is irrigated when there is a great need to supply irrigation water to Rabi crops.

Till some decades back, because of low population pressure and copious rainfall, the people of Assam were content to raise only one rainfed Kharif paddy crop. Since then there has been an excessive rise in population, owing mostly to heavy influx of immigrants since 1960. Moreover, rapid deforestation has led to an upset in the ecological balance, resulting in adverse changes in the rainfall pattern. Hence the need to create adequate irrigation facilities to make water available during the dry months of the year for popularisation of multiple cropping has gained an increasing urgency.

## 24.5 Source of Irrigation

There are two principal river systems in the state: the Brahmaputra and the Barak, which are the main sources of surface water. The Brahmaputra River system is the main source of surface water for irrigation. The river Brahmaputra with a total length of 2,900 km and a catchment area of 0.94 million square kilometres, enters into the north-eastern India and flows westwards through Assam. Many big and small tributaries meet the river at different places of its course. The tributaries of the Brahmaputra can be broadly classified into the northern-bank tributaries and the southern bank tributaries. The northern bank tributaries are mostly perennial in nature. The Brahmaputra River system has immense potential for development of surface water irrigation in the state.

The length of the Barak River from its origin upto the Assam--Bangladesh border is 560 km. It has numerous tributaries joining it from both north and south banks within the state of Assam. The total available annual runoff of the Barak and its tributaries within the state will be about 26.5 million acre feet. The total area of the Barak basin in the state is 6.96 lakh hectares (17.20 lakh acres). Thus the irrigation potential of the surface water resources is abundant to cover the whole Barak valley. Although there is an abundance of surface water resources in the state, because of the limitations imposed by the topographical characteristics of this region and reduced flow condition of the rivers in the winter months, it is necessary to explore the possibility of utilisation of ground water resources for irrigation in some areas within the state which cannot be provided with irrigation from surface water.

Besides, the Brahmaputra River basin has rich underground water resources owing to the occurrence of heavy rainfall and advantageous soil formation. The north bank districts of Brahmaputra valley [Barpeta, Nalbari, (North) Kamrup, Durrang and Sonitpur] overlie the northernmost piedmont belt. This belt is analogous to the 'Bhabar' belt of Ganga valley, and occurring close to the foothills is flawed to the south by a narrow Terai belt. The piedmont terraces are highly permeable in nature and ground water prevails at a depth of about 30 metres below the ground level.

**24.6 Development of Irrigation**

The irrigation achievement in the state up to the end of third plan was 46.627 thousand in the minor irrigation sector at an outlay of Rs 984.30 lakh. Thus up to the third plan the emphasis was only on the implementation of minor irrigation schemes. No major or medium irrigation scheme was constructed up to this period, as can be seen from Table 24.1.

According to the Irrigation Department, Assam, 1288 of government irrigation schemes of different categories, i.e., surface flow, surface lift and ground water lift, including Command Area Development/Water Course Schemes, had been completed up to 31st March, 1992 and work on 1,510 of such schemes were at different stages of progress. Some of these schemes are shown in Figure 24.2. The number of institutionally financed private shallow tube wells and flow lift points installed by the Assam State Minor Irrigation Development Corporation up to 1991--92 was 47,948.

Up to the end of the financial year 1990--91, irrigation had been extended to an aggregate area of 5.70 lakh hectares of cropped land. Of this area, government sponsored medium and minor irrigation schemes accounted for 1.49 lakh hectares and 2.77 lakh hectares, respectively. The remaining 1.44 lakh hectares were covered by privately sponsored shallow tube wells and surface lift irrigation schemes. The objective of the state government for the financial year 1991-92 was to bring an additional 90,600 hectares of cultivated land under irrigation map. Of this, minor and medium irrigation schemes of the government would cover 27,800 hectares and 26,800 hectares respectively. The remaining 36,000 hectares would be in the private sector.

However, up to 1991-92 Assam had already created facilities to bring 5.95 lakh hectares of the total cropped area under irrigation coverage (17.5 percent), through various completed and partially completed government major, medium and minor schemes as well as private small minor, medium and minor schemes as well as private small minor irrigation works installed through institutional finance schemes. According to a 1992 based estimate, out of the 27 lakh hectares of ultimate irrigation potential in the state, only 276,969 hectares (approx. 2.77 lakh hectares) were covered under irrigation schemes. Of the state's total geographical area of 78.438 lakh hectares, the gross cropped area was estimated at 34.6 lakh hectares. As against this, the target for the state's gross cropped area to be irrigated was set at 27 lakh hectares. Of this area an area of 20 lakh hectares was targeted to be irrigated with surface water and the remaining 7 lakh hectares with ground water.

Furthermore, of the targeted total 27 lakh hectares of cropped land to be irrigated, 10 lakh hectares were to be covered by major and medium irrigation projects and the remaining 17 lakh hectares were to be brought under minor irrigation projects.

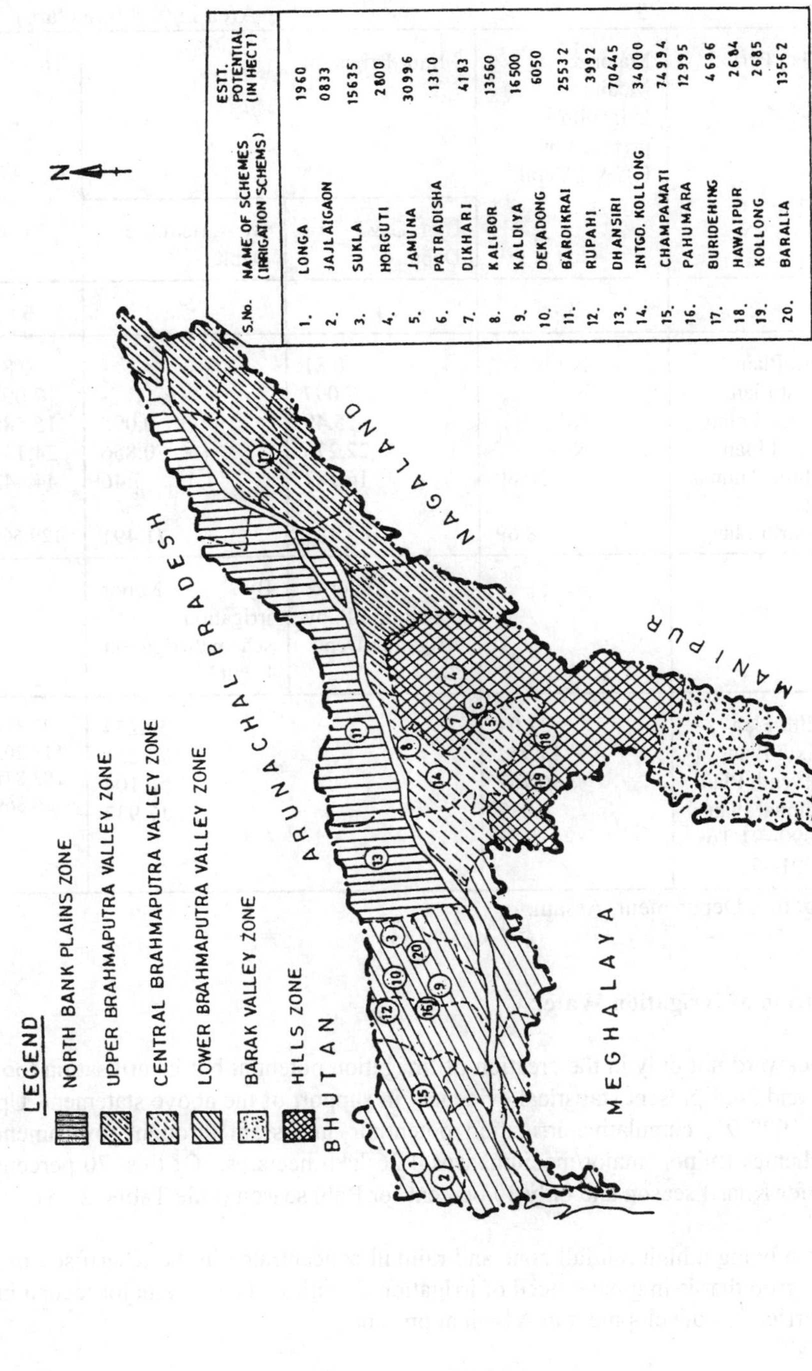


Fig. 24.2 Irrigation schemes in Assam.

Table 24. 3 The level of irrigation development in Assam during the plan period.
Area in (1000 hectare)

| Sl.No | Plan Period | Major & Medium Irrigation prepared by F.C.& I Deptt. | Minor Irrigation | | Total |
|---|---|---|---|---|---|
| | | | By F.C.&T Deptt. | By Agriculture Deptt. | |
| 1 | 2 | 3 | 4 | 5 | 6 |
| 1. | Pre-Plan | Nil | 0.81 | - | 0.81 |
| 2. | First Plan | Nil | 7.097 | - | 7.097 |
| 3. | Second plan | Nil | 15.49 | 0.098 | 15.588 |
| 4. | Third Plan | Nil | 22.276 | 0.856 | 24.132 |
| 5. | Three Annual Plans | 20.80 | 16.183 | 7.46 | 44.443 |
| 6. | Fourth Plan | 8.69 | 49.488 | 71.491 | 129.669 |
| | | | Major & Medium I.P. by Irrigation Deptt. | By Minor irrigation scheme/Irrigation Deptt. | |
| 7. | Fifth Plan | 27.04 | | 50.274 | 77.314 |
| 8. | Sixth Plan | 43.952 | | 74.251 | 118.203 |
| 9. | Seventh Plan | 46.767 | | 56.104 | 102.871 |
| 10. | Annual Plans: 1990--91 To 1991--92 | 8.932 | | 17.937 | 26.869 |

Source: Irrigation Department, Assam.

## 24.7 Utilisation of Irrigation Water

Assam is backward not only in the creation of irrigation potential but its utilisation too. Tables 24.4 and 24.5 present statistical evidence in support of the above statement. Up to the year 1990-91, cumulative irrigation potential in Assam through government irrigation schemes (minor, major/medium) was 4.26 lakh hectares. Of this, 70 percent was created for Kharif season and only 30 percent for Rabi season (vide Table 24.5).

In fact, Assam being a high rainfall zone and rainfall concentrated in the Kharif season, it is the Rabi crop that is in greater need of irrigation facilities. This is a major lacuna in the trend of irrigation development in Assam at present.

Table 24.4 Cumulative irrigation potential created in Assam.

Area in hectares

| | Minor Irrigation Govt. | Private | Total | Major &Medium Irrigation (Govt.) | Total cumulative Potential Created (govt.) | Total cumulative Potential Created |
|---|---|---|---|---|---|---|
| 1 | 2 | 3 | 4 | 5 | 6 | 7 |
| 1980--81 | 151,415 | --- | 151,415 | 57,784 | 209,199 | 209,199 |
| 1990--91 | 276,696 | 143,844 | 420,540 | 149,323 | 426,019 | 569,863 |

Source: Chief Engineer, Irrigation Department, Assam.

Table 24.5 Irrigation potential created up to 31-3-91 through Government irrigation schemes in Assam.

| Type | Kharif | Rabi | Total Area in ha |
|---|---|---|---|
| Minor | 190,299 | 81,840 | 272,139 |
| Major/Medium | 108,725 | 45,155 | 153,880 |
| Total | 299,024 | 126,995 | 426,019 |

Source: Irrigation Department, Assam.

The potential utilised up to 1990-91 by all types of Government irrigation schemes was 2.39 lakh hectares (Table 24.6) which is about 56 percent of the total potential created. 84 percent of the total potential utilised was in the Kharif season and only about 16 percent of it was utilised during Rabi and Pre-Kharif season.

Thus relative to the potential created through minor and major/medium irrigation projects, the utilisation of irrigation water in Assam is substantially low in Rabi and pre-Kharif season. About the considerable gap between the potential created and that utilised, the following facts may be worth consideration.

(i) Widely prevalent seepage and percolation loss of irrigation water;
(ii) Absence of adequate canal network and field channels;
(iii) Inadequate drainage system and waterlogging;
(iv) Poor maintenance;
(v) Lack of filling and land levelling;

(vi) Absence of consolidation of small and scattered holdings;
(vii) Untimeliness of the supply of irrigation water;
(viii) Defective alignment of canals;
(ix) Lack of steady and adequate supply of electricity especially in operation of electric lift irrigation schemes;
(x) Lack of communication facilities in the command area;
(XI) Lack of motivation among the farmers for utilisation of irrigation water to obtain optimum benefit from the project; and
(ix) Lack of co-ordination and co-operation among the concerned Government departments and agencies.

Table 24.6 Cumulative irrigation potential utilised by Government irrigation schemes in Assam up to 31-03-91.

| Schemes | POTENTIAL UTILISED | | | | | | Total Area in ha |
|---|---|---|---|---|---|---|---|
| | KHARIF | | | RABI AND PRE-KHARIF | | | |
| | Minor Irrig. | Major/Medium Irrig. | Total | Minor Irrig. | Major/ Medium Irrig. | Total | |
| Surface flow | 69,737 | 70,419 | 140,156 | 6,872 | 7,187 | 14,059 | 154,215 |
| Surface Lift | 32,877 | 17,600 | 50,477 | 14,590 | 3,742 | 18,332 | 68,809 |
| M.W. Lift | 10,515 | - | 10,515 | 5,558 | - | 5,558 | 16,073 |
| Mobile Diesel | 27 | - | 27 | 69 | - | 69 | 96 |
| Total: | 113,156 | 88,019 | 201,175 | 27,089 | 10,029 | 38,018 | 239,193 |

Source: Irrigation Department, Assam.

In this context it may not be out of place to reflect on the present position of irrigation development in the State. The three basic issues that arise pertaining to the present study relate to:

(i) the present policy of irrigation development;
(ii) the present pattern of distribution of irrigation facilities created amongst different categories of farmers; and
(iii) the present pace of development of irrigation potential. These issues will have far reaching impact on the goal towards achievement of self-sufficiency in food grains production in the State in the years to come. A general review has been undertaken to address these questions as follows.

Almost all the developments that have taken place in the irrigation sector in the State are based on the single purpose approach. They comprise of a diversion barrage or a pump installed on a barge along with a canal network. The above infrastructure does not have any provision for storage of irrigation water.

In the first decade and a half of the planning period when the land to man ratio in the State was favourable and there was abundant scope for extensive agriculture, this policy of irrigation planning did not present any constraint to agricultural development in general and food grains production in particular. Later, with mounting pressure of population on land and with the falling land to man ratio, the river diversion type of irrigation proved to be a major impediment to agricultural growth.

Today clearly the need for storage based irrigation projects is strongly felt. At the same time the inadequacy of the existing river diversion type of irrigation schemes in substantially augmenting agricultural production which has been stagnating over the years, is now better appreciated.

It is worthwhile mentioning the Jamuna Irrigation Scheme, the first medium irrigation project of Assam, where the irrigation potential created for Kharif crop is above 40,000 acres, its irrigation potential created for the Rabi season is only 400 acres, even though there exists a strong demand for irrigation water during the non-monsoon period. The constraint in the extension of irrigation facilities during the Rabi season may be ascribed to the absence of a storage dam reservoir, upstream of the existing irrigation headworks at Bakulia on the Jamuna River. The Jamuna irrigation headworks simply diverts a part of the available flow in the river.

Thus there is an imperative need to make a break from the existing policy of irrigation planning of river diversion schemes to storage-based irrigation projects as done elsewhere in the country to facilitate Rabi cultivation. Another major lacuna in the present irrigation policy of the state is that the ground water resources of the state have not been given their due importance. The state is rich in ground water resources. The ultimate irrigation potential from ground water, which was earlier pegged at 7 lakh hectares, has been revised to lakh hectares based on comprehensive data available now.

According to a report jointly prepared by the IBRD and IDA (1976): "About 2 billion cubic metres of ground water in the Brahmaputra basin are within 5 metres of the surface easily reached by shallow tube wells." The central Task Force on Agricultural Economy of Assam (March 1980) remarked" it is amply evident that the ground water resources particularly in the Brahmaputra Valley are immense, awaiting utilisation. As against these potentials, the progress with respect to ground water development in the state seems as if it has hardly begun".

From the progress reported in the Report on Clause Seven of the Assam Accord, from 1981--82 when the ground water development programme was taken up, about 43,000 shallow tube-wells were constructed up to the end of March 1990. In addition there were as many as 1,000 public deep wells in the state by 1990. From the view of the long term development of the state there will, however, be a need for storage dams at sites where adequate capacity of water for storage exists because of favourable geo-morphological formations. The storage irrigation projects involving dams and barrages are, however, highly capital intensive with relatively long gestation periods.

In Assam, it is observed that some of the bigger projects like Dhansiri and Champamati have suffered from unconscionable time and cost over-runs. This is, however, not an argument for avoiding such capacity intensive big projects. These big projects have certain inherent advantages such as extremely low recurring expenditure, creation of other inputs of development like hydel power generation, large scale flood moderation and improvement of navigation and, resulting lowering of transport cost. In short, the chief benefits of these projects accrue from their multi-purpose character which alone will make then economically viable.

Likewise, the ground water projects enjoy certain advantages, such as shorter gestation period, low initial capital investment and lower level of technical skill required for their maintenance. But the drawbacks of such projects relate to high operational cost and requirement of large scale uninterrupted power supply. For assured irrigation, the sole dependence on ground water may not be prudent as over-exploitation of ground water may lead to lowering the water table to uneconomical levels.

Moreover, the acute power shortage in Assam is a serious bottleneck in the state's development efforts. A disturbing fact to be remembered is that even after completion of seven five-year plans, the peak power availability in Assam is a little over 349 MW against a projected demand of 7,349 MW for 2000 A.D. This is likely to be even considerably higher at the end of the projected period of this study (2020 A.D.). The other option is to energise the diesel pumpsets, the use of which will considerably push up the running cost, thereby imposing a heavy burden on the state exchequer. There is also considerable malpractice (e.g., theft and misuse) associated with the use of diesel.

Above all, however high the capital intensity of storage dams may be, the benefits that will accrue from them will be available for generations to come, provided they are properly maintained.

## 24.8 Projection of Irrigation Potential

Adapting the date for the period 1979--80 to 1991--92, a projection for irrigation potential has been made up to 2020 A.D. The linear regression relation has been adopted for the purpose of reflecting the trend in the last decade. The annual rate of growth of irrigation potential depends on many complex factors, viz, government decisions and policy, construction period of projects and their associated delays, shortage of finances, project delays owing to natural calamities and unforeseen socio-political changes among others. As such, it is very difficult to adopt multiple regression approach accounting for the above stated factors. Since the present aim is to elicit a general assessment of the irrigation potential likely to be achieved keeping in view the aforesaid constraints, the linear regression relation has been adopted which would provide a fair idea of the likely situation in the projected period.

The projection has been made to review the probable rate of irrigation potential created in different years. The projected values of the irrigation potential likely to be created at the present pace of irrigation development are given in Table 24.7. A perusal of the projected values indicates that irrigation potential likely to be created at the present rate will fall considerably short of the total net area sown of 27 lakh hectares, even at the end of 2020 A.D. At the end of the projected period it will be only 16.12 lakh hectares. From the above finding it seems imperative to bring about a radical change in the present policy of irrigation development to accelerate its growth rate.

Table 24.7 Projection of irrigation area (potential created annually).

| Year | Area (ha) | Year | Area (ha) |
|---|---|---|---|
| 1990--91 | 558,561 | 2005--06 | 1085,521 |
| 1991--92 | 5,6,201 | 2006--07 | 1123,161 |
| 1992--93 | 633,841 | 2007--08 | 1160,801 |
| 1993--94 | 671,481 | 2008--09 | 1198,441 |
| 1994--95 | 709,121 | 2009--10 | 1236,081 |
| 1995--96 | 784,401 | 2010--11 | 1273,721 |
| 1996--97 | 784,401 | 2011--12 | 1311,361 |
| 1997--98 | 822,041 | 2012--13 | 1349,001 |
| 1998--99 | 859,681 | 2013--14 | 1386,641 |
| 1999--2000 | 871,921 | 2014--15 | 1424,281 |
| 2000--01 | 897,321 | 2015--16 | 1461,921 |
| 2001--02 | 934,961 | 2016--17 | 1499,561 |
| 2002--03 | 972,601 | 2017--18 | 1537,201 |
| 2003--04 | 1010,241 | 2018--19 | 1574,841 |
| 2004--05 | 1047,881 | 2019--20 | 1612,484 |

The sowing time, crop growing period, etc., have been adapted from the Agricultural Guide Book published by the Agriculture Department of Assam. The computer programme available in FAO (Food and Agricultural Organisation Paper No. 24, 1977), has been modified and implemented by Sharma and Soni, 6 working out the crop water requirement for 1 hectare area. The data required for estimation, viz., rainfall, wind velocity, percentage of sunshine, temperature, relative humidity, were collected from the Directorate of Economic and Statistics, Guwahati and the I.M.D., Guwahati. Subsequently, districtwise and cropwise gross irrigation requirements are estimated from the proposed cropping pattern in 2020 A.D. The cropping pattern (Table 24.8) is projected to meet the food requirement of the projected population by 2020 A.D.

The aggregated irrigation requirements for Assam till 2020 A.D. stand at 5.94 million ha-m. The required figure for the Brahmaputra basin is found out to be 1.95 million ha-m for Kharif and 3.25 million ha-m for Rabi while the total figure for the basin as a whole stands at 5.2 million ha-m.

## 24.9 Summary

There is no dearth of water resources in the state for the simple fact that the

Brahmaputra alone carries an average yield of about 60 million ha-m (420 million acre feet), which can be further supplemented by considerable ground water reserves. Therefore it can be concluded at this stage that the future irrigation water requirement in 2020 A.D. which will be available to support the proposed cropping pattern to bring about self-sufficiency in indigenous food grains output in general and rice in particular, is precariously balanced with the available water resources.

Table 24.8 Cropping pattern (2019-20).

| Kharif Crops | Cropping Intensity | Area (lakh ha) |
|---|---|---|
| Ahu Paddy | 25% | 6.75 |
| HYV Autumn Paddy | 40% | 10.80 |
| Normal Sali | 35% | 9.45 |
| HYV Winter Paddy | 35% | 9.45 |
| Jute | 10% | 2.75 |
| Total | 145% | 39.20 Lakh ha |
| Rabi Crops | | |
| Boro Paddy | 30% | 8.10 |
| Wheat | 25% | 6.75 |
| Mustard | 10% | 2.70 |
| Pulses | 10% | 2.70 |
| Perennial | | |
| Sugar-cane | 5% | 1.35 |
| Grand Total | 225% | 60.80 Lakh ha |

Even though it might appear to have surplus water resources in the state at present, but after three decades the state might run short of the requisite irrigation water to produce its own foodgrains, if inter-basin transfer of water is resorted to from the Brahmaputra, as talked about from time to time. Hence any programme for diversion of water from the Brahmaputra River should be considered only with regard to the possible consequences of such diversion in Assam on programmes of irrigation and agricultural development.

## References

Borthakur, B.C. and Upadhyay, L.P. (1980). "Take-off in agricultural production in Assam, Seminar Paper.

Champati, M., and Patnaik, S.C. (1985). "Utilisation of irrigation potentiality of major irrigation projects - a case study." Indian J. of Regional Science, Vol. XVII, No. 2.

Chakraborti, S. (1982). "In search of growth pattern for food grain production in India." Economic and Political Weekly, Review of Agriculture, December.

Charan, A.S. (1984). "Impact of irrigation on agriculture - a farm level analysis, Indian J. Agricultural Economics, Vol. 39, No. 3, July-Sept.

Dantwala, M. L. (1970). From stagnation to growth." Indian Economic J., Vol. 18, Oct.-Dec.

Desai, B.A., and Thingalaya, N.K. (1965). "Irrigation factor and yield variability in rice growing districts in India." Indian J. Agricultural Economics, Vol. 20, No. 3, July-Sept.

Sawant, S.D. (1975). "Extent of multiple cropping in irrigated and unirrigated areas in India - some implications for usefulness of irrigation statistics, Indian J. Agricultural Economics, Vol. 30, No. 2, Apr.-June.

Sawant, S.D. (1980). "Water requirements for agricultural use in Maharastra in 2000 A.D." Indian J. Agricultural Economics, Vol. 35, No. 1, Jan - Mar.

Saikia, P.D. and Bora, A.K. "Farmers response to improved agricultural practices in Cachar district." Agro Economic Research Centre, Jorhat.

Sharma, P. (1992). A study of irrigation development and its impact on food grains production in Assam. Ph.D thesis, Guwahati University, Guwahati.

# CHAPTER 25

# DEVELOPMENT OF POWER

P.C. SHARMA

It is argued that power development holds the key to the overall development of the northeastern region of India. Fortunately the main ingredients for such development exist abundantly in the region. International cooperation with Nepal, Bhutan, China and Bangladesh is essential to development of hydropower, the enormous potential of which is not being utilized. This chapter presents different aspects of power development in the northeast.

## 25.1 Introduction

After crossing the two waves of agricultural and industrial revolutions, the developed countries are in the third wave of science and technological revolution, i.e., they are now after efficiency rather than growth. It has been possible in U.S.A. to expand economic activities by 40 times during last 20 years, while holding down energy consumption to the 1973 level. Japan also was able to educe energy consumption relative to the gross domestic product (GDP) by 31% between 1973-83. In the changed context of the world, the developing countries have to grow fast and steadily at a high rate, with a balance between agriculture, and industry, as well as science and technology. The power sector of India, reeling under a resource crunch, therefore has to critically examine the energy paradigm.

Secondly, the north eastern (NE) region of India is chronically backward in spite of plentiful natural resources. Human resources play the most vital role in development. The region has a colourful ethnic and non-ethnic population representing a miniature India. But the political aspirations of the people are yet to consolidate into an economic whole. Water resource development is a regional subject. The Indian exercises in regionalisation have not succeeded so far. In the world of today, optimal development of India is a must. For the NE region to settle down to a peaceful life one also has to break its economic stagnation. This issue has been discussed in the third section.

The third 15 year national power plan has given the topmost priority to hydropower development. This has opened up opportunities for the region to develop its water resources to its best benefit for breaking the stagnation trap. Fourthly, the choice of hydel projects should be such as to boost the agricultural sector with cheap irrigation in a backward economy, as well as cheap water transport, flood control, etc.. The demand in the region being low, the evaluation of the bulk power is another important issue.



*V.P. Singh et al. (eds.), The Brahmaputra Basin Water Resources*, 436-472.

Providing a large industrial umbrella is yet another issue. These have been discussed in the fifth section.

Multipurpose projects are the unavoidable future of mankind. Changing patterns of land use and growth of population will stand in the way of their utilisation with delay. This aspect has been discussed in the sixth section.

With all these issues kept in mind, the approach should be to work out a development paradigm for the region vis-a-vis the country. The commercial interests in power generation may side track the issue of optimal utility of water resources. Similarly, the international commercial interests also may side track the issue of optimal utility of the NE regional resources. In the case of hydropower development in the north east, international co-operations with Nepal, Bhutan, and Bangladesh are essential. Thus a balanced strategy must be framed, keeping the long-term view in mind. These issues are discussed in the seventh section on hydropower generation.

## 25.2 The Energy Paradigm

The word 'change' can be conceived as a self-acting process where matter and energy interact incessantly. The principles of conservation of matter, energy, and both while in conversion of one to the other, guide the interactions. Since nothing is created nor destroyed, the process is qualitative in character, where change of forms only takes place. We call it development because we want conscious participation in it. Land and mines, water, the forests, and the human beings are the four basic resources of development. Man works as the catalyst and all of them change in the process.

Production can be defined as interaction between matter and energy. We can, therefore, visualize development as the process of social production for the optimal utilization of all the resource for the benefit of mankind. Since it is a continuous process, the social savings play a very vital role in development. Secondly, the forms of production are many but the use of energy is common for all of them. Therefore the pattern of energy use and its change can indicate the mode of social change.

During the period of 10 years from 1972-73 to 1982-83, the share of the commercial form of energy in India increased from 26% to 34% in the total energy use. The rural population made large use of the traditional forms, mostly in an inefficient manner, and exported these to the cities for livelihood as well. The use of commercial energy increased from 19.76 MTCE in 1953-54 to 101.5 MTCE in 1990-91. In it the industrial sector had a high consumption compared to the other sectors of agriculture, transport and services. Coal, gas and oil accounted for 83.4% in 1990-91, in the use of commercial energy. The growth of share of electricity was from 9.6% in 1970-71 to 17.6% in 1990-91. But here too, thermal generation from coal, gas and oil dominated the consumption against hydro-electricity, when 80% of the hydel potential just flows out to the sea. The trend of exploiting the fund resources and wasting the flow resources is an indication of bad planning, particularly when India is a net importer of energy.

The outlay for the power sector at 18.8% of the total for the seventh plan (1985--1990) was quite high but the results were poor. For 1990--95 it worked out to Rs 128,000 crores. India needed a high and steady growth of the economy in the changed global scenario, demanding much higher investment for power development. Therefore it becomes necessary to examine the energy paradigm in greater detail. The basic variables for working out the investment required for expansion of the installed capacity of electricity generation are: (i) unit cost of installed capacity; (ii) new capacity required; (iii) the growth rate of GDP; and (iv) the growth rate of electricity required for unit growth rate of GDP.

The issue of optimization of the investment raised a large number of questions. In respect of lowering the correlation factor between electricity growth rate and GDP growth rate, the need for energy efficient technology and optimization of the sectoral growth-rates become important. As regards optimization of the new capacity required, the important issues were: (i) reduction of TD loss; (ii) energy conservation; (iii) demand management; (iv) an integrated national grid; (v) capacity utilization; (iv) appropriate generation mix; (vii) decentralized generation at lower ends of the distribution system; and (viii) tapping the non-conventional sources of generation. All these issues are to be examined within the preview of two ruling questions: What should be the reliability levels for the supply of power and energy? What are the market demands of power and energy ?

We are then prepared to examine the first degree questions: What are the unit costs of the projects? How should we compromise with the GDP growth rate? But even this is not the final stage. The mammoth power supply industry can be conceived as one whole as the unifier of the various heterogeneous regions within a developed national grid. Sectoral profits and cross-subsidy are inherent in it, which create conflicting situations. But the final burden falls on the 75 million consumers using about 225 billion units of electricity in all the regions of the country in the 1991-92 condition. How should we bear with them about a rational power tariff, remembering also the prospective ones?

The Central Electricity Authority (CEA) had undertaken detailed studies on the national power development plan of India up to 2006-07, for the Planning Commission to finally supplement the overriding aspects. It could reveal, amongst many other facts, that India had followed a wasteful path of power development in the matter of hydro-thermal mix, which was the ruling one. The ratio had dwindled to 27% in 1992. In order to raise it to a level of 37% by 2007 A.D. about 52,000 MW of hydro-electric capacity should be commissioned in these 15 years and advance action may have to be taken for the 11th five year plan (2007--2012) for about 26,000 MW additional capacity, in order to maintain the trend[2].

Water resource development is a regional subject, rather than one for the states. In India the problem of regional development has eluded a solution. This appears to be the reason how shortsighted commerce of thermal power development, though wasteful as a whole, could dominate the power scenario. The same trend is visible in the matter of participation of the private sector, in the power development in India, where hydel development has yet to receive due importance.

Energy strategies in the developing countries have to deal with a totally different set of challenges from those in the developed world. Energy developments in the third world have to be viewed as an instrument of economic development. On the other hand, in several developed countries the problem is really one of bringing about an adjustment of what Paul Ehrlich describes as a state of 'mal-development', implying a pattern of consumption of goods and services which imposes a major net burden on our ecosystem and natural resources[3]. This explains the underlying trend of lower growth rates in the developed world, and higher growth rates in the developing world, with flow of investment as the common tie between the two.

India had paid more attention to foreign investment in the thermal power sector, as seen from the current power policy. The good sign is that there has been, of late, realization of it, as seen from the 26th report of the Parliamentary Standing Committee on Energy: The committee is not impressed by the response received from the private sector for setting up hydel projects. Only 26 schemes for a total capacity of 4,165 MW have been received so far. As against the desired hydro-thermal mix of 40:60 for providing peaking support, the present ratio is 27:73. Out of 130,000 MW of economically exploitable hydroelectric potential only 14 percent has been exploited so far. Non-availability of geological and hydrological data, which are reportedly being treated as 'classified', and inadequate incentive to hydel projects considering their special problems, are stated to be the reasons for private investors not evincing much interest and not making desirable progress in hydro-projects. The ministry of power has stated that in a further bid to encourage private investment a separated policy for hydel tariff is under consideration. The committee urges that this should be finalised early (The Economic Times, 1955).

## 25.3 Regional Development and the North Eastern Region

Development, as conceived by science, includes geographical space and historic time. Therefore we have to define a region which is the territorial unit for optimal development of all the resources in it. Optimal world development can be conceived as the complementary development of the regions. The federal area of a large country, conglomeration of countries, or of small states within a region can be conceived as regulatory forces at macro and micro levels. In India the development exercise had been started in 1950. Many models had been evolved till 1989. The latest one claims to have removed the shortcomings in the earlier models, but final results are awaited (Roy, 1992). Depressed regions get further depressed in the existing trend of development. Therefore the delay in regional economic planning had advisedly affected the NE region.

India had been divided into five electricity regions for review of the power development. These are the (1) Western, (2) Northern, (3) Southern, (4) Eastern, and (5) North Eastern regions. The total and the sectoral per capita consumption of electricity for 1991-92 are presented in Table 25.1 for supply of energy by utilities. The data show the glaring backwardness of the NE region in general, and the backwardness in the primary and the secondary sectors in particular.

As in the report of the Forest Survey of India, fuel wood consumption is in the highest per capita range of above 0.50 tonnes in the rural areas of Arunachal, Meghalaya, and Tripura, followed by Assam and Nagaland in the second highest range of 0.40 to 0.50 and Manipur in the high range of 0.25 to 0.40 tonnes. Amongst urban areas, Arunachal and Mizoram come in the highest range, followed by Tripura in the second highest range. In urban areas of Assam, Manipur, Meghalaya, and Nagaland, fuel wood consumption is between 0.25 to 0.40 tonnes per capita per annum which also is high.

Table 25.1 Sectoral per capita consumption of electricity in India: 1991--92 (utilities only).

| Regions | Population (millions) | (KWH) Domestic | (KWH) Comme--rcial | (KWH) *Industry *Public water *Traction *Miscellaneous | (KWH) Public Lighting | (KWH) Agriculture | (KWH) Total |
|---|---|---|---|---|---|---|---|
| Northern | 245.91 | 48.78 | 17.05 | 93.93 | 2.11 | 83.45 | 245.32 |
| Western | 190.31 | 55.39 | 18.39 | 182.32 | 2.73 | 98.83 | 357.66 |
| Southern | 199.20 | 46.17 | 13.13 | 127.27 | 2.82 | 83.31 | 272.70 |
| Eastern | 189.13 | 19.16 | 7.90 | 78.37 | 0.70 | 13.58 | 119.71 |
| North eastern | 31.89 | 15.54 | 7.26 | 45.87 | 1.08 | 2.04 | 71.79 |
| India | 856.44 | 41.86 | 14.24 | 116.11 | 2.06 | 68.37 | 242.45 |

Source of data: CEA

The pattern of commercial energy consumption, annual compounded growth rate of per capita consumption (PCC) of electricity from 1975--76 to 1992--93 for 17 years, and the CMIE infrastructure rating for 1992--93 are furnished for all the seven states of the NE region in Table 25.2. For Assam, the largest economy of the region, the PCC of commercial energy consumption itself is much lower than in the poor states and its share of electricity is still lower. The annual compounded growth-rate of PCC of electricity is 7.43 per cent against 7.57 per cent for the rich states indicating that the gap is widening.

The per capita consumption of power for agriculture, Industry (HV), industry (LV), and the rest of the consumer categories for 1991--92, in respect of the N.E. region of India, and the developed western region have been worked out in Table 25.3 along with the cross-subsidy burden.

One may note the glaringly low quantum of consumption and at the same time the highly distorted pattern of consumption. We may also note along with that every unit of HV industrial consumption in the NE region has to cross-subsidise 4.51 units of consumption in the other categories in spite of low agricultural consumption, whereas in the Western region it is only 1.667 units. Such a situation would not have arisen had there been regional planning for the removal of economic disparity.

Table 25.2 Pattern of commercial energy consumption and infrastructure rating.

| States | 1992--93 | | Electricity PCC (kwh) | | Annual compounded growth rate of PCC of electricity (%) | C.M.I.E.index of infra-structure 1992--93 |
|---|---|---|---|---|---|---|
| | PCC of Energy KgOE | SHARE OF ELE. | 1975--76 KWH | 1992--93 KWH | | |
| Arunachal | 130 | 34.9 | 7.2 | 128.3 | 18.46 | 44 |
| Assam | 84 | 38.8 | 26.9 | 91.0 | 7.43 | 93 |
| Manipur | 88 | 56.8 | 9.0 | 140.4 | 17.54 | 81 |
| Meghalaya | 121 | 46.7 | 31.3 | 159.4 | 10.05 | 65 |
| Mizoram | 90 | 49.6 | 4.3 | 126.1 | 21.99 | 63 |
| Nagland | 77 | 44.8 | 23.4 | 100.0 | 8.92 | 71 |
| Tripura | 55 | 54.1 | 7.9 | 83.9 | 14.91 | 63 |
| India | 232 | 50.6 | 98.4 | 330.6 | 7.39 | 100 |
| Rich states | 374 | 57.5 | 174.9 | 604.9 | 7.57 | 210 (Punjab) |
| Poor states | 178 | 41.3 | 64.8 | 207.6 | 7.09 | 44 (Arunachal) |

Source of data: centre for monitoring Indian economy (CMIE)

Table 25.3 Cross subsidy burden on industry (HV) for 1991--92.

| Regions | Industry HVKH (1) | Industry LVKH(3) | Agriculture | Others KWH(4) | Ratio = (1)/[(2)+(3)+(4)] |
|---|---|---|---|---|---|
| NE Region | 13.03 | 18.10 | 2.04 | 38.62 | 1:4.51 |
| India | 79.83 | 22.14 | 68.37 | 72.11 | 1:2.037 |
| Western Region | 134.13 | 28.11 | 98.93 | 96.59 | 1:1.667 |

There are examples of other damages too for neglecting the regional perspective. The load curves vary widely amongst the regions. As in 1991 data, the ratio between peak load and lean load was 1.37 with the minimum load of 8375 MW in the northern region,

1.28 with 9,320 MW in the western region, 1.2 with 7200 MW in the southern region and 1.34 with 3,200 MW in the eastern region. Against these, the ratio between the peak and the lean loads in the N.E region was as high as 2.3 and the minimum load was as low as 260 MW. A similar situation existed earlier too when a coal fired station away from pit head of capacity 240 MW was set up at Bongaigaon, ignoring the hydel alternative in the region. The result was very low PLF as well as energy and peaking shortages arising out of poor hydro-thermal co-ordination owing to techno-commercial problems of grid and tariff inadequacy, regional as well as inter-regional in nature. This had seriously handicapped load growth. It is essential that generation and tariff regime in the grid be homogenized regionally and inter-regional links be established.

As regards cross-subsidy burden on the H.V. industrial consumers of the NE region as per the example given, one can infer that for a healthy sustaining growth of the power sector, what matters is the development of linkages of power with agriculture and a strong industrial umbrella. Boosting agricultural demand and supplying power to it can trigger the growth of industry which can then increase the growth of power too, and this is the self-developing cycle.

Lastly, regional economic planning can contain human aberrations in development, too. It is participation of human beings which make the word 'change' purposeful. Unknowingly or with a wrong motive, man also can stimulate anti-development in the same process. Therefore the game of development is played by two teams of players opposing each other. It is not the business of science to question why and how this happens. But it can provide us with a central software containing the regional out look from the top to all levels down, so that the game is played truthfully.

As per the 14th Electric Power Survey (EPS) India will reach the PCC level of 586 KWH by 2007 A.D. Assuming that the NE region will reach this level by the same year and attain the same consumption structure, the annual compounded growth rate would have to be 17.88 per cent from the level of 2,289 MU in 1992 A.D up to 26,311 MU in 2007 A.D. It is a difficult job, yet it should be planned and pursued as per a befitting system model prepared for the purpose.

A decade ago CEA had suggested that the government and the Planning Commission should consider the location of energy intensive industries in the chronically backward NE region even assuming the intermediate transportation of raw materials and semi-process products, in view of cheap and bulk availability of power from the mega hydel projects in the region (CEA, 1985).

## 25.4 The Third National Power Development Plan

India embarked upon 15 year perspective planning for power in 1983. The first national power plan (NPP) covered the period from 1980 to 1995. The load forecast and supply reliability are the key issues for power planning. The Electric Power Survey (EPS), now the 14th in operation, provides the load forecast. Supply reliability expressed in terms of 'Loss of Load Probability' (LOLP) and energy not served (ENS) is decided by the

Planning Commission (PC).

The first NPP assumed high reliability at one percent LOLP. The second NPP, brought out in 1987, was with a lower reliability level corresponding to five percent LOLP, was considered enormous, demanding very high investment. This NPP covered up to the year 2000.

During the Sixth Plan (1980--85) only 14,228 MW additional capacity could be installed against the target of 19,666 MW. During the Seventh Plan (1985--90), the resource difficulty and environmental clearance hampered progress. As a result the target of 26,000 MW capacity addition was revised downwards to 22,245 MW. During 1990--95, then considered as the Eighth Plan, the target for capacity addition (CA) was 48,000 MW. But this also had to be revised downward to 38,369 MW, subject to the availability of funds.

In 1989--90 the peaking shortage of about 7,000 MW was more than 16 percent. The energy shortage of about 21 billion units was 8 percent. These conditions were damaging the economy. Studies revealed that these indicated a poor reliability level of 5.6 percent LOLP and 1.34 percent ENS on the average with regional variation. These indicated that power planning should have a higher reliability criterion, though not as high as that corresponding to 1 percent LOLP.

Thus caught in the dilemma of poor reliability and high investment, the Eighth Plan was deferred by two annual plans to 1992-97, whilst the third NPP covering the period of 15 years from 1992 was attempted with a 2 percent LOLP and ENS of less than 0.15 percent by 2007 AD together with other specified strategies to cut down CA to the workable minimum (CEA Report).

The optimal use of hydel capacity in the total is about 40% in India. The country started with 37.31% in 1947, touched 39.51% in 1959-60, rose to 50.62% in 1962--63, declined again to 40.02% in 1979--80 and continued further downward to reach the all time low of 27.05% by 1992--93.

The I.C. of the power sector stood at 69,874 MW in 1991--92. If the same trend of low hydro development continues, the CA required to meet the power demand during the 15 years would be 157,000 MW. If a higher hydro scenario is taken where the growth of hydel capacity has to be double the growth rate as existing now, the CA required will come down to 142,000 MW, resulting in a saving of 15,000 MW thermal capacity costing Rs 60,000 crores. There also will be savings in coal mining and coal transportation for 20 million tonnes from the requirement of 387 MT. In the transport sector rail transport had dominated road transport in a 60:40 proportion in the past. Of late the position has reversed to 40:60. Road transport is much costlier, the high hydro scenario will also indirectly help national savings on road transport.

The system load factors can be improved from the existing level of 65.20% to 72.38%, in the high hydro scenario. Also, an energy conservation by 10% each in domestic and agricultural sectors and 15% in the industrial sector, as well as a reduction of TD loss from 23% to 18% are possible. On achieving these, there will be further saving of 30,000 MW I.C costing Rs 120,000 crores. Over and above these, by making the national grid fully operational by 2007, about 10,000 MWIC costing Rs 40,000 crores can be saved.

It was, therefore, anticipated that following a high hydro scenario, demand management, energy conservation, TD loss improvement, and achieving a fully operational grid will cut down the required CA to 120,000 MW from the level of 157,000 MW, demanding a growth rate of 35,000 MW I. C per each 5-year plan from 1992 to 2007 A.D. But the strategy may not be fully successful. If there is a total failure, the reliability level will be as poor as LOLP 15% and ENS over 5%. It should be maintained at least in the level that existed during the Seventh Plan years. C.E.A, therefore, suggested a capacity addition between 120,000 MW and 130,000 MW.

Assuming the capacity addition as 125,000 MW, the total I.C. by 2007 A.D will be 194,874 MW from the 1992 level of 69,874 MW. As per the high hydro scenario, the capacity addition for hydro power was assumed as 52,000 MW. Keeping this the same, the hydel composition by 2007 A.D will reach the level of 37%. To this we have to add those for advance action so as to be operative by 2007--12 within the 11th plan.

## 25.5 Power Evacuation and Choice of Project

The basin-wise details of the estimated hydro potential in India are given in Table 25.4.

Table 25.4 Basin wise hydro potential.

| Basin / Rivers | MW at 60 % L.F |
|---|---|
| 1. Indus | 19,988.00 (23.8) |
| 2. Ganga | 10,715.00(12.7 |
| 3. Central Indian Rivers | 2,740.00 (3.3) |
| 4. West Flowing Rivers | 6,149.00 (7.3) |
| 5. East Flowing Rivers | 9,532.00 (11.3) |
| 6. Brahmaputra | 34,920.00 (41.5) |
| | 84,044.00 (100 ) |

The share of Brahmaputra basin in the total of all the basins is 41.5%. The status of development of these potentials regionwise, with state wise break up for the NE region, are furnished in Table 25.5. Out of the potential of Brahmaputra basin and some of the east flowing rivers, the hydro electric potential in the NE region comes to 31,857 MW at 60% load factor which is 38% of the total hydro potential of the country. In terms of I.C, it may be of the order of 55,000 MW. But only 0.99% of it has been developed. Considering work in progress it comes to 1.94% and assuming that those schemes cleared by CEA will also be cleared by PC, it will rise to 3.19% (Narasimhan and Singh, 1994).

In the world today, association as opposed to isolation is the key to development. The Economic Sub-commission for Asia--Pacific (ESCAP) conducted a study on the potential for a trans-country power grid between India, Nepal, Bhutan, and Bangladesh. At present India has bilateral arrangement only in respect of the 336 MW Chukha hydel project with Bhutan. Nepal, with a large potential of hydro-power of the order of 83,000 MW, is eager for an outlet to Bangladesh as well. At present the Indian initiative is limited to Karnali, Pancheswar, and Saptakoshi hydel projects in Nepal and 1,500 MW Samkosh project in Bhutan. The large chunk of hydro-power in the NE region cannot be developed without a large export outlet from this region. CEA has identified (i) difficult terrain, (ii) lack of demand, and (iii) non-availability of bulk transmission corridors, etc., as the three primary reasons for the extremely poor development of the hydel resources of the NE region. It is hoped that the Government of India will sort out the problem of power evaluation from the NE region with international cooperation among India, Bhutan, Nepal, and Bangladesh. We may note that land communication between the Indian mainland and the NE region is through a narrow chicken neck of about 10 km width.

It is a matter of concern for the environmentalist that in India, over 50,000 MW IC of thermal plants produce annually 200 billion units of electricity with an ejection of 55 million tonnes of fly ash, 200 million tonnes of $CO_2$, 1.6 million tonnes of $SO_2$ and 1.2 million tonnes of NO, besides producing three times the energy as heat into the atmosphere, consuming 144 million tonnes of coal, 1,000 million litres of oil and 3,000 billion litres of water over and above the diesel and coal consumed in transportation of coal. Against such odds, hydro electricity, a renewable form, is pollution free, saves C.A by about 50% against the thermal option, besides Rs150 crores savings of coal per 1,000 MW hydro IC and many other advantages. Nevertheless, 500 billion units of untapped hydro-power flows into the sea annually (Naidu, 1995).

A detailed project report of the Dihang dam project (20,000 MW) and Subansiri dam project (4,800 MW) had been submitted by the Brahmaputra Board in 1983[10]. The completion periods were 13 years and 9 years respectively. Dihang could reduce the peak floods of Brahmaputra by 1.75 meter at Pandu and 1.1 meter at Jogighopa. The contributions of the Subansiri project also are substantial, being 0.6 meter and 0.4 meter respectively. Besides power and flood control the project could offer the benefits of inland water transport as well as irrigation. But decisions on these projects are yet to come. The DPRs of Tipaimukh dam project (1,500 MW) in the Barak basin and Pagladiya dam project (5.5 MW) is the Brahmaputra basin also were submitted in 1988 and 1992, respectively, which have the irrigation and flood moderation benefits besides power, and were economical. Decisions on these two also are wanting. A long gestation period is often quoted as a handicap for hydel projects. The review of the Seventh Plan puts this figure at a 10- year average. Assuming no scope for cutting it down, this can be solved through advanced planning. The cost of hydro power generation is relatively low and so is the cost of operation and maintenance. Hydel energy's cost is fairly stable and is free from inflationary pressures of the economy. So advance planning is less problematic. But the bureaucratic handicap is highly formidable in India, which must be contained.

Assam is the major producer of food grains in the region. The Planning Commission estimated a gaping hole in agaiculture of Assam at Rs 700 crores per year to swell to Rs 3,500 crores within 1990--95[1]. Out of it a large part is owing to flood damage which was Rs 142.52 crores per year on average during 1980--88. The crop damages were 96.3% of it[12]. Poor irrigation is another factor for which agricultural productivity is poor. Only 23% of the rice cropped area is under irrigation in Assam against 95% in Tamilnadu. The yield of rice in Assam is 1.1 tonnes per hectare against 1.75 tonnes the Indian average and 2.96 tonnes per hectare in Tamilnadu. Experts believe that multi-purpose projects are needed to meet the large demand of irrigation. Electric irrigation is a burden on other power consumers. By itself it is a highly inefficient process. Measured in terms of coal for thermal generation and linking up coal mining to find the availability of water for irrigation through all the intermediate processes, 100 tonnes of coal might yield finally 5 tonnes coal equivalent of utility only. The terrain in the North East has wide variations of altitudes. Irrigation in the foot hills area from canal schemes appears feasible. Pumped storage will be necessary for energy banking as well as efficient operation of the EHV transmission system. Such storages may also be useful for canal irrigation, largely using the force of gravity.

In a study conducted at Roorkee University, a canal scheme in the north bank of the Brahmaputra from a pick up barrage, somewhere around Chauldhowaghat on Subansiri, skirting Arunachal Pradesh was suggested. Similarly, a south bank canal scheme from a barrage on Digang somewhere near Kobo, skirting Nagaland, the Karbi Anglong districts and Meghalaya, had been suggested.

Starting from the first pumped storage scheme for commercial use near Zurich in 1879, more than 300 pumped storage stations of capacity 83,561 MW are now operating in the world with 24,857 MW to be added to it soon. The first of its kind in India was of 700 MW capacity at Nagarjunasagar commissioned in 1970. Two schemes are in operation in South Africa which are used for pumping water from one basin to another (Verma and Krishnan, 1995).

The 'Jabo system of farming' comprising protected forest land on hill tops, a well planned water harvesting tank at the middle, and cattle yard and paddy fields towards the foot hills is very common in Nagaland. This is a progressive alternative to traditional shifting cultivation[15]. CEA has identified 10 schemes of pumped storage in the Barak region having a capacity of 16,900 MW total. Such stages may have potential for making the 'Jabo' system free from dependence on rain. CEA should investigate similar pumped storage reservoirs in the Brahmaputra basin too. The tributaries of Brahmaputra have sufficient variations in peak discharges. Floods in the tributaries were the causes of flood in Assam in 1986 and 1988. Pumped storage for transfer of flood water from one basin to the other may appear promising too, along with other utilities for the power system.

Land, water, forest, and human resources plus savings are the five categories involved in development. Since matter and energy interaction is the core, we conceive development as a process of social production. Therefore, if energy is assured, man is participative and other inputs of nature are available; generation of savings holds the key to development.

In the developed countries during 1991--92, the economically active population in agriculture varied between 2% in U.S.A. and U.K. to 6% in Italy. It was 12% in the erstwhile U.S.S.R. and 66% in China. For India and Bangladesh it was 66% and 67% respectively. Therefore the agricultural sector holds the key to generation of saving in the backward world.

The developed countries after the agricultural and the industrial revolution are in the third wave of the Science and Technology revolution. But the developing countries cannot jump into it, bypassing the earlier stages. This applies more particularly to the chronically backward areas of a developing country like the NE region, which is in a stagnation trap. The success of its development strategy as the exporter of hydroelectricity will depend on how best the other benefits of water resources can be extracted which are: flood control, irrigation and water transport. Therefore the choice of the projects should be for multi-purpose utility.

The state of Meghalaya in the region has surplus in power since its birth a couple of decades ago. But its economic stagnation continues. This is an example to show that mere availability of power may not be adequate to regenerate the economy of a backward region.

## 25.6 The Need for Multi-Purpose Projects

The development is a process of social production and reproduction involving man who causes the matter and energy interactions. The universal laws of conservation guide the interactions and all the resources within it change with development establishing a new balance every moment. When man can no longer stand the change, he says that eco-balance is lost. We can visualise the energy perspective of the world from this angle.

World Energy Consumption (1988) and World Population growth (0-1992) are furnished below:

World Energy Consumption (1998):

| Energy form | MTOE | % |
|---|---|---|
| Oil | 3,040 | 32.5 |
| Coal | 2,430 | 26.0 |
| Gas | 1,630 | 17.5 |
| Nuclear | 440 | 4.7 |
| Hydro | 540 | 5.8 |
| Biomass | 1,260 | 13.5 |
| Total | 9,340 | 100 |

Table 25.5 Status of hydro electric potential development as on 31-10-1993

| Region/ State | Hydro Potential MW at 60% L.F | Developed MW at 60% L.F | % Develo-ped | Under Development MW at 60%L.F | Under Development MW at 60%L.F | % Potential Developed and Under Development | Schemes Cleared by CEA. MW at 60% L.F | Schemes Cleared by CEA.% | Total % Developed Underdevel-opment CEA cleared |
|---|---|---|---|---|---|---|---|---|---|
| Nothern | 30,155.00 | 3,954.47 | 13.11 | 2,308.27 | 7.65 | 20.77 | 1,414.83 | 4.69 | 25.46 |
| Western | 5,679.00 | 1,763.67 | 31.06 | 1,585.00 | 27.91 | 5897 | 17.22 | 0.30 | 59.27 |
| Southern | 10,763.00 | 5,273.42 | 49.00 | 1,098.17 | 10.20 | 59.20 | 83.43 | 0.78 | 59.97 |
| Eastern | 5,590.00 | 969.33 | 15.53 | 725.90 | 12.99 | 28.52 | 487.78 | 8.73 | 37.25 |
| North Eastern | 31,857.00 | 315.50 | 0.99 | 303.88 | 0.95 | 1.94 | 396.17 | 1.24 | 3.19 |
| All India | 84,044.00 | 12,175.38 | 14.46 | 6,021.22 | 7.16 | 21.65 | 2,399.43 | 2.85 | 25.51 |
| 1. Arunachal Pradesh | 26,756.0 | 6.17 | 0.02 | 116.50 | 0.44 | 0.46 | 253.83 | 0.95 | 1.41 |
| 2. Assam | 351.00 | 105.00 | 29.91 | 97.50 | 27.78 | 57.69 | 0.00 | 0.00 | 57.69 |
| 3. Manipur | 1,176.00 | 73.17 | 6.22 | 2.00 | 0.17 | 6.39 | 45.83 | 3.90 | 10.29 |
| 4. | 1,070.00 | 121.67 | 11.37 | 0.00 | 0.00 | 11.37 | 0.00 | 0.00 | 11.37 |
| Meghalaya | 1,455.00 | 1.00 | 0.07 | 6.00 | 0.41 | 0.48 | 96.50 | 6.63 | 7.11 |
| 5. Mizoram | 1,040.00 | 0.00 | 0.00 | 81.88 | 7.87 | 7.87 | 0.00 | 0.00 | 7.87 |
| 6. Nagaland | 9.00 | 8.50 | 94.44 | 0.00 | 0.00 | 94.44 | 0.00 | 0.00 | 94.44 |
| 7. Tripura | | | | | | | | | |

At present we are increasing our numbers by 1.7 percent per year. That means the world population would double within the next forty years with incalculable consequences. In many developing countries increased food production is only possible through irrigation, whilst hydro-power is the only natural energy resource. Thus the amount of water drawn from freshwater sources worldwide which presently stands at an annual total of 3,500 $km^3$, can be expected to increase by roughly a further third by the end of the millennium, and the management of these resources will necessitate the provision of additional storage facilities.

World Population Growth:

| Year | Million |
|---|---|
| 0 | 250 |
| 1780 | 800 |
| 1800 | 1,000 |
| 1900 | 1,600 |
| 1950 | 2,815 |
| 1960 | 3,000 |
| 1975 | 4,000 |
| 1990 | 5,292 |
| 2000 | 6,280 |
| 2025 | 8,480 |

As can be seen from the pattern of energy consumption in the world, at present fossil fuels are used to satisfy three quarters of primary energy requirements worldwide. Given the foreseeable depletion of these resources, especially of oil and gas, plus the problem of the greenhouse effect and global warming, there is an urgent need to replace them with methods of energy production that do not release $CO_2$ into the atmosphere. In other words, in addition to the use of nuclear energy, which is itself controversial, we must concentrate as far as possible on renewable sources of energy. In short and medium terms, however, the only sources of renewable energy that will permit large scale exploitation are biomass and hydropower.

As a conservative estimate exploitable hydropower potential world wide totals over 12 million GWH of which only 17% is currently utilized. Flood protection will also continue to be a significant factor in dam construction, as long as 40% of all fatalities from natural catastrophes world-wide are caused by flooding.

The world hydro potential exploitable and developed (1989) are furnished below:

World hydro potential (1989) TGWH: Thousand GWH

| Continent | exploitable (TGWH) | developed (TGWH) | % |
|---|---|---|---|
| Africa | 1,335 | 41 | 3.07 |
| America | 4,454 | 929 | 20.86 |
| Asia | 4,244 | 383 | 9.02 |
| Australia | 203 | 37 | 18.23 |
| Europe | 872 | 485 | 55.62 |
| USSR | 1,100 | 230 | 20.91 |
| Total | 12,208 | 2,105 | 17.24 |

The logic in support of large multi-purpose projects given by Mr. W. Pircher, President of the International Commission on Large Dams as above, is very strong against the environmentalists, who stage a general warfare against the large multi-purpose projects. Population explosion is a bigger problem for the developing world. So water resources must have the fullest use. The potential yet to be tapped is high, in India it is 79%. The food problems are increasing, the NE region is an example. The problem of rehabilitation will rise with time. Small power utilization is not merely suboptimal from the standpoint of power generation, it may rob us of the advantage of irrigation, flood control and other benefits from many possible multipurpose schemes for ever. Thus there is no alternative to speedy large scale utilization of water resources. Rehabilitation and safety surely are the key issues, but these cannot be taken as a logic for the denial of large dam construction. In India the activists are also raising technical questions on dam safety in connection with the Tehri Project in the Himalayan belt. The questions are of general as well as a project-specific nature. Expert bodies should be in a position to carry with them the nonpartisan intelligentsia through technical discourses.

Isolation is a curse to development and the NE region suffers from it. Except by air, it is connected to the world through a narrow corridor of rail and road links with the Indian mainland. During the second world war road communication to Burma had been established, now not in operation. Access to the seaport of Chittagong by surface and to Calcutta by inland navigation along the Brahmaputra did not operate satisfactorily after the Indian partition, Sir M. Visresvaraya, an eminent engineer in the Central Water and Power Commission, India, in 1951 opined that "it would be technically possible to construct a direct navigable canal connecting West Bengal and Assam, and running entirely through Indian territory. The canal, which would be a still water canal, about 300 miles long .......... would be a long term solution to the problems of an adequate national transport link between Bihar and Assam to supplement the limited capacity of the Assam link" (Acharya, 1995). The Brahmaputra is one of the leading sediment transporting rivers of the world. At that time the possible canal siltation was cited as a problem.

After the partition of India inland water transport on the Brahmaputra stopped functioning; after the earthquake of 1959 it became inoperative owing to river siltation; now it is operating in a very small way, almost confined to ferrying services. In the mean time, with the increased flood intensity, the investigation of the Brahmaputra got an impetus. It was found that with the construction of large dam projects up stream of Brahmaputra, flood could be substantially contained and this also could restore inland water navigation effectively. Along with it the idea of Brahmaputra-Ganga canal also became promising.

For uniform motion against friction, one horse power can haul about 150 kg by road, 500 kg by rail, and 4000 kg by water. Therefore water transport is very cheap. The economy of the region can get a boost with a cheap transportation system. The proposed still water canal can be helpful not only for water transport, but also for water transfer when there is an excess, as well as for irrigation on the way. To mitigate the level differences, large scale pumping may be necessary, the power for which may be supplied from the proposed hydro projects.

As the geographical area of Assam is 7.54 mha, the flood-prone area averaged about 40% until 1988. But with the largest flood affected area going up in 1988, the flood prone area also went up to 62 percent. However, if the two hill districts of Karbi, Anglong and North Cachar are excluded to get an approximation of the plains of Assam, 77 percent, or over three quarters, becomes flood-prone and went under water in 1988. The available one-sixth of the plain area of Assam can go under water as many as four times in a decade.

The flood damage trends in Assam are furnished in Table 25.6. Affected area, affected crop area, affected population and area intensity, overall damage and crop damage all are high and are on the increase.

The Brahmaputra is among the high sediment carrying large rivers of the world. The sediment deposits are increasing at high rates after the 1950 earthquake. It was observed that the bed levels at Dibrugarh had increased by 3 meters and 6 meters, respectively. In the down stream area at Bahadurabad in Bangladesh, the rise indicated was from 11.9 meters in the 1950 decade to 13.4 meters in the decade of 1960. Before 1950 the danger level at Dibrugarh had been crossed only four times in 1931, 1938, 1942, and 1946 for 48 hours. But after 1954 more than once in a year, and for 16 to 20 days, floods crossed the danger level. As from expert assessment, between 1955 and 1979 the Brahmaputra had carried about 402 million tonnes of sediments per year at Pandu. More than 95% of this annual load was transported between May and October. The sediment load increased rapidly with increases in water discharge.

Studies by Goswami showed that between 1971 and 1979, 790 MCUM of sediments, which is about 70 percent of all the sediments brought into the Brahmaputra and all its tributaries, got deposited over the 607 km stretch from Ranaghat to Jogighopa.

Table 25.6 Flood damage trends in Assam.

| Period | Average area affected per annum (mha) | Average annual crop area affected (mha) | Crop area affected % | Average population affected per ann-um, (m) | Affected populati-on per hectare of flood area | Average total flood damages per year (Rs Crore) | Average Annual Crop damages as % of total flood damages |
|---|---|---|---|---|---|---|---|
| 1953--59 | 1.13 | 0.10 | 8.85 | 0.86 | 0.76 | 5.86 | 66.02% |
| 1960--69 | 0.75 | 0.16 | 21.33 | 1.52 | 2.03 | 7.57 | 92.33% |
| 170--79 | 0.87 | 0.18 | 20.69 | 2.00 | 2.3 | 15.18 | 89.33% |
| 1980--88 | 1.43 | 0.40 | 28.05 | 4.55 | 3.2 | 142.52 | 96.38% |

Source: Centre for Science and Environment

As regards high sedimentation, whether anthropogenic factors like deforestation and shifting cultivations, etc, were the dominant cause, or whether natural factors of geological origin, etc, were the dominant cause, the indications were strongly in favour of the latter. In October 1964, a study group was formed by the central government. It reported that the Brahmaputra from Chaikhowaghat to Dhubri had shoals up to a depth of 6.5 meters consisting of sand only. These shoals are, therefore, easily susceptible to erosion even at low velocities. Owing to these reasons, shoals disappear and then reappear in different flood waves (GOA, 1988). It was decided to take up dredging on an experimental basis in 1966. The first attempt was made during 1974--75 and the second during 1975--76. By and large it was unsuccessful.

With such a background it is logical to conclude that harnessing the Brahmaputra through multi-purpose projects is the way for a lasting solution of the flood problem in the Brahmaputra valley. The average annual runoff and sediment load of some of the major tributaries of the Brahmaputra are given in Table 25.7. The contributions of the Dihang, the Dibong, and the Lohit at the head of the Brahmaputra valley and the Subansiri which joins the main river about 150 km down stream of Kobo is about 66% of the average annual flow of the river at Pandu. The storage reservoirs provided are expected to achieve substantial flood moderation, making navigation possible for heavy steamers up to Dibrugarh and Pasighat. The surplus flows can also be transferred to the deficit Ganga basin with other additional advantages (Phukan, 1988).

## 25.7 Hydropower Generation

Now we have to frame a development paradigm from the energy paradigm. Our job is to break the development stagnation of the North East region. There are four issues involved in the task: (1) Mega generation from the multipurpose projects; (2) agriculture for the growth potency and industry for the growth protection; (3) growth of power demand; and (4) export of power to the Indian mainland and the neighbouring countries. The first three constitute the development cycle of the region in the proper sequential order. The fourth one is the ignition key for setting the development cycle of the region in motion, because without larger generation the multi-purpose benefits to the region will be missing, and without evaluation generation will be missing.

On the export front there are four important issues: (i) Import of hydro-electric power from Nepal and Bhutan; (ii) supply of power from the NE region to the export pool; (iii) supply, of power to the Indian mainland from the export pool; and (iv) export of power to Bangladesh. The development in these regards are yet to be transparent. International commerce should not handicap the development of the multipurpose mega-projects on which regional priorities exist. Secondly, all generations in the NE region, except small scale, should be pooled, irrespective of the ownership and the kind of generation, whether thermal or hydel for uninhibited all round regional development. The third guideline should be that a grid bottleneck must not stand in the way of export of large power from the NE region. Fourthly, the N.E. region should be a full participant in all decision-making, involving planning, project formulation, design, and operation of the entire evaluation scheme.

The other important issue is the growth of power demand in the regional Electric Power Survey (EPS) of CEA plays an important role here. The targets used are aimed at about 40% equalization of PCC with the average of the country. But the region used to miss even such low targets every time, with the result that subsequent targets used to be lowered substantially in the latter reviews. A comparison is provided in the Table 25.8.

For the year 1999--00, the fall is from 10142 MU to 8,483 MU in energy demand and from 2,399 MW to 1,867 MW in peak power demand. Development plans presume a certain level of human effort, the variation of which today does reflect in the expectations of tomorrow. Depression of electricity demand is primarily owing to a poor level of reliability of power in the region, as can be inferred from the chronic load shading. In the year 1992 (October) the shortages were 2.78% in energy and 11.85% in peaking power. In 1993 (October) these were 1.6% and 2.91%, respectively. In 1994 (May) the peak storage was 16%. The maximum load shedding for 1992 (October) was 156 MW and that for 1993 (October) was 178 MW. The annual growth rates of electricity sales in Assam are presented in Table 12.9. The unsteady and downward trends are alarming.

A comparison of the available I.C and peak demands of the region from 1993-94 up to 1998-99 are furnished in Table 25.10a and 25.10b. The NEREB under the CEA comprises of seven state utilities, seven state governments, the central hydro-power generating companies and the power grid corporation.

Only ongoing projects have been considered for capacity addition in this scenario, which are Lakwa extension (60 MW), Karbi Langpi (100 MW), Nuranang (6MW), Ranganadi (405 MW), Kopili Extension (100MW), and Kathalguri (2,91 MW). Secondly 12% annual rise per annum of peak load has been assumed for the estimate of peak from the level unsuppressed demand of 675 MW in 1993--94. The estimated demand is 1,190 MW only by 1988--99 against 1697 MW as per the 14th EPS, indicating that if the present state of poor supply reliability continues, there will be heavy depression of demand in the forthcoming 15th EPS.

Table 25.7 Some projects for flood moderation of the Brahmaputra.

| CLASS | NAME OF TRIBUTARY | Average annual runoff, (mhm) | Average annual sediment load (hm) | Installed power capacity (MW) |
|---|---|---|---|---|
| Main Stem | 1. Dihang (Siang) | 18.5 | 14,408 | 20,000 |
| North Bank | 2. Subansiri | 5.74 | 2,503 | 4,800 |
| do | 3. Ranganadi | 0.33 | 155 | 505 |
| do | 4. Dikrong | 0.31 | 338 | - |
| do | 5. Bargong | 0.15 | 178 | - |
| do | 6. Jia Bhoroli | 2.59 | 1,593 | 2,000 |
| do | 7. Dhansiri | 0.26 | 498 | - |
| do | 8. Puthinari | 0.19 | 198 | - |
| do | 9. Pagaldiya | 0.12 | 229 | 5.5 |
| do | 10. Pahumara | 0.14 | 28 | - |
| do | 11. Manas | 3.76 | 1,817 | 5,000 |
| do | 12. Aie | 0.59 | 1,576 | - |
| do | 13. Champawati | 0.21 | 88 | - |
| do | 14. Tipkai | 0.42 | 72 | - |
| do | 15. Sonkosh | 1.80 | 793 | 1,500 |
| do | 16. Gaurang | 0.21 | 108 | - |
| do | 17. Dibang | 3.91 | 2,620 | 2,500 |
| South bank | 18. Lohit | 4.66 | 4656 | 3,000 |
| do | 19. Noa Dihing | 0.36 | 351 | 100 |
| do | 20. Burhi Dihing | 1.36 | 1,247 | - |
| do | 21. Disang | 0.42 | 212 | - |
| do | 22. Dikhou | 0.42 | 206 | - |
| do | 23. Jhanji | 0.17 | 52 | 49 |
| do | 25. Dhansiri | 0.59 | 864 | 105 |
| do | 25. Kopili | 0.94 | 334 | - |
| do | 26. Kulsi | 0.23 | 30 | - |

Table 25.8 Power and energy forecast of CEA for the N.E. region.

| SURVEY REPORT | 1990--91 | 1992--93 | 1993--94 | 1995--96 | 1996--97 |
|---|---|---|---|---|---|
| 11th | 3,749 MU<br>775 MW | - | - | - | - |
| 12th | 3,389 MU<br>740 MW | 3,744 MU<br>814 MW | 5,550 MU<br>1,185 MW | 6,118 MU<br>1,302 MW | 9,013 MU<br>1,886 MW |
| 13th | - | - | 5,552 MU<br>1,322 MW | 6,260 MU<br>1,484 MW | 10,142 MU<br>2,391 MW |
| 14th | - | - | - | 5,503 MU<br>1,248 MW | 8,483 MU<br>1,867 MW |

Table 25.9 Annual growth rate of electricity consumption in Assam (1986-94).

| YEAR | TOTAL SALE (MU) | ANNUAL GROWTH % | INDUSTRY (MU) | ANNUAL GROWTH % |
|---|---|---|---|---|
| 1986--87 | 1,157.40 | - | 633.56 | - |
| 1987--88 | 1,250.51 | 8.0 | 695.78 | 9.82 |
| 1988--89 | 1,402.0 | 12.1 | 816.24 | 11.31 |
| 1989--90 | 1,528.52 | 9.0 | 850.90 | 4.25 |
| 1990--91 | 1,569.75 | 2.7 | 878.59 | 3.25 |
| 1991--92 | 1,584.28 | 0.9 | 789.29 | (-)10.16 |
| 1992--93 | 1,590.0 | 0.4 | 781.01 | (-)1.05 |
| 1993--94 | 1,611.0 | 1.3 | 819.03 | 4.87 |

We can see from Table 25.13 that the rise of lean load of Assam is about 15 MW per annum only which is alarming for hydro-thermal coordination. However, assuming this figure for estimation, the lean load in 1993--94 of the region may rise to, say, 345 MW by 1998--99. Assuming 60% as the minimum stable and economic load for thermal generation and all the hydel units shut down during the lean load period and also assuming 83% annual plant availability for the thermal units, we find that a large thermal capacity will need a shut down at least once, if not more, every day. The idle quantum of thermal IC would range from 97 MW in 1993-94 to 132 MW, 123 MW 123 MW, 299 MW, 364 MW and 339 MW up to 1998--99 for optimal utilization of which the lean load must increase sufficiently. Regarding the peak front, the thermal unit under operation will now deliver the full capacity totalling 450 MW, 475 MW, 500 MW, 525 MW, 550 MW and 575 MW for the years 1993-94 to 1998--99, respectively. Assuming a 90% availability of the hydel units, the additional peaking availability from the hydel sources would be 410 MW, 410 MW, 437 MW, 527 MW, 675 MW and 1,049 MW, respectively for the years 1993--94 to 1998--99. We have not assumed a system loss, but even when this is considered there will be a situation of surplus power without sufficient demand growth.

Table 25.10a Power demand and supply balance for the N.E. region (1993--99).

| Year | Hydel capa-city | Thermal capacity | Anticipated lean load from actual | Estimate of co-ordinated peak # (mw) | Anticip-ated peak from actual (mw) | Forecast of 14 th e.p.s. (mw) |
|---|---|---|---|---|---|---|
| 1993--94 | 455 | 547 | 270 MW* | 450T+410H = 860 | 675* | 968 |
| 1994--95 | 455 | 607 | 285 MW | 475MW+410H = 937 | 756 | 1,091 |
| 1995--96 | 485 | 623 | 300 MW | 500T+437H = 937 | 847 | 1,248 |
| 1996--97 | 585 | 824 | 315 MW | 525T+527H = 1052 | 948 | 1,355 |
| 1997--98 | 760 | 914 | 330 MW | 550T+684H = 1234 | 1062 | 1,538 |
| 1998--99 | 1,165 | 914 | 345 MW | 575T+1049H= 1624 | 1190 | 1,697 |

# See Explanation * actual

On the energy front the actual demand in 1991—92, assuming 23% loss, was 321.5 MW, on the average, continuous for the region. Assuming a 10% annual growth in energy demand, 65% PLF of thermal units, and 30% continuous output of the IC of hydro projects, energy availability has been worked out for the years from 1993-94 up to 1998--99 in the Table 12.10b.

We find that in spite of the depression of demand owing to the poor operation of the power system an energy surplus condition prevails in the grid except during 1994--95 and 1995--96 when there are marginal shortages of 3 MW and 30 MW. For other years, excess thermal capacities range from 11 MW in 1993--94 to 89 MW, 167 MW and 135 MW during 1996-97 to 1998--98, respectively. We may now note that the power scenario, as worked out, had to presume the running of the steam turbine units in the system during the low lean load periods, since these units should run continuously. Otherwise the system will be in artificial shortages of energy and power both. The fine coordination needed for this is failing every now and then and hence the shortages. NEREB is mostly confined to ad hoc administrative and commercial decisions, rather then techno-commercial coordination of the grid. Free flow of power in the grid is handicapped owing to the ad hoc decisions. The grid is too weak to ensure hydro-thermal co-ordination. Thirdly, the power commerce in the region has accommodated the unhealthy choice of hydel power against thermal power. Fourthly, the small isolated grid had no option but to ease out energy and power excess conditions without a link up with the national grid.

Table 25.10b Energy demand and supply balance for the N.E. region (1993--1999).

| Year | Availability of continuous hydel power (mw) | Availability of continuous thermal power (mw) | Availability of continuous total power (wm) | Actual demand of continuous power (mw) | Forecast of continuous power by 14th e.p.s (mw) |
|---|---|---|---|---|---|
| 1993--94 | 123 | 272 | 395 | 389 | - |
| 1994--95 | 123 | 302 | 425 | 428 | - |
| 1995--96 | 131 | 310 | 441 | 471 | 628 |
| 1996--97 | 158 | 410 | 566 | 518 | 704 |
| 1997--98 | 205 | 455 | 660 | 570 | 786 |
| 1998--99 | 315 | 455 | 700 | 625 | 874 |

The operational costs and the returns of the power utilities of the seven states in the region are furnished in Table 25.11(Gupta, 1994).

From the Table 25.11 we find that all the utilities in the region are in extremely bad shape. The average cost of operation ranges from 123.3 paise in the hydel energy surplus state of Meghalaya upto 303 paise in the thermal energy surplus state of Assam. Be it either for reasons within them or without, they failed to mend their houses. So they looked for opportunities to purchase cheap hydel power, the option for which was open for them from the central sector hydro-power generating corporations, NEEPCO and NHPC and the hydel stations of Meghalaya. The tendency indirectly favoured the continuance of energy surplus situation and, for that matter, that led to a lack of urgency about the development of the grid and the exercise of grid control as the consequences. The wrong policy of discriminating the cost of hydel and thermal energy on the ownership basis in a backward economy had seriously handicapped load growth in the NE region through the mechanism of poor quality of supply and depression of demand. The economy of Assam was the largest in the region. With 100 per cent thermal generation the ASEB utility of Assam was hard hit by this stagnation trap.

Unfortunately, it could not resist the death syndrome, but rather succumbed to it, as can be seen from Table 25.12, where the net imports of energy by Assam are furnished against its own generation for the period 1990-94 (Sharma, 1994--95).

During the period from 1990 to 1993 the generation of ASEB fell from 1206 MU to 940 MU, while net import increased from 831 MU to 1,239 MU, making the energy available in the grid between 1925 MU to 1999 MU. There were no major breakdowns in at least 3 × 60 MW units of BTPS As reported, only one unit was out owing to boiler explosion. But the PLE of BTPS was too poor even for the rest.

The composition of IC at the grid during 1993--94 was 455 MW hydro and 547 thermal, catering to the load of 675 MW. Of the thermal units 240 MW was coal based, 30 MW steam turbine with gas as burner fuel, and 60 MW steam turbine using oil, making the total of 330 MW. The lean load in the grid, occurring at different hours, was only 260 MW. The grid operating at low load had low system resilience. It was running in isolation for non-finalization of commercial terms with the eastern grid. Within the region the interstate tariff for import was different from export. The minimum loads during 1987, 1988, 1989 and 1990 may be seen from Table 25.13

Table 25.11 Cost of operation sale price of power in N.E. region: 1991-92.

| Name of utility | Average cost of operation | Average sale price | Short fall | Percentage of recovery of cost |
|---|---|---|---|---|
| State Electricity 1. Assam | 303 | 116 (As on 31.5.93) 240 * (As on 16.5.93) | 187 | 38 |
| 2. Meghalaya | 123.3 | 50.6 | 72.7 | 41 |
| Electricity Department 3. Arunachal | 171.8 | 45.0 | 126.8 | 26 |
| 4. Manipur | 138.8 | 52.50 | 86.3 | 38 |
| 5. Mizoram | 290.8 | 87.0 | 203.8 | 30 |
| 6. Nagaland | 202.3 | 110.0 | 92.3 | 54 |
| 7. Tripura | 165.15 | 53.0 | 112.15 | 32 |

The dominant peak used to be in the evening, with some seasonal variation in the timing. But the troughs, apart from being low, had seasonal as well as diurnal variations in the timings, demanding on-line grid control for hydro-thermal coordination. At least three timings were quite important: around 2 A.M., around 6 A.M., and around 2 P.M. The development strategies should have been to: (i) develop power generation as a regional subject; (ii) construct hydel projects rather than coal based thermal; (iii) develop the grid and take the umbrella of a better developed one, i.e, the eastern grid, for the release of the excess generation; and (iv) develop distribution and small generations as sub-regional units.

Table 25.12 Generation and import of energy by Assam (1990--1993).

| Year | Generation (mu) | Auxilliary consumption (mu) | Export (mu) | Import (mu) | Net import (mu) | Energy (mu) |
|---|---|---|---|---|---|---|
| 1990—91 | 1,206 | 112 | 60 | 891 | 831 | 1,925 |
| 1991—92 | 1,080 | 109 | 39 | 1,036 | 997 | 1,968 |
| 1992—93 | 1,068 | 98 | 78 | 1080 | 1,010 | 1,981 |
| 1993—94 | 940 | 80 | 87 | 1,226 | 1,239 | 1,999 |

The NE regional grid is in the early stage of formation in respect of a control system as well as the equipments. Over and above such expenses, large amounts of money have to be spent for the expansion of the transmission system too we can have an idea about it from Table 25.14.

Table 25.13 Peak and lean loads in ASEB system (1987--1989).

| Month year | March | | June | | September | | December | |
|---|---|---|---|---|---|---|---|---|
| | Peak (MW) | Lean (MW) | Peak (MW) | Lean (MW) | Peak (MW) | Lean (MW) | Peak (MW) | Lean (MW) |
| 1987 | 273 | 150 (2 P.M) | 230 | 162 (3 P.M) | 267 | 162 (7 A.M) | 310 | 155 (2 P.M 2A.M) |
| 1988 | 333 | 150 (2 P.M) | 267 | 175 (2 P.M) | 279 | 170 (6 A M) | 337 | 185 (1.30P.M) |
| 1989 | 346 | 158 (2 P.M) | 350 | 183 (3 P.M 7.A.M) | 350 | 186 (6 A.M) | 380 | 218.0 (3 A.M) |
| 1990 | 360.2 | 182 (2 P.M) (2 A.M) | 369.3 | 232 (4 P.M) | 366.8 | 228 (4 A.M) | 358.3 | 222.0 (3A.M |

Assuming that transmission lines above 90 KV are distance dependent, rather than load dependent, we are using the regional area as the quotient for the comparison of transmission need, and energy density as a measure of comparison for distribution need. On the basis of the Indian average, the region is lagging in the proportion of 27.06:6.26 for transmission and 27.23:20.45 for distribution.

At present PGCL, a central sector company, is engaged for grid management in the region, the expansion and operation costs of which are heavy. The corporation is now working as the agent for distribution of hydel power to the states of the region from NHPC and NEEPCO, the central sector companies, charging a landing cost of about 10 paise per unit of hydro electricity. The primary peak of the regional grid is to optimize the availability of energy and power for the region. The issues of inter-regional export or import also should be its responsibility. The regional grid becomes meaningless when confined to narrow goals of commerce for the central government. Let it acquire and assume the basic task of power marketing in the grid, irrespective of ownership or the kind of generation, from all the power stations in the grid for supply to the distribution sub-regions and for export.

Let us now sum up the regional problem and evolve a solution. Artificial shortage of energy and power in the grid, arising out of inadequate hydro-thermal co-ordination, causes depression of demand of energy and power in the grid. This is damaging the regional economy. The solution lies in treating regional generation as a whole. In doing so, the narrow commercial approach is avoided. Hydel power import of ASEB has increased by 408 MU between 1990--91 and 1993-94 causing ASEB to pay at least 4.08 crores in excess from 1991--92 upto 1993--94. But this is not the solution for breaking the stagnation trap in the energy front. Marketing of power from BTPS must be opened up and demand must grow in the region at a much higher rate. The regional grid has to assume the responsibilities for it and take over all the evacuation lines from all other power stations of the region in order to make pooled power available for the distribution areas of the region. The power stations should be given full autonomy. The distributing agencies should be responsible to the consumers for the tariff exceeding the cost of pooled generation cost at the grid. Since the energy density is low in the region, for catering to consumer expansion at low investment, small generators preferably from renewable sources, at the tail-end, in isolated areas, or injecting the power at intermediate voltages will be promising. Such generators may be left to the distributing agencies.

Table 25.14 Regional variations of transmission line and energy density: India --1991.

| Region | Area Thousand sq. km | Txn lines 220 KV to 400 | | Txn lines 90 KW to 132 KW | | Energy Density KWH/sq. km |
|---|---|---|---|---|---|---|
| | | Ckt kms | Per 1,000 sq km. | Ckt kms | Per 1,000 sq. km | |
| Northern | 1,010.72 | 24,944 | 25.68 | 22,614 | 22.37 | 59,686 |
| Western | 951.46 | 30,355 | 31.90 | 24,046 | 25.27 | 71,540 |
| Southern | 636.30 | 21,844 | 34.33 | 24,180 | 38.00 | 85,371 |
| Eastern | 433.69 | 10,499 | 25.21 | 14,065 | 32.43 | 52,204 |
| North Eastern | 225.09 | 1,410 | 6.26 | 4,604 | 20.45 | 8,975 |
| India | 3,287.26 | 89,052 | 27.06 | 89,509 | 27.23 | 63,166 |

The third vital issue is agricultural and industrial development, viz., regeneration of the regional economy. Man is part and parcel of nature and not someone standing outside it. Science conceives development as an optimizing process of resource utilization. The spatial unit for it is the region, which we have to define. But resource distribution is non-uniform throughout the world, governed by natural laws, where man has no hand. This compels us to question the kinds of resources, which are two: the flow resources, i.e., renewables such as water, sun, wind, etc., and fund resources such as petroleum, gas, coal, other minerals, etc.. There is also a third kind of resources such as forest, which man can use as a flow resource or as a fund resource (Munsi, 1984).

Man is a form of renewable resource, constantly in the flow like water. His development is secure with the flow resources of nature. Therefore regional identification of man on the basis of water resources is a valid logic. Conservation of minerals, coal, petroleum, gas, etc., are his nature culture. His behaviour with the forest resources should be such that eco-balance is maintained at all times. Seen from this angle the NE region of India comprising Arunachal, Assam, Manipur, Meghalaya, Mizoram, Nagaland, and Tripura, can be assumed as a natural economic region of India for macro development. The states may operate within it at the meso level, whereas the regions and sub-regions within them at the mini and the sub-mini levels of the economy. In all likelihood this will bring forth unity amongst the diversities of human culture from the bottom up, a process for breaking the stagnation trap of the economy. This appears to be the only way to stop the political anti-culture of either blaming the centre for the economic deprivations or praising the centre for giving alms, while all the varieties of political anti-culture have the common modus operandi of creating the diversity of the people, without creating the bonds of their unity.

The region has a vast potential for hydel power projects. Table 25.15 furnishes certain details for about 44,000 MW IC of such projects. In addition to that, a long list of medium hydro-power schemes also has been furnished as per Table 25.16, the total of which will run to about 10,512.6 MW at 60% L.F plus about 570 MW additional. In Arunachal Pradesh small and mini hydel schemes totalling 23.65 MW are in operation, and 48.95 MW are under implementation. Feasibility reports for about 40 small hydro projects with a total of 106.1 MW for the investment of Rs 44,240.33 lakhs are ready in hand.

In Table 25.17 the salient features of four hydro projects, Dihang, Subansiri, Tipaimukh, and Pagaldiya have been furnished. Out of the 7 million hectare meter of identified live storage capacity of the Brahmaputra basin, Dihang, and Subansiri will cover a live storage of 3.5 + 1 = 4.5 mhm, i.e 64% of the total. The project cost and generation cost are highly attractive along with irrigation and flood control benefits. Similarly, the Tipaimukh project of the Barak valley with live storage of 0.9 mhm also will have a high contribution in flood moderation and irrigation as well as cheap electricity. The Pagladiya scheme is for irrigation and flood moderation with its own power for the deep tube wells. However, it is an unpleasant fact that the time to decide such projects is long enough and comparable to the construction time of the projects themselves. Therefore rapid development projects from amongst the medium and the small schemes will have promise. Such schemes can be investigated from three angles: (i) As captive generation to existing industries; (ii) for injecting power into the existing distribution network at the lower ends of the grid or towards the tail-ends of the distribution network; and (iii) for power supply to area development schemes comprising of agro-industrial complexes.

## 25.8 Concluding Remarks

The chronically backward NE region is a drag on the optimal development of India. The region has a high endowment of fund and flow resources, which are underutilized. If the region is organized economically with the needed political authority and its isolation to the Indian main land and the world is removed, it can turn out to be a boon to the Indian economy. Harnessing the Brahmaputra water resources, in particular, can break its economic stagnation. The power utilities, the generating companies, the transmission network, and the grid should be reorganized into a regional whole. Load growth in the region should be induced under an integrated plan. A trans-national grid system should be developed, with the objective of evacuation of bulk power from the region and its export to the Indian mainland and the neighbouring countries, side by side, the multi-purpose projects should be taken up. In order to meet the immediate power need of the region under accelerated growth programme, the medium and the small hydel schemes should be developed particularly for load sharing applications.

## Bibliography and References

Acharya, S.K. (1995). The Vital Link Canal. The North East Times, April 9.
Alvin Toffler: The Third Wave.
CEA (1985). Twelfth Electric Power Survey.
CEA Report on Perspective of Power Development up to 2006-07.
Centre for Science and Environment. State of India's Environment: A citizens Report.
GOA (1988). Report of the Committee for Examining the Causes of Flood Occurred in Assam in May and June.
Gupta, A.K. (1994). Institutional Strengthening of SEBS / State Power Departments in North- Eastern States.
Munsi, S.K. (1984). India resource regions and regional disparity. People's Publishing House.
Naidu, B. S. K. Techno - economic Benevolence of Hydro. CBIP Journal.
Narasimhan, S.R. and Singh, R. (1994). Hydro-electric Development - Indian Scenario: CBIP Journal, January.
Pachauri, R.K. (1992). Population and Energy: Science, Population and Development, edited by Vasant Gowariker.
Phukan, S.N. (1988). An approach to flood Control in the Brahmaputra Valley. The Sentinel, September 3 and 4.
Pircher, W. 36,000 Large Dams-And Still More Needed.
Planning commission (1990). Economic Development of Assam. April.
Roy, B.K. (1992). Population, Regional Development and Regionalisation: Science, Population and Development, edited by Vasant Gowariker.
Sinha, C.P. (1995). I.E (I) A.S.C. Souvenir, Tenth W.R. Day.
Sharma, N. Strategy for Accelerated Irrigation Development: Profile of a Development Strategy for India's North East.
Sharma, P.C. (1995). Potential of multi-purpose Dihang Dam Project: The Assam Tribune, April 19.
Sharma, P.C. (1994-95). The Power Sector of Assam: Problems and Prospects: Industrial Prospects for Assam 1994-95. Corporate Communications: Souvenir.
TERI Energy Data Directory and Yearbook.

The Assam Tribune, July 15, 1995.
The Economic Times (1955). Suggestions of the Energy Panal.1-7-1955.
Verma, C.V.J. and Krishnan, K. (1994-95). On Pumped Storage Schemes. C.B.I.P Journal: 1994, October and 1995 Jan -March.

Table 25.15 Hydro power projects in the Brahmaputra basin of N.E. region.

| Name of Project | Average Annual Yield (mcum) | Dam Height above river bed level (m) | Storage Gross/Live (ha m) | Regulated Flow (cumecs) |
|---|---|---|---|---|
| 1.Umain-Umtru-Khri Basin.<br>--------------------<br>1.1.Umian | - | - | - | 12.20 |
| 1.2 Umian Stage - 1 | 260 | 73.20 | 18,272<br>-----------<br>14,346 | 6.40 |
| 1.3 Umian Stage - 2 | - | - | - | 6.40 |
| 1.4 Umian Stage -3 | 175 | 27.50 | 650<br>------<br>280 | 11.04 |
| .5 Umian Stage -4 | - | 43.40 | 180<br>-----<br>80 | 12.50 |
| 1.6 Umian Stage -5 | - | 25.00 | - | 20.00with Khri<br>♉13.00 |
| 1.7 Umian Stage -6 | - | 91.00 | 42,400<br>---------<br>25,000 | 43.00with Khri<br>♉33.00 |
| 2. Lower Umian Basin<br>2.1 Umswai | - | 39.50 | 35,758<br>--------<br>25,894 | 25.00 |
| 3. Barapani Basin | - | - | - | - |
| 3.1 Upper Barpani | - | 106.00 | 24,382<br>-------<br>18,518 | 12.00 |

Table 25.15 Hydro power projects in the Brahmaputra basin of N.E. region (contd.).

| Name of Project | Continuous Output.MW/MU (annual) | Installed Capacity (MW) | Remarks. |
|---|---|---|---|
| 1.Umain-Umtru-Khri Basin.<br>------------------<br>1.1.Umian | - | 341.2<br>11.2<br>(4×2.8) | in operation under MeSEB |
| 1.2 Umian Stage - 1 | 22.40/196.0 | 36.0<br>(4 ×9) | in operation under MeSEB |
| 1.3 Umian Stage - 2 | - | 18.0<br>(2 ×9) | in operation under MeSEB |
| 1.4 Umian Stage -3 | 13.40 | 60.0<br>(2 ×30) | in operation under MeSEB |
| 1.5 Umian Stage -4 | 14.70 | 60.0<br>(2 ×30) | in operation under MeSEB |
| 1.6 Umian Stage -5 | 8.00 with Khri♉13.00 | 36.0<br>(2 ×18) | under investigation of MeSEB. (Khri diversion will improve output ) |
| 1.7 Umian Stage -6 | 43.00with Khri♉ 59.00 | 120.0<br>(4 ×30) | under investigation of MeSEB. (Khri diversion will improve output ) |
| 2. Lower Umian Basin 2.1 Umswai | 49/429 | (2 ×60)<br>120.0 | under investigation of SEB. |
| 3. Barapani Basin | - | 220.0 | - |
| 3.1 Upper Barpani | 18/157 | 60.0<br>(2 ×30) | Under Investigation of ASEB |

Table 25.15 Hydro power projects in the Brahmaputra basin of NE region (contd.).

| Name of Project | Average Annual Yield (mcum) | Dam Height above river bed level (m) | Storage Gross/Live (ham) | Regulated flow (cumecs) |
|---|---|---|---|---|
| 3.2.Intermediate Borpani | - | 45.00 | - | 12.00 |

| | | | | |
|---|---|---|---|---|
| 3.3.Lower Borpani (Langpi) | - | 37.50 | 70/45 | 16.00 |
| 4.Kopili Basin<br>------------------<br>4. Kopili Stage1 | 1,604 | 53.55<br>and<br>28.00 | - | - |
| 4.2. Kopili Extension Phase-1 | - | - | - | 45.00 to 25.86 |
| 4.2. Kopili Extension Phase-2 | - | - | 63,457/59,753 (Kh) and 17,528/15,679 (Um) | - |
| 4.4. Lower Kopili | - | 73.00 | 39,259<br>--------<br>32,716 | - |
| 4.4. Amring | - | 7,000 | 41,975<br>--------<br>35,802 | 21.30 |
| 5. Dhansiri South Basin | - | - | - | - |
| 5.1 Doyang | 2227 | 96.00 | 150,000<br>----------<br>122,000 | 44.00 |
| 6. Noa-Dihing | 2732 | 99.00 | 38,000<br>--------<br>24,000 | 63.00 |

Table 25.15 Hydro power projects in the Brahmaputra basin of NE region (contd.).

| Name of Project | Continuous Output. MW/MU (annual) | Installed Capacity (MW) | Remarks. |
|---|---|---|---|
| 3.2.Intermediate Borpani | 18/157 | 60.00 $(2\times 30)$ | Investigation by A.S.E.B. |
| 3.3.Lower Borpani (Langpi) | 38/332 | 100.0 $(2\times 50)$ | Under Construction of A.S.E.B. |
| 4.Kopili Basin<br>------------------<br>4. Kopili Stage1 | 12/105<br>and<br>50/447 | 718.0<br>150.0<br>$2\times 25+2\times 50)$ | Khandong and Kopili P/H in operation by NEEPCO |
| 4.2. Kopili Extension Phase-1 | 95/835 | 35$(1\times 35)$ and | Under Construction by NEEPCO |

| | | | |
|---|---|---|---|
| | | 150(2 × 75) | |
| 4.3. Kopili Extension Phase-2 | 100/876 | 100and 150 | Planned by NEEPCO |
| 4.4. Lower Kopili | 40/350 | 100.0 | investigation by A.S.E.B. |
| 4.4. Amring | 15/130 | 33.0 (2 × 16.5) | Investigation by A.S.E.B. |
| 5. Dhansiri South Basin | - | 105.0 | - |
| 5.1. Doyang | 41/363 | 105.0 (3×35) | Under Construction by NEEPCO |
| 6. Noa-Dihing | 30/263 | 100.0 | Under Investigation of B.B. |

Table 25.16 Medium hydropower schemes of the Brahmaputra basin.

| Serial No | Name of Scheme | Head (m) | Power Out put at 60% L.F.(MW) |
|---|---|---|---|
| | Arunachal Pradesh. u Brahmaputra river system. | | |
| 1. | Lungchang | 170.3 | 13.3 |
| 2. | Tipang * | 142.1 | 50.5 |
| 3. | Yangman | 94.3 | 38.8 |
| 4. | Mpopnong * | 26.0 | 10.8 |
| | Lohit River System. | | |
| 5. | Diyung Dam* | 62.8 | 17.0 |
| 6. | Chemba | 140.0 | 5.8 |
| 7. | Toysu | 85.0 | 4.5 |
| 8. | Kalai | 310.0 | 1,122.7 |
| 9. | Hutong | 110.0 | 425.8 |
| 10. | Gimliang | 140.0 | 14.0 |
| 11. | Duliang | 100.0 | 11.0 |
| 12 | Raigam | 80.0 | 15.8 |
| 13. | Tidding -1 | 130.0 | 14.2 |
| 14. | Tidding -2 | 100.0 | 11.0 |
| | Dihang Dibang River System | | |
| 15. | Mihundon | 280.0 | 32.2 |
| 16. | Amulin | 120.0 | 60.8 |
| 17. | Emini | 150.0 | 65.2 |
| 18. | Agoline | 135.0 | 52.7 |
| 19. | Malinye | 240.0 | 74.8 |
| 20. | Attunli | 140.0 | 45.8 |
| 21. | Etalin | 440.0 | 681.2 |
| 22. | Emra - 1 | 200.0 | 60.8 |

| | | | |
|---|---|---|---|
| 23. | Emra - 2 | 520.0 | 180.3 |
| 25. | Elango | 360.0 | 37.3 |
| 25. | Yang Sang | 90.0 | 29.2 |
| 26. | Rigong | 170.0 | 50.8 |
| 27. | Mirak | 200.0 | 64.2 |
| 28. | Minnying | 200.0 | 56.0 |
| 29. | Pauk | 160.0 | 22.0 |
| 30. | Hoo | 230.0 | 39.8 |
| 31. | jarong | 115.0 | 45.5 |
| 32. | Hirong | 155.0 | 79.3 |
| 33. | Tato - 2 | 280.0 | 157.2 |
| 34. | Tato - 3 | 160.0 | 33.2 |
| 35. | Gameng | 100.0 | 16.3 |
| 36. | Naying | 240.0 | 218.7 |
| 37. | jeru * | 60.0 | 34.3 |
| 38. | Yapin | 105.0 | 60.3 |
| 39. | Pengging | 59.0 | 35.3 |
| 40. | Pasighat* | 220.0 | 36.0 |
| 41. | Sessari | 110.0 | 23.0 |

Table 25.16 Medium hydro-power schemes of the Brahmaputra (contd.).

| Serial No | Name of Scheme | Head (m) | Power Out put at 60% L.F.(MW) |
|---|---|---|---|
| | Subansiri River System | | |
| 42. | Oju - 1 | 275.0 | 680.3 |
| 43. | Oju - 2 | 365.0 | 912.7 |
| 44. | Niare | 220.0 | 560.6 |
| 45. | Naba | 200.0 | 514.2 |
| 46. | Tammu | 90.0 | 37.0 |
| 47. | Milli | 200.0 | 29.2 |
| 48. | Sape | 75.0 | 16.8 |
| 49. | Chomi | 120.0 | 35.2 |
| 50. | Chela | 95.0 | 32.2 |
| 51. | Nyapin | 140.0 | 14.0 |
| 52. | Hiya | 135.0 | 17.8 |
| 53. | Kurung Dam - 1* | 105.0 | 130.3 |
| 54. | Kurung Dam - 2* | 60.0 | 74.3 |
| 55. | Hegio | 120.0 | 166.0 |
| 56. | Tamen | 40.0 | 98.7 |
| 57. | Siri* | 47.7 | 20.7 |
| 58. | Tago - 1 | 290.0 | 10.7 |
| 59. | Tago - 2 | 75.0 | 6.3 |
| 60. | Yazali Storage * | 75.0 | 62.3 |
| 61. | Kimin | 155.0 | 14.8 |
| 62. | Par | 170.0 | 23.0 |
| 63. | Yazali Diversion | 313.4 | 270.3 |

| | | | |
|---|---|---|---|
| 64. | Dardu | 120.0 | 27.0 |
| | Total | | 3,754.2 |

Table 25.16 Medium hydro-power schemes of the Brahmaputra (contd.).

| Serial No. | Name of Scheme | Head (m) | Power Out put at 60% L.F.(MW) |
|---|---|---|---|
| 65. | Duimukh* | 81.0 | 111.3 |
| | Kameng River System | | |
| 66. | Chanda | 150.0 | 26.8 |
| 67. | Badao | 150.0 | 31.2 |
| 68. | Rebby | 160.0 | 7.8 |
| 69. | Para | 200.0 | 13.7 |
| 70. | Talong | 92.0 | 36.2 |
| 71. | Taley | 80.0 | 9.5 |
| 72. | Satuk | 130.0 | 18.2 |
| 73. | Kapak Leyak | 240.0 | 49.8 |
| 74. | Lachung | 110.0 | 10.8 |
| 75. | Phanchung | 200.0 | 22.8 |
| 76. | Utung | 295.0 | 51.0 |
| 77. | Nazong | 135.0 | 30.5 |
| 78. | Dibbin | 130.0 | 44.8 |
| 79. | Doom | 60.0 | 2.8 |
| 80. | Djrang | 80.0 | 4.0 |
| 81. | Them Lang | 100.0 | 11.8 |
| 82. | Khuitam | 120.0 | 15.5 |
| 83 | But | 100.0 | 14.0 |
| 84. | Mathithing | 135.0 | 19.0 |
| 85. | Bichom* Storage - 1 | 120.0 | 125.2 |
| 86. | Bichom - 2 | 130.0 | 134.7 |
| 87. | Pichang * | 136.0 | 20.5 |
| | Total | | 810.9 |

Table 25.16 Medium hydro-power schemes of the Brahmaputra (contd.).

| Serial No | Name of Scheme | Head (m) | Power Out put at 60% L.F. (MW) |
|---|---|---|---|
| 88. | Tarang Warang | 220.0 | 15.7 |
| 89. | Sepla | 140.0 | 12.0 |
| 90. | Chaurate | 90.0 | 8.2 |
| 91. | Papu* | 290.0 | 104.3 |
| 92 | Pasar* | 395.0 | 21.0 |
| 93. | Kimi | 335.0 | 356.5 |

| | | | |
|---|---|---|---|
| 94. | Bhareli Lft - 1 | 37.0 | 144.2 |
| 95. | Bhareli Lft - 1 | 48.0 | 198.8 |
| 96. | Pakke | 460.0 | 29.0 |
| 97. | Sepa | 140.0 | 26.5 |
| | ASSAM<br>Upper Brahmaputra<br>River System. | | |
| 98. | Dilli* | 52.0 | 21.8 |
| 99. | Disai Dam P.H.* | 26.5 | 10.2 |
| 100. | Kaliani* | 66.0 | 9.7 |
| | Kalang (Kopili) River | | |
| 101. | System | 37.8 | 19.0 |
| 102. | Jamuna Kam P.H. | 90.0 | 25.3 |
| | *langey | | |
| 103. | Schemes identified by | - | 2x3 MW I.C |
| 104. | ASEB | - | 2x1.5 MW I.C |
| 105. | Myntriang - 1 (Koplil basin) | - | 2x1.5 MW I.C |
| | Myntriang - 2 (Koplil basin) | | |
| | Lungit - 1(Kopili basin) | | |
| | Total | | 1001.2 |

Table 25.16 Medium hydro-power schemes of the Brahmaputra (contd.).

| Serial No | Name of Scheme | Head (m) | Power Out put at 60% L.F.(MW) |
|---|---|---|---|
| 106. | Lungit -2 (Kopili basin) | - | 2 × 1.5 MW I.C. |
| 107. | Dalaima (Kopili basin) | - | 3 × 2 MW I.C. |
| 108. | Jamuna (Kopili basin) | - | 2 × 1 MW I.C. |
| 109 | Jenum (Barak basin) | - | 2 × 2 MW I.C. |
| 110. | Sunani (Kopili basin) | - | 2 × 75 KW |
| 111. | Chapanala (Kopili basin) | - | 2 × 75 KW |
| 112. | Kalanga (Kopili basin) | - | 3 × 150 KW |
| 113. | Bar Dikorai Canal Drop | - | 3 × 1000 KW |
| | NAGALAND Upper Brahmaputra Basin | | |
| 114. | Dikhu Lft Dam | 30.5 | 81.6 |
| 115. | Dikhu Dam P.H.* | 117.1 | 310.5 |
| 116. | Yangnyu * | 152.0 | 90.2 |
| 117. | Jhanji Lft Dam | 26.0 | 12.3 |
| 118. | Jhanji Storage * | 121.3 | 49.0 |
| 119. | Diphupard * | 70.0 | 6.3 |
| 120. | Regma* | 132.0 | 8.0 |

| | | | |
|---|---|---|---|
| 121. | Doyang - IV | 64.0 | 12.0 |
| 122. | Doyang - III | 48.0 | 8.7 |
| 123. | Doyang - II | 54.0 | 9.3 |
| 125. | Doyang - I* | 62.5 | 10.7 |
| | Tezu River System. | | |
| 125. | Sakhai - I * | 48.0 | 5.3 |
| | Total | | 603.9 |

Table 25.16 Medium hydro-power schemes of the Brahmaputra (contd.).

| Serial No | Name of Scheme | Head (m) | Power Out put at 60% L.F. (MW) |
|---|---|---|---|
| 126. | Sakhai-II | 183.0 | 20.3 |
| 127. | Sakhai-III | 61.0 | 7.0 |
| 128. | Khuzami * | 77.0 | 21.5 |
| 129. | Rurrur * | 188.0 | 25.0 |
| 130. | Zungki * | 53.0 | 32.3 |
| 131. | Laruri | 82.0 | 52.0 |
| 132. | Tizu * | 230. | 243.0 |
| | MEGHALAYA | | |
| | South Flowing Rivers. | | |
| 133. | Sushen | 320. | 30.0 |
| 134. | Selim* | 475.0 | 37.5 |
| 135. | Umngot Storage * | 798.0 | 125.5 |
| 136. | Nongpodu | 100.0 | 16.2 |
| 137. | Umjaut | 320.0 | 9.5 |
| 138. | Umduma | 343.0 | 13.0 |
| 139. | Manhu | 358.0 | 16.2 |
| 140. | Nonglynogkien | 225.0 | 11.7 |
| 141. | Umrgi Storage | 338.0 | 23.3 |
| 142. | Rangmaw | 207.0 | 16.0 |
| 143. | Nongnam | 690.0 | 55.8 |
| 144. | Mowpat | 95.0 | 9.0 |
| 145. | Mawpat | 98.0 | 10.5 |
| 146. | Kynshi –II | 473.0 | 116.1 |
| | Total | | 890.4 |

Table 25.16 Medium hydro-power schemes of the Brahmaputra (contd.).

| Serial No | Name of Scheme | Head (m) | Power Out put at 60% L.F.(MW) |
|---|---|---|---|
| 147. | Umkram * | 83.0 | 7.3 |
| 148. | Kynshi - I * | 869.0 | 196.8 |
| 149. | Nongmawlar | 206.0 | 7.7 |
| 150. | Mawsyrpat | 274.0 | 11.7 |
| 151. | Mawthaba * | 89.0 | 26.3 |

| | | | |
|---|---|---|---|
| 152. | Mawblei Storage * | 438.0 | 44.2 |
| 153. | Simsang Dam P.H. * | 58.0 | 40.8 |
| 154. | Sidugiri | 122.0 | 12.7 |
| 155. | Amegam Storage * | 261.0 | 23.5 |
| | Lower Brahmaputra River System. | | |
| 156. | Umrina | - | 8.2 |
| | Kalang River System. | | |
| 157. | Sumer - I * | 165.0 | 7.5 |
| 158. | sumer - II | 81.0 | 3.7 |
| 159. | Umlanphang | 280.0 | 14.2 |
| | New Schemes of Brahmaputra Board. | | |
| 160. | Kulsi ( Assam ) | - | 25MW I.C. |
| 161. | Jadukata ( Meghalaya ) | - | 345MW I.C. |
| 162. | Someswari ( Meghalaya ) | - | 72MW I.C. |
| 163. | Umnigot ( Meghalaya ) | - | 100MW I.C. |
| | * STORAGE SCHEME | | 404.6 |
| | | | 10512.6 |

Table 25.17 Salient features of projects for which detailed project report submitted by the Brahmaputra Board.

| Projects features | Pagladiya Assam | Tipaimukh Manipur, Mizoram, Assam | Subansiri Arunachal Assam | Dihang Arunachal Assam |
|---|---|---|---|---|
| Latitude Longitude | 26°31'30" N 91°31' E | 24°14' N 93°1.3' E | 27°33'15" N 94°15'30" E | 28°11' N 95°14' E |
| Catchment (sq. km ) | 570 | 12,758 | 27,000 | 247,500 |
| Dam Height (m) | 29.0 | 162.8 | 240.0 | 296.0 |
| Installed Capacity (MW) | 5.5 (Foruse in D.T.W.) | 1,500 (6 × 250) | 4,800.0 (12 × 400) | 20,000.0 (40 × 500) |
| Annual Generation (GWH) | 88.27 | 3,609.0 | 17,200.0 | 58,975.0 |
| Irrigation Benefit (Ha) | 34,360 | 98,844 | - | 250,000 |
| Flood Benefit | 10,000 | 150,000 | 200,000 | 500,000 |

| | | | | |
|---|---|---|---|---|
| (Ha) | | | | |
| Area to be Submerged (Ha) | 5,550 | 31,100 | 19,300 | 49,000 |
| Population to be affected (Number ) | 13,895 | 2,232 | 7,500 | 35,000 |
| Cost of Project (Rs in Crores) | 337 (at 1992 price) | 1316 (at 1988 price) | 3938 (at 1988 Price) | 12,170 (at 1988 Price) |
| Cost of energy (Paise per KWH) | To be compared with ASEB Price | 35.0 paise at 1988 prices | 25.0 paise at 1988 prices | 27.0paiseat 1988 prices |
| Completion Time. D.P.R Submission Present State | 7 years 1992 under appraisal | 12 years 1988 final DPR ready | 9 years 1993 under re-examination | 13 year 1983 under re-examination |

# CHAPTER 26

# NAVIGATION DEVELOPMENT

R. M. DAS

Transport is a personal activity, a social service, and an industry which constitutes one of the most important activities of man in every stage of his civilisation. No nation could, indeed, afford to overlook the crucial role of transport in economic development at the national and regional levels as well as in the expansion of international trade. Today efforts have been made to explain regional imbalance in terms of infrastructure of which transport is a vital component. The impact and counter impact of each different type of transport facility may also be different in the same region from time to time.

Inland water transport has played an important role as the maiden transportation system throughout world. With the introduction of trains, automobiles, and the aeroplane, however, it has had to face serious competition. It has not merely survived but continues to play an important role in many advanced countries of the world, like U.S.A., Germany, the former U.S.S.R., etc., in spite of the fact that other modes of transport, such as rail and road, are far more developed in these countries as countries as compared to India. This is because of its inherent advantage as a low cost mode of transport. Many rivers in the world provide natural transportation routes for the movement of large quantities of goods.

River barges carry heavy cargo along the Illinois, Mississippi, Ohio, and Hudson rivers. The waterways of the United States transport nearly 10 percent of the nation's freight. In Europe many rivers are crowded with barges moving from country to country. In India the first steam propelled vessel sailed with passengers from Kulpi Road to Calcutta a distance of 80 kms on the Hooghly in 1823. A regular steamer service over the Brahmaputra between Calcutta and Assam was launched in 1863.

Unfortunately the development of inland water transport in India has suffered a setback for many reasons and there is a notion that, as a modern transport system, inland water transport has no place in the growth of the national economy. This impression is entirely wrong. The modern concept is that all modes of transport, viz., rail, road, air, and water have equal importance and each of these is to be considered on individual merit and as part of an integrated transport system. Inland water transport is likely to assume a much greater role in the Indian transport network in the near future.

The distribution of India's water resources is highly uneven. As much as 32 per cent of the country's water resources flow down the Brahmaputra and Barak rivers. India's navigable inland waterways extend nearly 14,500 km, comprising a variety of river



*V.P. Singh et al. (eds.), The Brahmaputra Basin Water Resources*, 473-519.

systems, canals, backwaters, creeks, and tidal inlets. Assam has a total of 1983 navigable waterways, with 2.53 density (sq. km waterways per 100 sq. km of area) of waterways.

A fresh look into the navigation potentiality in Assam has become much more important than ever before, in view of the rapidly mounting global energy crisis and its effect on the future economy of the region. The National Seminar on Inland Water Transport (January 19-20, 1984) has already highlighted its relevance in the context of the national economy.

## 26.1 Role of Inland Water Transport in Economic Development

Trade by inland water transport gave to the world its earliest civilisations in Mesopotamia and along the Indus, the Nile and the Hoang Ho. Apart from being one of the earliest modes of transport, the inland water transport is accepted as one of the most efficient modes of transport from the standpoint of energy consumption.

The economic development of a region or a modern state depends, amongst other things, on the development of transport systems as a whole. The major modes of transportation presently available in Assam comprises railways, roadways, airways, and inland waterways. The quantum of contribution to the transport sector, at present, by airways and waterways in Assam is insignificant. The major share of passenger and goods transport has been taken care of by rail and road transport. Viewed from the trend of development of these modes it is estimated that about 90% of inland passenger and freight traffic in 2000 AD will have to be met by road and rail. The problem is whether adequate road and rail transport facilities can be provided to meet such huge demands in the future at a reasonable cost to society and at the same time ensure safe, efficient, convenient, and pollution-free service to the public. This problem is made all the more complicated with the energy crisis generated world-wide, which undoubtedly is going to affect the future transportation scenario of the world. The transport sector consumes about 35 percent of (our) nation's requirement of petroleum products and this demand is increasing at the rate of 8 per cent per annum. With nearly 60 per cent of our nation's requisites being met by imports, every increase in the price of oil has eaten into the nation's foreign reserves significantly.

Keeping this in mind, the National Planning Commission recently initiated a study of the energy efficiency of various modes of transport, and from the view point of fuel efficiency, this study has shown that under Indian conditions the electrically run train is the best, and this is followed by diesel train, water transport (barge), diesel-truck, and then steam train, which is the least efficient mode of transport as far as fuel efficiency is concerned (Table 26.1).

Therefore inland water transport is one of the most efficient modes of transport from the standpoint of energy consumption. Economically, in fact, inland water transport is the cheapest mode of transport for certain kinds of traffic, both for long and short distances, particularly if the points of origin and destination are located on a water front and no transhipment is involved.

Table 26.1 Comparison of energy efficiency of various modes of transport.

| Mode of Transport Consumption | | Energy In BTU per Tonne M |
|---|---|---|
| 1. | Electric train | 100 |
| 2. | Diesel train | 170 |
| 3. | Inland Water Transport (Barge) | 330 |
| 4. | Diesel truck | 1,600 |
| 5. | Steam train | 2,800 |

(Source: Inland Waterways Authority of India)

The report of the National Transport Policy Committee (1980) contains an 'Inter-model cost comparison' based on RITES cost data for rail and road and TPP (Transport Policy Planning) cost data on Inland Water Transport for self propelled barges as presented in Table 26.2.

Table 26.2 Resource cost comparison, operation cost by railways road transport and inland waterways.

| | Cost in paise per tonne-me | | | | | | |
|---|---|---|---|---|---|---|---|
| Distance (in km) | Railways (Die-single line, operation, | | Road Transport Traction port Wagon long movement | | IWT (for self-propelled vessels) capacity at 75% load factor | | |
| | Coal | Fertiliser | Coal | Fertiliser | 500 | 1,000 | 1,500 |
| 1 | 2 | 3 | 4 | 5 | 6 | 7 | 8 |
| 50 | 23.6 | 23.4 | 27.0 | 35.0 | 13.7 | 10.3 | 9.4 |
| 100 | 14.6 | 14.3 | 21.0 | 19.0 | 9.1 | - | 5.8 |
| 300 | 8.6 | 8.3 | 15.0 | 17.4 | 6.1 | 4.1 | 3.4 |

Inland Water Transport has other inherent advantages as well. Firstly, whenever navigable waterways exist this mode can provide immediate accessibility to the people of the area, without requiring investment in trackage as in other modes of transport. Investment has, of course, to be made in vessels, and in improvement and maintenance of navigable channels and terminal facilities. Secondly, Inland Water Transport is a very labour intensive mode of transport. It generates more employment per rupee of investment than other mode, and that, too, benefits primarily the weaker sections of the community for whom this may be the only source of employment.

Notwithstanding these advantages, Inland Water Transport has gradually declined in its importance in the Assam region. The decline in its importance is owing to lack of investment in a scientific and planned manner in essential infrastructural facilities. Although this mode has some inherent disadvantages, e.g., movement, limited natural spatial accessibility, no direct point to point accessibility without transhipment, and finally non-availability of navigational channels throughout the year, still it can play an important role in the region's economy handling categories of freight and passenger which have no particular preference for quick movement within the limited natural spatial accessibility, between the points on the bank of the rivers and their territories by improving and maintaining the navigable channels for the whole year. The cost of improvement and maintenance of river navigable channels will be much more economical compared to required investments in trackage in other modes of transport.

## 26.2 The River Brahmaputra, its Morphology and Hydrology

The Brahmaputra ranks among the most important rivers of the world. It originates from the south west of Tibet from a glacier mass of the northernmost chains of the Himalayas called Mansarowar. It flows in Tibet for about 1,561 km under the name of the Tsangpo, almost in an easterly direction. It then takes a sudden turn towards south and enters India from the extreme north-east. It flows in the same direction for about 161 km under the name of the Dihang or Siang up to Pasighat. A few miles below Pasighat it receives the water of the Dibong and Luhit from the east and assumes the name of the Brahmaputra. It then turns towards southwest and flows in that direction for about 805 km, passing by the towns of Assam--Dibrugarh, Jorhat (Neamati), Tezpur, Guwahati, Goalpara, and Dhubri. A few miles below Dhubri it turns towards south and enters Bangladesh. It then flows about 225 km in the same direction with the name of the Meghna and joins the Ganga at Goalundo. It then flows, with the name Padma, in the southeast direction and falls into the Bay of Bengal, after a further course of about 225 km Thus the total course of the river Brahmaputra from its source to the mouth is about 2,977 km (Figure 26.1). The Brahmaputra River, the total course of which is 2,977 km long, includes two long navigable reaches separated from each other by falls while crossing through the Himalayan chain. The upstream reach, at an altitude of 3500 m on the high plateau of Tibet is 650 km long, the downstream reach, located in the low plain and the Bengal delta, is about 1300 km long (from Dibrugarh down to Calcutta).

The characteristics of the north and south bank tributaries of the Brahmaputra in Assam are conspicuously different. The north bank tributaries have bigger catchments, very steep slopes, carry heavy sediment loads and have sandy beds. In contrast, the south bank tributaries have flatter slopes, deep meandering channels almost from the foothills, beds and banks of fine alluvial soil and a comparatively low silt charge. The principal north bank tributaries are the Subansiri, Burai, Bargang, Bharali, Dhansiri, Puthimari, Pagladiya, Beki, Manas, Aie, Champabati, and the Gangadhar. The principal south bank tributaries are the Noa-Dihing, Burhi-Dihing, Disnag, Dihkou, Jhanji, Kopili, Kulsi, Krishnai, and the Jinary (Figure 26.2).

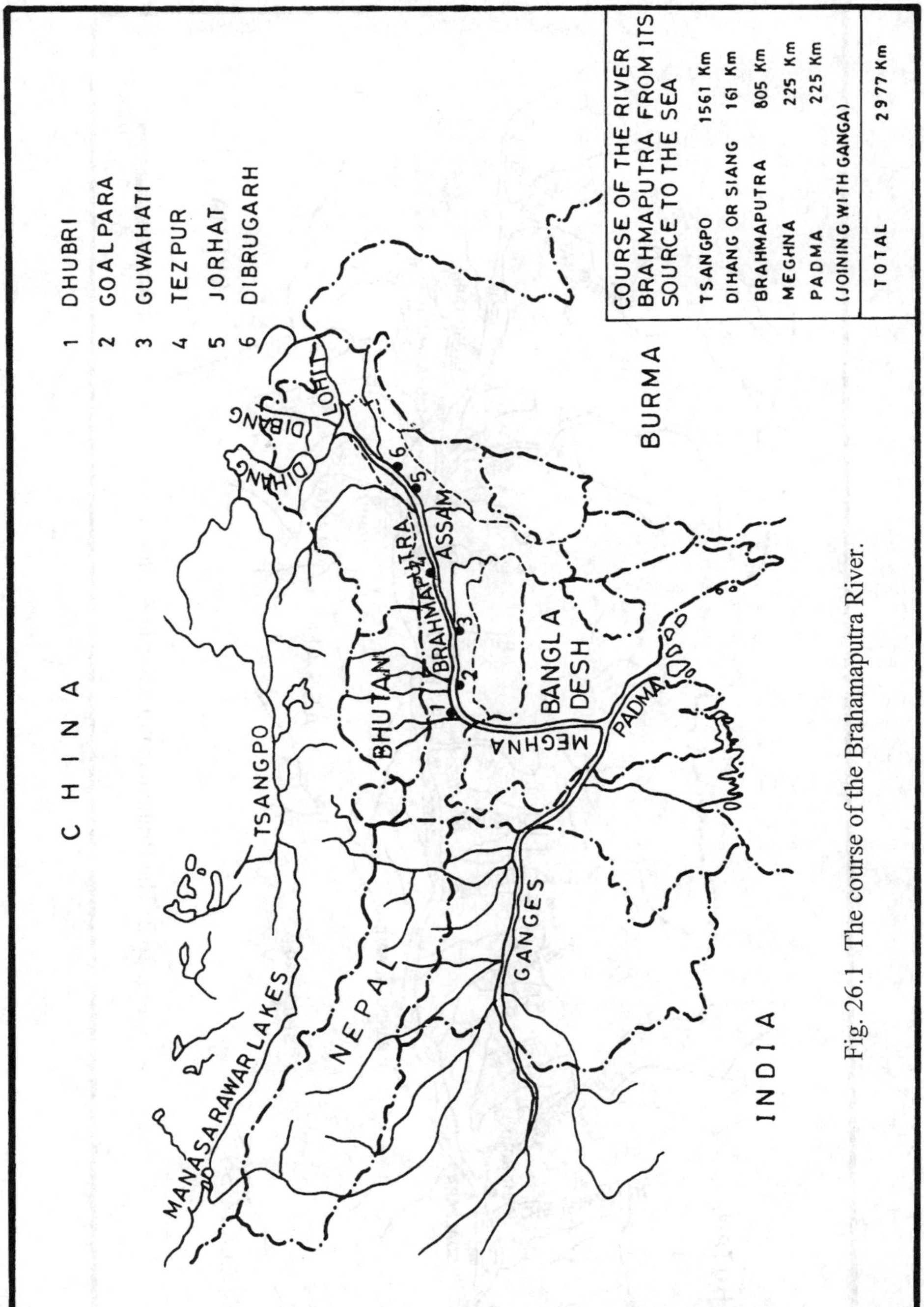


Fig. 26.1 The course of the Brahamaputra River.

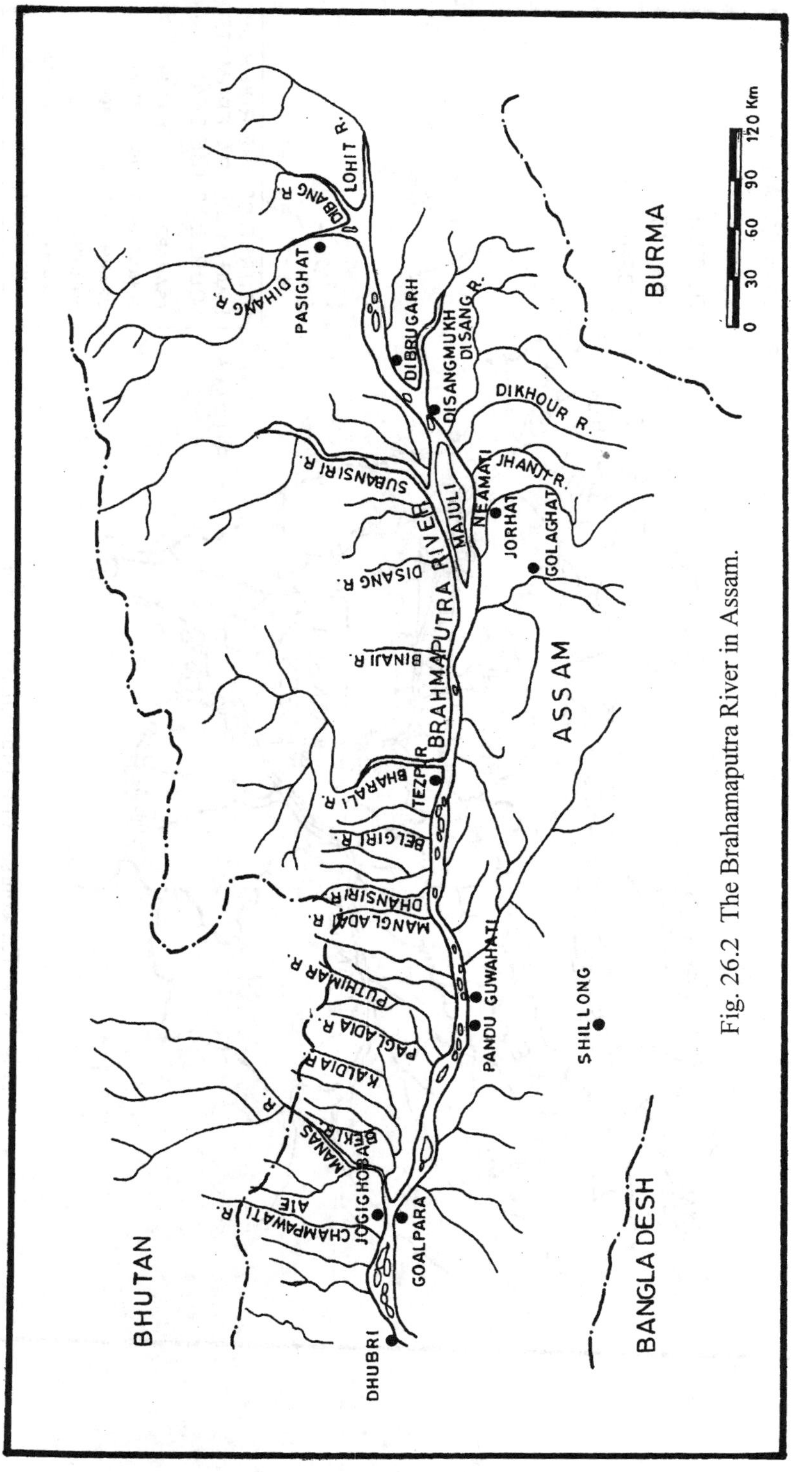


Fig. 26.2 The Brahamaputra River in Assam.

The average annual flow of the Brahmaputra at the point where it enters Bangladesh is of the order of 510,450 million cubic metres. The average annual flow contributed by the major tributaries is schematically shown in Figure 26.3. The Brahmaputra basin extends over an area of 580,000 sq. km in Tibet (China), Bhutan, India, and Bangladesh. The drainage area of the basin lying in India is 194,413 sq. km, nearly 5.9 % of the total geographical area of the country (India). The basin lies in the states of Arunachal Pradesh, Assam, Meghalaya, Nagaland, Sikkim, and West Bengal.

In India, the rainfall ranges from as high as 11,000 mm at Cherrapunji in Meghalya to as low as 215 mm in the vicinity of Jaisalmer in Rajasthan. The annual rainfall in the Brahmaputra valley ranges from 1,750 mm in Kamrup District to about 6,400 mm in the North Lakhimpur, rainfall in the valley is negligible from November to March, the average being 100 mm. Rainfall commences from April and is in full swing towards the end of May with the onset of monsoon.

### 26.2.1 GENERAL CHARACTERISTICS

The Brahmaputra River is a big alluvial river and the river is estimated to bring in nearly 800 million tonnes of sediment every year. Its discharges are influenced by the monsoon. The discharge variation between the maximum and minimum observed at Pandu (Guwahati) is 75,000 $m^3$/s. The dominant discharge during the monsoon, is about 41,000 $m^3$/s. The Brahmaputra River is of braided type. The channel cross-section being wide and shallow and non-uniformity in flow distribution only favour formation of islands and braiding pattern in the Brahmaputra River (Figure 26.4).

The river flow in Assam valley is controlled by isolated hills on either side of the river in certain reaches. In such reaches the river develops high velocities owing to constriction and the sediment picked up in the constriction debouches into vast delta type plains immediately downstream causing formation of high shoals and multiple channels. The river reaches immediately downstream of the Tezpur bridge, the Saraighat bridge at Guwahati, and the Palasbari, Gumi area about 35 kms downstream of Guwahati and Jogighopa can be cited as examples in the formation of local deltas typifying the braided nature of the river (Figure 26. 4 and 26.5).

Frequent earthquakes in the Assam region, particularly that of 1950, have caused changes in topographical features, drainage pattern, rise in sediment load, and river bed changes in the Brahmaputra basin. Studies of river plains indicate river exhibiting purely local meandering behaviour in some sections, flow guided by naturally controlled hills, superimposed by the braiding pattern and island formations.

The river slope of the Brahmaputra in its upper reaches is steep for a river of its size. Between the source and Kobo the slope is approximately 270 cm/km. This steep slope suddenly flattens when the river enters the Assam valley. Between Kobo and Dibrugarh the slope is 26 cm/km. Between Dibrugarh and Neamati the slope further flattens and is around 18 cm/km and between Guwahati and Dhubri the slope become 9.5 cm/km.

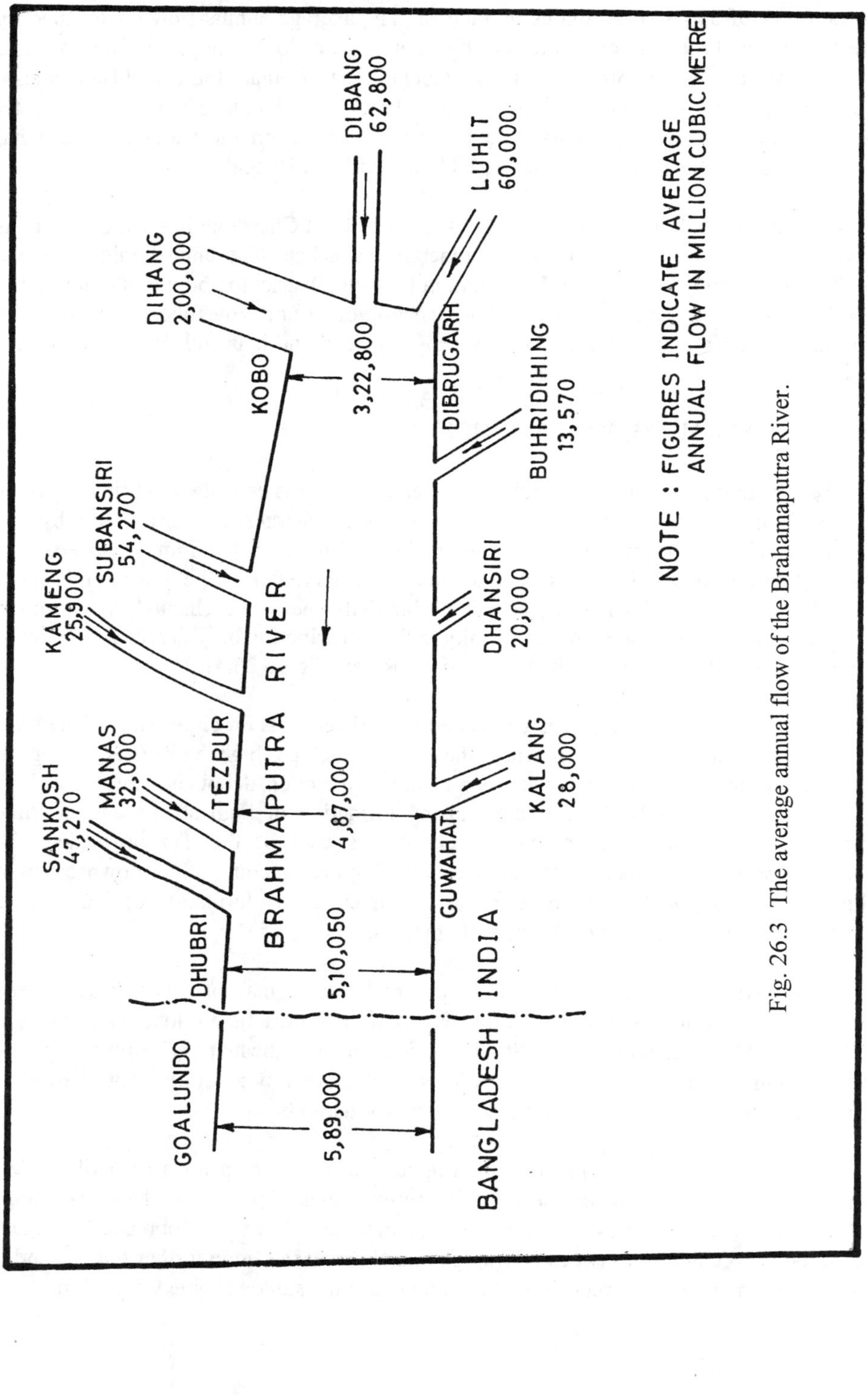


Fig. 26.3 The average annual flow of the Brahamaputra River.

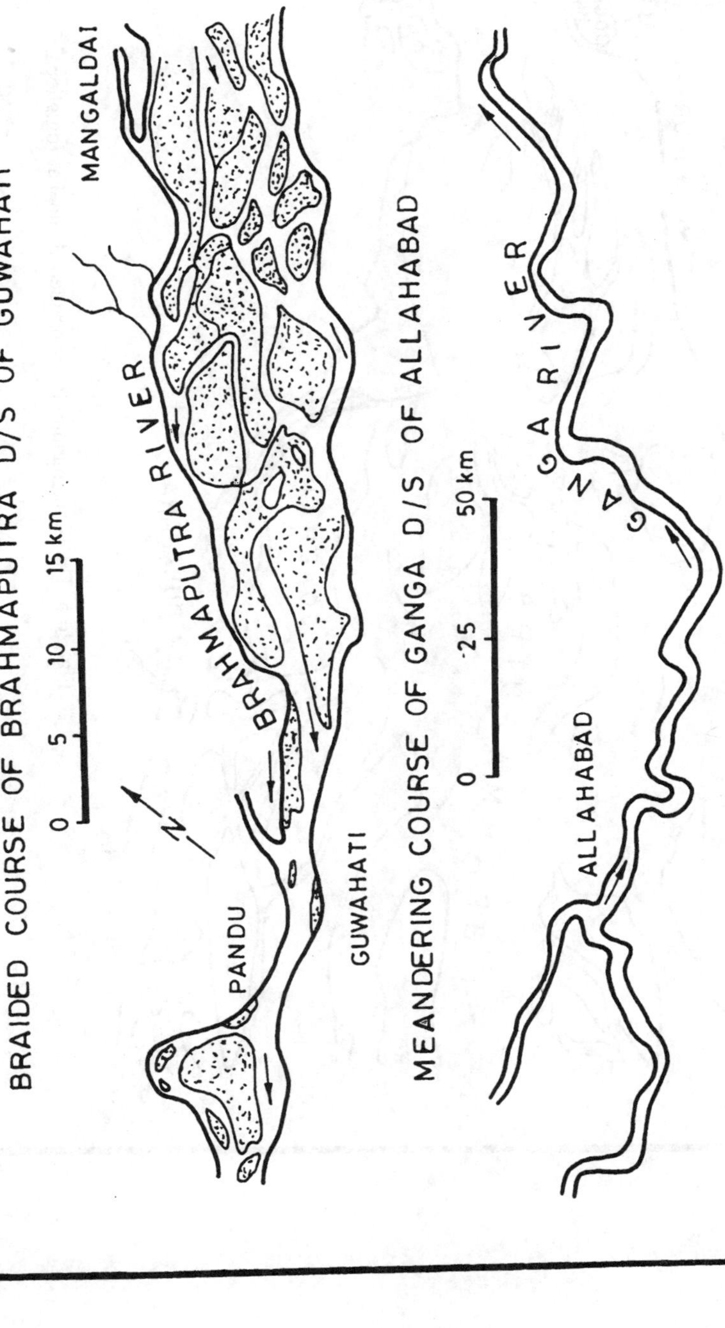


Fig. 26.4 The Brahamaputra River plan form.

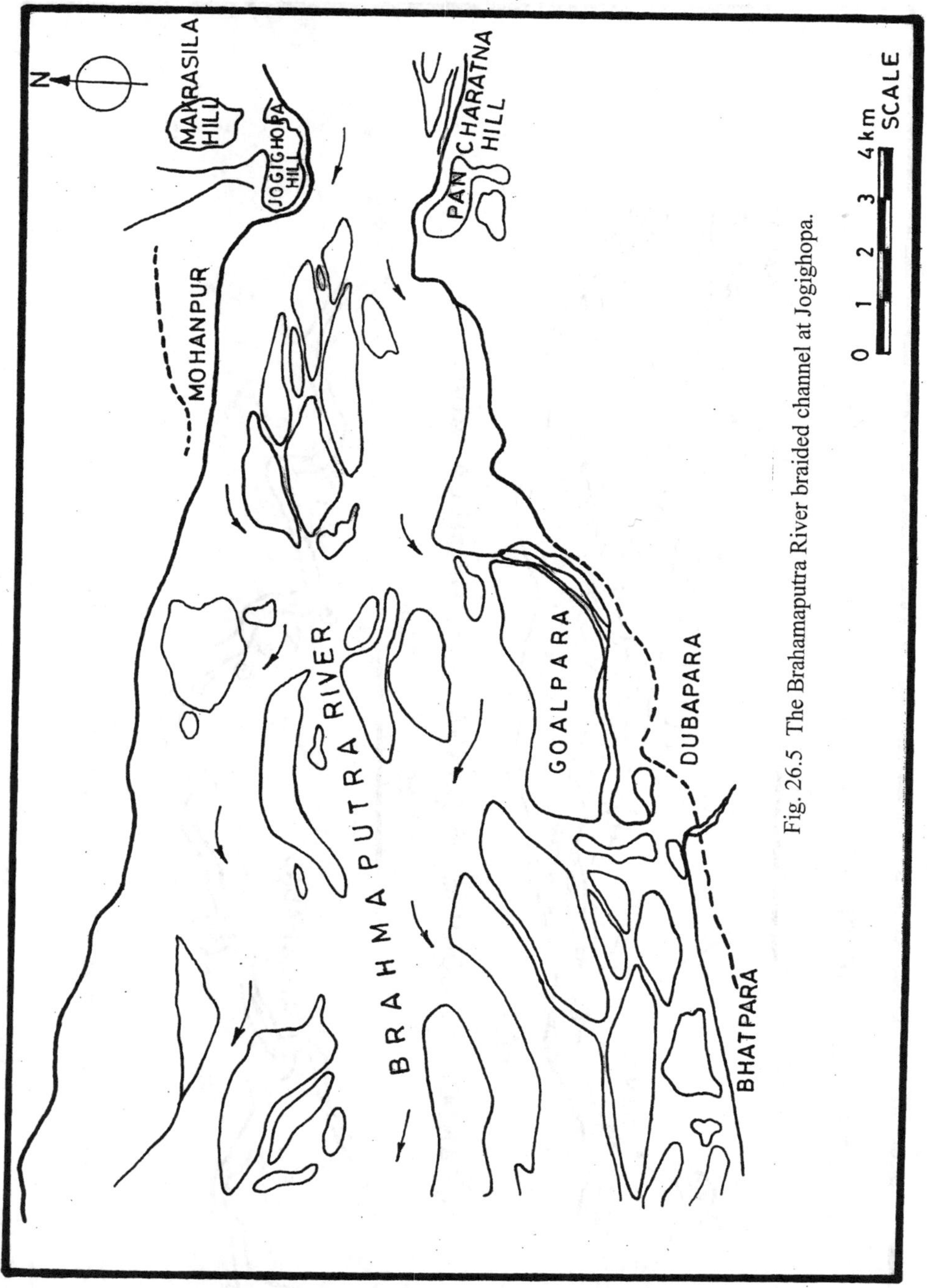


Fig. 26.5 The Brahamaputra River braided channel at Jogighopa.

## 26.2.2 MORPHOLOGICAL PARAMETERS RELEVANT TO NAVIGATION

The gauge discharge site of the Brahmaputra at Pandu (Guwahati) is the only site having historical records. Till recently this was the only discharge site covering the Indian territory. From the gauge hydrograph of the river, on 23 August 1962 the maximum discharge was observed at Pandu, corresponding to the maximum water level (MWL) of 49.33 m, as 72.794 $m^3/s$ (Figure 26.6). Figure 26.7 shows the corresponding high and low water profile at different gauge stations in the Brahmaputra valley, including Pandu in 1962. Figure 26.8 shows the high and low water profiles for the lowest year, namely, 1968. This data, shown in Figure 26.8, especially the low water level (LWL), is more relevant to navigation during the lean period. Figure 26.9 shows the discharge hydrograph of the river at Pandu for the year 1955 to 1958. This figure also indicates that moderate floods can be expected in the months of May, end of June, and in August.

It is seen that heavy sediment inflow of the order of 800 million tons annually, together with the very flat gradient of the river, are amongst the various factors contributing to its braided nature. This results in very shallow depth, multiple channels, wide flood plains, and formation of high shoals. All these factors are impediments for smooth navigation. Figure 26.10 shows the monthly discharge and silt load of the Brahmaputra at Pandu observed in 1962 when maximum flood passed through Pandu site. A perusal of this figure indicates that the fine sediment of size 0.075 mm and below is negligible compared to medium sediment of size between 0.075 mm and 0.2 mm and coarse sediment of size higher than 0.2 mm. This type of distribution of sediment is perhaps responsible for formation of shoals and the braided nature of the river. Figure 26.10 also indicates that maximum inflow does not necessarily mean maximum sediment inflow, as shown by the data for the months of June and August. In the case of the Ganga River, which is mostly of meandering type in the major reaches of the alluvial plain, nearly, 75% to 80% of the load is wash load of the size 0.075 mm and lower, although the total sediment is of the order of 1,600 million tons annually. These factors contribute towards the braided nature of the Brahmaputra River.

The data of mechanical analysis of bed material of the Brahmaputra is relevant for dredging. The details of analysis of bed material collected at different sites of the Brahmaputra are given in the Table 26.3. The record of maximum and minimum velocities observed along the various reaches covered by navigation is essential for the design parameters of the vessel, manueverability, etc.. However, except at Pandu the prototype data is not available. These (Pandu data) are reproduced in Table 26.4.

From a perusal of the above data it is observed that while 3.5 m/s is the maximum velocity that could be usually expected in the flood season, abnormally high velocity was observed in the month of August in 1959 and 1961 at 5.57 m/s and 5.85 m/s, respectively. These two observations could be perhaps owing to some localised constraint when the Saraighat bridge at Guwahati was still under construction. The construction of the Saraighat bridge was completed in October, 1962. Besides, the river just downstream of Pandu is constricted to a width of about 1.0 km owing to the presence of hills on either side, and hence velocities in this reach may be on the higher side. Hydraulic Model studies carried out for the proposed bridge at Jogighopa had indicated that 3.5 m/s velocity could be expected during the annual flood.

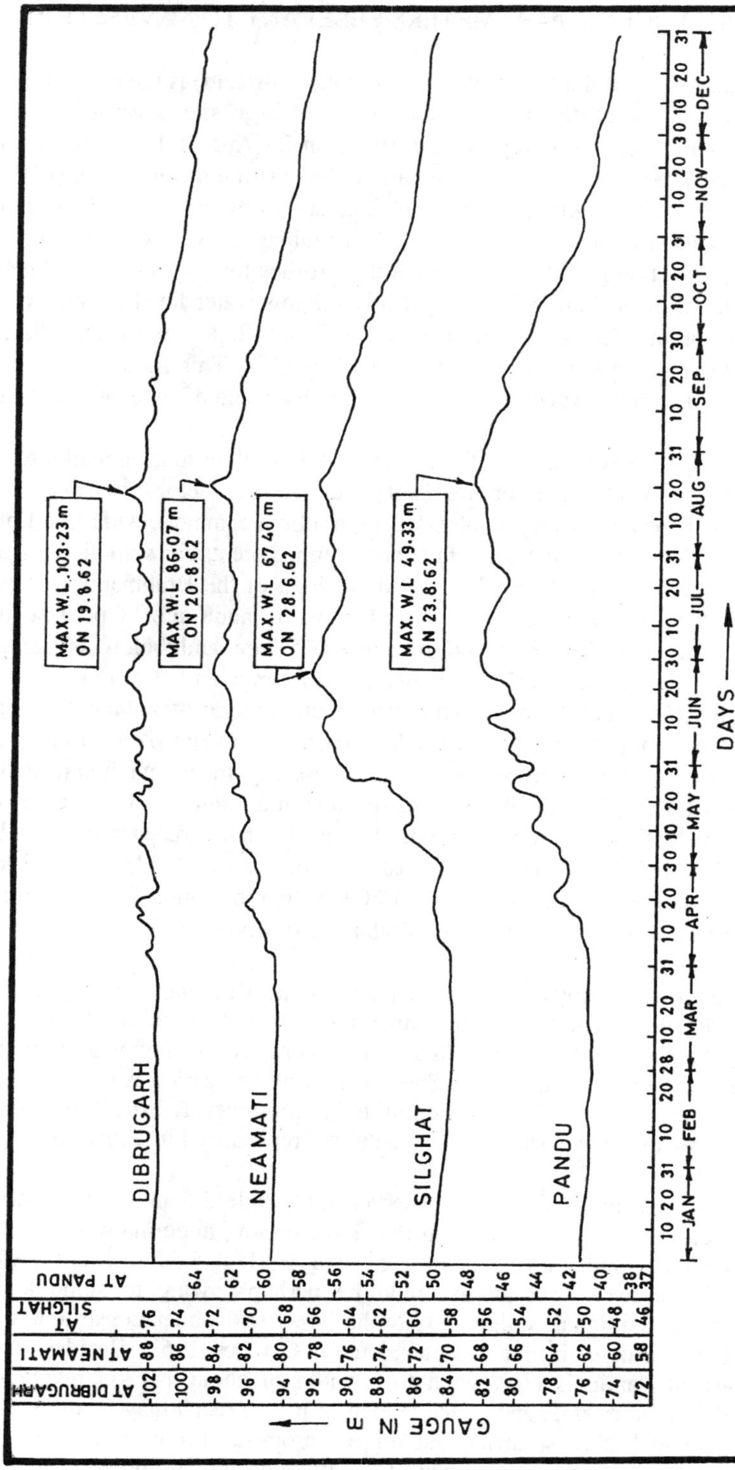


Fig. 26.6 The gauge hydrograph of the Brahamaputra River at Pandu (Guwahati) in 1962.

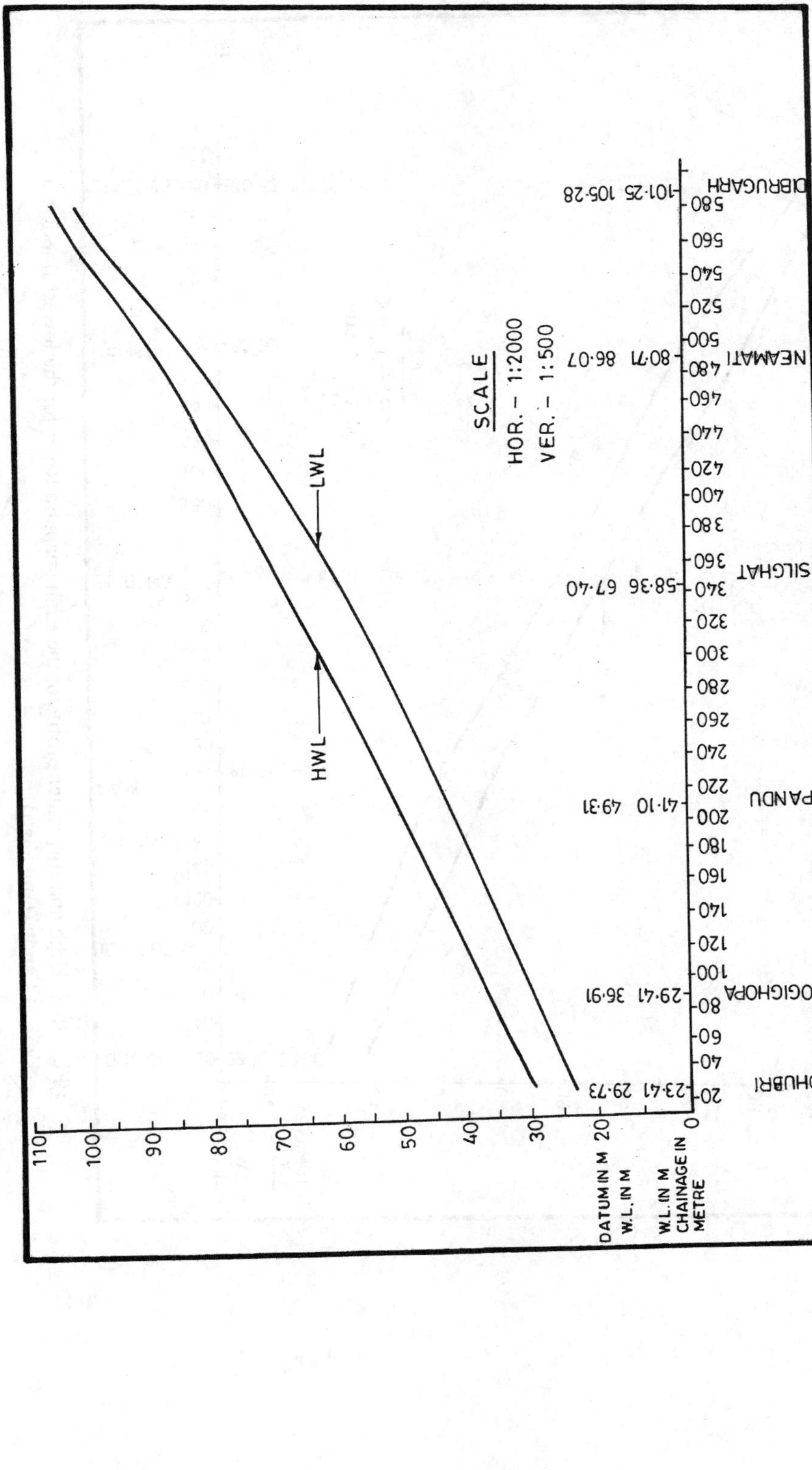


Fig. 26.7 High and low water profile of the Brahamaputra River in 1962 at Pandu (Guwahati).

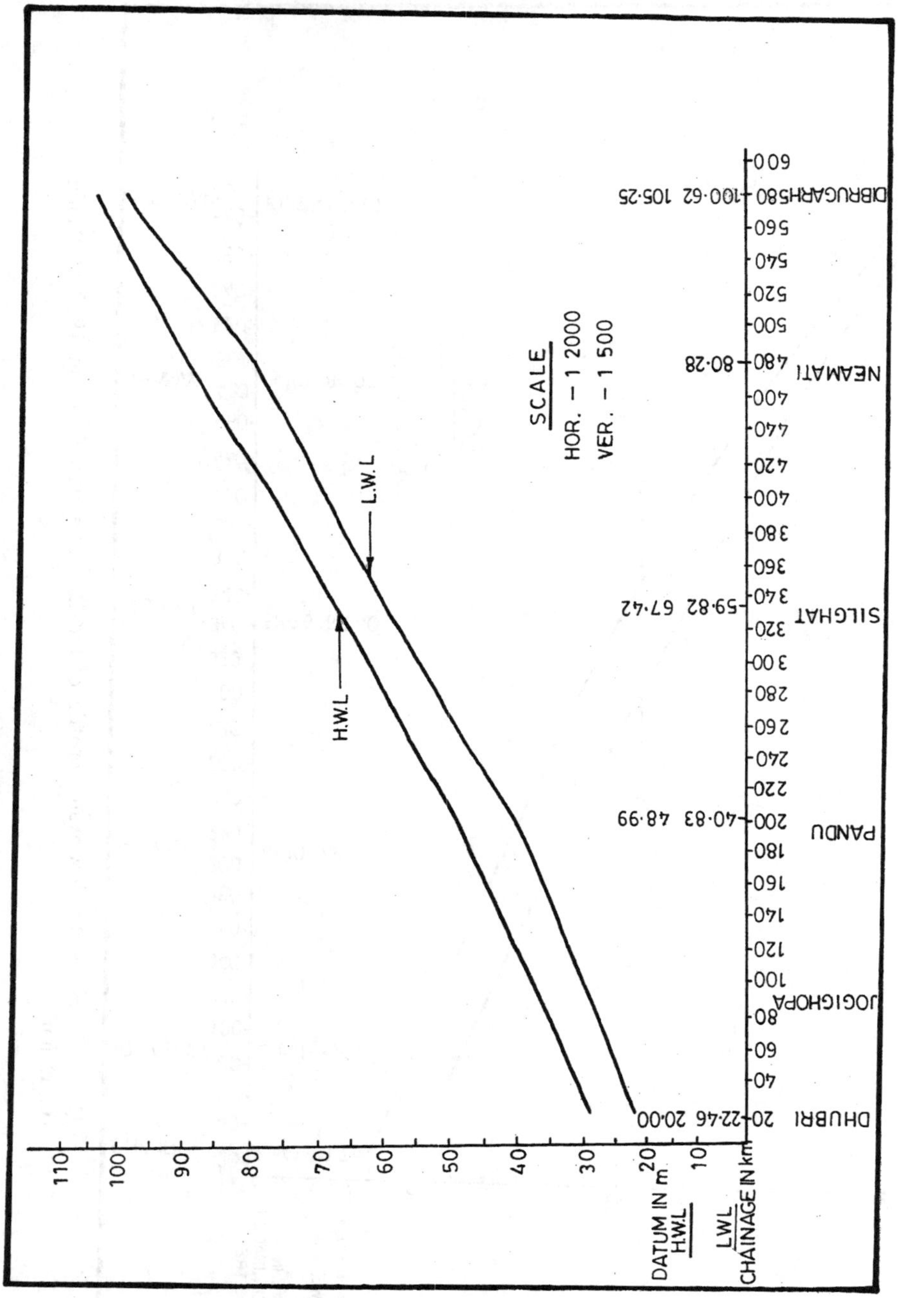


Fig. 26.8 High and low water profile of the Brahamaputra River for the lowest flood year 1968 at Pandu (Guwahati).

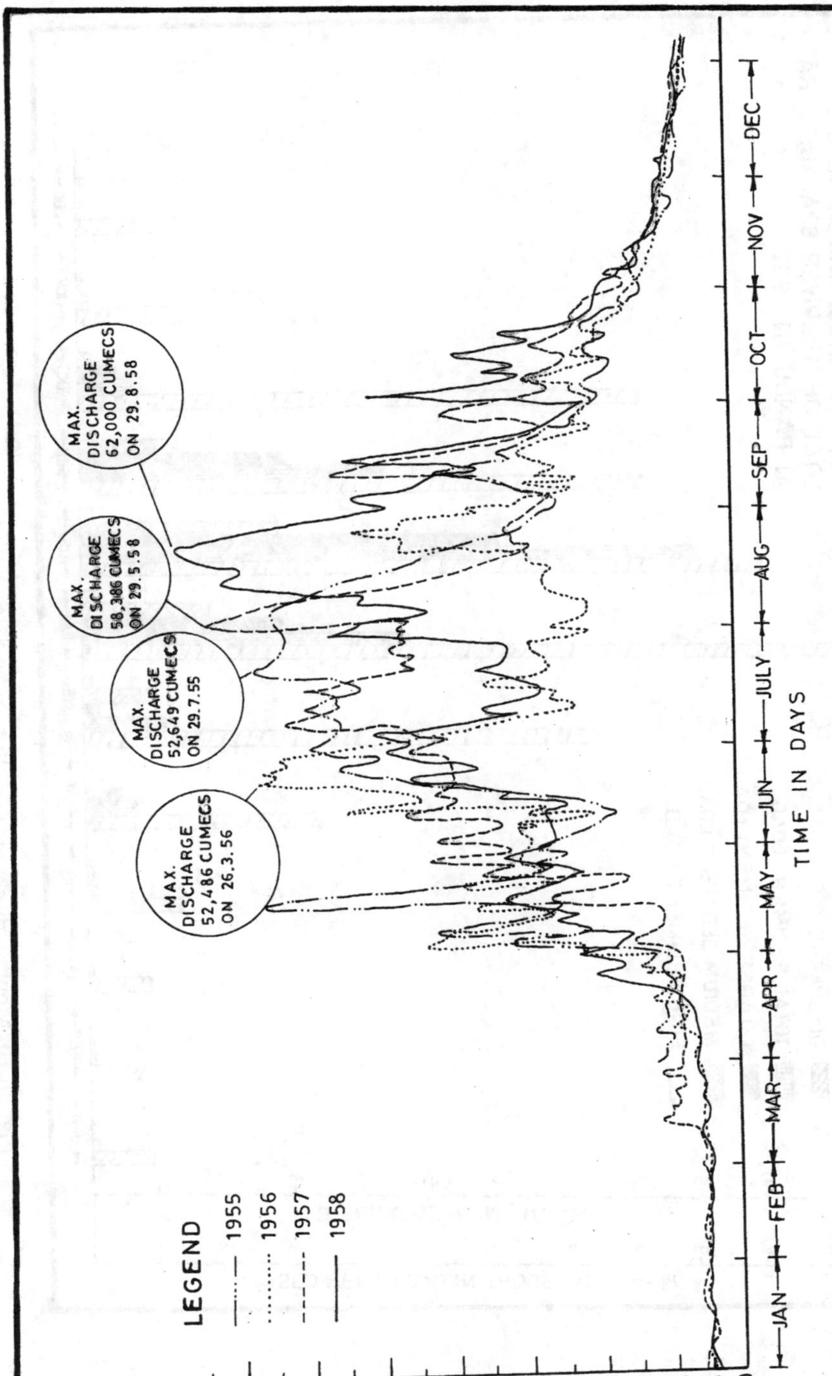


Fig. 26.9 Discharge hydrograph of the Brahamaputra River during 1955-1958 at Pandu.

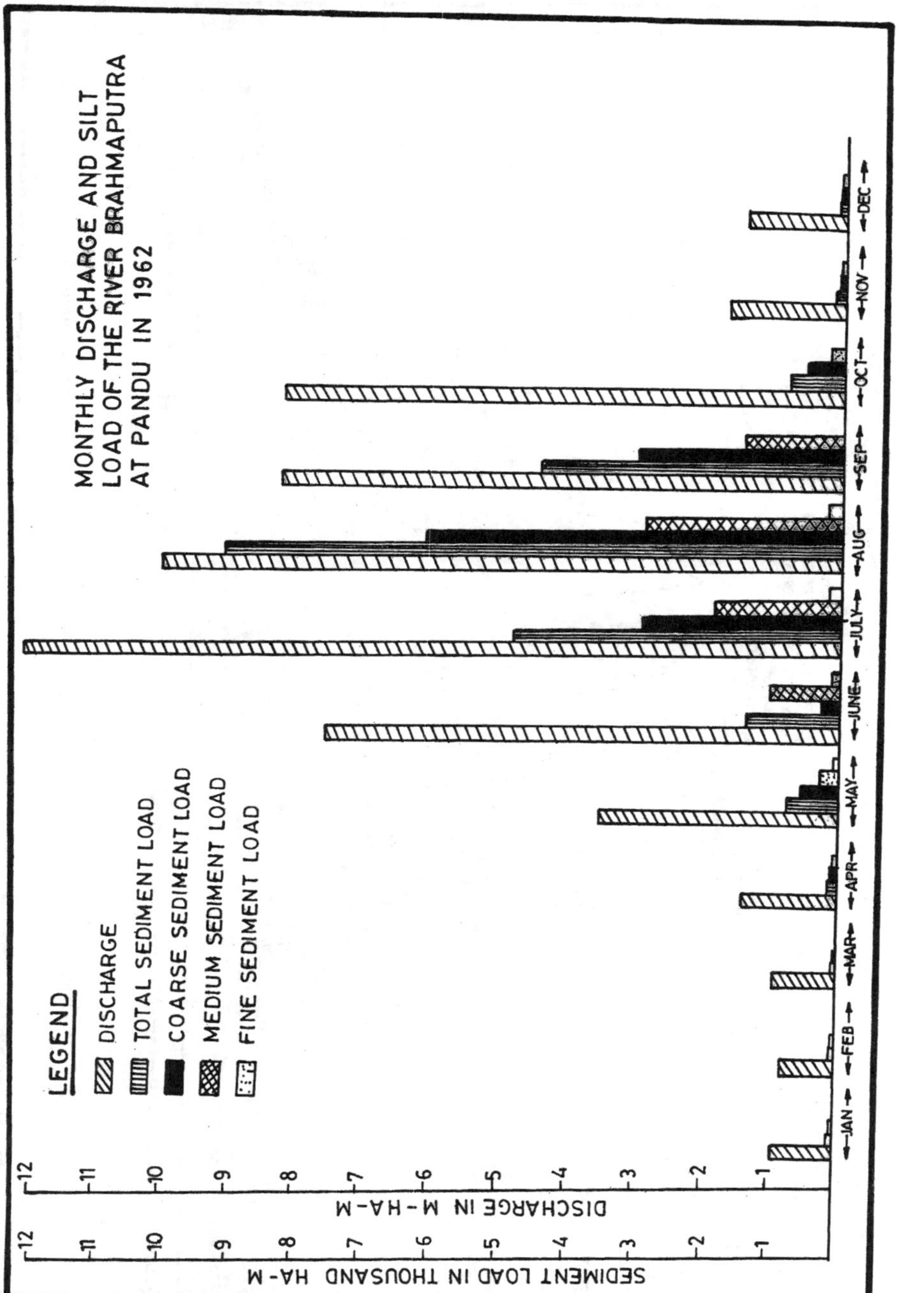


Fig. 26.10 Monthly discharge and silt load of the Brahamaputra River in 1962 at Pandu.

Table 26.3 Analysis of bed material of Brahmaputra River.

| S. N. | Site | Specific gravity | Classifi-cation | $D_{60}$ (mm) | $D_{50}$ (mm) | $D_{30}$ (mm) | $D_{10}$ (mm) | Uniformity coefficient | Remark |
|---|---|---|---|---|---|---|---|---|---|
| 1. | Sonarigaon | 2.65 | Sand | 0.23 | 0.22 | 0.17 | 0.12 | 2.0 | Poorly graded |
| 2. | Dibrugarh | 2.65 | Sand | 0.25 | 0.21 | 0.16 | 0.099 | 2.5 | -do- |
| 3. | Bhomora-guri, Tezpur | 2.65 | Sand | 0.30 | 0.20 | 0.15 | 0.10 | 3.0 | -do- |
| 4. | Bhuraband h Silghat | 2.66 | Sand | 0.30 | 0.25 | 0.20 | 0.10 | 3.0 | -do- |
| 5. | Pandu | 2.65 | Sand | 0.35 | 0.30 | 0.24 | 0.18 | 2.0 | -do- |
| 6. | Amingaon | 2.70 | Sand | 0.22 | 0.20 | 0.18 | 0.14 | 1.5 | -do- |
| 7. | Jogighopa | 2.70 | Sand | 0.20 | 0.18 | 0.15 | 0.09 | 2.0 | -do- |
| 8. | Dhubri | 2.68 | Sand | 0.18 | 0.16 | 0.12 | 0.07 | 1.6 | -do- |

In the absence of prototype data for other river reaches, it seems reasonable to assume that a maximum velocity of the order of 4 m/s owing to local concentration could be expected, which needs to be taken into consideration for design purposes.

## 26.3 Waterways Resources of Assam

Water is precious and is rated as the most important commodity available in nature next to air and light for sustenance of human, animal, and plant life as laid down in our scriptures:

" O EARTH, MY MOTHER
AIR (WIND) MY FATHER,
O FIRE (LIGHT) MY FRIEND :
WATER MY KINSMAN,
SPACE MY BROTHER,
HERE DO I BOW BEFORE YOU
WITH FOLDED HANDS"

Vairagyashataka of Bhartrahari, Sanskrit, 7th Century, AD

Table 26.4 Record of maximum and minimum Mean Velocities of the River Brahmaputra at Pandu (in metres per second).

| Months | | 1959 | 1961 | 1977 | 1978 |
|---|---|---|---|---|---|
| Jan. | Max. | 0.52 | 0.53 | 1.16 | 0.83 |
| Fed. | Min. | 0.39 | 0.47 | 0.83 | 0.73 |
| Mar. | Max | 0.54 | 0.53 | 1.10 | 0.86 |
| | Min. | 0.34 | 0.47 | 0.76 | 0.75 |
| Apr. | Max. | 1.19 | 1.35 | 1.49 | 1.28 |
| | Min. | 0.47 | 0.68 | 1.02 | 0.86 |
| May. | Max. | 2.21 | 2.74 | 2.05 | 1.76 |
| | Min. | 0.76 | 0.95 | 1.05 | 1.12 |
| June | Max | 3.14 | 2.50 | 2.61 | 3.63 |
| | Min. | 1.28 | 1.52 | 1.76 | 1.29 |
| July | Max. | 2.14 | 3.21 | 3.00 | 2.57 |
| | Min. | 1.55 | 1.50 | 2.00 | 1.70 |
| Aug. | Max. | 5.57 | 5.85 | 3.11 | 2.58 |
| | Min. | 1.61 | 1.84 | 2.55 | 1.63 |
| Sept. | Max. | 2.1 | 2.19 | 2.90 | 2.16 |
| | Min. | 1.36 | 1.24 | 1.65 | 1.60 |
| Oct. | Mix. | 1.50 | 1.53 | 2.40 | 1.84 |
| | Min. | 1.08 | 0.92 | 1.16 | 1.20 |
| Nov. | Max. | 1.15 | 0.97 | 1.14 | 1.33 |
| | Min. | 0.64 | 0.57 | 0.95 | 0.97 |
| Dec. | Max. | 0.66 | 0.64 | 1.12 | 1.02 |
| Min. | | 0.32 | 0.45 | 0.82 | 0.78 |

The water resource is undoubtedly acclaimed as one of the most vital national assets and holds the key to the nation's economic prosperity and stability. India is generously endowed with this gift of nature but this precious resource is unevenly distributed in time and space over its geographical dimensions. As in the recent assessment, the water resources potential of the river basins of India is 186.93 million hectare meters and the ground water potential is 45.22 million hectare metres, and, interestingly, as much as 32 per cent of the country's water resources flow down the Brahmaputra and Barak rivers. Therefore, judicious utilisation of surface and ground water for a variety of purposes, such as domestic, irrigation, energy, industry, navigation and other uses has become a challenging task for planners, and engineers, as well as political leadership of the country. In the regions where natural river navigation potentiality is available, the development for utilisation of water resources should not occur at the cost of navigation.

Navigable waterways are confined to only a few states in India. The share of inland waterways in the country's transport network, comprising roads, railways, coastal shipping, and inland waterways, is 1 percent, and the density is 0.44 (sq. km waterways navigable per 100 sq. kms of area). Assam's navigable inland waterways extend to 1,983 km out of India's total 14,544 km comprising 44 rivers, the longest being the Brahmaputra (720 kilometres from Dibrugarh to Dhubri), whose share is around 14 per cent of India's total waterways.

Table 26.5 Navigable length of important rivers in Assam.

| Rivers | By Boat (km) | By Mechanised Craft (km) |
|---|---|---|
| Brahmaputra River System (Assam) | | |
| Brahmaputra | 724 | 637 |
| Subansiri | 143 | 45 |
| Borak | 153 | 64 |
| Total | 1,020 | 746 |

(Source: Central Water Power Commission, Navigable Waterways of India, New Delhi, October 1961, p. 199).

Most waterways in India, however, suffer from navigational hazards such as shallow water and narrow width during dry seasons, siltation, and bank erosion. At present about half of the river length, which is 5,200 kilometres of major rivers and 485 kilometres of canals, is suitable for mechanised crafts. Presently the Ganga Bhagirathi-Hooghly River system, the Brahmaputra River system, the Godavari River system, and the West Canal in Kerala, appear to be very promising in the Project Report of Development of Inland Water Transport in National Waterways No.2 by the Inland Waterways Authority of India (1990). The important navigable rivers in Assam and the extent to which these are navigable by boats and mechanised crafts are shown in Table 26.5.

Steamers used to ply from Dhubri to Desangmukh (632 km) and feeder steamers from Desanmukh to Dibrugarh (64 km) till the earthquake of 1950, which caused considerable changes in the region of the Brahmaputra resulting in the deterioration of navigable channels, particularly in the upper reaches. All other rivers, except the Kallong (Nowgong), Dihing (Dibrugarh distance less than 160 km and some are navigable only during the rainy season. Table 6 indicates navigable waterways as per utilisation by different categories of boats and vessels in Assam in recent years.

Crews of the heydays of the river services in Assam, before 1950, used to say that navigable channel depth was 5 metres minimum all through the year and the maximum was up to 50 metres during the high flood time in the Brahmaputra River. The river bed level has come up considerably in the year 1950 owing to the earthquake, and thereafter year by year owing to heavy silt deposition in the absence of any regular scientific channelling work so far, and owing to falling of the average rainfall as well as in spurts of rainfall. It may be mentioned here that northern banks of the Brahmaputra basin carry soft alluvial soil, whereas the southern bank rivers carry old alluvium (harder) owing to the nature of geological formations. The rainfall observed in a wide area of Upper Assam in the last 56 years, from 1938 to 1993, is given in the Table 26.8.

Table 26.6 Navigable waterways in Assam.

| Description | Length in Kilometres |
|---|---|
| Navigable by large country boat | 2599 |
| Navigable by small country boat only | 1499 |
| Navigable by powered river vessels | 637 |
| Navigable by low draft powered feeder vessels only | 288 |
| State I.W.T. operation in Assam waters by their vessels | 760 |
| C.I.W.T.C. (Govt. of India undertaking) operating in Assam water by their vessels | 624 |

Source: Govt. of Assam, Transport Department.

The least available depth (draughts) in the Brahmaputra River in January and July of 1996, in sixteen recording stations, from Dhubri to Sadiya are given in Table 26.7.

### 26.4 River Transport Brahmaputra in Retrospect

Assam's earliest trade with the neighbouring provinces was mainly carried by river transport, the main ways to Bengal, Bihar, and Orissa being the Brahmaputra and the Ganges. As early as in 1839, when the Assam Company was formed, the company started its own fleet of country boats, and although a steamer was purchased at a cost of 13,000 pounds it was unsuccessfully tried on the Brahmaputra in 1842. The first steamer service in the Brahmaputra between Calcutta and Guwahati (1960 kilometres) was established by the Company in 1947. It is significant that the first steamboat to begin service on the Mississippi River (U.S.A.) was not much before the Brahmaputra. This was in 1,811. Yet it was irregular owing to the shortage of steamer; tea chests had to wait a long time at Guwahati for export. But, the steamer service was profitable, the receipt always exceeded the expenses in sixty trips in 1847, and the average earnings per trip were over Rs 5,000 as recorded by Dr. H.K. Barpujari in 'Assam in the days of the Company (1963)'. In 1856, the Government steamer service was extended from Guwahati to Dibrugarh.

Table 26.7 Depth (draught) readings of the Brahmaputra River, stationwise, in metres.

| Station | 1 | 2 | 3 | 4 | 5 | 6 | 7 | 8 | 9 | 10 | 11 | 12 | 13 | 14 | 15 | 16 |
|---|---|---|---|---|---|---|---|---|---|---|---|---|---|---|---|---|
| Jan. '96 | 2.3 | 2.8 | 2.3 | 1.8 | 2.0 | 2.0 | 2.9 | 2.0 | 2.3 | 2.1 | 0.9 | 1.5 | | | | |
| July | 3.0 | 5.0 | 2.8 | 2.6 | 6.0 | 2.7 | 3.0 | 3.0 | 2.5 | 2.4 | 2.0 | 2.2 | 2.0 | 2.3 | 1.5 | 1.8 |

(Source: IWAI, Guwahati)

| | | | |
|---|---|---|---|
| Remarks: | January '96 | Minimum at Dibrugarth | 1.5 |
| | | Minimum at Madarguir | 0.9 |
| | July '96 | Minimum at Paglam | 1.5 |

In 1861 Colonel Hopkins, the Commissioner, drew a gloomy picture of the isolation of Assam. In fact, it was the cultivation of tea which turned the course of history. In 1961 itself the Indian General Stream Navigation (IG SN) company started a regular service on the Brahmaputra with arrangements for transport of labour into Assam. In 1862 the River Steam Navigation (RSN) company also started operation with three steamers and three flats. By 1963 the IGSN company was more attracted to Cachar areas as compared to the Ganges trade.

In 1879 when the Eastern Bengal Railway introduced their services, both the passenger and goods traffic between Serajanju, Narayanganj, Dacca, and Cachar stations through railway tickets and documents, the IGSN and the RSN Companies faced challenges poised by rail. The I.G.S.N. Company, therefore, negotiated with the Eastern Bengal Railway for carriage of goods between certain river cum railway junctions by the vessels of the IGSN Company under railway documents. By 1882 the RSN Co. started a daily service (called the Assam mail Service) between Dhubri and Dibrugarh, by entering into an agreement with the Assam Government. In the same year the "Combined Cargo Service" was introduced by the IGSN Company for carrying goods between Assam and Goalundo and then shipment by rail to Calcutta for the advantage of avoiding the dangerous navigation through the Sunderbans. It is interesting to note that around 1882 river transport helped the railway construction in Assam in a big way.

Table 26.8 Rainfalls in a wide area of upper Assam for 56 years.

| Year | Rainfall (mm) | Year | Rainfall (mm) |
|---|---|---|---|
| 1938 | 3,900.41 | 1979 | 2,118.90 |
| 1940 | 2,807 | 1980 | 3,194.72 |
| 1945 | 2,187.92 | 1983 | 3,306.73 |
| 1950 | 2,623.57 | 1990 | 2,675.80 |
| 1955 | 3,057.91 | 1991 | 2,815.00 |
| 1960 | 2,790.80 | 1992 | 2,979.30 |
| 1965 | 3,073.15 | 1993 | 2,436.60 |
| 1970 | 3,104.96 | | |
| 1975 | 2,879.31 | | |
| 1978 | 2,014.96 | | |

Source: A Lab. in Upper Assam.

| | | |
|---|---|---|
| Remarks Highest in | 1938 | 3,900 mm |
| Lowest in | 1978 | 2,015 mm |

By 1885 the only railway in Assam was the Dibru-Sadiya Railway (Metre gauge). Then the introduction of the Assam Bengal Railway in 1895 which connected Assam with outside, the IGSN and RSN Companies faced challenges poised by rail. In 1899 these two companies were combined to form what was popularly known as the Joint Steamer Companies.

The river services had their heydays before the "Assam Link" brought the Assam State nearer to the rest of the country by rail in 1950. The Joint Steamer Companies were operating regular services between Calcutta and Sadiya throughout the year. In addition, they operated the following services on the feeder rivers:

(i) Subansiri River for a distance of 130 km;
(ii) Dihing River for a distance of 35 km;
(iii) Disang River for a distance of 75 km;
(iv) Dikhow River for a distance of 30 km;
(v) Dhansiri River for a distance of 40 km;
(vi) Kopili-Kolong River for a distance of 90 km.

The earthquake of 1950 caused considerable changes in the regime of the Brahmaputra, resulting in deterioration of the navigational channels, particularly in the upper reaches. However, the inland water transport between 1950 and 1965 continued to carry 80 per cent of tea and 90 per cent of jute, and considerable quantities of POL from Assam to Calcutta, and carried a maximum river traffic of 840,640 tonnes in 1951 and a minimum of 523,514 tonnes in 1962.

Unfortunately, after the outbreak of war in 1965 with erstwhile East Pakistan the Inland Water Transport in Assam received another setback. Even then the river traffic carried by the central Inland water Transport (established in 1966), after the liquidation of the joint Steamer Companies, was not very discouraging.

The Inland Water Transport Directorate (DIWT) was established by the Assam Government in 1959, in pursuance of the recommendations of the Gokhale Committee (1959) to look after the development of waterways, ferries, and for the training of Inland Water Transport personnel. Since then the directorate has been doing useful work, running as many as 66 ferry services in Assam (42 on the Brahmaputra and 24 on the Borak) as well running a commercial service within Assam, and with Bangladesh and Calcutta since 1974.

With the ferry service network Assam IWT has been carrying nearly about 60,000 passengers, 2,000 M.T. goods everyday by different ferry routes. It is not only the people of Assam who are benefitted by ferry operations of Assam I.W.T., but also people from adjoining states like Meghalayas, Nagaland, Manipur, and Arunachal Pradesh are also enjoying the benefit. Some of the major ferry services connecting Assam with the neighbouring states are as below:

The Assam IWT is also going to operate a new ferry service between Dhubri and Phulbari, connecting Assam and Meghalaya very soon. In recent times the inward and outward traffic movement in Assam, as well as the along side and cross traffic movement by both the CIWTC and the Assam IWT is about 140 thousand tonnes yearly. The major commodities which are being handled are cement, fertiliser, iron and steel, lubricating oil, tea, jute, forest products such as bamboo, wooden-sleeper and paddy. Out of these the CIWTC handles almost all the above-mentioned commodities, while machineries, cement, iron and steel, forest products and paddy are usually handled by the Assam DIWT The present cargo handling capacity of the commercial fleet is nearly 10,000 MT and the revenue earned by the Assam IWT is nearly 150 lakhs per annum.

| Ferry Service | State (s) Connected with |
|---|---|
| 1. Sadia-Saikhowa | Arunachal Pradesh, directly. |
| 2. Saikhowa-Puglum | -do- |
| 3. Neamati-Kamalabari Lohit -Khabalu | Arunachal Pradesh and Nagaland by shortest roadways. |
| 4. Dhansiri-Gamiri | -do- |
| 5. Jaleswar-Dhubri | Meghalaya by shortest roadways. |
| 6. Panchartna-Jogighopa | -do- |
| 7. Dhurbi-Fakirganj | -do- |

**26.5 Traffic Studies for Brahmaputra**

Assam first became a British Protectorate at the close of the First Burmese War in 1826. In 1826 Cachar (all the three Barak valley Districts at present) was annexed, in 1835 the Jaintia Hills were included in the East India Company's dominions, and in 1839 the Upper Assam was annexed to Bengal. In 1874 Assam was detached from Bengal and made a separate Chief Commissionership. On the partition of Bengal in 1905 it was united with the Eastern District of Bengal under a Lieutenant Governor. From 1912 the chief Commissionership of Assam was revived and in 1921 a Governorship was created. On the partition of India in 1947, the whole of the predominantly Muslim district of Sylhet was merged with East Bengal (now Bangladesh). Transport Development in Assam is conditioned, in addition to other factors, by the topography of the region.

Situated in the North Eastern Region of India, Assam is encompassed on three sides by foreign countries such as China on the North, Burma on the east, and Bangladesh in the southwest. As the state has a common border with Arunachal Pradesh, Nagaland, Manipur, Tripura, Meghalaya and Mizoram, all lines of communication to and from these areas have to pass through Assam. Furthermore, interestingly, the State is connected with the rest of India by a narrow strip of land (of about 50 kilometres) through West Bengal.

Assam has an area of 78,438 sq. km., accounting for 2.4 percent of the land surface of India. The State bifurcates itself into two physio-geographical regions, the plains and the hills. The plain region has an area of 49,485 sq. km (63%) and the hill regions have an area of 28,953 sq. km (37%). The total number of districts in the State is 23 with 78 district towns, out of which 58 towns are located in the Brahmaputra valley districts.

The population of the State was 224.14 lakhs in 1991. The decadewise growth of population since 1951 has been more than 10 percent higher in Assam than that of the country as a whole. Above 89 percent of the population of Assam live in rural areas against 75 per cent of the country. The State's average density of population is 286 persons per square kilometres (it is 240 in the Brahmaputra valley and 360in the Barak valley). In general, during the post-independence period, the growth of income has been sluggish and per capita income in real terms has remained practically stagnant.

## 26.6 Economic Scene

The major economic activity in Assam is basically confined to agriculture, including plantation and forestry. An analysis of the agricultural pattern of Assam shows that the State has some main crop regions. The Upper Brahmaputra valley is specialized in the cultivation of tea, rice, sugarcane, oilseed, and pulses. In the Lower Brahmaputra valley, where the crop pattern is more diversified, rice, oilseeds, pulses, sugarcane, jute, potato, and tobacco are grown. The Barak valley has the same production pattern as in the Brahmaputra valley. In the hill districts, rice, oilseeds, potato, and fruits are grown.

There is a total of 709 tea gardens in Assam (616 in the Brahmaputra valley and 93 in the Barak valley) and 86 per cent of them are in the Brahmaputra valley alone. The total production of tea in Assam was 399 million kg (356 million kg in the Brahmaputra valley, and 43 million kg in the Barak valley) in a total area of 216,117 hectares as on 31-12-95, which was more than 50% of the total tea produced (753 million kg) at 31-12-95 produced in the country.

So far as forest production is concerned, Assam possesses extensive forests and large areas of cultivatable waste. These are about seventy varieties of wood in Assam, which has the biggest bamboo reserve, yielding annually about 60,000 tonnes, contributing about 42 per cent of India's total. According to experts it can be increased ten fold. Moreover, almost the entire quantity extracted moves outside the state as raw materials for various products. Apart from agriculture and forest products, serviculture is also an important item in the economic activity of Assam. Out of a total 21995 inhabited villages in Assam, 6994 villages were declared as Serviculture villages.

In addition to resources such as agriculture and forests, Assam is full of other pastoral and mineral resources. It has reserves of 70.4 million ton (MT) of crude oil, 23 M cubic metres of mineral gas, 927.9 MT of coal, 300 MT of limestone, 73.80 MT of marble, 31 MT of iron ore, and about 90 per cent of the total sillimanite output of India. In fact Assam's sillimanite, a refractory mineral, is said to be the best in the world.

Although nature's bounties in the form of some important industrial raw materials have endowed the state with great possibilities, these have not been harnessed to the full extent, except in certain selected sectors such as tea, petroleum, and forest (wood) based industries.

Assam, with about 2.4 per cent of land surface of India and 2.7 per cent of India's population, accounts for only 1.8 per cent of country's gross industrial output, 1.3 per cent of the value added by manufacturing goods in the country, and 1.9 per cent of total employment in the registered factories.

## 26.7 The Brahmaputra and Its Hinterland

The Assam State comprises the hinterlands of the Brahmaputra River and Barak and the hill districts. The Brahmaputra valley, the hinterland of the Brahmaputra River, covers an area under 11 districts, around 63% of the whole of Assam, and the Brahmaputra River passes though the districts Dibrugarh, Lakhimpur, Sibsagar, Jorhat, Sonitpur, Darrang, Nowgaon, Kamrup, Nalbari, Barapeta, Goalpara, Abhayapuri, and Dhubri. A table comprising the Brahmaputra valley district, their areas, population, distance of river flow, etc., is presented hereunder for making the picture more clear.

Table 26.9 Population area, number of towns, river distances of the Brahmaputra valley Districts.

| Districts | Population '1000' (1991) | Area sq. km | Nos. of Towns | Distance of river flow (in km) | Distance of the farthest point from the river in the district (km) |
|---|---|---|---|---|---|
| 1 | 2 | 3 | 4 | 5 | 6 |
| Dibrugarh(E) | 2,004 | 7,171 | 11 | 171 | 111 |
| Lakhimpur (E) | 1,231 | 5,514 | 3 | 165 | 65 |
| Sibsagar | 908 | 2,668 | 5 | 55 | 87 |
| Jorhat | 1,699 | 6,353 | 5 | 164 | 120 |
| Sonitpur | 1,424 | 5,324 | 4 | 169 | 150 |
| Darrang | 1,299 | 3,481 | 3 | 86 | 53 |
| Nowgong(E) | 2,533 | 5,535 | 5 | 180 | 128 |
| Kamrup | 3,016 | 6,692 | 9 | 110 | 71 |
| Barpeta | 1,386 | 3,245 | 6 | 75 | 66 |
| Goalpara | 1,476 | 7,463 | 5 | 90 | 68 |
| Dhubri | 2,133 | 2,838 | 5 | 110 | 70 |
| Assam | 22,414 | 78,438 | 78 | E- Erstwhile | |

As may be seen from the Table 26.9 the maximum distance or river flow is 180 km in Nowgong district, whilst in Sibsagar it is only 55 km. There are about 41 river ferry services on the Brahmaputra between its banks. The minimum distance of these ferry routes is one km for Dukuchi-Bhukhuradia and the maximum is 100 km for Dibrugarh-Orium.

The ghatwise hinterland radius has been also attempted, irrespective of the districts, in the Brahmaputra valley districts, and also some of the remaining districts from where forest products, etc. are handled for trade to Calcutta and beyond, are given in Table 26.10.

## 26.7.1 IWT TRAFFIC MOVEMENT IN MAJOR HINTERLAND CLUSTERS OF THE BRAHMAPUTRA RIVER UP TO 2005 AD

In a recent project study Howe (India) Pvt. Ltd., the Central Water Power Research Station and Water and Power Consultancy Services (India) Ltd. have jointly made a commoditywise projection of the total transportation demand for the region on the basis of production tenders in the past, its planned target for the future and the post-consumption pattern vis-à-vis demand in future, including both private and public consumption. Out of the twenty one principal commodities being moved by road, and Inland Water Transport, thirteen commodities are identified as commodities which prima facie have been found suitable for movement by Inland Water Transport via the Brahmaputra River are presented in Table 26.11.

Table 26.10. Main river ghats and radii of the hinterland (km).

| River Ghat | Radius in km | River Ghat | Radius in km |
|---|---|---|---|
| Saikhowaghat | 45 | Tezpur | 40 |
| Sadia | 20 | Amingaon | 25 |
| Sibsagar | 40 | Pandu | 30 |
| Dibrugarh | 45 | Santali | 15 |
| Neamati | 45 | Jogighopa | 35 |
| Bokaghat | 35 | Goalpara | 25 |
| Mohanbari | 20 | Dhubri | 25 |
| Silghat | 20 | Fakirganj | 15 |

Some of the commodities are grouped with other related items of particular cargo, e.g., under coal coke is also included; under forest products imported logs are also included under POL, and SKO, RPC, lubricating oil, bitumen, paraffin and furnace oil, IDO, etc., are included.

The main focus of these estimates is on production and demand ratio of the commodities. For example, if demand for a commodity is more than its production the gap between demand and production is covered by the movement of goods either from within the state of from outside the state, including imports and vice-versa. For production estimates a thorough analysis has been made keeping in view (i) past performance vis a vis the targets and (ii) assessment of likely impact of various developmental and expansion programmes both in public and private sectors and policies of attainment of self sufficiency in food grains by the end of ninth plan period for balanced regional industries, etc., as embodied in the draft ninth plan documents.

The Inland Water Transport is the ideal mode of transportation for the over-Dimensional Cargo (ODC) required for industrial plants being set up on or near waterways of the Brahmaputra, owing to their size, shape and environmental impact. It is very different as well as highly as expensive to transport most of the ODC by roadways or railways since ODC must be dismantled with special efforts to suit the accommodation of road vehicles and railways.

## 26.8 Barak Valley and its Transport Economy

The Barak valley in Assam comprises three districts, viz., Cachar, Hailakandi, and Karimganj spread over an area of 6,922 square kilometre and having a population of 2491,000 (1991), with a population density of around 360 per square kilometre against 286 for Assam. The Barak valley shares 11 per cent of population of the whole of Assam.

The entire riverine area of the valley is very is rich in products, such as ginger, cotton, til jute, chillies, mustard seeds, sugar cane, pineapples, orange, timber and cane, etc., which are transported to Silchar for marketing. Daily necessities such as rice, pulses flour, mustard oil, kerosene, clothes, stationery articles, and building materials such as cement, brick, lime, steel, etc., are transported upstream by country boats.

Table 26.11 Commodity wise estimates of IWT traffic in national waterway No.2 (Sadia to Dhubri on Brahmaputra).

(In 1000 tonnes)

| S1. No. | Commodity | 1995-2000 | | 2000-2005 | |
|---|---|---|---|---|---|
| | | IWT Traffic | Total Traffic | IWT Traffic | Total Traffic |
| 1 | 2 | 3 | 4 | 5 | 6 |
| 1. | Foodgrains | 144 | 2,204 | 184 | 2,308 |
| 2. | Tea | 114 | 485 | 148 | 583 |
| 3. | Coal/Coke | 409 | 2,199 | 520 | 2,400 |
| 4. | POL | 400 | 2,414 | 509 | 2,470 |
| 5. | Iron & Steel | 155 | 306 | 147 | 338 |
| 6. | Cement | 165 | 705 | 211 | 761 |
| 7. | Jute | 108 | 316 | 138 | 356 |
| 8. | Fertilizer | 45 | 385 | 57 | 422 |
| 9. | Forest | 43 | 374 | 55 | 419 |
| 10. | Salt | 26 | 327 | 33 | 348 |
| 11. | Sulphur | 40 | 50 | 51 | 56 |
| 12. | Limestone | 380 | 750 | 485 | 850 |
| 13. | Stone Quartz | 250 | 250 | 319 | 319 |
| | | 2,239 | | 2,857 | |

There are 11 tea gardens around Silchar and Lakhimpur out of a total of 93 in the valley. The total tea production in the valley was, as on 31-12-95, 43 million kg which shares around 11 per cent of Assam's production (399 million kg. at 31-12-95). At present all the tea gardens send their tea to Silchar, then to other parts of the country by trucks. If mechanised a river transport service is introduced much of the traffic can be diverted to the river transport. As there is no mode of transport/conveyance military personnel also move up to Tipaimukh through the year by country boats.

There is no road between Silchar and Tipaimukh beyond Lakhimpur, about 35 km from Silchar. Eastern parts of the valley, the entire Lakhimpur block, and western portions of Manipur state presently depend mostly on the Barak River for their transportation requirements. Silchar is connected with Jowai, Agartala, Aizwal and Imphal by National Highways 44, 52, 54, and 53, respectively. But still, in the Barak valley a total of 181 villages out of a total of 1,575 villages, i.e., 11.5 percent, are at a distance of 16 km or more from a surfaced road. The present railway route kilometreage in villages is 186 and route length per thousand square kilometres of area is 26.8 km

"The Barak valley's topography does not favour the development of railways and road, Inland Water Transport thus plays a very important role in the economic development of the region." (Report of the Inland Water Transport Committee, Chaired by B. Bhagavati, M.L.A., Assam 1970, Government of India).

Up to now since 1984, the Assam Inland Water Transport Department restricted their service only to the Cachar and Hailakandi districts with a total of 24 (twenty four) ferry services mostly for passengers and goods across the rivers.

A cost benefit study of the scheme for running a commercial service on the Barak River between Kacharbali, a distance of 100 km, was made by the State Government. According to that study lakhs of people living in the hinterland of the area of operation would be benefitted by this service. This point was also recorded in the report of the said Bhagavati Committee.

## 26.9 Barak System---Catchment Area

The Barak River with its tributaries is the second largest river system of the North Eastern Region and is part of the Ganga-Brahmaputra-Meghna system. The river basin bounded by Barail, Patkai, and Lusai hills extends over an area of 41,723 sq. km. in the State of Assam (26,193 km.), Manipur, Meghalaya, Mizoram, Nagaland and Tripura in India and 751 sq. km. in Burma up to the Bangladesh border. It is situated on the route of the South West monsoon. During the monsoon period the intensity of rainfall can be as high as 250 mm per day and most of the heavy rainfall occurs in the short span of three months, viz., June, July and August.

### 26.9.1 THE BARAK

The Barak originates in Nagaland, a short distance from the Nagaland-Manipur Border.

After flowing through a hilly terrain in Manipur State, and then along the Manipur-Mizoram and Assam-Manipur borders for a distance of about 403 km, the river reaches Lakhimpur in Assam. From here it flows for a distance of 129 km towards the west in a zigzag course, through the plains of Cachar and Karimganj Districts of Assam till it reaches Bhanga, where the Barak bifurcates into two branches known as the Surma and Kushiyara and ultimately, these two branches enter into Bangladesh. Silchar town is situated on the left bank of the river Barak (Figure 26.11).

### 26.9.2 BARAK'S TRIBUTARIES

Throughout the reach from Lakhimpur to the Indo-Bangladesh Border a good number of tributaries have joined the Barak. The Chiri, the Madhura, the Jatinga on the right bank and the Sonai-Rukni, the Dhaleswar, Katakhal the Singla and the Longai on the left bank are worth mentioning (see Table 26.12 for details).

Table 26.12 Main tributaries of the Barak covering length and originating point.

| | River | Length before joining to Barak (in km) | Originates at |
|---|---|---|---|
| 1. | Cheri | 49 | Southern Slopes of N.C.Hills |
| 2. | Madhura | 39 | Foot hills of N.C.Hills. |
| 3. | Jatinga | 58 | N.C.Hills (Near Haflong) |
| 4. | Sonai-Rukni | 96 | Lusai Hills of Mizoram |
| 5. | Dhaleswari - Katakhal | 290 | Lusai Hills * (Note I) |
| 6. | Singla | 52 | Lusai Hills * (Note II) |
| 7. | Longai | - | Lusai Hills * (Note III) |

Note: 1: Dhaleswar is known as Kalangdong in the Lusai hills, and joins the Barak near Siyaltek Bazar. In the early 19th century a Cachar Raj cut a channel from the right bank of Dhaleswari for silting up Bakrbaor known as the Katakhal.

2: After 52 km from the origin falls into the Sonbeel, then after emerging from this Beel as the Kachua and flowing 20 km falls into the Kushiyara River at about 5 km to the east of the town of Karimganj.

3: Formerly, the Longai was connected to the Kushiyara Branch of the Barak within India by a channel called the Natikhal to facilitate water transport for the neighbouring tea gardens, and some time ago the channel was closed.

## 26.10 Features of River Barak System

The Barak River flows in a deep channel in narrow valleys and steep hills from its source to Lakhimpur where it enters the alluvial plain. From Lakhimpur onwards the river has a meandering course. This alluvial valley has a length of about 60 km along the valley axis, from Lakhimpur to Bhanga, where in the meandering length in between these reaches is 129 km.

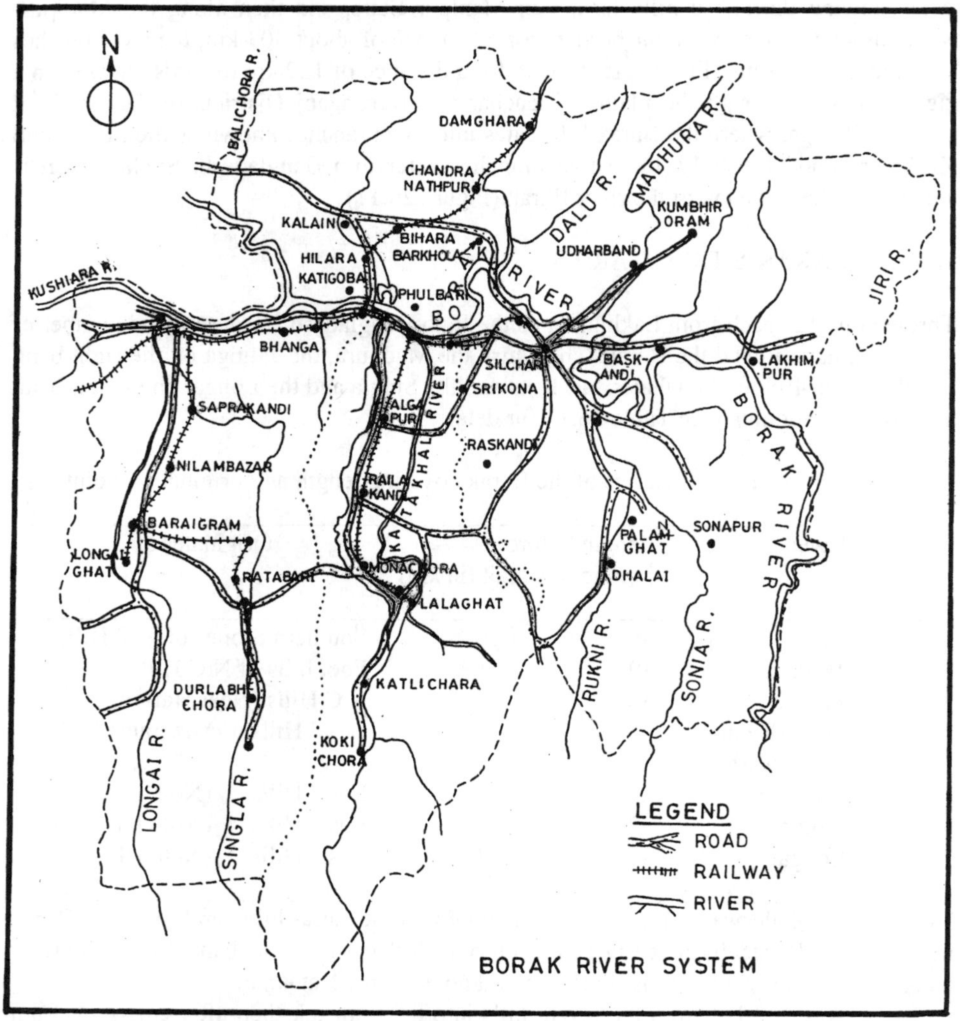


Fig. 26.11 the Barak River system.

Many cut-offs have taken place on the Barakin the past and they have left evidence of it in the from of ox-bow lakes, well detached from the river. These are locally known as 'Anua". One such cut-off is Ramnagar-Arua at Ramnagar Tarapur Shibbari area of Silchar town. The present tendency of the river is to revive the old course through this problematic zone. The phenomenon is further aggravated owing toa cut-off at a place called Dungripar located at about 20 km upstream of Silchar town in October 1992. This cut-off has reduced the length of the river by about 10 km.

Unlike other rivers in the State of Assam, the Barak and its tributaries are also characterised by the critical problem of failure of river banks in the shape of vertical settlement, etc. owing tosloughing and sliding at most of places. Such a problem is owing to peculiar properties of the soil along the river bank.

## 26.11 Hydrological Data and Topography of Barak Valley

The flood gradient of the Barak River is 1 in 21500 (the Master Plan of Barak subbasin). The region is subjected to very intense rainfall during the period from April to October. The highest annual rainfall recorded at Silchar E & D Colony Raingauge Station was 4194 mm in the year 1989 and the annual average rainfall is 3400. The highest recorded level of the Barak River was 21.84 m on 1st August 1989 at Annapurnaghat (Silchar) gauze/ discharge site and the maximum discharge was recorded as 6696 cumec on 14th June 1966 at the same observation site. The river Barak is highly meandering in nature for which it has changed its course at many places. Such process of changing of river course is still continuing. One such change of course took place earlier at Ramnagar-Tarapur-Shibbari area of Silchar town and the old abandoned course is known as Tarapur - Ramnagar Anua.

## 26.12 Climate of Barak Valley

The climate of Barak valley is not very congenial. The rainfall is heavy during the summer when the temperature is also quite high resulting in an oppressive humidity and hot summer. Again, winter is not as cold as in the hilly regions of the sub-basin. The maximum and minimum temperature for Silchar for 30 years for 1931 to 1960 are $39.4^0$ C and $5^0$ C.

## 26.13 Soils of Barak Valley

The soils of Barak- valley - Plains are alluvial and predominantly clayee isolated patches of a mixture of clay and sand in varying proportions. The soils of Tilla Langs (Hillocks) are, mostly covered with tea gardens, red and loamy under laid with shale.

## 26.14 River Transport in Barak in Retrospect

By 1863 the Indian General Steam Navigation (IGSN) Company was more attracted to the Barak valley as compared to the Ganges trade. In 1864 the East Bengal Railway and RSN Company, entering into an agreement, established a rail - cum steamer transport for traffic to the Barak valley and vice versa. A similar experiment was tried in relation to steamers for Dacca and the Barak valley when Port Canning, now in Bangladesh, was connected to Calcutta by the railways. (Griffiths, 1967).

By 1975 there was a fortnightly steamer service to Cachar from Calcutta, up to 1886, the steamer companies (IGSN & RSN) carrying cargo and passengers to Barak took considerable time. In 1887 the Companies, aided by a Government Subsidy, established a daily steamer service, with a powerful steamer, between Goalundo (now in Bangladesh, 40 km from Calcutta) and Silchar during the rainy season and between Goalundo and Frenchuganj in the cold weather, and which took almost half the time of the earlier service. Towards the end of the nineteenth century steamer transport dominated trade to and from the Barak region compared to boat transport during 1893-99 (Table 26.13).

Table 26.13 Import and export of Barak valley (1893- 94) (in mounds).

| Route | Import by | | Export by | |
|---|---|---|---|---|
| Between | Boat | Steamer | Boat | Steamer |
| Barakand Bengal (Excluding Calcutta) | 51,000 | 17,000 | 10,000 | - |
| Barakand Calcutta | - | 78,000 | 1,000 | 98,000 |
| Total: | 51,000 | 95,000 | 11,000 | 98,000 |

Source: Compiled from Assam Administration Reports and Report on the River Borne Trade of Assam.

In 1899 the IGSN and RSN Companies were combined to form a company which was popularly known as the Joint Steamer Companies. After the partition of 1947, the river transport had a special role in movement of traffic to and from the Barak valley district. It was the Barak valley which served as the origin and destination of the large boarding area, comprising Tripura and the Mizo Hills, now Mizoram, of the then Assam. The region, which had three lines of communication before partition, namely, the rail river rail route via Chandpur (now in Bangladesh,) the rail route via Santahar, Mymensing and Akhaura (all now in Bangladesh), and the rail route from Chittagong began to have a circuitous rail route after partition, and hence the dependence of the Barak region on the river route increased several fold. By 1954 the traffic from Calcutta to the Barak valley rose to 11 lakhs mounds as against 4 lakhs mounds in 1946. Similarly, riverborne exports to Calcutta rose to about 9 lakhs mounds in 1954 as against about 4 lakhs mounds in 1946. On the contrary, the volume of traffic by the Ganges Feeder Services to and from the Barak districts declined after independence from 1 lakh mounds in 1938 to about 0.5 lakhs mounds in 1954. (Source: Report on the Ganga- Brahmaputra water Transport Bord, 1954, pp 287-290). This was partly because of the diversion of traffic of rail transport owing to shorter distance and quicker transit by rail with stations in the Ganges Feeder Service route and avoidance of formalities of customs.

During the post-independence era the transport requirements of the Barak districts were much more than the capacity of the river services. In 1954 the RSN company informed the Karimganj Merchants' Association that the cargo from Calcutta was restricted to 21,000 mounds per week, of which 6,000 mounds per week was for stations above Karimganj ghat and the balance for Karimganj ghat. The booking of perishable cargo (such as onion) was wholly restricted for fear of deterioration in transit. The shortage of capacity of the river services to and from the Cachar area was conspicuous when the RSN Company advised the Karimganj Merchants association to offer as much of their goods for Karimganj ghat or via Karimganj ghat to upper section stations for onward carriage by rail.

The river service also was not perfect in the nineteen fifties. The correspondence between the Karimganj Merchants association and the RSN Company in 1955 shows that the movement of both inward and outward cargo during 1954-55 was very slow, and a consignment took one month to reach Karimgang from Calcutta instead of the normal period of ten days. Such a state of affairs was the reason for the withdrawal of bigger vessels from the line owing to channel's bad condition.

The J.S. companies rendered, on the whole, good services in view of the fact that except a partial grant made by the Government of Assam for bandalling work, the companies not receiving any subsidy for the development of water transport. (Source: Lakum, N.S.: Report on the Ganga - Brahmaputra water Transport Board, 1954, pp. 39.). In 1962 Mr. J.M. Parsons, an experienced officer of Macneill and Barry Ltd. Calcutta, took stock of the river's transport capacity and commented: "The Cachar route is served by 12 towing vessels and 40 dumb units with a total capacity of 7,400 tons and carrying 125,000 tons a year consisting solely of Joint Steamer Companies Services." (Source: Federation of India Chamber of Commerce: Report on Transport and Power Conference, New Delhi, September 20-22, 1962, P. 183).

So far as external trade is concerned, 1964 was the last year of glory or the steamer service in Assam, because the Indo-Pak hostilities of 1965 led to closure of the Indo-Pak river route, about 22 steamers and 40 fleets remained in Assam Waters. " River craft numbering 55 belonging to the RSN Companies and cargo in transit valued at about Rs. 3.19 crores were detained in East Pakistan" (Source: Lok-Sabha Secretariat, Lok- Sabha Debates, February 22, 1966). As in the case of interstate trade, the river transport has natural advantages, bercause navigable waterways of 153 km by boat and 64 km by steamer is available in the barak system.

## 26.15 I.W.T. Traffic Potential on The Barak System Projected up to 2005 Inter State

Food grains, fertilisers, iron, steel and cement are the major items for movement by IWT from Calcutta to the Barak valley and then to feed the State such as Mizoram, Tripura, and Manipur. It will be desirable to meet the fertilizer requirement of Barakvalley districts of Assam, Mizoram, Manipur and Nagaland with river connection from Calcutta port from cogitaic angle. An estimate of traffic of fertilizer to the valley are 0.11 lakhs tonnes in 1999-2000 and 0.124 lakh tonnes in 2004-2005.

Rich deposits of lime and dolomite in Meghalaya are being explored for distribution from Calcutta. About 0.5 lakhs tonnes per annum of minerals are expected to be moved by Inland Water Transport. Coal on the Calcutta Karimaganj route to the tune of 0.45 million tonne is expected by the year 2000. Other commodities identified for transportation of IWT in the valley and with the Calcutta port or Bangladesh are forest products, tea, jute, sulphur, bitumen, etc..

The future requirement of cement, iron and steel of the Barak valley Districts, Manipur, and Mizoram has been projected in Table 26.14.

## 26.16 Navigation Channel Design

There is a close relationship between the dimensions of navigable channel and the dimensions of vessels which are to use it. Therefore it is necessary to visualize and stipulate the length, breadth, and draft of the user vessels or combination of vessels when designing a channel for navigation. This may be termed the 'Design Vessel' for the purpose. The 'Design vessel' is a conceptual vessel which may or may not exit in reality - whose dimensions are to be taken as the basis for deciding the channel design parameters. Naturally such a design vessel must bear a close resemblance to a good cross-section of the larger vessels plying or expected to ply on the National Waterway (NW) No. 2 as well as on the waterways of the Barak system.

Table 26.14 Projected requirements of cement, iron, and steel of Barak valley districts and Manipur and Mizoram. (In tonnes)

| Barak valley | | | Manipur | | Mizoram | |
|---|---|---|---|---|---|---|
| Commodity | Year 1999-2000 | Year 2004-2005 | Year 1999-2000 | Year 2004-2005 | Year 1999-2000 | Year 2004-2005 |
| Iron & Steel | 325 | 500 | 9,650 | 12,900 | 10,400 | 14,800 |
| Cement | 55,750 | 71,500 | 63,000 | 76,950 | 36,000 | 57,000 |

| Route | Commodity suitable for IWT |
|---|---|
| 1. Calcutta-Karimganj: | Foodgrain, Cement, fertilzer, steel, Bitumen for the state of Tripura |
| 2. Karimganj-Calcutta: | Paper from H.P.C., Jute, Bamboo Products and Forest products of Tripura |

3. Calcutta-Badarpur: Foodgrain, cement for Mizoram, Industrial raw materials of H.P.C. bitumen and Steel for Mizoram, Pipes of ONGC.
4. Badarpur-calcutta: Paper of HPC, Coal (Meghalaya Origin), Forest products of Mizoram

In a recent project report of the Inland Waterways Authority of India (1990) the 'design vessel' dimensions are proposed as below, stagewise, for the purpose of channel design.

| | First stage (medium term future) | Second stage (long term future) |
|---|---|---|
| Length | 100 m | 100 m |
| Breadth | 12 m | 12 m |
| Draft (max.) | 2.25 m | 3.00-35 m. |

The main waterway parameters which have been taken into consideration while designing the navigation channel system are:

(i) depth of navigable channel (below datum/low water level);
(ii) width of navigable channel;
(iii) water current velocity;
(iv) ratio of wetted cross-sections of the channel and midship form of the design vessel ('n' ratio);
(v) whether two-way or one way navigation is to be permitted;
(vi) radius of curvature of channel bends;
(vii) vertical clearance under structures such as bridges;
(viii) vertical clearance under overhead power cables extending across channels, above HF; and
(ix) Horizontal clearance between vertical, fixed structures flanking navigation channels.

The 'design vessel' parameters which have a direct bearing on the above channel design aspect are:

- Max. Length
- Max. Width
- Max. Draft (Loaded)
- Max. Height above water line (W.L.) – (light condition)
- Full form of vessel and its midship-cross section.
- Speed while proceeding upstream.
- Squat of vessel at full speed.

It is evident that much research and development effort are required to optimise these numerous factors in order to arrive at the most efficient and economical design of channels. Unfortunately, little work of this nature has been done indigenously so far.

This is only to be expected since the renaissance of IWT is itself at its beginning in India.

However, it is known that a Government of India committee had recommended that a channel of 45 m width and 1.5 m depth should be the minimum National Waterways Standard. But the said aforementioned standard should not be taken as final and IWAI may make their recommendations on this independently, considering individual features of each and every waterway.

IWI has already, through a project report (1990), recommended channel dimensions for navigation on NW-No.2 (Brahmaputra) based only on technical consideration are as below:

| | 1st Stage | 2nd Stage | 3rd Stage |
|---|---|---|---|
| * Channel width, bottom | 7.5m | 125 m | 150 m |
| * Channel depth at lowest WL | 2.20 m | 2.5m | 3.0 m |
| * Radius of curvature inside | 660 m | 600 m | 600 m |
| * Flattening at curves | $L^2/2R$ | $L^2/2R$ | $L^2/2R$ |
| * 'n' ratio | 5 | 10 | 15 |
| * Minimum vertical clearance above HFL: under bridges | 10 m | 10 m | 10 m |
| * Minimum vertical clearance above HFL: under high voltage transmission cables | 19 m+ 1cm. for each addl KV above 110K.V. | | |

where L = Vessel (design) length in metres, and R = Radius of curvature of bend in metres.

Similar efforts for channel design for navigation on the Borak system have yet to be taken up, even on the basis of technical considerations. Channel design, based on technical considerations and assuming that the design vessel will be required to navigate the entire length of the waterway, may not go well for sound investment decision making. Economic considerations, such as volume of traffic on various sections (stage to stage), cost of channel development, etc., have to be regarded as vitally important for sound investment decision making.

Finally, the IWAI considering all relevant factors, the building up of the traffic to the projected levels which, in turn, is related inter alia to the time required for building the terminals---only a 45 m channel for vessels with 2m draft have been proposed for the immediate future and the cost of such a channel is estimated as the amount of Rs 28.7 crores (in 1990) for the first phase waterway development restricted to Dhubri to GAMIRI (upstream of Tezpur, with water depth 2.2 m). Furthermore IWAT's project report proposed for the next stage of development as below:

| | Stage | Channel | Channel Width | Channel Depth | Vessel Draft |
|---|---|---|---|---|---|
| 1. | First | Dhubri-Gamiri | 45 m | 2.2 m | 2.00 |
| 2. | Second | Dhubri-Sadiya | 45 m | 2.2 m | 2.00 |
| 3. | Third | Dhubri-Pandu | 74 m | 2.5 m | - |
| 4. | Fourth | Dhubri-Dibrugarh | 75 m | 2.5 m | - |

However, it is necessary to refer to the result of work done in some of the leading IWT countries such as the USA, France, Holland, Germany, the former U. S. S. R., etc. for cost effectiveness of channel development. The main Government authority in the USA on IWT is the US Army Corps of Engineers. In Europe there are several deliberative regulatory or advisory bodies connected with IWT. The more prominent of them are: The Permanent Association of the International Navigation Congress, Brussels (PIANC), the Central River Commission (CRC), the Conference of European Ministers of Transport (CEMT), the UN Economic Commission for Europe (ECE), etc., apart from various national authorities.

## 26.17 Channel Conditions and Navigation

The main artery of the states in the North Eastern Region of India, the Brahmaputra River receives a number of tributaries from north and south. Many of which are also reportedly navigable for varying distances from their mouths. The Subansiri, Dihang, Burhi-Dihing, and Dibang rivers are important among these (Figure 26.2).

The Brahmaputra River is internationally known as much for its instability as for its size and might. The Brahmaputra is a 'braided' type river and not an aggrading or meandering type unlike the course of the Ganga (Figure 26.3). The meandering type is the full and final stage of river development, whilst the other two are interim types which might be maintained as long as the same courses and effects remain unchanged. But in many instances when rivers are divided into two or more branches it is not possible for every branch to maintain a state of equilibrium in respect of sediment flow, resulting in the reduction of the cross sectional area in the branch where sediment load is in excess of the transportation capacity, and in turn decreases the discharge and accelerates further deposition. The other branch passing more discharge will be degraded or enlarged.

In the high discharge months floods and bank inundation is an annual event. During this period though channel depth for navigation is ample, identification channel route is problematic. Moreover, velocities of the water current as high as 8 to 10 km/hour and more possess obvious problems for navigation.

Often abrupt and violent shifts of channel at locations/alignment are unpredictable and this makes the marking of the channel to be navigated difficult. For the same reason the construction of fixed level/location terminals is also highly problematic. Because of all this, navigation and pilotage on the Brahmaputra is still more of an art than a science. The relatively shallow depths available in the low water season is one of the main constraints in developing water transport on the Brahmaputra, and it deprives the main advantage normally offered by water transport.

### 26.17.1 SOLUTIONS

To stabilize the river channels, reduce erosion, and check the formation of shifting shoals and sand-bars which hamper river transport, it is necessary to reduce the silt charge in the river. Also for uninterrupted transport with larger, and consequently more economical vessels, greater water depth in the river during periods of low flow has to be ensured. Studies have shown that both these requirements can be met to a large extent by creating storage reservoirs in the main river and tributaries. The four tributaries of the Brahmaputra, namely, the Dihang, 200,000 M cu m), the Dibang, (62,800 M cu.m) the Lohit (60,000 M cu.m) and the Subansiri, (54,270 M cu.m) taken together contributes about 77 per cent of the average annual yield of the Brahmaputra at Pandu (487,000 M cu.m) (Figure 26.3).

The reservoirs of Dihang and Subansiri (the project reports prepared and submitted to Government of India in 1983) the upstream of the Brahmaputra are likely to go a long way to boost the navigability of the river system by releasing definite quantities of the generated power, thereby augmenting the lean flow and the availability of draft during summer period.

It is understood that calculations indicate that the increase in draft during the lean season at Neamati, Tezpur and Pandu will be, respectively, as much as 2.0 m, 1.63 m and 1.08 m, if only the planned Dihang dam is constructed and 2.00 m., 2.0 m, 1.53 m, respectively, if both the Dihang and Subansiri dams are constructed. Therefore the construction of at least the Dihang dam is the real key to achieving the objective of developing the NW-2 inland navigation in a big way. This will minimise the need for costly dreadging and other conservancy measures such as bandalling which are only temporary and short-lived solutions.

However, in the interregum, which is bound to be of long duration, the navigation channel has to be improved and maintained through conservancy measures, river training, and dredging. Owing to the great changes that take place in the channel every year, calls for a major and continuous effort in hydrographic surveys, channel marking, etc., is thereby called for. The required navigation channels have been understood from the point of depth, width and curvature required by the 'Design Vessel' as envisaged in the section on 'Navigation Channel Design', based on the available data.

### 26.17.2 OTHER NAVIGATIONAL PROBLEMS

Apart from the main problems which affect navigation in rivers, viz., abrupt and violent shifts of channel, relatively shallow depths available in the low water season, other impediments to navigation that may be experienced on the Brahmaputra are as follows: velocities; eddies/ vortices/whirlpools, etc.; wave action; poor visibilities; and debris and snags.

26.17.3 VELOCITIES

The water currents attain quite high velocities in the Brahmaputra, especially in the monsoons. Quite high velocities, i.e., 3.5 to 4 m/s, occur for a few hours/days at a time almost every year in the month of June to August as observed by different agencies at various locations and seasons. Sometimes high velocities may occur in late May or early September. Records show that in August 1961 a high of 19.10 ft./s (i.e., 5.85 m/s) was reached by the current velocity at Pandu. In August 1959 it reached 18.30 ft/s (i.e., 5.59 m/s).

These velocities are indeed too high for vessels to navigate productively/safely whether proceeding down or upstream. Fortunately such high velocities usually do not sustain for more than a few hours---or worst a day or two at a time. However, the velocities are not uniform across the entire channel width. The advantage of this feature is normally used by the experienced and skilled river navigators and pilots. Moreover, velocities at Pandu are likely to be on the high side compared to most other reaches, as the river is relatively very narrow and confined at Pandu. Little can be done to reduce current velocities on the Brahmaputra in the monsoons. However, it will be useful to monitor velocities at several places and also to arrange for transmission of this information and associated warning when warranted to vessels before they reach an area of very high current, particularly in the vulnerable season.

Both maximum and average velocities generally decline with falling discharges, and river levels and may be quite benign, i.e., 1 to 2.5 m/s (2 to 4 knots) from mid-September or October to about mid-May. It will be worth/mentioning that since the Brahmaputra has a discharge of about 1.0 lakh cusecs at its lowest stage the average velocities to be contended with by vessels on this river tend to be higher in the lean season than on other comparable rivers in India. This factor may influence the bull shapes and towing-configurations to be used on the Brahmaputra.

26.17.4 EDDIES / VORTICES / WHIRLPOOLS

Eddies/vortices/whirlpool are bound to exist in the river, particularly during the high/flood stages and normally, they do not form a hazard to the navigation by modern/large vessels on the Brahmaputra. However, while plying a country-craft this factor cannot be ignored.

26.17.5 WAVE ACTION

Normally, waves of any significant size (height and period) do not build up in inland waters because wave size is a function of many factors, such as fetch, wind direction, velocity and duration, etc.. On a river, the fetch is rarely long enough owing to the turns and twists that its courses take as well as to the relative narrowness of the water course. However, in the case of the Brahmaputra in flood or at high stages, there may be an extensive area of water which may facilitate the building up of waves---particularly during gales and cyclones, sufficient in size to endanger the inland vessels which are required to have relatively small free board (compared to coastal or sea-going vessels).

Obviously, little can be done to mitigate wave action on the river. However, systems of weather warnings should be organised and properly implemented. Masters of vessels are, of course, expected to take suitable necessary action, including "heaving to" if required and feasible, as in the ordinary practice of seamen.

### 26.17.6 POOR VISIBILITY

Poor visibility owing to heavy rain and fog is a common problem faced by all navigators. On the Brahmaputra an additional cause of poor visibility is the occurrence of storms, owing to extensive sand-banks bared during the dry season, which may be whipped up by strong wings from March.

Exercising normal precautions and giving fog signals while underway are all that can be done in this regard. The authorities of the waterways may notify certain reaches which are prone to such sand storms.

### 26.17.7 DEBRIS AND SNAGS

The higher reaches of the Brahmaputra, i.e., from (beyond) Sadia to Dibrugarh and further downstream pose an additional problem of floating tree trunks and associated with flooding debris during season. It is understood that this is owing to trees uprooted from the banks of the tributaries being carried down by the flood waters into the main Brahmaputra. These tree trunks and snags may be visible or partially submerged, and thus can consitute a hazard to vessels through collisions and fouling of propellors. Similarly, such debris may ground on the bed and thus cause unexpected obstructions and dangers to navigation.

The way to deal with this menace is to remove its root cause. This is tantamount to deforestation of the flood prone areas on the banks of the Brahmaputra and its upstream tributaries. Risk of fouling of propellors may be mentioned by providing propellors guards.

## 26.18 Navigation Related Services and Infrastructure

### 26.18.1 HYDROGRAPHIC SURVEYS

Hydrographic survey is a basic investigation and the charts resulting from them is the essential information for the navigation by any kind of vessel. The hydrographic survey chart is put to many other uses such as for inducing model studies. So far the hydrographic survey of the entire length of the NW No. 2, i.e., from Dhubri to Sadia has not been completed.

The practice so far has been to follow the 'navigation channel' as indicated by local 'pilots'. This is something of a paradox, since it is only after the survey that the best channel can be identified.

26.18.2 RIVER NOTICES AND WARNINGS

The continuous changes taking place in the Brahmaputra have not only to be monitored but notified as soon as possible to all concerned, and particularly to those affected by it. The established method for this is the issue of river notices and warnings. The river notices may give channel changes in location or depths of shoals (together with chartlets when required), information regarding any activity which may call for special care in navigation, such as dredging, current observations, civil construction on the waterfront mandatory restrictions on navigation, i.e., draft reduction, speed controls, forbidding meeting or passing in certain areas, etc..

26.18.3 WEATHER AND STORM WARNINGS

The Meteorological Department of the Government of India (Met. Dept.) is already operating a fairly elaborate system for conveying weather-related information and warnings, both by radio and visually, in coastal waters. The Meteorological Department particularly, for storm warning signals, can issue the warnings for inland water navigation in cyclonic conditions. This can be done in two ways, storm warnings signals, as normally displayed in coastal waters at major and minor ports, should be displayed on flag staffs at various convenient points. In addition, the carrying of at least a single-band radio receiver on board all powered IWT vessels may be made a mandatory requirement by the authority as a prelude to making arrangement with All India Radio (AIR) and Met. Dept. for broadcasting special weather warnings (including extraordinary precipitation) for inland waterways traffic on Brahmaputra along the lines of special weather bulletins which are broadcast for fishing vessels and fisherman in coastal waters.

26.18.4 PILOTAGE

Pilotage service is essential for the safety of the vessels in all restricted waters and also of the restricted channels and the structures and infrastructure associated therewith. Pilotage would be of great value of NW2. In fact, pilots have been in use on the Brahmaputra for decades. The Waterway Authority may assume responsibility for provision of more efficient pilotage services throughout the length and breadth of NH2., by issuing Pilotage licenses to all those with acknowledged experience as pilots and serving presently CIWTC, DIWT, IWAI and others, in the first instance.

A system of holding (oral) examination, prescribing minimum educational and experience requirements, etc., may be established. Apart from good experience of navigating on the river Brahmaputra, good knowledge of the navigational aids used on the river and their significance, etc., will have to be learnt by licensed pilots.

## 26.18.5 RULES OF THE ROAD FOR NAVIGATION

In restricted water on the Brahmaputra River (NW-2) it is essential to have rules of movement, particularly when two or more vessels are using the same navigational route. Such rules of movement are called 'Rules of the Road (RORs)' and are now very well codified under the SOLAS conventions of International Maritime Organisation of the UN based in London. A proper knowledge of RORs and any special modifications thereof or local rules on NW-2 framed in future must be, inter-alia, a requirement for aspirants and possessors of pilotage licenses.

## 26.18.6 SALVAGE/RESCUE/ASSISTANCE

Sinking of vessel on NH No. 2 is reportedly rare. However, the authority(s) should have some infrastructure and facilities for dealing with emergencies, viz. collisions, groundings and fires on vessels. It will be useful for the Waterway Authority to procure and maintain powerful tug-cum-fire float, reasonably fast, fitted with all equipment for towing, fire-fighting and similar emergency duties. Good communications system including VHF, etc. will be desired on such vessels.

## 26.18.7 AIDS TO NAVIGATION

It is necessary that the navigation channel be marked in a clear and unambiguous manner so that its navigation can be done with confidence and relative ease. Such markers and various other devices for like use are commonly called aids to navigation. While visibility conditions may differ from time to time and season to season and as between day and night, a practical distance between two consecutive channel markers/buoys in riverine channels would be approximately, 2--3 km. This would ensure their being sighted and 'recognised' well in time by navigators in normal visibility conditions.

Owing to the entirely different environment and natural condition of inland waters particularly in non-tidal rivers and canals, the principal systems of maritime voyage, viz, the 'Cardinal' and 'Lateral' systems and channel markings have not been found to be suitable and indigenous marking systems have been developed in various countries through more or less informal processes. Obviously they are different from the systems as unified by International Conventions and they have not been internationalised except to a limited extent in Europe and that only recently.

With the progress of science and technology various radio aids to navigation have been developed. These include many types of Radio Beacons, Positioning systems, Radar, etc., Laser based aids to navigation are also available. However, these are much more expensive and can only be economically justified in conditions of heavy traffic or congested channels.

There are a number of special problems which mitigate against the adoption and use of conventional navigational aids to demarcate the navigational channel of the Brahmaputra. Some of these are mentioned below:

i) Water level fluctuations;
ii) High current velocities;
iii) Unstable channels.

After careful consideration of the problems and the various riverine navigation aids systems in use in countries where inland navigation is well established, it is felt the system developed by the European Economic Commission, known as 'Signs and signals on Inland Waterways' or 'SIGNI' could be advantageously adopted for the Brahmaputra, with such modifications as may be found desirable in actual practice. The 'SIGNI' system will ensure long lasting continuity. The system also has a comprehensive set of lights and their combinations for night navigation. It may be also mentioned that the use of the 'SIGNI' system has been recommended for Asia and Far East in the ESCAP seminar in 1980.

Lastly, if it is felt that the SIGNI system is too elaborate for the present state of IWT development on the Brahmaputra, the signs and signals which are useful or required at present need to be implemented. The SIGNI system is flexible enough to be introduced in parts or phases, and therefore the authority may suit its convenience in this respect.

26.18.8 OTHER AIDS

Two other types of aids to navigation may be provided on the Brahmaputra to facilitate navigation.

*26.18.8.1 Visual Information System for River Level*

Gauges and Semaphore gauges may be erected at places where the river channel is closed to the bank. High Semaphores may be erected which can indicate the rise of water level above a datum in a manner which is visible and readable for several km upstream and downstream. They are also convenient for display of other important visual signals such as storm signals, etc.. For ordinary guages on the Hooghly may be adopted for the Brahmaputra. Semaphores gauges, in use on, inter alia, in the Hooghly River are found to be very helpful by navigators.

*26.18.8.2 On-Board Navigational Aids/Equipments*

As traffic increases, the economic compulsions of night navigation will have full impact on on-board navigational aids/equipment for upgrading. These will comprise echo-sounders, radar and radio telephone links between ships and shore stations.

26.18.9 MOORING HOYS

The development and removal of buoys require special types of vessels, sometimes called morning hoys. These features, in particular, include: bow rollers and arrangements for handling lowering and hoisting heavy anchors, mooring chains and other tackle, fairly powerful mechanical or motorised winch/capstan(s); and a swinging derrick of say 2 to 3 tons to lay/pick buoys on an associated storage deck to keep the buoys, etc..

## 26.19 Conclusions

In 1950-51 Assam's per capita income was more than the per capita income of India. But the economic scenario of Assam has unfortunately been deteriorating since then and per capita income of Assam in 1993-94 was only Rs 5,916/ against Rs 6,929/ of India, which is 15 per cent less than the per capita income of India.

In a developing economy transport facilities should always be in excess of the immediate needs. An estimated 90 per cent of inland passenger and freight traffic in India in the beginning of the next century will have to be met by road and rail only. Foreseeing an energy crisis in the immediate near future, the National Planning Commission has given due importance to exploring the potentialities of Inland water Transport as well as transportation on a new canal system.

Assam possesses ample potentialities for the development of Inland Water Transport. Although it can still play a vital role in the regional economy, presently only a fraction of its potentialities have been explored and utilised. Therefore intensive efforts are necessary for the study of its technologies, management systems and practices.

The traffic movement prevailing on various river routes is not disheartening. It appears that if efforts are made in a consonant way even once, the flow of trade on the river routes in the Brahmaputra will tremendously increase, as narrated in the Detailed Project Report for Development of Inland Water Transport on National Waterway No. 2 (1990), since the Inland Water Transport system is one of the cheapest modes of Transport from the latest study and more appropriate to the Assam region with least pollution and providing more employment.

With the passing of the Inland Waterways Authority of India Act in 1985, the Inland Waterways Authority of India came into existence in 1986. This was a landmark in the History of the Development of Inland Waterways, as for the first time the Government decided to take on the full responsibility for the Development of navigation through mechanised vessels on the Inland Waterways of India. At present the Brahmaputra appears to be one of the very promising waterways and it has already been declared National Waterways No-2, from Dhubri to Sadia.

It is also high time to focus on the Barak valley to explore its natural navigational potentialities available for the benefit of the local economy of the valley. It may be noted that during the post-independence era, the transport requirement of Borak districts, for inter-state, were much more than the capacity of river services. Though, so far as trade of the Barak valley is concerned, 1964 was the last year of glory of 'steamer service' as the Indo-Pak hostilities of 1965 led to closure of the Indo-Pak river route for a period of years. However, after a decade the river traffic demand again became quite substantial. And the reason for non-development of international as well as inter-state Inland Water Transport Service is mainly the lack of development and maintenance of inland-waterways in the valley, and the present protocol system of international trade.

In view of the above, the early declaration of waterways of the Barak River as National

Waterways No. 4 will be justified, for the interest of proper development and maintenance of the same.

It is essential to maintain the river channel of Calcutta-Guwahati, Calcutta-Karimganj stretch for perennial navigation with night navigation facilities. It is also essential to improve the river channel from Badarpur to Panchgram to transport finished products of Hindustan Paper Corporation and thence to Calcutta throughout the year by the water route.

In view of gradually the improving relationship with Bangladesh, the Government may take necessary steps at the diplomatic level to explore the possibilities of establishing a river cum train service between Assam and Calcutta as well as between Assam and places of Bangladesh which were very viable in the pre-independence era to avoid complicated and hazardous navigation through the Sundarbans, as well as to reduce the distance to make it more economically sustainable.

It is expected that certain policy measures are to be adopted by the Government of India during the 9th Plan-period, 1997-2002, for the development of IWT in Assam. Liberal investment of the plan fund in the 9th plan is essential, unlike the 1st plan to 8th plan, where it suffered a setback compared to roadways and railways.

To attract the private entrepreneurs for investing in the IWT sector, IWT should be declared an industry like tourism. The IWT sector should be declared as a priority sector on a par with similar nascent industries to develop the backward region like North East Government as well as public service undertakings should come forward with long-term cargo assurance to the private entrepreneurs with bulk cargo in the river routes. Tax concessions for investment in the IWT sector for infrastructural development should be provided.

A regional institute in Assam for the development of waterways and management is highlighted. The existing regional Institute (NERIWALM) set up by NEC in Assam (at Tazpur) for NE Region for development of Water Resources and Management, otherwise may also be strengthened for action research and training.

Ship building subsidy as applicable to sea going vessels should be extended to Inland Water Vessels to relieve ship builders. Terminal facilities, with modern handling equipment are to be established at Karimganj, Badarpur, and Silchar. Speedy river craft are to be placed in operation for connecting the North Eastern States by main waterways, as well as their tributaries, with the rest of the country.

### 26.19.1 STANDARDISATION OF VESSEL/BOATS

Although the history of Inland Water Transport in the North East Region of India is more than a century old, till date, no standardisation for its vessels/boats has been adopted by any organisation. Standardisation of vessels has got a number of advantages, namely:

It economizes design and construction;
It economizes maintenance;
Also, it improves the overall economy of transportation.

Furthermore, it is necessary to visualize and stipulate the length, breadth, and draft of the user vessels or combination of vessels when designing the navigation channel. The fleet of Commercial Services of Assam, IWT and CIWTC may be of integrated tug--barge system. Self- propelled vessels should be reduced to a bare minimum. The barge sizes should be such that it facilitates the loading of containers of standard sizes as per ISO container economically.

The concept originating in the United States of a container carrier that could transport a valuable type of cargo, usually from door to door, with minimum handling but with speed and efficiency, captured the imagination of the transportation industry in the highly developed countries of Europe, and later in Japan.

The integrated tug—barge system composed of a leading barge with carefully shaped prow and propelling barge having a stern like that of a twin-screw towing boat, both with one square end. Between these two one or more double-square ended units are secured so as to form one barge. For designing the intermediate units one may follow the ISO Standards LASH (Lighter Aboard Ship) Barges, which are Ship borne Barge series. Each of these barges are single rectangular boxes with end symmetry such that they can be fitted flush with each other side by side.

Recently the Ministry of Surface Transport (India) has assigned with the Department of Ocean Engineering and Naval Architecture, Indian Institute of Technology, Kharagpur for a Research and Development Project proposal for 'Standard Hull Design for Mechanised Country Boat'. Welcoming the said development, it is expected more such research projects for standardisation of ferry and commercial vessels of both Assam IWT and CIWTC.

## 26.19.2 STANDARDISATION OF CHANNEL DESIGN AND DEVELOPMENT

All IWAI already has completed the design part channel for National Waterways No.2 from Sadia to Dhubri. Due importance should be given for its development and maintenance, phasewise, as a time bound project.

## 26.19.3 UTILISATION OF WIND POWER AS ADDITIONAL SOURCE OF PROPULSION

In a study by the author at IIT Kharagpur in 1984, it was found that wind characteristics over the Brahmaputra River are found favourable for introduction of wind propulsion system in addition to the heat engine propulsion in the Inland Vessels in Assam. Aerofoil fig system was proposed for the purpose. For an efficient control a microcomputer based control system is to be added to the rig-system.

Once the microcomputer is available on board the vessel, a variety of other jobs may be assigned to it such as wind direction sensing, wind velocity measurement, keeping track of actual distance covered, inventory control, and other computational aids to the navigation.

In view of the recent dwindling reserves of conventional fuels, recent price rises may get due importance in the immediate future for utilisation of non-conventional sources of energy in navigation, particularly, wind energy.

26.19.4 ALTERNATIVE MATERIALS FOR BOAT BUILDING

In fact, a majority of the river activities in Assam are still managed by wooden boats. The recent Supreme Court verdict banning the felling of trees for the interest of preserving ecological balance in the North East region of India has triggered the importance for finding alternative materials for boat building.

As such, the Central Government should look to entrusting another research project for the same. Lastly, planned integrated development of all modes of transport for effective economic and social improvement is a necessary, though not sufficient, condition. And water resource development needs that meets the needs of the present generation without compromising the scope of future generation to meet their own needs may alone be considered as the concept of sustainable development in water resources.

## References

Encyclopedia Americana: Vol: XXVII.
Medhi, S.B. 'Transport System And Economic Development in Assam.'
M' Cosh: Topography of Assam.
Shakespeare, L.W. (1929). History of Assam Rifles.
Barpujari, H.K. (1963). Assam in the days of Company (1963).
Griffiths, P.(1967). The History of the Indian Tea Industry (1967).
Gait, E. A History of Assam.
IWAIS (1990). Detailed Project Report for Development of Inland Water Transport on N.H. No. 2.
Government of India (1970). Report of the IWT Committee, Chaired by B. Bhagavati, M.L.A Assam.
Gokhale Committee Report on IWT (1951).
Report of the Committee of Transport Policy and Coordination (1974) and (1980).
Das, R.M. Theoretical Modeling of Cargo-Transportation in the River Brahmaputra and an Experimental study of utilisation of Wind-Power as Additional source of propulsion with Computer-Aided Mechanised- Sail in Inland-Water Vessel. A thesis, IIT, Kharagpur.

# CHAPTER 27

# TRANSPORT INFRASTRUCTURAL SCENARIO

S. B. MEDHI

The transport infrastructure in the Brahmaputra basin is gloomy at best. The need for modern and different modes of transport could not be greater. Without adequate transport facilities the northeastern region of India will remain backward and under-developed. This chapter reflects on the transport infrastructural scenario in the Brahmaputra basin and advances a mixture of suggestions to develop transport systems in the basin.

## 27.1 Introduction

While thinking about and writing this chapter, I am just looking at the Brahmaputra and distant blue hills on its banks from my house situated at the top of a hill on the bank of this mighty river. And several questions have flashed on my mental screen. What do we mean by basin? What are the tributaries that connect to this basin? What is the geography---area, population, climate, etc.---of this basin? What is the transport infrastructural scenario of this basin in the transport infrastructural gap? (I also find it interesting to look through the past history of at least the last 100 years to note the changes over time in understanding the present and future scenerios.) Our futuristic studies may become more realistic if we do not completely forget our past history.

A basin is a depressed area below the level of the surrounding terrain and it may be caused by reversal reasons of the globe crust, by upswelling of molten magma, often accompanied by volcanic action, or by prolonged erosion. A basin often tends to provide natural watersheds and as such they may embrace a substantial drainage area. Most of the river basins are caused by erosion and they generally become deep and broad as rains wash away the soil and currents eat into deeper levels.

## 27.2 Brahmaputra

The Brahmaputra, or Sri Lauhitya, is a sanskrit word meaning son of the God Brahma, the creator. In Assamese it is also called Lohit or Luit. This river is considered sacred by the Tibetans and Indians. The legendary account of the origin of the river is given in the KALIKA PURANA. Some salient points may be summarised as follows:

- Source of the river---the Kailas Range of the Himalaya mountains;



*V.P. Singh et al. (eds.), The Brahmaputra Basin Water Resources*, 520-534.

- Total length---2900 km up to the Bay of Bengal;

- Drainage capacity-about 935,490 to 1,600,000 sq. km carrying about 10 lakh tons of sediment per day;

- Entry point in Assam---(including 24 major tributaries) near Sadiya, joined by the Dibang and Luhit rivers;

- Length in Assam---about 800 km;

- Gradient of the river at different places---rising to more than 12,000 feet above sea level---in Assam it is about 370 feet above mean sea level (at the Himalayas it passes through a height of 25,440 ft.)

**27.3 Brahmaputra Basin**

The Brahmaputra basin was a meeting place for the Indo-Aryan and Indo-Mongoloid peoples. And the mighty AHOM kings, who were rulers in this basin in the 17$^{th}$ and 18$^{th}$ centuries, built communication systems, apart from temples and palaces. As such the basin was comparatively not sparsely populated even before 250 or 300 years.

The climate of the basin is different at different places. It is cold and dry in Tibet but hot, humid, and damp in Assam. The Brahmaputra is constantly changing its course at various places in the basin---and it has led to floods every year especially in the lower course---when we have heavy rainfall coupled with huge discharges of its tributaries. During the monsoon season the river has a discharge of more than 5 lakh cusecs and the Brahmaputra often floods to 8 km wide rising 30 to 40 ft, depositing sediments carried down from the mountains.

The valley has a fall of more than 300 feet within a distance of 450 miles (700 kms). The slope of the river above Dibrugarh is about 16 inches per mile and from Dibrugarhg to Tezpur it is 6 inches per mile. From Tezpur to Dhurbi border the velocity is less. This leads to sedimentation and many of our fertile areas are lost---apart from damages to crops during floods.

The basin has a marshy jungle, teak forest, commercial fisheries, wasteland, wetlands, etc. Some areas are famous for rice, jute, tea, and sugar cane production. But the proportion of fertile areas has gone down in view of the hydrological features of the river. It has been rightly estimated that the Brahmaputra's moving capacity is immense and as such it brings more water and sediment to the Ganges-Brahmaputra delta than does the Ganges, whose basin is more than 50 percent larger.

Despite such problems it has been rightly felt that the history and culture of the Assamese people are intimately connected (Brice, 1954) with this Brahmaputra River---as the Assamese have watered their fields with its life-giving water.

Today the Brahmaputra basin in Assam contains 6 districts on the north bank and 6 districts on the south bank. And the broadest part of the basin was found in the old Sibsagar (the present Jorhat) district and in the old Lakhimpur (the present Dibrugarh) district. In the Kamakhya temple (Nilachal) area the river is only about 800 yards broad and in the Dhubri area in the extreme west the basin widens with the width of the river.

## 27.4 Char Areas

One interesting fact of the basin is that it contains a number of river islands, called 'Chaparis" or "Chars". These almond-shaped river islands are formed out of suspended matter coming with river currents. But after some years these chaparis get covered with various needs and grasses making it habitable for people and cattle, etc.. Today Assam's Brahmaputra basin has about 2090 char areas totaling 2.40 lakh hectares, having 15 percent of Assam's population. The people occupying these char areas are mostly immigrants coming from Bangladesh (erstwhile East Bengal) and Nepal since about 1901 till today. The immigrants have produced various agricultural products in these char areas with their hard labor and often with improved inputs. As such transport needs of these char areas have increased about 200 times compared to 1901. The constant flow of immigrants to chars and nearby areas have led not only to extension of cultivation but also to increasing livestock farming to meet the needs of the consumers. With proper transport facilities, many of these agricultural products may be economically marketed in the entire state and neighbouring areas, as there is increasing demand now for vegetables, fruits, spices, etc., all over Northeast India.

## 27.5 Natural Resources

The basin has valuable stone quarries and is well known for water power (owing to a large number of water falls), forest resources, etc.. The soil is also fertile and suitable for growing a wide variety of crops. As already noted, floods and soil erosion have become a challenge if we note that now more than 500 villages of Assam are in the midst of the Brahmaputra River.

Areawise more than half of Assam can be said to be in the basin and about two thirds of the population resides in this basin. This looks surprising when we note that the basin also contains about 500 beels and 100 small rivers/streams. It is also interesting to note that the rate of growth of the population and the density of population is also the highest in the basin as compared to other areas of the state.

## 27.6 Production Pattern

We have noted that the main agricultural products of the basin are rice, tea, maize, jute, etc., and the major industries are very few. But tea, coal, petroleum, plywood, miscellaneous factories/industries (such as paper mill, sugar factory, jute mill, cement factory, fertilizer factory, rice mill) cotton and Silk Handloom industries, bell metal industry, handicrafts, etc., have produced an impact on the transport infrastructure of this basin, although many of these industries have grown up recently and a proper assessment of transport needs has not yet been made by the authorities concerned. The Brahmaputra basin is a land locked area and we want to open it to border trade and trade with other Northeast Eastern states but till today there is no proper input output analysis of the transport sector of the basin. As such, Majuli, the biggest riverine island in the world, is the worst sufferer.

## 27.7 Transport Scenario in the Past

While discussing the transport scenario I should restrict my analysis only to the Assam portion of the Brahmaputra basin and not to other areas beyond Assam, owing mainly to the paucity of data and reliable statistics. Even within the Brahmaputra basin in Assam, we can think of several sub-basins for looking into the real transport scenario.

Looking into the past history of transport, I find certain general observations which may be noted as follows:

- A gap existed in the past between the 'academic economist' and the business man or politician while there was some action on the transport front;
- The transport history of Assam area is one of slow development, but the pioneers showed foresight in seeking the best rather than the quickest solutions;
- Transport has shaped the economic development of Assam just as the geography of Assam has shaped its transport; we have failed to win over the natural barriers through innovative systems;
- Development depends not only on transport but on a complex set of inputs of which transport is only one, although it is a vital input;
- Without proper integration or coordination amongst all modes of transport the entire transport infrastructure of an area remains sub-optimal or below optimal, and this leads to either inadequate capacity with poor service quality or excess capacity (idle capacity) which is wasteful for poor or backward areas;
- The exact nature of the impact of transport infrastructure varies from region to region and industry to industry and time to time. Sometimes transport plays a passive (and not active) part. Again it may be a specific growth factor without any substitute.

27.7.1 RIVER OR WATER TRANSPORT

From very early times river transport was the main mode of transport in the Brahmaputra basin. In 1834, when the steamer services were introduced on the Ganges basin, the journey down stream from Goalpara (the westernmost district of Assam) to Calculatta took about 30 days and in the upward direction about 8 days more making it tedious and unsafe. The tea planters could not despatch tea by steamer for shortage of steamers, high rates, and lack of safety and were compelled to keep their country boats. (vide my book Transport System and Economic Development of Assam, p. 19). In 1844 the river journey from Gauhati to Sadiya by country boats took 6 weeks and that from Goalundo (Bengal) took 3 months or more. The steamer service in the Brahmaputra basin between Calcutta and Gauhati (960 km) was established by the Government in 1847 but it was irregular. In 1856, the steamer service was extended from Gauhati to Dibrugarh. Four years later the IGSN (the Indian General Steamers Navigation Company) being faced with the competition from the East Indian Railway in the Ganges basin, made a contract to run a pair of vessels every 6 weeks from Calcutta to Dibrugarh. In fact, the Brahmaputra basin was a backward area with a terrible sense of isolation for British investors and entrepreneurs. It was the cultivation of tea that turned the course of history. By 1873--75 the acreage under tea and yield in Assam was the maximum in the whole of India. By 1882 another company, the River Steam Navigation Company, entered into an agreement with Assam where navigation was maximum between Dhubri and Dibrugarh.

By 1880-81 the Brahmaputra basin had various government ships and vessels, totalling about 275, worked by private or Public Works Department for intrastate traffic movement. By this time, almost the entire quantity of cotton piece goods, woollen piece goods, liquors, drugs, rice, pulse, iron and other metals (worth about Rs 40 lakhs) moved into Assam by steamer. The country boats monopolized transport of salt, timber, lime and limestone, whereas in export of tea (worth about Rs 40 lakhs) and silk (for which the leading role, the export of timber from the Brahmaputra basin, is famous since ancient times), steamers employed about 600 boats, which belonged to Bengal.

Around 1882 river transport (the steamers of the Assam Railways and Trading Company) helped the railway construction in a big way. The pattern of development in the area is interesting, first, river transport helped the railway development, second, with railway development coal production became possible, and third, with coal production in the area, more space in the steamer became available for carrying other cargo to Assam as no coal was required to be imported from Bengal anymore.

Gradually the size, speed, and capacity of the steamers to handle goods and passenger traffic increased, and by 1889-90 the value of the trade of Assam with Calcutta (Bengal) was about 71 per cent of the whole trade of the province. And 94 percent of the inward and 95 percent of the outward traffic of the Brahmaputra valley was carried by the steamers (river services). In the Brahmaputra basin steamers played a more important role than boat transport both in imports and exports. The main imports by boat were kerosene, sugar, tobacco, dry fish, stone, lime, and rice, Usually the more bulky and less valuable items like paddy, raw cotton, mustard, timber, jute, hides, brass, til, shell, lac, etc., were exported by boat owing to its smaller transport cost. The more valuable items, such as opium, liquor, coal, cotton piece goods, pulses, rice, metal, leather, blankets, brass and iron ware, were handled by railways; and items such as ivory, tea, wax, rubber, and raw silk from the Brahmaputra basin, were handled by the steamer.

By 1900 the railways in trade stood only at about 3.4 percent and 6.5 percent---in terms of quantum and value, respectively. When we find that 70 percent of tea from Assam was exported by river transport we can safely conclude that the river transport enabled the tea industry to take root in Assam more that half a century before the advent of long---distance railway services.

By 1930 the steamer (river) routes in the Brahmaputra basin totalled about 700 miles and were more extensive than the system of metal roads, and were in total length only slightly exceeded by the railways. During the Second World War, the government diverted traffic to river routes, and steamer ghats experienced congestion owing to huge volumes of import traffic by river, After partition of India in 1947, the two steamer Companies (RSN and IGSN ) exercised a virtual monopoly of traffic (especially in case of tea and jute) as they could offer direct transportation between the Brahmaputra basin and Calcutta, which the railways could not. Before the completion of the Assam RAIL LINK in 1949 the river transport (steamers) carried about 80 percent of tea, 90 percent of jute, and large volumes of petroleum products to Calcutta till 1963-64, apart from moving large volumes of imports as the ASSAM LINK (RAILWAY) was not dependable owing to floods.

But with the Great Earthquake of 1950 changing the regimes of the river, navigational problems started. The main line river service had to be terminated about 70 kms downstream of the main tea producing areas in Dibdrugarh, Tinsukis, etc.. By 1953 many important feeder river services were also closed.

In fact, river transport played a dominant part when we look into the fact that up to 1953 the Brahmaputra basin had 510 registered steam vessels operating in the Calcutta--Assam route. Private companies played their part, as about 200 of these vessels belonged to the 7 steamer companies that handled about security and delivery arrangements. The J.S. Companies handled about 40 percent of total traffic of the Brahmaputra valley. This Company got only a partial grant from the Government of Assam for handling work, but did not get any subsidy for development of water transport, although the company kept surplus (excess) capacity of 150 million tonnes per year in both directions to meet emergency needs of the state.

We have learnt from history that tea gardens---even after Partition---preferred river services for several advantages such as less transit time, ease of reaching tea transit sheds and other warehousing establishments, and each of the flotillas of the JS Companies could carry an equivalent of 4 metre gauge goods trains (or 1(1/2) broad gauge goods train). We also learnt from history that water transport could save foreign exchange because tugs and barges were constructed in India, with an average life of about 50 years which is better than the life of railway rolling stock and road vehicles, and with less pollution too. Moreover, the capital and maintenance cost of water transport system was lower than those of the roads and railways in the Brahmaputra valley.

By 1959, the total traffic handling picture was as follows: railways, 35 percent; waterways, 50 percent; other modes, 15 percent. And 93 percent of tea and 90 percent of jute were exported by the river transport. In 1962 it was calculated that there was a surplus downward (export) capacity from Assam by river to the tune of about 200,000 tons per year owing to the seasonal nature of tea and jute crops. And it is important to note that by 1962 there was a shortage of overall transport capacity to the tune of 120,000 tons a year on account of input needs such as iron, cement, fertilizer, machinery, etc., for public and private projects.

In 1965 the Indo-Pak hostilities brought the river services to standstill. Moreover, the steamer vessels were then about 100 years old, and there was complete neglect in improvement of channel, terminal facilities, goodowns, etc. in the Brahmaputra valley. Gradually the carrying capacity decreased and the transit time by river increased, although there was a ready market for river transport.

Turning to the intra-state river transport, we find that Assam has 44 good and navigable rivers (such as Desang, Kullong, Drhing, etc.) and rivulets with good services. There were about 25 ferry crossings (12 with power driven boats) in the Brahmaputra and about 300 ferry crossings in smaller rivers---the most important being Pandu Amingaon (till the construction of the bridge in 1958). But the crossings were spaced at great distance---causing problems to all kinds of traffic. In addition, the frequency, regularity, and safety were matters of great concern. Nearmati, Bordutti, Gamiri, Bishnath, Tezpur, and Dhubri were important for outward traffic, while Neemati, Pandu and Gauhati Bazar Ghat handled higher volumes of inward traffic. The inter-state traffic by river stood at 85 percent as against intra-state traffic of about 15 percent in 1961. In the Brahmaputra basin about 70 lakh mounds of different kinds of commodities were carried by boats every year as per 1961 Central Government estimates. About 4 lakh passengers moved districtwise at major ferry crossings during the year around 1961.

With the winding up of the JS Companies in 1967, the CIWTC (Central Inland Water Transport Corporation) came into existence as a public undertaking and it had 20 steamers and 50 flats with carrying capacity up to 700 tons in each, but it was found that the CIWTC was running at a great loss for various reasons.

In Assam we find some thrust on river transport infrastructure only from the Third Five Year Plan. But even up to this 9$^{th}$ Plan we find haphazard plans, inadequate funds, lack of implementing capacity and mismanagement leading to various scams. The State Transport Department (and the State Government Inland Water Transport Directorate) have failed to play an effective part in building up a proper infrastructure for sustainable development in the Brahmaputra basin. The Government has also miserably failed to have riverside plans and Brahmaputra riverside development schemes for regenerating the environments and attracting tourists like the London Rivers Association and the Cardiff Bay Development Corporation in the UK. Such schemes would not only make the places beautiful but also protect the river embankment, create promenades, and riverside seating, and help in improving ghats and jetties and the maintenance of heritage structures. We may also have walkways, ornamental gardens, fountains, etc. which could generate even more revenues to the state with private sector participation.

The Economic Survey of Assam: 1997-98 is the latest report published by the State of Assam. The report mentions about Directorate of Inland Water Transport Department and the Central Inland Water Transport Corporation opening services in the basin on a commercial basis. The State Inland Water Transport Department claims that in 1995-96 it handled about 49,000 tonnes of cargo and 12,000 thousand passengers with 7 light vessels, 21 modern vessels, 20 mar boats, 69 single boats, 10 pantoons and 6 relief boats. The CIWTC's data are not available in view of its very poor service at the moment. The State Government-published data also indicates 78 items of inward and outward traffic handled by river transport in recent years. These items are rice, gram, pulses, wheat, maize, oil seeds, raw cotton, jute, fruit and vegetables, cattle---sheep, goat, animals of other sorts, raw skin, hide, bone, coal and coke, lime and limestone, manganese ore, iron ore, mineral oil, teak wood, kerosene, timber, sugar, gur, salt, tea, ground nut, mustard and coconut oils, gunny bags, cloth, cement, brick, tiles, iron and steel, machinery equipment (especially for the new refinery at Noomligarh), metal products, hemp, rubber, shoes, lac and shellac, tobacco raw, etc.. But the traffic figures are not rising at the expected rate owing to poor services. The government reports mention about 62 ferries, of which 40 are in the Brahmaputra basin for about the same number of services in various ghats (river crossing points). But modern vessels are few and most services are run by mar boats or single boats which are unsafe and have less carrying capacity. Some of the important points, such as Sadiya, Sissi, Neamati, Laokhowa (Tazpur), Rajaduar, Sualkuchi, Amingaon, and Fakirganj (Dhubri), need modern vessels as transport needs are about 4 to 5 times higher than the existing river transport capacity at the close of the 20th century. The ferry ghats must be improved for better and safer services with strict time schedules. Moreover, taking account of the peculiar topography of the Brahmaputra basin it is high time that we think of introducing hovercraft as water taxi services, as I had a pleasant experience while crossing the English channel from England to France by such means. Hovercraft have several advantages, viz., reduction of congestion on city roads and speed (covering 60 km say in 60 to 80 minutes), promotion of tourism, etc.. Moreover a bigger Hovercraft will also not cost as much compared to other vehicles. The operating cost is also less, as it does not require a jetty or berthing facility. A smooth landing place is sufficient for parking and its operation is not affected by the depth of water. It can ply over flood waters, marshy lands and dry river beds, saving transit time and anxiety. In the 21st century hovercraft have a tremendous future for creating a stable transport infrastructure in the Brahmaputra basin.

## 27.8 Road Transport Infrastucture

The development of roads and road transport services are also, to a great extent, the result of the efforts of the tea industry in the Brahmaputra basin. The Assam Trunk Road came into existence in 1866 (more than 134 years ago) but made some progress only after 1880 with the establishment of the Local Boards. However, we find that after partition in 1947 Assam had a very low road kilometreage compared to other states. Many villages in the basin were away from good roads and the length of roadways varied from region to region. The development of private and public road transport speeded up with the opening of the Brahmaputra bridge (first of its kind on this river) in the 1960s. Although the State and National Highways have touched many places in the

Brahmaputra basin the length and quality is very poor compared to those in other basins in other states in India. The private road transport companies, handling both inter-state as well as intra-state traffic, have been playing a more conspicuous role than the public road transport infrastructure in this basin. The Assam Road Transport Corporation (ASTC), a state public undertaking, handles mainly intra-state traffic, and now its position is very pitiful with a very few buses operating in a few areas of the basin with very low quality service and management, coupled with a high level of corruption with political persons as Heads of the Department and the Organisation. The Central Road Transport Corporation, a Central public undertaking, has also failed to contribute to creating a proper infra-structure owing to a number of reasons, including the lack of political will and mismanagement leading to corruption at various levels.

Assam has only 8572 km of surfaced roads out of a total length of 33,110 km, and of this length two thirds (66 percent) are in the Brahmaputra basin. There are wide regional/district variations in length and quality, marked by missing links, weak and incomplete bridges, etc. The condition of the interior's village roads is beyond description, and even bullock carts find them difficult to tread on at various times. At the same time, the total number of motor vehicles is increasing year by year, reaching a mark of about 5 lakhs by 2000 AD. Motor cycles and scooters are increasing at a high rate, followed by public carriers and private cars. In view of the increasing number of trucks, truck traffic congestion and pollution have reached a dangerous level in many places of the basin. Another noticeable point is that there are fewer registered vehicles in Assam, indicating two things: firstly, there is a lot of bridge traffic coming from outside of Assam and passing through the basin to go to neighboring northeastern states; and secondly, the Assam State is a loser because the taxes of these vehicles are paid in the originating states and Assam as such does not have much revenue to spend on road repair and maintenance owing to heavy bridge traffic, unless some funds come from the central government.

The Brahmaputra basin could have had an excellent road transport infrastructure with 8130 km of route kilometreage of the Assam State Road Transport Corporation, which has a fleet of 731 (including 641 buses). But this corporation does not operate trucks to earn from the huge freight traffic, and even out of 641 buses only about 50 are operating with any semblance of a standard service and efficiency. As such the public has lost confidence in it and most of the passenger traffic for interstate and interstate routes is diverted either to private bus transport or to nearby state government corporation buses. In particular the West Bengal State Corporation buses have already captured sizeable traffic from the Brahmaputra basin even with old buses, as their services are better than those of the Assam Transport Corporation. Despite their old and unsafe buses with irregular and heavily congested (overloaded) passenger services, private bus companies are earning huge profits, taking advantage of the poor role played by the Assam State Government Transport Corporation buses. The consumers (passengers) are compelled to pay higher charges with very poor quality services in the interior areas of the Brahmaputra basin, and these are the areas where State buses should have created the right infrastructure, as they had resources and huge staff totaling about 6000.

Apart from poor passenger services by private and government road transport system we find a poor and inadequate goods transport system in the basin, as heavy trucks cannot move in view of weak bridges and worse roads (mostly kutcha or unmetalled). The mini-trucks moving inside the basin for import and export traffic charges freight rate as they will, and accidents are very common in view of the lack of strict government road rules and licencing system. Moreover, many vehicles are aged, creating a high level of pollution. The people have no option but to hire or use these, as there is a dearth of requisite number of carriers in view of the high transport needs in the feeder roads. The old bullock carts have now almost vanished and these are not preferred nowadays in view of longer transit times, except in compelling situations. Added to this is the problem of insurgents of various types who extort huge sums of money from private truck operators at various points in their trips. This leads to higher cost of transport apart from anxiety and uncertainty, as these risks are not insurable in the basin. It appears to me that the only solution for having a proper road transport infrastructure in a insurgent-ridden area is to start an efficient truck transport operation at a large scale through the Assam State Road Transport Corporation, in addition to improving the operation and management of this corporation for more and better passenger services in the 21$^{st}$ century.

The transport infrastructure of this basin is over-dependent on the National Highways as this basin has only 2 percent of the total National Highways of India but over this 2 percent length we find movement of about 40 percent of goods and passengers of the country. There is a need for more State Highways and rural roads, and these should be served by both state corporation buses and private buses in order to have a situation of market competition in the interest of consumers. There is a gap in the road transport sector to the tune of 100 percent, especially in the interior rural areas of the Brahmaputra basin.

## 27.9 Railways

In May 1882 the first metre gauge locomotive passed between Dibrugarh and Jaipur Road, and in 1883 the Jorhat Provincial Railway was opened with 2 ft. gauge to start railway history in the Brahmaputra basin. By 1885 the Dibru Sadiya Railway (metre gauge) came into being. But it was the opening of the Assam Bengal Railway in 1895 which connected the Brahmaputra basin with the outside world. The tea gardens also had many garden transways (18 inches to metee gauge) totaling about 330 miles, which served as very useful feeders to the Assam Bengal Railway and the river transport. By 1902 another railway---the Eastern Bengal Railway---came as far as Dhubri. After 18 years in 1920 the Chaparmukh Silghat Railway started its services. Then in 1942 the Assam Bengal Railway (1,133 miles) was amalgamated with the Eastern Bengal Railway to form the Bengal and Assam Railway, and it has a history of carrying refugee traffic, 3,000 per day, coming from Burma. We also find that in 1883 the Jorhat Provincial Railway (40 kms.) was opened as a state enterprise with a 2 ft. gauge system. Similarly in 1894 we had the Tezpur-Balipara Railway (2 ft. 6 inches gauge) just to serve the tea gardens.

A close analysis indicates that the railways in the early period of development were nothing more than feeders to the river transport, which played the dominant part in both the export and import trade in the Brahmaputra basin. In 1904 the railways were also linked to the port of Chittagong to have an additional port like Calcutta for export of tea. The railway facilitated the development of the area traversed by them. At first the Dibru Railway handled more traffic, but later on the Assam Bengal Railway became the dominant carrier in the basin. By the middle of the 1940s the railways invested the largest amounts of capital as compared to other transport services, viz., steamer and road.

After partition the railway link to and from Assam was cut off and its capacity was also limited for various reasons. The real development of railway infrastructure started after 1962 with the completion of the Brahmaputra bridge and the extension of the railway line to the farther northeast near Arunachal Pradesh, connecting vast isolated virgin areas. 1965 and 1966 were notable years from the point of new railway extension, as 220 km were opened to traffic during this time.

To the railway engineer the Brahmaputra is a masculine river, as it not only splits Assam along its length but also cuts off seven sister states (of Northeast India) from the whole country. As it flows in the most unpredictable manner, the 1.30 km. long first bridge over the Brahmaputra was a challenge to railway engineers. But the basin was able to have a second road bridge (3 km. long) near Tezpur in 1987. A third bridge over the Brahmaputra was also completed in April 1998 (for a 2.30 km rail cum road line) at Jogighopa as a part of the latest line connecting Guwahati with Jogighopa. The total cost of the bridge was Rs 300 crores, and today it connectes NH No. 31 B on the north bank and NH No. 37 on the south bank to provide an alternative link to Guwahati via South bank.

Today the total length of railway route in Assam is about 2,435 kms (out of which about 531 kms are under broad gauge). The rail transport infrastructure varies from place to place in the Brahmaputra basin. Many areas are still untouched by the railways and some important feeder/branch lines of British days are closed, despite their viability. The basin is also at a low scale in regards to broad gauge (only 22 per cent of the total length), electrification, modernization for safety and various other facilities for all types of traffic. Although some new rail services have been started from this basin to various cities of the country, the quality and quantity of services are questionable, mainly because of the style of management, which is neither customer or traffic friendly. The NF Railway is incurring losses owing to diversion of goods traffic to roadways in the basin mainly because of poor quality services. The railways are a natural long distance carrier for all traffic of this basin, and they have the potential market with ever expanding traffic of all types. But the railway infrastructure has not come up to mark to help all round development of the Brahmaputra basin. The rail--road coordination and the old agency system is totally forgotten here. The services are also irregular often because of insurgency and BANGHS (forced holidays for the public). The short distance road transport can like the interior areas of the basin with the railways to have a better infrastructure---as railways cannot be built everywhere in view of cost and topography. As such, the out agency system and rail--road coordination becomes urgent. But it is not yet planned by the railways or the state government.

As a result we do not have a proper rail infrastructure in the basin. The rail capacity is far below the needs for passenger and goods traffic in this basin. It is difficult to get suitable wagons in time even today. Mismanagement has also led to theft, damage, pilferage, and delay in transit in goods traffic. Moreover, many railway projects are progressing slowly, and timely completion is a big question. I made a calculation to find that the rail transport capacity is not commensurate with the requirements of both agricultural and industrial sector of the basin. Apart from increasing rail services with new lines, another most important task is the improvement of operational efficiency so that existing capacities may be optimally utilized, by avoiding restrictions, and starting uneconomic lines with innovative ideas, such as tram cars, quick disposal of claims, better use of line capacities, improving loads per train with better wagon usage, improving the net tonne kilometers per engine hour, assuring punctuality, avoiding accidents and starting better repair facilities in the local basin area, etc..

## 27.10 Air Transport

The Brahmaputra basin needed air services for urgent traffic, especially in emergency situations to save transit time. Even before 1949, i.e., immediately after the partition of the country, air companies carried traffic to the tune of 150,000 pounds per day from Calcutta to Guwahati, the most important airport in the basin. The main imports were consumer goods, medicines, machinery, and the main exports were orange, tea, supari, and tejpata. By 1954, the traffic increased to 750,000 pounds per day, about 5 times within 5 years. As imports to the Assam basin was greater than exports the bulk freight rates were higher for incoming traffic (almost double).

In 1959 the Government of India started air services with the Indian Airlines Corporation (IAC) for passenger and freight traffic. But it was found that the losses that incurred were about Rs 21 lakhs per year mainly owing to the competition between IAC and private operators which carried about 30 to 40 percent of the total freight in 1959. But the private operators rendered valuable services in difficult areas of the basin like Jorhat, Mohanbari, and North Lakhimpur.

Several areas of the basin are covered by airports like Guwahati, Dibrugarh, Tezpur, North Lakhimpur, Jorhat etc., and traffic carried has been increasing at a very high rate, with some new carriers in addition to IAC. But the cost is also increasing and the quality of services are often questioned by all concerned. There is a wide gap in air services as well. We need small aircraft to link all small towns of the Brahmaputra basin as early as possible. In such a case all parts of Assam can be reached within one hour. Moreover, old good airports, such as Rupshi airport (in Dhubri district), should be developed which would cost little to fetch the potential heavy traffic of the vast uncovered area with very bad alternate transport services.

Recent State Governments reports on air traffic indicate that our inward traffic is about three times more than the outward traffic in 1995-96, and our traffic with Calcutta is the heaviest. We need a quick action plan and implementation for better air services. It is a pity that the Borjhar airport (Guwahati) has been elevated to an international airport and Rs 150 crore had been earmarked in the Central budget about 3 years ago. But till today nothing is done to make it international. The runway has not been extended and till today there is no night landing facility in this Guwahati/Borjhar airport. The safety and security system is also much below national standards. Often the NRIs from Assam and foreign tourists coming and going from Assam do not get confirmed seats despite their best efforts months ahead. It is a pity that the airport also does not have night accommodation facilities for various kinds of needy passengers. It is unfortunate that traffic is vast but facilities of a reasonable standard are lacking.

## 27.11 Pipelines

During the Second World War, pipelines had a glorious record because petroleum was a major item in the movement demand in Assam's Brahmaputra basin area. Between 1943 and 1945 two American pipelines were laid down from Calcutta to China following rail and road routes. These projects were notable and great but little has been written by anyone about these projects so far. Pipelines have reduced the strain on other modes of transport and have stopped large scale empty movement of rail tankers apart from being a least costly mode of transporting crude oil. In fact, pipeline is competitive with inland water transport as well for bulk movement.

In 1964 we had the major pipeline for transporting crude oil from the Naharkatiya and Moran oil fields (which produced about 3 million tonnes crude oil annually) to refineries at Gauhati and Barauni. This 1,157 km pipeline was constructed by Oil India Ltd. to transport 2.75 million tonnes crude oil per year---that is 7,550 tonnes a day. The other two crude pipelines are Naharkatiya--Digboi and Lakua--Moran. Amongst the two PRODUCT pipelines, one is Digboi--Tinsukia and the other is Gauhati--Siliguri (420 km) commissioned in 1964 to transport white products, such as motor spirit, superior kerosene, high speed diesel, light diesel oil, etc., of the refinery. There are also pipelines to transport natural gas from Naharkatiya to Namrup and other new areas owned by the Assam Gas Company, a state public undertaking of Assam. The several advantages noted above indicate the importance of this mode of transport in the basin.

There is further scope for more pipelines in the Brahmaputra basin if we can plan to supply gas to domestic consumers for cooking purposes immediately as done in Delhi, Bombay, and other places. We have the natural gas and we have the ready market. What is needed is a plan and implementation as this will generate large scale and constant employment in the Brahmaputra basin in addition to supplying gas reliably to consumers to minimize anxieties, uncertainties, and accidents from gas cylinders. Many gas cylinders contain spurious materials owing to large-scale corruption. In my mind this should be a priority scheme because we have been wasting gas every day by simple burning, losing huge amounts of money for several years.

## 27.12 Ropeways

It is sad to note that we have dismantled the Goalpara Tramway (a heritage item) near Kachugaon in the present Kokrajhar district. It was built in 1901 and its total length increased to 20 miles. The idea was to move timber from Goalpara forests to EB Railway. By 1934 its length increased to 72 km. Similarly we find the Singri-Panchnoi River Tramway in Darrang district, which was managed by steamer companies.

These short tramways were in fact ready alternatively to ropeways because ropeways were not constructed quickly, despite their several advantages. In West Bengal and Bihar we see more than 28 ropeways with an installed capacity of 2,500 tonnes per hour. But till today Assam has no ropeway although there are strong reasons for construction. It was calculated that the cost of transportation goes down by about 80 percent in the case of ropeways in difficult areas where roads, railways, etc. cannot be built. Construction of ropeways would also generate employment, in addition to helping a safe and quicker movement of high value perishable items. The ropeways can also be linked with the bus and railway stations in selected places to provide an integrated and coordinated transport infrastructure.

## 27.13 Concluding Remarks

Let me pinpoint the following facts here after a quick study of the transport infrastructure scenario of the Brahmaputra basin.

- The area of the Brahmaputra basin has remained almost constant. In fact, there is degradation of soil, reduction of fertility, and increase in environmental pollution in the area with the growth of time. But there is a rapid increase in pollution resulting in heavy demand for both interstate as well as interstate transport facilities. Almost two-thirds of the entire population of Assam is now concentrating on this basin for various reasons, economic, social, and cultural etc..

- Most of the additional traffic is now easily diverted to road transport, causing a huge pressure on roads and inadequate road transport system. This has also led to huge ecological imbalance, as already noted in the entire basin. New roads are being constructed at the cost of paddy fields and valuable forests in various areas of the basin.

- Our short and long term policy should be to divert traffic to river transport as far as possible. The river transport infrastructure needs urgent attention for future development. Owing to the rise of river beds many areas of the basin will be under floods in the next few years and road and rail transport will not be able to save the situation. The quality of roads is already very bad. The cost of repair will be immense. The breached railway line to Dhemaji is still unrepaired after the last 9 months or so. It is only in name, the river transport development will be comparatively cheaper. There will also be less environmental pollution. Hovercraft have a great future in this basin.

- There is a need for better quality services in all existing modes of transport. Facility and quality are two important words for transport infrastructure in this basin.

- The State Government has failed to give due importance to river and road transport. The Assam State Government Corporation and the Inland Water

Transport Department must be managed efficiently with new business concepts with proper management, as huge resources are lying idle in these two organisations. Many of the 100 government bus stations have a deserted look today in the entire Brahmaputra basin. These two organisations have become SICK and a revitalisation scheme has become urgent along with a serious implementation and action plan with right persons at the top. Politician heads as chairmen would not produce any results, as we see from Assam's recent history of the last 50 years.

- Mere privatization will not help. We need competition between both private and public (government) enterprises in transport infrastructure. If the state cannot run certain areas, these should be leased out to good private parties under certain conditions.

- All modes of transport should be properly integrated to have an integrated system with proper coordination to avoid wastages.

- Incomplete projects under plans must be completed in time and there must be concurrent evaluation of all on-going transport projects.

- Banks and private financial institutions must come forward for investment funds in the transport sector, so that private parties may take up new innovative projects like ropeways, hovercraft, etc.

- Laws/Acts and Rules relating to all modes of transport must be modified or ammended to suit the needs of the present traffic for sustainable development in the 21st Century. For this we need a very strong political will.

- Lastly, there must be a political will to put the right person in right job to run/head the transport organisations as these cannot be managed optimally by politicians with no knowledge or expertise.

The Brahmaputra basin has a great potential and it must be developed in a right manner in a right time so that it becomes a leading growth area for the entire north east India and thus inspire the entire country.

## References

Government of Assam. Dispur-Various Five Year Plans.

Government of Assam (1997). Various Published Reports, viz., Statistical Handbook of Assam, 1997 (latest).

Mehdi, S.B. (1978). "Transport System and Economic Development in Assam", (424 pages) : Publication Board, Guwahati, Assam, India.

# CHAPTER 28

# FLOOD MANAGEMENT

A.K. MITRA

Flooding in the Brahmaputra basin is an annually occurring problem. Thus far, only limited success has been achieved in mitigating flood disasters and the ensuing damage. This chapter discusses flood management using planned measures.

## 28.1 Introduction

Floods and their disastrous impact on the plains adjoining the rivers and on the drainage channels have vexed civilisation since time immemorial and constitutes one of the major natural calamities faced by India almost every year resulting in substantial loss of life, large scale damage to property, disruption of community lifelines, besides entailing untold misery to the millions.

The North Eastern Region of India comprising the states of Assam, Arunachal Pradesh, Meghalaya, Manipur, Mizoram, Tripura, and Nagaland is a huge sprawling land mass made up of extensive countless hills and mountainous terrain that rises in the north to the snow-capped heights of the Himalaya, and is the playground of the mighty river Brahmaputra regarded as one of the largest rivers of the world and its tributaries. This region accounts for nearly 40% of the total water resources of the country and this bounty of natural resources is still unharnessed. The tentative assessment of this dynamic resource in the North Eastern part of India is about 60 million hectare meters. The flood scenario of the Assam state and the ravages brought by the recurrent floods cannot be expressed in words. The recurrent and chronic flood havoc created by the mighty Brahmaputra and Borak rivers is the major deterrent to the development of this region. For Assam the Brahmaputra and Borak rivers represent the very essence of the life of its people as rivers which contain so much of promise and which can yield such rich dividends if only we can properly plan and utilise its enormous potentials.

Complete immunity from flood or absolute 'Flood Control' is utopian in concept and economically not viable. The concept of flood management aims at taking such planned measures as will ensure profitable and economic utilisation of the flood plains and the water resources for the benefit of the mankind, while simultaneously ensuring that during the periods of high flood there is no severe damage to the extent possible.



*V.P. Singh et al. (eds.), The Brahmaputra Basin Water Resources*, 535-559.

## 28.2 The Brahmaputra River System

The Brahmaputra River characterized by high seasonal variability in flow, sediment transport and configuration is a unique river system of the world that drains such diverse environments as the cold, dry plateau in Tibet formed of highly erodible deposited soil, the steep rain- drenched slopes of the Himalayas, the landlocked alluvial plains in Assam and vast deltaic flat in Bangladesh. The Brahmaputra River originates in a great glacier mass in the Kailash range of Himalayas south of lake Gunkyud in South West Tibet at an elevation of 5,150 m. The Brahmaputra River, known as the 'Tsangpo' in Tibet, the Siang or Dihang in Arunachal Pradesh, and the Jamuna in Bangladesh, is one of the biggest rivers of the world (see Table 28.1). The Brahmaputra basin extends over an area of 580,000 sq. km up to its confluence with the Ganga in Bangladesh.

Out of the drainage area of 580,000 sq. km, 293,000 sq. km lie in Tibet (China) and 2,40,000 sq. km in India and Bhutan, and 47,000 sq. km. in Bangladesh. The Brahmaputra River traverses 1,685 km in Tibet, 918 km in India, and 337 km in Bangladesh. It is a braided river in its entire length in Assam (Figure 28.1). Its bed and bank are built of fine alluvial soil with rock outcrops at several places. The river flows through alluvial plains from Pasighat to Dhubri. During its journey through the alluvium plains it changes its channel courses and forms large sand islands. Whilst in the dry season it breaks into numerous channels of various magnitudes, during rainy season it becomes a sheet of water up to a width of 15 km submerging all sand islands almost all throughout its length from Pasighat to Dhubri (Figure 28.2).

The width of the Brahmaputra Valley is confined between the foothills of the Himalayas and the Borak range to only 80--90 km. Natural vegetation in the Brahmaputra basin varies with altitude from topical evergreen and mixed deciduous forest within the valley and foothills, to alpine meadows and steppes in the higher ranges and in Tibet, about 209 km of the Brahmaputra valley is forested.

The Brahmaputra valley in Assam is underlain by recent alluvium approximately 200 m--300 m thick, consisting of clay, silt, sand, and pebbles. Within this alluvium range from Sadia to Dhubri there are about 40 tributaries on its north bank and 20 tributaries on its south bank. The prominent tributaries on the north bank are the Subansiri, Jiya-Bharali, Bornadi, Puthimari, Pagladia, Beki, Mans, etc., and some on the south bank are Burhi Dihing, Dikhow, Disang, Dhansiri, Korili, etc..

The catchment of the Brahmaputra River, excluding the Tibetan portion, forms an integral part of the monsoonal regime of south east Asia. The rainfall averages 2,300 mm annually with a variability of 15% to 20% (Figure 28.3). There is a marked spatial variation in the distribution of precipitation over the catchment. The Himalayan sector receives 5,000 mm of rainfall per year. About 85% of the total rainfall takes place during the months of May to September, and as a result during that period the Brahmaputra remains persistently in its high stage. This persistent high stage of the Brahmaputra along with 4--5 flood waves during the monsoon is the main cause of flooding and drainage congestion of the Valley (Figure 28.4).

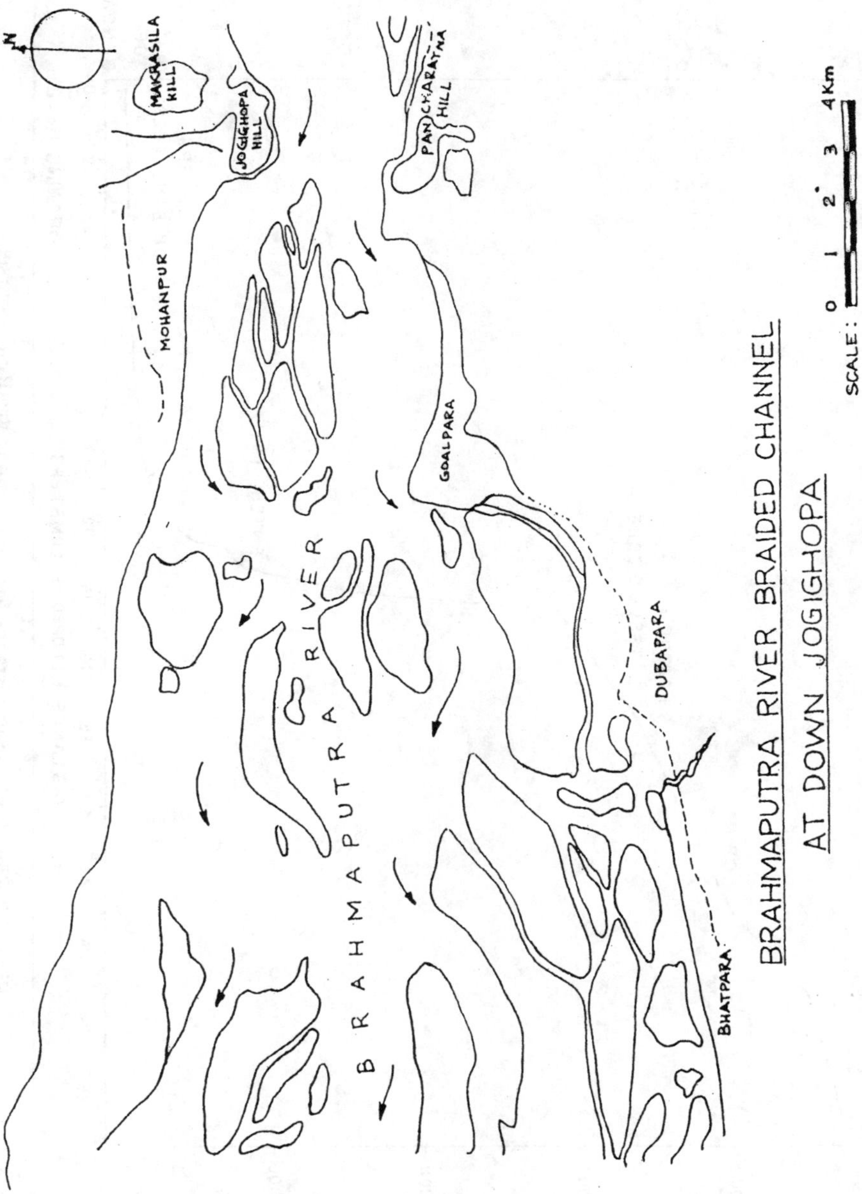


Fig. 28.1 The Brahmaputra River as a braided channel down Jogighopa.

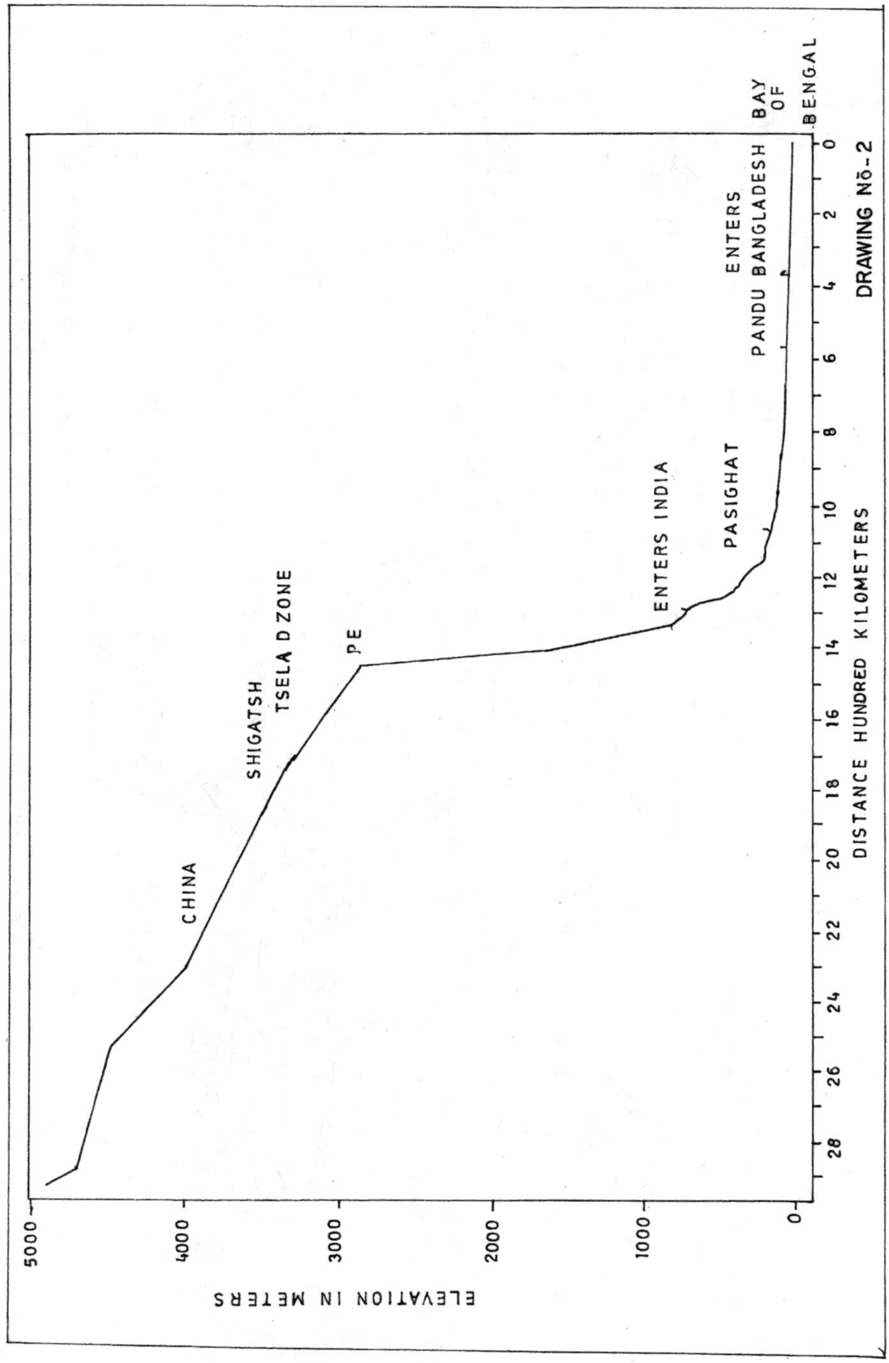


Fig. 28.2 Elevation versus distance for the Brahmaputra River course.

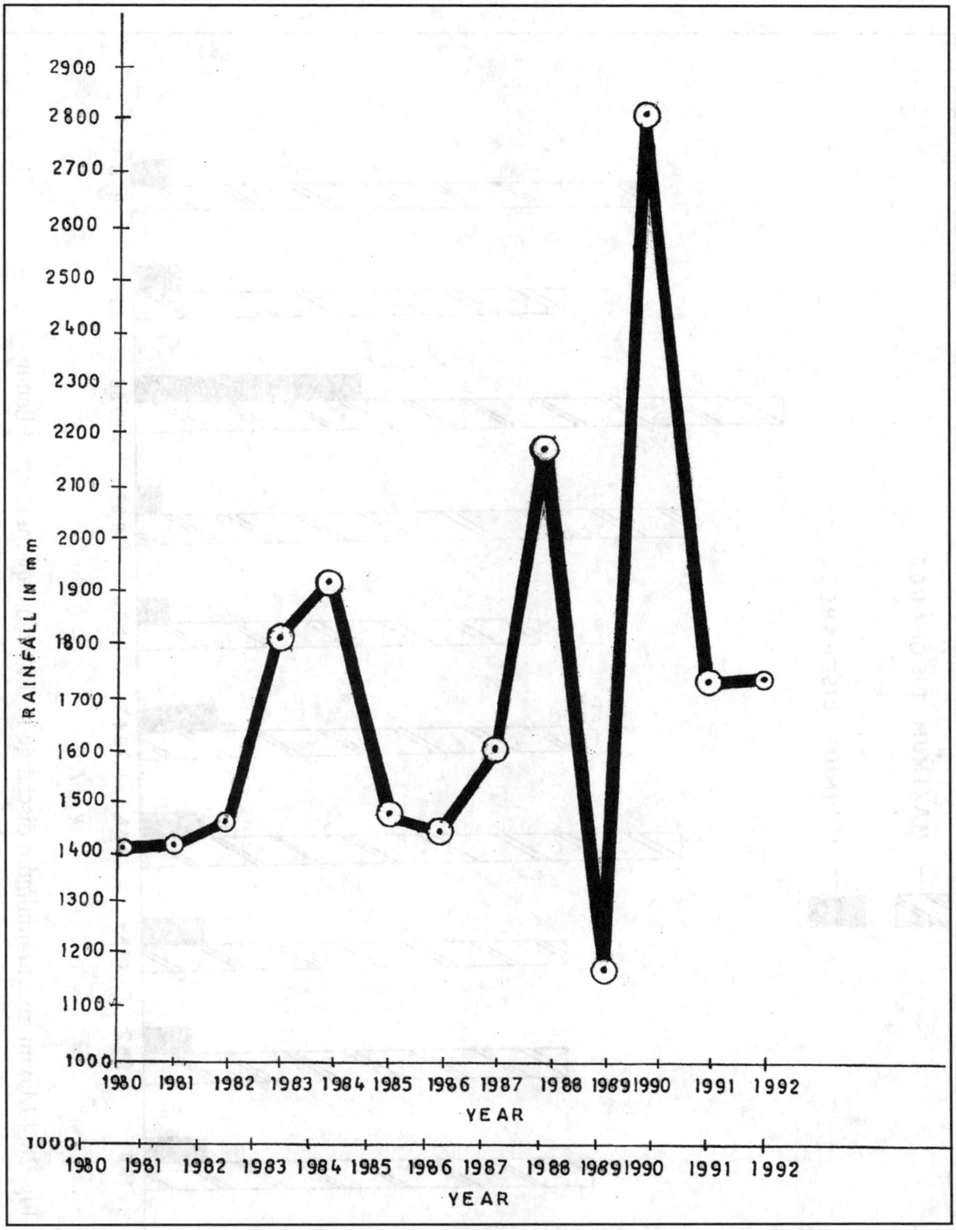


Fig. 28.3 Annual rainfall of the Guwahati area in Assam.

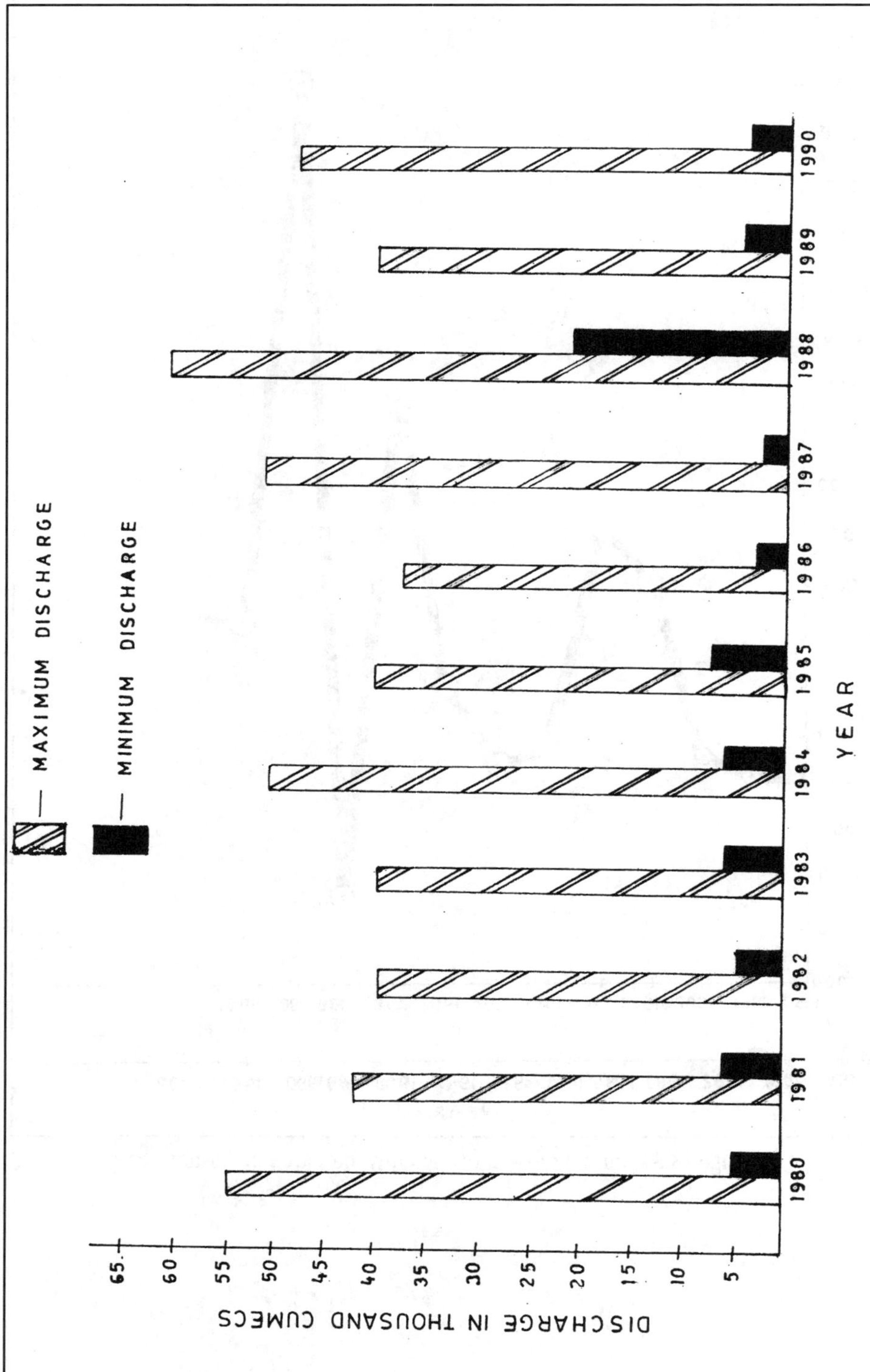


Fig. 28.4 Maximum and minimum discharge of the Brahmaputra River at Pandu.

Table 28.1 Nodal points on the Brahmaputra.

| S.No. | Site | Width at the nodal point | Width upstream of nodal point | Width downstream of the nodal point |
|---|---|---|---|---|
| 1 | 2 | 3 | 4 | 5 |
| 1. | Murkong selek (627) | 4.8 | 11.7 | 9.6 |
| 2. | Near Disangmukh (522) | 5.10 | 11.0 | 10.0 |
| 3. | Downstream of Jhanzimukh(495) | 3.75 | 8.6 | 8.4 |
| 4. | Downstream of Dhansiri North (425) | 4.0 | 9.6 | 7.5 |
| 5. | Downstream of Dhasirimukh at Gamiri(420) | 4.4 | 11.0 | 5.2 |
| 6. | Upstream of Tezpur (335) | 3.6 | 5.5 | 12.3 |
| 7. | At Pandu near Guwahati (198) | 1.2 | 11.0 | 7.5 |
| 8. | Pancharatna (85) | 2.4 | 8.0 | 10.5 |

## 28.3 River Morphology and Configuration

The morphology and behaviour of the Brahmaputra River in Assam undergoes drastic changes in response to variations in the flow regime and pattern of sediment transport and deposition in the river following the seasonal rhythm of the monsoon and the freeze, thaw, etc. of the Himalayan snow. The Brahmaputra provides a classic example of a braided river with multiple channels twinning around numerous mid-channels as well as lateral sand bars that are locally known as CHARS. The bankline of the Brahmaputra for the most part is extremely unstable. Besides, its wide alluvial channel is marked by continuous shift of the thaw, and the drastic transportation in the geometry and location of sand bars. Shear failures in the upper bank materials, that seem to be by far the most widespread made by bank failures in the case of the Brahmaputra, are caused either by undercutting of the upper bank materials by currents during high flows producing an over-hanging cantilevered block which eventually fails or by over-steepening of the bank materials owing to migration of the thalweg closure to the bank during falling stages of the river. High moisture content, low proportions of clay, and good sorting of the bank materials make them highly susceptible to erosion by the river. Within the 720 km long Assam section of the river there are several stretches, where the rate of bankline migration is extremely high and the change in channel configuration too drastic. In such actively eroding sections the rate of bankline recession may be of the order of several hundred meters in the courses of a single year, creating thereby an extremely serious hazard for scores of people living in the densely populated and intensely cultivated adjoining flood plains.

The catchment of the Brahmaputra River, being in a seismic zone, the great earthquake of 1950 seriously disturbed the whole river system. Since then the suspended load of the river had increased appreciably. Apart from its own sediment load (average annual yield 800 million ton) it is also fed by the sediment yield from its tributaries. Under these circumstances the discharge of the river loses its dynamic equilibrium and causes erosion all along its bank. Thus the main factors of erosion are excess sediment charge

traversing its valley in a series of deep and narrow threat followed by broad and shallow aggrading reaches. There are several nodal points all along its course where the velocity is increased and deep scouring takes place, if the upstream and downstream of the nodal points where it discharges its silt load causing severe erosion. Brief details of these constrictions or nodal points are furnished in Table 28.1.

The study group set up by Government of India in 1964 to assess the problem of erosion of the Brahmaputra after a detailed study concluded that between Kobo and Dhubri the left bank has undergone erosion over a length of about 350 km, whereas on the right bank the same has occurred over a length of 230 km.

The Brahmaputra is one of the most sediment-charged rivers of the world. Amongst the largest rivers of the world, it is second only to the Yellow River in China in the amount of sediment transported per unit of drainage area. The volume of sediment load (suspended) carried by Brahmaputra, based on measurement at Pandu, varies from 36,609 ha m (1958) to 1487 ha m (1973) per year.

**28.4 Flooding and Major Flood Events**

The problem of flooding is endemic in many parts of India. Nowhere is the problem more acute and the possible solutions more complex than in the Brahmaputra Valley. On average, flooding affects an area of 1 million hectares in Assam compared to 9 million hectares in the country as a whole every year.

The main reasons for frequent floods in the North Eastern Region are the narrowness of the valleys, the high rainfall, the heavy encroachment of the flood plains and the increasing pressure of population. The seismic activity in this part of the country is a continuous process. Earthquakes of magnitude 5 or thereabout on the Richter scale are quite frequent. The severe earthquake of 1897 (8.7 Richter scales) and 1950 (8.6 Richter scale) had seriously disturbed the drainage of the Brahmaputra valley. The acutely braided nature of the Brahmaputra, coupled with soil and sand strata of its banks, competitively steeper slopes and high sediment load, etc., is the main cause of its excessive erosive activity in the valley. The unstable hill slopes with frequent seismic activity coupled with the practice of Jhuming cultivation in some of the upper catchment area is responsible for the high sediment load in the rivers/streams and the consequent acute erosion problems in most of the reaches of Brahmaputra and Borak.

During the last few years increasing flood problems and consequent flood damage have created havoc in both valleys and have put the people at large to untold miseries. The floods of 1987, 1988, 1992 and 1995 have been particularly severe in the Brahmaputra Valley. The entire embankment system was put to maximum strain resulting in 102 breaches and cuts in 1987 and 220 in 1988. Acute bank erosion has continued unabated.

A tentative assessment of flood damage in Assam during the decades 1953 to 1995 tabulated below reflects an extremely gloomy picture of the state, and such colossal damage caused by recurrent floods in Assam have throttled the process of basic agrarian economic development for years on end.

Table 28.2 Tentative assessment of flood damage of Assam during 1953 to 1995.

| | | Total | Maximum |
|---|---|---|---|
| (i) | Area affected in million hectares | 41.66 | 3.82 (1988) |
| (ii) | Population affected in million (Nos) | 98.10 | 10.47 (1987) |
| (iii) | Damage to crops in million hectares | 5.08 | 1.13 (1988) |
| | value in term of rupees | 3,288.31 | 334.10 (1988) |
| (iv) | Damage to houses | | |
| | (a) Total Nos. | 3327,189 | 4,998 (1988) |
| | (b) Value in terms of rupees | 296.80 | 103.92(1988) |
| (v) | Cattle lost | 431,537 | 108,913 (1987) |
| (vi) | Human lives lost | 1,724 | 226 (1988) |
| (vii) | Damage to public utilities in rupees (crores) | 832.42 | 225.82 (1988) |
| (viii) | Total damage to crops, houses and public utilities in terms of million rupees | 44,175.30 | 6,638.40 |

Most of the valley areas of the rivers in the north eastern region are heavily encroached upon by the increasing population, resulting in flood problems of a continuous nature. Natural occurrences, like floods with heavy rainfall in any region, cannot be effectively controlled but the damage on account of them can certainly be reduced. Flood damage is a complex phenomenon being the result of socio-economic activity of the human beings in the flood plains of a river. Since completely staying away from the flood plains may neither be possible nor desirable, mankind should be willing to regulate its developmental activities in the flood plains so that the flood damage could be kept to a minimum. It is necessary to recognize that by adoption of appropriate flood management practices, both structural and non-structural, flood hazard and damage could be minimised. Therefore, the northeastern states have practised mainly the structural methods of flood management within the constraints of the availability of the resources during the different plan periods. In Assam, the following structural methods have so far been executed which could afford a reasonable protection to 16.00 lakh hectares of flood prone areas out of a total flood prone area of 31.50 lakh hectares.

(i) Embankment .................................. 4,500 Kms.
(ii) Anti-erosion works.......................... 600 Nos.
(iii) Drainage channel ........................... 850 Nos.
(iv) Major sluices ................................. 85 Nos.

With the launching of the National Policy on Flood Control in 1954, the tempo of construction of embankments had increased rapidly, and with the result that the length of embankments increased from 6000 km in 1954 to 15,675 km in 1990 besides improvement of 30,857 Km of drainage channel. In both the Brahmaputra and Borak valleys, the construction of embankments has been used primarily as the first and only step of effective flood management so far, and there is hardly any scope of further construction of new embankments. Since 1955 the Brahmaputra River had experienced seven major floods, discharge of which ranges from 55,092 cumecs to 72,800 cumecs.

The years corresponding to the major flood events along with corresponding discharges at Pandu (near Guwahati) are given in Table 28.3.

Table 28.3 Floods at Pandu.

| Year | Discharge (cumecs) |
|---|---|
| 1957 | 57,733 |
| 1958 | 61,320 |
| 1960 | 57,985 |
| 1962 | 72,800 |
| 1966 | 57,570 |
| 1980 | 55,092 |
| 1988 | 61,015 |

( Historic Floods in the World )

The disastrous floods of 1954 and the tragedy and sufferings that followed in the wake of these floods prompted the need for tackling the flood problem on the basis of national emergency. The Union Minister for Planning and Irrigation and Power placed before parliament the reports "Floods India-Problems and Remedies" and "The floods in the Country". These statements brought out for the first time the national policy regarding the problems of floods and the remedies in a comprehensive manner.

## 28.5 Measures Taken in the Field of Management

The outline of the programme and the nature of the remedial measures as summarised in the statements of 3rd September 1954 were to undertake a programme divided into three phases, namely, (a) immediate, (b) short term, and (c) long term. The immediate phase, extending over a period of two years, was to cover immediate works for the protection of towns, and also construction of some embankments in the most vulnerable reaches. The second phase, which was to coincide with the commencement of the second plan, envisages flood control measures, such as construction of embankments, channel improvements, raising of villages, and protection of towns. In the long term phase further works, which included creation of storage reservoirs, were contemplated to stabilise the benefits of the works constructed earlier and also to give reasonable protection to more areas.

In order to implement the above programme it was felt necessary to set up a suitable machinery both at the state and central levels. The Central Flood Control Board was formed in 1954 and was entrusted the following responsibilities:

(1) to lay down principles and policies in connection with floods and flood control measures;

(2) to consider and approve the master plans for flood control which may be submitted by the state/river commissions;

(3) to consider and approve specific schemes of flood control works

which may be submitted to them for approval; and

(4) to arrange for necessary assistance in connection with investigations, planning, and execution of flood control works.

The State Flood Control Boards, headed by the respective Chief Ministers, were also constituted and their responsibilities were:

(1) to approve the long range plans and specific schemes for flood control measures and their priorities;

(2) to assess the flood problem in the State to deal with policy matters concerning floods, flood relief, and measures for flood control and protection;

(3) to consider and approve the arrangements and programme for carrying out surveys, investigation and collection of data for flood control works;

(4) to watch the progress of implementation of the approved schemes and to issue suitable directives to the department concerned;

(5) to approve measures proposed for dealing with any emergent situation; and

(6) to examine and approve proposals for an effective flood warning system/for the State,

**28.6 Achievements**

Since the enunciation of the National Flood Control Programme in 1954, the State Flood Control Board, assisted by its technical Advisory Committee initially guided the state in carrying out investigations, and in the formulation and implementation of flood control programmes, subsequently, considering the seriousness and complexity of the flood problem, the Brahmaputra Flood Control Commission was set up in 1971, for the preparation of a comprehensive plan of flood control and its implementation in a systematic manner. Based on the recommendations of the various committees, State Irrigation Ministers Conference, and experts appointed from time to time, the Brahmaputra Flood Control Commission has prepared an outline of a comprehensive plan for the Brahmaputra basin.

The types of flood management works so far adopted have been mainly embankments, construction of drainage channels, town protection works, and erosion control works. Opinions have differed from State to State and from time to time in the twenties and thirties of the 20th century regarding effectiveness of embankment as a flood control measure. Opinion in Assam, as in other flood prone states in the country, was against the construction of embankments, but gradually it veered round in its favour, particularly after 1954 when the National Flood Control Policy was formulated by the Government of India for a programme in three phases, viz., the immediate, the short, and the long term. During the last three decades, the Government of Assam has completed the following works as a part of a short-term measure of flood management.

BRAHMAPUTRA VALLEY:

| | | | |
|---|---|---|---|
| i) | Embankments | | |
| | a) | Brahmaputra: | 984 km |
| | b) | Tributaries: | 2,568 km |
| ii) | Antierosion Schemes: | | 421 Nos. |
| iii) | Drainage Channel: | | 600 km |
| iv) | Sluices: | | 56 Nos. |

BARAK VALLEY:

| | | |
|---|---|---|
| i) | Embankments | |
| a) | Barak: | 214 km |
| b) | Tributaries: | 524 km |
| ii) | Anti erosion Scheme: | 81 |
| iii) | Drainage Channel: | 247 km |
| iv) | Sluices : | 29 |

A review of the achievements since the launching of the 1954 programme, will lead to a conclusion that some of the programme of immediate, short, and long measures still remain to be completed. The initial programme of completing those measures in a period of 12 years was perhaps too ambitious. The difficult terrain of the Brahmaputra basin and insufficient hydrological and geological data made investigations of multi-purpose reservoirs both difficult and time consuming. Nevertheless, it cannot be denied that the flood control measures implemented so far have afforded a reasonable protection to an area of 16.35 lakh ha out of the total flood-prone area of 31.50 lakhs ha. But for the protection works, which were taken up in 1954--55, Dibrugarh town would have ceased to exist.

## 28.7 Bank Protection and River Training Works

It is observed that the magnitude of works of this category which are usually taken up to counteract the attack of the river on the banks or embankments for affording protection to the vulnerable areas liable to damage, is increasing almost in all the states during recent years. These works do not generally provide a long lasting remedy to the situation but usually transfer the problem to other reaches of the river and require periodical maintenance owing to recurring damage during the floods. As in the case of the embankments, maintenance of these works in damaged locations is again subject to availability of funds, and more often only priority works get the needed attention and that too, generally not to the desired level, in view of the inadequacy of allotted funds.

Erosion by the Brahmaputra and its tributaries has been causing concern to both the people and the administration. At some places a few kilometres of the bank, along with villages, habitations, fertile lands, and roads have been eroded by the river. The reaches of the Brahmaputra where erosion is acute at present are: downstream of Goalpara; Palashbari-- Gumi, Moriahola; and Nathala on the south bank and Mathmara, Majuli, Sonarighat, Tezpur, and Mukalmua on the north bank.

The Brahmaputra has braided channels in most of its traverse through the plains of Assam. Owing to excessive sediment charge there is constant shifting of these channels and formation of sand shoals. The flood discharge and sediment charge of its tributaries also influence the formation of these channels. This process of constant shifting of the river has been continuing through ages. The river has a tendency to shift laterally, as can be observed clearly from the changes in the outfalls of its various tributaries, particularly on the south bank. The south bank tributaries, near their outfalls normally flow parallel to the Brahmaputra. The lateral movement of the river in a southerly direction has the effect of shifting outfalls of these tributaries eastward. For instance Dibrugarh, as its name implies, was situated on the Dibru River. The confluence of the Dibru and the Brahmaputra used to be about 6.5 km downstream of the town in the later part of the last century. Now it is about 13 km upstream of the town. Besides this lateral movement, the river attacks its banks sporadically through erosion. There is a considerable loss of land each year through this process of erosion. The extent of loss owing to erosion in the valley varies from year to year, depending on the severity of floods.

The high sediment load which the Brahmaputra carries is mainly owing to the frequent seismic disturbances of low magnitude besides the disastrous earthquakes that occur from time to time in the North Eastern Region, which geologically is the most unstable part of India. One of the most disastrous earthquakes in history occurred on the 15th August, 1950, which caused widespread devastation throughout Upper Assam, particularly in the Arbor and Mishmi hills of Arunachal Pradesh, and in parts of Lakhimpur and Sibsagar districts of Assam. There were large scale landslides in these hills. These brought enormous quantities of sediment and debris into the rivers. The Government of Assam had taken up bank protection works at a number of affected sites in the last 30--35 years. The important works are at Dibrugarh, Jorhat (Kokilamukh), Tezpur, Guwahati, Soalkuchi, Palasbari--Gumi, Goalpara, and Dhubri. There are a large number of works of smaller sizes in the various tributaries of the Brahmaputra, on the Barak river and its tributaries. The works at Dibrugarh, Jorhat, Soalkuchi and Dhubri were done on the basis of model tests by the Central Water and Power Research Station, Pune (India). The work at the Palasbari--Gumi site was taken in the nature of a prototype for tests. On the basis of the evaluation study taken up on this prototype test, works on other important chronic sites would be planned. The Assam Government had also taken up a number of river training schemes with a combination of measures like stone bed bars, bamboo porcupines, permeable screens with timber bellies in the tributaries and obtained very good results. The type of works were also done in Pagladiya, Puthimari and many other tributaries of the Brahmaputra, and it has been noticed that the north bank tributaries respond very well to these measures. Even in side channels of the Brahmaputra with shallow depth these measures are satisfactory. The measures are inexpensive and do not shift the problem to other sites.

## 28.8 The Performance of the Palasbari--Gumi Anti-Erosion Scheme

The Palasbari--Gumi area, downstream of Guwahati, was being subjected to persistent erosive activity of the Brahmaputra for which the embankments had to be retired repeatedly in the past. After careful consideration an anti-erosion scheme for protection of Palasbari--Gumi area from erosion, prepared by the Government of Assam has been approved by the Government of India for Rs 12.94 crores (revised to Rs 15.30 crores). The major portion of the work was executed by the Government of Assam during 1985--86. The main work comprises 7 land spurs connected to the main embankment which tie the bunds. The scheme covers a reach of about 8 km protecting the Palasbari-Gumi area. The scheme was accepted and approved by Government of India as an experimental prototype study for utilising experience gained from the performance of this scheme in formulating an anti-erosion scheme for protecting other affected reaches by the Brahmaputra River. For studying the prototype behaviour of this scheme a monitoring Committee was also constituted by the Government of India.

The performance of the anti-erosion works, as executed under this scheme, was observed during the floods of 1986, 1987 and 1988. Whilst the overall performance of the anti-erosion works during the high flows could be termed satisfactory, the direct attack in the embankment zone between the spurs owing to inclined flows through the low patches between the island (which normally develop during the falling stages of the floods) threatens the base of the spurs and the connected/embankments tie bunds. The locations where these direct attacks will develop and threaten the embankment/spurs remain uncertain. Constant vigil, both during the flood and the lean season, is required to be kept to reduce the damage affected by this direct current and to protect the base of the spurs. Certain effective measures would have to be evolved to see that these direct attacks do not develop and threaten the spurs and the tie bunds. Some of the measures that could be thought of are silt inducing devices at the mouth of such channels, and anti-erosion in the form of full apron for protecting half the length of tie bund from the upstream armour zone of the downstream spur. These and other measures would have to be firmed up on the basis of the experience gained during the next few years for replication.

Apart from the progress made on the flood protection measures, flood forecasting has received considerable attention after the recommendations of the Ministers Committee of Floods and Flood Relief 1970 for its extension to vulnerable regions. Flood forecasting units have been set up at Guwahati and Dibrugarh with control rooms at Guwahati, Silchar, Nalgari, Goalpara, Barpeta Road, Dibrugarh, Jorhat, Pasighat, and Tezpur. These control rooms report the actual river levels as well as the forecasts alongwith daily rainfall covering the Brahmaputra and Diburgarh, Neamati, Guwahati, Goalpara, and Dhubri. The flood forecasting units also cover the following tributaries, the Pagladiya, the Dihang, the Subansiri, the Burhi Dihing, the Manas, the Beki, the Puthimari, the Kopili, and the Borak. In Assam there exist 23 of flood forecasting stations (Figure 28.5).

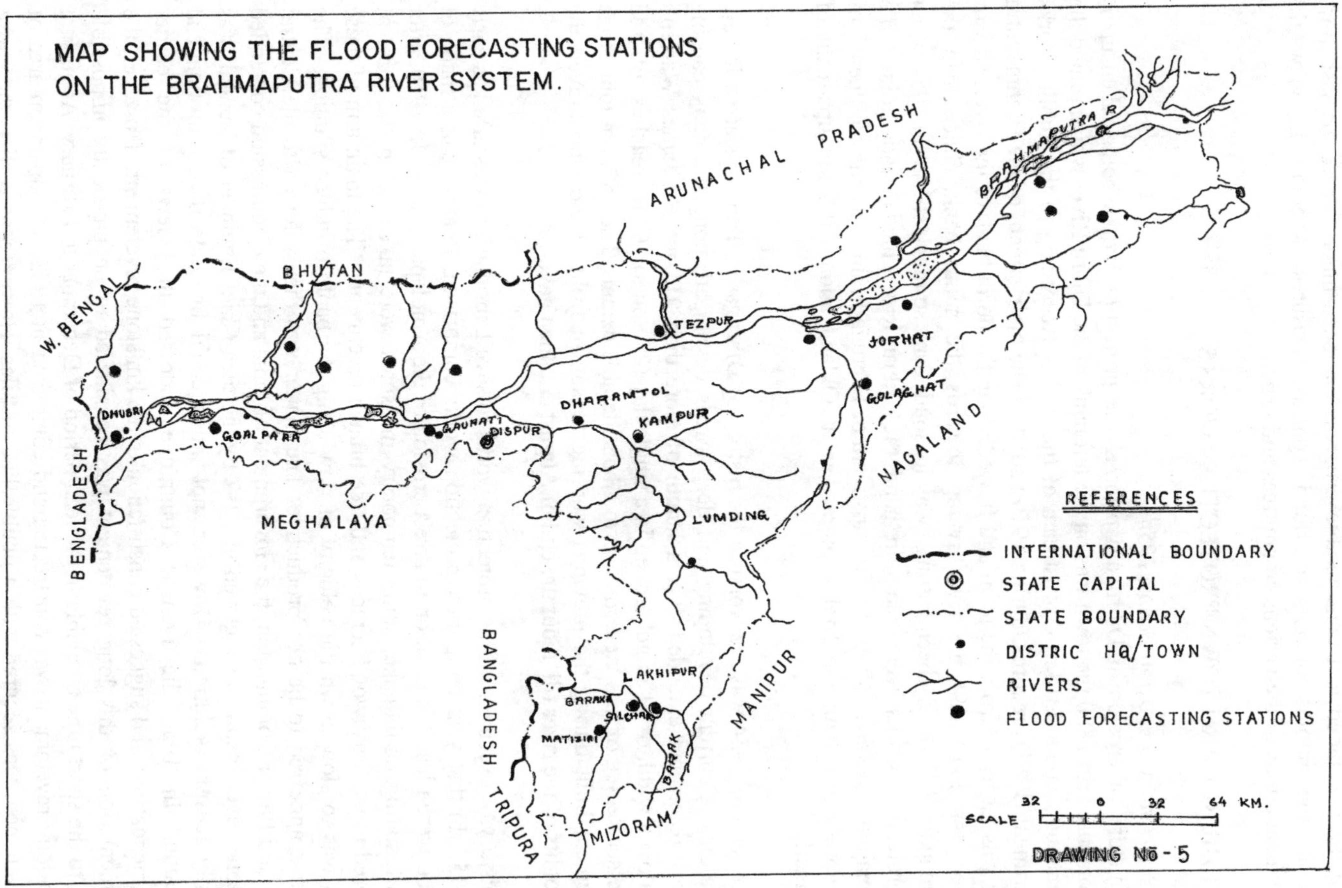


Fig. 28.5 Flood forecasting stations of the Brahmaputra River system.

It may be concluded that considerable progress has been made during the last thirty years in the implementation of flood control measures, keeping in view the recommendations of the various committees and experts.

## 28.8.1 DIBRUGARH TOWN PROTECTION WORKS

### *28.8.1.1 History o Erosion Before 1954*

The history of erosion at Dibrugarh prior to the protection works being taken up is phenomenal. In 1880 the town was on the left bank of the Dibru River which joined the Brahmaputra about 5.0 km downstream of the town. According to the Public Works Department (PWD) records, during 1910 and 1911 the main stream of the Brahmaputra River was far away from the town and flowed along the north bank of the river. It was in the year 1913 that the river swung towards the south bank. Since than the Brahmaputra has slowly been cutting away its south bank so much so that in 1954 when the protection works were taken in hand the Dibru River had already joined the Brahmaputra at Naijan about 4.0 km upstream of Dibrugarh. In 1957 the confluence of the Dibru River with the Brahmaputra was at Okland about 10.0 km upstream of Dibrugarh.

The erosion of 1954 was the worst in the history of Dibrugarh. Trouble started in July with a strong current developing and deep swirls developed along the bank causing severe erosion and extended over a length of more than 915 m. At James Warren's bungalow a width of 31 m got eroded in a single day (24 hours). Other reaches suffered to a lesser degree of loss, nevertheless to a considerable extent. The 457.3 m long stone revetement built during the previous working season in front of the club braved the onslaughts of the river till about the end of August and then gave way.

The history of execution of the protection works to save Dibrugarh town started as early as 1935. In that year temporary protection works comprising floating cages made of bamboo were placed in position to arrest erosion of the Amolapatty area. In 1950, after the devastating earthquake, some temporary measures were taken up in the form of anchored trees, brushwood screens and tree branch revetments. The main aim of these works was to slow down the velocity of flow along the bank to induce siltation. The measures appeared to be non-productive. The matter was then referred to the Central Water and Power Commission, the Government of India and the Commission concluded that stone revetement, as suggested in 1937 by the then Chief Engineer of Bengal, was the only feasible solution. However, not much could be done owing to financial constraints. In 1952 the Iyengar Committee endorsed the views of the earlier recommendations and suggested construction of a 6 km stone revetement. However, the implementation of this long revetement was considered to be physically impossible owing to the short time available in the lean period of December to February. As such, a 450 m long revetement was constructed and completed in 1953. The upstream area of the revetement was secured with a double line of piles. However, as earlier, the entire revetment was outflanked in September 1954 and erosion continued unabated. In 1954 the situation was so grave that the then Prime Minister, Pt. Jawaharlal Nehru, took personal interest in the matter and visited the site and the Central Water and Power Commission was entrusted with the planning and designing of the works. The Central

Water & Power Research Station (CWPRS), Pune, carried out model studies and recommended certain protection measures.

The major schemes have been further divided into three broad headings, viz., Original works, Extension of Downstream Protection Works--Phase I, and Extension of Downstream Protection works Phase--II.

*28.8.1.2 Original Works*

To derive a permanent solution model tests were conducted by CWPRS, Pune, and accordingly protection works were adopted to cover a stretch of 8 km. The work was taken in hand from the year 1954 and comprised the following:

(i) Seven stone spurs at an average interval of 760 m. Whilst the spur No.11 (kahai) was about 120 m long, spur No. VII was about 30 m long. All other spurs were 60 m long. The spurs were constructed projecting into the river from the bank. Furthermore to kill eddies between the stone spore, a series of timber/pile spore were also constructed;

(ii) In the most vulnerable stretch near the court building, a 1300 m long stone revetement was constructed;

(iii) To abate the spilling of flood water into the town, a 9.6 m long earthen dyke was constructed in the year 1955. The dyke that commenced from the high grounds west of the Naijan Bridge was taken along the river bank in the town area and ended in south Amolapatty. Two subsidiary dykes were constructed to protect the George Institution and Mohonghat;

(iv) With a view to giving relief to the town area from drainage congestion owing to the construction of the dyke, a 14 km long drain was constructed in 1956-57 through the town from Palton Bazar to Bogibeel, outfalling into the Brahmaputra River.

*28.8.1.3 Performance of Protection Works*

The protection works have stood well even though they had to bear the attack of the flood through successive monsoons. The maintenance works of replenishing stone and other repairs to the spurs, including timber spurs and bank revetements, have been carried out from time to time. Whilst the stone spurs are in good shape the timber spurs are mostly found to be damaged. The stone spurs have helped in diverting the current away from the southern bank. Whilst in the initial stage there was erosion on the downstream side of the town of Dibrugarh, the erosion has now shifted to upstream of Maijan.

Boulders used for stone spurs are transported from a distance of about 500 km. Materials such as soil cement blocks HDPE cage, filled up with lean mortar or ash bricks could be a substitute for stone boulders. It will be worth while to work out the economics of alternative materials.

*28.8.1.4 Performance of Dykes*

The average top elevation of the embankment of Dibrugarh is 107.19 m against HFL of El 106.32 attained in 1988. The top surface of the embankment from Kahai (stone spur No. 11) to Mohanaghat is concrete paved. The rest of the portion is not pucca but is motorable. The side slopes are 2:1 u/s and 3:1 d/s. There is habitat all along the toe of the embankment, and at places the downstream slope has been encroached. The existing section of the embankment appears inadequate to withstand the flood fury of the Brahmaputra River. Considering the floods of 1988, the embankment needs raising and strengthening with proper slope to account for the hydraulic gradient line. The earth for the dyke is transported by trucks, as no earth is available in its vicinity. It may be worthwhile to ascertain the suitability of the silt/earth from the river bed which may be available during working seasons after the monsoon.

*28.8.1.5 Performance of Drainage*

The existing main drain is not functioning properly owing to stagnation and congestion at the outfall. The capacity of the drainage channel in the Bazar area has decreased owing to dumping of debris/garbage. The inadequate functioning of the drain results in spilling of drainage water, causing inundation in the surroundings. Also there is no proper drainage system in the recently developed southern sector of the town. Suitable plans have therefore to be drawn up to keep the existing drain clear of silt, garbage, and other encroachments, to consider pumping options to evacuate the drainage water from the drain into the Brahmaputra River at one or more locations and to provide a suitable drainage channel(s) for drainage of the southern part of the town. The first stretch of the drainage index would need review and the designs to be modified accordingly. A network of drainage channels for disposal of drainage water through different outfalls may need consideration.

In conclusion, it can be stated that after the construction of spurs and revetements the main current of the southern channel has been pushed towards the north, thereby more or less arresting the erosion of the bank in the town area. The dyke system is functioning well to cut off the entry of flood waters into the Dibrugarh town and its suburbs. However, the river flows, its behaviour, and the existing protection works, have to be continuously monitored and suitable timely remedial measures taken from time to time to prevent erosion and spill of water into the area and its suburbs.

## 28.9 Erosion at Gumi-Kalatoli Area of the Brahmaputra Dyke and Measures Adopted

### 28.9.1 CHARACTERISTICS OF THE PROBLEM

Bank erosion is the most serious problem and critical phenomenon associated with the embankment system in this reach of the Brahmaputra River. This river, after passing through a narrow constricted reach (nodal reach) right from Chandrapur in the upstream to 1.0 km downstream of the Saraighat Bridge (Agiathuri Hill), enters the alluvial plains and fans out abnormally. The average width of the river in this reach (from Gumi to Kalatoli) of the Brahmaputra River varies from 15.00 km to 20.00 km, with innumerable land chars in between. The river here maintains relatively low velocity, discharges its excessive sediment load, and tends to increase its waterway resulting in a severe bank erosion problem. Frequent reorientation of the sand chars after each wave of flood changes the flow pattern abruptly and causes an acute bank erosion problem. Moreover, there is a constant shifting tendency of the Brahmaputra River towards its south bank, particularly in this reach. This lateral movement of the river attacks the bank sporadically through active erosion. The soil content in this reach of the Brahmaputra River is such that they resist erosion. There is bank erosion also by sloughing of the saturated bank. Owing to these there has been a severe bank erosion problem in this Alikash--Borakat reach of the Brahmaputra River for a long and there has been considerable loss of land each year owing to this continuous process of bank erosion.

## 28.10 Past Efforts for Management of the Problems

The Brahmaputra Dyke on the left bank from Gumi to Kalatoli (from ch. 22.0 km to ch. 40.00 km) was constructed during the year 1954--55. Owing to the active erosion of the Brahmaputra River the embankment had to be retired several times from 1981 to 1988. Although this reach of B/dyke from ch. 30th km to ch. 37th km (Alikash--Majortoop Borakhat) was subject to continuous/serious bank erosion since the nineteen sixties, the erosion in the reach has become more pronounced/acute/active since 1985. In 1985 the erosion in the reach took a serious turn and a breach occurred in the B/dyke from ch.32.500 m to ch. 33.150 m owing to bank erosion. Subsequently, a retirement linking the ch. 31.50 km to 37.60 km (for a length of 6.10 km) was constructed during 1985--86 at a distance of 800 m from the eroded river bank. Erosion continued unabated in the reach from ch. 31.0 km to ch. 34.0 km during 1986 and a vast strip of land was engulfed by the river. Initially some palliative measures in the form of laying or dumping boulder-filled cages were taken up in the 1st one kilometre of the newly constructed retirement to arrest further encroachment but later on the rate of erosion was so alarming that no further attempts were made to arrest the erosion. In the meantime the stone Bull head no. 5 along with 12 timber dampeners located just in the upstream of Borakhat (part of Alikash Protection Works) were outflanked by the turbulent current, and the erosion was subsequently migrated towards downstream, which virtually threatened the very existence of the newly constructed retirement. The depth of erosion varied from 200 m to 620 m with the maximum scour depth of 12.00 m for a reach of about 3.0 km leaving a berm of only 175 m to the newly constructed retirement.

In order to save the newly constructed retirement as well as the entire area from the erosion of the Brahmaputra River an anti-erosion scheme consisting of construction of 20 of timber Dampers armoured with a boulder apron at an estimated cost of Rs 150.00 lakhs was submitted to the Central Water Commission (CWC) for their approval, after it was duly recommended by the State Technical Advisory Committee in its ninth meeting.

Just before the commencement of flood of 1987, as a remedial measure some boulder-filled cages were laid on the river bank in the most vulnerable reaches to arrest the erosion. But in the flood of July 1987 the temporary protective measures of boulder-filled cages were damaged by outflanking and after a channel being developed joining with the old borrow pit, and the channel was flowing only 15.0 to 20.0 m away from the river side of the retirement too. In order to save the retirement from imminent breach by erosion, construction of 5 stone deflectors was immediately taken up in this reach in addition to launching of percumines, tree bushes, boulder filled cages, boulder filled sausages, etc.. But in the flood of September 87 these deflectors were washed away by the turbulent current and the embankment was again breached at ch. 2543 m of the retirement. The breach caused by erosion was closed later on by constructing a retirement from ch.1,300 m to ch. 4300 m in the year 1987--88.

Before obtaining the formal approval from the Central Water Commission for the execution of major anti-erosion schemes, the erosion in the reach before the flood of 1988 again became very severe and boulder filled cages (in four rows of two layers) were laid on the eroded bank to hold the line in position/to reduce the rate of erosion. This reduced the erosion rate satisfactorily and the retirement was saved from the imminent breach, leaving only a berm of 15.00 m. The unprecedented flood of August, 1988, caused a breach at ch. 36.0 km of B/dyke, a little downstream of the proposed location of the dampeners, had changed the configuration of the entire reach considerably.

The work of construction of 20 timber dampeners were taken up in January 1989 and was completed in March 1990. Owing to the change of river configuration in this reach the locations of the dampeners have been shifted towards upstream along with structural modifications.

## 28.11 The Anti-Erosion Project

### 28.11.1 SCOPE OF THE PROJECT

As is evident from the past records, the rate of erosion in this reach (of the Brahmaputra River from Gumi to Kalatoli-Broakhat being at the middle of dyke system) is abnormally high. An area of approximately 102.0 sq. km/41.00 sq. km. has already eroded away by the erosive activity of the Brahmaputra River in this reach (Gumi to Kalatoli during the period from 1911 to 1976 to 1988 (Drawing No.) and if the erosion continued unabated, in addition to the huge loss of land, there would be a loss of public and private properties worth crores of rupees in the near future. There was also every likelihood of merging of the Brahmaputra River with the Kulsi which is also flowing almost parallel to the Brahmaputra at an average distance of only two to three km and creation of a big island of about sq. km extending up to Nagarbera, the outfall of river Kulsi at Brahmaputra. It was also apprehended that the proposed Kamakhaya-Jogihopa B.C. Railway Line, which was to pass through this locality, would come under the erosion activity of the river and would also disrupt communication through the National Highway 37 in the long run. Over and above this, the rehabilitation of a large number of families would also create a serious problem to the state exchequer.

In view of this chronic and serious erosion problem of this reach of the Brahmaputra River, the Flood Control Department, Government of Assam, had recommended taking up the construction of this Anti-erosion scheme as a remedial measures to contain further migration of the bank line. The approximate area benefited from the scheme/project was about 44.251 bighas (5924.00 ha ). The benefit cost ratio thus calculated/worked out to be 1.23:1.

### 28.11.2 STRUCTURAL COMPONENTS

The anti-erosion scheme, thus, recommended by the State Technical Advisory Committee and accepted by the Government of India, consists of construction of 20 timber dampeners armoured with boulder apron. The detailed specifications are given below:

a) Each dampener contains 38 of Sal pile of 20 to 25 cm diameter in two rows, 10 m below the excavated level of foundation trenches and 1.50 m to 1.90 above the excavated level. The length of each dampener is 42.0 m and the spacing of dampeners is 125.0 m.

## 28.12 Intermediate Flood Management Proposals

The State Government's proposal to take up the following measures to improve the flood management infrastructure created during the last four decades:

I. Raising and strengthening of embankments in reaches where there is inadequate section and to provide impervious blanket, clay core, etc., to reduce seepage etc.;

II. To provide an adequate number of sluices to remove drainage congestion behind embankments along with an improvement of existing drainage channels;

III. To provide raised platforms along the embankments in vulnerable reaches prone to breaches/drainage congestion. The spacing of the raised platforms could be a 1 km interval. Depending on the density of population the spacing may be changed;

IV. To take up adequate and appropriate river training measures where embankments are prone to bank erosion;

V. To initiate model tests in the reaches of the Brahmaputra where there is chronic bank erosion and also at nodal points, protection measures and anti-erosion schemes need to be taken up to protect prosperous towns and villages and to stabilise embankment system;

IV. To take up integrated watershed management schemes to reduce silt charge of the river system.

## 28.13 Proposal for Long-Term Measures

In the Master Plan prepared by the Brahmaputra Board, it has been indicated that the persisting high stages of the Brahmaputra and the Barak owing to 85% of the rainfall taking place in the rainy season is the main cause of flooding and drainage congestion problem of the Brahmaputra and Barak Valleys in Assam. It has also been indicated that the key to a long term effective and reasonably permanent solution to the flood problem in the two valleys lies in construction of some large storage reservoirs in the upper reaches of the main rivers and some of their tributaries. The storage reservoirs have been identified in the Master Plan. The total quantity of live storage identified in the Brahmaputra basin is 7.09 million ha m. Out of this the capacity of the reservoir to be formed in the Dihang project is 3.5 million ha m and that of Subansiri is 1.00 million ha m. Thus the total capacity of Dihang and Subansiri reservoir combined will be 4.5 million ha m which constitute 64% of the total storage identified in the Brahmaputra basin. Computer studies have indicated that with Dihang and Subansiri dam in place there will be substantial moderation of the peak flood in the entire valley, and the flood and drainage congestion problem will be mitigated to a large extent. Besides, these two projects will generate a huge quantity of bulk hydel power. Installed capacity of Dihang is 20,000 MW and that of Subansiri is 4,800 MW. The Pagladiya Dam project with irrigation, flood control and power components will go a longway for a permanent solution of the flood problem of the lower Assam region

Besides being a solution of the flood and drainage congestion problems, these storage reservoirs, along with integrated watershed treatment, will also reduce the high silt charge brought down by these two rivers to the Brahmaputra and bring stability to the river. With the reduction of the silt the erosion problem in the valley could be contained. The river will then be amenable to conventional treatment of river training. Similarly, in the Barak basin 1.7 m ha m of live storage has been identified. Out of this the Tipaimukh dam project alone will have a live storage capacity of 0.9 m ha m. This project is an important component of the Barak Master Plan. The project envisages construction of a 160 M high rock fill dam at Tipaimukh and will be able to absorb a 100 year flood. It will mitigate the flood and drainage congestion problem substantially. It will also produce a huge block of cheap hydel power. The installed capacity is 1,500 MW. The estimated cost of the project is 28,990 million rupees at the 1995 price level.

## 28.14 Hydropower Generation

The hydropower potential of the country is estimated at 84,000 MW, out of which more than half of the potential is located in the North Eastern Region. The identified potential in the Brahmaputra basin is 41,000 MW and in the Barak basin 3,000 MW. The potential so far developed is 340 MW and 820 MW is being developed which constitute only 2.6%. Unfortunately, even in the country as a whole, hydropower development has been very limited. The installed capacity of hydropower now is only 17600 MW. Out of the potential of 44,000 MW in the region, the Brahmaputra Board has already prepared detailed project reports of three large projects having a total installed capacity of 26,300 MW. A fortunate circumstance is that these large projects will contribute in a big way to mitigating the chronic flood problem of the two valleys. It has also been established that these projects are techno-economically sound.

## 28.15 Future Programme and Financial Projection for the 21st Century

On the basis of various studies the recommendations of High Level Committees, and recommendations of the Master Plan of the Brahmaputra, an ambitious flood management programme up to 2050 A.D. combining both short term and long term measures may be summarised as below with tentative financial projection at the 1995 price level for the North Eastern region, and Assam State in particular:

A. SHORT TERM MEASURES

| | Nos. of scheme | | Approx. cost (Million Rs.) |
|---|---|---|---|
| 1. | Raising strengthening of embankment. | 3,122K.M. | 13,000 |
| 2. | New embankments | 365 Km. | 1,000 |
| 3. | Bank protection works. | 375 Km | 2,500 |
| 4. | Drainage schemes | | |
| | a) Drainage channel | 357 Km | 500 |
| | b) Sluices | 43 Km | 220 |
| 5. | Raised platform | 76 Nos. | 160 |
| 6. | Investigation and data collection. | 55 years | 1,100 |
| | Total | | 18,480 |

B. LONG TERM MEASURES:

| | Nos. of scheme | | Approx. cost (Million Rs.) |
|---|---|---|---|
| 7. | Pagladiya dam project | 1 No. | 4,800 |
| 8. | Subansiri dam project | 1 No. | 42,000 |
| 9. | Tipaimukh dam project | 1 No. | 28,990 |
| 10. | Dehang Dam project | 1 No. | 250,000 |
| Total | | | 325.790 |

The amount required under short term measures is Rs 18,480 millions and under long term measures is Rs 325,790 millions. The above flood management programme is costly and has a long gestation period requiring huge financial investments.

Considering the inherent limitations of the short term structural measures and huge capital cost, and the uncertainty of the long term measures such as creation of reservoirs on the main rivers Brahmaputra and Barak as well as on tributaries, the approach to an effective flood management in Assam should be a blended strategy of structural and non-structural methods. This is exclusively important and needs to be implemented by assigning due priority to a systematic programme of improving the existing measures to provide maximum benefits to the flood prone areas.

### 28.15.1 NON-STRUCTURAL MEASURES

In view of the cost effectiveness of non-structural measures and speedier implementation, the main thrust is now on non-structural flood management measures. The non-structural measures which are being implemented are:

A. Modifying the susceptibility to flood damage

i. Flood plain management.
ii. Flood proofing including disaster preparedness and response planning.
iii. Flood forecasting and warning.

B. Modifying the less burden

i. Disaster Relief.

ii. Flood fighting including public health measures.

iii. Flood insurance.

## 28.16 Conclusions

Flood management is a complex phenomenon and requires a multi-disciplinary approach for the prevention of flood damage. Owing to the stupendous cost and long gestation periods of various long-term structural methods of flood management, there is a need for advocating an integrated approach to flood management. Flood plain zoning and flood forecasting and warning constitute the main non-structural measures. In the North Eastern Region, particularly in Assam, flood forecasting and warning is being effectively managed in some vulnerable areas but not much progress has been achieved in flood plain zoning. The structural and non-structural aspects would provide an integrated approach to the future flood management programmes. Whilst structural aspects would keep the floods away from the people, the non-structural aspects are aimed at keeping away from the floods. Therefore a judicious mix of structural and non-structural aspects would optimise benefits which would accrue from structural measures. Whilst the physical flood protection works, like embankments and dykes, etc., will continue to be aligned carefully, the emphasis should be on non-structural measures for the minimisation of the losses by flood. Reduction of flood losses to a reasonable degree and appropriate exploitation of water resources of the region will foster socio-economic development for a smooth passage into the 21st century.

# CHAPTER 29

# REMOTE SENSING AND GIS IN FLOOD MANAGEMENT

V. BHANU MURTHY AND V. HARI PRASAD

Space technology is finding its niche in water resources and disaster management, including flood management. Satellites offer information about areas which are beyond easy human reach. They also provide vast amounts of data. This chapter discusses the role of remote sensing and GIS in flood management in the Brahmaputra basin.

## 29.1 Introduction

Floods have been causing severe damage in flood plains throughout history. Attempts have been constantly made to reduce the damages through the use of various types of flood prevention techniques. India is the worst flood-affected country in the world after Bangladesh (Agarwal and Narian, 1991) and accounted for one fifth of the global death count owing to floods. About 40 million hectares (mha), or nearly 1/8$^{th}$ of India's geographical area, is flood-prone. The total area affected annually is about 8 m ha. The annually affected crop area is about 3.5 m ha and could be as high as 10 m ha in the worst years. Flood control works executed consist of mainly construction of new embankments, drainage channels, afforestation to save 546 towns and raising of 4700 villages above the flood level (Ramamurthy, 1989).

Human activity, however, tends to concentrate in flood-liable areas which are often convenient and attractive locations for settlement and other economic endeavours resulting in greater flood damages. If losses owing to floods in terms of life and property are to be minimized, the solution is not merely to provide relief measures but also to undertake necessary measures encompassing a wide rage of activities, namely, long and short-term prediction, prevention, warning, monitoring and relief. This involves the interaction amongst different government and private agencies on the one hand, and the people of country facing the disaster, on the other hand, in making use of information and carrying out the above tasks.

The chronic flood-prone river basins in India are the Ganga and the Brahmaputra. These Himalayan rivers flowing down the hills cause flood problems in Uttar Pradesh, Bihar, West Bengal, and Assam owing to high discharges concentrated during the monsoon months (June to September) and large volumes of sediment are carried to the plains. Drainage problems also arise concurrently if floods are prolonged and the outfalls of major drainage arteries get blocked. On account of frequent changes in river courses and braiding of channels, erosion of river banks also assumes importance as one of the



*V.P. Singh et al. (eds.), The Brahmaputra Basin Water Resources*, 559-577.

major problems associated with floods. These often get aggravated with the occurrence of cyclones, which generally are accompanied by heavy rainfall and high winds. In north Bengal the tributaries of the Brahmaputra rise owing to flash floods and deposition of coarse sediments which cause river spills owing to inadequate carrying capacity of the river channels.

Permanent protection at all the flood-prone areas and for all magnitudes of floods by structural means may be neither feasible nor economical. Furthermore, by controlling the river by structural measures, such as embankments, the flood damage may be minimized to some extent, but on the other hand, flood waters are going waste and silt may be deposited on the river bed, reducing its channel capacity. At present the country is suffering with floods in one part (owing to vast amounts of water) and in another part with drought (owing to lack of water). Therefore, the engineers' attention is being focused on various methods to mitigate floods to reduce damages and at the same time to best utilize flood waters. Flood control has to be viewed within the broad context of the economic and social development in the country, since seventy percent of the population is dependent on agriculture. An overall comprehensive approach is therefore required for obtaining the best possible utilisation of land and water resources for optimum production.

Space technology has made a substantial contribution in every aspect of flood management, such as preparedness, prevention and relief (Rao, 1994). The advantages of the information acquired by satellite remote sensing are those of wide area, periodicity, and spectral characteristics, and especially the easiness of comparing the data before and after flood disasters. While satellite communications have significant potential for real-time dissemination of information and early warning, earth observation satellites enable continuous monitoring of atmospheric as well as surface parameters contributing to the phenomenon. These considerable advances in the operational use of space technology towards management and mitigation of floods offer viable technological solutions to these problems. Although several countries, including a few developing nations, are using satellite technology to address these problems in their respective regions, many developing countries are yet to have the necessary infrastructure and trained manpower to absorb this high technology for direct problem solving. This chapter discusses the state of art of satellite technology in flood management.

## 29.2 The Brahmaputra Valley

The Brahmaputra is one of the largest rivers in the world. It originates from a glacier east of Mansarowar, and after traversing 1,625 km in Tibet enters India near Tuting in Arunachal Pradesh. Its course within India is 918 km. A little downstream of Dhubri the river enters the plains of Bangladesh, through which it flows for 258 km before joining the Ganges (Padma) at Goalundo. The combined stream then flows to meet the Mehgna (called the Barak higher up) which falls into the Bay of Bengal. Its catchment area in Tibet is 293,000 sq. km, that in India and Bhutan 240,000 sq. km and in Bangladesh 47,000 sq. km. At Pandu near Guwahati the river carried a maximum discharge of 72,794 cumecs on 23.8.1992 and a minimum discharge of 1.757 cumecs on

20th February 1968. The average dry season discharge is 4420 cumecs. The average annual flow (1956--62) of the river near its confluence with the Ganges at Gaolundo in Bangladesh is 61.4 million ha.m The average annual yield as observed at Pandu is 49.43 million ha.m. Although two third of its length and more than half of the catchment are in Tibet, about two third contribution of its flow is from the drainage area in India.

During its course in the Assam valley from Kobo to Dhubri, the river is joined by about 20 important tributaries on its north bank and about 13 on its south bank. Some of the north bank tributaries originate from the snow-clad Himalayas and others from lower Himalayas. The south bank tributaries originate from the hill ranges of Mehgalaya, Nagaland, and Arunachal Pradesh. The distinct features of the drainage area of the north bank are more intensive rainfall and less stable soil with steep land slopes. Consequently, these rivers carry large amounts of silt and bring flash floods.

The Brahmaputra River is braided and unstable in its entire reach in the Assam Valley. The instability of the river is mainly attributed to high sediment charge, steep slopes, and transverse gradient. Apart from these, the entire area is in a highly seismic zone and receives earthquake shocks of severe intensity periodically. This also is a major cause of the unstable character of the river. The drainage network of the lower Assam and upper Assam were considerably disturbed by the major earthquake of 1897 and 1950.

## 29.3 Hydrometeorology

The meteorological conditions in the catchment of the Brahmaputra in Tibet and in India are different and lie in different climatic zones. The zone in Tibet comprises plateau regions and that in India comprises hilly ranges and plain areas of the Assam Valley. The drainage area in India is also subjected to cyclonic storms originating in the Bay of Bengal, particularly during the pre-monsoon period of May and June and the later part of September, besides regular monsoon from June to September.

The mean annual rainfall over the whole catchment, excluding Tibet and Bhutan is about 2,300 mm. The mean annual rainfall over the sub catchments varies widely from 2,590 mm in Dihang catchment to 1,735 mm in Kopili sub-catchment. In the northern part, monsoon rainfall accounts for less than 50 per cent of the annual rainfall, the pre-monsoon around 35 per cent, and winter rain around 10 per cent. In the rest of the catchment towards the south, 60 to 70 percent of thew rainfall occurs during the monsoon period and 20 to 30 percent in the pre-monsoon season and a small amount during winter.

The basic requirement for planning and developing and management of water resources is a correct assessment of the availability of water from precipitation and its distribution in time and space in different parts of the basin. It is also necessary to collect other hydrological data such as gauge, discharge and sediment flow of the main river and its important tributaries. A network of rain gauge stations has been established and is being improved progressively. There are, at present, 336 ordinary rain gauges and 113 self-recording rain gauges. However, to conform to the standards of the World

Meteorological Organisation (WMO) an additional 636 ordinary raingauges and 67 self-recording rain gauges have to be established. Some gauge observation sites at various places were established as far back as 1907--08. The gauge discharge and silt observation sites were set up and are being maintained from around 1955. The first discharge observation site was, however, started at Pasighat in 1949. The gauge discharge sites on a number of tributaries have also been established and data are being collected since 1956-57. Studies on flood frequency, silt movement, river aggradation or degradation, and annual yield have been made from the available data. It is necessary to continue to observe the hydrometeorological and hydrological data and continue to make detailed studies in order to understand different river characteristics and to formulate proposals. These data are required to plan various projects for harnessing water resources for flood control, development of hydropower, extension of irrigation and navigation.

The existing network for hydrometeorological and hydrological observations on the main river and its tributaries should be further extended and strengthened. All the sites should be provided with modern equipment for observations and telecommunication.

## 29.4 Geology

A study of major geological formations, geomorphology, land slopes and seismicity is a pre-requisite for planning and development of water resources. The geological successions in the Shillong Plateau and those in the Naga-Patkai ranges are at variance with those seen in the Himalayan catchment in the northern part. The basement constituents of the main Shillong Plateau are made up of high grade metamorphites, gneisses, and granites. These are overlain by low grade meta-sediments of the Shillong group with granite emplacements.

On the northern side of the Brahmaputra Valley the formations generally comprise alluvial plains abutting against the Siwalik ranges of the Himalayas. In the Lohit area on the extreme eastern side, the Siwalik belts are not exposed and the alluvial deposits are by and large directly abutting against the meta sediments followed by rock outcrops of different characters. It is, however, felt that the basement of the entire basin is made up of the type of metamorphites of Shillong Plateau and Mikir Hills. The geomorphology of the basin has also been studied from the LANDSAT imagery which reveal the following distinct zones:

1. Himalayan zone including the Mishmi block;
2. Naga-Lushai-Paktal zone including the Tripura block;
3. Meghalaya uplands including the Mikir Hills; and
4. Himalayan fore deep inclusive of the Brahmaputra and Surma valleys.

The geological history of the Brahmaputra valley indicates that the Brahmaputra is a very young river and its present configuration took shape only during the Pleistocene and recent times. The erosive power of the river is still great and it has been adopting many changes in its course. Since the Brahmaputra River is not stable a study of its geomorphology and its behaviour may be taken up in a systematic manner. A programme for such studies will have to be drawn in consultation with the Geological Survey of India.

Landslides are a natural phenomenon of degradation in hilly areas of this basin. The process is accelerated by torrential rains, steep hill slope and earthquakes. Rainwater easily percolates through the joints, fractures, and foliation in the rocks making the soil loose which gradually slumps down.

The entire basin area is known for its high seismicity, and there have been a large number of occurrences of major earthquakes in the past from 1920 to 1980 with distribution as under:

| Magnitude (Richter scale) | No. of occurrences |
|---|---|
| 5 to 6 | 270 |
| 6 to 7 | 167 |
| 7 to 8 | 15 |
| 8 and above | 3 |
| Total | 455 |

Major earthquakes in the area have occurred in the years 1865, 1869, 1897, 1908, 1937, and 1950. A limited number studies have so far been carried out with regard to seismicity in the region. Steps have since have taken to set up a few MEQ stations. A network of such stations should be planned and regular observations taken for planning and design of various projects and for stabilising hillslopes.

Some important features of the flow regime of the Brahmaputra River in Assam in general, and the flood regime in particular through an analysis of the flow data covering the last 40 years in general and the decade 1971--1980 in particular, is described here. Several aspects of the basin environment that have a considerable bearing on the flow regime of the river is also described here. As a gigantic fluvial system with unique characteristics of flow, sediment transport, and channel configuration, the Brahmaputra River is close to being a virgin river when it comes to fluvial research. While Coleman (1969) and Bristow (1987) examined the fluvial processes and sedimentation of the river in Bangladesh, Goswami (1983, 1985a. 1985b) studied different aspects of floods in the Brahmaputra River in Bangladesh. Like the Hwang-Ho River prior to the 1950s or the Mississippi River before the 1930s, the Brahmaputra today is a veritable source of disaster affecting millions of people living in its valley.

Floods comprise a significant natural hazard in the Brahmaputra basin in India. These high-magnitude episodic events are capable of performing gigantic geomorphic activities and bring about dramatic consequences to the landscape (Wolman and Gerson, 1978; Gupta, 1988). Management of flood waters and reduction of the hazard has been the major concern of water resource scientists and engineers. The technological challenges inherent in major flood management efforts, economic benefits derived from them, their social justifications and environmental implications are some of the vital considerations involved in such endeavours. Flood studies in the tropics, including India, are greatly impaired by the inadequacy of long-term flood records and lack of sophistication in both measurements and analytical procedures. The flooding rivers in the humid tropics are extremely dynamic features of the natural environment. There is an urgent need to enhance our understanding and expertise on these river regimes. Some of the developing countries of humid tropics, such as India, where most disastrous flood events recur rather too frequently, flood management is identified as the key factor of economic and socio-cultural transformation. Unless properly planned, designed, and implemented, the flood management activities may create complex social and environmental problems. These impacts are difficult to measure and their assessment requires databases longer and more extensive than normally available. The colossal mass of water draining through this massive system flows unharnessed into the Bay of Bengal. Although the importance of taming this river and exploiting its resources seems to have become a part of popular consciousness in recent times, no headway has really been made so far in this regard. It is imperative that we know the river system well enough before going in a big way to manipulate its regime through structural measures. Fluvial research on the river is, therefore, as much a matter of high academic interest as of great practical significance. The present study, despite the constraint of an inadequate database, is expected to provide a glimpse of the fluvial regime and flood flow panorama of the river against the backdrop of its unique basin environment.

## 29.5 Environmental Setting of the Basin

The Brahmaputra basin represents a unique physiographic setting vis-a-vis the eastern Himalayas, a powerful monsoon rainfall regime under wet humid conditions, fragile geological base, and active seismicity. It covers an area of 580,000 km$^2$ of which 50.5% lie in China, 33.6% in India, 7.8% in Bhutan and 8.1% in Bangladesh. The river originates in a great glacier mass in the Kailas Range of the Himalayas, south of Gunkyud Lake in the southwest. Tibet at an elevation of 5,300 m and flows through China, India, and Bangladesh for a total distance of 2,880 km, before emptying into the Bay of Bengal through a joint channel with the Ganga. Flowing eastward for 1,100 km on the Tibetan Plateau, the Brahmaputra, known there as Tsangpo, enters a deep narrow gorge at Pe (3,500 m), then continues southward across the east--west trending ranges of the Himalayas, viz., the Greater Himalayas (average elevation 6,000 m), Middle Himalayas (3,000-5,000 m) and sub-Himalayas (1,000-2,000 m) before debouching onto the Assam plain near Pasighat (155 m). The different geo-ecological zones represented by these Himalayan ranges through which the river cuts its course have distinctive assemblages of topographical, geological, climatological and floral characteristics (Karan and Iijima, 1985). The gradient of the Brahmaputra is as steep as 4.3 to 16.8 m/km in the gorge section upstream of Pasighat, but near Guwahati it is as flat as 0.1 m/km (Goswami, 1985a). In Assam the Brahmaputra, with an average channel width of 8 km, flows in a highly braided channel marked by the presence of numerous sandbars and islands.

The Brahmaputra Basin is underlain, for the most part, by very young and unweathered sedimentary formations, with the result that the river carries mainly fine sand and silt with very little clay. The sub-Himalayan ranges bordering the Brahmaputra Valley on the north consist mainly of Tertiary sandstones and are marked by the presence of many raised, relatively young terraces (Gansser, 1964). The Middle Himalayas are underlain by lower Gondwana deposits comprising shales, slates, and phyllites. The Greater Himalayas consist primarily of granites and gneisses. Further to the north, the trans-Himalayas of Tibet (average elevation 4,500 m) are made up of sedimentary rocks of the Palaeozoic to Eocene ages (Wadia, 1968). The Patkai Naga ranges bordering the Brahmaputra Valley on the east and southeast (average elevation 1,000 m) consist of Tertiary formation riddled with numerous active faults. The highlands to the south comprising the Meghalaya plateau and Mikir Hills (elevation 600--1,800 m) are made up of gneisses and schists of the Precambrian age. The Brahmaputra Valley in Assam is a tectono-sedimentary basin, 720 km long and 80-90 km wide, underlain by recent alluvium, approximately 200--300 m thick. The flood plain of the Brahmaputra exhibits an array of alluvial features including natural levees, crevasses---splay deposits, pointbars, meander-scrolls, ox-bow lakes, channel bars and flood basins (GSI, 1977). The channel of the river is characterized by rapid aggradation, dramatic channel and excessive bankline recession (Goswami, 1985a). Because of the presence of the natural levee of the Brahmaputra, several tributaries in their lower reaches tend to flow parallel to the mainstream until they find an opening to traverse it. This has resulted in the formation of swamps along the lower parts of tributaries. In the foothill region the sudden change in gradient of the rivers has led to the deposition of coarse alluvial materials in large coalescing alluvial fans and braided channels.

The Brahmaputra Valley and its adjoining hill ranges are seismically very unstable. The effects of the earthquakes of 1897 and 1950, measuring 8.7 on the Richter scale, included extensive landslips on the Himalayan hillslopes, subsidence and fissuring of the ground in the valley, and changes in the course and configuration of several tributary rivers as well as the mainstream (Oldham, 1899; Poddar, 1952). The seismic activity in the region appears to have a great impact on the fluvio-sedimentary regime of the Brahmaputra and its tributaries. The effect of the 1950 earthquake on the highest and lowest flood levels of the Brahmaputra will highlight this. It is evident that the lowest water level rose by about 3m as a result of the earthquake.

The Brahmaputra basin in India lies within the monsoon rainfall regime receiving an annual rainfall of about 230 cm. There is a marked variability in the distribution of precipitation over the catchment. For example, the Jiadhol catchment on the north east receives 410 cm of rainfall while the Kopili catchment in the south central part receives only 175 cm. Rainfall in the Himalayan sector amount to 500 cm per year with the lower ranges receiving more than the highest areas. For example, the annual rainfall at Dibrugarh in the eastern part of the valley is 285 cm, whereas at Pasighat, located in the foothill region, it is 507 cm, and at Tuting, located further up in the Himalayas, it is 274 cm. Monsoon rains from June to September account for 60-70 percent of the annual rainfall. These rains that contribute a large portion of runoff in the Brahmaputra and its tributaries are primarily controlled by the position of a belt of depression called the monsoon trough extending from northwest India to the head of the bay of Bengal. In the course of its north--south oscillations in summer, when this axis moves to the foothills of the Himalayas heavy, precipitation is caused in Assam and adjoining highlands. The pre-monsoon caused primarily by depressions moving from the west and by the local conventional storms. Precipitation in the Meghalaya plateau to the south of the valley is of the order of 1,000 cm per year.

The natural vegetation of the Brahmaputra Basin is marked by variation according to altitude evergreen and mixed deciduous forests within the valley and foothills to alpine meadows and steppes in the higher ranges in Tibet. About 59% of the Brahmaputra Basin is forested (Goswami, 1997). In Assam, the percentage of forest cover is 20.56 (ARSAC, 1990). Large tracts of mixed subtropical forests are found in the foothills region and in the reserved forests within the valley. In the flood-prone riverine areas of the valley deciduous trees such as simul (Boombax Malabaricum) and Koroi (Alibizzia Procera) grow, whereas low and grasses with Saccharum Spontanium as the main component are found on the river banks, active flood plains and sandbanks. Deforestation in the Himalayan catchments owing to wanton logging operations and widespread shifting cultivation assumes alarming proportion in recent times thereby accentuating the problems of flood and erosion in the valley.

Soils in the lower Himalayan ranges developed on Tertiary sandstones are friable and shallow consisting mainly of sands with admixtures of cobbles and boulders. These soils have a lower threshold of erosion. The Brahmaputra valley is made up of alluvial soil formed on recent river deposits called 'new alluvium'. There are a few isolated pockets of Pleistocene deposits called 'old alluvium' within the valley and along the foothill region. In the hilly areas, especially to the south of the Brahmaputra, laterites and red loams are found.

From the above discussion it is evident that steep slopes, high monsoonal rainfall, and snowmelt from Himalayan glaciers, easily erodible soils in the catchment, and tectonic instability of the region, together with anthropogenic factors combine to provide the Brahmaputra its unique flow regime.

## 29.6 Flow Regime of the Brahmaputra

The flow regime of the Brahmaputra responds to the seasonal rhythm of the monsoons and freeze--thaw cycle of the Himalayan snow. It is ranked fourth among the largest rivers of the world with an average annual discharge of 19,830 $m^3$ $s^{-1}$ at mouth (Table 29.1). The rates of discharge per unit drainage area for yield of the Brahmaputra basin rivers also among the highest in the world. The yield of the Brahmaputra from the catchment above Pandu, Assam, is 0.0306 $m^3$ $s^{-1}$ $km^2$. For the three major tributaries, viz., the Subansiri, Jia Bharali and Manas rivers, the rates are 0.076, 0.086, and 0.023 $m^3$ $s^{-1}$ $km^2$ respectively (Table 29.2). These are very high rates compared to the figures for other major rivers of the world (Leopold et al 1964, pp. 76). High monsoonal rainfall in the upper catchments and the steep gradient of the rivers are considered to be the major factors responsible for the high rates of unit discharge which in turn help generate the high sediment yield from the basins and contributes significantly towards causing drainage congestion in the valley.

Table 29.1 Ranking of the Brahmaputra among large rivers of the World in terms of discharge and sediment yield.

| River | Average Discharge at mouth $10^3 m^3 s^{-1}$ | River | Sediment yield (tons km$^{-2}$yr$^{-1}$) |
|---|---|---|---|
| Amazon | 99.15 | Yellow | 1,40.3 |
| Congo | 39.66 | Brahmaputra (At Baha durabad, Bangladesh) | 1,128 |
| Yangtze | 21.80 | Brahmaputra (At Pandu, Assam) | 804 |
| Brahmaputra | 19.83 | Irrawaddy | 616 |
| Hwang Ho | 19.83 | Yangtze | 246 |
| Yamser | 17.39 | Mekong | 214 |
| Mississippi | 17.30 | Onnoco | 212 |
| Onnoco | 17.00 | Colorado | 212 |
| Lena | 15.49 | Missouri | 159 |
| Parana | 14.90 | Amazon | 46 |
| Irrawadly | 13.56 | Indus | 103 |
| Ob | 12.49 | Mississippi | 64 |
| Ganga | 11.67 | Nile | 37 |

Source: Goswami (1985a, 1989)

Table 29.2 Water and sediment yields of selected tributaries of the Brahmaputra River, Assam.

| River | Drainage Area (km$^2$) | Water yield (m$^3$s$^{-1}$ km$^2$) | Sediment yeild (tons km$^2$ yr$^{-1}$) |
|---|---|---|---|
| Brahmaputra at | | | |
| Tsela 1D', Zang (China) | 191,222 | 0.0105 | 100 |
| Pasighat (India) | 244,700 | 0.0231 | 340 |
| Pandhu (India) | 500,000 | 0.0306 | 804 |
| Bahadurabad (Bangladesh) | 580,000 | 0.0331 | 1128 |
| Dibang | 12,120 | 0.1066 | 3765 |
| Lohil | 22,077 | 0.0709 | 1960 |
| Subansin | 27,400 | 0.0756 | 959 |
| Jia Bharali | 11,300 | 0.0858 | 4721 |
| Pathuman | 1,787 | 0.0403 | 2887 |
| Paglodia | 383 | 0.1087 | 1883 |
| Manas | 36,300 | 0.0232 | 1581 |
| Bu | 4,923 | 0.0788 | 1129 |
| Desong | 3,950 | 0.00382 | 622 |
| Dta | 10,240 | 0.0184 | 379 |
| Kopili | 13,556 | 0.0182 | 230 |

Source: Goswami (1955a and 1989).

## 29.7 Present Satellite Systems

Indian Remote Sensing Satellites are being used for obtaining information on flood inundated areas and estimation of flood damages. The IRS series (IRS-1B, IRS-P2, IRS-1C, IRC-P3 and IRS-1D) which collect information in optical region of the electro magnetic spectrum at different spatial resolution (Table 29.3) provide valuable information on flood situation.

Table 29.3 Details of Indian remote sensing satellite (IRS).

| Satellite | Sensor | Spatial Resolution in meters | Revisit period days |
|---|---|---|---|
| IRS-1B | LISS-I<br>LISS-II | 72.5<br>36.25 | 22<br>22 |
| IRS-P2 | LISS-II | 36.25 | 24 |
| IRS-P3 | WiFS | 188 | 5 |
| IRS-1C | WiFS | 188 | 5 |
| IRS-1D | LISS-III<br>PAN | 23.25<br>5.8 | 24<br>24 |

WIFS sensor, which has a wide swath of about 800 km and a repetivity of five days, proves to be very effective in monitoring floods in the entire basin. The 5-day repetivity of WIFS data results in about 25 coverage over any specific area during flood months (June to September). Monitoring of different waves of floods, extent of inundation during the flood wave, duration of floods, etc., which will form critical inputs for basinwise planning for flood management. The WIFS, LISS-III and PAN sensors will be providing the much needed close temporal coverage, large area synoptic viewing of the entire basin and more detailed view of ground conditions (Thiruvengadachari et al. 1996). The suite of these sensors on board IRS-1C and 1D is expected to enhance effective flood management in the flood-prone areas of the country. In addition to the IRS satellites, the NRSA has been receiving data from Landsat 5 Thematic Mapper which has 30 m resolution and 16 days repetivity, and microwave data from European Remote Sensing Satellite (ERS) which has a C-band SAR sensor on board.

## 29.8 State of the Art in Satellite Remote Sensing Applications

### 29.8.1 GLOBAL EXPERIENCE

Many countries are utilizing remote sensing data for flood management. For example, the Mississippi River flood in the USA was investigated using remotely sensed data from ERTS-1 (Landsat 1) during 1973 (Deutsch and Ruggles, 1974; Moore and North, 1974). It was concluded that Landsat data are extremely useful as a regional tool for flood management. Furthermore, a study of Landsat images acquired during different times showed apparently the levee system along the Mississippi River which was providing adequate protection, whereas flooded areas on the tributaries indicated locations where new and additional works might be necessary. Such an evaluation was not possible before the comprehensive regional coverage from satellite became available. A change detection analysis is used to study normal and flood flow conditions of rivers (Philipson and Hatker, 1981; Vehara et al. 1983). Furthermore, the agriculture pattern difference in the flood plain was studied which was indicative of flood-prone areas. Never before was such a complete picture of flooding observed as by using satellite data. In Australia Landsat data were utilized to define the spatial pattern of flooding and identifying flood ways based on the tonal variations observed owing to different sediment concentration on Darling River in New South Wales (Green et al. 1983). Landsat images were acquired for different magnitudes of floods of Black River in the USA and the areal extent of flooding was correlated with in-situ river discharges and a relationship between the area inundated and discharge was developed (Phillipin, et al. 1980).

Multi-temporal satellite data are used along with Digital Elevation Model (DEM) to identify the flood inundation areas including flooding under vegetation canopies. Furthermore, for the 1993 Mississippi flood a comprehensive assessment of devastation by floods was made. Each land use class affected by floods was estimated and the quantitative risk owing to floods was assessed for different types of land use classes (EOSAT, 1993). Subsequent developments in satellite remote sensing provided with microwave data which has all weather capability and which has high potential in monitoring floods under monsoon clouds. Airborne X and L band SAR images have been used to map the floods of Red River in the USA. L band SAR data from SIR-B at different 'look-angles' were used to delineate flooded forests in the Sundarban areas (Lowry et al. 1979; Imhoff and Gesch, 1990).

China is operationally using airborne microwave data along with satellite data for monitoring floods in the Yellow and Yangtze Rivers. A comprehensive data as well as data from automated ground observation stations (like water level information) to processing stations have been obtained. Digital elevation models (DEMs) are developed to process the remotely sensed data along with the ground data using the disaster evacuation model for determining the flood progression, damage assessment and for planning relief measures (Shuhu, 1991).

Enhancement techniques and digital procedures have been developed to overlay pre-flood, during flood, and post-flood images for detecting the changes so as to assess flood damage. NOAA data were also being used for mapping large floods, though the resolution is coarse. However, regional level information on the extent of flooding can be obtained (Berg et al. 1979, Pramanik 1988).

## 29.8.2 INDIAN EXPERIENCE

Optical and microwave data from IRS, Landsat and ERS satellite are being used to map and monitor flood events over the entire country in near real-time on operational mode and the information is being furnished to departments concerned so as to assist in organising necessary relief measures and to make a reliable assessment of flood damages. The WiFS data from IRS-1C and 1D have shown a promising future in monitoring floods more effectively in view of its large swath and high repetivity.

*29.8.2.1 Near Near-Time Flood Mapping and Monitoring*

Whenever a flood occurs in the country the first and foremost activity that will be taken up with top priority is relief, for which a flood inundation map is a vital input. It can be said that satellite remote sensing is the most powerful tool for preparing a flood inundation map in near real-time which can be effectively used for relief management. Furthermore, with the availability of multiple satellite data from the IRS SERIES, Landsat 5, and ERS it is now possible to monitor the flood situation in a particular basin where multiple flood waves have occurred. It can also provide approximate duration of flooding, provided that daily water level/discharge information is available. The flood inundation map helps the decision makers to make a scientific assessment and for better management of relief activities.

For the last single decade major floods which occurred in the country were mapped in nearly real time using satellite remote sensing and provided the information to the departments concerned. The utility of the satellite remote sensing has been operationally demonstrated for mapping the flood inundation areas. During the flood season a constant watch has been kept on the flood situation in the country through new media, TV/radio report etc., and satellite data were screened for cloud cover covering flood-affected areas. Even partially cloud-free data are acquired, analyzed and interpreted and flood maps are prepared in nearly real time. During the 1993 flood season, for example, attempts were made in advance to procure ERS-1 C-band SAR which has got cloud penetration, covering flood-affected areas and flood maps were prepared and ERS-1 SAR data were extensively used for mapping the floods in near real-time on operation level.

*29.8.2.2 Flood Damage Assessment*

Based on satellite data acquired during flood, the pre-flood, and post flood, along with ground information, flood damage is being estimated. Digital overlays have been prepared in GIS (Geographical Information System) environment pertaining to topographical, hydrological, and flood plain land cover/land use information. The flood boundaries derived from the analysis of satellite data were intercepted with the GIS overlays. Based on the duration of flooding, the magnitudes of floods, number of flood waves, area affected, type of land use features, etc., flood damage is estimated.

*29.8.2.3 Mapping of Post-Flood River Configuration and Existing Flood Control Works*

Most of the flood-prone rivers in India change their courses frequently after every flood wave, attacking different locations at different times. Hence it becomes necessary to understand the behaviour of the river and its latest configuration so as to plan the flood control measures effectively. At the same time, it is equally important to monitor the existing flood control structures from time to time is observed to avoid breaches in view of frequent changes in river configuration. To assist the planners in providing the latest, accurate, multi-temporal data, high resolution satellite data were used to map the river configuration and existing flood control works. Every year, for selected flood plane rivers, post-flood satellite data were acquired, enhanced, and enlarged and geometrically corrected for positional accuracy and used for mapping the configuration of the river after flood waves have passed, along with flood control embankments and major spurs. This information can be effectively used to identify the vulnerable river reaches and the status of the flood control embankments/spurs so that necessary measures can be taken accordingly to avoid breaches.

*29.8.2.2.4 Identification of Erosion Prone Areas*

Every year the banks of the flood-prone rivers are subjected to erosion, eating away fertile land and at times villages on the banks. In order to provide bank protection works, vulnerable areas subjected to bank erosion along the rivers have to be monitored. Satellite data acquired at different time intervals have been used to identify bank erosion, and maps have been prepared for protection works and town protection works.

*29.8.2.5 Identification of Drainage Congested Areas*

It is observed quite frequently that flood waters are being stagnated, even for months, after the flood wave has passed, owing to various reasons, causing extensive damage. Satellite data acquired during floods of different magnitudes were compared with the pre-flood situation and the post-flood situation, in the basin and areas badly affected by drainage congestion were delineated to facilitate planning for making drainage improvement works.

*29.8.2.6 Flood Disaster Impact Minimization*

One of the major goals of flood control and management is to improve flood forecasting. The use of remote sensing techniques can increase the accuracy of forecast and considerably increase the warning time of forecasts. An example of the lower Godavari basin is given in the following paragraphs to demonstrate the utility of the satellite data for flood forecasting.

The main goal of the project is to improve flood forecasting in the lower Godavari basin and development of the `Spatial Flood Warning System' in the Bhadrachalam and Paloncha Divisions of Khammam District of Andhra Pradesh, covering an area of about 625 km$^2$. The lower Godavari basin covers an area of about 175,000 km$^2$ between Sriram Sagar Project at the upstream and Dhoweleswarm Barrage at the downstream. The principal tributaries joining the main stream in this reach are Pranahita (75,000 km$^2$) and Indravathi (40,000 km$^2$). The contribution from the two tributaries is about 75% of the total inflow of the basin. The Bhadrachalam Division is unprotected, wherein floods of the average magnitude have occurred eight times in a span of forty years.

At present, forecasts are issued by the Central Water Commission using conventional rainfall-runoff models. The accuracy of forecasts is around 65%--70%, whilst the warning time of forecasts is between six and twelve hours. The poor performance can be attributed to the high spatial variability of rainfall not captured by ground measurements and lack of spatial information on the catchment characteristics of the basin, such as current hydrological landuse/cover, spatial variability of soils, etc. Incorporation of remote sensing inputs such as satellite derived rainfall estimates, current hydrological landuse/cover, soil information, etc. in a rainfall-runoff model which is of spatially distributed type and capable of being operated in a GIS environment is being addressed in this project.

The flood forecasts are at present being issued in terms of the rise in water level and persistence and not in terms of the likely spatial impact. The state government authorities (Commissioner of Relief, Government of Andhra Pradesh) interpret the flood water level in terms of likely impact based on the past experience and draw up flood contingency plans, such as dissemination of forecasts to residents, evacuation measures, food supply for people of marooned villages, etc.. However, in the absence of definitive and objective information regarding flood plains likely to be affected by the incoming floods, these measures, based on the historical experience, are found to be inadequate. The development of a 'Spatial Flood Warning System', with a comprehensive database including DEM generated using Differential Global Positioning System (DGPS), hydraulic/hydrologic modelling capabilities and a Decision Support System (DSS) for appropriate relief response is being addressed in cooperation with the Office of Principal Secretary (Panchayathi Raj), Rural Development and Relief Department, the Government of Andhra Pradesh.

## 29.9 Flood Risk Zone Mapping

### 29.9.1 CONCEPT OF FLOOD RISK MAPPING

A risk map is a special topographic map in which the hypothetical flood characteristics are represented graphically. Flood risk maps mean more than just a map for one specific flood event. Risk mapping is the basic tool and starting point of any regional intervention policy for flood control. Flood risk maps can be used for several purposes.

1. They provide the basic initial information for land use planning.
2. They allow correct development for new urban areas.
3. The cost of flooding and risk reduction benefits can be adequately evaluated by using these maps.
4. The feasibility of non-structural flood control measures, such as flood-proofing can be correctly assessed.
5. They can form the basis for any type of insurance plan.
6. Flood risk maps serve as a logical base for investment planning and priority setting, mainly for non-structural measures, and
7. Last, but not least, flood risk maps increase the public awareness of risk.

Flood risk ignorance is usually responsible for most of the flood damage. Flood risk maps are essential tools for land use planning flood-prone areas. The basic criteria for mapping are usually chosen according to flood return periods. A flood risk map is considered as a preliminary, yet necessary, initial step for all regional development policies.

### 29.9.2 APPROACH

Floods have many features interesting enough to be reproduced on maps, either for definition of priorities, structural intervention, or assessment of the damage likely to occur. The first property of floods that is interesting to map is flood frequency. This is usually done by stating the flood return period. The simplest and the most frequent case is to draw flood contour maps for a set of occurrence probabilities. This usually fixes a series of nested bands as the flood return period increases.

Flood duration is also a very important characteristic, mainly for agricultural damage evaluation. Many crops, and especially trees, can stand one day underwater without serious damage. However, beyond that duration, agricultural damage increases severely. Hydrological characterization requires delineation of the flood hydrograph, not just the peak. These flood characteristics mentioned above can be numerically defined.

Zoning may be classified as a 'Prohibition Zone' wherein generally new building development shall be prohibited and the use of land confined to agriculture or recreational purpose. Beyond the prohibited area there is a flood fringe zone which does not contribute substantially to the downstream movement of flood water, as the water moves very slowly there. This may be called the 'Restrictive Zone'. In this zone, though it is not necessary to prohibit the building activities, development should still be regulated to prevent excessive flood damage. The third zone can be the 'Warning Zone' which covers the area liable to inundation by the largest flood which may occur in the river. Zoning requires regular revision where and when large-scale flood control measures are implemented.

Flooding risk analysis is important as a preliminary step for river-basin planning with respect to decisions on land use and flood-reduction strategies. The methodology refers to a large-scale study for a basin where river training and land use are already at an advanced stage.

### 29.9.3 ROLE OF REMOTE SENSING

One way to prepare flood risk zone maps is the hydrological and hydraulic approach, in which the area inundated under peak flood level is determined using mathematical formulae and stream channel geometry by theoretical calculation of the required flood plain to pass a given discharge and associated flood elevation. This approach demands extensive field surveys and complex analysis, which is time consuming. Furthermore, the latest information on flood plain, flood duration, river configuration, etc. is required to be incorporated in the risk map, which is a complex task using conventional methods. In this context satellite remote sensing plays a very important role.

Remote sensing can provide information on flood inundated areas for different magnitudes of floods so that the extent of flooding can be related to flood magnitude. The duration of flooding can be estimated in view of multiple coverage of the same area within 2--3 days by satellites. High resolution satellites data provide information on the flood plain and details of flood control works. The extent of inundation for specific flood return periods can be estimated. Using close contour information, the inundation extent for a given water level elevation can be estimated, which is a vital input for risk zone mapping.

## 29.10 A Case Study

### 29.10.1 THE 1993 FLOODS IN THE BRAHMAPUTRA BASIN

Severe floods occurred in the Brahmaputra basin during 1993. Most of the optical satellite data were found to be cloud covered. Though a special arrangement with European Space Agency, ERS-1, SAR data acquisition was programmed and acquired, covering flood-affected areas in parts of the Brahmaputra basin. SAR data were acquired on June 29, July 4, and July 7, 1993 and analyzed along with available partially cloud-free optical data.

### 29.10.2 ANALYSIS OF SAR DATA

Using medium filtering techniques, speckle has been removed and the data have been piecewise linearly stretched for better delineation of the land--water boundary. The data were geometrically corrected for positional accuracy. Digital values over as set of cross-sections in the image were taken in order to fix the thresholds for classification of flood inundated areas. Accordingly, the classification was carried out and flood boundaries were obtained and the inundation extent was estimated in near real-time and information was passed on to the user departments. Owing to the continuous acquisition of satellite data it was possible to map the multiple waves of floods that have occurred in the Brahmaputra basin. The flood inundation extent for each magnitude of flood was calculated.

For assessing flood damages, various digital overlays were developed containing information on land use/cover, topographical details, crop detail etc., using GIS for the Marigaon District of Assam. Satellite acquired during pre-flood and post-flood were examined to understand the situation in district, prior to the flood and the effect of the flood wave that passed. The delineated flood boundaries from SAR data were intercepted with the GIS overlays and flood damages in terms of villages marooned, submerged roads/ railways affected owing to floods was estimated. The flood damages were estimated at sub-district level (Table 29.4).

### 29.10.3 INTEGRATION OF ERS-1 SAR DATA WITH OPTICAL DATA

The ERS-1 SAR data acquired during floods were integrated with pre-flood IRS data as one of the channel and a false colour composite (FCC) was made. This has highlighted the flood extent with respect to normal flow configuration of the river, area affected, submerged and non-submerged crop areas distinctly (Figure 29.1). In order to identify the affected roads and villages SPOT data, acquired during the pre-flood, were merged with ERS-1 SAR flood data. The SPOT data were edge enhanced and edges were extracted.

This was used as a channel along with SAR and SPOT PLA data and a composite image was made in which the marooned villages and the roads were highlighted. For the first time ERS-1 SAR data were extensively used, perhaps only in India. For monitoring floods and assessing flood damage in nearly real time. ERS-1 SAR data, along with other satellite data, increased the repetitive coverage of the flood affected areas. Owing

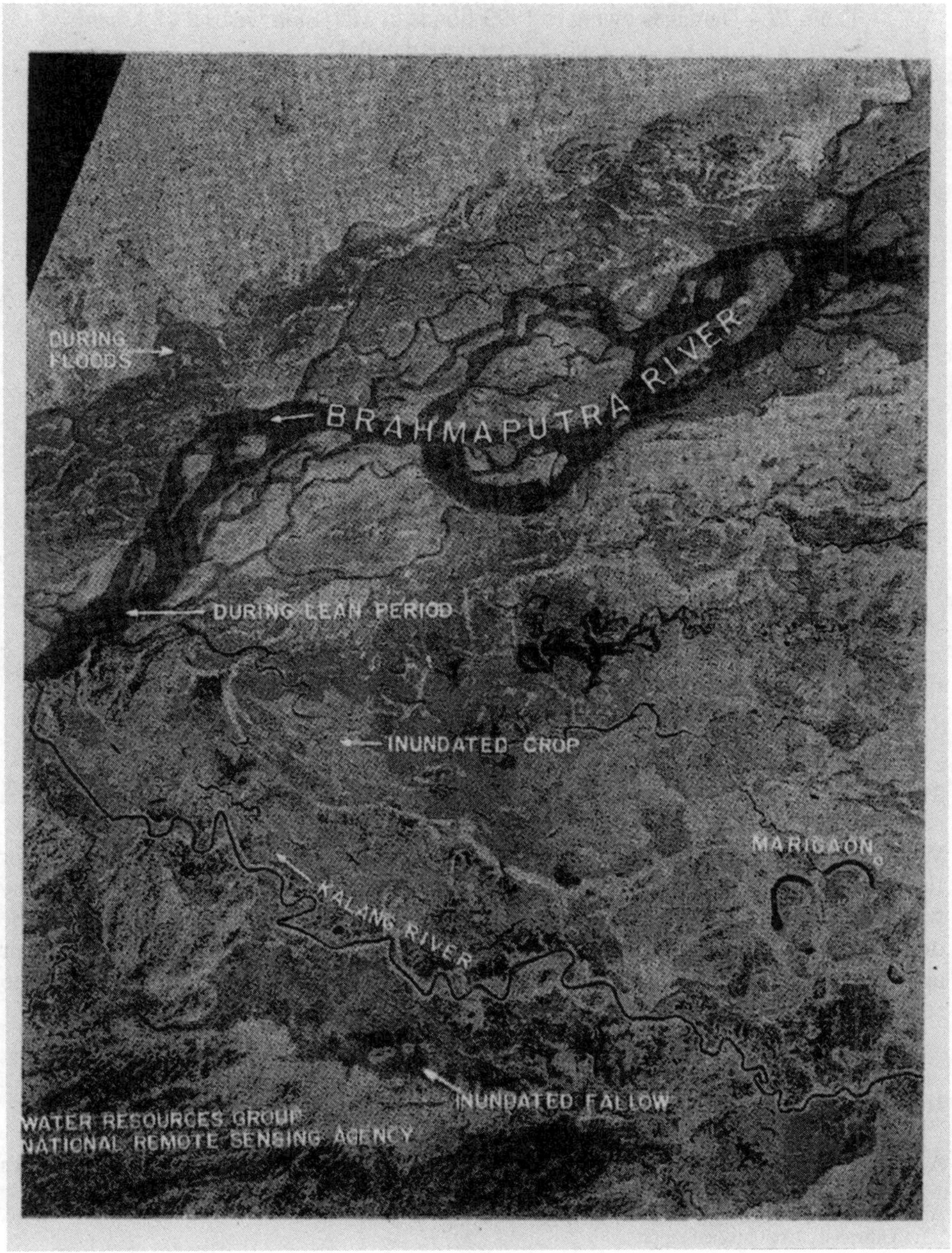


Fig. 29.1 Enhanced multi-satellite image showing flood situation in the Marigaon District (Assam) on July 7, 1993, based on ERS (flood) and IRS (pre-flood) data.

to the absence of clouds in SAR images, the flood damage assessment and estimation of the flooding extent was improved. Better results were achieved when ERS-1 SAR data were used in conjunction with other optical satellite data.

Table 29.4 Damages owing to 1993 floods in Marigaon District of Assam (sub-district wise details).

| Serial Number | Sub-district/Mouza | Inundated area km | Affected crop area km |
|---|---|---|---|
| 1. | Mayang | 83.00 | 22.68 |
| 2. | Pakaria | 85.11 | 16.18 |
| 3. | Bokani | 66.32 | 13.99 |
| 4. | Nij Gaghua | 66.60 | 60.08 |
| 5. | Manaha | 53.76 | 49.11 |
| 6. | Gobha | 21.69 | 18.92 |
| 7. | Uttar Khola | 62.38 | 32.92 |
| 8. | Tetelia | 40.87 | 37.00 |
| 9. | Marigaon | 31.87 | 31.27 |
| 10. | Bhuragaon | 43.45 | 29.53 |
| 11. | Lahorighat | 47.17 | 23.96 |
| 12. | Mairabari | 15.77 | 1.41 |
| 13. | Mikirbheta | 34.46 | 31.38 |
| 14. | Dandua | 26.35 | 25.68 |
| 15. | Chairabahi | 24.44 | 22.94 |
| 16. | Silpukhuri | 27.92 | 16.10 |
| | Total | 731.06 | 432.15 |

## 29.11 Concluding Remarks

With rapid advances occurring in space and information technologies, opportunities for flood management have increased tremendously. Early warning systems are an easy way of averting a flood disaster. It is argued that these technologies must be disseminated through educational curricula so there is little resistance to their application in field. The power of these technologies is almost limitless and is here to stay. There will be few areas which will not be affected by them.

## References

Agarwal, A. and Narian, S. (1991). Floods, flood plains and Environmental myths. State of India's Environment -3. Centre for Science and Environment, New Delhi, India.

ARSAC (1990). Landuse/Landcover maps and reports of Assam Remote Sensing Application Centre, Guwahati. Unpublished report.

Berg, C.P. Wiesnet, D.R. and Matson, M. (1979). Assessing the Red River of the north 1978 flooding from NOAA satellite data. Satellite Hydrology.

Deutsch, M. and Ruggles, F. (1974). Optical data projected applications of the ERTS-1 imagery covering the 1973 Mississippi river valley floods. Water Resources Bulletin, V 10, pp 1023-1039.

EOSAT (1993). Landsat Users Notes, V.8.
Green, A.A. Whitehouse, G. and OutHet, D. 1983 Causes of Flood stream lines observed on Landsat Images and their use as indicators o flood ways. International Journal of Remote Sensing, V. 4.
Imhoff, M.L. and Gesch, D.B. (1990). The derivation of a sub-canopy digital terrain model of flooded forest using SAR. Photogrammetric Engineering and Remote sensing, v. 56.
Lowry, R.T., Langham, E.J. and Mundry, N. (1979). A preliminary Analysis of SAR Mapping of the Manitoba flood March, 1979. Satellite Hydrology.
Moore, G.K. and North, G.W. (1974). Flood inundation in the south eastern United States form aircraft and satellite imagery. Water Resources bulletin, v. 10.
Philipson, W.R. and Hafker W.R. (1981). Manual versus digital Landsat analysis of delineating river flooding. Photogrammetric Engineering and Remote Sensing, v.47.
Phillipin, W.R., McLeester, J.N. and Hafker, W.R. (1980). Development of a model for estimating the extent of river flooding with satellite and in situ data. Unpublished project report submitted to the office of Water and Research Technology, Washington, D.C.
Pramanik, M. (1988). Role of remote sensing technology in flood studies. Seminar on Floods in Bangladesh, Dhaka.
Rammamurthy, A.S. (1989). Space and flood management. International Astronautical Federation 40th Congressm, Malaga, Spain.
Rao, U. R. (1994). Floods and droughts. Symposium on Space applications for disaster prevention, Warning, Mitigation and relief, Un committee on use of outer space, Vienna.
Shuhu, C. (1991). Chaina - The Flood Control. Remote Sensing Information System and its applications in China. Remote Sensing Technology Pllicatiuon Centre, Chaina.
Thiruvengadachari, S., Jonna, S., Hakeem, A., Raju, P.V., Bhanumurthy, V., Rao, G.S. Paul, P.R. Sankar, E.S. and Jayaseelan, A.T. (1996). Improved Water Management: The IRS-C contribution, Current Science, v 70. Pp 589-599.
Vehara, S., Sato, T., Tsuchiya, K. and Yamura, Y. (1983). Detection of Flooded area in Hokkaido based on Landsat MSS data. XVII International Symposium on Remote Sensing of Environment, Michigon.

# CHAPTER 30

# SEISMO-TECTONICS AND EARTHQUAKE DESIGN PARAMETERS

AMITA SINVHAL AND VIPUL PRAKASH

Recognizing the importance of seismicity in the Brahmaputra basin, this chapter considers design parameters which can be employed for construction of safer or earthquake-resistant buildings and hydraulic works. It also reflects on the future of building codes.

## 30.1 Introduction

North east India is one of the most seismically active regions of the world. This region has been home to two of the world's most destructive earthquakes. This chapter begins with a brief description of their effects. The geological features which cause such devastating earthquakes are then described, followed by the probable cause of earthquakes in and around the Brahmaputra basin. A site near Dibrugarh is chosen for illustrating the use of empirical formulae for estimating the possible values of peak ground accelerations from some representative earthquake scenarios. The next section describes the extent to which the present building codes consider the seismic hazard and the way in which the effects of the earthquake's shaking of the ground are considered for the design of buildings. The next two sections describe the goals and principles of earthquake-resistant design of buildings. Finally, a very brief introduction describes the direction in which future building codes are evolving.

## 30.2 A Profile of Earthquakes in Northeast India

Two great earthquakes of magnitude greater than 8 within a short span of only 53 years in 1897 and in 1950, have devastated north east India. Each of these two earthquakes devastated a very large part of the Brahmaputra basin and produced all possible earthquake effects on the terrain, the built environment, and transient phenomena, which any single earthquake has ever produced. The spectacular ground effects produced by these two earthquakes included surface faulting, topographic changes and formation of fissures. These gave rise to sand and water fountains, subsidence and elevation of ground, extensive landslides along the Brahmaputra and its many tributaries, damming of rivers, leading to change in the course of rivers and post-earthquake floods. Buildings, roads, railways, bridges, and telephone lines were damaged to varying degrees in a very large area. These earthquakes are discussed briefly in this chapter. The Srimangal earthquake of 1918 ($M > 7.6$) and the Dhubri



*V.P. Singh et al. (eds.), The Brahmaputra Basin Water Resources*, 578-610.

earthquake of 1930 (M > 7.0) also caused large scale damage. The epicentral location of more than 1,300 known earthquakes in northeast India is shown in Figure 30.1. These make northeast India one of the most seismically active regions of the world.

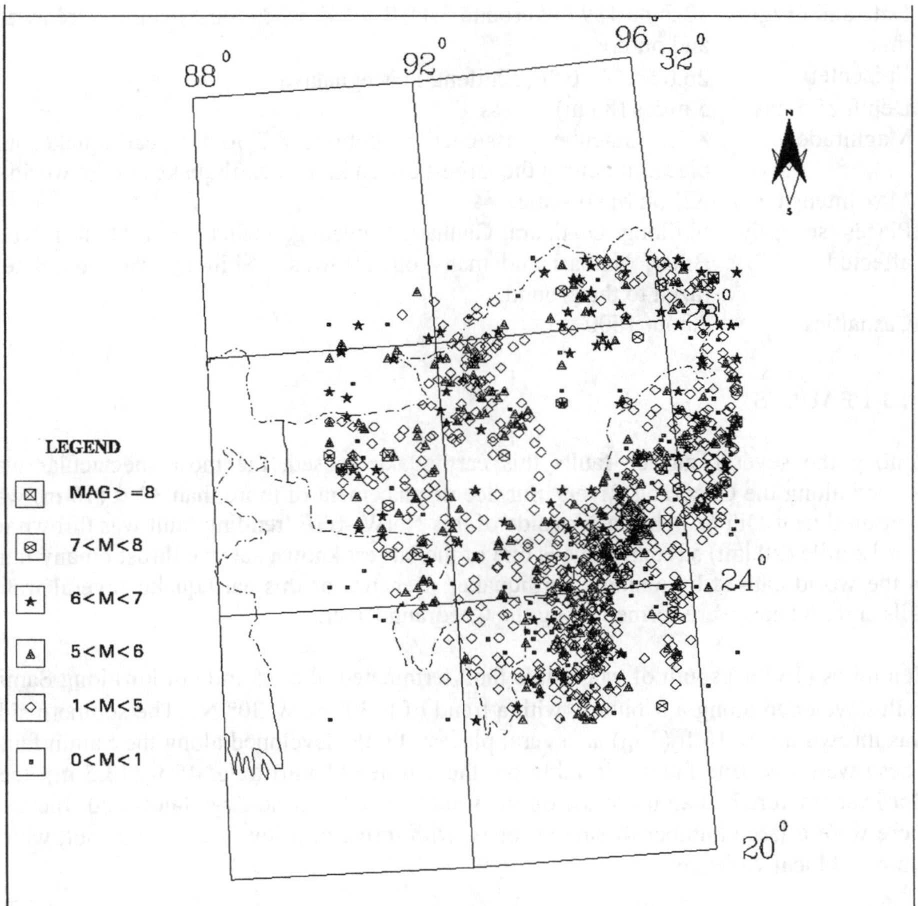


Figure 30.1 Epicentral map of north east India and contiguous regions, between latitudes 22° N to 29° N and longitudes 90° E to 97° E. Data from the USGS earthquake data-base, which shows more than 1300 epicenters, including historical earthquake data. Dotted lines show international boundaries.

## 30.3 The Assam Earthquake of 1897

This was probably one of the world's largest earthquakes that ever occurred in recent history. Oldham (1899) authored a valuable scientific memoir on this earthquake, believed to be the first book totally devoted to a single earthquake. Most of the following material has been collected from this memoir. The earthquake devastated the Shillong plateau and affected the whole of north east India and was felt in an area of approximately 1,750,000 sq. miles (4,550,000 sq. km). Serious damage to masonry buildings occurred in more than 145,000 square miles (377,000 sq. km); or, if the area from which no reports were procurable is included, the area affected would rise to about 160,000 sq. miles (416, 000 sq km). Salient features of this earthquake are as follows:

| | |
|---|---|
| Date and Origin time | 12 June 1897 at around 5.15 PM IST when most people were awake and outdoors. |
| Epicenter | 26.0 °N 91.0 °E, Shillong in Meghalaya |
| Depth of focus | 5 miles (8 km) or less |
| Magnitude | 8.7. Gutenberg assigned magnitude 8.7 to this earthquake, and placed it among the largest dozen known earthquakes in the world. |
| Max Intensity | XII on MMI scale |
| Places severely affected | Shillong, Goalpara, Gauhati, Nowgong, Sylhet, Tura, Dhubri, Kuch Bihar, Nalbari and many other towns. Shillong was completely razed to the ground. |
| Casualties | About 1600 |

### 30.3.1 FAULTS

Among the several surface faults this earthquake caused, the most spectacular was formed along the Chedrang River. Surface displacement of more than 32 ft (10 m) was measured near Dilma. The eastern side of this NNW--SSE trending fault was thrown up in a 12 mile (20 km) stretch. This is one of the largest known surface thrust on any fault in the world caused by a single earthquake. Because of this earthquake several water falls and 30 lakes were formed along the Chedrang River.

Ten miles (16 km) south of where this fault terminated, the 2.5 mile (4 km) long Samin fault developed along a tributary, with a trend of E 30° S, W 30° N. The southern side was thrown up by 10 ft (3 m) at several places. Pools developed along the Samin fault. These were two true faults. In addition, the 7 mile (11 km) long, 45 ft (13.5 m) deep Bordwar fracture had an up-throw on the south. Besides these large faults and fractures there were a large number of smaller ones, with thrust of a few inches to a foot, which indicated local violence.

It is interesting to note that compared to this the moderate sized Latur earthquake (M = 6.3) of 1993 in central India which killed over 10,000 people had an oblique displacement of 0.26 m over soil along a 30° southerly dipping plane. This was 2.5 km northwest of the Killari village in an area of 1,600 sq. m of fallow land (Pande et. al, 1995; Sinvhal et. al, 1994).

### 30.3.2 FISSURES

Wide gaping fissures opened in all directions in the alluvial plains of the Brahmaputra, spouting solid columnar fountains of water and sand, one-meter high. The outpouring sand filled beds of rivers, tanks, and wells. This disturbed the drainage system and caused extensive damage. This phenomenon was extensive in Goalpara and Kamrup districts, the western part of Darrang, Nowgogn, Sylhet, north Cachar, Rangpur,

Dinajpur, Rajshahi, Maldah, Purnea, Pabna, Bogra, Maimansingh and Dacca. In Assam fissures were reported from as far away as the Lakhimpur district, but were confined west of Subansiri River. These were also formed in the alluvium of Dhaneswari (Dhansiri) valley. In other words, the area over which fissures were formed was about 400 miles (650 km) from east to west and 350 miles (560 km) from north to south. The extreme limits from which these were reported were Sibsagar in the east and Bihar in the west, which is nearly 600 miles (960 km) in an ENE--WSW direction. In the north-south direction these were formed between the Terai regions of Nepal and Midnapur, a distance of about 300 miles (480 km).

### 30.3.3 LAKES AND POOLS

Lakes and pools were formed as a result of faulting and owing to the change of gradient of drainage channels. These were extensive in the north eastern part of the Garo hills, and to the south and south east of the Chedrang and Samin faults.

### 30.3.4 FLOODS

All along the Brahmaputra the floods of 1897 were extensive and the water level rose higher than it had done so before. Unusual flooding could be attributed to several factors such as subsidence of land, filling of river channels and shaking down of high riverbanks. Floods extended over vast tracts of land around Shillong, to the west of Garo hills and in Goalpara and Sylhet. Borpeta (26° 20', 91° 03') was the worst effected region.

### 30.3.5 LANDSLIDES

Gigantic landslides occurred in the Assam hills wherever the hill was high, narrow from side to side, fractured, and the slope was steep. Oldham described hillsides so denuded of soil that bedrock stratification was exposed. Landslides caused large scale surface distortions. Conspicuous landslides developed along the southern edge of the Garo and Khasi hills, and near the Paniathit River. In the Mahadeo valley the watercourse was modified. Sylhet, Cherrapunji, and Tura also witnessed landslides. Everywhere hillsides facing the valley were stripped bare from crest to base. Landslides were common east of the 91° meridian. The eastern limits of landslides were the north Cachar hills. No landslips occurred in the Mikir hills further to the northeast.

Landslides were more widespread in the Himalayas north of the Brahmaputra than in the south of the Brahmaputra valley. These maximized in and around Goalpara. Tezpur was the eastern limit; and Bhutan, Sikkim and Darjiling marked the western limit.

### 30.3.6 ACCELERATIONS

Shaking of the ground was so strong that Oldham concluded the maximum acceleration to exceed gravity at several places in the vast epicentral tract. Several posts shot out of their holes and boulders were lifted out of the ground vertically upwards, without cutting the edges of their former seats, indicating that vertical accelerations exceeded gravity. Eyewitnesses reported seeing pebbles bouncing on the ground 'like peas on a drumhead'. A 30 ft (10 m) length of wave is reported to have had a height (amplitude) of one foot. The maximum amplitude of horizontal vibration was as much 7 inches (17.5 cm), the period being one second, giving a response acceleration of 0.7 g at T = 1.0 seconds (EQ 98 - 10).

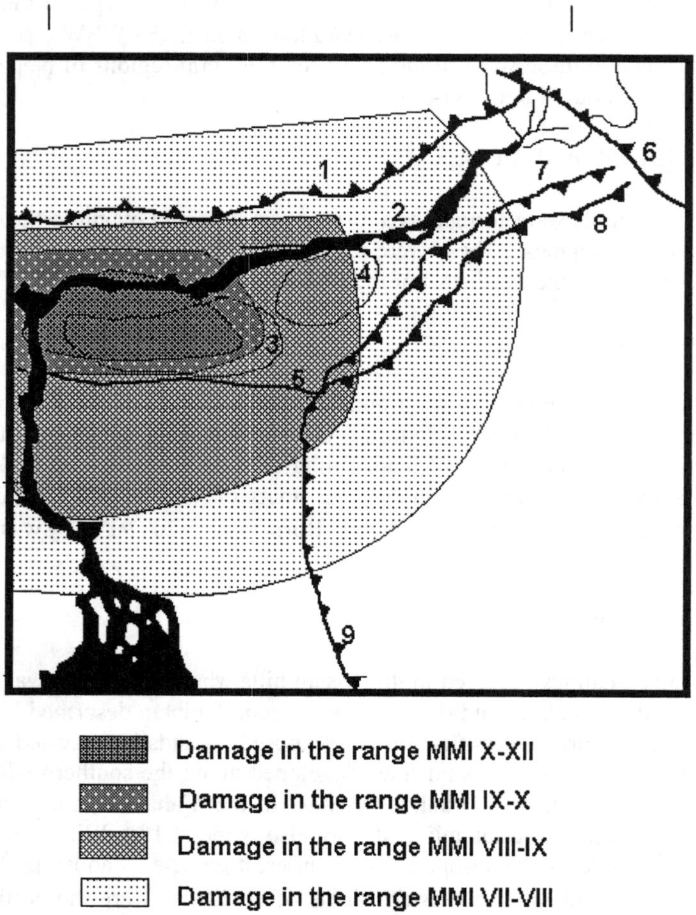


Figure 30.2 Iso-seismal map for the great Assam earthquake of 1897. (Redrawn after Oldham, 1899). Simplified tectonics around the Brahmaputra are redrawn after Balakrishnan, (1997). 1 Himalayan frontal arc, 2 River Brahmaputra, 3 Shillong plateau, 4 Mikir hills, 5 Dauki fault zone, 6 Mishmi and Lohit thrust zone, 7 Naga thrust, 8 Disang thrust zone, 9 southward continuation of the Naga--Disang thrust zone.

## 30.3.7 SURFACE DISTORTIONS

Changes in level were observed in far off places on the Shillong plateau, such as at Mairang (25°34'N, 91°42'E) and Tura (25°29'N, 90°16'E). The description at Mairang thus goes: "Before the earthquake I was informed that only a short stretch of the road was visible, where it rounded a spur at about 3 miles off; the crest of an intervening ridge hiding the road before it came out round the next spur. Now a much longer stretch of the road is visible, and it can be seen rounding the next spur, where I was positively informed the road could not be seen before the earthquake."

A military station in the hills of Tura regularly exchanged heliograph signals with Rowamari, which was on the River Brahmaputra. Before the earthquake it was possible to do this from a certain spot by a grazing ray over an intervening hill. After the earthquake there was no difficulty at all, and instead of Rowamari being just visible over the hilltops, a broad stretch of plains east of the Brahmaputra could be seen. Re-surveys after the earthquake confirmed extensive changes, but since all monuments were in the disturbed area no reference points could be assumed to be fixed and details of warping could not be derived.

At Tura ground deformations were conspicuous and included landslides, liquefaction, and formation of cracks, fissures, ridges, and furrows. Some hills sank bodily while many others rose, or changed positions. A classical example is that of heliograph signaling. Houses built on steep hill slopes or at the base of cliffs were destroyed by landslides. Large crater-like pits (diameter of 60--140 ft) were formed in the ground. Large cracks (1 mile long, 2--3 ft wide, and 16 ft deep) developed at several places. Fountains ejected sand, water, coal, peat, resin, semi-petrified timber, and 6--7 ft high through these fissures. Because of liquefaction houses sank bodily into the ground, and for some only the roof was visible. The canal bed was heaved up by 20 ft, and so was the bed of the Jinjiram River, but not to the same level. Both settled down after the earthquake, though not to their normal level. Several water bodies, such as 'bhils' also silted up. Embankments sank 2-3 ft for long distances. Large stones were thrown up in the air (of 20" diameter, 10" thickness), 3 ft away and were turned upside down. Stone abutments of bridges over the Ronkon and Mafilkol were shaken down. An 80 ft long timber bridge was heaved up to a height of 8 ft above its proper level at the twenty-second mile of the Mankachar road. These effects were not peculiar to Tura alone but extended to the entire Shillong plateau, in places such as Dhubri, Goalpara, Gauhati, Shillong, Cherrapunji and Dilma.

Permanent changes accompanying the earthquake were not confined to one spot, but extended over the northern part of the Assam hills for a distance of 100 miles from east to west. Surface faulting was most conspicuous in this region. These changes were most conspicuous west of the 91° meridian.

### 30.3.8 PROBABLE CAUSE OF EARTHQUAKE

The original disruption started on a thrust fault, which initiated several aftershocks along branch faults. Aftershocks were closely connected with subsequent movements of these faults and helped in locating these. Oldham believed the existence of several foci in a 300 km by 80 km tract. According to Oldham, the NNW–SSE trending Chedrang fault is considered the causative lineament for the 1897 earthquake.

## 30.4 Assam Tibet Earthquake of 1950

This was an equally disastrous earthquake (M = 8.7), which affected the remotest parts of north east India and the contiguous regions of Burma, China and Tibet. The damage owing to this earthquake was even more devastating than that of the 1897 earthquake. It had profound effects downstream of the Brahmaputra River and its many tributaries in upper Assam. Salient features of this earthquake are as follows:

| | |
|---|---|
| Date and Origin time | 15 August 1950, Evening, at around 7.40 pm, (Pramanik and Mukerji, 1953.) |
| Epicenter | ➢ 28° 40' N 96° E, in the Dibang valley (Poddar, 1953).<br>➢ 28.7°N, 96.6°E, Rima, on the China--Tibet border, in the Mishmi block, outside the Indian border.<br>➢ The epicenter of the earthquake, determined by the seismographic recordings in India and other countries, was about 200 miles north of Sadiya, in mountainous country on the north--east border of Assam (now in Arunachal Pradesh). |
| Depth of focus | ➢ 15--25 km (Banerji, 1953).<br>➢ 14 km (Pramanik and Mukerjee, 1953) |
| Magnitude | ➢ 8.7, Instrumental<br>➢ 8.6 estimated from records from Pasadena and Strasbourg (Banerji, 1953) |
| Iso-seismal maps | Various workers prepared these.<br>➢ Ray (1953) based the map on reports published between August 15 and September 3, 1950. The twelve point MMI scale was used.<br>➢ Poddar (1953) and Tandon (1953) used the ten point Rossi Forel Scale (RF)<br>All isoseismals were elongated along the Brahmaputra valley and indicate that the damage experienced at Dibrugarh was the highest in the scale, XII on the MMI scale and X on the RF scale. |
| Places severely affected | The Abor and Mishmi hills in the northern portion of upper Assam were the worst affected, together with the central Brahmaputra valley. Severe damage was reported in Lakhimpur and the northern part of the Sibsagar districts. Dibrugarh, Jorhat, Sadiya, Kebo Pasighat road, Lohit valley road, and Saikhowa Bongloi road and 70 villages were destroyed. |
| Felt area | Estimated area over which extensive and heavy damage occurred was 300 sq. miles. The estimated area in Assam, which suffered minor damage, was 30,000 sq. miles.<br>The earthquake was felt over an area greater than one million sq. km in India, China, Burma and Bangladesh, in the south western quadrant with respect to the epicenter. Assuming that the violence of the earthquake was felt similarly in the other three quadrants, it may reasonably be estimated that the earthquake was felt in an area exceeding 4 million sq. km. |
| Casualties | 1,526, in a sparsely populated area. Of these 952 were in the |

| | |
|---|---|
| | Mishmi and Abor Hills, 80% of the rest were owing to floods in the Subansiri. Therefore post-earthquake floods caused more casualties than did the earthquake. The affected region was sparsely populated and the earthquake occurred when the people were awake, otherwise the casualties would have been much higher. |
| Epicentral Accelerations | 0.2 g for rock formations, 0.4 g for alluvium, 1.0 g on top of very tall structures (Pramanik and Mukerjee, 1953). |
| Seismic Moment (Molnar, 1990) | ➢ $M_0 = 6\times10^{21}$ Nm, on a 200 km long thrust fault, gently dipping towards the north. The average slip on a thrust fault 100 km wide would have been 9 m.<br>➢ $M_0 = 2\times10^{21}$ Nm, on a NNW trending right lateral strike slip fault, about 400 km long. |

### 30.4.1 LANDSLIDES

Gigantic landslides scarred the hills of northeast Assam, disrupting the drainage of innumerable streams; inundation of which swept the countryside for months after the quake. Changes in the main drainage lines of the Brahmaputra were reported. Rivers rose high after the earthquake bringing down sand, mud, trees, and all kinds of debris. This was largely owing to the change of topography, which was brought about by enormous landslides in the Mishmi and Abor hills. Pilots who flew over the meizo-seismal area reported great changes in topography owing to landslides on the rivers Dihang, Dibang, Subansiri, Tidding, Lohit, Burhi Dihing and other tributaries of the Brahmaputra. Artificial lakes and dams on these rivers, caused by massive landslides, burst and caused havoc by flooding the area downstream. The natural dam across the Subansiri burst four days after the earthquake and 20 ft (6 m) high waves swept away villagers and caused 532 deaths. This caused more damage to life and property than the earthquake itself.

## 30.5 Why is the Region Around the Brahmaputra Repeatedly Devastated by So Many Earthquakes?

The answer is provided by the theory of plate tectonics. This theory postulates that the entire surface of the earth is made up of several large and small plates. These plates are in constant motion with respect to each other and the margins of these plates are prone to earthquakes. Convection currents inside the earth are responsible for plate movement. The convection currents generated are of two types – ascending and descending. These currents cause the movements of plates either towards each other (convergent) or slowly force them apart (divergent). Thus when two plates converge mountain ranges are formed and uplifted; and when they diverge the two adjoining plates move away from each other. The Himalayan mountain ranges were formed by the convergence of the Indian and Eurasian plates.

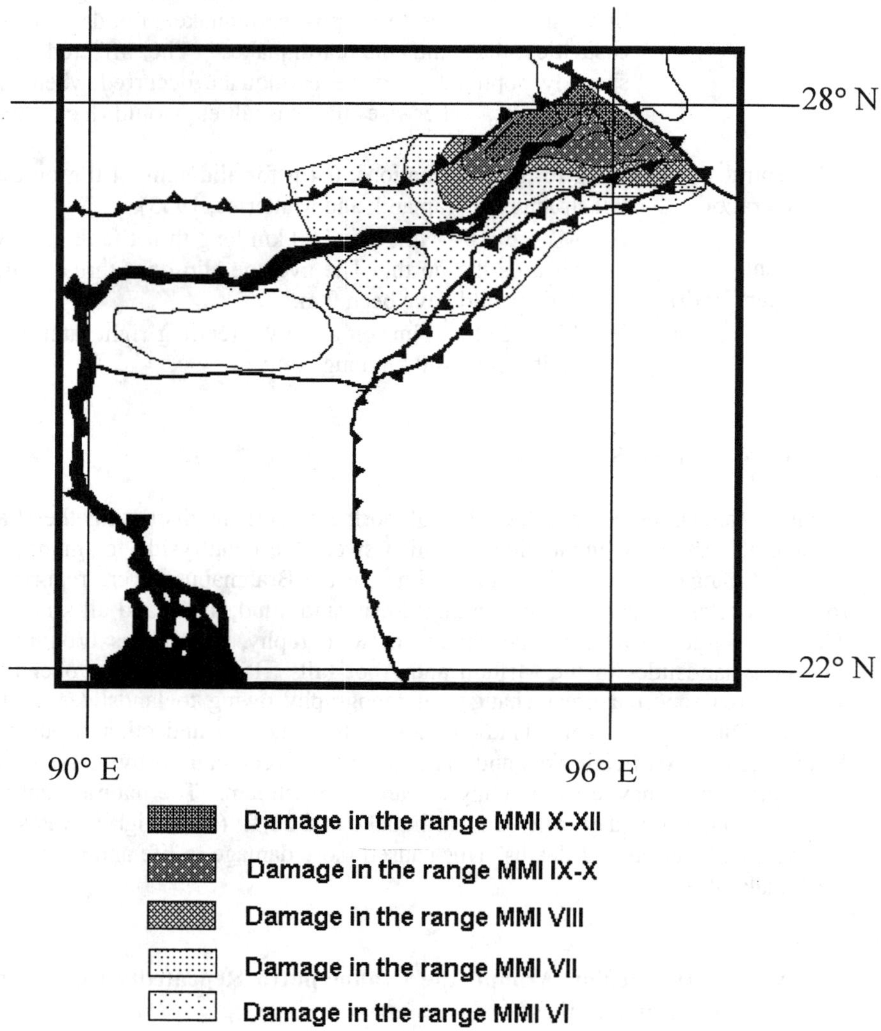


Figure 30.3 Iso-seismal map for the great Assam earthquake of 1950. (Redrawn after Poddar, 1953). Simplified tectonics around the Brahmaputra is the same as shown in figure 2.

Sieches: Seiches were observed as far away as Norway and England.

All the present day continents once formed a single super continent called Pangea. 135 million years ago they started splitting, first into two large continents---the northern one was called Laurasia and the southern one was called Gondwanaland. The present day land mass of India was part of Gondwanaland. Gradually the two large continents further split. India separated as an island from Gondwanaland and slowly moved northward, closing the huge Tethys Sea. In the meantime part of Laurasia formed the present day Eurasia.

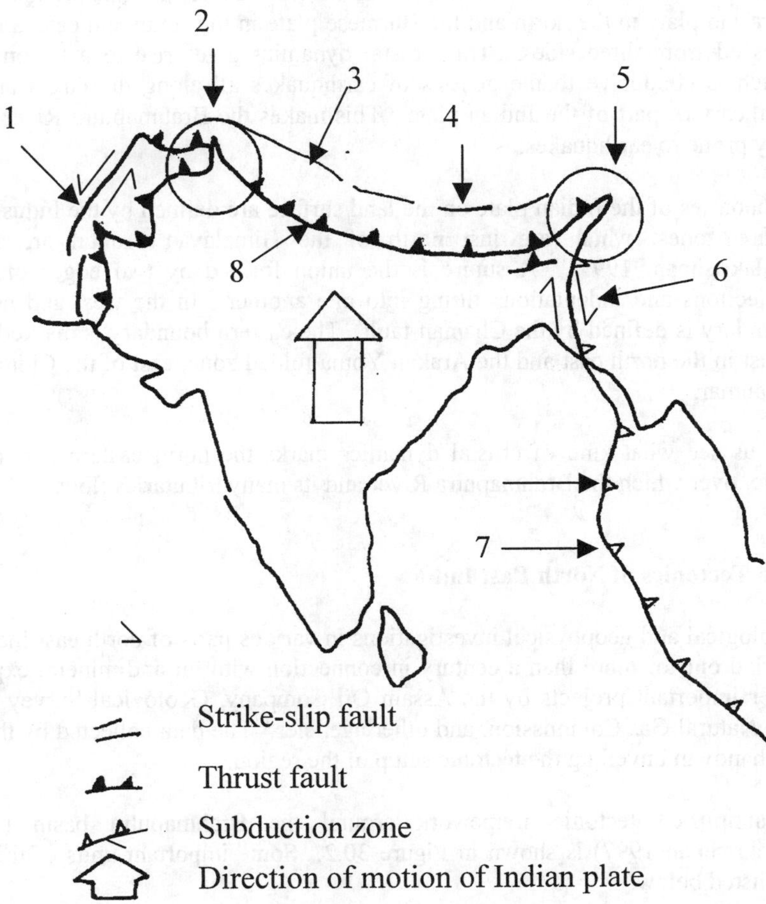


Figure 30.4 Map showing simplified boundaries of part of the Indian plate (modified and redrawn after Balakrishnan, 1997; Yeats et. al, 1997). 1 Chaman fault, 2 North western syntexis, 3 Indus suture zone, 4 Tsangpo suture zone, 5 North eastern syntexis, (Lohit and Mishmi thrust zone), 6 Arakan Yoma folded zone, 7 subduction zone, 8 Himalayan frontal arc.

India collided with Eurasia, first near the present day Ladakh, about 60 million years ago. Then the rest of the island closed in with the Eurasian land mass, the north eastern part of India being the last to do so. The northward push crumpled, the intervening landmass to form the Himalayan ranges. Remnants of the Tethys sea exist as fossils in the Himalayan ranges. The northern edge of the moving Indian land mass is marked by two very sharp knee bends (or syntexial bends), marked by the Nanga Parbat in the west (in north Kashmir) and Namche Barwa in the east. The eastern bend is the point where the Brahmaputra takes a very sharp U- bend and then enters India.

The north eastern part of the northward moving Indian plate is wedged between the Eurasian plate in the north and the Burmese plate in the south and east, and is therefore stressed from three sides. This crustal dynamics gives rise to a tectonic framework which is conducive to the genesis of earthquakes all along the edges and within the northeastern part of the Indian plate. This makes the Brahmaputra River and its basin very prone to earthquakes.

Boundaries of the Indian plate on the land surface are defined by the Indus and Tsangpo suture zones, which are just north of the Himalayan Frontal arc in the north (Balakrishnan, 1997). A suture is the union formed by two edges of a plate with projections and indentations fitting into one another. In the west and north west the boundary is defined by the Chaman fault. The eastern boundary is marked by the Lohit thrust in the north east and the Arakan Yoma folded zone, east of the Chittagong hills in Myanmar.

Let us see what kind of crustal dynamics marks the north eastern part of the Indian plate, over which the Brahmaputra River and its many tributaries flow.

## 30.6 Tectonics of North East India

Geological and geophysical investigations in various parts of north east India have been carried out for more than a century in connection with oil and mineral exploration and other important projects by the Assam Oil Company, Geological Survey of India, Oil and Natural Gas Commission, and other agencies. The data collected by these agencies are handy in unveiling the tectonic setup of the region.

A simplified tectonic framework around the Brahmaputra basin (mostly after Balakrishnan 1997) is shown in Figure 30.2. Some important units of this framework are listed below:

- The Himalayan Frontal Arc.
- The Shillong Plateau.
- The Haflong–Disang–Naga thrust zone.
- The east Himalayan Syntexis, (a term used to include both pure melting and assimilation of country rock) which includes the Mishmi and Lohit thrust zone.
- The Assam basin.
- The Indo Myanmar arc which includes the Arakan Yoma folded zone in the north and the Andaman and Nicobar islands in the south.

### 30.6.1 THE HIMALAYAN FRONTAL ARC

This arc borders the upper Brahmaputra valley on the north and north east. The Main Boundary fault and the foothill thrusts are part of this arc. The latter forms the lowermost structural feature of the Arunachal Himalayas and gives rise to the neo-tectonic activity along this arc.

### 30.6.2 THE SHILLONG PLATEAU

This plateau together with the smaller Mikir massif, represents an uplifted block of Archaean to Pre-Cambrian age. It includes the Garo, Khasi, Jaintia and Mikir hills. The existence of a criss-cross pattern of faults as well as shattered belts in the Shillong plateau and Mikir hills bear testimony to the great stress suffered by these areas, which may be owing to the probable northward and eastward movement of the Indian plate. A very high shear stress occurs in this area, which could be a precursor of high seismic activity. The Upper Brahmaputra basin and the Shillong plateau are part of the Indian peninsula. The limits of the Shillong plateau are demarcated by several major faults on all sides as follows:

- The western limit of this plateau is marked by the NNW--SSE trending Dhubri fault (west of Tura), which separates the Indian peninsular from the folded regions in the east.
- The east–west trending Dauki Fault extends till Haflong in the east (25° N), south of Shillong. The Shillong plateau has moved 250 km eastward along this strike slip fault (Evans,1964). The fault shows vertical tectonics (Auden 1972), the throw being 17 km (Hiller and Elahi, 1988). The Dauki fault also marks the southern limit of the Shillong plateau. The Dauki fault zone can be considered a major lineament associated with the earthquake of 1897. The great earthquake of 1897 could have been caused by movements along the Dauki fault or the Chedrang and Samin faults, features oblique to the Dauki fault.
- Towards north is the east--west trending Mikir fault. A deep trough exists between this fault and the Main Boundary Fault (MBF) and the Main Central Thrust (MCT) of the Himalayas, which occurs a little further to the north. A fault along the Brahmaputra River separates the Shillong plateau from the MBF. Both the Dauki and the Brahmaputra faults are cut off on the west by the Dhubri fault. The MCT, MBF, the Brahmaputra fault and the Dauki fault form a system of parallel east west trending faults.
- The alluvium of the Kopili valley, which shows a pronounced NW--SE trend, and the Naga thrust, marks the eastern margin of the plateau. The Shillong plateau is cut off by the Naga and other thrusts on the east. These thrusts also cut off the Dauki, Brahmaputra and Mikir faults at the surface. The subsurface extension of these faults indicates that the Shillong plateau is seen to plunge under the thrusts and terminates in a fault parallel to the Naga thrust. The Dauki fault also extends eastwards under the thrust. This separates the N--S trending Tripura hills from the NNE--SSW trending hills of the Arakan Yoma. It is an apparent extension of the peninsular mass of the Shillong plateau.

### 30.6.3 THE HAFLONG–DISANG–NAGA THRUST ZONE

This zone is south east of the upper Assam valley and these merge near Haflong. This is a possible subduction zone along the India--Myanmar plate boundary. In a subduction zone two plates collide with each other and one plate goes down beneath the other. The Indian plate is going beneath the Myanmar plate. The Naga thrust and the Disang thrust have a pronounced NE--SW trend and are believed to be associated with deep-seated basement faults. The Naga thrust and Haflong--Disang thrust converge at depth and then probably merge into each other in the east. This zone has undergone less tectonic deformation during Quaternary times (0--2 million years). The south east extension of these NE--SW trending thrusts merge with the eastward extension of the Dauki fault zone near Haflong. These thrusts were tectonically active in the late Tertiary and sub-recent times. This thrust zone is assumed to dip 25° towards the south east and east.

### 30.6.4 THE EAST HIMALAYAN SYNTEXIS

This marks the north eastern boundary of the Indian plate. It is characterized by the NW--SE trending Mishmi block and is popularly known as the Lohit Himalayas. The Mishmi thrust separates the Mishmi block from the sedimentary basins of the Brahmaputra valley, to its west. The NW--SE trending and the high angle (75° towards the northeast) Lohit thrust denotes the eastern margin of the Himalayan Syntexis. Mishmi and Lohit thrusts are amongst the youngest thrusts in the region. These can be considered as potential seismic source zones. The great earthquake of 1950 originated in this region.

### 30.6.5 THE ASSAM BASIN

This basin, with Tertiary sediments in the Brahmaputra valley, is surrounded on all sides by the above mentioned tectonic units. Between the Arunachal Himalayas and the Haflong—Disang--Naga thrust zone the river Brahmaputra drains through Assam in an ENE--WSW direction. The valley is covered with alluvium with only an occasional presence of gneissic rocks and sedimentary hillocks along its course. It is possible that the course of the Brahmaputra in the middle and upper reaches is controlled by basement faults.

### 30.6.6 THE INDO MYANMAR ARC

This arc separates the Indian plate from the adjoining Myanmar plate. The Indian plate subducts below the Myanmar plate in the east. The line of contact at the surface is the Arakan--Yoma, east of the Chittagong hills. This continues southwards as the Sunda trench and the arc formed by the Andaman and Nicobar Islands. These islands indicate the downward buckling of the Indian crust. The Chindwin and Irrawady basins, the volcanic arc and the Andaman sea are thrust over the Indian plate. The volcanic arc consists of the extinct core of Taugthoulon, several craters near Vuntho, craters and intrusives near Manywa, the extinct volcanoes Mt Popa and the dolerite sills and dykes of Pegu Yoma. The arc continues seawards and contains the active volcanoes of Narcondam and Barren islands.

The complex tectonics of the region has given rise to many major faults around the Brahmaputra river such as the Main Boundary fault and the Main Central Thrust in the north, Lohit and Mishmi thrust in the north east, Haflong-Disang-Naga-Lushai thrusts in the southeast, and several more bounded within the region described by the above three, such as the Dauki fault, Jamuna fault, Sylhet fault, Barapani shear zone, Dudhnoi fault,

Kalyani shear zone, Bomdilla lineament, Dhansiri Kopili fault, Anterkhet fault, the two Hinge zones, North Mikir fault, East Barail fault, Bengal basin margin fault, south edge of Bengal shelf, east Andaman fault and Sagaing fault.

### 30.7 Probable Cause of Earthquakes in and Around the Brahmaputra Basin

The Himalayan frontal arc is convex to the south. The Indo Myanmar arc is convex to the west. These two arcs meet in the form of a semicircle around the Himalayan syntexis. These arc systems are a measure of the enormous stresses in operation in the region, which is also the spot where the great earthquake of 1950 is believed to have originated. The Himalayan arc is involved in a continent to continent collision, whereas the Indo Myanmar arc is involved in a subduction process (Dewey and Bird, 1970; Seeber and Ambruster, 1981).

Major earthquakes in the north eastern region of India are associated with the Himalayan Frontal Arc, the Eastern Himalayan Syntexis, the chain of faults in the Indo--Myanmar fold belt and major faults that encompass the Shillong plateau. These are potential seismic sources in this region. Because of the very high seismicity of north east India, the possibility of damaging earthquakes recurring is large. The brief description of the damage caused by the earthquakes of 1897 and 1950 is only a grim reminder of what can go wrong in the region. Shallow focus earthquakes occur along these thrusts. However, south east of the Naga Disang thrust zone, towards the Manipur--Myanmar border region and the Arakan Yoma folded belt, the epicentres are more concentrated and the foci are at greater depths. It shows that moving from Naga Disang thrust to the Arakan Yoma belt there is a gradual increase in depth of focus from shallow to intermediate. The India Myanmar border region along the Arakan Yoma folded belt represents a subduction zone of the Indian plate, which under-thrusts below the Myanmar region.

Fault plane solutions for events in northeast India show thrusting with significant strike slip movements. In the Arunachal Himalayas, near the Main Boundary fault, pressures are oriented almost along the strike of the fault while along the Indo-Myanmar region the pressures (compression) act at right angles or acute angles to the mountains. This was reported by Choudhury and Srivastava (1976) on the basis of data which came from observatories in the USSR, China and India.

Fitch (1970) has shown that the focal mechanism solutions of earthquakes in northeast India were consistent with the under-thrusting of the Indian plate towards the north beneath the Himalayas. For the Indo Myanmar belt, for earthquakes of intermediate depth, the tension axes were oriented down the dip of the descending plate. These solutions were consistent with the postulated plate tectonic model of the region. However, Ichikawa et al. (1972), Tandon and Srivastava (1975) do not find a consistent relationship between the focal mechanism solution and plate tectonics. According to them the compression and tension axes, whose plunges were shallow, are nearly parallel and perpendicular to the mountain ranges in northeast India.

Because of such a complex mix of tectonics and earthquakes, more devastating earthquakes are inevitable in this region. Also, since the hydroelectric potential of the

Brahmaputra is tremendous, therefore this mighty river will be harnessed by hydroelectric projects and high dams (EQ 83-9, EQ 83-17, EQ 83-26, EQ 84-2) at several places. The development activity will lead to more construction activity such as rail and road bridges, super-thermal power plants (EQ 84-4), oil fields (EQ 88-4), refineries and industrial complexes, multistoried buildings (EQ 83-16, EQ 83-15, EQ 84-11, EQ 84-1, EQ 84-19, EQ 84-22, EQ 90-11, EQ 90-14) etc. It is imperative that their planning and design should take into account the earthquake potential of the region and safety factors be introduced at the design stage.

## 30.8 Seismotectonic Implications for a Site Near Dibrugarh -- A Case Study

A rail-cum-road bridge is proposed to be constructed at Bogibil in Dibrugarh district (EQ Srivastava, 1984-85). The site lies on the Indian plate in the seismically vulnerable east Himalayan syntexis. It is surrounded by several known major regional faults and thrusts, which are capable of generating great or major earthquakes. These are as follows.

1. **The Dauki fault**. It is situated more than 300 km south west of the project site and is considered the major lineament associated with the 1897 Assam earthquake.
2. The eastward continuation of the Dauki fault is the Haflong fault, which extends towards the north east into the Naga--Disang thrust zone.
3. The **Mishmi and Lohit thrust zone**, which supported the great Assam earthquake of 1950, is about 110 km north east of the project site.
4. The **Main Boundary Fault** and other faults and thrusts associated with it are 25 km northwest of the site. This fault is considered capable of generating great earthquakes. It is believed that the great Kangra earthquake of 1905 in Himachal Pradesh originated on the westward extension of this fault.
5. Besides these known mega-faults and thrusts other major and minor blind and capable faults may also exist. This unknown factor would further complicate the seismotectonic picture around the site. The Brahmaputra River probably has its course aligned along such lineaments. In the absence of any definite information hypothetical earthquakes are often considered for evaluating ground motion characteristics like accelerations. Earthquakes of various magnitudes, with their epicenters in the alluvial planes of Assam, have also been observed. These are probably related to **unknown faults** in the basement. Therefore an unknown fault, capable of generating an earthquake in the immediate environs of any project site, should also be considered for design purposes.

Based on the simplified tectonic framework around the Brahmaputra River discussed earlier, the earthquake history and potential of the area, and implications of these to any project site in the area, the following earthquake parameters may be adopted for evaluating peak ground accelerations at the project site, using several magnitude—distance--acceleration--attenuation relationships. These are given in Table 30.1.

The possible source zones and the kind of earthquakes they can generate are as follows.

1. An earthquake of magnitude 8.7 at a depth of 33 km at a distance of about 300 km southeast of the site, originating in the Dauki fault zone. This corresponds to the 1897 earthquake.

2. An earthquake of magnitude 8.7 at a depth of 33 km at a distance of about 110 km northeast of the project site, originating in the NW--SE trending Mishmi and Lohit thrust zone. This corresponds to the 1950 earthquake.
3. An earthquake of magnitude 7.5 with a depth of focus 20 km, occurring at a distance of about 60 km from the project site, on the Haflong--Disang--Naga thrust zone.
4. An earthquake of magnitude 7.5 with depth of focus 20 km occurring at a distance of about 25 km northwest of the project site, on the Main Boundary Fault.

Besides these four scenarios, other earthquake scenarios may also be visualized for unknown faults. Four such scenarios are considered in Table 30.1. This table shows the computed peak ground accelerations (PGA) by considering three of the many formulae currently in vogue. Different formulae give a very large range of PGA for the same seismic parameters. These formulae do not consider many aspects of the problem, including: (i) the nature of the causative fault, such as its length, depth, dip, strike and its rupture during an earthquake; and (ii) soil conditions at the site.

The project site lies in seismic zone V of the current seismic zoning map of India (IS:1893–1984). This is the zone of the severest seismicity and corresponds to the seismic intensity IX on the Modified--Mercalli Intensity (MMI) scale. During the great earthquake of 1950, the project site was assigned an even higher intensity, MMI X+. As the site is of strategic importance a detailed seismic assessment is necessary.

## 30.9 Consideration of Seismic Hazard to Buildings in Present Building Codes

Earthquakes pose a number of potential hazards to building structures. Earthquake ground shaking, fault movements, ground failure (liquefaction, differential compaction, and landslides), and flooding can all damage buildings and lead to their collapse, and all these effects have been amply demonstrated in the Brahamputra basin during the 1897 Assam earthquake and during the 1950 Assam--Tibet earthquake. Out of these hazards, building codes usually address only the hazard resulting from earthquake's shaking of the ground. The other hazards, i.e., owing to fault movements, ground failure, and flooding also pose a significant threat to the safety of life and can result in immense economic losses. These hazards can sometimes be mitigated by selecting less hazardous sites and by adopting remedial measures. Although usually not addressed in building codes, these hazards should also be explicitly considered in the planning and design process.

The earthquake-induced ground shaking has higher intensity near the epicenter and this intensity usually reduces with distance from the epicenter. The earthquake-induced ground shaking can be resolved into one vertical and two horizontal (longitudinal and transverse) components. The intensity of the vertical component diminishes more rapidly with distance from the epicenter. In building codes it has been customary to take the intensity of the vertical component as either one-half or two-thirds of the intensity of the horizontal components. However, close to the epicenter the intensity of the vertical component is comparable with the intensity of the horizontal components and may be taken equal.

Table 30.1 Peak ground accelerations owing to potential sources.

| S.N | Seismic Source | Earthquake Parameters | | | Peak Ground Accelerations | | |
|---|---|---|---|---|---|---|---|
| | | Magnitude | Distance (km) | Depth (km) | McGuire 1977* | Abrahamson & Litehiser 1989 | Campbell 1989 |
| 1 | Dauki Fault zone (corresponding to 1897 earthquake). | 8.7 | 300 | 33 | 0.07 g | 0.03 g | 0.05 g |
| 2 | Mishmi and Lohit Thrust zone (corresponding to the 1950 earthquake) | 8.7 | 110 | 33 | 0.20 g | 0.08 g | 0.14 g |
| 3 | Haflong Naga Disang Thrust | 7.5 | 60 | 20 | 0.17 g | 0.09 g | 0.10 g |
| 4 | Main Boundary fault | 7.5 | 25 | 20 | 0.30 g | 0.17 g | 0.17 g |
| 5 | Unknown faults associated with the Brahmaputra | 7.5 | 20 | 10 | 0.39 g | 0.18 g | 0.23 g |
| | | 7.5 | 10 | 10 | 0.482 g | 0.24 g | 0.31 g |
| | | 7.0 | 20 | 10 | 0.28 g | 0.15 g | 0.17 g |
| | | 7.0 | 10 | 10 | 0.350 g | 0.20 g | 0.23 g |

*McGuire's formula (1977) is valid for a hypocentral distance greater than 15 km.

The vertical component of the earthquake-induced ground shaking causes vertical loads and the horizontal components cause horizontal loads in a building. These loads are proportional to ground acceleration and also to the weight (mass) of the building in accordance with Newton's Second Law of Motion. Buildings usually have sufficient strength against vertical loads, because they are conservatively designed for the ever-present gravity loads owing to the weight of the building. However, buildings are usually weak and have low strengths against horizontal loads, which occur only rarely during earthquakes and wind storms. For this reason, buildings are generally more vulnerable to the horizontal components of the earthquake-induced ground shaking. Building codes specify that the total horizontal load, H, in any direction be calculated by a formula of the following type.

$$H = C_h W \tag{30.1}$$

in which W is the total weight of the building and $C_h$ is a coefficient for horizontal seismic load. $C_h$ may depend on several factors, such as the seismic zone for the site and the level of earthquake considered; the type of the soil profile at the site; the type of the foundation proposed; the importance of the structure; and the fundamental period (T), the damping ratio ($\xi$), and the behavioral characteristics of the lateral load resisting system of the structure.

In India, as per IS:1893--1984 Indian Standard Criteria for Earthquake Resistant Design, $C_h$ is given by the following expression:

$$C_h = \beta I F_0 \frac{S_a}{g} K \qquad (30.2)$$

in which: $\beta =$ the soil-foundation system factor ranging between 1.0 for good soil-foundation types and 1.5 for poor soil-foundation types; $I =$ the importance factor ranging between 1.0 for ordinary buildings (most buildings) and 1.5 for important buildings (essential facilities like hospitals, fire-brigade, etc.; and buildings where large loss of life is a possibility such as cinema halls, etc.); $F_0 =$ the seismic zone factor ranging from 0.05 for the Zone I (the least seismic zone) to 0.4 for Zone V (the most seismic zone); $S_a / g =$ the average acceleration coefficient which is a function of vibration time period, $T$, and damping ratio, $\xi$, of the structure. This function is called the design acceleration response spectra. For the most commonly adopted damping ratio of 5 %, it has a maximum value of 0.2, which corresponds to 1-storey to 6-storey buildings having a time period ranging from 0.1 to 0.6 seconds; and $K =$ the performance factor depending on the structural framing system and brittleness or ductility of construction. Its value ranges from 1.0 for moment resistant ductile frames with or without reinforced concrete shear walls to 1.6 for frames with masonry infills and non-ductile reinforced concrete frames.

As per IS:1893--1984, the whole of north east India, including the Brahmaputra basin, lies in the most severe seismic zone, zone V; and for this zone the value of $C_h$ ranges between 0.08 (= 1.0×1.0×0.4×0.2x1.0) to 0.288 (= 1.5×1.5×0.4×0.2×1.6). The lower value of 0.08 is for ordinary buildings with ductile frames having good soil-foundation types; and, the higher value of 0.228 is for important buildings with non-ductile frames and founded on relatively poor soil-foundation types. The ground motions corresponding to these values of $C_h$ are underestimates of those that can be expected and this is made clear by clause 0.7 of IS:1893--1984 which states:

*"It is pointed out that structures will normally experience more severe ground motion than the one envisaged in the seismic coefficient specified in this standard. However, in view of the energy absorbing capacity available in the inelastic range, ductile structures will be able to resist such shocks without much damage. It is, therefore, necessary that ductility must be built into the structures since brittle structures will be damaged more extensively."*

### 30.10 Goals of Earthquake Resistant Design of Buildings

The low values of seismic loads specified in building codes follow from the realization that it is economically not feasible to design ordinary buildings to remain linear elastic (i.e., undamaged) in a rare major earthquake. It is common to design for seismic loads that are a fraction, perhaps as low as $1/12$ to $1/4$ (i.e., 8% to 25%) of those corresponding to elastic response (Chopra 1995, Paulay 1992, Penelis 1997, Priestley et al 1996). This design practice is at variance with the design practice for other loads such as owing to gravity or wind where the loads are conservatively overestimated. The performance objectives implied for earthquake resistant design for ordinary buildings are as follows:

- No damage to either structural or non-structural components in frequent minor earthquakes.
- Limited non-structural damage, but no structural damage to structural components in occasional moderate earthquakes.
- No collapse (i.e., the building should still be able to carry its weight so that the loss of life may not occur) in a rare major earthquake. The building may be damaged beyond economic repair and may have to be demolished after the earthquake.

Thus buildings are expected to exhibit linear elastic behavior in minor earthquakes, slightly nonlinear behavior in moderate earthquakes, and highly nonlinear behavior in a major earthquake. Owing to the dearth of computer programs for conducting nonlinear analysis and lack of a broad consensus on how such programs are to be used, the present building codes usually specify seismic loads corresponding to minor earthquakes for design using linear analysis techniques. A satisfactory performance in rare major earthquakes can be expected only if a building's nonlinear behavior under a rare major earthquake is similar to its linear behavior under a minor earthquake. This is usually possible if simple principles of earthquake resistant design are followed.

### 30.11 Principles of Earthquake Resistant Design of Buildings

Some of these principles can be stated in brief as follows.

1. **Building should be simple**: Inspections of earthquake damage demonstrate that the simpler the building the better the behavior, all other parameters being the same. There are two main reasons for this. First, it is easier to understand the linear and nonlinear seismic behavior of a simple building compared to a complex one. Second, it is easier to understand, formulate in drawings, and construct simple structural details than complicated ones.

2. **Building should be symmetric in both plan directions**: A lack of symmetry causes the floors of a building to twist about vertical axes during the action of seismic lateral loads. These twists cause outer columns and walls to undergo much greater deformation compared to the interior ones. The torsion effects (twisting) are difficult to assess properly and can be very destructive. IS:1893--1984 therefore recommends that buildings having plans with shapes like L, T, E and Y, etc., shall preferably be separated into rectangular parts.

3. **Building should be regular in plan and elevation**: In a regular building there is a uniform and continuous distribution of strength, stiffness and ductility. A lack of regularity leads to stress-concentration and damage near the locations of irregularity. In shaking a building an earthquake will relentlessly seek out every structural weakness. These weaknesses are usually created by sharp changes in strength, stiffness and ductility. A common enough example is provided by buildings in which the walls at the ground level are removed to provide space for parking or shops. In these buildings the ground storey has much lower stiffness and strength compared to the upper storeys and thus forms a **soft storey**. In such buildings failure takes place in the soft storey.

4. **Building should be lightweight** (avoid unnecessary masses): The seismic forces ($H = \alpha_h W$) produced during an earthquake increase as the weight ($W$) of building increases. Therefore the use of building materials with a high strength to weight ratio is preferable. Thus the best construction material is good quality seasoned timber for low rise buildings; and, structural steel for high rise buildings. Reinforced concrete and reinforced masonry are next best. Unreinforced stone masonry and unreinforced brick masonry are among the least desirable and should not be used in highly seismic regions.

5. **Structure should have sufficient initial stiffness** to avoid significant deformations and damage to adjacent non-structural components (such as unreinforced masonry partitions) under minor and moderate earthquakes.

6. **Structure should be sufficiently ductile** (stability of strength with deformation) so that it is less sensitive to the effects of sudden failure of interacting non-structural components under a rare major earthquake. Structural elements attract seismic lateral loads in proportion to their stiffness. When a reinforced concrete frame is designed to resist lateral loads and unreinforced masonry walls are placed within this frame; then walls which are very stiff compared to frame members, attract most of the seismic lateral loads. Unfortunately, unreinforced walls tend to be weak and brittle. Their failure leads to sudden transfer of their load to the columns. This sudden transfer is equivalent to a gradual transfer of twice the load. Consequently columns, too, fail immediately after failure of the brittle walls. Therefore, it is important that sufficient ductility is built into the frame members so that they can absorb the sudden transfer of load from adjacent non-structural walls. This is normally achieved by the provision of closely spaced ties at the ends of beams and columns (IS:4326--1993 Earthquake Resistant Design and Construction of Buildings, IS:13920--1993 Ductile Detailing of Reinforced Concrete Structures subjected to Seismic Forces). It is also important that the walls be reinforced so that their load carrying capacity reduces gradually rather than suddenly on failure.

7. **Components resisting seismic loads must be effectively tied together** so that they can collectively resist the load. An isolated wall is strong for lateral loads along its plane, but it is very weak for lateral loads perpendicular to it and can topple easily. When cross walls are tied to it at right angles, then these walls can resist large lateral loads in any direction. The tying can be accomplished by providing reinforced concrete floors and roof. It can also be accomplished at intermediate levels by providing reinforced concrete bands at the plinth level, sill

level, lintel level and roof level. These bands need to be continuous at junctions. More bands are needed if the mortar used is weak and in high seismic regions.

The seismic lateral loads amplify greatly from the base to the top of a building. Water tanks, chimneys and parapet walls on the roof are particularly vulnerable. These must be effectively secured in their place to prevent their fall. These principles and many other good engineering practices must be scrupulously observed in order to justify the low values of seismic lateral loads used in design. Buildings designed for building code level seismic loads without regard to the principles of earthquake resistant design and good engineering practice often prove fatal during a moderate or major earthquake.

## 30.12 Consideration of Seismic Hazard in Forthcoming Building Codes

Owing to the availability of the computer programs DRAIN-2DX DRAIN-3DX and DRAIN-BUILDING (Prakash and Powell, 1993a,b,c) for static and dynamic nonlinear analyses of buildings, efforts to develop a broad consensus on nonlinear analysis techniques for design and evaluation of structures for seismic loads corresponding to moderate and major earthquakes are continuing. More precise specification of minor, moderate, and major earthquakes and the corresponding performance targets are being attempted. Thus future building codes are expected to specify several levels of earthquake ground shaking with their different expected performance levels (targets).

## 30.13 Earthquake Levels Specified in ATC-40

The Applied Technology Council (ATC), USA, in its report ATC-40: Seismic Evaluation and Retrofit of Concrete Buildings (1996), defines the following three levels (SE, DE and ME) of earthquake ground shaking.

- **The Serviceability Earthquake** (SE): The SE is defined probabilistically as the level of ground shaking that has a 50% chance of being exceeded in a 50 year period. The SE has a mean return period of approximately 75 years and represents a frequent level of ground shaking that is likely to be felt during the life of the structure. This level of earthquake ground shaking is typically about 0.5 times the level of ground shaking of the Design Earthquake (DE).

- **The Design Earthquake** (DE): The DE is defined probabilistically as the level of ground shaking that has a 10% chance of being exceeded in a 50-year period. The DE has a mean return period of approximately 500 years and represents an infrequent level of ground shaking that may occur during the life of the structure.

- **The Maximum Earthquake** (ME): The ME is defined deterministically as the maximum level of earthquake ground shaking which may ever be expected at the structure site within the known geological framework, or the ground motion with a 5% chance of being exceeded in a 50 year period. The ME has a return period of approximately 1,000 years and is intended to represent an upper bound on the level of ground shaking that could be reasonably expected to occur at the site. This level of ground shaking is typically about 1.25 to 1.5 times the level of ground shaking of the DE.

## 30.14 Performance Levels Specified in ATC-40 for Ordinary Buildings

Corresponding to the three levels of earthquake ground shaking (SE, DE, and ME), ATC-40 and ATC-34: A Critical Review of Current Approaches to Earthquake Resistant Design (1995), recommend the following basic (minimum) **performance levels** (targets) for new construction.

- After the Serviceability Earthquake (SE) the structure must remain **operational** and ready for immediate occupancy. Only very limited structural and non-structural damage may occur and the basic vertical and horizontal load resisting systems of the building must retain nearly all their pre-earthquake characteristics and capacities. Although minor disruption and cleanup should be expected, all equipment and machinery should be working. As described earlier, the SE represents a frequent level of ground shaking that is likely to be felt many times during the life of the structure, and therefore to meet this objective the structure must remain essentially elastic under SE.

- After the Design Earthquake (DE) significant damage to the structure may occur but some margin against either total or partial structural collapse must remain to ensure **safety** of **life**. Major structural components must not become dislodged threatening safety of life. Whilst injuries during the DE may occur, the risk of life-threatening injury from the structural damage must be very low. It should be expected that extensive structural repairs would be necessary prior to reuse of the structure. To meet this objective the structure must have an adequate combination of strength and ductility.

- After a Maximum Earthquake (ME) the structure may be on the verge of experiencing partial or total collapse, but must be **structurally stable**. Substantial damage to the structure, including significant degradation in the stiffness and strength of the lateral load resisting system, may occur. But all significant components of the vertical load resisting system must continue to carry the gravity loads. To meet this objective the structure must have an adequate combination of strength and ductility.

The performance levels for essential facilities and hazardous facilities may be more stringent.

## 30.15 Determination of Elastic Horizontal Seismic Load Coefficient $C_{hE}$ in ATC-40

As per ATC-40 the value of the horizontal seismic load coefficient $C_{hE}$ corresponding to SE, DE and ME depends upon Site Soil Characteristics and Site Seismicity Characteristics as follows.

(a) **Site Soil Characteristics:** Each site is assigned a Soil Profile Type based on Table 30.2.

For sites in the Brahmaputra basin the soil profile may correspond to types $S_B$, $S_C$, $S_D$, $S_E$ or $S_F$. Soil profile types $S_C$, $S_D$ or $S_E$ are likely to be applicable for most sites. Sites with soil profile type $S_F$ are better avoided.

(b) **Site Seismicity Characteristics:** Site Seismicity characteristics are based on the Seismic Zone Factor for the site, the proximity of the site to active seismic sources, and the site's soil profile characteristics.

Table 30.2 Soil profile types.

| Soil Profile Type | Soil Profile Description | Average Soil Properties for Top 100 feet of Soil Profile | | |
|---|---|---|---|---|
| | | Shear Wave Velocity, $\overline{v}_s$ (feet per second) | Standard Penetration Test, $\overline{N}$ (or $N_{CH}$ for cohesionless soil layers) (blows/foot) | Undrained Shear Strength, $\overline{s}_u$ (psf) |
| $S_A$ | Hard Rock | $\overline{v}_s > 5{,}000$ | Not Applicable | |
| $S_B$ | Rock | $2{,}500 < \overline{v}_s < 5{,}000$ | Not Applicable | |
| $S_C$ | Very Dense Soil and Soft Rock | $1{,}200 < \overline{v}_s < 2{,}500$ | $\overline{N} > 50$ | $\overline{s}_u > 2{,}000$ |
| $S_D$ | Stiff Soil Profile | $600 < \overline{v}_s < 1{,}200$ | $15 < \overline{N} < 50$ | $1{,}000 < \overline{s}_u < 2{,}000$ |
| $S_E$ | Soft Soil Profile | $\overline{v}_s < 600$ | $\overline{N} < 15$ | $\overline{s}_u < 1{,}000$ |
| $S_F$ | Very Soft Soil Profile | Soil Requiring Site Specific Evaluation | | |

(i) Seismic Zone Factor: Table 30.3 gives the values of the Seismic Zone Factor for sites in US.

Table 30.3 Seismic zone factor Z.

| Zone | 1 | 2A | 2B | 3 | 4 |
|---|---|---|---|---|---|
| $Z$ | 0.075 | 0.15 | 0.20 | 0.30 | 0.40 |

Zone 4 is for sites with the severest seismicity in the state of California. Considering the earthquake data and seismo-tectonic implications of the geological features, the Brahamputra basin is arguably more seismic compared to sites in California, and can be assigned the severity of Zone 4 with $Z = 0.4$.

(ii) Near Source Factor: Each site is assigned near source factors, $N_A$ and $N_V$, according to tables 4 and 5.

Considering that in the Brahmaputra region there have been two great earthquakes (M > 8) within a span of 50 years and a large of number of M7+ earthquakes and a SR > 20 mm/year, this region corresponds to Seismic Source Type A.

Table 30.4 Seismic source type.

| Seismic Source Type | Seismic Source Description | Seismic Source Definition | |
|---|---|---|---|
| | | Maximum Moment Magnitude, $M$ | Slip Rate, $SR$ (mm/year) |
| A | Faults that are capable of producing large magnitude events and which have a high rate of seismic activity | $M \geq 7.0$ | $SR \geq 5$ |
| B | All faults other than types A and C | Not Applicable | Not Applicable |
| C | Faults that are not capable of producing large magnitude events and that have a relatively low rate of seismic activity | $M < 6.5$ | $SR < 2$ |

Table 30.5 Near source factors, $N_A$ and $N_V$

| Seismic Source Type | Closest Distance to Known Seismic Source | | | | | | | |
|---|---|---|---|---|---|---|---|---|
| | < 2 km | | 5 km | | 10 km | | ≥ 15 km | |
| | $N_A$ | $N_V$ | $N_A$ | $N_V$ | $N_A$ | $N_V$ | $N_A$ | $N_V$ |
| A | 1.5 | 2.0 | 1.2 | 1.6 | 1.0 | 1.2 | 1.0 | 1.0 |
| B | 1.3 | 1.6 | 1.0 | 1.2 | 1.0 | 1.0 | 1.0 | 1.0 |
| C | 1.0 | 1.0 | 1.0 | 1.0 | 1.0 | 1.0 | 1.0 | 1.0 |

As per ATC-40 the shape and the intensity of the earthquake response spectra is determined by the values of Seismic Coefficients $C_A$ and $C_V$, which depend on the soil type and the product $ZEN_A$ or $ZEN_V$, where $E$ is taken as 0.5 for Serviceability Earthquake, 1.0 for Design Earthquake and 1.25 (Zone 4 sites) or 1.50 (Zone 3 sites) for the Maximum Earthquake. Table 30.6 gives the values of $ZEN_A$ or $ZEN_V$ for sites in the Brahmaputra basin.

Table 30.6 Values of $ZEN_A$ or $ZEN_V$ for sites in Brahmaputra basin.

| Earthquake Level | Closest Distance to Known Seismic Source | | | | | | | |
|---|---|---|---|---|---|---|---|---|
| | < 2 km | | 5 km | | 10 km | | ≥ 15 km | |
| | $ZEN_A$ | $ZEN_V$ | $ZEN_A$ | $ZEN_V$ | $ZEN_A$ | $ZEN_V$ | $ZEN_A$ | $ZEN_V$ |
| *SE* | 0.30 | 0.40 | 0.24 | 0.32 | 0.20 | 0.24 | 0.20 | 0.20 |
| *DE* | 0.60 | 0.80 | 0.48 | 0.64 | 0.40 | 0.48 | 0.40 | 0.40 |
| *ME* | 0.75 | 1.00 | 0.60 | 0.80 | 0.50 | 0.60 | 0.50 | 0.50 |

Seismic coefficients $C_A$ and $C_V$ for Soil Profile types $S_C$, $S_D$ and $S_E$ are computed corresponding to the values of $ZEN_A$ or $ZEN_V$ according to Tables 30.7 and 30.8.

Table 30.7 Computation of seismic coefficient $C_A$

| Soil Profile Type | Shaking Intensity, $ZEN_A$ | | | | | |
|---|---|---|---|---|---|---|
| | = 0.075 | = 0.15 | = 0.20 | = 0.30 | = 0.40 | > 0.40 |
| $S_B$ | 0.075 | 0.15 | 0.20 | 0.30 | 0.40 | 1.0 ($ZEN_A$) |
| $S_C$ | 0.09 | 0.18 | 0.24 | 0.33 | 0.40 | 1.0 ($ZEN_A$) |
| $S_D$ | 0.12 | 0.22 | 0.28 | 0.36 | 0.44 | 1.1 ($ZEN_A$) |
| $S_E$ | 0.19 | 0.30 | 0.34 | 0.36 | 0.36 | 0.9 ($ZEN_A$) |

Table 30.8 Computation of Seismic Coefficient $C_V$

| Soil Profile Type | Shaking Intensity, $ZEN_V$ | | | | | |
|---|---|---|---|---|---|---|
| | = 0.075 | = 0.15 | = 0.20 | = 0.30 | = 0.40 | > 0.40 |
| $S_B$ | 0.075 | 0.15 | 0.20 | 0.30 | 0.40 | 1.0 ($ZEN_A$) |
| $S_C$ | 0.13 | 0.25 | 0.32 | 0.45 | 0.56 | 1.4 ($ZEN_V$) |
| $S_D$ | 0.18 | 0.32 | 0.40 | 0.54 | 0.64 | 1.6 ($ZEN_V$) |
| $S_E$ | 0.26 | 0.50 | 0.64 | 0.84 | 0.96 | 2.4 ($ZEN_V$) |

By using Tables 30.6, 30.7 and 30.8 the computed values of $C_A$ and $C_V$ for *SE, DE* and *ME* as a function soil profile type and closest distance to known seismic source are given in Table 30.9.

Using the values of $C_A$ and $C_V$, the horizontal seismic coefficient, $C_{hE}$, for elastic response and 5 % damping ratio is computed by the following equations.

$$T_S = \frac{C_V}{2.5\ C_A} \qquad (30.3), (30.4)$$

$$T_A = 0.2\ T_S$$

$C_{hE} = C_A$ at Period, $T = 0.0$.

$$C_{hE} = C_A\left(1.0 + 1.5\frac{T}{T_A}\right) \text{ for Periods, } 0.0 \le T \le T_A \text{ seconds.} \tag{30.5}$$

$$C_{hE} = 2.5\ C_A \text{ for Periods, } T_A \le T \le T_S \text{ seconds.} \tag{30.6}$$

$$C_{hE} = \frac{C_V}{T} \text{ for Periods, } T_S \le T < 3.0 \text{ seconds.} \tag{30.7}$$

The values of $C_{hE}$ for a site with soil profile type $S_C$ and located at a distance ≥ 15 km are plotted as a function of time period, *T*, in Figure 30.5.

Figure 30.5 shows that the maximum value of horizontal seismic coefficient for elastic response is 1.25, 1.0 and 0.6 for ME, DE, and SE, respectively. These high values imply that for elastic response a building must be designed for horizontal loads of the same order as gravity loads. This is neither practical nor economical for ordinary structures, and provides the reason for existence of a distinct philosophy for seismic design, as discussed earlier.

Table 30.9 The values of $C_A$ and $C_V$ for sites in the Brahmaputra basin.

| Earthquake Level | Closest Distance to Known Seismic Source | | | | | | | |
|---|---|---|---|---|---|---|---|---|
| | < 2 km | | 5 km | | 10 km | | ≥ 15 km | |
| | $C_A$ | $C_V$ | $C_A$ | $C_V$ | $C_A$ | $C_V$ | $C_A$ | $C_V$ |
| For Soil Profile Type $S_B$ | | | | | | | | |
| *SE* | 0.30 | 0.40 | 0.24 | 0.32 | 0.20 | 0.24 | 0.20 | 0.20 |
| *DE* | 0.60 | 0.80 | 0.48 | 0.64 | 0.40 | 0.48 | 0.40 | 0.40 |
| *ME* | 0.75 | 1.00 | 0.60 | 0.80 | 0.50 | 0.60 | 0.50 | 0.50 |
| For Soil Profile Type $S_C$ | | | | | | | | |
| *SE* | 0.33 | 0.56 | 0.28 | 0.47 | 0.24 | 0.37 | 0.24 | 0.32 |
| *DE* | 0.60 | 1.12 | 0.48 | 0.90 | 0.40 | 0.67 | 0.40 | 0.56 |
| *ME* | 0.75 | 1.40 | 0.60 | 1.12 | 0.50 | 0.84 | 0.50 | 0.70 |
| For Soil Profile Type $S_D$ | | | | | | | | |
| *SE* | 0.36 | 0.64 | 0.31 | 0.56 | 0.28 | 0.46 | 0.28 | 0.40 |
| *DE* | 0.66 | 1.28 | 0.53 | 1.02 | 0.44 | 0.77 | 0.44 | 0.64 |
| *ME* | 0.83 | 1.60 | 0.66 | 1.28 | 0.55 | 0.96 | 0.55 | 0.80 |
| For Soil Profile Type $S_E$ | | | | | | | | |
| *SE* | 0.36 | 0.96 | 0.35 | 0.86 | 0.34 | 0.72 | 0.34 | 0.64 |
| *DE* | 0.54 | 1.92 | 0.43 | 1.54 | 0.36 | 1.15 | 0.36 | 0.96 |
| *ME* | 0.68 | 2.40 | 0.54 | 1.92 | 0.45 | 1.44 | 0.45 | 1.20 |

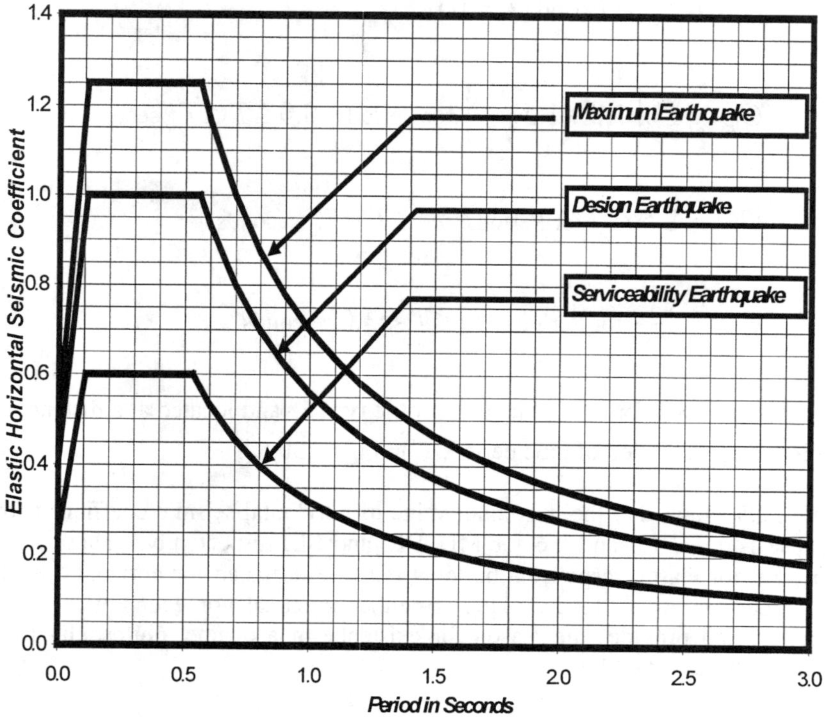


Fig. 30.5 5% Damped Elastic Horizontal Seismic Coefficient for Serviceability Earthquake (SE), Design Earthquake (DE) and Maximum Earthquake (ME) for Site with Soil Profile Type Sc and located atleast 15 km from Known Seismic Source

## 30.16 Conclusions

The Brahmaputra basin lies in one of the most seismically active regions of the world and has been home to two great earthquakes (in 1897 and 1950). The seismo-tectonic environment, the nature of ground motions and the response spectrum, that may be reasonably considered for design have been discussed in this chapter. For construction of any engineering structure, such as dams, hydroelectric projects, rail and road bridges, oil fields, refineries, industrial complexes, multi-storied complexes, consideration of the possible seismic effects cannot be ignored. These considerations will lead to a safer built environment in case of future earthquakes in the region.

## 30.17 Acknowledgments

The authors would like to thank the Head of the Department, Department of Earthquake Engineering, University of Roorkee, Roorkee, Uttaranchal, for providing all necessary facilities for preparation of this chapter.

**Appendix**

**MODIFIED MERCALLI INTENSITY SCALE:** This comprehensive scale describes the intensity of earthquakes precisely. The main definitions are given in IS code 1893--1984, pages 57--62. It first describes the type of building and then classifies the damage it has suffered.

**a) Type of Structure (Buildings)**

| | |
|---|---|
| Structure A | Buildings in field stone, rural structures, houses, clay houses. |
| Structure B | Ordinary brick buildings, buildings of large block and prefabricated type, half timbered structures, buildings in natural hewn stone. |
| Structure C | Reinforced buildings, well built wooden structures. |

**Definition of Quantity**

| | |
|---|---|
| Single, few | About 5 percent |
| Many | About 50 percent |
| Most | About 75 percent |

**b) Classification of Damage to Buildings**

| | |
|---|---|
| Grade 1 | **Slight damage**. Fine cracks in plaster; fall of small pieces of plaster. |
| Grade 2 | **Moderate damage**. Small cracks in walls; fall of fairly large pieces of plaster, pantiles slip off; cracks in chimneys; parts of chimney fall down. |
| Grade 3 | **Heavy damage**. Large and deep cracks in walls; fall of chimneys. |
| Grade 4 | **Destruction**. Gaps in walls; parts of buildings may collapse; separate parts of the building lose their cohesion; inner walls collapse. |
| Grade 5 | **Total damage**. Total collapse of buildings. |

**c) Intensity Scale:**

I **Not noticeable:** The intensity of the vibration is below the limit of sensibility; the tremor is detected and recorded by seismographs only.

II **Scarcely noticeable (very slight):** Vibration is felt only by individual people at rest in houses, especially on upper floors of buildings.

III **Weak, partially observed:** The earthquake is felt indoors by a few people, outdoors only in favourable circumstances. The vibration is like that caused by the passing of a light truck. Attentive observers notice a slight swinging of hanging objects, somewhat more heavily on upper floors.

IV **Largely observed:** The earthquake is felt indoors by many people, outdoors by few. Here and there people awake, but no one is frightened. The vibration is like that caused by the passing of a heavily loaded truck. Windows, doors and dishes rattle. Floors and walls creak. Furniture begins to shake. Hanging objects swing slightly. Liquids in open vessels are slightly disturbed. In standing motor cars the shock is noticeable.

V **Awakening:** a) The earthquake is felt indoors by all, outdoors by many. Many sleeping people awake. A few run outdoors. Animals become uneasy. Buildings tremble throughout. Hanging objects swing considerably. Pictures knock against walls or swing out of place. Occasionally pendulum clocks stop. Unstable objects may be overturned or shifted. Open doors and windows are thrust open and slam back again. Liquids spill in small amounts from well filled open containers. The sensation of vibration is like that owing to heavy object falling inside the buildings.
b) Slight damage in buildings of type A are possible.
c) Sometimes a change in the flow of springs.

VI **Frightening:** a) Felt by most indoors and outdoors. Many people in buildings are frightened and run outdoors. A few persons lose their balance. Domestic animals run out of their stalls. In a few instances dishes and glassware may break, books fall down. Heavy furniture may possibly move and small steeple bells may ring. b) Damage of Grade 1 is sustained in single buildings of Type B and in many of Type A. Damage in a few buildings of Type A is of Grade 2. c) In a few cases cracks up to widths of 1 cm are possible in wet grounds; in mountains occasional landslips; changes in the flow of springs and in the level of well water are observed.

VII **Damage to buildings:** a) Most people are frightened and run outdoors. Many find it difficult to stand. The vibration is noticed by persons driving motor cars. Large bells ring. (b) In many buildings of Type C damage of grade 1 is caused; in many buildings of Type B the damage is of grade 2. Most buildings of Type A suffer the damage of Grade 3, a few of grade 4. In single instances land slips of roadways on steep slopes; cracks in roads; seams of pipelines damaged, cracks in stone walls.

VIII **Destruction of buildings:** a) Fright and panic; also persons driving motor cars are disturbed. Here and there branches of trees break off. Even heavy furniture moves and partly overturns. Hanging lamps are damaged in part. b) Most buildings of Type C suffer damage of grade 2, and a few of grade 3. Most buildings of Type B suffer damage of grade 3, and most buildings of Type A suffer damage of grade 4. Many buildings of Type C suffer damage of grade 4. Occasional breaking of pipe seams. Memorials and monuments move and twist. Tombstones overturn. Stone walls collapse. c) Small land slips in hollows and on banked roads on steep slopes; cracks in ground up to widths of several centimetres. Water in lakes becomes turbid. New reservoirs come into existence. Dry wells refill and existing wells become dry. In many cases changes in the flow and level of water.

**IX** **General damage to buildings:** a) General panic; considerable damage to furniture. Animals run to and fro in confusion and cry out. b) Many buildings of Type C suffer the damage of grade 3, a few of grade 4. Many buildings of Type B show damage of grade 4; a few of grade 5. Many buildings of Type A suffer damage of grade 5. Monuments and columns fall. Considerable damage to reservoirs; underground pipes partly broken. In individual cases railway lines are bent and roadways damaged. c) On flat land an overflow of water, sand and mud is often observed. Ground cracks to widths of up to 10 cm on slopes and river banks more than 10 cm; furthermore, a large number of slight cracks in the ground; fall of rock, many land slides and earth flows: large waves in water. Dry wells renew their flow and existing wells dry up.

**X** **General destruction of buildings:** a) Many buildings of Type C suffer damage of grade 4, a few of grade 5. Many buildings of Type B show damage of grade 5; most of Type A have destruction of grade 5; critical damage to dams and dykes and severe damage to bridges. Railway lines are bent slightly. Underground pipes are broken or bent. Road paving and asphalt show waves. b) In the ground cracks up to widths of several centimetres, sometimes up to 1 metre. Parallel to water courses broad fissures occur. Loose ground slides from steep slopes. From river banks and steep coasts considerable landslides are possible. In coastal areas displacement of sand and mud; change of water level in wells; water from canals, lakes, rivers, etc, thrown on land. New lakes occur.

**XI** **Destruction:** a) Severe damage even to well built buildings, bridges, water dams and railway lines; highways become useless; underground pipes destroyed. b) Ground considerably distorted by broad cracks and fissures, as well as by movement in horizontal and vertical directions; numerous land slips and falls of rock. The intensity of the earthquake requires to be investigated specially.

**XII** **Landscape changes:** a) Practically all structures above and below the ground are greatly damaged or destroyed. b) The surface of the ground is radically changed. Considerable ground cracks with extensive vertical and horizontal movements are observed. Fall of rock and slumping of river banks over wide areas, lakes are dammed; waterfalls appear, and rivers are deflected. The intensity of the earthquake requires to be investigated specially.

Intensity I is assigned to an event which is felt by a few only under exceptionally favourable conditions, e.g., by people at rest in high rise buildings. Intensity VI means that the earthquake was felt by all at that particular place. Not much damage is associated with it, except that there may be some cases of fallen plaster, etc. Intensities X to XII are very severe cases of destruction where earthquake safety cannot be ensured in traditional buildings. However, for large crucial structures such as dams, nuclear power plants, power houses, bridges, etc. earthquake safety measures can be incorporated at an increased expenditure of only 5%--10% of the project cost. Intensity XII represents a catastrophe in which nearly all man made structures collapse and objects are thrown up in the air. The intermediate intensities vary and describe various degrees of shaking.

## References


Abrahamson, N.A., and J.J. Litehiser (1989). Attenuation of Vertical Peak Acceleration, BSSA, v. 79, No. 3, 549-580p.

ATC – 34 (1995). A Critical Review of Current Approaches to Earthquake Resistant Design, Applied Technology Council, 555 Twin Dolphin Drive, Suite 550, Redwood City, California 94065, 94p.

ATC – 40 (1996). Seismic Evaluation and Retrofit of Concrete Buildings, Applied Technology Council, 555 Twin Dolphin Drive, Suite 550, Redwood City, California 94065, 520p.

Auden, J.B. (1972). Review of the Tectonic Map of India, Mem. Geol Soc India, 13, 101, 107.

Balakrishnan, T.S. (1997). Major Tectonic Elements of the Indian Subcontinent and Contiguous Areas: A Geophysical View, Mem 38, Geological Society of India,155p.

Banerji, S.K. (1953). The Origin and the Nature of the Disturbance Produced by the Assam Earthquake of August 15, 1950 and its Aftershocks, p 11 – 15, *in* M.B.R Rao (Ed), A Compilation of Papers on the Assam Earthquake of August 15, 1950, Pub No. 1, The Central Board of Geophys, Calcutta, Govt. of India.

Campbell, K.W. (1989). Near Source Attenuation of Peak Horizontal Acceleration, BSSA, v. 71, 2039-2070 p.

Choudhury, H.M. and H.N. Srivastava (1976). Seismicity and Focal Mechanism of Some recent Earthquakes in Northeast India and Neighbourhood, Amali di Geofisica, v XXIX, No 1-2, p41 – 57.

Chopra, A.K. (1995). Dynamics of Structures – Theory and Applications in Earthquake Engineering, Prentice Hall, 750p.

Dewey, J.F. and Bird, J.M. (1970). Mountain belts and the new global tectonics, Journ Geophys. Res., V. 75, p 2625 – 2647.

EQ 83 – 9, Design Earthquake Parameters for Dihang project site in Arunachal Pradesh, Department of Earthquake Engineering, University of Roorkee, Roorkee, Unpublished report.

EQ 83 –16 (1983). Earthquake Engineering Studies, Design Earthquake Parameters for Earthquake Resistant Design of Rail-cum-Road Bridge Across Brahmaputra at Dibrugarh, Assam, Department of Earthquake Engineering, University of Roorkee, Roorkee, 39p. Unpublished report.

EQ 83 – 17, Design Earthquake Parameters for Ranganadi Project site in Arunachal Pradesh, Department of Earthquake Engineering, University of Roorkee, Roorkee, Unpublished report.

EQ 83 – 25, Design Earthquake Parameters for bridge site across Barak river, Department of Earthquake Engineering, University of Roorkee, Roorkee, Unpublished report.

EQ 83 – 26, Design Earthquake Parameters for Dhaleswari dam site in Arunachal Pradesh, Department of Earthquake Engineering, University of Roorkee, Roorkee, Unpublished report.

EQ 84 – 1, Earthquake Parameters for Design of Road bridge at Pasighat site, Arunachal Pradesh, Department of Earthquake Engineering, University of Roorkee, Roorkee, Unpublished report.

EQ 84 – 2, Design Earthquake Parameters for Khuga dam site in Arunachal Pradesh, Department of Earthquake Engineering, University of Roorkee, Roorkee, Unpublished report.

EQ 84 – 4, Design Earthquake Parameters for Borgalai Super Thermal Power Plant site in Assam, Department of Earthquake Engineering, University of Roorkee, Roorkee, Unpublished report.

EQ 84 – 19, Design Earthquake Parameters for Brahmakund Bridge site, Arunachal Pradesh, Department of Earthquake Engineering, University of Roorkee, Roorkee, Unpublished report.

EQ 84 – 22, Design Earthquake Parameters for Doyang Bridge site, Arunachal Pradesh, Department of Earthquake Engineering, University of Roorkee, Roorkee, Unpublished report.

EQ 88 – 4, Design Earthquake Parameters for Digboi oil Field, Assam, Department of Earthquake Engineering, University of Roorkee, Roorkee, Unpublished report.

EQ 93 – 03, Seismic Parameters for Damwe and Kameng Hydro electric Project site, Arunachal Pradesh, Department of Earthquake Engineering, University of Roorkee, Roorkee, Unpublished report.

EQ 98 –10 (1998). Seismic Ground Motions for Design of Rail-cum-Road Bridge Across River Brahmaputra at Bogibil, Department of Earthquake Engineering, University of Roorkee, Roorkee, 82p. Unpublished report.

Evans, P. (1964). The Tectonic Framework of Assam, Jour Geol Soc India, v 5, p 4 - 34.

Fitch, T.J. (1970). Earthquake Mechanism in the Himalayan, Burmese and Andaman Regions and Continental Tectonics in Central Asia, JGR, 75, 2699.

Geological Survey of India (1974). Geology and Mineral Resources of the States of India, Part IV- Arunachal Pradesh, Assam, Manipur, Meghalaya, Mizoram, Nagaland and Tripura (Calcutta GSI, Miscellaneous Publications.)

Gutenberg, B. (1956). Great earthquakes 1896 - 1903, Trans Am Geopys Union, v 37, p608 –614.

Hiller, K. and Elahi, K. (1988). Structural growth and hydrocarbon entrapment in the Surma basin, Bangladesh, *In*: .Wagner, H.C., Wagner, L.C., Wang, F.F.H. and Wong, F.L., Petroleum Resources of China and related subjects: Houston, Texas, Circum-Pacific Council for energy and mineral resources, Earth Science Series, No 10, p657 - 669.

Ichikawa, M., H.N. Srivastava, and J. Drakapoulos (1972). Focal Mechanism of Earthquakes Occurring in and Around the Himalayan and Burmese Mountain Belts, Meteorology and Geophysics, 23, 149 – 162.

IS: 1893-1984, (Fourth Revision) Indian Standard Criteria for Earthquake Resistant Design of Structures, Bureau of Indian Standards, New Delhi, India.

IS: 4326-1993, Indian Standard Code of Practice for Earthquake Resistant Design and Construction of Buildings, Bureau of Indian Standards, New Delhi, India.

IS: 13920-1993, Indian Standard Code of Practice for Ductile Detailing of Reinforced Concrete Structures subjected to Seismic Forces, Bureau of Indian Standards, New Delhi, India.

McGuire, R.K. (1977). Seismic Design Spectra and Mapping Procedures using Hazard Analysis Based Directly on Oscillator Response, Earthquake Engineering and Structural Dynamics, v 5, No. 3, p 211-234.

Molnar, P. (1990). A Review of the Seismicity and the Rates of Active Underthrusting, Journ Himalayan Geol, v1, No 2, p 131 – 154.

Naeim, F. (1989). The Seismic Design Handbook, Van Nostrand, Reinhold, 450p.Oldham, R.D. (1899). Report on the Great Earthquake of 12th June 1897, Memoirs of the Geological Survey of India, v XXIX, 379p.

Pande, P., S.K. Gupta, N.V. Venkataramn and B. Venkataramn (1995). Terrain changes consequent to the Killari earthquake of 30$^{th}$ September, 1993, p215 - 220, *in* Geological Survey of India Special Publication, No 27, 261p, Eds M. Ramakrishnan, K.D. Viswanatham, B.S.R. Murty, and L Harendranath.

Paulay, T., Priestley, M.J.N. (1992). Seismic Design of Reinforced Concrete and Masonry Buildings, John Wiley & Sons, 744p.

Penelis, G.G., Kappos, A.J. (1997). Earthquake Resistant Concrete Structures, E & FN Spon, 572p.

Poddar, M.C. (1953). A Short Note on the Assam Earthquake of Aug 15, 1950, p38 – 42, *in* M.B.R Rao (Ed), A Compilation of Papers on the Assam Earthquake of August 15, 1950, Publication No. 1. The Central Board of Geophysics; Calcutta, Govt. of India.

Prakash, V. and Powell, G.H. (1993). DRAIN-2DX Version 1.10 – Dynamic Response Analysis of Inelastic 2D Building Structures, NISEE/Computer Applications, Department of Civil Engineering, University of California, Berkeley, USA

Prakash, V. and Powell, G.H. (1993). DRAIN-3DX Version 1.10 – Dynamic Response Analysis of Inelastic 3D Building Structures, NISEE/Computer Applications, Department of Civil Engineering, University of California, Berkeley, USA

Prakash, V. and Powell, G.H. (1993). DRAIN-BUILDING Version 1.10 – Dynamic Response Analysis of Inelastic Building Structures, NISEE/Computer Applications, Department of Civil Engineering, University of California, Berkeley, USA

Pramanik, S.K., and Mukherjee, S.M. (1953). The Assam Earthquake of 1950, p26-34, in M.B.R Rao (Ed), A Compilation of Papers on the Assam Earthquake of August 15, 1950, Publication No. 1. The Central Board of Geophysics; Calcutta, Govt. of India.

Priestley, M.J.N., Seible, F., Calvi, G.M. (1996). Seismic Design and Retrofit of Bridges, John Wiley & Sons, 686p.

Raju, A.T.R. (1968). Geological Evolution of Assam and Cambay Tertiary Basins of India, Am Assoc. of Petro Geologists Bull, v 52, No 12, p 2422 – 2437.

Ray, S. (1953). Isoseismals for the Great Assam Earthquake of August 15, 1950, in M.B.R. Rao, A Compilation of Papers on the Assam Earthquake of August 15, 1950, Publication No. 1. The Central Board of Geophysics; Govt. of India, Calcutta.

Satellite Imageries: Geocoded Subscene Sat – ID : IRS – 1B L2B 2 Map ID : 83I11 PO14 – R48, Band : 4, Acquired 14 Feb 95, 10:02:58 POL CC S44 – A151 IND between 94° 30'E – 94° 45'E and 27°15'N - 27°30'N, IRS DPS I GT – 73687 29 Jul 1996, 08 : 30 : 30, ISRO/ NRSA, GT:73685, 30 Jul 96.

Satellite Imageries: Geocoded Subscene Sat – ID : IRS – 1B L2B 2 Map ID : 83I15 PO14 – R48, Band : 4, Acquired 14 Feb 95, 10:02:58 POL CC S44 – A151 IND between 94° 45'E – 95° 00'E and 27°15'N - 27°30'N, IRS DPS I GT – 66965 19 Feb 1996, 04 : 28 : 46, ISRO/ NRSA, GT: 66965, 22 Feb 96.

Seeber, L. and Ambruster, J.G. (1981). Great detachment earthquakes along the Himalayan Arc and llong term forecasting, in Simpson, D.W. and Richards, P.G., (eds.), Earthquake Prediction: An international Review, Maurice Ewing Series, Amer. Geophys. Union, Washington, 259 – 277.

Sinvhal, A., P.R. Bose and R.N.Dubey (1994). Damage report for the Latur-Osmanabad Earthquake of September 30, 1993, Bull Ind Soc Earthquake Tech, v 31, no 1, 15 - 54p.

Srivastava, V.B. (1984-85). Geotechnical Report on Road-cum-Rail Bridge over Brahmaputra near Dibrugarh town, Dibrugarh Distt., Assam. Submitted to and reported in annexure 2 the document of Northeast Frontier Railway and RITES titled, "Detailed Investigation Report for Road-cum-Rail Bridge across river Brahmaputra at Bogibil near Dibrugarh Volume 2, Annexures and Estimates," January 1998.

Tandon, A.N. (1953). The very great earthquake of Aug 15, 1950, p 80 – 89, in Rao, M.B.R., (1953), A Compilation of Papers on the Assam Earthquake of August 15, 1950, Publication No. 1. The Central Board of Geophysics; Govt. of India.

Tandon, A.N., and H.N. Srivastava (1975). Focal mechanism of Some recent Himalayan earthquakes and regional Plate tectonics, BSSA, v 65, p 963 – 970.

Yeats, R.S., Sieh, K. and Allen, C.R. (1997). The geology of Earthquakes, Oxford University press, New York, 568p.

# Water Science and Technology Library

1. A.S. Eikum and R.W. Seabloom (eds.): *Alternative Wastewater Treatment*. Low-Cost Small Systems, Research and Development. Proceedings of the Conference held in Oslo, Norway (7–10 September 1981). 1982 ISBN 90-277-1430-4
2. W. Brutsaert and G.H. Jirka (eds.): *Gas Transfer at Water Surfaces.* 1984 ISBN 90-277-1697-8
3. D.A. Kraijenhoff and J.R. Moll (eds.): *River Flow Modelling and Forecasting*. 1986 ISBN 90-277-2082-7
4. World Meteorological Organization (ed.): *Microprocessors in Operational Hydrology.* Proceedings of a Conference held in Geneva (4–5 September 1984). 1986 ISBN 90-277-2156-4
5. J. Němec: *Hydrological Forecasting.* Design and Operation of Hydrological Forecasting Systems. 1986 ISBN 90-277-2259-5
6. V.K. Gupta, I. Rodríguez-Iturbe and E.F. Wood (eds.): *Scale Problems in Hydrology.* Runoff Generation and Basin Response. 1986 ISBN 90-277-2258-7
7. D.C. Major and H.E. Schwarz: *Large-Scale Regional Water Resources Planning*. The North Atlantic Regional Study. 1990 ISBN 0-7923-0711-9
8. W.H. Hager: *Energy Dissipators and Hydraulic Jump*. 1992 ISBN 0-7923-1508-1
9. V.P. Singh and M. Fiorentino (eds.): *Entropy and Energy Dissipation in Water Resources*. 1992 ISBN 0-7923-1696-7
10. K.W. Hipel (ed.): *Stochastic and Statistical Methods in Hydrology and Environmental Engineering.* A Four Volume Work Resulting from the International Conference in Honour of Professor T. E. Unny (21–23 June 1993). 1994
    10/1: Extreme values: floods and droughts ISBN 0-7923-2756-X
    10/2: Stochastic and statistical modelling with groundwater and surface water applications ISBN 0-7923-2757-8
    10/3: Time series analysis in hydrology and environmental engineering ISBN 0-7923-2758-6
    10/4: Effective environmental management for sustainable development ISBN 0-7923-2759-4
    Set 10/1–10/4: ISBN 0-7923-2760-8
11. S.N. Rodionov: *Global and Regional Climate Interaction: The Caspian Sea Experience.* 1994 ISBN 0-7923-2784-5
12. A. Peters, G. Wittum, B. Herrling, U. Meissner, C.A. Brebbia, W.G. Gray and G.F. Pinder (eds.): *Computational Methods in Water Resources X.* 1994 Set 12/1–12/2: ISBN 0-7923-2937-6
13. C.B. Vreugdenhil: *Numerical Methods for Shallow-Water Flow*. 1994 ISBN 0-7923-3164-8
14. E. Cabrera and A.F. Vela (eds.): *Improving Efficiency and Reliability in Water Distribution Systems.* 1995 ISBN 0-7923-3536-8
15. V.P. Singh (ed.): *Environmental Hydrology.* 1995 ISBN 0-7923-3549-X
16. V.P. Singh and B. Kumar (eds.): *Proceedings of the International Conference on Hydrology and Water Resources* (New Delhi, 1993). 1996
    16/1: Surface-water hydrology ISBN 0-7923-3650-X
    16/2: Subsurface-water hydrology ISBN 0-7923-3651-8

16/3: Water-quality hydrology ISBN 0-7923-3652-6
16/4: Water resources planning and management ISBN 0-7923-3653-4
Set 16/1–16/4 ISBN 0-7923-3654-2

17. V.P. Singh: *Dam Breach Modeling Technology.* 1996 ISBN 0-7923-3925-8
18. Z. Kaczmarek, K.M. Strzepek, L. Somlyódy and V. Priazhinskaya (eds.): *Water Resources Management in the Face of Climatic/Hydrologic Uncertainties.* 1996 ISBN 0-7923-3927-4
19. V.P. Singh and W.H. Hager (eds.): *Environmental Hydraulics.* 1996 ISBN 0-7923-3983-5
20. G.B. Engelen and F.H. Kloosterman: *Hydrological Systems Analysis.* Methods and Applications. 1996 ISBN 0-7923-3986-X
21. A.S. Issar and S.D. Resnick (eds.): *Runoff, Infiltration and Subsurface Flow of Water in Arid and Semi-Arid Regions.* 1996 ISBN 0-7923-4034-5
22. M.B. Abbott and J.C. Refsgaard (eds.): *Distributed Hydrological Modelling.* 1996 ISBN 0-7923-4042-6
23. J. Gottlieb and P. DuChateau (eds.): *Parameter Identification and Inverse Problems in Hydrology, Geology and Ecology.* 1996 ISBN 0-7923-4089-2
24. V.P. Singh (ed.): *Hydrology of Disasters.* 1996 ISBN 0-7923-4092-2
25. A. Gianguzza, E. Pelizzetti and S. Sammartano (eds.): *Marine Chemistry.* An Environmental Analytical Chemistry Approach. 1997 ISBN 0-7923-4622-X
26. V.P. Singh and M. Fiorentino (eds.): *Geographical Information Systems in Hydrology.* 1996 ISBN 0-7923-4226-7
27. N.B. Harmancioglu, V.P. Singh and M.N. Alpaslan (eds.): *Environmental Data Management.* 1998 ISBN 0-7923-4857-5
28. G. Gambolati (ed.): *CENAS. Coastline Evolution of the Upper Adriatic Sea Due to Sea Level Rise and Natural and Anthropogenic Land Subsidence.* 1998 ISBN 0-7923-5119-3
29. D. Stephenson: *Water Supply Management.* 1998 ISBN 0-7923-5136-3
30. V.P. Singh: *Entropy-Based Parameter Estimation in Hydrology.* 1998 ISBN 0-7923-5224-6
31. A.S. Issar and N. Brown (eds.): *Water, Environment and Society in Times of Climatic Change.* 1998 ISBN 0-7923-5282-3
32. E. Cabrera and J. García-Serra (eds.): *Drought Management Planning in Water Supply Systems.* 1999 ISBN 0-7923-5294-7
33. N.B. Harmancioglu, O. Fistikoglu, S.D. Ozkul, V.P. Singh and M.N. Alpaslan: *Water Quality Monitoring Network Design.* 1999 ISBN 0-7923-5506-7
34. I. Stober and K. Bucher (eds): *Hydrogeology of Crystalline Rocks.* 2000 ISBN 0-7923-6082-6
35. J.S. Whitmore: *Drought Management on Farmland.* 2000 ISBN 0-7923-5998-4
36. R.S. Govindaraju and A. Ramachandra Rao (eds.): *Artificial Neural Networks in Hydrology.* 2000 ISBN 0-7923-6226-8
37. P. Singh and V.P. Singh: *Snow and Glacier Hydrology.* 2001 ISBN 0-7923-6767-7
38. B.E. Vieux: *Distributed Hydrologic Modeling Using GIS.* 2001 ISBN 0-7923-7002-3

39. I.V. Nagy, K. Asante-Duah and I. Zsuffa: *Hydrological Dimensioning and Operation of Reservoirs*. Practical Design Concepts and Principles. 2002 ISBN 1-4020-0438-9
40. I. Stober and K. Bucher (eds.): *Water-Rock Interaction*. 2002 ISBN 1-4020-0497-4
41. M. Shahin: *Hydrology and Water Resources of Africa*. 2002 ISBN 1-4020-0866-X
42. S.K. Mishra and V.P. Singh: *Soil Conservation Service Curve Number (SCS-CN) Methodology*. 2003 ISBN 1-4020-1132-6
43. C. Ray, G. Melin and R.B. Linsky (eds.): *Riverbank Filtration*. Improving Source-Water Quality. 2003 ISBN 1-4020-1133-4
44. G. Rossi, A. Cancelliere, L.S. Pereira, T. Oweis, M. Shatanawi and A. Zairi (eds.): *Tools for Drought Mitigation in Mediterranean Regions*. 2003 ISBN 1-4020-1140-7
45. A. Ramachandra Rao, K.H. Hamed and H.-L. Chen: *Nonstationarities in Hydrologic and Environmental Time Series*. 2003 ISBN 1-4020-1297-7
46. D.E. Agthe, R.B. Billings and N. Buras (eds.): *Managing Urban Water Supply*. 2003 ISBN 1-4020-1720-0
47. V.P. Singh, N. Sharma and C.S.P. Ojha (eds.): *The Brahmaputra Basin Water Resources*. 2004 ISBN 1-4020-1737-5